NOISE AND VIBRATION CONTROL ENGINEERING

NOISE AND VIBRATION CONTROL ENGINEERING
PRINCIPLES AND APPLICATIONS

SECOND EDITION

Edited by

**István L. Vér
and
Leo L. Beranek**

JOHN WILEY & SONS, INC.

This book is printed on acid-free paper. ∞

Copyright © 2006 by John Wiley & Sons, Inc. All rights reserved.

Published by John Wiley & Sons, Inc., Hoboken, New Jersey.
Published simultaneously in Canada.

No part of this publication may be reproduced, stored in a retrieval system, or transmitted in any form or by any means, electronic, mechanical, photocopying, recording, scanning, or otherwise, except as permitted under Section 107 or 108 of the 1976 United States Copyright Act, without either the prior written permission of the Publisher, or authorization through payment of the appropriate per-copy fee to the Copyright Clearance Center, Inc., 222 Rosewood Drive, Danvers, MA 01923, (978) 750-8400, fax (978) 750-4470, or on the web at www.copyright.com. Requests to the Publisher for permission should be addressed to the Permissions Department, John Wiley & Sons, Inc., 111 River Street, Hoboken, NJ 07030, (201) 748-6011, fax (201) 748-6008, e-mail: permcoordinator@wiley.com.

Limit of Liability/Disclaimer of Warranty: While the publisher and author have used their best efforts in preparing this book, they make no representations or warranties with respect to the accuracy or completeness of the contents of this book and specifically disclaim any implied warranties of merchantability or fitness for a particular purpose. No warranty may be created or extended by sales representatives or written sales materials. The advice and strategies contained herein may not be suitable for your situation. You should consult with a professional where appropriate. Neither the publisher nor author shall be liable for any loss of profit or any other commercial damages, including but not limited to special, incidental, consequential, or other damages.

For general information on our other products and services please contact our Customer Care Department within the United States at (800) 762-2974, outside the United States at (317) 572-3993 or fax (317) 572-4002.

Wiley also publishes its books in a variety of electronic formats. Some content that appears in print may not be available in electronic books. For more information about Wiley products, visit our web site at www.wiley.com.

Library of Congress Cataloging-in-Publication Data:

Noise and vibration control engineering : principles and applications / edited by Istvan L. Ver.—2nd ed.
    p. cm.
  Includes bibliographical references and index.
  ISBN-13 978-0-471-44942-3
  ISBN-10 0-471-44942-3
  1. Noise control. 2. Vibration. 3. Soundproofing. I. Vér, I. L. (István L.), 1934–
  TD892.N6512 2005 2006
  620.2′3—dc22
                                                                                          2005001372

Printed in the United States of America.

10 9 8 7 6 5 4 3 2 1

CONTENTS

Preface		vii
Contributors		ix
1.	**Basic Acoustical Quantities: Levels and Decibels** *Leo L. Beranek*	1
2.	**Waves and Impedances** *Leo L. Beranek*	25
3.	**Data Analysis** *Allan G. Piersol*	43
4.	**Determination of Sound Power Levels and Directivity of Noise Sources** *William W. Lang, George C. Maling, Jr., Matthew A. Nobile,* and *Jiri Tichy*	71
5.	**Outdoor Sound Propagation** *Ulrich J. Kurze* and *Grant S. Anderson*	119
6.	**Sound in Small Enclosures** *Donald J. Nefske and Shung H. Sung*	145
7.	**Sound in Rooms** *Murray Hodgson* and *John Bradley*	181
8.	**Sound-Absorbing Materials and Sound Absorbers** *Keith Attenborough* and *István L. Vér*	215
9.	**Passive Silencers** *M. L. Munjal, Anthony G. Galaitsis* and *István L. Vér*	279
10.	**Sound Generation** *István L. Vér*	345

vi CONTENTS

11. Interaction of Sound Waves with Solid Structures 389
István L. Vér

12. Enclosures, Cabins, and Wrappings 517
István L. Vér

13. Vibration Isolation 557
Eric E. Ungar and Jeffrey A. Zapfe

14. Structural Damping 579
Eric E. Ungar and Jeffrey A. Zapfe

15. Noise of Gas Flows 611
H. D. Baumann and *W. B. Coney*

16. Prediction of Machinery Noise 659
Eric W. Wood and James D. Barnes

17. Noise Control in Heating, Ventilating, and Air Conditioning Systems 685
Alan T. Fry and *Douglas H. Sturz*

18. Active Control of Noise and Vibration 721
Ronald Coleman and Paul J. Remington

19. Damage Risk Criteria for Hearing and Human Body Vibration 857
Suzanne D. Smith, *Charles W. Nixon* and *Henning E. Von Gierke*

20. Criteria for Noise in Buildings and Communities 887
Leo L. Beranek

21. Acoustical Standards for Noise and Vibration Control 911
Angelo Campanella, *Paul Schomer* and *Laura Ann Wilber*

Appendix A. General References 933

Appendix B. American System of Units 935

Appendix C. Conversion Factors 939

Index 943

PREFACE

The aim of this edition continues to be the presentation of the latest information on the most frequently encountered noise and vibration problems. We have endeavored to introduce new chapters and to update those chapters where the field has advanced. New or fully rewritten chapters are Sound Generation, Noise Control in Heating, Ventilating, and Air Conditioning Systems, Active Control of Noise and Vibration, Sound-Absorbing Materials and Sound Absorbers, Outdoor Sound Propagation, Criteria for Noise in Buildings and Communities, and Acoustical Standards for Noise and Vibration Control. Substantial new information has been added to Passive Silencers. All other chapters have been reviewed for timeliness.

Worldwide, there has been increased interest in noise and vibration control. Much of this interest has been generated by the expanding activities in the countries of the European Union and the Far East. Workshops on the latest developments in global noise policy are being held annually—the latest in Prague (Czech Republic) in 2004 and in Rio de Janeiro (Brazil) in 2005. There are signs of expanded interest in noise policy in the United States. Consumers are demanding quiet to a greater degree, the best example being the improved quiet interiors of automobiles. Other consumer products with greater noise control are already following or are likely to follow. Manufacturers must be alert to the increased competitiveness of imported products.

We are particularly indebted to John Bradley, Richard D. Godfrey, Colin H. Hanson, William W. Lang, George C. Mailing, Jr., Howie Noble, Robert Preuss, and Paul Schomer, who rendered help and criticism during the preparation of this book, and to the many individuals and organizations who gave us permission to use copyrighted technical information. We are also indebted to Acentech, Inc., BBN Technologies, Inc., Mueller-BBM Corporation and Harris Miller, Miller & Hanson, Inc., for encouraging members of their senior technical staff to contribute chapters and for assistance in many ways. It gives us special pleasure to acknowledge the help of Kathy Coleman and Jane Schultz, who assisted in the preparation of many chapters, and to Robert L. Argentieri, Fred Bernardi, and Robert H. Hilbert our editors at John Wiley & Sons, for their effective help and guidance in the production of this book.

ISTVÁN L. VÉR
LEO L. BERANEK

January 2005

CONTRIBUTORS

Grant S. Anderson, Harris Miller, Miller & Hanson Inc. Burlington, Massachusetts

Keith Attenborough, University of Hull, Hull, United Kingdom

James D. Barnes, Acentech, Inc., Cambridge, Massachusetts

H. D. Baumann, Consultant, Rye, New Hampshire

Leo L. Beranek, Consultant, Cambridge, Massachusetts

John Bradley, National Research Council of Canada, Ottawa, Canada

Angelo Campanella, Consultant, Columbus, Ohio

Ronald Coleman, BBN Systems and Technologies Corporation Cambridge, Massachusetts

W. B. Coney, BBN Systems and Technologies Corporation Cambridge, Massachusetts

Alan T. Fry, Consultant, Colchester, Essex, United Kingdom

Anthony G. Galaitsis, BBN Systems and Technologies Corporation, Cambridge, Massachusetts

Murray Hodgson, University of British Columbia, Vancouver, Canada

Ulrich J. Kurze, Mueller-BBM GmbH, Planegg near Munich, Germany

William W. Lang, Consultant, Poughkeepsie, New York

George C. Maling Jr., Consultant, Harpswell, Maine

M. L. Munjal, Indian Institute of Science, Bangalore, India

Donald J. Nefske, Vehicle Research Laboratory, General Motors R&D Center, Warren, Michigan

Charlers W. Nixon, Consultant, Kettering, Ohio, USA

Matthew A. Nobile, Consultant, IBM Hudson Valley Acoustics Laboratory, Poughkeepsie, New York

Allan G. Piersol, Piersol Engineering Company, Woodland Hills, California

Paul J. Remington, BBN Technologies Corporation, Cambridge, Massachusetts

Paul Schomer, Consultant, Champaign, Illinois

Shung H. Sung, Vehicle Development Research Laboratory, General Motors R&D Center, Warren, Michigan

Susanne D. Smith, Air Force Research Laboratory, Wright-Patterson Air Force Base, Ohio

Douglas H. Sturz, Acentech, Inc., Cambridge, Massachusetts

Jiry Tichy, The Pennsylvania State University, University Park, Pennsylvania

Eric E. Ungar, Acentech, Inc., Cambridge Massachusetts

István L. Vér, Consultant in Acoustics, Noise, and Vibration Control, Stow, Massachusetts

Henning E. von Gierke, Consultant, Yellow Spring, Ohio

Laura Ann Wilber, Consultant, Wilmette, Illinois

Eric W. Wood, Acentech, Inc., Cambridge, Massachusetts

Jeffrey A. Zapfe, Acentech, Inc., Cambridge, Massachusetts

CHAPTER 1

Basic Acoustical Quantities: Levels and Decibels

LEO L. BERANEK

Consultant
Cambridge, Massachusetts

1.1 BASIC QUANTITIES OF SOUND WAVES

Sound Waves and Noise

In the broadest sense, a *sound wave* is any disturbance that is propagated in an elastic medium, which may be a gas, a liquid, or a solid. Ultrasonic, sonic, and infrasonic waves are included in this definition. Most of this text deals with sonic waves, those sound waves that can be perceived by the hearing sense of a human being. *Noise* is defined as any perceived sound that is objectionable to a human being. The concepts basic to this chapter can be found in references 1–7. Portions are further expanded in Chapter 2.

Sound Pressure

A person who is not deaf perceives as sound any vibration of the eardrum in the audible frequency range that results from an incremental variation in air pressure at the ear. A variation in pressure above and below atmospheric pressure is called *sound pressure*, in units of pascals (Pa).* A young person with normal hearing can perceive sound in the frequency range of roughly 15 Hz (hertz) to 16,000 Hz, defined as the normal audible frequency range.

Because the hearing mechanism responds to sound pressure, it is one of two quantities that is usually measured in engineering acoustics. The normal ear is most sensitive at frequencies between 3000 and 6000 Hz, and a young person can detect pressures as low as about 20 μPa, which, when compared to the normal atmospheric pressure (101.3×10^3 Pa) around which it varies, is a fractional variation of 2×10^{-10}.

*One pascal (Pa) = 1 newton/meter squared (N/m^2) = 10 dynes/cm^2

1

Pure Tone

A pure tone is a sound wave that can be represented by the equation,

$$p(t) = p_0 \sin(2\pi f)t \qquad (1.1)$$

where $p(t)$ is the instantaneous, incremental, sound pressure (above and below atmospheric pressure), p_0 is the maximum amplitude of the instantaneous sound pressure, and f is the frequency, that is, the number of cycles per second, expressed in hertz. The time t is in seconds.

Period

A full cycle occurs when t varies from zero to $1/f$. The $1/f$ quantity is known as the period T. For example, the period T of a 500-Hz wave is 0.002 sec.

Root-Mean-Square Amplitude

If we wish to determine the mean value of a full cycle of the sine wave of Eq. (1.1.) (or any number of full cycles), it will be zero because the positive part equals the negative part. Thus, the mean value is not a useful measure. We must look for a measure that permits the effects of the rarefactions to be added to (rather than subtracted from) the effects of the compressions.

One such measure is the root-mean-square (rms) sound pressure p_{rms}. It is obtained, first, by squaring the value of the sound pressure disturbance $p(t)$ at each instant of time. Next the squared values are added and averaged over one or more periods. The rms sound pressure is the square root of this time average. The rms value is also called the *effective value*. Thus

$$p_{rms}^2 = \tfrac{1}{2} p_0^2 \qquad (1.2)$$

or

$$p_{rms} = 0.707 p_0 \qquad (1.3)$$

In the case of nonperiodic sound pressures, the integration interval should be long enough to make the rms value obtained essentially independent of small changes in the length of the interval.

Sound Spectra

A sound wave may be comprised of a pure tone (single frequency, e.g., 1000 Hz), a combination of single frequencies harmonically related, or a combination of single frequencies not harmonically related, either finite or infinite in number. A combination of a finite number of tones is said to have a *line spectrum*. A combination of an infinite (large) number of tones has a *continuous spectrum*. A continuous-spectrum noise for which the amplitudes versus time occur with a

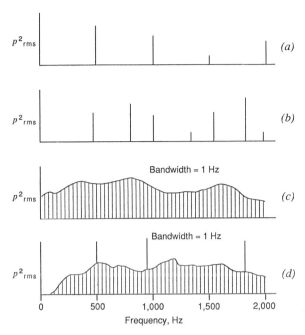

FIGURE 1.1 Mean-square sound pressure spectra: (*a*) harmonically related line spectrum; (*b*) inharmonically related line spectrum; (*c*) continuous power spectral density spectrum; (*d*) combination line and continuous power spectral density spectrum (complex spectrum).

normal (Gaussian) distribution is called *random noise*. Three of these types of noise are shown by the frequency spectra in Figs. 1.1*a*–*c*. A combination of a line and a continuous spectrum, called a complex spectrum, is shown in Fig. 1.1*d*.

Regardless of which type of sound wave is considered, when propagating at normal sound pressure amplitudes (to avoid nonlinearity) in air over reasonably short distances (so that sound attenuation in the air itself, which becomes significant at frequencies above 1000 Hz, can be neglected), the waveform is unchanged. Thus, a violin heard at a distance of 30 m sounds the same as at 5 m, although it is less loud.

Sound Intensity

The second quantity commonly measured in engineering acoustics is *sound intensity*, defined as the continuous flow of power carried by a sound wave through an incrementally small area at a point in space. The units are watts per square meter (W/m^2). This quantity is important for two reasons. First, at a point in free space, it is related to the total power radiated into the air by a sound source and, second, it bears at that point a fixed relation to the sound pressure.

Sound intensity at a point is directional (a vector) in the sense that the position of the plane of the incrementally small area can vary from being perpendicular to the direction in which the wave is traveling to being parallel to that direction. It has its maximum value, I_{max}, when its plane is perpendicular to the direction of travel. When parallel, the sound intensity is zero. In between, the component of I_{max} varies as the cosine of the angle formed by the direction of travel and a line perpendicular to the incremental area.

Another equation, which we shall develop in the next chapter, relates sound pressure to sound intensity. In an environment in which there are no reflecting surfaces, the sound pressure *at any point in any type of freely traveling* (plane, cylindrical, spherical, etc.) wave is related to the maximum intensity I_{max} by

$$p_{rms}^2 = I_{max} \cdot \rho c \quad Pa^2 \tag{1.4}$$

where p_{rms} = rms sound pressure, Pa (N/m²)
ρ = density of air, kg/m³
c = speed of sound in air, m/s [see Eq. (1.7)]
N = force, N

Sound Power

A sound source radiates a measurable amount of power into the surrounding air, called *sound power*, in watts. If the source is nondirectional, it is said to be a *spherical sound source* (see Fig. 1.2). For such a sound source the measured (maximum) sound intensities at all points on an imaginary spherical surface centered on the acoustic center of the source are equal. Mathematically,

$$W_s = (4\pi r^2) I_s(r) \quad W \tag{1.5}$$

where $I_s(r)$ = maximum sound intensity at radius r at surface of an imaginary sphere surrounding source, W/m²
W_s = total sound power radiated by source in watts, W (N · m/s)
r = distance from acoustical center of source to surface of imaginary sphere, m

A similar statement can be made about a line source; that is, the maximum sound intensities at all points on an imaginary cylindrical surface around a *cylindrical sound source*, $I_c(r)$, are equal:

$$W_c = (2\pi r l) I_c(r) \quad W \tag{1.6}$$

where W_c = total sound power radiated by cylinder of length l, W
r = distance from acoustical centerline of cylindrical source to imaginary cylindrical surface surrounding source

Inverse Square Law

With a spherical source, the radiated sound wave is spherical and the total power radiated in all directions is W. The sound intensity $I(r)$ must decrease with

BASIC QUANTITIES OF SOUND WAVES 5

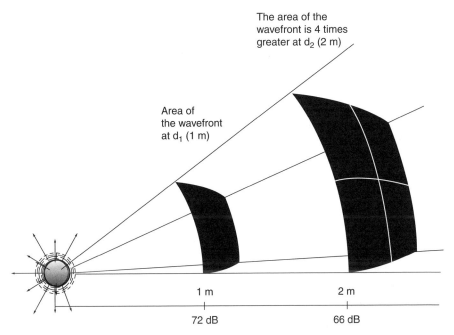

FIGURE 1.2 Generation of a one-dimensional spherical wave. A balloonlike surface pulsates uniformly about some equilibrium position and produces a sound wave that propagates radially away from the balloon at the speed of sound. (Courtesy of Ref. 8.)

distance in proportion to the sound pressure squared, that is $W/4\pi r^2$ [Eq. (1.5)]. Hence the name inverse square law. This is illustrated by the diagram in Fig. 1.2, where at a distance of 1 m the area of the wave front shown is a square meters and at 2 m it becomes $4a$ square meters. To preserve sound energy, the same amount of power flows through the larger area; thus Fig. 1.2 diagrammatically shows the intensity decreasing by a factor of 4, or, as shown later, 6 dB, from 72 to 66 dB.

Particle Velocity

Consider that the surface of the spherical source of Fig. 1.2 is expanding and contracting sinusoidally. During the first half of its sinusoidal motion, it pushes the air particles near its surface outward. Because air is elastic and compresses, the pressure near the surface will increase. This increased pressure overcomes the inertia of the air particles a short distance away and they move outward. That outward movement causes the pressure to build up at this new distance and, in turn, pushes more removed particles outward. This outward movement of the disturbance takes place at the speed of sound.

In the next half of the sinusoid, the sinusoidally vibrating spherical source reverses direction, creating a drop of pressure near its surface that pulls nearby

6 BASIC ACOUSTICAL QUANTITIES: LEVELS AND DECIBELS

air particles toward it. This reverse disturbance also propagates outward with the speed of sound. Thus, at any one point in space, there will be sinusoidal to-and-fro movement of the particles, called the *particle velocity*. Also at any point, there will be sinusoidal rise and fall of sound pressure.

From the basic equations governing the propagation of sound, as is shown in the next chapter, we can say the following:

1. In a *plane wave* (approximated at a large distance from a point source) propagated in free space (no reflecting surfaces) the sound pressure and the particle velocity reach their maximum and minimum values at the same instant and are said to be in phase.
2. In such a wave the particles move back and forth along the line in which the wave is traveling. In reference to the spherical radiation discussed above, this means that the particle velocity is always perpendicular to the imaginary spherical surface (the wave front) in space. This type of wave is called a longitudinal, or compressional, wave. By contrast, a transverse wave is illustrated by a surface wave in water where the particle velocity is perpendicular to the water surface while the wave propagates in a direction parallel to the surface.

Speed of Sound

A sound wave travels outward at a rate dependent on the elasticity and density of the air. Mathematically, the *speed of sound* in air is calculated as

$$c = \sqrt{\frac{1.4 P_s}{\rho}} \quad \text{m/s} \tag{1.7}$$

where P_s = atmospheric (ambient) pressure, Pa
ρ = density of air, kg/m³

For all practical purposes, the speed of sound is dependent only on the absolute temperature of the air. The equations for the speed of sound are

$$c = 20.05\sqrt{T} \quad \text{m/s} \tag{1.8}$$

$$c = 49.03\sqrt{R} \quad \text{ft/s} \tag{1.9}$$

where T = absolute temperature of air in degrees Kelvin, equal to 273.2 plus the temperature in degrees Celsius
R = absolute temperature in degrees Rankine, equal to 459.7 plus the temperature in degrees Fahrenheit

For temperatures near 20°C (68°F), the speed of sound is

$$c = 331.5 + 0.58°C \quad \text{m/s} \tag{1.10}$$

$$c = 1054 + 1.07°F \quad \text{ft/s} \tag{1.11}$$

Wavelength

Wavelength is defined as the distance the pure-tone wave travels during a full period. It is denoted by the Greek letter λ and is equal to the speed of sound divided by the frequency of the pure tone:

$$\lambda = cT = \frac{c}{f} \quad \text{m} \tag{1.12}$$

Sound Energy Density

In standing-wave situations, such as sound waves in closed, rigid-wall tubes, rooms containing little sound-absorbing material, or reverberation chambers, the quantity desired is not sound intensity, but rather the *sound energy density*, namely, the energy (kinetic and potential) stored in a small volume of air in the room owing to the presence of the standing-wave field. The relation between the *space-averaged sound energy density D* and the *space-averaged squared sound pressure* is

$$D = \frac{p_{av}^2}{\rho c^2} = \frac{p_{av}^2}{1.4 P_s} \quad \text{W} \cdot \text{s/m}^3 \ (\text{J/m}^3 \text{ or simply N/m}^2) \tag{1.13}$$

where p_{av}^2 = space average of mean-square sound pressure in a space, determined from data obtained by moving a microphone along a tube or around a room or from samples at various points, Pa²

P_s = atmospheric pressure, Pa; under normal atmospheric conditions, at sea level, $P_s = 1.013 \times 10^5$ Pa

1.2 SOUND SPECTRA

In the previous section we described sound waves with line and continuous spectra. Here we shall discuss how to quantify such spectra.

Continuous Spectra

As stated before, a continuous spectrum can be represented by a large number of pure tones between two frequency limits, whether those limits are apart 1 Hz or thousands of hertz (see Fig. 1.3). Because the hearing system extends over a large frequency range and is not equally sensitive to all frequencies, it is customary to measure a continuous-spectrum sound in a series of contiguous frequency bands using a sound analyzer.

Customary bandwidths are one-third octave and one octave (see Fig. 1.4 and Table 1.1). The rms value of such a filtered sound pressure is called the one-third-octave-band or the octave-band sound pressure, respectively. If the filter bandwidth is 1 Hz, a plot of the filtered *mean-square* pressure of a *continuous-spectrum*

8 BASIC ACOUSTICAL QUANTITIES: LEVELS AND DECIBELS

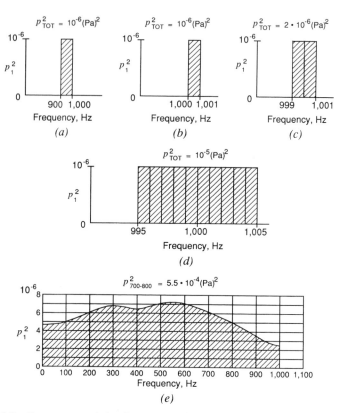

FIGURE 1.3 Power spectral density spectra showing the linear growth of total mean-square sound pressure when the bandwidth of noise is increased. In each case, p_1^2 is the mean-square sound pressure in a band 1 Hz wide.

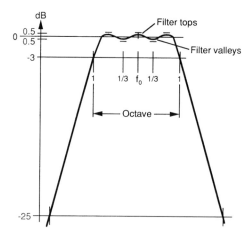

FIGURE 1.4 Frequency response of one manufacturer's octave-band filter. The 3-dB down points for the octave band and for a one-third-octave-band filter are indicated.

TABLE 1.1 Center and Approximate Cutoff Frequencies (Hz) for Standard Set of Contiguous-Octave and One-Third-Octave Bands Covering Audio Frequency Range[a]

	Octave			One-Third Octave		
Band	Lower Band Limit	Center	Upper Band Limit	Lower Band Limit	Center	Upper Band Limit
12	11	16	22	14.1	16	17.8
13				17.8	20	22.4
14				22.4	25	28.2
15	22	31.5	44	28.2	31.5	35.5
16				35.5	40	44.7
17				44.7	50	56.2
18	44	63	88	56.2	63	70.8
19				70.8	80	89.1
20				89.1	100	112
21	88	125	177	112	125	141
22				141	160	178
23				178	200	224
24	177	250	355	224	250	282
25				282	315	355
26				355	400	447
27	355	500	710	447	500	562
28				562	630	708
29				708	800	891
30	710	1,000	1,420	891	1,000	1,122
31				1,122	1,250	1,413
32				1,413	1,600	1,778
33	1,420	2,000	2,840	1,778	2,000	2,239
34				2,239	2,500	2,818
35				2,818	3,150	3,548
36	2,840	4,000	5,680	3,548	4,000	4,467
37				4,467	5,000	5,623
38				5,623	6,300	7,079
39	5,680	8,000	11,360	7,079	8,000	8,913
40				8,913	10,000	11,220
41				11,220	12,500	14,130
42	11,360	16,000	22,720	14,130	16,000	17,780
43				17,780	20,000	22,390

[a]From Ref. 6.

sound versus frequency is called the *power spectral (or spectrum) density spectrum*. Narrow bandwidths are commonly used in analyses of machinery noise and vibration, but the term "spectral density" has no meaning in cases where pure tones are being measured.

The mean-square (or rms) sound pressure can be determined for each of the contiguous frequency bands and the result plotted as a function of frequency (Fig. 1.3*e*).

Bandwidth Conversion. It is frequently necessary to convert sounds measured with one set of bandwidths to a different set of bandwidths or to reduce both sets of measurements to a third set of bandwidths. Let us imagine that we have a machine that at a point in space produces a mean-square sound pressure of $p_1^2 = 10^{-6}$ Pa^2 in a 1-Hz bandwidth between 999 and 1000 Hz (Fig. 1.3a). Now imagine that we have a second machine the same distance away that radiates the same power but is confined to a bandwidth between 1000 and 1001 Hz (Fig. 1.3b). The total spectrum now becomes that shown in Fig. 1.3c and the total mean-square pressure is twice that in either band. Similarly, 10 machines would produce 10 times the mean-square sound pressure of any one (Fig. 1.3d).

In other words, if the power spectral density spectrum in a frequency band of width Δf is flat (the mean-square sound pressures in all the 1-Hz-wide bands, p_1^2, within the band are equal), the total mean-square sound pressure for the band is given by

$$p_{\text{tot}}^2 = p_1^2 \frac{\Delta f}{\Delta f_0} \quad \text{Pa}^2 \tag{1.14}$$

where $\Delta f_0 = 1$ Hz.

As an example, assume that we wish to convert the power spectral density spectrum of Fig. 1.3e, which is a plot of $p_1^2(f)$, the mean-square sound pressure in 1-Hz bands, to a spectrum for which the mean-square sound pressure in 100-Hz bands, p_{tot}^2, is plotted versus frequency. Let us consider only the 700–800-Hz band. Because $p_1^2(f)$ is not equal throughout this band, we could painstakingly determine and add together the actual p_1^2's or, as is more usual, simply take the average value for the p_1^2's in that band and multiply by the bandwidth. Thus, for each 100-Hz band, the total mean-square sound pressure is given by Eq. (1.14), where p_1^2 is the average 1-Hz band quantity throughout the band. For the 700–800-Hz band, the average p_1^2 is 5.5×10^{-6} and the total is $5.5 \times 10^{-6} \times 100$ Hz $= 5.5 \times 10^{-4}$ Pa^2.

If mean-square sound pressure levels have been measured in a specific set of bandwidths such as one-third-octave bands, it is possible to present accurately the data in a set of wider bandwidths such as octave bands by simply adding together the mean-square sound pressures for the component bands. Obviously, it is not possible to reconstruct a narrower bandwidth spectrum accurately (e.g., one-third-octave bands) from a wider bandwidth spectrum (e.g., octave bands.) However, it is sometimes necessary to make such a conversion in order to compare sets of data measured differently. Then the implicit assumption has to be made that the narrower band spectrum is continuous and monotonic within the larger band. In either direction, the conversion factor for each band is

$$p_B^2 = p_A^2 \frac{\Delta f_B}{\Delta f_A} \tag{1.15}$$

where p_A^2 is the measured mean-square sound pressure in a bandwidth Δf_A and p_B^2 is the desired mean-square sound pressure in the desired bandwidth Δf_B.

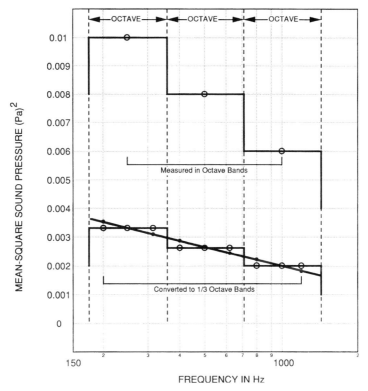

FIGURE 1.5 Conversion of octave-band mean-square sound pressures into third-octave-band mean-square sound pressures. Such conversion should be made only where necessary under the assumption that the third-octave-band mean-square pressure levels decrease monotonically with band midfrequency. The sloping solid curve is the assumed-correct converted spectrum.

As an example, assume that we have measured a sound with a continuous spectrum using a one-octave-band filterset and have plotted the intensity for the contiguous bands versus the midfrequency of each band (see the upper three circles in Fig. 1.5). Assume that we wish to convert to an approximate one-third-octave-band spectrum to make comparisons with other data possible. Not knowing how the mean-square sound pressure varies throughout each octave band, we assume it to be continuous and monotonic and apply Eq. (1.15). In this example, approximately, $\Delta f_B = \frac{1}{3} \Delta f_A$ for all the bands (see Table 1.1), and the mean-square sound pressure in each one-third-octave band with the same midfrequency will be one-third that in the corresponding one-octave band. The data would be plotted versus the midfrequency of each one-third-octave band (see Table 1.1 and Fig. 1.5). Then the straight, sloping line is added under the assumption that the true one-third-octave spectrum is monotonic.

12 BASIC ACOUSTICAL QUANTITIES: LEVELS AND DECIBELS

Complex Spectra

The mean-square sound pressure resulting from the combination of two or more pure tones of different amplitudes p_1, p_2, p_3 and *different frequencies* f_1, f_2, f_3 is given by

$$p^2_{\text{rms}}(\text{total}) = p_1^2 + p_2^2 + p_3^2 + \cdots \qquad (1.16)$$

The mean-square sound pressure of two pure tones of the *same frequency* but different amplitudes and phases is found from

$$p^2_{\text{rms}}(\text{total}) = p_1^2 + p_2^2 + 2p_1 p_2 \cos(\theta_1 - \theta_2) \qquad (1.17)$$

where the phase angle of each wave is represented by θ_1 or θ_2.

Comparison of Eqs. (1.16) and (1.17) reveals the importance of phase when combining two sine waves of the same frequency. If the phase difference $\theta_1 - \theta_2$ is zero, the two waves are in phase and the combination is at its maximum value. If $\theta_1 - \theta_2 = 180°$, the third term becomes $-2p_1 p_2$ and the sum is at its minimum value. If the two waves are equal in amplitude, the minimum value is zero.

If one wishes to find the mean-square sound pressure of a number of waves all of which have different frequencies except, say, two, these two are added together according to Eq. (1.17) to obtain a mean-square pressure for them. Then this mean-square pressure and the mean-square pressures of the remainder of the components are summed according to Eq. (1.16).

1.3 LEVELS[6]

Because of the wide range of sound pressures to which the ear responds (a ratio of 10^5 or more for a normal person), sound pressure is an inconvenient quantity to use in graphs and tables. This is also true for the other acoustical quantities listed above. Early in the history of the telephone it was decided to adopt logarithmic scales for representing acoustical quantities and the voltages encountered in associated electrical equipment.

As a result of that decision, sound powers, intensities, pressures, velocities, energy densities, and voltages from electroacoustic transducers are commonly stated in terms of the logarithm of the ratio of the measured quantity to an appropriate reference quantity. Because the sound pressure at the threshold of hearing at 1000 Hz is about 20 µPa, this was chosen as the fundamental reference quantity around which the other acoustical references have been chosen.

Whenever the magnitude of an acoustical quantity is given in this logarithmic form, it is said to be a *level* in *decibels* (dB) *above* or *below* a zero *reference level* that is determined by a *reference quantity*. The argument of the logarithm is always a ratio and, hence, is dimensionless. The level for a very large ratio, for example the power produced by a very powerful sound source, might be given with the unit *bel*, which equals 10 dB.

Power and Intensity Levels

Sound Power Level. Sound power level is defined as

$$L_W = 10 \log_{10} \frac{W}{W_0} \quad \text{dB re } W_0 \tag{1.18}$$

and conversely

$$W = W_0 \text{ antilog}_{10} \frac{L_W}{10} = W_0 \times 10^{L_W/10} \quad \text{W} \tag{1.19}$$

where W = sound power, W (watts)
  W_0 = reference sound power, standardized at 10^{-12} W

As seen in Table 1.2, a *ratio* of 10 in the power W corresponds to a *level difference* of 10 dB regardless of the reference power W_0. Similarly, a ratio of 100 corresponds to a level difference of 20 dB. Power ratios of less than 1 are allowable: They simply lead to negative levels. For example (see Table 1.2), a power ratio of 0.1 corresponds to a level difference of -10 dB.

Column 4 of Table 1.2 gives sound power levels relative to the standard reference power level $W_0 = 10^{-12}$ W in watts.

Some sound power ratios and the corresponding sound-power-level differences are given in Table 1.3. We note from the last line that the sound power level for the *product of two ratios* is equal to the *sum of the levels* for the two ratios. For example, determine L_W for the quantity 2×4. From Table 1.3,

TABLE 1.2 Sound Powers and Sound Power Levels

Radiated Sound Power W, watts		Sound Power Level L_W, dB	
Usual Notation	Equivalent Exponential Notation	Relative to 1 W	Relative to 10^{-12} W (standard)
100,000	10^5	50	170
10,000	10^4	40	160
1,000	10^3	30	150
100	10^2	20	140
10	10^1	10	130
1	1	0	120
0.1	10^{-1}	-10	110
0.01	10^{-2}	-20	100
0.001	10^{-3}	-30	90
0.000,1	10^{-4}	-40	80
0.000,01	10^{-5}	-50	70
0.000,001	10^{-6}	-60	60
0.000,000,1	10^{-7}	-70	50
0.000,000,01	10^{-8}	-80	40
0.000,000,001	10^{-9}	-90	30

TABLE 1.3 Selected Sound Power Ratios and Corresponding Power-Level Differences

Sound Power Ratio W/W_0, R	Sound-Power-Level Difference[a] $10 \log W/W_0$, L_W (dB)
1000	30
100	20
10	10
9	9.5
8	9.0
7	8.5
6	7.8
5	7.0
4	6.0
3	4.8
2	3.0
1	0.0
0.9	−0.5
0.8	−1.0
0.7	−1.5
0.6	−2.2
0.5	−3.0
0.4	−4.0
0.3	−5.2
0.2	−7.0
0.1	−10
0.01	−20
0.001	−30
$R_1 \times R_2$	$L_{W_1} + L_{W_2}$

[a] To the nearest 0.1 dB.

$L_W = 3.0 + 6.0 = 9.0$ dB, which is the sound power level for the ratio 8. Similarly, L_W for a ratio of 8000 equals the sum of the levels for 8 and 1000, that is, $L_W = 9 + 30 = 39$ dB.

Sound Intensity Level. Sound intensity level, in decibels, is defined as

$$\text{Intensity level} = L_I = 10 \log \frac{I}{I_{\text{ref}}} \quad \text{dB re } I_{\text{ref}} \quad (1.20)$$

where I = sound intensity whose level is being specified, W/m²
I_{ref} = reference intensity standardized as 10^{-12} W/m²

Sound power levels should not be confused with intensity levels (or with sound pressure levels, which are defined next), which also are expressed in decibels. Sound power is a measure of the *total* acoustical power radiated by a source in watts. Sound intensity and sound pressure specify the acoustical "disturbance" produced at a point removed from the source. For example, their levels depend on the distance from the source, losses in the intervening air path, and room effects (if indoors). A helpful analogy is to imagine that sound power level is related to the total rate of heat production of a furnace, while either of the other two levels is analogous to the temperature produced at a given point in a dwelling.

Sound Pressure Level

Almost all microphones used today respond to sound pressure, and in the public mind, the word *decibel* is commonly associated with sound pressure level or A-weighted sound pressure level (see Table 1.4). Strictly speaking, sound pressure level is analogous to intensity level, because, in calculating it, pressure is first squared, which makes it proportional to intensity (power per unit area):

$$\text{Sound pressure level} = L_p = 10 \log \left[\frac{p(t)}{p_{\text{ref}}} \right]^2$$

$$= 20 \log \frac{p(t)}{p_{\text{ref}}} \quad \text{dB re } p_{\text{ref}} \quad (1.21)$$

where p_{ref} = reference sound pressure, standardized at 2×10^{-5} N/m² (20 μPa) for airborne sound; for other media, references may be 0.1 N/m² (1 dyn/cm²) or 1 μN/m² (1 μPa)
$p(t)$ = instantaneous sound pressure, Pa

Note that L_p re 20 μPa is 94 dB greater than L_p re 1 Pa.

As we shall show shortly, $p(t)^2$ is only proportional to sound intensity if its mean-square value is taken. Thus, in Eq. (1.21), $p(t)$ would be replaced by p_{rms}.

The relations among sound pressure levels (re 20 μPa) for pressures in the meter-kilogram-second (mks), centimeter-gram-second (cgs), and English systems of units are shown by the four nomograms of Fig. 1.6.

1.4 DEFINITIONS OF OTHER COMMONLY USED LEVELS AND QUANTITIES IN ACOUSTICS

Analogous to sound pressure level given in Eq. (1.21), *A-weighted sound pressure level* L_A is given by

$$L_A = 10 \log \left[\frac{p_A(t)}{p_{\text{ref}}} \right]^2 \quad \text{dB} \quad (1.22)$$

where $p_A(t)$ is the instantaneous sound pressure measured using the standard frequency-weighting A (see Table 1.4).

TABLE 1.4 A and C Electrical Weighting Networks for Sound-Level Meter[a]

Frequency, Hz	A-Weighting Relative Response, dB	C-Weighting Relative Response, dB
10	−70.4	−14.3
12.5	−63.4	−11.2
16	−56.7	−8.5
20	−50.5	−6.2
25	−44.7	−4.4
31.5	−39.4	−3.0
40	−34.6	−2.0
50	−30.2	−1.3
63	−26.2	−0.8
80	−22.5	−0.5
100	−19.1	−0.3
125	−16.1	−0.2
160	−13.4	−0.1
200	−10.9	0
250	−8.6	0
315	−6.6	0
400	−4.8	0
500	−3.2	0
630	−1.9	0
800	−0.8	0
1,000	0	0
1,250	+0.6	0
1,600	+1.0	−0.1
2,000	+1.2	−0.2
2,500	+1.3	−0.3
3,150	+1.2	−0.5
4,000	+1.0	−0.8
5,000	+0.5	−1.3
6,300	−0.1	−2.0
8,000	−1.1	−3.0
10,000	−2.5	−4.4
12,500	−4.3	−6.2
16,000	−6.6	−8.5
20,000	−9.3	−11.2

[a] These numbers assume a flat, diffuse-field (random-incidence) response for the sound-level meter and microphone.

Average sound level $L_{\text{av},T}$ is given by

$$L_{\text{av},T} = 10 \log \frac{(1/T) \int_0^T p^2(t)\, dt}{p_{\text{ref}}^2} \quad \text{dB} \quad (1.23)$$

where T is the (long) time over which the averaging takes place.

DEFINITIONS OF OTHER COMMONLY USED LEVELS AND QUANTITIES IN ACOUSTICS

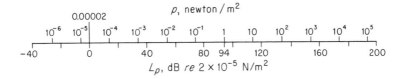

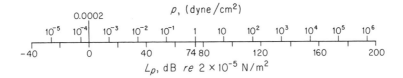

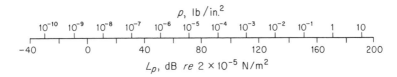

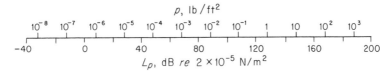

FIGURE 1.6 Charts relating L_p (dB re 20 µPa) to p in N/m²(Pa), dyn/cm², lb/in.², and lb/ft². For example, 1.0 Pa = 94 dB re 20 µPa.

Average A-weighted sound level $L_{A,T}$ (also called L_{eq}, *equivalent continuous A-weighted noise level*) is given by

$$L_{A,T} = L_{eq} = 10 \log \frac{(1/T) \int_0^T p_A^2(t) \, dt}{p_{ref}^2} \quad \text{dB} \quad (1.24)$$

The time T must be specified. In noise evaluations, its length is usually one to several hours, or 8 h (working day), or 24 h (full day).

Day–night sound (noise) level L_{dn} is given by

$$L_{dn} = 10 \log \frac{1}{24} \left[\frac{\int_{07:00}^{22:00} p_A^2(t) \, dt}{p_{ref}^2} + \frac{\int_{22:00}^{07:00} 10 \, p_A^2(t) \, dt}{p_{ref}^2} \right] \quad \text{dB} \quad (1.25)$$

where the first term covers the "daytime" hours from 07:00 to 22:00 and the second term covers the nighttime hours from 22:00 to 07:00. Here, the nighttime noise levels are considered to be 10 dB greater than they actually measure. The A-weighted sound pressure p_A is sampled frequently during measurement.

A-weighted sound exposure $E_{A,T}$ is given by

$$E_{A,T} = \int_{t_1}^{t_2} p_A^2(t)\, dt \quad \text{Pa}^2 \cdot \text{s} \tag{1.26}$$

This equation is not a level. The term $E_{A,T}$ is proportional to the energy flow (intensity times time) in a sound wave in the time period T. The period T starts and stops at t_1 and t_2, respectively.

A-weighted noise exposure level $L_{EA,T}$ is given by

$$L_{EA,T} = 10 \log\left(\frac{E_{A,T}}{E_0}\right) \quad \text{dB} \tag{1.27}$$

where E_0 is a reference quantity, standardized at $(20\ \mu\text{Pa})^2 \cdot \text{s} = (4 \times 10^{-10}\ \text{Pa})^2 \cdot \text{s}$. However, the International Organization for Standardization standard ISO 1999: 1990-01-5, on occupational noise level, uses $E_0 = (1.15 \times 10^{-5}\ \text{Pa})^2 \cdot \text{s}$, because, for an 8-h day, $L_{EA,T}$, with that reference, equals the average A-weighted sound pressure level $L_{A,T}$. The two reference quantities yield levels that differ by 44.6 dB. For a single impulse, the time period T is of no consequence provided T is longer than the impulse length and the background noise is low.

Hearing threshold for setting "zero" at each frequency on a pure-tone audiometer is the standardized, average, pure-tone threshold of hearing for a population of young persons with no otological irregularities. The standardized threshold sound pressure levels at the frequencies 250, 500, 1000, 2000, 3000, 4000, 6000, and 8000 Hz are, respectively, 24.5, 11.0, 6.5, 8.5, 7.5, 9.0, 8.0, and 9.5 dB measured under an earphone. An audiometer is used to determine the difference at these frequencies between the threshold values of a person (the lowest sound pressure level of a pure tone the person can detect consistently) and the standardized threshold values. Measurements are sometimes also made at 125 and 1500 Hz.[7]

Hearing impairment (hearing loss) is the number of decibels that the permanent hearing threshold of an individual at each measured frequency is above the zero setting on an audiometer, in other words, a change for the worse of the person's threshold of hearing compared to the normal for young persons.

Hearing threshold levels associated with age are the standardized pure-tone thresholds of hearing associated solely with age. They were determined from tests made on the hearing of persons in a certain age group in a population with no otological irregularities and no appreciable exposure to noise during their lives.

Hearing threshold levels associated with age and noise are the standardized pure-tone thresholds determined from tests made on the hearing of individuals who had histories of higher than normal noise exposure during their lives. The average noise levels and years of exposure were determined by questioning and measurement of the exposure levels.

Noise-induced permanent threshold shift (NIPTS) is the shift in the hearing threshold level caused solely by exposure to noise.

1.5 REFERENCE QUANTITIES USED IN NOISE AND VIBRATION

American National Standard

The American National Standards Institute has issued a standard (ANSI S1.8-1989, Reaffirmed 2001) on "Reference Quantities for Acoustical Levels." This standard is a revision of ANSI S1.8-1969. The authors of this book have been surveyed for their opinions on preferred reference quantities. Table 1.5 is a combination of the standard references and of references preferred by the authors. The two references are clearly distinguished. All quantities are stated in terms of the International System of units (SI) and in British units.

Relations among Sound Power Levels, Intensity Levels, and Sound Pressure Levels

As a practical matter, the reference quantities for sound power, intensity, and sound pressure (in air) have been chosen so that their corresponding levels are interrelated in a convenient way under certain circumstances.

The threshold of hearing at 1000 Hz for a young listener with acute hearing, measured under laboratory conditions, was determined some years ago as a sound pressure of 2×10^{-5} Pa. This value was then selected as the reference pressure for sound pressure level.

Intensity at a point is related to sound pressure at that point in a free field by Eq. (1.14). A combination of Eqs. (1.4), (1.20), and (1.21) yields the sound intensity level

$$L_I = 10 \log \frac{I}{I_{\text{ref}}} = 10 \log \frac{p^2}{\rho c I_{\text{ref}}}$$

$$= 10 \log \frac{p^2}{p_{\text{ref}}^2} + 10 \log \frac{p_{\text{ref}}^2}{\rho c I_{\text{ref}}}$$

$$L_I = L_p - 10 \log K \quad \text{dB re } 10^{-12} \text{ W/m}^2 \quad (1.28)$$

where $K = \text{const} = I_{\text{ref}} \rho c / p_{\text{ref}}^2$, which is dependent upon ambient pressure and temperature; quantity $10 \log K$ may be found from Fig. 1.7, or,

$K = \rho c / 400$

The quantity $10 \log K$ will equal zero, that is, $K = 1$, when

$$\rho c = \frac{p_{\text{ref}}^2}{I_{\text{ref}}} = \frac{4 \times 10^{-10}}{10^{-12}} = 400 \text{ mks rayls} \quad (1.29)$$

We may also rearrange Eq. (1.28) to give the sound pressure level

$$L_p = L_I + 10 \log K \quad \text{dB re } 2 \times 10^{-5} \text{ Pa} \quad (1.30)$$

TABLE 1.5 Reference Quantities for Acoustical Levels from American National Standard ANSI S1.8-1989 (Reaffirmed 2001) and As Preferred by Authors

Name	Definition	SI	British
		Preferred Reference Quantities	
Sound pressure level (gases)	$L_p = 20 \log_{10}(p/p_0)$ dB	$p_0 = 20~\mu\text{Pa} = 2 \times 10^{-5}~\text{N/m}^2$	$2.90 \times 10^{-9}~\text{lb/in.}^2$
Sound pressure level (other than gases)	$L_p = 20 \log_{10}(p/p_0)$ dB	$p_0 = 1~\mu\text{Pa} = 10^{-6}~\text{N/m}^2$	$1.45 \times 10^{-10}~\text{lb/in.}^2$
Sound power level	$L_W = 10 \log_{10}(W/W_0)$ dB	$W_0 = 1~\text{pW} = 10^{-12}~\text{N} \cdot \text{m/s}$	$8.85 \times 10^{-12}~\text{in.} \cdot \text{lb/s}$
	$L_W = \log_{10}(W/W_0)$ bel	$W_0 = 1~\text{pW} = 10^{-12}~\text{N} \cdot \text{m/s}$	$8.85 \times 10^{-12}~\text{in.} \cdot \text{lb/s}$
Sound intensity level	$L_I = 10 \log_{10}(I/I_0)$ dB	$I_0 = 1~\text{pW/m}^2 = 10^{-12}~\text{N/m} \cdot \text{s}$	$5.71 \times 10^{-15}~\text{lb/in.} \cdot \text{s}$
Vibratory force level	$L_{F0} = 20 \log_{10}(F/F_0)$ dB	$F_0 = 1~\mu\text{N} = 10^{-6}~\text{N}$	$2.25 \times 10^{-7}~\text{lb}$
Frequency level	$N = \log_{10}(f/f_0)$	$f_0 = 1~\text{Hz}$	$1.00~\text{Hz}$
Sound exposure level	$L_E = 10 \log_{10}(E/E_0)$ dB	$E_0 = (20~\mu\text{Pa})^2 \cdot \text{s} = (2 \times 10^{-5}~\text{Pa})^2 \cdot \text{s}$	$8.41 \times 10^{-18}~\text{lb}^2/\text{in.}^4$

The quantities listed below are not officially part of ANSI S1.8. They either are listed there for information or are included here as the authors' choice.

Name	Definition	SI	British
Sound energy level given in ISO 1683:1983	$L_e = 10 \log_{10}(e/e_0)$ dB	$e_0 = 1~\text{pJ} = 10^{-12}~\text{N} \cdot \text{m}$	$8.85 \times 10^{-12}~\text{lb} \cdot \text{in.}$
Sound energy density level given in ISO 1683:1983	$L_D = 10 \log_{10}(D/D_0)$ dB	$D_0 = 1~\text{pJ/m}^3 = 10^{-12}~\text{N/m}^2$	$1.45 \times 10^{-16}~\text{lb/in.}^2$
Vibration acceleration level	$L_a = 20 \log_{10}(a/a_0)$ dB	$a_0 = 10~\mu\text{m/s}^2 = 10^{-5}~\text{m/s}^2$	$3.94 \times 10^{-4}~\text{in./s}^2$
Vibration acceleration level in ISO 16831983	$L_a = 20 \log_{10}(a/a_0)$ dB	$a_0 = 1~\mu\text{m/s}^2 = 10^{-6}~\text{m/s}^2$	$3.94 \times 10^{-5}~\text{in./s}^2$
Vibration velocity level	$L_v = 20 \log_{10}(v/v_0)$ dB	$v_0 = 10~\text{nm/s} = 10^{-8}~\text{m/s}$	$3.94 \times 10^{-7}~\text{in./s}$
Vibration velocity level in ISO 1683:1983	$L_v = 20 \log_{10}(v/v_0)$ dB	$v_0 = 1~\text{nm/s} = 10^{-9}~\text{m/s}$	$3.94 \times 10^{-8}~\text{in./s}$
Vibration displacement level	$L_d = 20 \log_{10}(d/d_0)$ dB	$d_0 = 10~\text{pm} = 10^{-11}~\text{m}$	$3.94 \times 10^{-10}~\text{in.}$

Notes: Decimal multiples and submultiples of SI units are formed as follows: 10^{-1} = deci (d), 10^{-2} = centi (c), 10^{-3} = milli (m), 10^{-6} = micro (μ), 10^{-9} = nano (n), and 10^{-12} = pico (p). Also J = joule = W · s(N · m), N = newton, and Pa = pascal = 1 N/m². Note that 1 lb = 4.448 N.

Although some international standards differ, in this text, to avoid confusion between power and pressure, we have chosen to use W instead of P for power; and to avoid confusion between energy density and voltage, we have chosen D instead of E for energy density. The symbol lb means pound force.

In recent international standardization $\log_{10}$ is written lg and $20 \log_{10}(a/b) = 10 \lg(a^2/b^2)$, i.e., "20" is never used.

REFERENCE QUANTITIES USED IN NOISE AND VIBRATION 21

TABLE 1.6 Ambient Pressures and Temperatures for Which ρc(Air) = 400 mks rayls

	Ambient Pressure		Ambient Temperature T	
p_s, Pa	m of Hg, 0°C	in. of Hg, 0°C	°C	°F
0.7×10^5	0.525	20.68	−124.3	−192
0.8×10^5	0.600	23.63	−78.7	−110
0.9×10^5	0.675	26.58	−27.0	−17
1.0×10^5	0.750	29.54	+30.7	+87
1.013×10^5	0.760	29.9	38.9	102
1.1×10^5	0.825	32.5	94.5	202
1.2×10^5	0.900	35.4	164.4	328
1.3×10^5	0.975	38.4	240.4	465
1.4×10^5	1.050	41.3	322.4	613

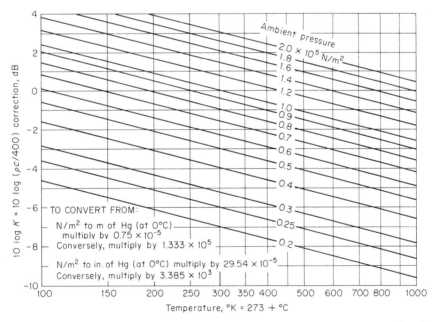

FIGURE 1.7 Chart determining the value of $10 \log(\rho c/400) = 10 \log K$ as a function of ambient temperature and ambient pressure. Values for which $\rho c = 400$ are also given in Table 1.6.

In Table 1.6, we show a range of ambient pressures and temperatures for which $\rho c = 400$ mks rayls. We see that for average atmospheric pressure, namely, 1.013×10^5 Pa, the temperature must equal 38.9°C (102°F) for $\rho c = 400$ mks rayls. However, if $T = 22$°C and $p_s = 1.013 \times 10^5$ Pa², $\rho c \approx 412$. This yields a value of $10 \log(\rho c/400) = 10 \log 1.03 = 0.13$ dB, an amount that is usually not significant in acoustics.

Thus, for most noise measurements, we neglect $10 \log K$ and in a free progressive wave let

$$L_p \approx L_I \tag{1.31}$$

Otherwise, the value of $10 \log K$ is determined from Fig. 1.7 and used in Eq. (1.28) or (1.30).

Under the condition that the *intensity is uniform over an area S*, the sound power and the intensity are related by $W = IS$. Hence, the sound power level is related to the intensity level as follows:

$$10 \log \frac{W}{10^{-12}} = 10 \log \frac{I}{10^{-12}} + 10 \log \frac{S}{S_0}$$

$$L_W = L_I + 10 \log S \quad \text{dB re } 10^{-12} \text{ W} \tag{1.32}$$

where S = area of surface, m^2
$S_0 = 1$ m^2

Obviously, only if the area $S = 1.0$ m^2 will $L_W = L_I$. Also, observe that the relation of Eq. (1.32) is not dependent on temperature or pressure.

1.6 DETERMINATION OF OVERALL LEVELS FROM BAND LEVELS

It is necessary often to convert sound pressure levels measured in a series of contiguous bands into a single-band level encompassing the same frequency range. The level in the all-inclusive band is called the *overall level* L(OA) given by

$$L_p(\text{OA}) = 20 \log \sum_{i=1}^{n} 10^{L_{pi}/20} \quad \text{dB} \tag{1.33}$$

$$L_p(\text{OA}) = 10 \log \sum_{i=1}^{n} 10^{L_{li}/10} \quad \text{dB} \tag{1.34}$$

The conversion can also be accomplished with the aid of Fig. 1.8. Assume that the contiguous band levels are given by the eight numbers across the top of Fig. 1.9. The frequency limits of the bands are not important to the method of calculation as long as the bands are contiguous and cover the frequency range of the overall band. To combine these eight levels into an overall level, start with any two, say, the seventh and eighth bands. From Fig. 1.8 we see that whenever the difference between two band levels, $L_1 - L_2$, is zero, the combined level is 3 dB higher. If the difference is 2 dB (the sixth band level minus the new level of 73 dB), the sum is 2.1 dB greater than the larger (75 + 2.1 dB). This procedure is followed until the overall band level is obtained, here, 102.1 dB.

DETERMINATION OF OVERALL LEVELS FROM BAND LEVELS   23

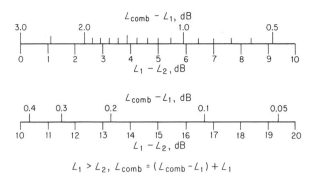

$L_1 > L_2$, $L_{\text{comb}} = (L_{\text{comb}} - L_1) + L_1$

FIGURE 1.8 Nanogram for combining two sound levels L_1 and L_2 (dB). Levels may be power levels, sound pressure levels, or intensity levels. Example: $L_1 = 88$ dB, $L_2 = 85$ dB, $L_1 - L_2 = 3$ dB. Solution: $L_{\text{comb}} = 88 + 1.8 = 89.8$ dB.

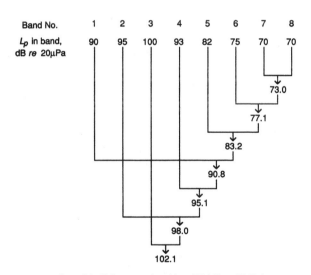

Overall L_p (8 frequency bands) ≈ 102 (dB re 20μPa)

FIGURE 1.9 Determination of an overall sound pressure level from levels in frequency bands (see also Fig. 1.10).

If we set $L_{\text{comb}} - L_1 = A$, then A is the number to be added to L_1 (the larger) to get L_{comb}.

It is instructive to combine the bands in a different way, as is shown in Fig. 1.10. The first four bands are combined first; then the second four bands are combined. The levels of the two wider band levels are then combined. It is seen that the overall level is determined by the first four bands alone. This example points up the fact that characterization of a noise by its overall level may be completely inadequate for some noise control purposes because it may ignore a large

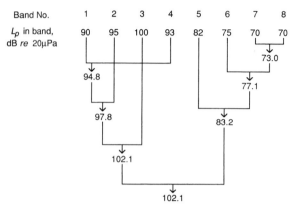

FIGURE 1.10 Alternate determination of an overall sound pressure level from levels in frequency bands (see also Fig. 1.9).

portion of the frequency spectrum. If the data of Fig. 1.10 represented a genuine noise control situation, the 102.1 overall level might be meaningless for some applications. For example, the sound pressure levels in the four highest bands might be the cause of annoyance or interference with speech communication, as is discussed in Chapter 19.

Finally, it should be remembered that in almost all noise control problems, it makes no sense to deal with small fractions of decibels. Rarely does one need a precision of 0.2 dB in measurements, and quite often it is adequate to quote levels to the nearest decibel.

REFERENCES

1. L. L. Beranek, *Acoustics*, Acoustical Society of America, Melville, NY, 1986.
2. M. Moeser, *Engineering Acoustics*, Springer, New York, 2004.
3. D. A. Bies and C. H. Hansen, *Engineering Noise Control*, 2nd ed., E&FN Spon, New York, 2002.
4. P. M. Morse and K. U. Ingard, *Theoretical Acoustics*, Princeton University Press, Princeton, NJ, 1987.
5. L. E. Kinsler, A. R. Frey, A. B. Coppens, J. B. Sanders, and L. Kinsler, *Fundamentals of Acoustics*, 4th ed., Wiley, New York, 1999.
6. L. L. Beranek, *Noise and Vibration Control*, rev. ed., Institute of Noise Control Engineering, Poughkeepsie, NY, 1988.
7. L. L. Beranek, *Acoustical Measurements*, rev. ed., Acoustical Society of America, Melville, NY, 1988.
8. Charles M. Salter Associates, *Acoustics*, William Stout, San Francisco, 1998.

CHAPTER 2

Waves and Impedances

LEO L. BERANEK

Consultant
Cambridge, Massachusetts

2.1 THE WAVE EQUATION[1]

Sound waves must obey the laws of physics. For gases these include Newton's second law of motion, the gas law, and the law of conservation of mass. Combined, these equations produce the wave equation that governs the behavior of sound waves regardless of the surroundings in which they occur.

Equation of Motion

The equation of motion (also called the force equation) is obtained by the application of Newton's second law to a small volume of gas in a homogeneous medium. Imagine the small volume of gas to be enclosed in a packet with weightless flexible sides and assume that there is negligible drag (friction) between particles inside and outside the packet. Suppose that this small volume exists in a part of the medium where the sound pressure p [actually $p(t)$] increases at the space rate of

$$\text{grad } p = \mathbf{i}\frac{\partial p}{\partial x} + \mathbf{j}\frac{\partial p}{\partial y} + \mathbf{k}\frac{\partial p}{\partial z} \tag{2.1}$$

where $\mathbf{i}$, $\mathbf{j}$, and $\mathbf{k}$ are unit vectors in the x, y, and z directions, respectively. Obviously, **grad** p is a vector quantity.

The difference between the forces acting on the sides of the packet is a force $\mathbf{f}$ equal to the rate at which the force changes with distance times the incremental dimensions of the box:

$$\mathbf{f} = -\left[\mathbf{i}\left(\frac{\partial p}{\partial x}\Delta x\right)\Delta y\ \Delta z + \mathbf{j}\left(\frac{\partial p}{\partial y}\Delta y\right)\Delta x\ \Delta z + \mathbf{k}\left(\frac{\partial p}{\partial z}\Delta z\right)\Delta x\ \Delta y\right] \tag{2.2}$$

Note that a positive gradient causes the packet to accelerate in the negative direction.

Division of both sides of the equation by $\Delta x \, \Delta y \, \Delta z = V$ gives the force per unit volume acting to accelerate the box,

$$\frac{\mathbf{f}}{V} = -\text{grad } p \tag{2.3}$$

By Newton's law, the force per unit volume of Eq. (2.3) equals the time derivative of the momentum per unit volume of the box. Because the box is a deformable packet, the mass inside is constant. Hence,

$$\frac{\mathbf{f}}{V} = -\text{grad } p = \frac{M}{V}\frac{D\mathbf{q}}{Dt} = \rho'\frac{D\mathbf{q}}{Dt} \tag{2.4}$$

where $\mathbf{q}$ is the average vector velocity of the gas in the packet, ρ' is the average density of the gas in the packet, and $M = \rho'V$ is the total mass of the gas in the packet.

The partial derivative D/Dt is not a simple one but represents the total rate of the change of velocity of the particular bit of gas in the packet regardless of its position that is, because its position changes when a sound wave hits it:

$$\frac{D\mathbf{q}}{Dt} = \frac{\partial \mathbf{q}}{\partial t} + q_x \frac{\partial \mathbf{q}}{\partial x} + q_y \frac{\partial \mathbf{q}}{\partial y} + q_z \frac{\partial \mathbf{q}}{\partial z} \tag{2.5}$$

where q_x, q_y, and q_z are the components of the vector particle velocity $\mathbf{q}$.

If $\mathbf{q}$ is small enough, the rate of change of momentum of the particles in the box can be approximated by the rate of change of momentum at a fixed point $D\mathbf{q}/Dt \doteq \partial \mathbf{q}/\partial t$, and the instantaneous density ρ' can be approximated by the average density ρ. Then

$$-\text{grad } p = \rho \frac{\partial \mathbf{q}}{\partial t} \tag{2.6}$$

Gas Law

At audible frequencies, the wavelength of a sound wave is long compared to the spacing between air molecules, so that expansions and contractions at two different parts of the medium occur so rapidly that there is no time for heat exchange between points of differing instantaneous pressures. Hence, the compressions and expansions are adiabatic. From elementary thermodynamics,

$$PV^\gamma = \text{const} \tag{2.7}$$

where γ for air, hydrogen, oxygen, and nitrogen equals 1.4. If we let $P = P_s + p$ and $V = V_s + \tau$, where P_s and V_s are the undisturbed pressure and volume of the packet, we get, for small values of incremental pressure p and incremental volume τ,

$$\frac{p}{P_s} = \frac{\gamma \tau}{V_s} \tag{2.8}$$

The time derivative of Eq. (2.8) yields

$$\frac{1}{P_s}\frac{\partial p}{\partial t} = -\frac{\gamma}{V_s}\frac{\partial \tau}{\partial t} \tag{2.9}$$

Continuity Equation

The continuity equation is a statement that the mass of the gas in the deformable packet is constant. Thus the change in the incremental volume τ depends only on the divergence of the vector displacement ξ:

$$\tau = V_s \text{ div } \xi \tag{2.10}$$

or

$$\frac{\partial \tau}{\partial t} = V_s \text{ div } \mathbf{q} \tag{2.11}$$

where $\mathbf{q}$ is the instantaneous (vector) particle velocity.

Wave Equation in Rectangular Coordinates

The three-dimensional wave equation is given by combining Eqs. (2.6), (2.9), and (2.11) and setting

$$c^2 = \frac{\gamma P_s}{\rho} \tag{2.12}$$

which yields

$$\nabla^2 p = \frac{1}{c^2}\frac{\partial^2 p}{\partial t^2} \tag{2.13}$$

where

$$\nabla^2 p = \frac{\partial^2 p}{\partial x^2} + \frac{\partial^2 p}{\partial y^2} + \frac{\partial^2 p}{\partial z^2} \tag{2.14}$$

The one-dimensional wave equation is simply

$$\frac{\partial^2 p}{\partial x^2} = \frac{1}{c^2}\frac{\partial^2 p}{\partial t^2} \tag{2.15}$$

We could also have eliminated p in the combination of the three equations and retained $\mathbf{q}$, in which case we would have had

$$\nabla^2 \mathbf{q} = \frac{1}{c^2}\frac{\partial^2 \mathbf{q}}{\partial t^2} \tag{2.16}$$

2.2 SOLUTIONS TO THE ONE-DIMENSIONAL WAVE EQUATION

General Solution

The general solution to Eq. (2.15) is the sum of two terms,

$$p(x,t) = f_1\left(t - \frac{x}{c}\right) + f_2\left(t + \frac{x}{c}\right) \quad \text{Pa} \tag{2.17}$$

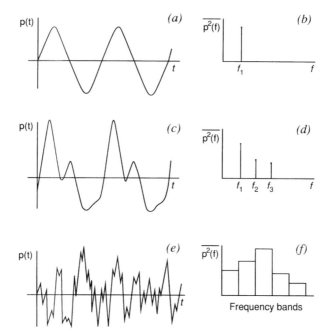

FIGURE 2.1 (*a, c, e*) Forms of time function f_1 or f_2 of Eq. (2.17). Corresponding spectra are shown on the right. (*b, d*) Line spectra. (*f*) Complex spectrum (very large number of tones in each band).

where f_1 and f_2 are arbitrary functions. As we shall illustrate shortly, the first term represents an outgoing wave and the second term a backward-traveling wave. The functions f_1 and f_2 represent the shapes of the two sound waves being propagated. Examples of typical time histories and spectra of $p(t)$ at a fixed location are given in Fig. 2.1. We also recognize c as the speed of sound in air.

Outwardly Traveling Plane Wave

An apparatus for producing an outward-traveling plane wave is shown in Fig. 2.2. A piston at the left moving sinusoidally generates a sound wave that travels outward in the positive x direction and becomes absorbed in the anechoic termination so that no reflected wave exists. Equation (2.17) becomes

$$p(x, t) = f_1\left(t - \frac{x}{c}\right) = P_R \cos k(x - ct) \quad \text{Pa} \quad (2.18)$$

where P_R is the peak amplitude of the sound pressure.

Let us choose the space and the time origins, as shown by the left-hand sine wave in Fig. 2.2, so that P_R has its maximum value at $x = 0$ and $t = 0$. After a time t_1, the wave will have traveled a distance $x_1 = ct_1$. Similarly for $x_2 = 2x_1 = 2ct_1$.

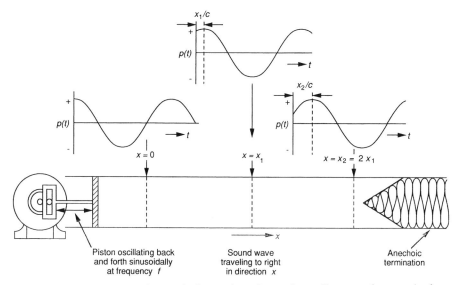

FIGURE 2.2 Apparatus for producing a plane forward-traveling sound wave. A plane wave generated by the piston at the left travels to the right and is absorbed by the anechoic termination. The three waves at the top give the variation in sound pressure with time at the three points indicated, $x = 0$, $x = x_1$, and $x = x_2 = 2x_1$.

Figure 2.3 shows a set of four spatial timeshots taken at $t = 0$, $\frac{1}{4}T$, $\frac{1}{2}T$, $\frac{3}{4}T$, where T is the time period of the piston machine. Each shows the sound pressure over a spatial extent of one wavelength, $\lambda = c/f = cT$. The 20 vertical lines along each snapshot enable one to observe the spatial variation of sound pressure at a given point at the four different times.

Snapshot (*a*) represents the pressure-versus-distance relation for times $t = 0, T, 2T, 3T, \ldots, nT$. The maximum value $+P_R$ exists at $x = 0$. Because the wave is periodic in space, the maximum value must also occur at $x = \lambda, 2\lambda, 3\lambda, \ldots$.

Snapshot (*b*) shows the sound pressure a quarter of a period, $\frac{1}{4}T$, later, that is, the wave in (*a*) has moved to the right a distance equal to $\frac{1}{4}\lambda$ to become the wave in (*b*). Similarly for (*c*) and (*d*). To convince yourself that the wave is traveling to the right, allow your eyes to jump successively from (*a*) to (*b*) to (*c*) to (*d*) and note that the peak, $+P_R$, moves successively to the right.

Wavenumber. The cosine function of Eq. (2.18) repeats its value every time the argument increases 2π radians (360°). From the definition of wavelength, $\lambda = c/f = cT$, we can write this periodicity condition as

$$\cos[k(x + \lambda - ct)] = \cos[k(x - ct) + 2\pi] \qquad (2.19)$$

so that $k\lambda = 2\pi$ and $k = 2\pi/\lambda$ radians per meter. We see that the meaning of the parameter k, called the *wavenumber*, is a kind of "spatial frequency."

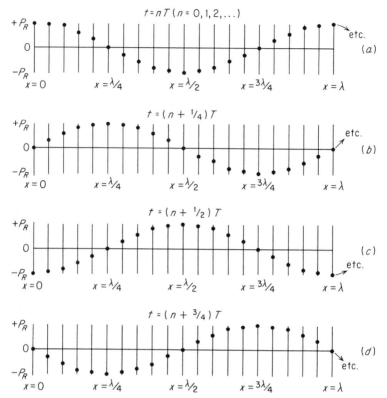

FIGURE 2.3 Graphs showing sound pressure in a plane free-progressive wave traveling from left to right at 20 equally spaced axial locations at four instants of time t. The wave is produced by a source at the left and travels to the right with the speed c. The length of time it takes a wave to travel a distance equal to a wavelength is called the period T. Forward-traveling wave: $p(x, t) = P_R \cos k(x - ct)$; $k = 2\pi/\lambda = 2\pi/(cT) = 2\pi f/c = \omega/c$.

The argument of the cosine in Eq. (2.18) may be written in any one of the following ways:

$$k(x - ct) = \frac{2\pi}{\lambda}(x - ct) = 2\pi f\left(\frac{x}{c} - t\right) = 2\pi \left(\frac{x}{\lambda} - \frac{t}{T}\right)$$

$$= \frac{2\pi x}{\lambda} - 2\pi f t = kx - \omega t$$

From Eq. (2.19) we can write the equations for the snapshots of Fig. 2.3 as

(a) $\quad t = nT \quad\quad p = P_R \cos \dfrac{2\pi x}{\lambda}$

(b) $\quad t = \left(n + \tfrac{1}{4}\right)T \quad\quad p = P_R \cos\left(\dfrac{2\pi x}{\lambda} - \dfrac{\pi}{2}\right)$

(c) $t = \left(n + \tfrac{1}{2}\right)T$ $p = P_R \cos\left(\dfrac{2\pi x}{\lambda} - \pi\right)$

(d) $t = \left(n + \tfrac{3}{4}\right)T$ $p = P_R \cos\left(\dfrac{2\pi x}{\lambda} - \dfrac{3\pi}{2}\right)$

where $n = 0, 1, 2, 3, \ldots$.

Root-Mean-Square Sound Pressure

As discussed in Chapter 1, a measure of the strength of a sound wave is needed that avoids the problem of directly averaging over a cycle. Such an average is zero. The measure that has been standardized is the rms sound pressure p_{rms}. Its magnitude is 0.707 times the peak value P_R [Eq. (2.18)]. The rms value is also called the *effective value*.

Particle Velocity

From Eq. (2.6) we may derive a relation between sound pressure p and particle velocity u, where u is the component of $\mathbf{q}$ in the x direction. In one dimension,

$$-\frac{\partial p}{\partial x} = \rho \frac{\partial u}{\partial t} \qquad (2.20)$$

Substitution of Eq. (2.18) into Eq. (2.20) gives

$$-u = \frac{1}{\rho}\int \frac{\partial p}{\partial x}\,dt = \frac{-P_R}{\rho c}\cos\, k(x - ct) \qquad (2.20\text{a})$$

or

$$u = \frac{p}{\rho c} \qquad (2.21)$$

where $\rho =$ time-averaged density of air, $= 1.18$ kg/m^3 for normal room temperature $T = 22°C$ (71.6°F) and atmospheric pressure $P_s = 0.751$ m (29.6 in.) Hg

$c =$ speed of sound [see Eqs. (1.5)–(1.8)], which at normal temperatures of 22°C equals 344 m/s (1129 ft/s)

$\rho c = 406$ mks rayls (N · s/m^3) at normal room temperature and pressure; at other T's and P_s's, ρc is found from Fig. 1.7

Intensity

A freely traveling progressive sound wave transmits energy. We define this energy transfer as the *intensity* **I**, the energy that flows through a unit area in unit time. The units are watts per square meter (N/m · s). It has its maximum value, I_{max}, when the plane of the unit area is perpendicular to the direction in which the wave is traveling. Intensity, analogous to electrical power, equals the time average of the product of sound pressure and particle velocity,

$$I_{max} = \overline{p \cdot u} \quad \text{W/m}^2 \ (\text{N/m} \cdot \text{s}) \tag{2.22}$$

For the wave of Figs. 2.2 and 2.3,

$$I_{max} = \lim_{T \to \infty} \frac{1}{T} \int_0^T \frac{P_R^2}{\rho c} \cos^2 k(x - ct) \, dt \tag{2.23}$$

where $I(\theta) = I_{max} \cos \theta$, θ being the angle between the direction of travel of the wave and a line perpendicular to the plane of the unit area through which the flow of sound power is being determined.

Let $T = \infty$ be a time long enough that I_{max} has reached its asymptotic value within experimental error.

Because the time average of the cosine is zero,

$$I_{max} = \frac{P_R^2}{2\rho c} = \frac{p_{rms}^2}{\rho c} \quad \text{W/m}^2 \tag{2.24}$$

where p_{rms} is the square root of the mean (time) square value of $p(t)$, as can be demonstrated by finding $I\rho c$ from Eq. (2.24) and W is watts.

Backward-Traveling Plane Wave

A backward-traveling plane wave may be produced by interchanging the source and termination of Fig. 2.2. The wave now travels in the $-x$ direction and is described by

$$p(x, t) = P_L \cos k(x + ct) \quad \text{Pa} \tag{2.25}$$

Comparison of Eqs. (2.18) and (2.25) show that, if the two variables x and ct are separated by a negative sign, the wave travels in the positive direction, and if the two variables are separated by a positive sign, the direction reverses.

The four "snapshots" of Fig. 2.4 illustrate the backward-traveling wave. Allowing your eyes to jump from (a) to (b) to (c) to (d) and following the movement of $+P_L$ from right to left convinces one that this is true.

One-Dimensional Spherical Wave

Sound Pressure. The equation for the sound pressure associated with a free-progressive, spherically traveling sound wave, produced as shown in Fig. 1.2 in Chapter 1, is

$$p(r, t) = \frac{A}{r} \cos k(r - ct) \quad \text{Pa} \tag{2.26}$$

where A is an amplitude factor with dimension newtons per meter. Because the sign between r and ct is negative, the wave is traveling outward in the positive

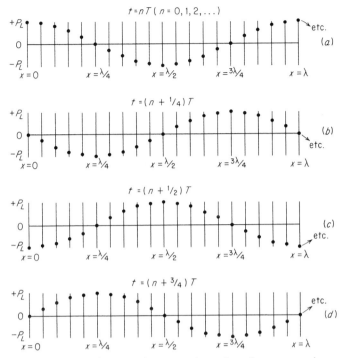

FIGURE 2.4 Graphs showing sound pressure in a plane free-progressive wave traveling from right to left at 20 equally spaced axial locations at four instants of time t. The wave is produced by a source at the right (or by being reflected from a boundary at the right) and travels to the left with a speed c. The period T is defined as for Fig. 2.3. Backward-traveling wave: $p(x,t) = P_L \cos[k(x+ct)]$; $k = 2\pi/\lambda = 2\pi/(cT) = 2\pi f/c = \omega/c$.

r direction. In a spherical wave, the pressure amplitude is inversely proportional to the radial distance r.

Particle Velocity. The particle velocity for a spherical wave, using Eq. (2.20a) with r substituted for x, is

$$u(r,t) = -\frac{1}{\rho} \int \left[\frac{A}{r} k \sin k(r-ct) + \frac{A}{r^2} \cos k(r-ct) \right] dt$$

$$= -\frac{1}{\rho} \left[\frac{-kA}{kcr} \cos k(r-ct) - \frac{A}{r^2 kc} \sin k(r-ct) \right]$$

or

$$u(r,t) = \frac{A}{\rho c r} \cos k(r-ct) \left[1 + \frac{1}{kr} \tan k(r-ct) \right] \quad (2.27)$$

For large values of kr,

$$u(r,t) \doteq \frac{p(r,t)}{\rho c} \qquad k^2 r^2 \gg 1 \quad (2.28)$$

34 WAVES AND IMPEDANCES

For very small values of kr, $k^2r^2 \ll 1$,

$$u(r,t) \doteq \frac{A}{k\rho cr^2} \sin k(r-ct)$$

$$\doteq \frac{p(r,t)}{\rho ckr} \angle 90° \text{ re phase of } p(r,t) \quad (2.29)$$

Equations (2.27) and (2.29) show that as one approaches the center of a spherical source, the sound pressure and particle velocity become progressively more out of phase, approaching 90° in the limit.

Intensity. For a freely traveling spherical wave, Eq. (2.26) states that for *all values of r*, the sound pressure varies as $1/r$. Because $u = p/\rho c$ [Eq. (2.28)] for the case $k^2 r^2 \gg 1$, we can write for all values of r that

$$I_{max} \text{ at radius } r = \overline{p \cdot u} = \frac{p_{rms}^2 \text{ at } r}{\rho c} \quad \text{W/m}^2 \quad (2.30)$$

2.3 SOUND POWER OUTPUT OF ELEMENTARY RADIATORS

Monopole (Radiating Sphere)

The total power W_M radiated from a simple source, a pulsating sphere, called a monopole, is given in Table 2.1, where

where $\hat{Q}_s$ = source strength, $= 4\pi a^2 \hat{v}_r$, m³/s
$\hat{v}_r$ = peak value of velocity of sinusoidally pulsating surface, m/s
a = radius of pulsating sphere, m
$k = 2\pi f/c$, m⁻¹
ρc = characteristic impedance of gas, 406 mks rayls (N · s/m³) for air at normal room temperature and atmospheric pressure

Dipole (Two Closely Spaced Monopoles)[2]

By definition, two monopoles constitute a dipole when $(kd)^2 \ll 1$ and when they vibrate 180° out of phase. A dipole has a figure-eight radiation pattern, with minimum radiation in the direction perpendicular to a line connecting the two monopoles. The total sound power, W_D, radiated is given in the second row of Table 2.1, where $\hat{Q}_s$, $\hat{v}_r$, a, k, and ρc are as given above and

where d = separation between monopoles, m

Oscillating Sphere[3,4]

An oscillating sphere is defined as a rigid sphere moving axially, back and forth, around its rest position. The total power, W_{OS}, radiated is given in the third row of Table 2.1, ρc and k are as defined for a monopole, and

where $\hat{v}_x$ = peak back-and-forth velocity, m/s
a = radius of oscillating sphere, m

TABLE 2.1 Sound Power Output of Elementary Radiators

Source Type	Source Behavior 180° Phase Difference	Sound Power Output, W	Auxiliary Expressions
Monopole		$W_M = \dfrac{\rho c k^2}{8\pi(1 + k^2 a^2)} \hat{Q}_s^2$	$\hat{Q}_s = 4\pi a^2 \hat{v}_r$
Dipole		$W_D = \dfrac{\rho c k^4 d^2}{12\pi} \hat{Q}_s^2$	d = distance between monopoles
Oscillating sphere		$W_{OS} = \dfrac{2\pi \rho c k^4 a^6}{3(4 + k^4 a^4)} \hat{v}^2$	$\hat{v}_x$ = peak sinusoidal vibration velocity
Baffled piston		$W_{BP} = \dfrac{\rho c k^2}{4\pi} \hat{Q}^2 \; ka \ll 1$ $W_{BP} = \dfrac{\rho c}{2\pi a^2} \hat{Q}^2 \; ka \gg 1$	$\hat{Q} = \pi a^2 \hat{v}_x$ where $\hat{v}_x$ is as above

Baffled Piston[5]

An axially vibrating diaphragm in an infinite plate is called a baffled piston. The total power radiated to one side of the rigid wall, W_{BP}, is given in the fourth row of Table 2.1, both for the case of a piston whose radius is small compared to a wavelength, $ka \ll 1$, and vice versa, $ka \gg 1$, where

where $\hat{v}_x$ = peak axial velocity of piston, m/s
        a = radius of piston, m

The other quantities are as defined above.

2.4 INTERFERENCE AND RESONANCE

The sound pressure and incremental density in a sound wave are generally very small in comparison with the equilibrium values on which they are superposed. This is certainly true for speech and music waves. As a result, it is possible in such acoustical situations to determine the effect of two sound waves in the same space by simple linear addition of the effects of each sound wave separately. This is a statement of the principle of *superposition*.

In the previous section we presented a series of spatial snapshots for plane sound waves traveling to the right (Fig. 2.3) and to the left (Fig. 2.4). According to the principle of superposition, the effect of the sum of these two waves will be the sum of their effects, which we can see graphically by adding Figs. 2.3 and 2.4. The result is shown in Fig. 2.5, where we have set the amplitude of the forward-traveling wave P_R equal to the amplitude of the backward-traveling wave P_L.

The interference of the two waves has produced a surprising change. No longer does the sound pressure at one place occur to the right or to the left of that place at the next instant. The wave no longer travels; it is a standing wave. We see that at each point in space the sound pressure varies sinusoidally with time, except at the points $x = \frac{1}{4}\lambda$ and $\frac{3}{4}\lambda$, where the pressure is always zero. The maximum value of the pressure variation at different points is different, being greatest at $x = 0$, $x = \frac{1}{2}\lambda$, and $x = \lambda$. The sound pressures at the points between the points $x = \frac{1}{4}\lambda$ and $x = \frac{3}{4}\lambda$ always vary together, that is, increase or decrease in phase. At the same times the sound pressures for the points to the left of $x = \frac{1}{4}\lambda$ and to the right of $x = \frac{3}{4}\lambda$ decrease or increase together (in phase). Thus all pressures are in time phase in the standing wave, but there is a space difference of phase of 180° between the sound pressures at the points at $x = 0$ and $x = \frac{1}{2}\lambda$.

Remembering that $P_R = P_L = P$, that is, that the amplitude of the wave traveling to the right is equal to the amplitude of the wave traveling to the left, we find that the sum of the two waves is

$$p(x,t) = P \cos[k(x - ct)] + P \cos[k(x + ct)]$$
$$= 2P(\cos kx)(\cos 2\pi ft) \quad \text{Pa} \qquad (2.31)$$

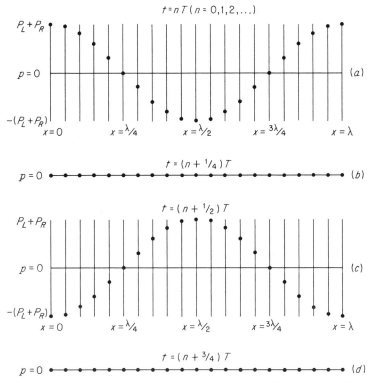

FIGURE 2.5 Graphs giving sound pressure in a plane standing wave at 20 places in space at four instants of time t. The wave is produced by two sources equal in strength at the right and left of the graphs (or the right source is a perfectly reflecting boundary that sends back a wave of equal amplitude to that produced by the source on the left). Standing wave: $p(x, t) = 2P(\cos kx)(\cos 2\pi f t)$, where k is the wavenumber.

From Eqs. (2.18), (2.20), and (2.31) we see very clearly the differences between a standing and a traveling wave. In a traveling wave distance x and time t occur as a sum or difference in the argument of the cosine. Hence, for the traveling wave, by adjusting both time and distance (according to the speed of sound) in the argument of the cosine, we can always keep the argument and thus the magnitude of the cosine the same. In Eq. (2.31) distance and time no longer appear together in the argument of a single cosine. So the same sound pressure cannot occur at an adjacent point in the space at a later time.

Standing waves will exist in any regular enclosure. In a rectangular room, for example, three classes of standing waves may exist (see Chapter 6). One class includes all waves that are perpendicular to one pair of opposing walls, that is, that travel at grazing incidence to two pairs of walls, the $(n_x, 0, 0)$, $(0, n_y, 0)$, $(0, 0, n_z)$ modes of vibration. A second class travels at grazing incidence to only one pair of walls, the $(n_x, n_y, 0)$, $(n_x, 0, n_z)$, $(0, n_y, n_z)$ modes of vibration. A third class involves all walls at oblique angles of incidence, the (n_x, n_y, n_z) modes of vibration. Each free-standing wave in an acoustical space is called a normal mode of vibration, or simply a resonance. The frequencies at which resonant standing waves can exist are related to the separation between the reflecting surfaces. For example, the lowest frequency for a resonant standing wave in a one-dimensional system consisting of two rigid parallel walls is given by

$$f = \frac{c}{2d} \text{ Hz} \tag{2.32}$$

where f = lowest frequency for resonant standing wave, Hz
c = speed of sound, m/s
d = distance separating two reflecting surfaces, m

Resonant standing waves can also exist at every integral multiple of this frequency. That is to say,

$$f = \frac{nc}{2d} \text{ Hz} \tag{2.33}$$

where n is an integer $1, 2, 3, \ldots$.

2.5 IMPEDANCE AND ADMITTANCE

Reference to Eqs. (2.18) and (2.21) reveals that the magnitudes of sound pressure and particle velocity are directly proportional to each other. Also, in the special case of plane-wave sound propagation, the time dependence of sound pressure is exactly the same as the time dependence of particle velocity, and at any point in the wave there is no phase difference between the two quantities. Thus, in a plane sound wave the ratio of sound pressure to particle velocity at all instants of time is a constant equal to ρc.

In general, however, for linear (small-signal) acoustical phenomena in the steady state, there is a difference in the time functions of sound pressure and particle velocity, leading to a phase difference of one relative to the other. Thus, at any point the particle velocity may lead or lag the sound pressure. In many situations, both the ratio of the magnitudes and the relative phase may be functions of frequency.

In several of the chapters that follow, it is convenient in acoustical design to avoid separate consideration of steady-state sound pressure and steady-state particle velocity (or other quantities that may be derived from them, such as force and volume velocity) and instead to deal with either one and with their complex ratio, as defined below.

Complex Notation

The foundations for the designation of steady-state signals with the same frequency but different phases by complex notation are expressed in the identities

$$|A|\cos(\omega t + \theta_1) \equiv \text{Re } \overline{A}e^{j\omega t} \tag{2.34}$$

where $|A| \equiv$ amplitude of cosine function
$\theta_1 \equiv$ phase shift at time $t = 0$
Re $\equiv$ "real part of"
$j \equiv \sqrt{-1}$

and

$$\overline{A} \equiv A_{\text{Re}} + jA_{\text{Im}} = |A|\ e^{j\theta_1} \tag{2.35}$$

$$|A| \equiv \sqrt{A_{\text{Re}}^2 + A_{\text{Im}}^2} \tag{2.36}$$

$$\theta_1 \equiv \tan^{-1}\left(\frac{A_{\text{Im}}}{A_{\text{Re}}}\right) \tag{2.37}$$

We note also that

$$e^{j\theta} \equiv \cos\theta + j\sin\theta \tag{2.38}$$

so that

$$A_{\text{Re}} \equiv |A|\cos\theta \tag{2.39}$$

$$A_{\text{Im}} \equiv |A|\sin\theta \tag{2.40}$$

These equations say that a cosinusoidal time-varying function, given by the left-hand side of Eq. (2.34), can be represented by the real-axis projection of a vector of magnitude $|A|$ given by Eq. (2.36), rotating at a rate ω radians per second. The angle θ_1 is the angle of the vector (in radians) relative to the positive real axis at the instant of time $t = 0$.

We might therefore express a time-varying steady-state sound pressure or force by

$$\overline{A}e^{j\omega t} = |A|\ e^{j\omega t}e^{j\theta_1} \tag{2.41}$$

Also, a time-varying steady-state velocity or volume velocity might be expressed by

$$\overline{B}e^{j\omega t} = |B|\ e^{j\omega t}e^{j\theta_2} \tag{2.42}$$

Definitions of Complex Impedance

Complex Impedance Z. In general, complex impedance is defined as

$$\overline{Z} \equiv \frac{\overline{A}}{\overline{B}} = \frac{|A|\ e^{j(\omega t+\theta_1)}}{|B|\ e^{j(\omega t+\theta_2)}} = \frac{|A|\ e^{j\theta_1}}{|B|\ e^{j\theta_2}} = |Z|\ e^{j\theta} \tag{2.43}$$

and

$$\overline{Z} \equiv R + jX = \sqrt{R^2 + X^2}\ e^{j\theta} = |Z|\ e^{j\theta} \tag{2.44}$$

where $\overline{Z}$ = complex impedance as given above
A = steady-state pressure or force

B = steady-state velocity or volume velocity
$|Z|$ = magnitude of complex impedance
θ = phase angle between the time functions A and $B, = \theta_1 - \theta_2$
R, X = real and imaginary parts, respectively, of complex impedance $\overline{Z}$
R = resistance, Re Z
X = reactance, Im $\overline{Z}$

Often $|A|$ and $|B|$ are taken to be the rms values of the phenomena they represent, although, if so taken, a factor of $\sqrt{2}$ must be added to both sides of Eq. (2.34) to make them correct in a physical sense. Whether amplitudes or rms values are used makes no difference in the impedance ratio.

Complex impedances are of several types, according to the quantities involved in the ratios. Types common in acoustics are given below.

Acoustical Impedance Z_A. The acoustical impedance at a given surface S is defined as the complex ratio of (a) sound pressure averaged over the surface to (b) volume velocity through it. The volume velocity $U = uS$. The surface may be either a hypothetical surface in an acoustical medium or the moving surface of a mechanical device. The unit is N · s/m^5, also called the mks acoustical ohm. That is,

$$Z_A = \frac{p}{U} \quad \text{N · s/m}^5 \text{ (mks acoustical ohms)} \tag{2.45}$$

Specific Acoustical Impedance Z_s. The specific acoustical impedance is the complex ratio of the sound pressure at a point of an acoustical medium or mechanical device to the particle velocity at that point. The unit is N · s/m^3, also called the mks rayl. That is,

$$Z_s = \frac{p}{u} \quad \text{N · s/m}^3 \text{ (mks rayls)} \tag{2.46}$$

Mechanical Impedance Z_M. The mechanical impedance is the complex ratio of the force acting on a specific area of an acoustical medium or mechanical device to the resulting linear velocity through or of that area, respectively. The unit is the N · s/m, also called the mks mechanical ohm. That is,

$$Z_M = \frac{f}{u} \quad \text{N · s/m (mks mechanical ohms)} \tag{2.47}$$

Characteristic Resistance ρc. The characteristic resistance is the ratio of the sound pressure at a given point to the particle velocity at that point in a free, plane, progressive sound wave. It is equal to the product of the density of the medium and the speed of sound in the medium (ρc). It is analogous to the characteristic impedance of an infinitely long, dissipationless transmission line. The unit is the N · s/m^3, also called the mks rayl. In the solution of problems in this book we shall assume for air that $\rho c = 406$ mks rayls, which is valid for a temperature of 22°C (71.6°F) and a barometric pressure of 0.751 m (29.6 in.) Hg.

Normal Specific Acoustical Impedance Z_{sn}. At the boundary between air and a denser medium (such as a porous acoustical material) we find a further definition necessary, as follows: When an alternating sound pressure p is produced at the surface of an acoustical material, an alternating velocity u of the air particles is produced through the surface. The to-and-fro motions of the air particles may be at any angle relative to the surface. The angle depends both on the angle of incidence of the sound wave and on the nature of the acoustical material. For example, if the material is porous and has very low density, the particle velocity at the surface is nearly in the same direction as that in which the wave is propagating. By contrast, if the surface were a large number of small-diameter tubes packed side by side and oriented perpendicular to the surface, the particle velocity would necessarily be only perpendicular to the surface. In general, the direction of the particle velocity at the surface has both a normal (perpendicular) component and a tangential component.

The normal specific acoustical impedance (sometimes called the unit-area acoustical impedance) is defined as the complex ratio of the sound pressure p to the *normal component* of the particle velocity u_n at a plane, in this example at the surface of the acoustical material. Thus

$$Z_{sn} = \frac{p}{u_n} \quad \text{N} \cdot \text{s/m}^3 \text{ (mks rayls)} \tag{2.48}$$

Definition of Complex Admittance

Complex admittance is the reciprocal of complex impedance. In all ways, it is handled by the same set of rules as given by Eqs. (2.34)–(2.44). Thus, the complex admittance corresponding to the complex impedance of Eq. (2.43) is

$$\overline{Y} \equiv \frac{\overline{B}}{\overline{A}} = \frac{|B| \, e^{j\theta_2}}{|A| \, e^{j\theta_1}} = |Y| \, e^{j\phi} \tag{2.49}$$

where $|Y| = 1/|Z|$
$\phi = -\theta$

The choice between impedance and admittance is sometimes made according to whether $|A|$ or $|B|$ is held constant during a measurement. Thus, if $|B|$ is held constant, $|Z|$ is directly proportional to $|A|$ and is used. If $|A|$ is held constant, $|Y|$ is directly proportional to $|B|$ and is used.

REFERENCES

1. W. J. Cunningham, "Application of Vector Analysis to the Wave Equation," *J. Acoust. Soc. Am.*, **22**, 61 (1950); R. V. L. Hartley, "Note on 'Application of Vector Analysis to the Wave Equation,' " *J. Acoust. Soc. Am.*, **22**, 511 (1950).
2. H. Kuttruff, *Room Acoustics*, Halstead, New York, 2000.

3. M. Moeser, *Engineering Acoustics*, Springer-Verlag, New York, 2004.
4. L. Cremer, *Vorlesungen uber Technische Akustik*, Springer-Verlag, New York, 1971, Section 2.5.2.
5. L. L. Beranek, *Acoustics*, Acoustical Society of America, Melville, NY, 1986, Sections 4.3 and 7.5.

CHAPTER 3

Data Analysis

ALLAN G. PIERSOL

Piersol Engineering Company
Woodland Hills, California

For studies related to noise and vibration control engineering, the analysis of measured acoustical noise and/or vibration data may be accomplished with a number of goals in mind. The most important of these goals can be divided into four broad categories: (a) an assessment of the severity of an environment, (b) the identification of system response properties, (c) the identification of sources, and (d) the identification of transmission paths. The first goal is commonly accomplished using one-third-octave-band-level calculations, or perhaps frequency-weighted overall level measurements, as described in Chapter 1. The other goals often require more advanced data analysis procedures. Following a brief discussion of the general types of acoustical and vibration data of common interest, the most important data analysis procedures for accomplishing goals (b)–(d) are outlined, and important applications of the results from such analyses are summarized. Because of the broad and intricate nature of the subject, heavy use of references is employed to cover details.

3.1 TYPES OF DATA SIGNALS

Acoustical and vibration data are commonly acquired in the form of analog time history signals produced by appropriate transducers (details of acoustical and vibration transducers and signal conditioning equipment are available from the data acquisition documents listed in the Bibliography and the literature published by acoustical and vibration measurement system manufacturers). The signals are generally produced with the units of volts but can be calibrated into appropriate engineering units (g, m/s, Pa, etc.) as required. From a data analysis viewpoint, it is convenient to divide these time history signals into two broad categories, each with two subcategories, as follows:

1. Deterministic data signals: (a) steady-state signals; (b) transient signals.
2. Random data signals: (a) stationary signals; (b) nonstationary signals.

Deterministic Data

Deterministic data signals are those for which it is theoretically feasible to determine a mathematical equation that would predict future time history values of the signal (within reasonable experimental error), based upon a knowledge of the applicable physics or past observations of the signal. The most common type of deterministic signal, called a periodic signal, has a time history $x(t)$ that exactly repeats itself after a constant time interval T_p, called the period of the signal; that is,

$$x(t) = x(t \pm T_p) \tag{3.1}$$

The most common sources of periodic acoustical and vibration signals are constant-speed rotating machines, including propellers and fans. Ideally, such signals would have only one dominant frequency, allowing them to be represented by a simple sine wave. However, it is more likely that the periodic source will produce a complex signal that must be described, using a Fourier series representation,[1] by a collection of harmonically related sine waves, as illustrated in Fig. 3.1. In any case, periodic signals are called steady state because their average properties (mean value, mean-square value, and spectrum) do not vary with time.

There are steady-state acoustical and vibration signals that are not rigorously periodic, for example, the data produced by a collection of independent (unsynchronized) periodic sources, such as the propellers on a multiengine propeller airplane. Such nonperiodic steady-state signals are referred to as almost periodic. Most nonperiodic deterministic signals, however, are also not steady state; that is, their average properties change with time. An important type of time-varying

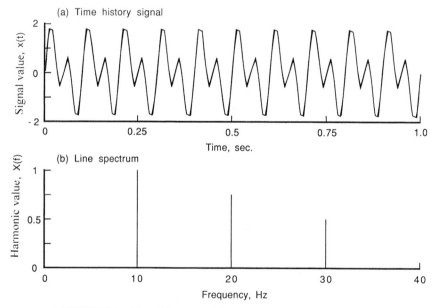

FIGURE 3.1 Time history and line spectrum for periodic signal.

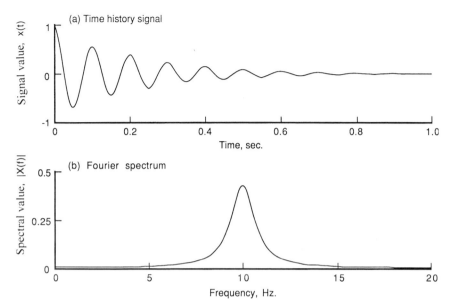

FIGURE 3.2 Time history and Fourier spectrum for deterministic transient signal.

signal is one that begins and ends within a reasonable measurement time interval. Such signals are called transient signals. Examples of deterministic transient signals include well-controlled impacts, sonic booms, and aircraft landing loads. Such data can be described, using a Fourier integral representation,[1] by a continuous spectrum, as illustrated for an exponentially decaying oscillation in Fig. 3.2.

Random Data

Random acoustical and vibration data signals may be broadly defined as all signals that are not deterministic, that is, where it is not theoretically feasible to predict future time history values based upon a knowledge of the applicable physics or past observations. In some cases, the border between deterministic and random signals may be blurred. For example, the pressure field produced by a high-speed fan at its blade passage rate with a uniform inflow would be deterministic, but turbulence in the inflow would introduce a random property to the pressure signal. In other cases, the data will be more fully random in character (sometimes called "strongly mixed"). Examples include the pressure fields generated by fluid dynamic boundary layers (flow noise), the loads produced by atmospheric turbulence, and the acoustical noise caused by the exhaust gas mixing from an air blower. These sources of acoustical and vibration signals cover a wide frequency range and have totally haphazard time histories, as illustrated in Fig. 3.3.

When the mechanisms producing random acoustical or vibration data are time invariant, the average properties of the resulting signals will also be time invariant. Such random data are said to be stationary. Unlike steady-state deterministic data, stationary random data must be described by a continuous spectrum.

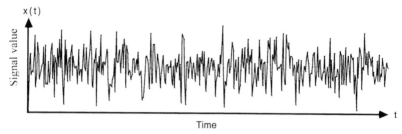

FIGURE 3.3 Time history for stationary random signal.

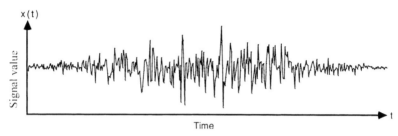

FIGURE 3.4 Time history for nonstationary random signal.

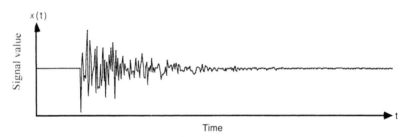

FIGURE 3.5 Time history for stochastic transient signal.

Furthermore, because of the probabilistic character of the data, the measurement of the spectrum and all other signal properties of interest will involve statistical sampling errors that do not occur in the analysis of deterministic signals. These statistical errors will be summarized later.

If the average properties of random signals vary with translations in time, the signals are said to be nonstationary. An illustration of nonstationary data is shown in Fig. 3.4. Although there is a well-developed theoretical methodology for the analysis of arbitrary nonstationary signals,[1] the analysis procedures often require more data than are commonly available and further involve extensive and complex computer calculations. An exception is a special class of nonstationary signals, called stochastic transients, that begin and end within a reasonable measurement time interval. Common sources of stochastic transients are hard impact loads and pyrotechnic devices. An illustration of a stochastic transient signal is shown in Fig. 3.5. Such data can be analyzed by procedures similar to those used to describe deterministic transients discussed earlier, except now there will be a statistical sampling error problem that must be addressed.

3.2 MEAN AND MEAN-SQUARE VALUES

The most rudimentary measures of any steady-state or stationary data signal are the mean value and the mean-square (ms) value (or variance), which provide single-valued descriptions of the central tendency and dispersion of the signal. Mean and ms values of signals can be measured using either digital computations[1] or analog instruments.[2] Hence, appropriate algorithms are presented for both digital and analog analysis procedures.

Mean Values

For acoustical and vibration signals, the mean value is commonly zero, simply because most transducers used for acoustical pressure and vibration acceleration measurements do not sense static values. However, if a transducer is used that senses static values and the central tendency of the signal is of interest, the mean value for a time history measurement, $x(t)$, of duration T_r is computed by

$$m_x = \frac{1}{T_r} \int_0^{T_r} x(t)\, dt \qquad (3.2a)$$

For digital data with a sampling interval of Δt, $x(t) = x(n\,\Delta t)$, $n = 1, 2, \ldots, N$, and the mean value is computed by

$$m_x = \frac{1}{N} \sum_{n=1}^{N} x(n\,\Delta t) \qquad (3.2b)$$

The mean value m_x in Eq. (3.2) has units of volts and is essentially the quantity computed by a direct current (DC) voltmeter. If the data signal is periodic, the calculation in Eq. (3.2) will be accurate as long as T_r (or $N\,\Delta t$) is an integer multiple of the period T_p. For random data, the calculation will involve a statistical sampling error that is a function of T_r (or N) and the spectral characteristics of the signal,[1] to be discussed later.

Mean-Square Values

The ms value of a steady-state or stationary acoustical or vibration signal $x(t)$ [or a digitized signal $x(n\,\Delta t)$] is defined by

$$w_x = \frac{1}{T_r} \int_0^{T_r} x^2(t)\, dt = \frac{1}{N} \sum_{n=1}^{N} x^2(n\,\Delta t) \qquad (3.3)$$

The ms value w_x in Eq. (3.3) has the units of volts squared, which is proportional to power or power per unit area, and hence w_x is often referred to as the overall "power" or "intensity" of the acoustical or vibration signal. If the mean value of the signal is zero, the ms value in Eq. (3.3) is equal to the variance of $x(t)$, defined as

$$s_x^2 = \frac{1}{T_r} \int_0^{T_r} [x(t) - \mu_x]^2 \, dt = \frac{1}{N} \sum_{n=1}^{N} [x(n \, \Delta t) - \mu_x]^2 \quad (3.4)$$

where μ_x is the true mean value of the signal (m_x as $T_r \to \infty$). The positive square roots of the quantities defined in Eqs. (3.3) and (3.4), $w_x^{1/2}$ and s_x, are called the rms value and the standard deviation, respectively, of the signal $x(t)$. Again, $w_x^{1/2} = s_x$ if the mean value of the signal is zero, as will be assumed henceforth. The value s_x in Eq. (3.4), or $w_x^{1/2}$ in Eq. (3.3) when $\mu_x = 0$, is essentially the quantity measured by a true rms voltmeter (not to be confused with an alternating current (AC) voltmeter that employs a linear rectifier calibrated to measure the correct rms value for a sine wave and reads about 1 dB low when measuring random noise). As for mean-value calculations, the ms value calculation in Eq. (3.3) will be precise for periodic data if T_r is an integer multiple of the period T_p. For random data, however, there will be a statistical sampling error that is a function of T_r (or N) and the spectral characteristics of the signal,[1] to be discussed later.

Weighted Averages

The averaging operation indicated in Eqs. (3.2)–(3.4) is a simple linear sum of prior values over a specific time interval. This type of average, referred to as an unweighted or linear average, is the natural way one would average any set of discrete data values and is the simplest way for a digital computer to calculate an average value. However, some acoustical and vibration data analyses are still performed using analog instruments, where a relatively expensive operational amplifier is required to accomplish an unweighted average. Hence, the averaging operation in analog instruments is commonly weighted so that it can be accomplished using inexpensive passive circuit elements.[2] The most common averaging circuit used by analog instruments is a simple low-pass filter consisting of a series resistor and shunt capacitor (commonly called an RC filter) that produces an exponentially weighted average. For a ms value estimate (assuming the mean value is zero), the exponentially weighted average estimate is given by

$$w_x(t) = \frac{1}{K} \int_0^t x^2(\tau) \exp\left(-\frac{t-\tau}{K}\right) d\tau \quad (3.5)$$

where $K = RC$ is the product of the numerical values of the resistance in ohms and capacitance in farads used in the averaging circuit. The term K has units of time (seconds) and is referred to as the *time constant* of the averaging circuit. An exponentially weighted averaging circuit provides a continuous average value estimate versus time, which is based on all past values of the signal. It follows that after starting the average calculation, a period of time must elapse before the indicated average value is accurate. As a rule of thumb, when averaging a steady-state or stationary signal, at least four time constants ($t > 4K$) must elapse to obtain an average value estimate with an error of less than 2%. The error in question is a bias error. For the analysis of random data signals, there will also be a statistical sampling error, to be discussed later.

Running Averages

When the acoustical or vibration data of interest have average properties that are time varying (nonstationary data), "running" time averages are often used to describe the data. For the case of a ms value estimate, this could be accomplished by executing Eq. (3.3) repeatedly over short, contiguous time segments of duration $T \ll T_r$. Exponentially weighted averaging is particularly convenient for the computation of running averages because it produces a continuous average value estimate versus time. However, a near-continuous estimate can also be generated using an unweighted average (as is more desirable for digital data analysis) by simply recomputing an average value every data-sampling interval Δt rather than only at the end of the averaging time T. An illustration of a running average for the rms value of typical nonstationary acoustical data measured near an airport during an aircraft flyover is shown in Fig. 3.6 (the measurement in this illustration is a frequency-weighted rms value called *perceived noise level*).

The basic requirement in the computation of a running average is to select an averaging time T (or averaging time constant K) that is short enough not to smooth out the time variations of the data property being measured but long enough to suppress statistical sampling errors in the average value estimate at any time (assuming the data are at least partially random in character). Analytical procedures for selecting the averaging time T that provides an optimum compromise between the smoothing and statistical sampling errors have been formulated,[1] but trial-and-error procedures coupled with experience will usually provide adequate results. Also, many of the simpler acoustical and vibration

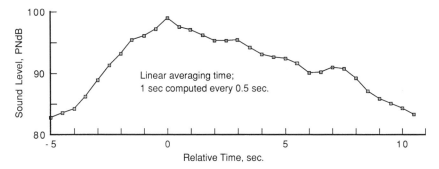

FIGURE 3.6 Running average for weighted rms value of aircraft flyover noise. (Courtesy of Acoustic Analysis Associates, Inc., Canoga, Park, CA.)

measurement instruments used in the field have fixed averaging times ("fast" and "slow" averaging circuits) built into the instrument.[2]

Statistical Sampling Errors

As mentioned earlier, there are no fundamental errors (other than instrument and calibration errors) associated with the calculation of mean and ms values of periodic data signals, assuming the averaging time is an integer multiple of the period of the signal. For random data signals, however, there will be a statistical sampling error due to the fact that the record duration and averaging time can never be long enough to cover all unique signal values. It is convenient to describe this statistical sampling error in terms of a normalized standard deviation of the resulting estimate, called the normalized random error (also called the coefficient of variation). The normalized random error of an estimate, $\hat{\theta}$, for a signal property θ is defined as

$$\epsilon_r[\hat{\theta}] = \frac{\sigma[\hat{\theta}]}{\theta} \tag{3.6}$$

where $\sigma[\cdot]$ denotes the standard deviation as defined in Eq. (3.4) with $T_r \to \infty$ and the hat denotes an estimate. The interpretation of the normalized random error is as follows. If a signal property θ is repeatedly estimated with a normalized random error of, say, 0.1, then about two-thirds of the estimates, $\hat{\theta}$, will be within ±10% of the true value of θ.

The normalized random errors of the mean, ms, and rms value estimates are summarized in Table 3.1.[1] In these error equations, B_s is a measure of the spectral bandwidth of the data signal, defined as[1]

$$B_s = \frac{w^2}{\int_0^\infty G^2(f)\,df} \tag{3.7}$$

where w is the ms value and $G(f)$ is the auto (power) spectrum of the data signal, to be addressed later. The error formulas in Table 3.1 are approximations that are valid for $\epsilon_r \leq 0.20$. Note in Table 3.1 that the random error for mean-value estimates is dependent on the true mean (μ_x) and standard deviation (σ_x) of the signal as well as the bandwidth (B_s) and averaging time (T or K). However, the

TABLE 3.1 Normalized Random Errors for Mean, ms, and rms Value Estimates

	Normalized Random Error, ϵ_r	
Signal Property	Linear Averaging with Averaging Time T	RC Weighted Averaging with Time Constant K
Mean value, m_x	$\sigma_x/[\mu_x(2B_sT)^{1/2}]$	$\sigma_x/[\mu_x(4B_sK)^{1/2}]$
ms value, w_x	$1/(B_sT)^{1/2}$	$1/(2B_sK)^{1/2}$
rms value, $w_x^{1/2}$	$1/(4B_sT)^{1/2}$	$1/(8B_sK)^{1/2}$

random error for ms and rms value estimates is a function only of the bandwidth and averaging time.

Synchronous Averaging

Periodic vibration and acoustical data signals produced by rotating machinery (including propellers and fans) are sometimes contaminated by additive, extraneous noise such that the measured signal is $x(t) = p(t) + n(t)$, where $p(t)$ is the periodic signal of interest and $n(t)$ is the noise. In such cases, the signal-to-noise ratio of the periodic signal can be strongly enhanced by the procedure of synchronous averaging,[3] where the data record is divided into a collection of segments $x_i(t)$, $i = 1, 2, \ldots, q$, each starting at exactly the same phase angle during a period of $p(t)$. The collection of segments can then be ensemble averaged to extract $p(t)$ from the extraneous noise as well as other periodic components that are not harmonically related to $p(t)$ as follows:

$$p(t) \approx \frac{1}{q} \sum_{i=1}^{q} x_i(t) \qquad (3.8)$$

Synchronous averaging is illustrated in Fig. 3.7 for the pressure field generated in the plane of the propeller on the sidewall of a propeller airplane powered by a reciprocating engine. With 1000 ensemble averages (accomplished in less than 2 min), the procedure extracts the propeller pressure signal cleanly from the engine and boundary layer turbulence noise.

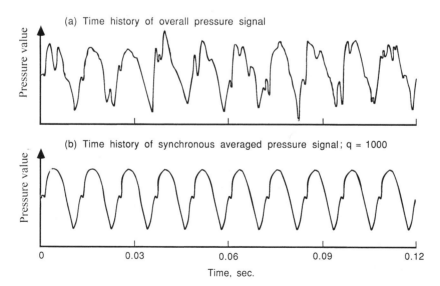

FIGURE 3.7 Original and synchronous averaged time histories for pressure measurement on sidewall of propeller airplane.

The primary requirement for synchronous averaging is the ability to initiate new records at a desired instant during a period of $p(t)$. This is most effectively accomplished using a trigger signal that is a noise-free indicator of the phase during each period of $p(t)$. For rotating machines, a noise-free trigger signal is commonly obtained using either an optical detector or a magnetic pulse generator on the rotating element of the machine. The time base accuracy of the trigger signal determines the accuracy of the magnitude of the resulting synchronous averaged signal; that is, time base errors in the trigger signal cause a reduction in the indicated signal amplitude with increasing frequency. The signal-to-noise ratio enhancement for the synchronous averaged signal is given in decibels by $10 \log_{10} q$, where q is the number of segments used in the ensemble-averaging operation[3].

3.3 SPECTRAL FUNCTIONS

The ms value of a digital signal constitutes an overall measure of the "power" or "intensity" represented by the measured quantity, but much more useful information is provided by a frequency decomposition of the signal values. As noted in Chapter 1, the computation of rms values in one-third-octave bands is widely used for the frequency analysis of acoustical data and sometimes for vibration data as well. However, the more advanced signal-processing techniques needed for system, source, and path identification problems require the computation of frequency spectra with a much finer resolution than one-third-octave bands. Furthermore, if the data are random in character, a frequency analysis in terms of power quantities per unit frequency (hertz) greatly facilitates the desired evaluations of the data signals.

Prior to 1965, most analysis of acoustical and vibration data signals, including the calculation of frequency spectra, was accomplished by analog instruments. Highly resolved frequency spectra were generally computed using narrow-band width analog filters, often employing mechanical elements such as resonant crystals or magnetostrictive devices.[2] Of course, narrow-bandwidth frequency spectra could also be computed, even at that time, on a digital computer using Fourier transform software, but the computations were time consuming and expensive because of the large number of data values needed to represent wide-bandwidth acoustical and vibration signals. In 1965, this situation changed dramatically with the introduction of an algorithm for the fast computation of Fourier series coefficients[4] that reduced the required computer calculations by several orders of magnitude. Various versions of this algorithm have since come into wide use and are generally referred to as fast Fourier transform (FFT) algorithms. The vast majority of all current narrow-bandwidth spectral analysis of acoustical and vibration data signals are performed using FFT algorithms.

The FFT Algorithm

Using lowercase letters for functions of time and uppercase letters for functions of frequency, the Fourier transform of a time history signal, $x(t)$, which is measured over the time interval $0 \leq t \leq T$, is defined for all frequencies (both positive and negative) by

$$X(f, T) = \int_0^T x(t) e^{-j2\pi ft} \, dt \qquad (3.9a)$$

In terms of a digital time series of N data values where $x(t) = x(n \, \Delta t)$, $n = 0, 1, \ldots, N - 1$ (starting the indexing at $n = 0$ is helpful in maintaining consistent relationships between time and frequency functions), the Fourier transform may be written as

$$X(f, T) = X(k \, \Delta f, N) = \Delta t \sum_{n=0}^{N-1} x(n \, \Delta t) \, \exp(-j2\pi fn \, \Delta t) \qquad (3.9b)$$

where the spectral components are generally complex valued and are defined only at N discrete frequencies given by

$$f_k = k \, \Delta f = \frac{k}{N \, \Delta t} \qquad k = 0, 1, 2, \ldots, N - 1 \qquad (3.9c)$$

The finite Fourier transform defined in Eq. (3.9), when divided by T (or $N \, \Delta t$), essentially yields the conventional Fourier series coefficients for a periodic function under the assumption that the time history record $x(t)$ is one period (or an integer multiple of one period) of the periodic function being analyzed. The Fourier components are unique only out to $k = \frac{1}{2}N$, that is, out to the frequency $f_k = 1/(2 \, \Delta t)$, commonly called the Nyquist frequency, f_N, of the digitized signal. The Nyquist frequency is that frequency for which there are only two sample values per cycle and, hence, aliasing will initiate.[1] Comparing the digital result in Eq. (3.9b) with the analog formulation in Eq. (3.9a), the first $\frac{1}{2}N + 1$ Fourier coefficients, from $k = 0$ to $k = \frac{1}{2}N$, define the spectral components at nonnegative frequencies, while the last $\frac{1}{2}N - 1$ Fourier coefficients, from $k = \frac{1}{2}N + 1$ to $k = N - 1$, essentially define the spectral components at negative frequencies.

The details of the various FFT algorithms are fully documented in the literature (see refs. 1 and 4 and the signal analysis documents listed in the Bibliography). It is necessary here only to note a few basic characteristics of the most common algorithm used for acoustical and vibration work, which is commonly referred to as the *Cooley–Tukey algorithm* in recognition of the authors of the 1965 paper[4] that initiated the wide use of such algorithms:

1. It is convenient to restrict the number of data values for each FFT to a power of 2, that is, $N = 2^p$, where values of $p = 8$ to $p = 12$ are commonly used.
2. The fundamental frequency resolution of the Fourier components will be $\Delta f = 1/(N \, \Delta t)$.
3. The Nyquist frequency where aliasing will initiate, denoted by f_N, occurs at the $k = \frac{1}{2}N$ Fourier component, that is, $f_N = 1/(2 \, \Delta t)$.

4. The first $\frac{1}{2}N+1$ Fourier components up to the Nyquist frequency are related to the last $\frac{1}{2}N-1$ components above the Nyquist frequency by $X(k) = X^*(N-k)$, $k = 0, 1, 2, \ldots, N-1$, where the asterisk denotes complex conjugate.
5. The Fourier components defined only for positive frequencies (called a one-sided spectrum) are given by $X(0)$, $X(N/2)$, and $2X(k)$; $k = 1, 2, \ldots, N/2 - 1$.

Line and Fourier Spectral Functions

For deterministic signals that are periodic, a frequency decomposition or spectrum of the signal is directly obtained by computing the Fourier series coefficients of the signal over at least one period of the signal using an FFT algorithm. Assuming the mean value equals zero, the one-sided spectra of Fourier components for a periodic signal, $p(t)$, are given by

$$P(f) = \frac{2X(f, T)}{T} \qquad f > 0 \text{ or } k = 1, 2, \ldots, \frac{N}{2} - 1 \qquad (3.10)$$

where $X(f, T)$ is as defined in Eq. (3.9). The Fourier component magnitudes, $|P(f)|$, are usually plotted versus frequency in the form of a discrete-frequency spectrum (often called a *line spectrum*), as illustrated earlier in Fig. 3.1b. Of course, each Fourier component is a complex number that defines a phase as well as a magnitude (the phases for the Fourier components in Fig. 3.1 are all zero). However, the phase information is generally retained only in those applications where there may be a need to reconstruct the signal time history or determine peak values.

Aliasing. Because of the aliasing problem inherent in digital spectral analysis,[1] it is important to assure there are no spectral components in the data signal being analyzed above the Nyquist frequency [$f_N = 1/(2 \Delta t)$] for the analysis. This can be guaranteed only when the analog signal is low-pass filtered prior to the digitization to remove any spectral components that may exist in the signal above f_N. The low-pass filters employed to accomplish this task are commonly referred to as *antialiasing* filters and should always be used for all spectral analysis.

Leakage Errors and Tapering. Ideally, in the anaysis of periodic data, the spectral computations should be performed over an exact integer multiple of one period to avoid truncation errors in the Fourier series calculation. In practice, however, the computations are often terminated at a time that is convenient for the FFT algorithm and not related to the exact period of the data. The resulting truncation error leads to a phenomenon called side-lobe leakage[1,5] that can severely distort the desired results. To suppress this leakage, it is common to taper the measured time history signal in a manner that forces the values at the start and finish of the measurement to be zero, so as to eliminate the discontinuity between the beginning and ending data values. Numerous tapering functions (often called "windows") have been proposed over the years,[5] but one of the earliest and still most widely used is the *cosine-squared* taper (commonly called the *Hanning* window) given by

$$u_h(t) = 1 - \cos^2 \frac{\pi t}{T} \qquad 0 \le t \le T \qquad (3.11)$$

The FFT is then performed on the signal, $y(t) = x(t)u_h(t)$, rather than directly on the original measured signal, $x(t)$. Leakage suppression can also be accomplished by equivalent operations in the frequency domain.

For deterministic signals that are not periodic and further have a well-defined beginning and end (transients), a spectrum of the signal is directly obtained by computing the Fourier transform of the signal over the entire duration of the signal, again using an FFT algorithm (to obtain a one-sided spectrum, the actual computation is $2X(k\,\Delta f)$, $k = 1, 2, \ldots, N/2 - 1$. The Fourier spectrum magnitude is plotted as a continuous function of frequency, as illustrated previously in Fig. 3.2*b*. Similar to the spectra for periodic signals, there is also a phase function associated with Fourier spectra, but it is generally retained only if the signal time history is to be reconstructed or peak values are of interest. As long as the FFT computation is performed over a measurement duration that covers the entire duration of the transient event, there is no side-lobe leakage problem in the analysis.

As a final point on transient signal analysis, it should be mentioned that transient data signals, particularly those produced by short-duration mechanical shocks, are often analyzed by a technique called the *shock response spectrum*,[6] which essentially defines the peak response of a hypothetical collection of single-degree-of-freedom mechanical systems to the transient input. The shock response spectrum can be a valuable tool for assessing the damaging potential of mechanical shock loads on equipment but is not particularly useful for noise and vibration reduction applications.

Auto (Power) Spectral Density Functions

The autospectral density function (also called the "power" spectral density function) provides a convenient and consistent measure of the frequency composition of random data signals. The autospectrum, denoted by $G_{xx}(f)$, is most easily visualized as the ms value of the signal passed through a narrow-bandpass filter divided by the filter bandwidth, as illustrated in Fig. 3.8. In equation form,

$$\hat{G}_{xx}(f) = \frac{1}{T\,\Delta f} \int_0^T x^2(f, \Delta f, t)\,dt \qquad (3.12)$$

where $x(f, \Delta f, t)$ denotes the signal passed by the narrow-bandpass filter with a center frequency f and a bandwidth Δf. To obtain the exact autospectral density function, the operations in Fig. 3.8 would theoretically be carried out in the limit as $T \to \infty$ and $\Delta f \to 0$ such that $T\,\Delta f \to \infty$. It is clear from Fig. 3.8 and

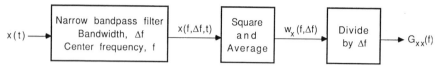

FIGURE 3.8 Autospectral density function measurement by analog filtering operations.

Eq. (3.12) that the units of the autospectral density function are volts squared per hertz.

The operations shown in Fig. 3.8 represent the way autospectra were computed by analog instruments[2] prior to the introduction of FFT algorithms and the transition to digital data analysis procedures. Today, with the ready availability of FFT hardware and software, the autospectral density function (at positive frequencies only) is estimated directly by[1]

$$G_{xx}(f) = \frac{2}{n_d T} \sum_{i=1}^{n_d} |X_i(f, T)|^2 \qquad f > 0 \qquad (3.13)$$

where $X_i(f, T)$ is the FFT of $x(t)$ computed over the ith data segment of duration T, as defined in Eq. (3.9), and n_d is the number of disjoint (statistically independent) data segments used in the calculation. To obtain the exact autospectral density function, the operations in Eq. (3.13) would theoretically be carried out in the limit as $T \to \infty$ and $n_d \to \infty$. As will be seen later, the number of averages n_d determines the random error in the estimate, while the segment duration T for each FFT computation determines the resolution and, hence, a potential bias error in the estimate. The collection of disjoint data segments needed to estimate a statistically reliable autospectrum is usually created by dividing the total available measurement duration T_r into a sequence of contiguous segments of duration T, as illustrated in Fig. 3.9. It follows that $n_d = T_r/T = \Delta f \, T_r$, often referred to as the *BT product* of the estimate. It can be shown[1] that Eq. (3.13) is equal to the result in Eq. (3.12) when the appropriate limits are imposed, that is, when $T \to \infty$ and $\Delta f \to 0$ in Eq. (3.12) and $T \to \infty$ and $n_d \to \infty$ in Eq. (3.13). Note that the autospectral density function is always a real number (there is no phase information associated with autospectra).

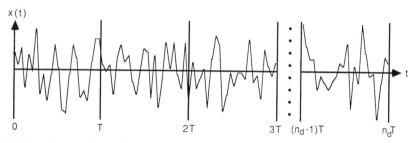

FIGURE 3.9 Subdivision of measured time history into n_d contiguous segments.

A number of grooming operations are commonly employed to enhance the quality of autospectral density estimates. A few of the more important ones are as follows (see the noted references for details).

Antialiasing Filters. As for periodic data analysis, to avoid aliasing, it is important that the random signal being analyzed have no spectral values above the Nyquist frequency, $f_N = 1/(2\,\Delta t)$. Hence, the analog signal must always be low-pass filtered prior to digitization to suppress any spectral content that may exist above f_N.[1]

Tapering Windows. Since random data signals essentially have an infinite period, there will always be a truncation error associated with the selected segment duration T. Hence, tapering operations (windows) are commonly used in random signal analysis to suppress the side-lobe leakage problem. Of the numerous available tapering functions,[5] the cosine-squared (Hanning) window defined in Eq. (3.11) is the most widely used.

Overlapped Processing. Although tapering operations on segments of the measured signal are desirable to suppress leakage, they also increase the bandwidth of the effective spectral window associated with the analysis.[5] If it is desired to maintain the same spectral window bandwidth with tapering that would have been achieved without tapering, the segment duration T for the analysis must be increased. However, assuming the total duration of the measurement T_r is fixed, this will reduce the number of disjoint averages n_d and increase the random error of the spectral estimates. This increase in random error can be counteracted by computing the spectrum with overlapped segments, rather than contiguous segments.[7,8] A 50% overlap is commonly used in such cases.

Zoom Transforms. As discussed earlier, FFT algorithms are usually implemented with a fixed number of data points. Hence once a desired upper frequency limit for an analysis (the Nyquist frequency f_N) has been chosen, the resolution of the analysis, Δf, as defined in Eq. (3.9c), is also fixed. Situations often arise when the desired upper frequency limit and frequency resolution are not compatible with the number of data points used by the FFT computation. In these cases, a finer resolution for a given value of f_N can be achieved using computation techniques referred to as *zoom transform* procedures.[1] The most common zoom transform techniques employ a complex demodulation calculation that essentially segments the frequency range of the signal into contiguous bands that are then analyzed separately.

Cross-Spectral Density Functions. The solution of acoustical and vibration control problems involving random processes is often facilitated by the identification of a linear dependence (correlation) between two measurements at different locations. The basic parameter that defines the linear dependence between two measured random signals, $x(t)$ and $y(t)$, as a function of frequency is the cross-spectral density function, which is estimated (at positive frequencies only) by[1]

$$G_{xy}(f) = \frac{2}{n_d T} \sum_{i=1}^{n_d} X_i^*(f, T) Y_i(f, T) \qquad f > 0 \qquad (3.14a)$$

where $X_i(f, T)$ and $Y_i(f, T)$ are the FFTs of $x(t)$ and $y(t)$, respectively, computed over the ith simultaneous data segments of duration T, n_d is the number of disjoint records used in the calculation, and the asterisk denotes complex conjugate. As for autospectra, the exact cross-spectral density function would be obtained in the limit as $T \to \infty$ and $n_d \to \infty$. All of the computational considerations and grooming procedures discussed for autospectra apply to cross-spectra as well. Unlike the autospectrum, however, the cross-spectrum is generally a complex number that includes both magnitude and phase information and, hence, may be denoted in complex polar notation as

$$G_{xy}(f) = |G_{xy}(f)| e^{j\theta_{xy}(f)} \qquad (3.14b)$$

Coherence Functions

For many applications, it is more convenient to work with a normalized version of the cross-spectral density function, called the coherence function (sometimes called coherency squared), which is defined as[1]

$$\gamma_{xy}^2(f) = \frac{|G_{xy}(f)|^2}{G_{xx}(f) G_{yy}(f)} \qquad (3.15)$$

The coherence function is a real-valued quantity bounded by zero and unity, that is,

$$0 \leq \gamma_{xy}^2(f) \leq 1 \qquad (3.16)$$

where a value of zero means there is no linear dependence and a value of unity means there is a perfect linear dependence between the signals $x(t)$ and $y(t)$ at the frequency f. A coherence value that is less than unity at one or more frequencies is usually indicative of one of the following situations[9]:

1. Extraneous noise is present in the measurements.
2. The frequency resolution of the spectral estimates is too wide.
3. The system relating $y(t)$ to $x(t)$ has time-dependent parameters.
4. The system relating $y(t)$ to $x(t)$ is not linear.
5. The output $y(t)$ is due to other inputs besides $x(t)$.

By carefully designing an experiment to minimize the first four possible reasons for a low coherence, the fifth reason provides the basis for a powerful procedure to identify acoustical noise and/or vibration sources. Specifically, if it is known that a constant-parameter linear system exists between a source and a receiver location, the source signal is measured with an adequate signal-to-noise ratio, and the spectra of the source and receiver signals are estimated with an adequate frequency resolution, then the coherence function defines the fractional portion of the receiver signal autospectral density that is due to the measured source signal. This is the basis for the coherent output power relationship, which is discussed and illustrated in Section 3.5.

Statistical Sampling Errors

There are no statistical sampling errors associated with the calculation of spectra for periodic signals, assuming the averaging time is an integer multiple of the period of the signal. The same is true of the calculation of Fourier spectra for deterministic transient signals, assuming the averaging time is longer than the transient. The calculation of spectral density quantities for random signals, however, will involve a random sampling error, as discussed previously in Section 3.2. First-order approximations for these random errors in autospectra, cross-spectra, and coherence function estimates are summarized in Table 3.2.[1] The random errors are presented in terms of the normalized random error (coefficient of variation) defined in Eq. (3.6), except for estimates of the cross-spectrum phase where the random error is given in terms of the standard deviation of the estimated phase angle in radians.

Beyond the random errors, there is also a bias error problem in the estimation of spectral density functions that occurs at peaks and valleys in the estimates. This bias error is caused by the finite-resolution bandwidth used for the calculations. For auto- and cross-spectral density magnitude estimates, the bias error is approximated in normalized terms by[1,9]

$$\epsilon_b[\hat{G}(f)] = \frac{b[\hat{G}(f)]}{G(f)} = -\frac{1}{3}\left(\frac{\Delta f}{B_r}\right)^2 \tag{3.17}$$

where $b[\cdot]$ denotes the bias error incurred by estimating $G(f)$ by its biased value $\hat{G}(f)$, Δf is the frequency resolution of the analysis, and B_r is the half-power-point bandwidth of a spectral peak in either $G_{xx}(f)$ or $|G_{xy}(f)|$ at that frequency. There is no general bias error equation for coherence function estimates, but error relationships have been formulated for special cases.[10]

TABLE 3.2 Normalized Random Errors for Autospectra, Cross-Spectra, and Coherence Function Estimates

Signal Property	Normalized Random Error ϵ_r or Standard Deviation σ_r
Autospectral density function, $G_{xx}(f)$	$\epsilon_r = 1/n_d^{1/2}$
Cross-spectral density magnitude, $|G_{xy}(f)|$	$\epsilon_r = 1/[n_d \gamma_{xy}^2(f)]^{1/2}$
Cross-spectral density phase, $\theta_{xy}(f)$	$\sigma_r = [1 - \gamma_{xy}^2(f)]^{1/2}/[2n_d \gamma_{xy}^2(f)]^{1/2}$
Coherence function, $\gamma_{xy}^2(f)$	$\epsilon_r = [1 - \gamma_{xy}^2(f)]/[0.5 n_d \gamma_{xy}^2(f)]^{1/2}$

3.4 CORRELATION FUNCTIONS

Certain noise and vibration control problems that involve relatively wide bandwidth random data signals are beast addressed using time domain signal-processing procedures, as opposed to the frequency domain spectral analysis techniques discussed in the previous section. The basic calculation of interest is the correlation function between two random data signals, $x(t)$ and $y(t)$, which is estimated by

$$R_{xy}(\tau) = \frac{1}{T - \tau} \int_0^{T-\tau} x(t) y(t + \tau) \, dt \qquad (3.18a)$$

where τ is a time delay. In digital notation,

$$R_{xy}(r \, \Delta t) = \frac{1}{N - r} \sum_{n=1}^{N-r} x[n \, \Delta t] y[(n + r) \, \Delta t] \qquad (3.18b)$$

where r is a lag number corresponding to a time delay of $r \, \Delta t$. The general quantity estimated in Eq. (3.18) is called the cross-correlation function between the signals $x(t)$ and $y(t)$. For the special case where $x(t) = y(t)$,

$$R_{xx}(\tau) = \frac{1}{T - \tau} \int_0^{T-\tau} x(t) x(t + \tau) \, dt \qquad (3.19)$$

is called the autocorrelation function of $x(t)$. Note that for $\tau = 0$, the autocorrelation function is simply w_x, the ms value of the signal. In both Eqs. (3.18) and (3.19), the estimated quantities will become exact in the limit as the averaging time $T \to \infty$. For finite values of T, there will be a random sampling error in the estimates, to be discussed later.

The correlation function is related to the spectral density function through a Fourier transform,[1]

$$G_{xy}(f) = 2 \int_{-\infty}^{\infty} R_{xy}(\tau) e^{-j2\pi f \tau} \, d\tau \qquad (3.20)$$

Equation (3.20), often called the *Wiener–Khinchine relationship*, is the basis for computing correlation functions in practice. Specifically, the spectral density function is first computed by the FFT procedures outlined in Section 3.3. An inverse Fourier transform of the spectral density function is then computed to obtain the correlation function. Due to the remarkable efficiency of the FFT algorithm, this approach requires substantially fewer calculations than that needed to compute Eq. (3.18b) directly. However, due to the *circular effects* associated with the FFT algorithm, a number of special operations are needed to obtain correct results, as detailed in reference 1.

Correlation Coefficient Function

For many applications, it is more convenient to work with the normalized cross-correlation function between $x(t)$ and $y(t)$, called the correlation coefficient function, which is given by (assuming the mean value is zero)

$$\rho^2(\tau) = \frac{R_{xy}^2(\tau)}{R_{xx}(0)R_{yy}(0)} = \frac{R_{xy}^2(\tau)}{w_x w_y} \quad (3.21)$$

The correlation coefficient function (sometimes called the squared correlation coefficient function) is similar to the coherence function, defined in Section 3.3, in that it is a real-valued quantity bounded by zero and unity, that is,

$$0 \le \rho_{xy}^2(\tau) \le 1 \quad (3.22)$$

where a value of zero means there is no linear dependence and a value of unity means there is a perfect linear dependence[1] between $x(t)$ and $y(t)$ at the time displacement τ. Hence, the correlation coefficient function is interpreted much like the frequency domain coherence function discussed in Section 3.3, except the correlation coefficient function applies to the entire frequency range of the two signals while the coherence function applies to specific frequencies. Also, from Eq. (3.14b), time delay information in the correlation coefficient function is related to the phase information in the cross-spectral density function by

$$\theta(f) = 2\pi f \tau \quad (3.23)$$

Hence, the phase of the cross-spectrum can be valuable for extracting time delay information when the time delay is a function of frequency.[11]

Statistical Sampling Errors

When applied to random data signals, the computation of correlation functions will involve a statistical sampling error. In terms of a normalized random error defined in Eq. (3.6), the error in a cross-correlation estimate can be approximated by[1]

$$\epsilon_r[\hat{R}_{xy}(\tau)] = \left(\frac{1 + 1/\rho_{xy}^2(\tau)}{2B_s T_r}\right)^{1/2} \quad (3.24)$$

where $\hat{R}_{xx}(\tau)$ is an estimate of $R_{xx}(\tau)$, T_r is the total measurement duration over which the computations are performed, and B_s is the smallest statistical bandwidth for the two data signals, as defined in Eq. (3.7).

3.5 DATA ANALYSIS APPLICATIONS

The applications for signal analysis in noise and vibration studies are extensive and can become quite elaborate.[9] However, as mentioned in the introduction to this chapter, there are three specific application areas of special interest for noise and vibration control problems: (a) the identification of system response properties, (b) the identification of excitation sources, and (c) the identification of transmission paths.

Identification of System Response Properties

The control of noise and vibration is often facilitated by the determination of gain factors between excitation sources and receiver location responses. The fundamental measurement of interest here is the frequency response function (sometimes called the transfer function) between the two points of interest. Given an excitation source signal $x(t)$ and a simultaneously measured response signal $y(t)$, the frequency response function between the source and receiver signal is given by[1]

$$H_{xy}(f) = \frac{G_{xy}(f)}{G_{xx}(f)} \tag{3.25a}$$

where the auto- and cross-spectral density functions are as defined in Eqs. (3.13) and (3.14), respectively. The frequency response function is generally a complex-valued quantity that is more conveniently expressed in complex polar notation as

$$H_{xy}(f) = |H_{xy}(f)| \, e^{i\phi xy(f)} \tag{3.25b}$$

where the magnitude function $|H_{xy}(f)|$ is the gain factor and the argument $\phi_{xy}(f)$ is the phase factor between $x(t)$ and $y(t)$. In the more advanced applications, such as normal-mode analysis,[12] both the gain and phase factor are needed. In many elementary applications, however, only the gain factor may be of interest.

The normalized random error in frequency response magnitude (gain factor) estimates is approximated by[1]

$$\epsilon_r[|\hat{H}_{xy}(f)|] \approx \frac{[1 - \gamma_{xy}^2(f)]^{1/2}}{[2n_d \gamma_{xy}^2(f)]^{1/2}} \tag{3.26}$$

where $\hat{H}_{xy}(f)$ is an estimate of $H_{xy}(f)$, $\gamma_{xy}^2(f)$ is the coherence function between the source and receiver signals, and n_d is the number of disjoint averages used to compute the autospectra and cross-spectra from which the gain factor is calculated. The random error in frequency response phase estimates is the same as given for the phase of cross-spectral density estimates in Table 3.2.

Like coherence function estimates, the random error in a gain factor estimate approaches zero as the coherence function approaches unity, even for a small number of averages in the spectral density estimates. Hence, if the coherence function is large, the gain factor can be estimated with greater accuracy than the spectral density estimates used in its computation.

There are several sources of bias errors in gain factor estimates,[1,9] but the most significant is due to the frequency resolution bias error in the spectral density functions used to compute the gain factor, as given by Eq. (3.17). As a rule of thumb, if there are at least four spectral components between the half-power points of peaks in the spectral data, that is, if $\epsilon_b[G(f)] < 0.02$ in Eq. (3.17), then the bias error in the gain factor estimate should be negligible.

As an illustration of the application of gain factor estimates, consider the experiment illustrated in Fig. 3.10, involving two vibration measurements made on a

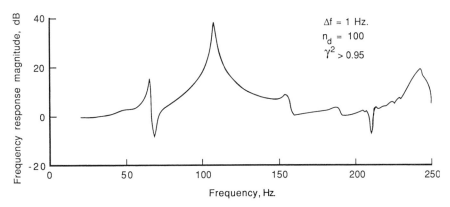

FIGURE 3.10 Gain factor estimate for component in space vehicle payload.

simulated spacecraft payload during a vibration test. One of the measurements is near the mounting point of the payload, and the other is on a critical payload element where vibration may adversely affect the payload performance. The gain factor clearly reveals a frequency region (around 110 Hz) where vibration at the mounting point is greatly magnified at the critical element of concern, due to a strong normal-mode response (resonance) of the payload at this frequency. It follows that efforts to reduce the vibration should be concentrated in this frequency region.

Identification of Periodic Excitation Sources

The identification of periodic acoustical and vibration excitations can usually be accomplished by a straightforward narrow-bandwidth spectral analysis plus a knowledge of the rpm of all rotating machinery producing the acoustical noise and/or vibration. This is illustrated in Fig. 3.11, which shows the spectrum for the vibration on the floor of a microelectronics manufacturing facility with extensive

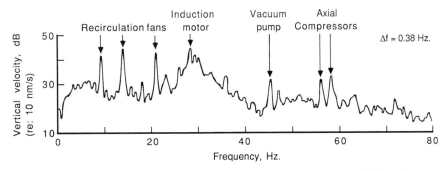

FIGURE 3.11 Fourier spectrum of floor vibration in microelectronics facility. (Courtesy of BBN Laboratories, Inc., Canoga Park, CA.)

air conditioning equipment. It is seen that the more intense spectral peaks in the vibration data can be directly identified with specific rotating machines simply through a knowledge of the rotational frequencies. Of course, this approach cannot separate the vibration contributions of two rotating machines that operate at exactly the same frequency. However, as long as there is some difference in the rotational speed between two machines, say, Δg hertz, the contributions of the machines can theoretically be separated by a spectral analysis with a frequency resolution Δf that is less than Δg. In practice, it is desirable to have

$$\Delta f = \frac{1}{T} < \frac{\Delta g}{3} \qquad (3.27)$$

where T is the segment duration for the spectral computation. Since the signals are periodic or almost periodic, there is no random error associated with the resulting spectral estimates.

Identification of Random Excitation Sources

When two or more sources of an acoustical and/or vibration environment are random in character and further cover essentially the same frequency range, the identification of the contributions of the individual sources is more difficult. If possible, the contribution of each source of excitation should be identified by turning off all but one of the sources so that the effects of each source can be measured individually. If this is not possible, then the principle of coherent output power might be applied. Specifically, measure the acoustical or vibration signal produced at the source location for each of a collection of q suspected sources, denoted by $x_i(t)$, $i = 1, 2, \ldots, q$. For each individual source signal, simultaneously measure the receiver signal $y(t)$ and compute the coherence function between the ith source and receiver signals using Eq. (3.15). Under proper conditions, the autospectrum of the receiver signal due solely to the contribution of the ith source signal is given by

$$G_{y:i}(f) = \gamma_{iy}^2(f) G_{yy}(f) \qquad (3.28)$$

where $G_{y:i}(f)$ reads "the portion of the autospectrum of the receiver signal that is due only to the source signal $x_i(t)$." Equation (3.28) is called the coherent output power relationship. If used properly, it can be a powerful tool for separating the contributions of various possible sources of random excitation in acoustical noise and vibration problems. The primary requirements for the proper application of the coherent output power relationship are as follows[9]:

1. The candidate sources of excitation (or the responses in the immediate vicinity of the sources) must be measured accurately and with negligible measurement noise. However, since the coherence function is dimensionless, any type of transducer (pressure, velocity, acceleration, or displacement) can be used for the source measurements as long as it generates a signal that has a linear relationship with the excitation phenomenon.

2. The candidate sources must be statistically independent, and there must be no interference (crosstalk) in the measurement of any one source due to energy propagating from the other sources; that is, the coherence functions among the measured source signals must all be zero.
3. There must be no significant feedback or nonlinear effects between the candidate source and receiver signals.

The normalized random error for coherent output power measurements is approximated by[1]

$$\epsilon_r[\hat{G}_{y:i}(f)] \approx \frac{[2 - \gamma_{iy}^2(f)]^{1/2}}{[n_d \gamma_{iy}^2(f)]^{1/2}} \quad (3.29)$$

where $\hat{G}_{y:i}(f)$ is an estimate of $G_{y:i}(f)$, $\gamma_{iy}^2(f)$ is the coherence function between the ith source signal and the receiver signal, and n_d is the number of disjoint averages used to compute the autospectra and cross-spectra from which the coherent output is calculated. It is clear from Eq. (3.29) that the number of disjoint records (and hence the total measurement duration, $T_r = n_d T$) required for an accurate coherent output power calculation will be substantial when $\gamma_{iy}^2(f) \ll 1$, as will commonly occur if there are numerous independent sources contributing to $y(t)$.

A serious bias error can occur in coherent output power calculations due to time delays between the source and receiver signals that may arise when there is a substantial distance between the source and receiver measurement positions[1,9,10,13] or the measurements are made in a reverberant environment[9,14] Since the two measurements are usually recorded and analyzed on a common time base, time delays between the source and receiver signals will cause a portion of the received signal to be uncorrelated with the source signal. These time-delay-induced bias errors can be suppressed by the use of precomputation delays or the selection of an appropriately long block duration T in the data analysis, as detailed in references 9, 13, and 14.

To illustrate the coherent output power calculation, consider the experiment outlined in Fig. 3.12, where a panel section excited by a broadband random vibration source radiates acoustical noise to a receiver microphone. The autospectrum of the radiated noise, $y(t)$, as seen by the receiver microphone with no other

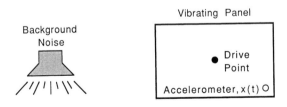

FIGURE 3.12 Acoustical source identification experiment with radiating panel in background noise.

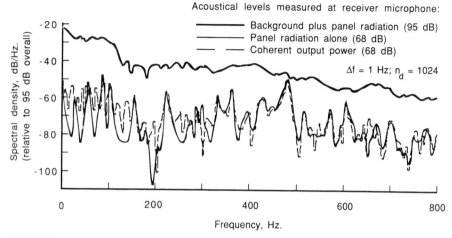

FIGURE 3.13 Overall and coherent output power spectra for radiating panel in background noise. (From Ref. 9 with the permission of the authors.)

noise sources present, is shown by the thin solid line in Fig. 3.13. Statistically independent background noise is now introduced by a speaker to produce acoustical energy with an overall value that is about 500 times more intense (27 dB higher) than the panel-radiated noise. The autospectrum of the total received microphone signal due to both the panel radiation plus the background noise is also shown in Fig. 3.13, by the heavy solid line. Finally, the coherent output power between the microphone and an accelerometer mounted on the panel is computed with the background noise present. This result is shown by the dashed line in Fig. 3.13. It is seen that the coherent output power calculation extracts the autospectrum of the radiated panel noise, as measured by an accelerometer, from the intense background noise with reasonable accuracy at most frequencies. The reason the procedure works in this example is that the radiated noise from the panel has a linear relationship with the panel motion measured by the accelerometer. Furthermore, since the panel is driven from only one point, the vibration response at any one point on the panel is representative of the vibration at all points.

Identification of Propagation Paths

Another analysis of great importance in acoustical noise and vibration control problems is the identification of the physical path or paths by which energy from a source of excitation travels to a receiver location. For those cases involving broadband random energy that propagates in a nondispersive manner (with a frequency-independent propagation speed), such as airborne noise, the identification of propagation paths can often be accomplished by a cross-correlation analysis. Specifically, assume a source signal $x(t)$ propagates in a nondispersive manner through r paths to produce a receiver signal $y(t)$. For simplicity, further assume the propagation paths have uniform (frequency-independent) gain factors denoted by H_i, $i = 1, 2, \ldots, r$. It follows that

$$y(t) = H_1 x(t - \tau_1) + H_2 x(t - \tau_2) + \cdots + H_r x(t - \tau_r) \quad (3.30)$$

where τ_i, $i = 1, 2, \ldots, r$, are the propagation times through each of the paths. Then, from Eqs. (3.18) and (3.19),

$$R_{xy}(\tau) = H_1 R_{xx}(\tau - \tau_1) + H_2 R_{xx}(\tau - \tau_2) + \cdots + H_r R_{xx}(\tau - \tau_r) \quad (3.31)$$

In words, the cross-correlation function between the source and receiver signals will be a series of superimposed autocorrelation functions, each associated with a nondispersive propagation path and centered on a time delay equal to the propagation time along that path. Referring to Eq. (3.20), if the source signal has a wide bandwidth, these autocorrelation peaks will decay rapidly when τ deviates from τ_i and will be sharply defined in the cross-correlation estimate, as illustrated in Fig. 3.14. Noting that the propagation time for each path is the ratio of distance to propagation speed, the physical path associated with each correlation peak can usually be identified from a knowledge of the length of the path and the propagation speed of the nondispersive waves in the medium forming the path. Finally, the portion of the receiver signal ms value that propagated through a specific path is proportional to the square of the magnitude of the correlation peak associated with that path. The normalized random error associated with the estimate of $R_{xy}(\tau)$ in Eq. (3.31) is given by Eq. (3.24).

For a cross-correlation analysis to be effective in identifying different nondispersive propagation paths, several requirements must be met, as follows.

1. The source $x(t)$ must be a broadband random signal.
2. The propagation paths must have reasonably uniform gain factors.
3. The propagation time through each path must be different from all other paths.

As a rule of thumb,[9] the difference in the propagation times through any pair of paths must be $\Delta t > 1/B_s$, where B_s is the spectral bandwidth of the receiver

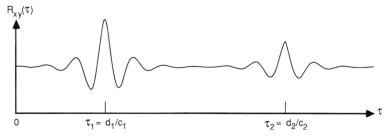

FIGURE 3.14 Cross-correlation function between source and receiver signals with two nondispersive propagation paths.

signal, as defined in Eq. (3.7). As the spectral bandwidth becomes small, the peaks in the cross-correlation function spread, and the ability to identify peaks representing individual propagation paths diminishes. There are a number of other signal-processing operations that will enhance the ability to detect individual propagation paths between two signals when the bandwidth is not wide.[9] Also, the use of envelope functions generated by Hilbert transforms can further enhance such detections.[1] Nevertheless, in the limiting narrow-band case where the source signal is a sine wave (or any periodic function), the identification of individual propagation paths cannot be achieved by any signal-processing procedure no matter how large a difference there is between the propagation times through the various paths.

To illustrate this application of cross-correlation analysis to a propagation path identification problem, consider the experiment shown in Fig. 3.15, which involves a speaker that produces acoustical noise with a bandwidth of approximately 8 kHz. Two microphones are used to measure the noise, one located in front of the speaker and another located 0.68 m from the speaker. There is a wall behind the receiver microphone that causes a back reflection producing a second path between the source and receiver microphones with a length of 1.7 m. The computed cross-correlation function between the source and receiver signals is shown in Fig. 3.16. It is seen that two maxima appear in the cross-correlation estimate at 2 and 5 ms. Noting that the speed of sound in air at room

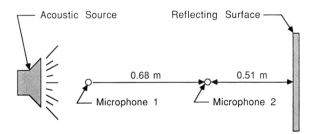

FIGURE 3.15 Acoustical propagation path experiment with back reflection.

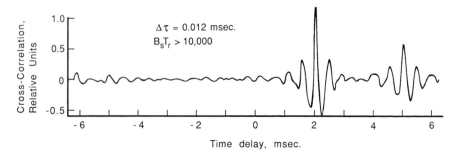

FIGURE 3.16 Cross-correlation function between two microphone signals with back reflection. (From Ref. 9 with the permission of the authors.)

temperature is about 340 m/s, these two peaks clearly identify the direct path and the back reflection. The magnitude of the correlation peak corresponding to the back reflection is only about 40% of the magnitude of peak corresponding to the direct path, as would be expected due to spherical spreading loss; that is, the reflected path is 2.5 times longer than the direct path and, hence, should be about 8 dB lower in level.

The cross-correlation analysis procedure often works well in multipath acoustical problems where the propagation is in the form of longitudinal waves that are nondispersive. In structural vibration problems, there may also be some nondispersive longitudinal wave propagation, but most vibratory energy in structures propagates in the form of flexural waves, which are dispersive,[15] that is, the propagation velocity is a function of frequency. Also, flexural waves strongly reflect and/or scatter at locations where there are changes in either the material properties or the geometry of the structural path. These facts greatly complicate the detection of individual propagation paths in multipath structural vibration problems. Nevertheless, meaningful results can sometimes be obtained through the judicious use of bandwidth-limited cross-correlation analyses[1,9] if the structural paths are reasonably homogeneous.

REFERENCES

1. J. S. Bendat and A. G. Piersol, *RANDOM DATA: Analysis and Measurement Procedures*, 3rd ed., Wiley, New York, 2000.
2. L. L. Beranek, *Acoustical Measurements*, rev. ed., Acoustical Society of America, Woodbury, NY, 1988.
3. H. Himelblau et al., *Handbook for Dynamic Data Acquisition and Analysis*, IEST-RP-DTE012.1, Institute of Environmental Sciences and Technology, Rolling Meadows, IL, 1994.
4. J. W. Cooley and J. W. Tukey, "An Algorithm for the Machine Calculation of Complex Fourier Series," *Math. Computat.*, **19**, 297–301 (1965).
5. F. J. Harris, "On the Use of Windows for Harmonic Analysis with the Discrete Fourier Transform," *Proc. IEEE*, **66**(1), 51–83, (1978).
6. S. Rubin, "Concepts in Shock Data Analysis," in C. M. Harris and A. G. Piersol (Eds.), *Harris' Shock and Vibration Handbook*, 5th ed., McGraw-Hill, New York, 2002.
7. P. D. Welch, "The Use of Fast Fourier Transforms for the Estimation of Power Spectra: A Method Based on Time Averaging Over Short, Modified Periodograms," *IEEE Trans. Audio Electroacoust.*, **AU-15**(2), 70–73 (1967).
8. A. H. Nuttall, "Spectral Estimates by Means of Overlapped Fast Fourier Transformed Processing of Windowed Data," NUSC TR-4169, Naval Underwater Systems Center, New London, CT, October 1971.
9. J. S. Bendat and A. G. Piersol, *Engineering Applications of Correlation and Spectral Analysis*, 2nd ed. Wiley, New York, 1993.
10. H. Schmidt, "Resolution Bias Errors in Spectral Density, Frequency Response and Coherence Function Estimates," *J. Sound Vib.*, **101**(3), 347–427 (1985).

11. A. G. Piersol, "Time Delay Estimation Using Phase Data," *IEEE Trans. Acoust. Speech Signal Proc.*, **ASSP-29**(3), 471–477 (1981).
12. R. J. Allemang and D. L. Brown, "Experimental Modal Analysis," in C. M. Harris and A. G. Piersol (Eds.), *Harris' Shock and Vibration Handbook*, 5th ed., McGraw-Hill, New York, 2002.
13. M. W. Trethewey and H. A. Evensen, "Time-Delay Bias Errors in Estimating Frequency Response and Coherence Functions from Windowed Samples of Continuous and Transient Signals," *J. Sound Vib.*, **97**(4), 531–540 (1984).
14. K. Verhulst and J. W. Verheij, "Coherence Measurements in Multi-Delay Systems," *J. Sound Vib.*, **62**(3), 460–463 (1979).
15. L. Cremer, M. Heckl, and E. E. Ungar, *Structure-Borne Sound*, 2nd ed., Springer-Verlag, New York, 1988.

BIBLIOGRAPHY

Brigham, E. O., *The Fast Fourier Transform and Its Applications*, Prentice-Hall, Englewood Cliffs, NJ, 1988.

Crocker, M. J., *Handbook of Acoustics*, Wiley, New York, 1998.

Doebelin, E. O., *Measurement Systems: Application and Design*, 5th ed., McGraw-Hill, New York, 2004.

Harris, C. M., *Handbook for Acoustic Measurements and Noise Control*, 3rd ed., McGraw-Hill, New York, 1991.

Harris, C. M., and A. G. Piersol, *Harris' Shock and Vibration Handbook*, 5th ed., McGraw-Hill, New York, 2002.

Hassall, E. R., and K. Zaven, *Acoustic Noise Measurements*, 5th ed., Bruel & Kjaer, Naerum, Denmark, 1989.

Mitra, S. K., and J. F. Kaiser, *Handbook for Digital Signal Processing*, Wiley, New York, 1993.

Newland, D. E., *Random Vibrations, Spectral and Wavelet Analysis*, 3rd ed., Wiley, New York, 1993.

Oppenheim, A. V., and R. W. Schafer, *Discrete-Time Signal Processing*, 2nd ed., Prentice-Hall, Englewood Cliffs, NJ, 1999.

Smith, S. W., *The Scientist and Engineer's Guide to Digital Signal Processing*, 2nd ed., California Technical Publishing, San Diego, CA, 2002. (Online: http://www.dspguide.com/pdfbook.htm)

Wirsching, P. H., T. L. Paez, and H. Ortiz, *Random Vibrations, Theory and Practice*, Wiley, New York, 1995.

Wright, C. P., *Applied Measurement Engineering*, Prentice-Hall, Englewood Cliffs, NJ, 1995.

CHAPTER 4

Determination of Sound Power Levels and Directivity of Noise Sources

WILLIAM W. LANG

Noise Control Foundation
Poughkeepsie, New York

GEORGE C. MALING, JR

Harpswell, Maine

MATTHEW A. NOBILE

IBM Hudson Valley Acoustics Laboratory
Poughkeepsie, New York

JIRI TICHY

The Pennsylvania State University
University Park, Pennsylvania

4.1 INTRODUCTION

Noise control can be considered as a system problem, the system containing three major parts: the source, the path, and the receiver.[1]

In any noise control problem, sound energy from the source or sources travels over a multiplicity of paths, both in solid structures and in air, to reach the receiver—an individual, a group of people, a microphone or other instrument, or a structure that is affected by the noise. Three action words are associated with the source–path–receiver model: *emission, transmission*, and *immission*. Sound energy that is emitted by a noise source is transmitted to a receiver where it is immitted. Transmission is treated in Chapters 5–7, 10, and 11. Immission is treated in Chapter 19.

Sound pressure level is the physical quantity usually used to describe a sound field quantitatively because the ear responds to sound pressure. A sound-level meter may be used to readily measure the sound pressure level at the location in the sound field occupied by the receiver. Therefore, the preferred descriptor of immission is the sound pressure level in decibels. However, the sound pressure level by itself is not a satisfactory quantity to describe the strength of a noise source (emission) because the sound pressure level varies with distance from the source and with the acoustical environment in which the source operates.

Two quantities are needed to describe the strength of a noise source, its *sound power level* and its *directivity*. The sound power level is a measure of the total sound power radiated by the source in all directions and is usually stated as a function of frequency, for example, in one-third-octave bands. The sound power level is then the preferred descriptor for the emission of sound energy by noise sources. The sound power level is usually expressed in decibels.*

The directivity of a source is a measure of the variation in its sound radiation with direction. Directivity is usually stated as a function of angular position around the acoustical center of the source and also as a function of frequency. Some sources radiate sound energy nearly uniformly in all directions. These are called nondirectional sources (see Fig. 4.1). Generally, such sources are small in size compared to the wavelength of the sound radiated. Most practical sources are somewhat directional (see Fig. 4.2); that is, they radiate more sound in some directions than in others. Measures of source directivity are presented in Section 4.12.

From the sound power level and directivity, it is possible to calculate the sound pressure levels produced by the source in the acoustical environment in which it operates. This is not an easy task, however, because the resulting sound

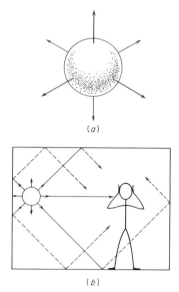

FIGURE 4.1 Sound source that radiates uniformly in all directions: *(a)* sound source in free space; *(b)* same source in enclosure showing reflections from interior surfaces. Solid lines show the direct sound; dashed lines show the reflected (reverberant) sound.

---

*Some industries have adopted the use of the bel, where 1 bel = 10 dB, as the unit of sound power level to distinguish clearly between sound pressure level in decibels and sound power level in bels.

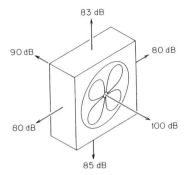

FIGURE 4.2 Sound source with directive radiation into free space. This behavior is typical of equipment noise.

pressure levels at points in the room depend not only on the characteristics of the source but also on the characteristics of the room itself. The sound energy from the source is reflected, absorbed, and scattered by the room boundaries, and some energy is transmitted through these boundaries to adjacent spaces. Thus, the same source could produce quite different sound pressure levels in different rooms or environments. This underscores the difference between *emission* and *immission* and illustrates why sound pressure level is not a good descriptor for noise emission from a source. Sound fields outdoors and in rooms are discussed in detail in Chapters 5–7.

A source may set a nearby surface into vibration if it is rigidly attached to that surface, causing more sound power to be radiated than if the source were vibration isolated. Both the operating and mounting conditions of the source therefore influence the amount of sound power radiated as well as the directivity of the source. Nonetheless, the sound power level alone is useful for comparing the noise radiated by machines of the same type and size as well as by machines of different types and sizes; determining whether a machine complies with a specified upper limit of noise emission; planning in order to determine the amount of transmission loss or noise control required; and engineering work to assist in developing quiet machinery and equipment.

4.2 SOUND POWER LEVELS OF SOURCES

Even though the sound power produced by a noisy machine is only a very small fraction of the total mechanical power that the machine produces, the range of sound powers produced by sources of practical interest is enormous—from less than a microwatt to megawatts. Shaw[2] has estimated the radiation ratio of a wide variety of noise sources.

A single-number descriptor for noise source emission is obtained when the sound power as a function of frequency is weighted using the A-frequency weighting curve. The result is the A-weighted sound power (Table 1.4). The

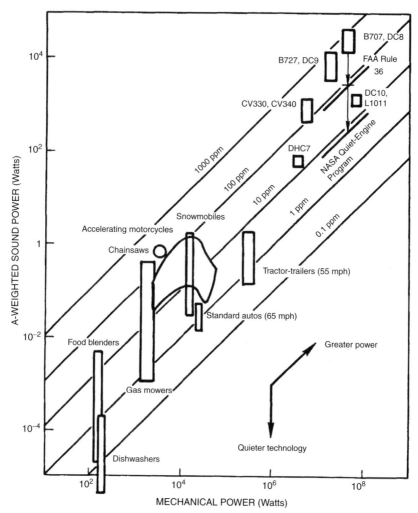

FIGURE 4.3 Estimated values of A-weighted sound power versus mechanical power for various machines. The diagonal lines are lines of constant mechanoacoustical efficiency (sound power per mechanical power) in parts per million. The line labeled FAA Rule 36 approximates 1975 noise levels for new aircraft designs, while the goal of a research program in progress at the time the figure was published is labeled NASA Quiet-Engine Program.

radiation ratio is the ratio of the A-weighted sound power level of the source and the mechanical power.

Estimates of the conversion ratio are shown in Fig. 4.3. The diagonal lines in the figure give conversion ratios ranging from 10^{-3} to 10^{-7}.

The sound powers of sources of practical interest cover a range of more than 12 orders of magnitude. Hence, it is convenient to express a sound power on a logarithmic scale using an internationally agreed-upon sound power, 10^{-12} W (watts), as the reference for the logarithm (see Chapter 1). The A-weighted sound power level* in decibels is defined as

$$L_{WA} = 10 \log\left(\frac{W_A}{W_0}\right) \qquad (4.1)$$

where W_A = A-weighted sound power
L_{WA} = A-weighted sound power level, dB
W_0 = reference sound power, internationally agreed upon as 10^{-12} W

Consequently

$$L_{WA} = 10 \log W_A + 120 \quad \text{dB re } 10^{-12} \text{ W} \qquad (4.2)$$

The A-weighted sound power level (L_{WA}) is usually expressed in decibels but may be expressed in bels. Since there is an order-of-magnitude difference between sound power levels in bels (emission) and sound pressure levels in decibels (immission), the ambiguity of expressing both emission and immission values in decibels is avoided. Use of this convention is particularly important in dealing with the public. In many countries, people who are unfamiliar with the technical details of acoustics are unable to distinguish between different quantities that are expressed in decibels. Nonetheless, noise control engineers and other practitioners in the field usually find it convenient to express sound power levels in decibels.

Example 4.1. A sound source radiates an A-weighted sound power of 3 W. Find the A-weighted sound power level in decibels.

Solution

$$L_{WA} = 10 \log 3 + 120$$
$$= 4.8 + 120 = 124.8 \text{ dB re } 10^{-12} \text{ W}$$

A level is a dimensionless quantity, and it is imperative that the reference be stated to avoid confusion.

4.3 RADIATION FIELD OF A SOUND SOURCE

Near Field, Far Field, and Reverberant Field

Since the sound power emitted by a source must be determined by measurement of a field quantity such as sound pressure or sound intensity (the sound energy

---

*In the past, the A-weighted sound power level was sometimes designated by the term *noise power emission level*, abbreviated NPEL. This usage is no longer common.

flowing through a unit area in a unit time), it is important to understand the radiation field of a sound source when it is placed in various acoustical environments. The character of the radiation field of a typical noise source usually varies with distance from the source. In the vicinity of the source, the particle velocity is not necessarily in the direction of propagation of the sound wave, and an appreciable tangential velocity component may exist at any point. This is the near field. It is characterized by appreciable variations of the sound pressure with distance from the source along a given radius, even when the source is in a free unbounded space commonly referred to as a free field. Moreover, in the near field, the sound intensity is not simply related to the mean-square value of the sound pressure.

The distance from the source to which the near field extends is dependent on the frequency, on a characteristic source dimension, and on the phases of the radiating parts of the surface of the source. The characteristic dimension may vary with frequency and angular orientation. It is difficult, therefore, to establish limits for the near field of an arbitrary source with any degree of accuracy. It is often necessary to explore the sound field experimentally.

In the far field, the sound pressure level decreases by 6 dB for each doubling of the distance from the source, provided that either the source is in free space (no boundaries to reflect the sound) or the reverberant field has not yet been reached (see Fig. 4.4). In this free-field part of the far field, the particle velocity is primarily in the direction of propagation of the sound wave.

If the source is radiating inside an enclosure, fluctuations of sound pressure with position are observed in the reverberant part of the far field, that is, in the region

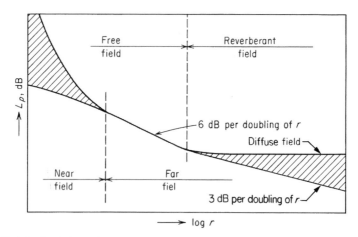

FIGURE 4.4 Variation of sound pressure level in an enclosure along a radius r from a typical noise source. The free field, reverberant field, near field, and far field are shown. The free far field indicates the region where the sound pressure level L_p decreases at the rate of 6 dB for each doubling of distance from the acoustical center of the source, although this region is often very short. In the reverberant far field, the sound pressure level in a highly reverberant room is constant. The lower edge of the shaded region is typical of sound fields in furnished rooms of dwellings and offices.

where the waves reflected from the boundaries of the enclosure are superimposed upon the direct field from the source (see Fig. 4.4). In a highly reverberant room in the region where the direct sound pressure from the source is considerably smaller than the contribution from the reflected sound, the sound pressure level reaches a value essentially independent of the distance from the source. This region may approximate an ideal *diffuse field* in which reflected energy propagates equally in all directions and the sound energy density is uniform.

As discussed in more detail in Chapter 7, in furnished rooms in dwellings and offices where the sound field is neither a free field nor a diffuse field, the sound pressure level decreases by about 3 dB for each doubling of distance from the source.

4.4 SOUND INTENSITY, SOUND POWER, AND SOUND PRESSURE

Sound Intensity and Sound Power

The sound energy (in joules, $1\ \text{J} = 1\ \text{N} \cdot \text{m}$) or the sound power (in joules per second or watts) radiated into free space by a source spreads out over a larger and larger area as the sound wave travels outward from the source. As a result, the sound intensity (in watts per square meter) and the sound pressure (in pascals) at a point in the sound field decreases as the distance from the source increases. Sound intensity, defined in Section 2.2 as the amount of sound energy flowing through a unit area per unit time, is a vector quantity $\mathbf{I}$ having magnitude $I = |\mathbf{I}|$ and direction $\hat{\mathbf{r}}$ pointing in the direction of sound energy propagation. If a closed surface that surrounds the source is selected, the sound power radiated by the source can be calculated from the following integral:

$$W = \int_S \mathbf{I} \cdot \mathbf{dS} \tag{4.3}$$

where W = sound power, W
$\mathbf{I}$ = time-average sound intensity vector, W/m^2
$\mathbf{dS}$ = infinitesimal element of surface area (a vector oriented normal to surface)
S = area of closed surface that surrounds source

Expanding the dot product in Eq. (4.3) allows the surface integral to be written in terms of scalar quantities as follows:

$$W = \int_S \mathbf{I} \cdot \mathbf{dS} = \int_S (|\mathbf{I}|\hat{\mathbf{r}}) \cdot (|\mathbf{dS}|\hat{\mathbf{n}})$$
$$= \int_S (I\ dS) \times (\hat{\mathbf{r}} \cdot \hat{\mathbf{n}}) = \int_S I\ dS \cos\theta = \int_S I_n\ dS \tag{4.4}$$

where $I_n = I \cos\theta$ is the component of sound intensity *normal* to the surface at the location of $\mathbf{dS}$; dS is the magnitude of the elemental surface area vector;

$\hat{\mathbf{r}}$ is the unit vector in the direction of sound propagation; $\hat{\mathbf{n}}$ is the unit vector normal to the surface; and θ is the angle between $\hat{\mathbf{r}}$ and $\hat{\mathbf{n}}$. It should be noted that the maximum value of the intensity is equal to its magnitude, $I_{\max} = I$, and is attained in the direction $\hat{\mathbf{r}}$; in general, components of the intensity vector, such as I_n (or components in the Cartesian directions $\hat{\mathbf{x}}$, $\hat{\mathbf{y}}$, or $\hat{\mathbf{z}}$) are less than $I_{\max}$.

The integral may be carried out over a spherical or hemispherical surface that surrounds the source. Other regular surfaces, such as a parallelepiped or a cylinder, are also used in practice, and, in principle, any closed surface can be used. If the source is nondirectional and the integration is carried out over a spherical surface having a radius r and centered on the source, sound intensity and sound power are related by

$$I(\text{at } r) = I_n(\text{at } r) = \frac{W}{S} = \frac{W}{4\pi r^2} \quad \text{W/m}^2 \quad (4.5)$$

where I = magnitude of intensity on the surface (at radius r)
  I_n = normal component of intensity on the surface (at radius r)
  W = sound power, W
  S = area of spherical surface, $= 4\pi r^2$, m^2
  r = radius of sphere, m

In general, a source is directional, and the sound intensity is not the same at all points on the surface. Consequently, an approximation must be made to evaluate the integral of Eq. (4.4). It is customary to divide the measurement surface into a number of subsegments each having an area S_i and to approximate the normal component of the sound intensity on each surface subsegment. The sound power of the source may then be calculated by a summation over all of the surface subsegments:

$$W = \sum_i I_{n_i} S_i \quad \text{W} \quad (4.6)$$

where I_{n_i} = normal component of sound intensity averaged over ith segment of area, W/m^2
  S_i = ith segment of area, m^2
  i = number of segments

Equation (4.6) may be expressed logarithmically as

$$L_W = 10 \log \sum_i S_i \times 10^{L_{I_i}/10} \quad (4.7)$$

where L_W = sound power level, dB re 10^{-12} W
  S_i = area of the ith segment, m^2
  L_{I_i} = normal sound intensity level averaged over the ith area segment, dB re 10^{-12} W/m^2 (where the subscript n has been omitted for simplicity)

When each of the subareas S_i on the measurement surface has the same area, S_ε, Eq. (4.7) reduces to

$$W = \sum_{i=1}^{N} I_{n_i} S_i = S_\varepsilon \sum_{i=1}^{N} I_{n_i} = \left(\frac{1}{N}\sum_{i=1}^{N} I_{n_i}\right) \times (N \times S_\varepsilon) = \overline{I}_n \times S \quad \text{W} \quad (4.8)$$

where $\overline{I}_n = (1/N)\sum_{i=1}^{N} I_{n_i}$, average normal sound intensity over the measurement surface, W/m²
S = total area of the measurement surface, m²

Equation (4.8) may be expressed logarithmically as

$$L_W = 10 \log \frac{\overline{I}_n}{I_0} + 10 \log \frac{S}{S_0} \quad (4.9)$$

$$L_W = L_I + 10 \log \frac{S}{S_0} \quad (4.10)$$

where L_W = sound power level, dB re 10^{-12} W
L_I = normal sound intensity level, dB re 10^{-12} W/m²
S = area of measurement surface, m²
$S_0 = 1$ m²
I_0 = reference sound intensity, internationally agreed upon as 10^{-12} W/m²

Equation (4.10) is usually used to determine the sound power level of a source from the sound intensity level except when the source is highly directional. For directional sources, the subareas of the measurement surface may be selected to be unequal and Eq. (4.7) should be used.

Free-Field Approximation for Sound Intensity

From Eq. (2.24) or (2.30), for the far field of a source radiating into free space, the magnitude of the intensity at a point a distance r from the source is

$$I(\text{at } r) = \frac{p_{\text{rms}}^2(\text{at } r)}{\rho c} \quad \text{W/m}^2 \quad (4.11)$$

where ρc = characteristic resistance of air (see Section 2.5), N · s/m³, equal to about 406 mks rayls at normal room conditions
p_{rms} = rms sound pressure at r, N/m²

Strictly speaking, this relationship is only correct in the far field of a source radiating into free space. Good approximations to free-space, or "free-field," conditions can be achieved in properly designed anechoic or hemi-anechoic rooms, or outdoors. Hence, Eq. (4.11) is approximately correct in the far field of a source over a reflecting plane provided that the space above the reflecting plane remains essentially a free field at distance r (see Section 4.3). Even if the free field is not perfect and a small fraction of the sound is reflected from the walls and ceiling of the room, an "environmental correction" may be introduced to allow valid measurements to be taken in the room. Environmental corrections are introduced in Section 4.9. The relations below are widely used in standards that determine the sound power of a source from a measurement of sound pressure level.

If a closed measurement surface is placed around a source so that all points on the surface are in the far field and the intensity vector is assumed to be essentially normal to the surface so that $I = I_n$ at all points on the surface, then Eqs. (4.7) and (4.11) can be combined to yield

$$W = \frac{1}{\rho c} \sum_i p_i^2 S_i \qquad (4.12)$$

where p_i = average rms sound pressure over area segment S_i, N/m²

We may express Eq. (4.12) logarithmically as

$$L_W = 10 \log \sum_i S_i \times 10^{L_{pi}/10} - 10 \log D \qquad (4.13)$$

where
L_W = sound power level, dB re 10^{-12} W
L_{pi} = sound pressure level over the ith area segment, dB re 2×10^{-5} N/m²
S_i = area of ith segment, m²
$10^{L_{pi}/10} = p_i^2/p_{ref}^2$
$p_{ref} = 2 \times 10^{-5}$ N/m²
$D = \rho c W_0/p_{ref}^2 = \rho c/400$

and values for $10 \log D$ may be found from Fig. 1.7; at normal temperatures and pressures, the $10 \log D$ term is negligible.

Note that in Eq. (4.13), if the A-weighted sound power level in bels were being computed, the constant 10 in front of the logarithm would disappear, and the result (with $D = 1$) would be

$$L_{WA} = \log \sum_i S_i \times 10^{L_{pAi}/10} \qquad (4.14)$$

When each subsegment S_i has an equal area and $10 \log D = 0$, in analogy with Eqs. (4.8)–(4.10), Eq. (4.13) can be expressed as

$$L_W = \langle L_p \rangle_S + 10 \log \left(\frac{S}{S_0}\right) \qquad (4.15)$$

where L_W = sound power level, dB re 10^{-12} W
$\langle L_p \rangle_S$ = sound pressure level averaged on a mean-square basis over the measurement surface (surface sound pressure level, dB re 2×10^{-5} N/m^2)
S = area of the measurement surface, m^2
$S_0 = 1$ m^2

Equation (4.15) is usually used to determine the sound power level of a source from the sound pressure level except when the source is highly directional. For directional sources, the subareas of the measurement surface may be selected to be unequal; Eq. (4.13) should be used (usually with 10 log $D = 0$).

Hence, the sound power level of a source can be computed from sound pressure level measurements made in a free field. Equations (4.13) and (4.15) are widely used in standardized methods for determination of sound power levels in a free field or in a free field over a reflecting plane.

Sound Power Determination in a Diffuse Field

The sound power level of a source can also be computed from sound-pressure-level measurements made in an enclosure with a diffuse sound field because in such a field the sound energy density is constant; it is directly related to the mean-square sound pressure and, therefore, to the sound power radiated by the source. The sound pressure level in the reverberant room builds up until the total sound power absorbed by the walls of the room is equal to the sound power generated by the source. The sound power is determined by measuring the mean-square sound pressure in the reverberant field. This value is either compared with the mean-square pressure of a source of known sound power output *(comparison method)* or calculated directly from the mean-square pressure produced by the source and a knowledge of the sound-absorptive properties of the reverberant room *(direct method)*. Depending on the method used, either Eq. (4.16) or Eq. (4.18) is used to determine the sound power level of the source in a diffuse field.

Diffuse sound fields can be obtained in laboratory reverberation rooms. Sufficiently close engineering approximations to diffuse-field conditions can be obtained in rooms that are fairly reverberant and irregularly shaped. When these environments are not available or when it is not possible to move the noise source under test, other techniques valid for in situ determination of sound power level may be used and are described later in this chapter.

All procedures described in this chapter apply to the determination of sound power levels in octave or one-third-octave bands. The techniques are independent of bandwidth. The A-weighted sound power level is obtained by summing (on a mean-square basis) the octave-band or one-third-octave-band data after applying the appropriate A-weighting corrections. A-weighting values are listed in Table 1.4.

Nonsteady and impulsive noises are difficult to measure under reverberant-field conditions. Measurements on such noise sources should be made either under free-field conditions or using one of the techniques described in Chapter 18.

82 DETERMINATION OF SOUND POWER LEVELS AND DIRECTIVITY OF NOISE SOURCES

4.5 MEASUREMENT ENVIRONMENTS

Three different types of laboratory environments in which noise sources are measured are found in modern laboratories: anechoic rooms (free field), hemi-anechoic rooms (free field over a reflecting plane), and reverberation rooms (diffuse field). In an anechoic room, all of the boundaries are highly absorbent, and the free-field region extends very nearly to the boundaries of the room. Because the "floor" itself is absorptive, anechoic rooms usually require a suspended wire grid or other mechanism to support the sound source, test personnel, and measurement instruments. A hemi-anechoic room has a hard, reflective floor, but all other boundaries are highly absorbent. Both anechoic and hemi-anechoic environments are used to determine the sound power level of a source, but the hemi-anechoic room is clearly more practical for testing large, heavy sources. The sound power level is derived from Eq. (4.7) if the sound intensity on a surface surrounding the source is measured directly or from Eq. (4.13) if the sound pressure level is measured on a surface surrounding the source and located in the far field.

In a reverberation room, where all boundaries are acoustically hard and reflective, the reverberant field extends throughout the volume of the room except for a small region in the vicinity of the source. The sound power level of a source may be determined from an estimate of the average sound pressure level in the diffuse-field region of the room coupled with a knowledge of the absorptive properties of the boundaries.

The sound pressure field in an ordinary room such as an office or laboratory space that has not been designed for acoustical measurements is neither a free field nor a diffuse field. Here the relationship between the sound intensity and the mean-square pressure is more complicated. Instead of measuring the mean-square pressure, it is usually more advantageous to use a sound intensity analyzer that measures the sound intensity directly (see Section 4.10). By sampling the sound intensity at defined locations in the vicinity of the source, the sound power level of the source can be determined. Equation (4.10) is used to determine the sound power levels from the sound intensity levels. If the subareas of the measurement surface are unequal, Eq. (4.7) may be used.

4.6 INTERNATIONAL STANDARDS FOR DETERMINATION OF SOUND POWER USING SOUND PRESSURE

ISO Standards

The International Organization for Standardization (ISO) has published a series of international standards, the ISO 3740 series,[3-13] which describes several methods for determining the sound power levels of noise sources. Table 4.1 summarizes the applicability of each of the basic standards of the ISO 3740 series. The most important factor in selecting an appropriate noise measurement method is the ultimate use of the sound-power-level data that are to be obtained.

TABLE 4.1 Description of the ISO 3740 Series of Standards for Determination of Sound Power

ISO Standard	Test Environment	Criteria for Suitability of Test Environment	Volume of Source	Character of Noise	Limitation on Background Noise	Sound Power Levels Obtainable	Optional Information Available
ISO 3741 Precision (Grade 1)	Reverberation room meeting specified requirements	Room volume and reverberation time to be qualified	Preferably less than 2% of room volume	Steady, broadband, narrow-band, or discrete frequency	$\Delta L \geq 10$ dB, $K_1 \leq 0.5$ dB	A weighted and in one-third-octave or octave bands	Other frequency-weighted sound power levels
ISO 3743-1 Engineering (Grade 2)	Hard-walled room	Volume ≥ 40 m^3, $\alpha \leq 0.20$	Preferably less than 1% of room volume	Any, but no isolated bursts	$\Delta L \geq 6$ dB, $K_1 \leq 1.3$ dB	A weighted and in octave bands	Other frequency-weighted sound power levels
ISO 3743-2 Engineering (Grade 2)	Special reverberation room	70 m$^3 \leq V \leq 300$ m^3, 0.5 s $\leq T_{nom} \leq 1.0$ s	Preferably less than 1% of room volume	Any, but no isolated bursts	$\Delta L \geq 4$ dB, $K_1 \leq 2.0$ dB	A weighted and in octave bands	Other frequency-weighted sound power levels
ISO 3744 Engineering (Grade 2)	Essentially free field over a reflecting plane	$K_2 \leq 2$ dB	No restrictions; limited only by available test environment	Any	$\Delta L \geq 6$ dB, $K_1 \leq 1.3$ dB	A weighted and in one-third-octave or octave bands	Source directivity; SPL as a function of time; single-event SPL; other frequency-weighted sound power levels

(*continued overleaf*)

TABLE 4.1 (*continued*)

ISO Standard	Test Environment	Criteria for Suitability of Test Environment	Volume of Source	Character of Noise	Limitation on Background Noise	Sound Power Levels Obtainable	Optional Information Available
ISO 3745 Precision (Grade 1)	Anechoic or hemi-anechoic room	Specified requirements; measurement surface must lie wholly inside qualified region	Characteristic dimension not greater than 1/2 measurement surface radius	Any	$\Delta L \geq 10$ dB, $K_1 \leq 0.5$ dB	A weighted and in one-third-octave or octave bands	Same as above plus sound energy level
ISO 3746 Survey (Grade 3)	No special test environment	$K_2 \leq 7$ dB	No restrictions; limited only by available test environment	Any	$\Delta L \geq 3$ dB, $K_1 \leq 3$ dB	A weighted	Sound pressure levels as a function of time
ISO 3747 Engineering or Survey (Grade 2 or 3)	Essentially reverberant field in situ, subject to stated qualification requirements	Specified requirements	No restrictions; limited only by available test environment	Steady, broadband, narrow-band, or discrete frequency	$\Delta L \geq 6$ dB, $K_1 \leq 1.3$ dB	A weighted from octave bands	Sound pressure levels as a function of time

Notes: ΔL is the difference between the sound pressure levels of the source-plus-background noise and the background noise alone; K_1 is the correction for background noise, defined in the associated standard; V is the test room volume; α is the sound absorption coefficient; T_{nom} is the nominal reverberation time for the test room, defined in the associated standard; K_2 is the environmental correction, defined in the associated standard; SPL is an abbreviation for "sound pressure level."

The principal uses of sound-power-level data include the development of quieter machines and equipment, compliance testing of products in production, acoustical comparisons of several products that may be of the same or different type and size, and preparing acoustical noise declarations for the general public.

In making a decision on the appropriate measurement method to be used, several factors should be considered: (a) the size of the noise source, (b) the movability of the noise source, (c) the test environments available for the measurements, (d) the character of the noise emitted by the noise source, and (e) the grade (classification) of accuracy required for the measurements. The methods described in this chapter are consistent with those of the ISO 3740 series. A set of standards with the same objectives is available from the ANSI or the Acoustical Society of America (ASA).

4.7 DETERMINATION OF SOUND POWER IN A DIFFUSE FIELD

Characteristics of Reverberation Rooms

Measurements of the sound power level of a device or machine may be performed in a laboratory reverberation room (see Section 7.8). The determination of the sound power level of a noise source in such a room is based on the premise that the measurements are performed entirely in the diffuse (reverberant) sound field (see Fig. 4.4). In the reverberant field, the average sound pressure level is essentially uniform, although there are fluctuations from point to point, and it is related to the sound power radiated by the source [see (Eq. 4.18)]. Information in regard to the directivity of the source cannot be obtained in a diffuse sound field.

The minimum volume of the room depends on how low in frequency valid measurements are to be taken. For example, if measurements in the 100-Hz one-third-octave band are desired, the minimum volume is recommended to be 200 m^3 in ISO 3741. On the other hand, if measurements are only needed down to the 200-Hz band, 70 m^3 will be adequate. Maximum room volume is constrained by the adverse effects of air absorption and the ability to make valid high-frequency measurements; a volume less than 300 m^3 is generally recommended. The equipment being tested should have a volume no greater than 2% of the room volume. The absorption coefficient of the surface that is closest to the equipment being evaluated (usually the floor) should not exceed 0.06, and the remaining surfaces of the room should be highly reflective, such that the reverberation time (in seconds) is greater than the ratio of volume V (m^3) to surface area S (m^2): $T_{60} > V/S$. In addition to these requirements on room volume and absorption, ISO 3741 includes requirements for background noise levels; temperature, humidity, and atmospheric pressure; instrumentation and calibration; installation and operation of the source under test; minimum distance between microphone and source; and performance requirements for the reference sound source, if used.

If the room is to be used to measure equipment that has discrete-frequency components in its noise emissions, it is often necessary to use additional microphone positions and additional source positions, and ISO 3741 includes detailed procedures to determine whether or not such positions are required. These procedures must be followed for each source tested, or, alternatively, the reverberation room itself can be "qualified" for the measurement of discrete-frequency components (see below). The need for additional source positions may be reduced by adding low-frequency sound absorption to the room or through the use of rotating diffusers.

Room Qualification

The ISO 3741 standard contains two detailed procedures for the qualification of reverberation rooms for the determination of sound power levels, one for qualifying the room for the measurement of broadband sound (Annex E) and one for qualifying the room for the measurement of discrete-frequency components (Annex A). For sources of broadband sound, the room is qualified by using a reference sound source placed at different positions in the room and determining the standard deviation of the measured space-averaged sound pressure levels in the room for each position. A minimum of six positions of the source are required, each with specified constraints on the distance between them, the distance from walls, and the distance from the microphone. The room is qualified according to ISO 3741 for the measurement of broadband sound if the standard deviation does not exceed the values given in Table 4.2.

The qualification of reverberation rooms for the measurement of noise that contains discrete-frequency components in the spectrum is complicated and time consuming. However, this only has to be performed once, and the benefit can be great. If the chamber is qualified for the measurement of discrete tones, there is no longer a need to perform initial tests to determine the number of microphone and source positions for each source under investigation.

Essentially, a "calibrated" loudspeaker is placed in the reverberation room and is driven by a series of discrete-frequency tones in each one-third-octave band. For example, there are 22 frequencies spaced 1 H apart in the 100-Hz one-third-octave band and 23 frequencies spaced 5 H apart in the 500-Hz octave band. The average sound pressure level in the room is determined at each frequency and corrected for the loudspeaker response (previously determined in a hemi-anechoic

TABLE 4.2 Qualification Requirements for Reverberation Room Used for Measurement of Broadband Noise Sources

Octave-Band Center Frequencies, Hz	One-Third-Octave-Band Center Frequencies, Hz	Maximum Allowable Standard Deviation, dB
125	100–160	1.5
250, 500	200–630	1.0
1000, 2000	800–2500	0.5
4000, 8000	3150–10,000	1.0

Source: ISO 3741: 1999.

Note: Annex A of ISO 3741 contains detailed procedures for the discrete-frequency qualification.

TABLE 4.3 Qualification Requirements for Reverberation Room Used for Measurement of Narrowband Noise Sources

Octave Band Center Frequency, Hz	One-Third-Octave-Band Center Frequency, Hz	Maximum Allowable Standard Deviation, dB
125	100–160	3.0
250	200–315	2.0
500	400–630	1.5
1000, 2000	800–2500	1.0

room). The room is qualified according to ISO 3741 if the standard deviation of the level in each one-third-octave band does not exceed the values given in Table 4.3.

Experimental Setup

An array of fixed microphone positions or a single microphone that traverses a path (often circular) in the reverberant room may be used to determine the average sound pressure level in the reverberant field. The number of fixed microphone positions, N_M, required depends on the results of an initial series of sound-pressure-level measurements using six positions. If the standard deviation of these initial measurements, s_M, is less than or equal to 1.5, then the original six microphone positions will suffice (i.e., the noise is essentially broadband). If $s_M > 1.5$, it is assumed that the source emits discrete tones and a larger number of microphone positions is usually required to obtain an adequate sampling of the sound field. In this case, N_M could range anywhere from 6 to 30 depending on the frequency and the magnitude of s_M. When a traversing microphone is used, the path length l must be at least $(\lambda/2)N_M$, where λ is the wavelength of sound at the lowest midband frequency of interest. The microphone path or array should be positioned in the room so that no microphone position is within a minimum distance d_{min} of the equipment being evaluated. The distance d_{min} is determined differently in the comparison method and the direct method for determination of sound power. These requirements are discussed below.

If the noise source under test is typically associated with a hard floor, wall, edge, or corner, it should be placed in a corresponding position in the reverberation room. Otherwise, it should be placed no closer than 1.5 m from any wall of the room. The source should not be placed near the geometric center of the room since in that location many of the resonant modes of the room would not be excited. For rectangular reverberation rooms, the source should be placed asymmetrically relative to the boundaries. ISO 3741 gives further information for special source locations and installation conditions.

Near the boundaries of the room and close to other reflecting surfaces such as stationary or rotating diffusers, the sound field will depart from the ideal state of diffusion. ISO 3741 prescribes the following conditions on the microphone positions used during the measurements:

When using fixed microphone positions:

1. No position shall be closer than 1.0 m from any surface of or within the room.
2. No position shall be closer than d_{min} from the source (defined below).
3. The distance between microphone positions shall be at least $\lambda/2$, where λ is the wavelength of sound at the lowest midband frequency of interest.

When using one or more continuous microphone traverses:

1. No point on the traverse shall be closer than 1.0 m from any surface of the room.
2. No point on the traverse shall be closer than 0.5 m from any surface of a rotating diffuser.
3. No position shall be closer than d_{min} from the source (defined below).
4. The microphone traverse should not lie in any plane within 10° of a room surface.
5. The length of the traverse shall be at least $l \geq 3\lambda$, where λ is the wavelength of sound at the lowest midband frequency of interest.
6. If multiple traverses are used, the minimum distance between their paths shall be at least $\lambda/2$.

Comparison Method

The procedure for determining the sound power level of a noise source by the comparison method[6] requires the use of a reference sound source (see Fig. 4.5 and ref. 13) of known sound power output. Using microphone positions that meet the above requirements, the procedure is essentially as follows:

1. With the equipment being evaluated at a suitable location in the room, determine, in each frequency band, the average sound pressure level (on a mean-square basis) in the reverberant field using the microphone array or traverse described above.
2. Replace the source under test with the reference sound source and repeat the measurement to obtain the average level for the reference sound source.

The sound power level of the source under test, L_W, for a given frequency band is calculated as

$$L_W = L_{Wr} + (\langle L_p \rangle - \langle L_p \rangle_r) \qquad (4.16)$$

where L_W = one-third-octave-band sound power level for source being evaluated, dB re 10^{-12} W

FIGURE 4.5 Reference sound source, Brüel and Kjær Type 4204. (Courtesy of Brüel and Kjær, Inc.)

$\langle L_p \rangle$ = space-averaged one-third-octave-band sound pressure level of source being evaluated, dB re 2×10^{-5} N/m^2

L_{Wr} = calibrated one-third-octave-band sound power level of reference source, dB re 10^{-12} W

$\langle L_p \rangle_r$ = space-averaged one-third-octave-band sound pressure level of reference sound source, dB re 2×10^{-5} N/m^2

To make certain that the reverberant sound field predominates in the determination of the average sound pressure level, the minimum distance $d_{\min}$ between the microphone(s) and the equipment being evaluated should be at least

$$d_{\min} = 0.4 \times 10^{(L_{Wr}-L_{pr})/20} \tag{4.17}$$

where L_{Wr} and L_{pr} are as defined for Eq. (4.16). [Note that although Eq. (4.17) states the actual requirement, ISO 3741 also recommends that the constant be 0.8 instead of 0.4 to ensure that the microphone is in the reverberant field.]

Direct Method

The direct method does not use a reference sound source. Instead, this method requires that the sound-absorptive properties of the room be determined by measuring the reverberation time in the room for each frequency band. Measurement of T_{60} is described in ISO 3741.

With this method, the space-averaged sound pressure level for each frequency band of the source being evaluated is determined as described above for the comparison method. The sound power level of the source is found from[4]

$$L_W = \overline{L_p} + \left\{ 10 \log \frac{A}{A_0} + 4.34 \frac{A}{S} + 10 \log \left(1 + \frac{S \times c}{8 \times V \times f}\right) \right.$$

$$\left. -25 \log \left[\frac{427}{400} \sqrt{\frac{273}{273 + \theta}} \times \frac{B}{B_0} \right] - 6 \right\} \quad \text{dB re } 10^{-12} \text{ W} \quad (4.18)$$

where L_W = band sound power level of sound source under test
$\overline{L_p}$ = band space-averaged sound pressure level of sound source under test, dB re 2×10^{-5} N/m²
A = equivalent absorption area of the room, = $(55.26/c)(V/T_{\text{rev}})$, m²
V = room volume, m³
T_{rev} = reverberation time for particular band
A_0 = reference absorption area, 1 m²
S = total surface area of room, m²
V = room volume, m³
f = midband frequency of measurement, Hz
c = speed of sound at temperature θ, = $20.05\sqrt{273 + \theta}$, m/s
θ = temperature, °C
B = atmospheric pressure, Pa
B_0 = 1.013×10^5 Pa
V_0 = 1 m³
T_0 = 1 s

To make certain that the reverberant sound field predominates in the determination of the average sound pressure level, the minimum distance $d_{\min}$ between the microphone(s) and the equipment being evaluated should be at least[4]

$$d_{\min} = 0.08 \sqrt{\frac{V/V_0}{T/T_0}} \quad \text{m} \quad (4.19)$$

where V and T_{rev} are as defined above. [Note that although Eq. (4.19) states the actual requirement, ISO 3741 also recommends that the constant be 0.16 instead of 0.08 to ensure that the microphone is in the reverberant field.]

Example 4.2. Assume a room at temperature 21.4°C with a volume of 200 m³, a surface area $S = 210$ m², and a reverberation time at 100 Hz of 3 s. The space-averaged sound pressure level $\langle L_p \rangle$ in the diffuse field with a given machine operating is 100 dB. Find the sound power level for this machine. Assume a discrete-frequency spectrum and an atmospheric pressure of 1000 mbars.

Solution Use Eq. (4.18) to determine the sound power. The wavelength corresponding to 100 Hz is 3.44 m:

$$c = 20.05\sqrt{273 + 21.4} = 344 \text{ m/s}$$

$$A = \frac{55.26}{344}\left(\frac{200}{3}\right) = 10.7 \text{ m}^2$$

The sound power level is

$$L_W = 100 + \left\{ 10 \log \frac{10.7}{1} + 4.34 \frac{10.7}{210} + 10 \log \left(1 + \frac{210 \times 344}{8 \times 200 \times 100}\right) \right.$$

$$\left. -25 \log \left[\frac{427}{400}\sqrt{\frac{273}{273 + 21.4}}\right] - 6 \right\}$$

$$= 100 + (10.3 + 0.22 + 1.62 - 0.3 - 6)$$

$$= 106.4 \text{ dB re } 10^{-12} \text{ W}(100 \text{ Hz mean frequency})$$

4.8 DETERMINATION OF SOUND POWER IN A FREE FIELD USING SOUND PRESSURE MEASUREMENTS

Determination of the sound power level produced by a device or a machine may be performed in a laboratory anechoic or hemi-anechoic room. Alternatively, a hemi-anechoic environment may be provided at an open-air site above a paved area, distant from reflecting surfaces such as buildings, and with a low background noise level. This environment approximates a large room with sound-absorptive treatment on ceilings and walls with the equipment under test mounted on the hard, reflecting floor. Detailed information on anechoic and hemi-anechoic rooms is presented in Section 7.9.

The determination of the sound power level radiated in an anechoic or a hemi-anechoic environment is based on the premise that the reverberant field is negligible at the positions of measurement for the frequency range of interest. Thus, the total radiated sound power may be obtained by a spatial integration, over a hypothetical surface that surrounds the source, of the component of sound intensity normal to the surface of the source [see Eq. (4.3)]. When all points of the measurement surface are in the far field of the source, the magnitude of the intensity can be assumed to be equal to $p^2/\rho c$ [see Eq. (4.11)]. With the additional assumption that the magnitude of the intensity is equal to the normal component of sound intensity (i.e., the direction of sound propagation is essentially normal to the surface at the measurement points), the relationships given in Eqs. (4.12) and

(4.13) can be used to determine the sound power level of a source from simple measurements of sound pressure level. This is the basis for two key international standards for the determination of sound power levels of noise sources, ISO 3744[8] and ISO 3745,[9] discussed below. Essentially, a measurement surface is chosen and microphone positions are defined over this surface. Sound-pressure-level measurements are taken at each microphone position for each frequency band, and from these, the sound power levels are computed.

Selection of a Measurement Surface

The international standards discussed below allow a variety of measurement surfaces to be used; some are discussed here. In selecting the shape of the measurement surface to be used for a particular source, an attempt should be made to choose one where the direction of sound propagation is approximately normal to the surface at the various measurement points. For example, for small sources that approximate a point source, the selection of a spherical or hemispherical surface may be the most appropriate—the direction of sound propagation will be essentially normal to this surface. For machines in a hemi-anechoic environment that are large and in the shape of a box, the parallelepiped measurement surface may be preferable. For "tall" machines in a hemi-anechoic environment having a height much greater than the length and depth, a cylindrical measurement surface[14,15] may be the most appropriate.

Measurement in Hemi-Anechoic Space

The sound power determination in a hemi-anechoic space may be performed according to ISO 3744[8] for engineering-grade accuracy or according to ISO 3745[9] for precision-grade accuracy. ISO 3744 is strictly for hemi-anechoic environments, while ISO 3745 includes requirements for both hemi-anechoic and fully anechoic environments. These standards specify requirements for the measurement surfaces and locations of microphones, procedures for measuring the sound pressure levels and applying certain corrections, and the method for computing the sound power levels from the surface-average sound pressure levels. In addition, they provide detailed information and requirements on criteria for the adequacy of the test environment and background noise, calibration of instrumentation, installation and operation of the source, and information to be reported. Several annexes in each standard include information on measurement uncertainty and the qualification of the test rooms.

In terms of allowable measurement surfaces, ISO 3744 currently specifies the hemisphere and the parallelepiped,* while ISO 3745 explicitly defines only the hemisphere and the sphere. However, ISO 3745 includes a clause for "other microphone arrangements" and cites as an example papers describing the cylindrical measurement surface.[14,15] If a hemisphere is used to determine the location

---

*At the time of this writing, a revision to ISO 3744 was under consideration, which included the cylindrical measurement surface as well.

of the microphone positions, it has its center on the reflecting plane beneath the acoustical center of the sound source. To ensure that the measurements are carried out in the far field, the radius of the hemisphere for ISO 3744 measurements should be equal to at least two "characteristic" source dimensions and generally not less than 1.0 m. For ISO 3745, the requirements are slightly more stringent with the radius being at least twice the *largest* source dimension or three times the distance of the acoustical center of the source from the reflecting plane (whichever is larger) and at least $\lambda/4$ of the lowest frequency of interest. No microphone position can lie outside of the region qualified for measurements (the free-field region), and this requirement generally prevents microphones from being located too close to the walls. Outdoors, atmospheric effects are likely to influence the measurements if the radius of the test hemisphere is much greater than about 15 m, even in favorable weather.

Figure 4.6 gives an array of 20 microphone positions that may be used for measurements according to ISO 3744[8] for sources that emit predominantly broadband sound. The concern when measuring in a hemi-anechoic room is that reflections of sound from the hard floor may cause far-field interference at the microphone positions. This becomes more problematic when the noise emissions contain discrete tones; a small change in the position of the microphone may result in large variations in measured sound pressure level in those bands containing the tones. Therefore, for sources that emit discrete tones, ISO 3744 specifies a different microphone array, one having a greater distribution in the vertical direction. For precision measurements made in a hemi-anechoic environment according to ISO 3745,[9] a microphone array of at least 20 positions, each with a different vertical height, is required. The coordinates of this array are given in Table 4.4. Procedures are specified in both standards for determining whether or not the number of microphones is sufficient and for defining additional positions, if required. Note that the simultaneous deployment of a very large number of microphones may result in a situation where the support structures of these microphones become effective scatterers of the direct sound at high frequencies and can introduce substantial errors. Scanning the sound field with a single microphone can eliminate this error.

A parallelepiped array of microphone positions is frequently used for the determination of sound power levels for box-shaped machines and equipment. Detailed requirements for the selection of microphone positions and the criteria for the sufficiency of the number of microphones are given in ISO 3744.[8] A minimum of nine positions is required, but this number rapidly increases as the size of the source under test increases. The basic nine-position parallelepiped arrangement is illustrated in Fig. 4.7. As can be seen, this array is limited in its sampling in the vertical direction and so should be used with caution if the source is directional or emits discrete tones.

Another convenient measurement surface standardized in at least one industry test code[16] is the cylindrical microphone array illustrated in Fig. 4.8. The use of this array facilitates the measurement of tall-aspect-ratio sound sources such as data-processing equipment installed in racks. Since the array is usually

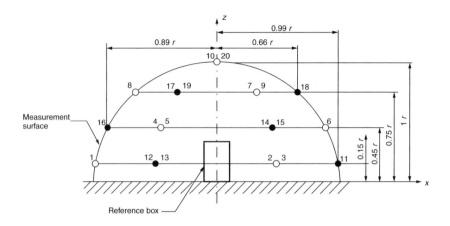
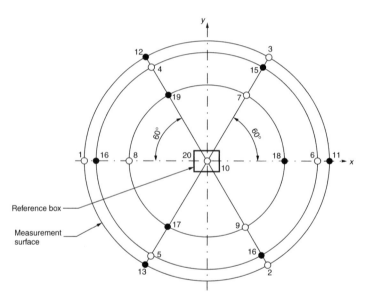

FIGURE 4.6 Microphone positions for a hemispherical measurement surface according to ISO CD 3744 (N1497). (Courtesy of the International Organization for Standardization, Geneva, Switzerland.)

implemented using continuously traversing microphones and a sufficient number of vertical heights, the accuracy is generally improved over that of the parallelepiped.[14,15]

The general procedure for determining the sound power level of the source according to either ISO 3744 or ISO 3745 can be summarized as follows:

1. The test room is set up and checked out and all environmental conditions are recorded.

TABLE 4.4 Twenty Microphone Positions on Surface of Hemisphere As Defined in ISO 3745

Position Number	x/r	y/r	z/r
1	−1.00	0	0.025
2	0.50	−0.86	0.075
3	0.50	0.86	0.125
4	−0.49	0.85	0.175
5	−0.49	−0.84	0.225
6	0.96	0	0.275
7	0.47	0.82	0.325
8	−0.93	0	0.375
9	0.45	−0.78	0.425
10	0.88	0	0.475
11	−0.43	0.74	0.525
12	−0.41	−0.71	0.575
13	0.39	−0.68	0.625
14	0.37	0.64	0.675
15	−0.69	0	0.725
16	−0.32	−0.55	0.775
17	0.57	0	0.825
18	−0.24	0.42	0.875
19	−0.38	0	0.925
20	0.11	−0.19	0.975

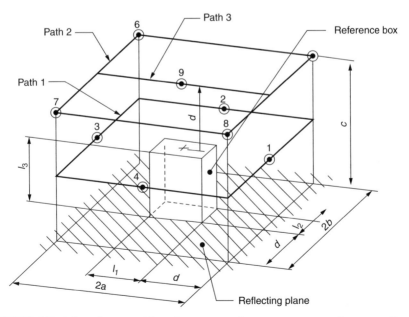

FIGURE 4.7 Microphone positions for a rectangular measurement surface according to ISO CD 3744 (N1497). (Courtesy of the International Organization for Standardization, Geneva, Switzerland.)

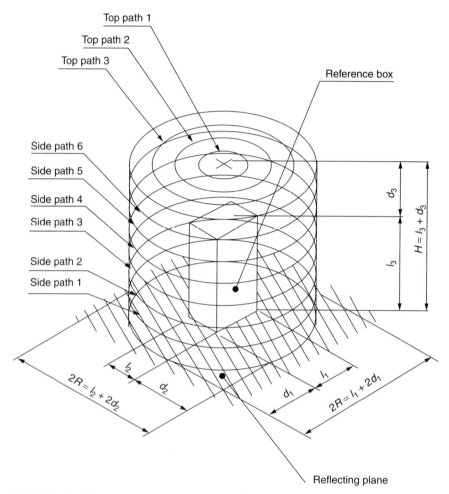

FIGURE 4.8 Microphone positions for a cylindrical measurement surface according to ISO CD 3744 (N1497). (From ISO CD 9997: 1999 Amended, Revision of Annex B (N1479). Courtesy of the International Organization for Standardization, Geneva, Switzerland.)

2. The microphones and instrumentation are calibrated.
3. The sound source is installed.
4. The measurement surface is selected and the microphones are set up in their proper positions.
5. The sound source under test is operated under specified conditions and sound pressure levels are measured at each microphone position for each frequency band of interest.
6. The sound source is turned off and the background noise levels are measured.

7. The sound pressure data are corrected for background noise and for environmental conditions (see below), if necessary.
8. The surface average band sound pressure levels are calculated.
9. The band sound power levels are calculated from the latter, taking into account corrections for meteorological conditions.

Equation (4.15) is used to determine the sound power level of the source when the subareas of the measurement surface associated with each microphone position are all equal. For highly directional sources, the standards require to increase the number of subareas of the measurement surface in the region of high directivity and to measure the sound pressure levels associated with each subarea. The sound power level can then be determined by using Eq. (4.13) for unequal subareas. The correction for background noise, denoted K_1, is specified in terms of the difference between the sound pressure levels measured with and without the source running, ΔL_i, as

$$K_1 = -10 \log(1 - 10^{-0.1 \Delta L_i}) \quad \text{dB} \qquad (4.20)$$

Measurement in Anechoic Space

In certain instances, the sound power level of a noise source must be determined in a totally free field, without the influence of a reflecting plane beneath the source. The reflecting plane not only causes constructive and destructive interference patterns in the far field—especially pronounced when the noise emissions contain discrete frequency components—but also may affect the radiated sound power of the source itself, especially at low frequencies.[17] Thus for critical measurements of tonal sources, measurements of noise sources that in normal use are not mounted over a hard surface, measurements of directivity, and other specialized measurements, a fully anechoic test environment may be desirable. In this case, the space-averaged mean-square sound pressure level is generally determined over a spherical measurement surface, and measurements are made according to ISO 3745.[9] Most of the procedures and requirements discussed above for hemi-anechoic measurements apply here also, with the exception of the measurement surfaces.

An array of at least 20 microphone positions is specified in ISO 3745, defined over the spherical surface by the Cartesian coordinates given in Table 4.5. If requirements on the sufficiency of the number of microphones given in ISO 3745 are not met, then the number of positions must be doubled, to 40. If the sound source is highly directional, the number of subareas in the regions of high directivity should be increased, as mentioned above. When measurements are made in an anechoic space according to ISO 3745, the sound power level of each frequency band is computed according to either Eq. (4.15) for equal subareas or Eq. (4.13) for unequal subareas. For the precision measurements of ISO 3745, however, two constants are included (instead of the single constant D), one to correct for meteorological conditions and one to "normalize" the measurements to specified "reference" meteorological conditions.

4.9 ENVIRONMENTAL CORRECTIONS AND DETERMINATION OF SOUND POWER LEVEL

In the previous sections, the only corrections applied for the determination of the average sound pressure level on the measurement surface were for the presence of background noise. However, if reflections from the room surfaces affect the measured surface sound pressure levels in any of the frequency bands of interest, a second correction may be required to account for these reflections. This so-called environmental correction, denoted K_2, is computed for both A-weighted values and individual frequency bands and is subtracted directly from the measured sound pressure level (after corrections for background noise are applied). The ISO 3744 standard[8] generally limits the environmental correction to a maximum of 2 dB. The correction may be determined in several ways:

1. By comparing the calibrated sound power level of a reference source, L_{Wr}, with the measured sound power level of the same source in the room, L_W. The environmental correction is then computed as $K_2 = L_W - L_{Wr}$.

TABLE 4.5 Twenty Microphone Positions on Sphere As Defined in ISO 3745

Position Number	x/r	y/r	z/r
1	−1.00	0	0.05
2	0.49	−0.86	0.15
3	0.48	0.84	0.25
4	−0.47	0.81	0.35
5	−0.45	−0.77	0.45
6	0.84	0	0.55
7	0.38	0.66	0.65
8	−0.66	0	0.75
9	0.26	−0.46	0.85
10	0.31	0	0.95
11	1.00	0	−0.05
12	−0.49	0.86	−0.15
13	−0.48	−0.84	−0.25
14	0.47	−0.81	−0.35
15	0.45	0.77	−0.45
16	−0.84	0	−0.55
17	−0.38	−0.66	−0.65
18	0.66	0	−0.75
19	−0.26	0.46	−0.85
20	−0.31	0	−0.95

2. By first determining the "equivalent sound absorption area" of the room, A, and computing the environmental correction as $K_2 = 10 \log[1 + 4(S/A)]$ decibels, where S is the area of the measurement surface. There are two methods currently specified in ISO 3744 for determining the value of A: a method using the measured reverberation time in the room and the so-called two-surface method. There is also a third method under consideration for standardization, a direct method using a calibrated reference sound source. The latest version of ISO 3744 should be consulted for details of using these methods.

3. By using the "approximate" method for determining A, which is then used only for determining an A-weighted value of the environmental correction, $K_{2A} = 10 \log[1 + 4(S/A)]$. Here the mean sound absorption coefficient α is estimated from Table 4.6.[10] The equivalent sound absorption area A is then calculated from $A = \alpha S_V$, where S_V is the total area of the boundary surfaces of the test room (walls, floor, ceiling) in square meters.

Reference should also be made to ISO 3746[10], a survey-grade standard for the determination of A-weighted sound power in rooms where the environmental correction may range up to 7 dB. This standard may be useful for taking measurements in situ, that is, when the source cannot be moved into a laboratory environment. The uncertainty of the A-weighted sound power level determined according to ISO 3746 is greater than that obtained when ISO 3744 or ISO 3745 is used. The average A-weighted sound pressure level is determined either on the parallelepiped measurement surface of Fig. 4.9 or the hemispherical measurement surface of Fig. 4.10, each shown with the minimum number of microphones required. Table 4.7 gives the coordinates for the four key microphone positions

TABLE 4.6 Approximate Values of Mean Sound Absorption Coefficient α

Mean Sound Absorption Coefficient α	Description of Room
0.05	Nearly empty room with smooth hard walls made of concrete, brick, plaster, or tile
0.1	Partly empty room; room with smooth walls
0.15	Room with furniture; rectangular machinery room; rectangular industrial room
0.2	Irregularly shaped room with furniture; irregularly shaped machinery room or industrial room
0.25	Room with upholstered furniture; machinery or industrial room with a small amount of sound-absorbing material on ceiling or walls (e.g., partially absorptive ceiling)
0.35	Room with sound-absorbing materials on both ceiling and walls
0.5	Room with large amounts of sound-absorbing materials on ceiling and walls

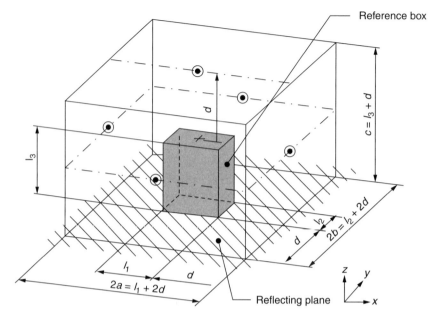

FIGURE 4.9 Microphone positions for a rectangular measurement surface according to ISO 3746. (Courtesy of the International Organization for Standardization, Geneva, Switzerland.)

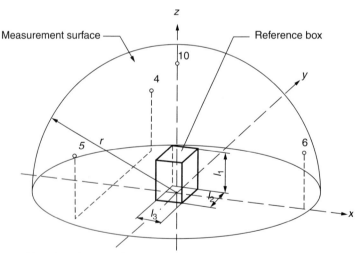

○ Key microphone positions

FIGURE 4.10 Microphone positions for a hemispherical measurement surface according to ISO 3746. (Courtesy of the International Organization for Standardization, Geneva, Switzerland.)

TABLE 4.7 Microphone Positions on Hemisphere for Survey Measurements According to ISO 3746

Microphone Position	Microphone Positions (All Heights, $z = 0.6R$)		
	x/r	y/r	z/r
4	−0.45	0.77	0.45
5	−0.45	−0.77	0.45
6	0.89	0	0.45
10	0	0	1.0
14	0.45	−0.77	0.45
15	0.45	0.77	0.45
16	−0.89	0	0.45
20	0	0	1.0

Note: The key microphone positions are 4, 5, 6, and 10. Additional microphone positions are 14, 15, 16, and 20.

shown in Fig. 4.10 for the hemisphere, along with the four additional positions that might be required under certain circumstances.

The measured sound pressure levels are first corrected for background noise according to Eq. (4.20), but for ISO 3746 only the A-weighted value, K_{1A}, is needed. The environmental correction, K_{2A}, is determined using methods similar to those mentioned above for ISO 3744, including the approximate method. The value of K_{2A} should not exceed 7 dB (or, the ratio S/A should be less than or equal to 1). The A-weighted sound power level is finally calculated according to Eq. (4.15).

4.10 DETERMINATION OF SOUND POWER USING SOUND INTENSITY

The fundamental procedures for the determination of sound power from sound intensity are formulated in Section 4.4. The sound intensity is measured over a selected surface enclosing the source. In principle, the integral over any surface totally enclosing the source of the scalar product (dot product) of the sound intensity vector and the associated elemental area vector provides a measure of the sound power radiated directly into the air by all sources located within the enclosing surface.

The precision of sound power determination based on sound intensity when the value of the intensity is calculated from measurements of sound pressure in a free field strongly depends on a number of factors. These include source type, measurement area related to source size, and presence of standing waves. The relationship between sound intensity and sound pressure in Eq. (4.11) is valid only in the far field of the free waves radiated by a source. Experience with the measurements on real sources has revealed that errors as large as 10 dB can occur if the selection of the measurement surface is unsuitable or if standing waves affect the measured sound pressure.

In recent years, methods and techniques for direct measurement of sound intensity have been developed. This direct measurement has made it possible to determine the sound power of a variety of source configurations and to make measurements in environments which are not suitable for use in sound power determinations based on measurement of sound pressure. These include sound power determination from measurements in the near field of large sources on test stands, elimination of the adverse effects of standing waves when measuring in enclosures, partial sound power determination of parts of the source with the entire source operating, and the analysis of the source behavior by finding the areas of sound radiation and absorption.

Other acoustical quantities are, by definition, based on sound power. These quantities include sound transmission coefficient, sound absorption coefficient, and radiation efficiency. Either the direct or indirect method described here can be used to determine the sound power for these purposes.

Sound intensity is defined as the sound power that propagates perpendicularly through a unit area. The sound power radiated by a source can be calculated using Eq. (4.5). Sound intensity can be calculated from the product of the sound pressure and particle velocity at a given field point. Because both of these quantities are functions of time, the sound intensity calculated from this time-dependent product is called instantaneous intensity. However, in noise control, it is more practical to work with time-averaged quantities so that acoustic intensity for periodic sound of frequency f and period $T = 1/f$ is defined at point r by

$$\mathbf{I}(r) = \frac{1}{T} \int_0^T p(r,t) \mathbf{u}(r,t) \, dt \qquad (4.21)$$

where $\mathbf{u}$ is the particle velocity and p is the sound pressure. The sound intensity $\mathbf{I}$ is a vector describing the time-averaged flow of power per unit area in watts per square meter normal to the intensity direction. This is typically applicable to the tonal components of the noise spectrum. The random components of the noise spectra usually consist of stationary noise and the length of the averaging time T is, in principle, related to the required measurement precision. In most situations, the sound power is determined in octave bands or one-third-octave bands, and the averaging time depends on the filter response—as specified in the instrumentation standards.

To determine the sound intensity, both the pressure and the particle velocity have to be measured. While good pressure microphones are available, precision measurement microphones for the particle velocity are not available. Recently, prototypes of velocity microphones that can measure all three components of the velocity vector by means of a hot-wire technique have been developed.[18] However, the standards for sound power measurements are based on a two-microphone (pressure microphones) technique for the velocity determination. This is based on Euler's equation, which links the particle velocity with the sound pressure gradient. This equation has the form

$$\mathbf{u} = -\int_0^t \frac{1}{\rho} \nabla p \, dt \qquad (4.22)$$

where ρ is the density of the medium, ∇p is the pressure gradient, and t is the integration time, which, as before, depends on the time function defining the noise. When measuring sound power, we are usually interested in the intensity vector component that is normal to the measurement surface. In this case, the gradient of p is replaced by $\partial p/\partial x$, where x is the direction of the normal to the measurement surface. The measurement of this gradient is approximated by $\partial p/\partial x \cong \Delta p/\Delta x \cong (p_1 - p_2)/\Delta x$. The pressure difference $p_1 - p_2$ is measured using two microphones spaced a distance Δx apart, which must be much smaller than the wavelength of sound. Figure 4.11 shows a two-microphone probe used to measure the pressure gradient. The magnitude of the measured sound intensity vector component $\hat{I}_x$ in the direction x is determined from

$$\hat{I}_x = -\frac{1}{2\rho \, \Delta x} E \left\{ [p_2(t) + p_1(t)] \int_0^t [p_2(t) - p_1(t)] \, dt \right\} \qquad (4.23)$$

where E is the expected value representing the time averaging and the pressures are on the two microphone diaphragms spaced a distance Δx apart. Figure 4.12 shows a block diagram of an intensity measurement instrument which calculates the intensity component $\hat{I}_x$ using Eq. (4.23).

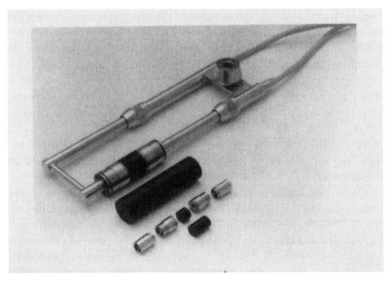

FIGURE 4.11 Sound intensity probe showing two $\frac{1}{2}$-in.-diameter microphones separated precisely by a 1.2-cm spacer. Just beneath is a 5-cm spacer. Below are two $\frac{1}{4}$-in. microphones separated by a 0.6-cm spacer. (Courtesy of Brüel and Kjær, Inc.)

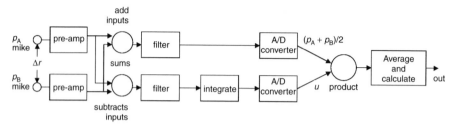

FIGURE 4.12 Block diagram of an analog/digital sound intensity analyzer using octave-band filters. Alternatively, the preamplifier outputs could be suitably amplified and converted to digital format. Then all further processing would be done digitally.

The instrument is also provided with filters with selectable bandwidths to be able to measure the sound power in frequency bands and also to calculate the total power with A-frequency weighting, as required by some standards. Because sound intensity is often used to locate the areas of sound energy radiation or absorption from the surface of a noise source, it is also very practical to measure the sound intensity in very narrow bands. This is conveniently accomplished by using a dual-channel FFT analyzer. The sound intensity measurement is based on a Fourier transform of Eq. (4.23):

$$\hat{I}_x(\omega) = \frac{1}{\omega \rho \, \Delta x} \text{Im}[S_{p_1 p_2}(\omega)] \qquad (4.24)$$

where $\text{Im}[S_{p_1 p_2}(\omega)]$ is the imaginary part of the cross-spectra of the output from the two microphones as measured by the FFT analyzer. This measurement is particularly suitable to analyze the sound power radiation from sources with pronounced line spectra. All sound sources which operate with mechanical periodicity such as rotating engines, fans, vehicles, and similar equipment usually have strong line spectra. The output of the FFT analyzer is usually connected to a computer, which executes the desired postprocessing. The most common calculations are the energy in the lines, in octave or one-third-octave bands, and the total energy, either linear or with A-frequency weighting, as required by some standards. To determine the energy in a band, at least 10 lines of the FFT analysis are generally needed.

The precision of intensity measurements depends on many factors, which can be summarized into two groups: the precision of the instrumentation and the precision of the sampling of the radiated intensity and subsequent calculation of the total radiated power.

There are two standards that define instrumentation requirements for intensity measurements: International Electrotechnical Commission (IEC) 1043[19] and ANSI 1.9–1996.[20] In principle, the intensity measured by an intensity meter should be the same as the intensity in a plane wave measured by a pressure microphone and calculated from the equation $I = p^2/\rho c$. An intensity meter consists of an intensity probe and a processor. The standards define allowable tolerances for both the probe and the processor for class 1 and 2 intensity instruments.

The probe shown in Fig. 4.11 consists of two pressure microphones separated by a spacer of a length selected by the user. The sound pressure is calculated from the arithmetic average of the microphone pressures. As shown in Eq. (4.23), the particle velocity is approximated by the pressure gradient, which depends on the difference of the microphone pressures. The selected microphone distance must be small enough to avoid a bias error at high frequencies. Figure 4.13 shows the bias error as a function of frequency and microphone distance. This type of bias error cannot be corrected.

Another bias error can occur at low frequencies if the microphones are too close together and not sufficiently phase matched. Figure 4.14 shows the bias error for a 0.3° microphone phase mismatch. The lower the frequency, the greater is the error. Increasing the microphone distance will decrease this error, but a large distance will cause, as mentioned above, a finite distance bias error. Fortunately, modern processors and signal processing can compensate for the phase mismatch error. Essential details are provided in the standards. Due to the existence of these two different bias errors, when the frequency range of the measured noise is large, the measurement usually needs to be repeated using two different microphone separations.

Another important quantity defining the useful frequency range of the instrument is the dynamic capability L_d, defined as

$$L_d = \theta_{pIR} - K \qquad (4.25)$$

where θ_{pIR} is the pressure minus residual intensity index and $K = 7$ for a 1-dB allowable measurement error. The importance and usefulness of L_d are shown and

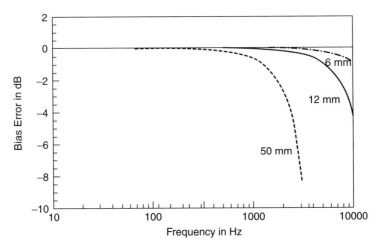

FIGURE 4.13 Intensity probe bias error as a function of frequency with different microphone distances as a parameter.

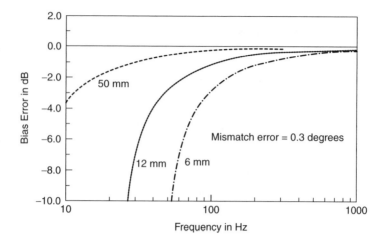

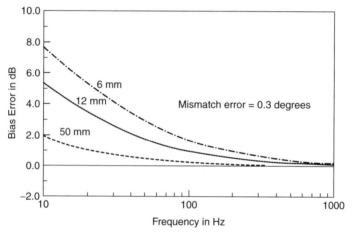

FIGURE 4.14 Intensity probe relative bias error as a function of frequency for different microphone distances as a parameter and mismatch error of 0.3° (plus and minus).

explained in Fig. 4.15. The pressure–residual intensity index θ_{pIR} can be obtained by placing both microphones in a small cavity in which the sound field is excited by an external source. In this way, both microphones are exposed to an identical sound pressure level L_p and the residual intensity level L_{IR} is measured using the intensity meter (usually a FFT analyzer). Ideally, the measured residual intensity should be zero, but due to the microphone mismatch, noise, and measurement instrumentation channel mismatch and other factors, L_{IR} is finite (see Fig. 4.15).

Both L_p and L_I are measured for an actual noise source and the pressure minus intensity index $L_{pI} = L_p - L_I$ is calculated. The intersection of the dynamic capability curve with L_I determines the lowest usable frequency. A phase mismatch will increase L_{IR} and shift the lowest usable frequency higher.

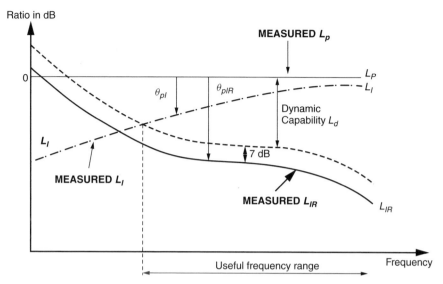

FIGURE 4.15 Dynamic capability of a sound intensity measurement system for 1-dB error as obtained from the pressure–intensity index. The values of L_p and L_I measured with the microphone in a small cavity are used to determine the residual pressure minus intensity index $\theta_{pIR} = L_p - L_{IR}$. The values of L_p and L_I on an actual noise source are used to determine the pressure minus intensity index, $\theta_{pI} = L_p - L_I$.

The pressure minus intensity index $L_{pI} = L_p - L_I$ obtained from the measurement of the actual noise source is an important quantity that characterizes at the measurement point the combination of the source properties and the sound field properties, particularly the effects of the wall reflections. The standards for the sound power measurements and other literature[21] provide essential details.

A typical machine (e.g., an engine) usually operates under conditions different from those under which it would be tested in a reverberant or anechoic room. Moreover, the radiation impedance "seen" by the source may be different from that in a controlled acoustical environment. Hence, its radiated power may be somewhat dependent on how it is mounted and on the proximity of surrounding surfaces.

The intensity technique permits, in most situations, measurement of the sound power of a source of any size operating in its natural environment or on a test stand. In many situations, the sound power radiated from parts of a source can be measured while the whole source is operating. In addition, the intensity technique has become an analytical tool to determine the sound power radiation from different regions of the surface of the source so that the areas of major power radiation can be found. These tasks require an extensive general knowledge of the characteristics of the sound fields, near and far, from the source, wave interference, reflected wave fields, and diffuse sound fields. The details of these important subjects can be found in the literature.[21,22]

Standards which use the direct measurement of sound intensity have been prepared by the ISO and are based on extensive research as well as practical experience gained from measurements. Part 1 of ISO 9614[23] is based on the sampling of sound intensity at discrete points. The intensity probe is moved from point to point and held at each point for a time sufficiently long to make the temporal averaging error small. Part 2 of the same standard uses a scanning technique. The intensity probe is moved over a prescribed path with a sufficiently low speed to satisfy the temporal averaging criterion. The measurement surface is subdivided into smaller sampling areas to secure the required uniformity in the motion of the hand-held probe. Part 3 of the same standard defines the conditions for using the scanning technique for more exact measurements than in Part 2. In practice, the intensity probe is held perpendicular to the measurement surface so that the scalar product converts into an algebraic product.

One of the principal advantages of using the intensity method in field situations is that the results are, in principle, independent of sources outside the measurement surface. The sound energy of external sources propagates through the measurement surface without contributing to the measured power of the sources within the surface—provided that no sound energy due to external sources is absorbed within the surface. Similarly, sound waves reflected from room boundaries or standing waves, unless they are too strong, do not affect the results of a sound power measurement. The standards are applicable to stationary sources located in a nonmoving medium. Because of instrumentation limitations, the frequency range is generally limited to the one-third-octave bands from 50 Hz to 6.3 kHz. A-frequency weighted data are calculated from one-third-octave-band data in this frequency range or from octave-band levels in the frequency range 63 Hz–4 kHz. The correction factors are given in Table 1.4.

Determination of the measurement uncertainty is an important component of the measurements. All standards define the methods and procedures for its determination. Before starting the measurement, the acoustical environment has to be examined for extraneous intensity, wind, gas flow, vibrations, and temperature. The next step consists of the calibration and field check of the instrumentation as specified in IEC 1043[19] or ANSI 1.9–1996.[20] The selection of the measurement surface is important. This is usually performed in two steps. First, an initial surface is selected and an initial measurement performed. The results are tested using a set of indicators that define the characteristics of both the pressure and intensity fields on the measurement surface.[23,24] If the values of these "field indicators" as specified in the standard are not satisfactory, the steps above have to be modified and the measurement repeated. Important field indicators are defined below.

The initial measurement surface is usually selected following the shape of the source at a distance greater than 0.5 m, unless that position is over an area that radiates an insignificant proportion of the sound power of the source under test. The selection of the number of measurement points depends on the shape, segments, and size of the measurement surface. A minimum of 10 points must be selected (greater number obviously leads to better precision, particularly at higher frequencies). If the source is large, one point per square meter of the measurement surface is usually selected, provided that the total number is not less than 50. If extraneous sound penetrates into the measurement area, the number of measurement points must be increased.

After both the sound pressure and the sound intensity are measured at all points, the results are tested by the field indicators in all frequency bands. Depending on the outcome from these indicators, the number and distribution of the measurement points and the distance of the measurement surface from the source may have to be changed. The standards provide tables and flow charts that define the actions to be taken.

The purpose of the indicators is to ensure a sufficient precision of the measurement. The statistical distributions of both the sound pressure and the sound intensity over the measurement surface depend on the source shape and its environment, primarily standing waves caused by sound reflections. Therefore, a general formula for the measurement error does not exist and the error must be determined experimentally. Figure 4.16 shows the flow diagram for the implementation of ISO 9614-1.

The field indicators as defined in ISO 9614 are as follows: F_1, temporal variability of the sound field; F_2, surface pressure–intensity; F_3, negative partial power; and F_4, field nonuniformity. The field indicators require measuring both the sound pressure and intensity. If the criteria for the indicators are not satisfied, the measurement arrangements must be modified. This concerns mainly the change of distance from the measurement surface and increase of the number of measurement points. The flow diagram indicates the actions to be taken. Table 4.8 provides detailed information on these actions.

Indicator F_1 checks the stationarity of the intensity field from several short time-average estimates of the sound intensity at one point of the measurement surface. Its value should be less than 0.6. This should assure that the source operation is steady and the environmental effects are not time variable.

Indicator F_2 is calculated from pressure square averages over the measurement area, converted into a pressure level L_p. Similarly, the intensity level L_I is determined from the arithmetic average of the sound intensities in individual points, all taken with positive sign, irrespective of the power flow out or the measurement surface at a particular point. Indicator F_2 must be smaller than the dynamic capability indicator L_d as defined by Eq. (4.25) and shown in Fig. 4.15 in order to keep the error caused by the instrumentation less than 1 dB for $K = 7$. This indicator is particularly important at low frequencies, as it is apparent in Fig. 4.15.

Indicator F_3 is similar to indicator F_2 except that the intensity level L_I is determined from intensity values with respect to its sign, which means that I is negative at the points where the sound power flows into the measurement surface. This can be caused by extraneous sources or strong reflections due to source environment. Thus $F_3 - F_2$, which is supposed to be less than 3 dB, is linked to the ratio of the sound power radiated out of the measurement surface to the sound power entering the measurement surface. Because both bias and

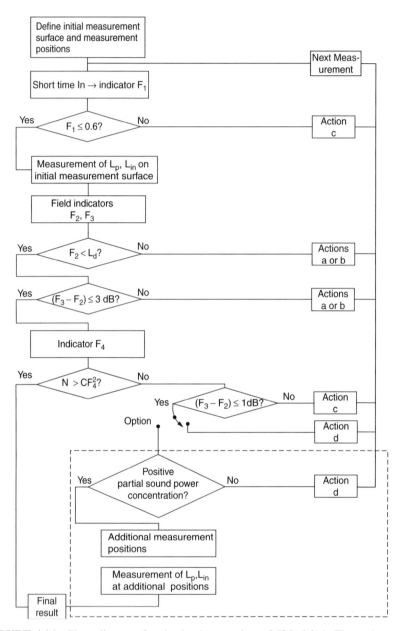

FIGURE 4.16 Flow diagram for the implementation of ISO 9614. The action codes defined in the center column of this figure are defined in Table 4.8. Calculation of the required number of measurement points requires a factor, C, which is defined in Table 4.9. The path enclosed in dashed lines represents an optimal procedure designed to minimize the number of additional measurement positions required on the initial measurement surface. (Courtesy of the International Organization for Standardization, Geneva, Switzerland.)

TABLE 4.8 Actions to Be Taken to Increase Grade of Accuracy of Determination

Criterion	Action Code[a]	Action
$F_1 > 0.6$	e	Take action to reduce the temporal variability of extraneous intensity, or measure during periods of less variability, or increase the measurement period at each position (if appropriate)
$F_2 > L_d$ or $(F_3 - F_2) > 3$ dB	a	In the presence of significant extraneous noise and/or strong reverberation, reduce the average distance of the measurement surface from the source to a minimum average value of 0.25 m; in the absence of significant extraneous noise and/or strong reverberation, increase the average measured distance to 1 m
	or b	Shield measurement surface from extraneous noise sources or take action to reduce sound reflections toward the source
Criterion 2 not satisfied and $1 \text{ dB} \leq (F_3 - F_2) \leq 3$ dB	c	Increase the density of measurement positions uniformly to satisfy criterion 2
Criterion 2 not satisfied and $(F_3 - F_2) \leq 1$ dB, and the procedure of ISO 9614 Section 8.3.2 either fails or is not selected	d	Increase average distance of measurement surface from source using the same number of measurement positions or increase the number of measurement positions on the same surface

[a] See Fig. 4.16.

random errors depend on $F_3 - F_2$, the 3-dB criterion satisfies the requirement to keep the bias errors low.

Indicator F_4 is the spatial variance of sound intensity measured at discrete points normalized to the average. The signs of intensity values are considered. This indicator reflects the variability of the power flow over the measurement surface. The higher is F_4, the more measurement points are needed. The number of the measurement points N is given by $N > C \times F_4^2$, where C is a factor which depends on frequency and required precision grade, as shown in Table 4.9.

Part 2 of ISO 9614 defines the measurement of both the sound pressure and intensity by scanning. The measurement area is subdivided into usually plane segments over which the probe is moved ("scanned") perpendicularly to the surface so that spatial averages of the measured quantities are obtained. The recommended moving pattern and speed are specified in the standard. Experimental evidence indicates that the results using the scanning technique are generally more precise than using point measurements.

TABLE 4.9 Values for Factor C

Octave Band Center Frequencies, Hz	One-Third-Octave-Band Center Frequencies, Hz	C Precision (Grade 1)	C Engineering (Grade 2)	C Survey (Grade 3)
63–125	50–160	19	11	
250–500	200–630	29	19	
1000–4000	800–5000	57	29	
	6300	19	14	
A weighted[a]				8

[a] 63 Hz–4 kHz or 50 Hz–6.3 kHz.

Part 3 of ISO 9614 is also based on the scanning technique. The requirements for the measurements and the tolerances to be satisfied are more strict than in Part 2.

In addition to the cited international standards, the ANSI has developed a standard, ANSI S12.12-1992,[25] which mirrors, in its fundamental concept and measurement procedures, the ISO standards. This standard contains a greater number of field indicators and provides more details on the measurement procedures. The selection of any of these standards depends on the product to be measured, the purpose of the sound power determination, and commercial criteria.

ECMA International has also issued a standard for determining sound power from sound intensity using a scanning technique.[26] The standard is intended for use with computer and business equipment.

After substantial experience with the determination of sound power via sound intensity has been achieved, it is expected that these standards will be revised.

4.11 SOUND POWER DETERMINATION IN A DUCT

The most common application of in-duct measurements is to determine the sound power radiated by air-moving devices. The sound power level of a source in a duct can be determined according to ISO 5136[27] from sound-pressure-level measurements, provided that the sound field in the duct is essentially a plane progressive wave, using the equation

$$L_W = L_p + 10 \log \frac{S}{S_0} \qquad (4.26)$$

where L_W = level of total sound power traveling down duct, dB re 10^{-12} W
L_p = sound pressure level measured just off centerline of duct, dB re 2×10^{-5} N/m^2
S = cross-sectional area of duct, m^2
$S_0 = 1$ m^2

The above relation assumes not only a nonreflecting termination for the end of the duct opposite the source but also a uniform sound intensity across the duct. At frequencies near and above the first cross resonance of the duct, the latter assumption is no longer satisfied. Also, when following the measurement procedures of ISO 5136, several correction factors are incorporated into Eq. (4.26) to account for microphone response and atmospheric conditions.

Equation (4.26) can still be used provided L_p is replaced by a suitable space average $\langle L_p \rangle$ obtained by averaging the mean-square sound pressures obtained at selected radial and circumferential positions in the duct or by using a traversing circumferential microphone. The number of measurement positions across the cross section used to determine $\langle L_p \rangle$ will depend on the accuracy desired and the frequency. (See ref. 27, Section 6.2.)

In practical situations, reflections occur at the open end of the duct, especially at low frequencies. The effect of branches and bends must be considered.[28] When there is flow in the duct, it is also necessary to surround the microphone by a suitable windscreen (see Chapter 14). This is necessary to reduce turbulent pressure fluctuations at the microphone, which can cause an error in the measured sound pressure level.

4.12 DETERMINATION OF SOURCE DIRECTIVITY[29,30]

Most sources of sound of practical interest are directional to some degree. If one measures the sound pressure level in a given frequency band a fixed distance away from the source, different levels will generally be found for different directions. A plot of these levels in polar fashion at the angles for which they were obtained is called the *directivity pattern* of the source. A directivity pattern forms a three-dimensional surface, a hypothetical example of which is sketched in Fig. 4.17. The particular pattern shown exhibits rotational symmetry about the direction of maximum radiation, which is typical of many noise sources. At low frequencies, many sources of noise are nondirectional—or nearly so. As the frequency increases, directivity also increases. The directivity pattern is usually determined in the far (free) field (see Fig. 4.4). In the absence of obstacles and reflecting surfaces other than those associated with the source itself, L_p decreases at the rate of 6 dB per doubling of distance.

Directivity Factor

A numerical measurement of the directivity of a sound source is the directivity factor Q, a dimensionless quantity. To understand the meaning of the directivity factor, we must first compare Figs. 4.17 and 4.18. We see in Fig. 4.18 the directivity pattern of a nondirectional source. It is a sphere with a radius equal in length to L_{pS}, the sound pressure level in decibels measured at distance r from a source radiating a total sound power W. The sources of Figs. 4.17 and 4.18 both radiate the same total sound power W, but because the source of Fig. 4.17

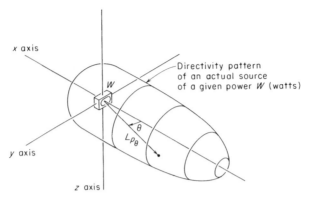

FIGURE 4.17 Directivity pattern of a noise source radiating sound power W into free space. A particular sound pressure level L_p is shown as the length of a vector terminating on the surface of the directivity pattern at angle θ. Sound pressure levels were measured at various angles θ and at a fixed distance r from the actual source in free space.

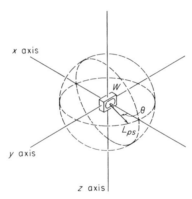

FIGURE 4.18 Spherical directivity pattern of a nondirectional source radiating acoustic power W into free space. At all angles θ and distance r, the sound pressure level equals L_{pS}, where $L_{pS} = 10 \log(10^{12} \times W/4\pi r^2)$.

is directional, it radiates more sound than that of Fig. 4.18 in some directions and less in others.

To derive a directivity factor Q, we must assume that the directivity pattern does not change shape regardless of the radius r at which it is measured. For example, if L_p at a particular angle is 3 dB greater than at a second angle, the 3-dB difference should be the same whether r is 1, 2, 10, or 100 m. This can only be determined in the far field of a source located in anechoic space.

The directivity factor Q_θ is defined as the ratio of (1) the mean-square sound pressure p_θ^2 [(N/m²)²] at angle θ and distance r from an actual source radiating W watts to (2) the mean-square sound pressure p_s^2 at the same distance from a nondirectional source radiating the same acoustic power W. Alternatively, Q_θ

is defined as the ratio of the intensity in the direction of propagation (W/m²) at angle θ and distance r from an actual source to the intensity at the same distance from a nondirectional source, both sources radiating the same sound power W. Thus

$$Q_\theta = \frac{p_\theta^2}{p_s^2} = \frac{I_\theta}{I_s} = \frac{10^{L_{p\theta}/10}}{10^{L_{pS}/10}} \quad \text{(dimensionless)} \quad (4.27)$$

or

$$Q_\theta = 10^{(L_{p\theta} - L_{pS})/10} \quad (4.28)$$

where $L_{p\theta}$ = sound pressure level measured a distance r and an angle θ from a source radiating power W into an anechoic space (see Fig. 4.17)
L_{pS} = sound pressure level measured at a distance r from a nondirectional source of power W radiating into anechoic space (see Fig. 4.18)

Note that Q_θ is for the angle θ at which $L_{p\theta}$ was measured and that L_{pS} and $L_{p\theta}$ are for the same distance r.

Directivity Index

The directivity index (DI) is simply defined as

$$\text{DI}_\theta = 10 \log Q_\theta \quad \text{dB} \quad (4.29)$$

or

$$\text{DI}_\theta = L_{p\theta} - L_{pS} \quad (4.30)$$

Obviously, a nondirectional source radiating into spherical space has $Q_\theta = 1$ and $\text{DI} = 0$ at all angles θ.

Relations between $L_{p\theta}$, Directivity Factor, and Directivity Index

The sound pressure level for the nondirectional source of Fig. 4.18 is

$$L_{pS} = 10 \log \frac{p^2 \text{ at distance } r}{4 \times 10^{-10}} \quad \text{dB} \quad (4.31)$$

From Eqs. (4.5) and (4.11) and taking the quantity $D = \rho c W_0 / p_{\text{ref}}^2$ to be small, L_{pS} is given by (see Fig. 4.18)

$$L_{pS} = 10 \log \frac{W \times 10^{12}}{4\pi r^2} \quad \text{dB} \quad (4.32)$$

From Eqs. (4.28) and (4.32), we find that

$$L_{p\theta} = 10 \log \frac{WQ_\theta \times 10^{12}}{4\pi r^2} \qquad (4.33)$$

where W is in watts and r is in meters. In logarithmic form

$$L_{p\theta} = L_W + \text{DI}_\theta - 20 \log r - 11 \text{ dB} \qquad (4.34)$$

where r is the distance from the acoustical center of the source in meters.

Determination of Directivity Index in Spherical Space

The directivity index DI_θ of a sound source in free space at angle θ and for a given frequency band is computed from

$$\text{DI}_\theta = L_{p\theta} - \langle L_p \rangle_S \qquad (4.35)$$

where $L_{p\theta}$ = sound pressure level measured at distance r and angle θ from source, dB
 $\langle L_p \rangle_S$ = sound pressure level averaged over test sphere of radius r (and area $4\pi r^2$) centered on and surrounding source

Determination of Directivity Index in Hemispherical Space

The directivity index DI_θ of a sound source on a rigid plane at angle θ and for a given frequency band is computed from

$$\text{DI}_\theta = L_{p\theta} - \langle L_p \rangle_H + 3 \text{ dB} \qquad (4.36)$$

where $L_{p\theta}$ = sound pressure level measured distance r and angle θ from source, dB
 $\langle L_p \rangle_H$ = sound pressure level of space-averaged mean-square pressure averaged over a test hemisphere of radius r (and area $2\pi r^2$) centered on and surrounding source

The 3 dB in this equation is added to $\langle L_p \rangle_H$ because the measurement was made over a hemisphere instead of a full sphere, as defined in Eq. (4.37). The reason for this is that the intensity at radius r is twice as large if a source radiates into a hemisphere as compared to a sphere. That is, if a nondirectional source were to radiate uniformly into hemispherical space, $\text{DI}_\theta = \text{DI} = 3$ dB.

Determination of Directivity Index in Quarter-Spherical Space

Some pieces of equipment are normally associated with more than one reflecting surface, for example, an air conditioner standing on the floor against a wall. The power level of noise sources of this type may be measured with those surfaces in place. This is done best in a test room with anechoic walls but with one hard wall forming an "edge" with the hard floor. The general considerations of the preceding paragraphs apply here as well. One determines the sound pressure level averaged over the quarter-sphere, $\langle L_p \rangle_H$, and also determines $L_{p\theta}$ as before. The directivity index is given by

$$\text{DI}_\theta = Lp_\theta - \langle L_p \rangle_Q + 6 \text{ dB} \tag{4.37}$$

REFERENCES

1. R. H. Bolt and K. U. Ingard, "System Considerations in Noise-Control Problems," in C. M. Harris (Ed.), *Handbook of Noise Control*, 1st ed., McGraw-Hill, New York, 1957, Chapter 22.
2. E. A. G. Shaw, "Noise Pollution—What Can Be Done?" *Phys. Today*, **28**(1), 46 (1975).
3. ISO 3740, *"Acoustics—Determination of Sound Power Levels of Noise Sources—Guidelines for the Use of Basic Standards,"* International Organization for Standardization, Geneva, Switzerland, 2000.
4. ISO 3741, *"Acoustics—Determination of Sound Power Levels of Noise Sources Using Sound Pressure—Precision Methods for Reverberation Rooms,"* International Organization for Standardization, Geneva, Switzerland, 1999.
5. ISO 3741, *"Acoustics—Determination of Sound Power Levels of Noise Sources Using Sound Pressure—Precision Methods for Reverberation Rooms,"* International Organization for Standardization, Geneva, Switzerland, 1999. Correction 1:2001.
6. ISO 3743-1, *"Acoustics—Determination of Sound Power Levels of Noise Sources—Engineering Methods for Small, Movable Sources in Reverberant Fields—Part 1: Comparison Method for Hard-Walled Test Rooms,"* International Organization for Standardization, Geneva, Switzerland, 1994.
7. ISO 3743-2, *"Acoustics—Determination of Sound Power Levels of Noise Sources Using Sound Pressure—Engineering Methods for Small, Movable Sources in Reverberant Fields—Part 2: Methods for Special Reverberation Test Rooms,"* International Organization for Standardization, Geneva, Switzerland, 1994.
8. ISO 3744, *"Acoustics—Determination of Sound Power Levels of Noise Sources Using Sound Pressure—Engineering Method in an Essentially Free Field over a Reflecting Plane,"* International Organization for Standardization, Geneva, Switzerland, 1994.
9. ISO 3745, *"Acoustics—Determination of Sound Power Levels of Noise Sources Using Sound Pressure—Precision Methods for Anechoic and Hemi-Anechoic Rooms,"* International Organization for Standardization, Geneva, Switzerland, 2003.
10. ISO 3746, *"Acoustics—Determination of Sound Power Levels of Noise Sources Using Sound Pressure—Survey Method Using an Enveloping Measurement Surface over a Reflecting Plane,"* International Organization for Standardization, Geneva, Switzerland, 1995.
11. ISO 3746, *"Acoustics—Determination of Sound Power Levels of Noise Sources Using Sound Pressure—Survey Method Using an Enveloping Measurement Surface over a Reflecting Plane,"* International Organization for Standardization, Geneva, Switzerland, 1995. Correction 1:1995.
12. ISO 3747, *"Acoustics—Determination of Sound Power Levels of Noise Sources Using Sound Pressure—Comparison Method in Situ,"* International Organization for Standardization, Geneva, Switzerland, 2000.

13. ISO 6926, *"Acoustics—Requirements for the Performance and Calibration of Reference Sound Sources Used for the Determination of Sound Power Levels,"* International Organization for Standardization, Geneva, Switzerland, 1999.
14. M. A. Nobile, B. Donald, and J. A. Shaw, "The Cylindrical Microphone Array: A Proposal for Use in International Standards for Sound Power Measurements," Proc. NOISE-CON 2000 (CD-ROM), paper 1pNSc2, 2000.
15. M. A. Nobile, J. A. Shaw, and R. A. Boyes, "The Cylindrical Microphone Array for the Measurement of Sound Power Level: Number and Arrangement of Microphones," Proc. INTER-NOISE 2002 (CD-ROM), paper N318, 2002.
16. ISO 7779, *"Acoustics—Measurement of Airborne Noise Emitted by Information Technology and Telecommunications Equipment (Second Edition),"* International Organization for Standardization, Geneva, Switzerland, 1999.
17. R. V. Waterhouse, "Output of a Sound Source in a Reverberation Chamber and Other Reflecting Environments," *J. Acoust. Soc. Am.*, **30**, 4–13 (1958).
18. Hans-Elias de Bree, *"The Microflown Report,"* Amsterdam, The Netherlands, www.Microflown.com, April 2001.
19. IEC 1043, *"Instruments for the Measurement of Sound Intensity—Measurement with Pairs of Pressure Sensing Microphones,"* International Electrotechnical Commission, Geneva, Switzerland, 1993.
20. ANSI S1.9, *"Instruments for the Measurement of Sound Intensity,"* Acoustical Society of America, Melville, NY, 1996.
21. F. J. Fahy, *Sound Intensity*, 2nd ed., E&FN SPON, Chapman & Hall, London, 1995.
22. J. Adin Man III and J. Tichy, "Near Field Identification of Vibration Sources, Resonant Cavities, and Diffraction Using Acoustic Intensity Measurements," *J. Acoust. Soc. Am.*, **90**, 720–729 (1991).
23. ISO 9614, *"Determination of Sound Power Levels of Noise Sources Using Sound Intensity Part 1: Measurement at Discrete Points"—1993; "Part 2: Measurement by Scanning"—1994; "Part 3: Precision Method for Measurement by Scanning"—2000*, International Organization for Standardization, Geneva, Switzerland.
24. F. Jacobsen, "Sound Field Indicators: Useful Tools." *Noise Control Eng. J.*, **35**, 37–46 (1990).
25. ANSI S12.12, *"Engineering Method for the Determination of Sound Power Levels of Noise Sources Using Sound Intensity,"* Acoustical Society of America, Melville, NY, 1992.
26. ECMA-160: *"Determination of Sound Power Levels of Computer and Business Equipment using Sound Intensity Measurements; Scanning Method in Controlled Rooms,"* 2nd edition, ECMA International, Geneva, Switzerland, December 1992. A free download is available from www.ecma-international.org.
27. ISO 5136, *"Acoustics—Determination of Sound Power Radiated into a Duct by Fans and Other Air Moving Devices—In-Duct Method,"* International Organization for Standardization, Geneva, Switzerland, 2003.
28. P. K. Baade, "Effects of Acoustic Loading on Axial Flow Fan Noise Generation," *Noise Control Eng. J.*, **8**(1), 5–15 (1977).
29. L. L. Beranek, *Acoustical Measurements*, Acoustical Society of America, Woodbury, NY, 1988.
30. L. L. Beranek, *Acoustics*, Acoustical Society of America, Woodbury, NY, 1986.

CHAPTER 5

Outdoor Sound Propagation

ULRICH J. KURZE

Mueller-BBM GmbH
Planegg near Munich, Germany

GRANT S. ANDERSON

Harris Miller Miller, & Hanson Inc.
Burlington, Massachusetts

5.1 INTRODUCTION

This chapter deals with the description and prediction of sound due to sources in an outdoor environment. Specifically, it deals with the propagation path from source to receiver. Propagation is affected by geometrical spreading, ground effects including reflection and refraction due to temperature and wind speed vertical gradients, attenuation from intervening barriers, general reflections and reverberation, atmospheric absorption, and attenuation from intervening vegetation.

5.2 GENERAL DISCUSSION

The atmosphere is in constant motion due to wind and sun at amplitudes that are large compared to the amplitudes of sound–particle velocity. This constant motion results in considerable distortion of sound waves and considerable variability of propagation conditions. Ever since the careful observations and first scientific modeling of outdoor sound propagation by O. Reynolds, Lord Rayleigh, and Lord Kelvin in the nineteenth century,[1] numerous experimental and mathematical studies have provided detailed understanding of the effects of mechanical and thermal turbulence, humidity (including fog), boundary conditions at the ground surface, and obstacles such as trees, walls, and buildings in the propagation path.

Unfortunately, much of the vast amount of information published on outdoor sound propagation in scientific papers is not relevant to the practical control of noise from recreational and industrial facilities or from road, rail, and air traffic. Relevance is determined by the rating criteria required by governmental agencies for assessment of community environments, such as described in the ISO 1996 series and various governmental regulations. Relevant criteria most often are given in the form of two acoustical descriptors: the equivalent continuous and the average maximum A-weighted sound pressure level, L_{eq} (as defined in Chapter 2) and $L_{A,max}$ (e.g., during a single-vehicle pass-by).

For that purpose, mainly average favorable conditions for sound propagation (plus the frequency of such conditions) need to be considered. Unfavorable conditions, which result in a large and uncertain range of low sound pressure levels, play a minor role in Europe.

For example, when favorable conditions occur during 50% of the time and unfavorable conditions result in at least 5-dB lower levels, then the equivalent continuous level

$$L_{eq} = 10 \lg \left(\frac{50}{100} 10^{L_{fav}/10} + \frac{50}{100} 10^{L_{unfav}/10} \right) \text{ dB}$$
$$< L_{fav} - 3 \text{ dB} + 10 \lg(1 + 10^{-5/10}) \text{ dB} = L_{fav} - 1.8 \text{ dB} \quad (5.1)$$
$$> L_{fav} - 3 \text{ dB}$$

lies some 2–3 dB below the average maximum level L_{fav} that occurs under favorable propagation conditions. It never lies outside that range, independent of the actual distribution of levels L_{unfav} for unfavorable propagation conditions.

The techniques in this chapter concentrate on such average favorable conditions for sound propagation to receiver positions about 4 m above the ground (first story above the ground story or higher). Additional techniques will be needed in governmental jurisdictions that require assessment of somewhat lower levels received 1.5 m above the ground—as is common within the United States and Canada.

In general, the level of outdoor sound decays with increasing distance between source and receiver. This geometrical divergence is most important for attenuation near the source, while meteorological conditions dominate attenuation further away.

Barriers, buildings, and hills that interrupt direct propagation from source to receiver are most important for excess attenuation. Of less importance are the effects of atmospheric absorption, porous ground (with receivers above 3–5 m), trees, single reflections from buildings, and reverberation in forests, valleys, and street canyons. Even though such effects are well understood, their computation requires input that is often not available in engineering practice—such as the spatial distribution of relative humidity or the effective flow resistivity of the ground between source and receiver. For planning purposes, engineering estimates or conventions have to be employed, rather than detailed models, to account for such effects.

In a homogeneous atmosphere at rest, geometrical divergence of sound from point sources is described by rigorous solutions of the wave equation in spherical coordinates. In these solutions, sound travels on a straight course in all directions at the same speed.

In reality, however, sound paths are not straight, because sound speed varies with temperature and wind velocity, mainly as a function of height above the ground. Underwater sound propagation is similarly dependent on height, due to the variation of salt content with height. To account for refraction of sound under water, various specialized mathematical models have been developed: ray theory, the spectral method or fast-field program (FFP), the normal mode, and the parabolic equation (PE).[2] Except for the normal-mode model, which is based on essentially two-dimensional fields between the bottom and the surface of an ocean, all these models have been proposed for airborne sound as well.

In addition, for airborne sound propagation the particular effects of ground impedance can be computed with a boundary element model (BEM)—although that model is limited to a nonrefracting (homogeneous) atmosphere. Also in the acoustical literature is a Meteo-BEM (presently limited to linear sound speed profiles) and a generalized-terrain PE. The latter is limited to axis-symmetric cases, just like the PE, but is also applicable to terrain profiles with moderate slope.[3]

In spite of these advanced models, the much-simpler ray theory has now gained engineering importance, exclusively. In the past, ray theory was dominant because of excessive computation-time requirements for FFP- and PE-like models. At present, the comparison of results obtained with FFP (within the limits of reliable atmospheric input data involving the Monin-Obukhov boundary layer theory) has shown no particular advantage to FFP over ray theory. Instead, the two models show surprising consistency for downwind conditions up to a distance of 2000 m.[4,5]

At a reception point outdoors, only the A-weighted overall sound pressure level is considered in most practical cases. Further calculation of sound transmission into buildings, which requires information about the spectral distribution, is limited to special problems of building acoustics. Consequently, older engineering estimates neglect the frequency dependence of propagation losses. Instead, they assume a single frequency band that represents the attenuation of A-weighted overall sound pressure from sources with typical spectra. Due to increased computer capabilities, advanced engineering models now include calculations in frequency bands. Except for distant propagation to receiver heights of 1.5 m, full-octave bands are sufficient—both for the precision obtainable and the requirement for traceable details. One-third-octave bands are needed for special cases only.

Regulations that allow various calculation procedures with frequency bands of different widths have a great disadvantage. The simpler procedure is not always on the safe side, and so users quickly learn to choose the procedure that yields "better" results, from their point of view. To avoid ambiguous results of this type, one should follow well-established governmental regulations and conventions—such as ISO 9613-2[6] in Europe and the regulations of funding authorities in the United States and Canada.

5.3 SOURCES

Point Sources at Rest

Historically, acousticians distinguished between point, line, and area sources. Modern computer capabilities allow for the exclusive use of large numbers of point sources to approximate lines and areas. Such approximation is officially considered satisfactory in Europe, but not in the United States, where federal regulations generally require use of line source algorithms and computer programs. Such line source programs eliminate the possibility that gaps in noise barriers, for example, will be ignored by computations when the approximating point sources are not sufficiently dense along the roadway.

European computer programs automatically break up extended sources into sufficiently small elements, which can be described as point sources. For this purpose, it is generally sufficient that

- the largest dimension of a source element be less than half the distance between source and receiver,
- the sound power be about equally distributed over the source element, and
- about the same propagation conditions exist from all points on the source element to the receiver.

The last requirement pertains to effects of the ground and obstacles in the propagation path. If they are relevant, the permissible difference in height above the ground between two source elements is typically less than 0.3 m. Whether a particular computer program follows these rules needs to be carefully determined by test cases.

Point sources are described by

- the location (x,y,z) of the center,
- the sound power level L_W in frequency bands (relative to 1 pW), and
- the directivity index D_I in frequency bands (in one or two dimensions).

Preferred frequency bands are octave bands with center frequencies from 63 to 8000 Hz. Lower frequencies may be important—for example, in the vicinity of jet engine test cells. They need special consideration. Higher frequencies are subject to strong and variable atmospheric attenuation. They can be neglected outdoors.

When available, one should use measured sound spectra and directivity of specific noise sources. If these are not available but the A-weighted overall sound power level is known, Table 5.1 contains an estimate of the octave-band spectra applicable to a surprising number of sound sources—such as roadway, railway and aircraft traffic, rifle fire, muffled diesel engines, and many industrial noise sources. For comparatively large and slow sources or for substantial vibration damping, the spectrum may be shifted to lower frequencies by one octave. On the other hand, comparatively small and fast-moving sources with little vibration

TABLE 5.1 Typical Unweighted and A-Weighted Octave-Band Source Spectra Relative to A-Weighted Overall Sound Power

Octave-Band Center frequency, Hz	63	125	250	500	1000	2000	4000	8000
Unweighted: $L_{W,\text{oct}} - L_{WA}$, dB	−2	1	−1	−3	−5	−8	−12	−23
A weighted: $L_{WA,\text{oct}} - L_{WA}$, dB	−28	−15	−10	−6	−5	−7	−11	−24

damping may have a spectrum that is shifted to higher frequencies by one octave. When the frequency band around 500 Hz is taken as an equivalent for the attenuation of A-weighted overall sound, this simplifying assumption is relative to one octave band below the average maximum of A-weighted octave-band noise.

The directivity index is normalized so that the average value of $10^{D_I/10}$ in all directions is unity. In many cases of rotational symmetry, the directivity index is sufficiently described in one dimension. Examples are smoke stacks and gunfire. For receivers on level ground, the directivity index is often described in the horizontal plane only. A practical example is the directivity of sound from aircraft engines during ground run-up tests. For turbo-jet aircraft on the ground, a typical dependence of directivity on angle and frequency is plotted in Fig. 5.1. Highest values occur in the direction $\phi = 120°$ from the forward direction and in the 1000-Hz frequency band.

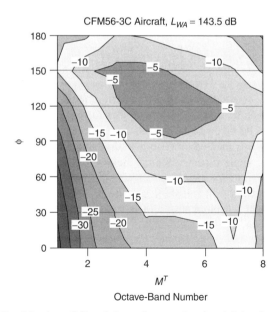

FIGURE 5.1 Combination of directivity and octave-band weighting for a turbojet engine with high bypass ratio at takeoff power setting. Octave-band number 2 is centered at 125 Hz and number 8 at 8000 Hz. Data are for a CFM-56-3C aircraft engine with $L_{WA} = 143.5$ dB.

The basic relation between the sound pressure level L_p at a distance d from a source and the sound power level L_W of the source is given by

$$L_p(d, \phi) = L_W + D_I(\phi) + D_\Omega - A(d, \phi) \qquad (5.2)$$

where $A(d, \phi)$ = transfer function due to effect of all elements in propagation path; attenuating or enhancing effects are discussed in Section 5.4
$D_I(\phi)$ = directivity index
D_Ω = index that accounts for sound propagation into solid angles less than 4π steradians

When the sound power output of a source is determined from outdoor measurements of the sound pressure level $L_p(d, \phi)$ in various directions ϕ at a certain distance d from the source, it is important to apply the same transfer function $A(d, \phi)$ from the sound power level to the sound pressure level as is applied in the opposite direction.

The description of sound emission in terms of the sound power level generally includes the assumption that the source radiates into free space (solid angle $\Omega = 4\pi$). Instead, when the source is located on the ground, and therefore the solid angle of radiation is 2π, two equivalent assumptions can account for the ground: (1) an additional incoherent image source beneath the ground, with the same sound power, or (2) a correction of the sound power level L_W by $D_\Omega = 3$ dB in Eq. (5.2). In general, when the source is located at a height h_S and the receiver at a height h_R above reflecting ground, the correcting index is

$$D_\Omega = 10 \lg \left[1 + \frac{d^2 + (h_S - h_R)^2}{d^2 + (h_S + h_R)^2} \right] \text{ dB} \qquad (5.3)$$

from geometrical consideration of incoherent image sources.[6] Furthermore, in some cases the ground effect A_{gr} from source to receiver, which is part of $A(d)$ in Eq. (5.2), must be taken into account.

When the source is located close to a wall or in a corner formed by two walls, sound pressure at some distance would include reflections from these walls as well. Similar to Eq. (5.3), the effect of these reflections is a level difference of about $D_\Omega = 6$ dB for a wall and about 9 dB for a corner. Alternatively—and necessarily for sound-absorbing walls—image sources may be considered separately.

Moving Sources

For traffic noise, it is common practice to consider the motion of sources along straight lines. Compared to continuous sources at rest, moving sources yield

- variable sound pressure levels at a stationary receiver,
- variable pitch of tonal components at a stationary receiver (Doppler effect), and
- different radiation due to different acoustical loading of the surrounding air.

The last of these effects is well described by theory, but in practice either included in the overall description of sound emission or neglected at low velocities.

The variation in pitch, from a higher pitch during vehicle approach to a lower pitch after passage, is determined by the factor $(1 + M)/(1 - M)$, in which $M = V/c$ is the Mach number resulting from the vehicle velocity V and the sound velocity c. For road traffic, this multiplicative factor roughly corresponds to one-third of an octave, but for high-speed maglev trains, it may reach a full octave. In addition, for maglev trains, the sinusoidal excitation of the track, which is due to the groove-passage frequency of magnets over the long-stator, results in a broadband maximum noise at receivers close to the track—when radiated frequencies shift upward from some parts of the track and downward from other parts (see Fig. 5.2). Pure tones from the vibration signal of the magnets are not audible during train passage.

For moving vehicles of any type, the variation in sound pressure level during approach and pass-by is generally not taken into account, except for its energy mean and, in some cases, its maximum value. An integrating sound-level meter is used to determine the total sound energy and then to report the pass-by's sound exposure level (SEL), which is the level of the integrated sound energy referred to a time interval of 1 s and is also called the single-event level. Since the direction of sound approach to a receiver from a long, straight track is equally distributed horizontally over 180°, Eq. (5.2) can be used without its directivity index to convert the SEL into a sound power level. Spreading this sound power over the distance the source travels in 1 h results in a sound power per unit length of track—the strength of a line source for one vehicle per hour. For further calculations, the line source is broken up into straight-line segments in the United States and Canada or into point sources in Europe.

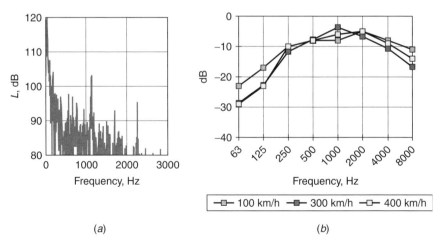

FIGURE 5.2 (*a*) Narrow-band spectrum of sound pressure level measured close to the magnets of a maglev train at 355 km/h. (*b*) A-weighted octave-band spectrum relative to A-weighted overall sound pressure level at some distance from a maglev train.

Although this procedure is normally applied to individual road and rail vehicles, the emission from the homogeneous section of a long train is usually evaluated differently. In such cases, the time integral of the squared sound pressure may be limited, without significant energy loss, to just when the center part of the homogeneous section passes by. This integral describes the sound emission from the homogeneous section in the same way as the SEL describes it for the entire train. Consequently, the same procedure can be used here (distribute the sound power over the distance the source travels in one hour) to determine the sound power per unit length of track for the train section selected. Sound power contributions from rolling noise of different vehicles are then added together. Contributions from sources at different heights need to be considered as separate line sources.

5.4 ELEMENTS IN PROPAGATION PATH

Overview

The transfer function of sound from a source to a receiver is determined by the sum of all attenuations along a particular ray path and by the contribution from all paths of direct and reflected sound. Attenuations account for spherical spreading A_{div}, ground effect A_{gr}, diffraction by barriers A_{bar}, partial reflections A_{refl}, atmospheric absorption A_{atm}, and miscellaneous others A_{misc}. They are described in the following.

Sound Propagation in Homogeneous Free Space over Ground

In the geometrical, high-frequency approximate solution of the Helmholtz equation for sound pressure p,

$$\Delta p + \left(\frac{2\pi f}{c}\right)^2 p = -\delta^2(\mathbf{r} - \mathbf{r}_S) \tag{5.4}$$

propagation of sound with frequency f from a point source at $\mathbf{r} = \mathbf{r}_S$ is described by ray theory. Rapid variations in phase are distinguished from slow variations in amplitude due to geometrical spreading and energy loss mechanisms. The ray trajectories are perpendicular to the surface of constant phase, which forms the wave front. The direction of average energy flux follows that of the trajectories. The amplitude of the field at any point can be obtained from the density of rays.

At short ranges, it is reasonable to assume straight rays. In free space, spherical spreading over a distance d results in an attenuation of

$$A_{div} = 10 \lg \frac{4\pi d^2}{d_0^2} \quad \text{dB} \tag{5.5}$$

where $d_0 = 1$ m. An observer above partially reflecting ground receives not only the direct ray but also a ground reflection, as sketched in Fig. 5.3. For simplicity,

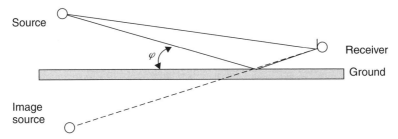

FIGURE 5.3 Ground reflection of a straight ray.

the plane-wave reflection coefficient

$$R_p = \frac{Z_s \sin\varphi - Z_0}{Z_s \sin\varphi + Z_0} \quad (5.6)$$

of a locally reacting surface of the ground, with impedance Z_s, may be used to determine the magnitude/phase effect of the reflection. The characteristic wave impedance of air, $Z_0 = \rho_0 c_0$, is generally very small compared to the magnitude of the ground impedance Z_s. Therefore, the reflection coefficient is usually about 1, except for very small angles φ—that is, when $|Z_s| \sin\varphi \ll Z_0$ and the reflection coefficient is about -1. Under the latter condition, the direct ray destructively interferes with the ground reflection and causes relatively low sound pressure close to the ground. In essence, the source and image source form a dipole, causing little radiation parallel to the ground. When the source is much closer to the ground than the receiver—e.g., for road traffic noise—the ground effect near the source can be attributed to a vertical radiation characteristic of the source. According to measurements over grassland near the shoulder of highways, the directivity index $D_I(\phi)$ of A-weighted road traffic noise drops from 0 dB at an elevation angle of 15° to about -5 dB at 0°. Generalized per reciprocity for arbitrary source and receiver heights, this relation converts to a reduction in overall A-weighted sound level[6]:

$$A_{\text{gr},D} = \left[4.8 - \frac{h_{\text{av}}}{d}\left(34 + \frac{600 \text{ m}}{d}\right)\right] \text{ dB} > 0, \quad (5.7)$$

where d is the source–receiver distance and

$$h_{\text{av}} = \tfrac{1}{2}(h_S + h_R) \quad (5.8)$$

is the average height of the sound ray from a source at height h_S to a receiver at height h_R.

Within a receiver distance of less than 200 m, this description is quite consistent with calculations of the Ontario noise regulation[7] from the equation

$$A_{\text{gr},O} = 10G \lg\frac{r}{15 \text{ m}} \text{ dB} > 0 \quad (5.9)$$

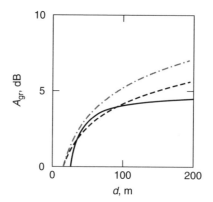

FIGURE 5.4 Excess attenuation A_{gr} over grassland as a function of distance d for an average height of 2 m: (—) according to Eq. (5.7); (- - -) according to Eq. (5.9) with $G = 0.5$; (-·—) according to Eqs. (5.9) and (5.10).

with

$$0 \le G \equiv 0.75\left(1 - \frac{h_{av}}{12.5 \text{ m}}\right) \le 0.66 \qquad (5.10)$$

for an average height of 2 m (or a receiver height of 4 m for a source close to the ground), as shown in Fig. 5.4.

More precise models have been developed to account for the curvature of wave fronts using a spherical reflection factor and possibly for an extended reaction of soft ground (e.g., fresh-fallen snow).[8] Such models explain a pronounced interference dip of sound pressure for sources and receivers within 2 m above the ground. This dip typically occurs in octave bands centered at 250 and 500 Hz. A dominant parameter of these more precise models is the effective flow resistivity of the ground. United States and Canadian agencies generally require the application of such models to properly account for this dip to receivers 1.5 m above the ground (the required receiver height). Note that this dip is caused by interference, not energy absorption at the ground.

However, near-complete destructive interference of two rays requires about equal amplitude and opposite phases. Diffuse rather than specular ground reflection, plus phase distortion due to thermal and wind turbulence in the propagation path, reduces or inhibits such near-complete interference, especially at large distances. Predicted ground attenuation A_{gr} of more than 20 dB for average lawn surface is unrealistic.

From an engineering point of view (especially in Europe), this interference effect is of little use. Some sound-level reduction may be experienced on a terrace surrounded by a lawn. But neighbors to industrial plants cannot be protected from sound impinging on their bedroom windows (4 m or more above the ground) by interfering reflections from a substantially lower ground. In Europe, planning of new roadways or commercial activities does not depend on the ground conditions of adjacent property, as it does in the United States and Canada.

Refraction of Sound in Inhomogeneous Atmosphere

Very important for outdoor sound propagation is the inhomogeneity of the atmosphere. The normal state of air is one of "convective equilibrium," in which the sound velocity c varies with temperature and wind velocity as a function of height above the ground. Higher temperature close to the ground results in higher sound velocity there and therefore in curvature of sound rays toward the sky. A component of wind in the direction of sound propagation causes refraction of sound rays toward the ground, because wind speed always increases with height. Linear wind and temperature profiles result in rays following a catenary curve that can be approximated by a circular arc with radius

$$R = \frac{1}{a \cos \phi} \tag{5.11}$$

as already described by Lord Rayleigh,[1] where a denotes the sound speed gradient due to vertical temperature and wind gradients. In the Nordic countries of Europe, measurements of wind and temperature are specified at heights of 0.5 and 10 m above level ground, expressly to determine a from the relation[9]

$$a = \frac{10^{-3}}{3.2 \text{ m}} \left(\frac{0.6 \, \Delta T}{1°C} + \frac{\Delta u}{1 \text{ m/s}} \right) \tag{5.12}$$

where $\Delta T = T(10 \text{ m}) - T(0.5 \text{ m})$ is the difference in temperature and $\Delta u = u(10 \text{ m}) - u(0.5 \text{ m})$ is the difference in wind speed component at those two heights.

For negative values of a, causing upward refraction, there is a limiting arc between source and receiver at a distance

$$D = \sqrt{\frac{2}{|a|}} \left(\sqrt{h_S} + \sqrt{h_R} \right) \tag{5.13}$$

that just grazes the ground (see Fig. 5.5). Beyond this distance, sound can reach the receiver not along a ray but only by diffraction. The limiting arc determines the boundary of an acoustical shadow zone. For source and receiver height $h_S = 0$ m and $h_R = 4$ m, respectively, and a moderate gradient $a = -10^{-4}$ m^{-1}, this distance $D = 282$ m. Stronger gradients reduce the distance, but typically not below 100 m. Consequently, meteorological conditions are often neglected for sound propagation over less than 100 m. To ensure reliable measurements of industrial noise at a minimum height of 4 m, the Nordic countries specify a minimum value $a = -10^{-4}$ m^{-1} at distances from 50 to 200 m and $a > -10^{-4}$ m^{-1} at larger distances.

Most noticeable is the case of temperature inversion ($a > 0$), which happens on calm days during dawn and dusk when dampness covers the ground and sunshine is restricted to the upper layers of the atmosphere. With temperature inversion, traffic noise is trapped in the lowest layer by downward refraction

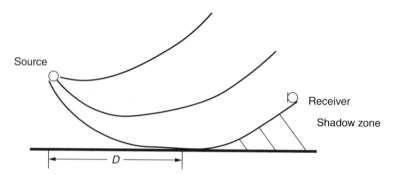

FIGURE 5.5 Sound rays and shadow formation for negative values of a.

and repeated reflections from the ground. As a result, sound can be heard over very large distances, due to divergence in only two (rather than three) dimensions—that is, no divergence upward. Even stronger effects can be observed for sound propagation over lakes and wide rivers.

Opposite effects occur from sound propagation over hot spots, as is well known from the poor quality of communication over an open fire. Ground surfaces warmed up by the sun or by industrial facilities may cause upward refraction and scattering of sound out of straight rays, both effects resulting in excess attenuation.

The combined effects of ground and micrometeorology have been studied in detail for the development of a heuristic model for outdoor sound propagation.[10] This model accounts for

- ray theory for the fastest path in air from source to receiver;
- one or more ground reflections during sound propagation downwind or for temperature inversion conditions, the number depending on the magnitude of the positive sound speed gradient;
- diffraction of sound into the shadow zone for a negative sound speed gradient;
- spherical reflection coefficient for refracted sound rays incident on the ground, depending on the effective flow resistivity of the ground; and
- reduced coherence of the various contributions to the sound pressure at the receiver, due to different travel times of these components and turbulence of the atmosphere.

Data calculated from this model have been compared to data measured over grassland at relatively low receiver positions with the following results:

- High excess attenuations calculated for negative gradients a at distances of more than 400 m have not been observed. Up to 1600 m, the average attenuation values were limited to approximately 20 dB in the frequency band centered at 630 Hz.

- For positive gradients a, average measured excess attenuations were limited to a few decibels and did not increase significantly with distance from 400 to 1600 m in the frequency bands below 200 Hz and above 630 Hz. Calculated values of excess attenuation came out higher.
- A pronounced excess attenuation in the frequency bands from 200 to 1000 Hz was both calculated and measured for low wind and a small positive gradient a.

A much simpler model for the ground effect on sound propagation outdoors has been adopted in ISO 9613-2[6]:

- Downwind (or a positive gradient a) is considered exclusively.
- The ground is either hard ($G = 0$) or porous ($G = 1$). If only a fraction of the ground is porous, G takes on a value between 0 and 1, the value being the fraction of the region that is porous.
- Three regions for ground reflection are distinguished: (1) the source region stretching $30 h_S$ (maximum) toward the receiver, (2) the receiver region stretching $30 h_R$ (maximum) toward the source, and (3) if the source and receiver regions do not overlap, a middle region in between.
- Similar to the angular parameter involved in Eq. (5.7), a parameter

$$q = \left(1 - 60 \frac{h_{\text{eff}}}{d}\right) \text{dB} > 0 \quad (5.14)$$

is introduced to account for reflections in the middle region by a ground attenuation equal to $-3q$ in the octave band centered at 63 Hz and equal to $-3q(1 - G_m)$ in all the higher octave bands up to 8 kHz center frequency, where G_m is the value of G in the middle region. Together with the assumption that sound is radiated into a half-space with solid angle 2π, there is a 6-dB increase in sound pressure level due to correlated ground reflection at 63 Hz, independent of ground porosity, and at higher frequencies for hard ground. Over porous ground, a 3-dB increase at higher frequencies is due to uncorrelated reflections.
- For the source and the receiver region, $q = 0.5$ accounts for the splitting of regions and G_m is replaced by G_S or G_R, where the indices S and R refer to the source and receiver regions, respectively. In addition, the following attenuations are taken into account for both regions in four octave bands:

Midband Frequency, Hz	Attenuation Contribution, dB	
125	$3Ge^{-[(h-5\text{m})/2.9\text{m}]^2}\left(1 - e^{-d/50\text{m}}\right)$	
	$+5.7Ge^{-(h/3.3\text{m})^2}\left(1 - e^{-(d/600\text{m})^2}\right)$	(5.15)
250	$8.6Ge^{-(h/3.3\text{m})^2}\left(1 - e^{-d/50\text{m}}\right)$	
500	$14Ge^{-(h/1.5\text{m})^2}\left(1 - e^{-d/50\text{m}}\right)$	
1000	$5Ge^{-(h/1.05\text{m})^2}\left(1 - e^{-d/50\text{m}}\right)$	

Largest attenuation occurs over porous ground in the octave band with center frequency 500 Hz. Consequently, the simplifying assumption that this frequency band is representative for the attenuation of A-weighted overall sound may not be conservative. Instead, it could actually underestimate the sound pressure level at the receiver.

The attenuation linearly increases with distance d up to about $d = 50$ m and then approaches a limit, except for the low-frequency band around 125 Hz, where a further extending ground wave is assumed.

Equations (5.15) show an exponential decay of attenuation with the square of receiver height. For the octave band at 500 Hz, the normalizing height in that band's equation is 1.5 m, roughly two wavelengths of sound. At 1000 Hz, this normalizing height is 1.05 m. The ratio of height to wavelength is similar in the relevant octave bands around 250 and 1000 Hz, indicating that the model is based on physical considerations together with experimental findings.

Despite essential simplifications compared to the heuristic model, the procedure described in ISO 9613-2 for octave-band calculations of ground attenuation is often replaced in practice by the procedure of Eq. (5.7) for A-weighted overall sound, mainly because ground factors G are often ambiguous.

The application of Eq. (5.15) to road and rail traffic noise with low source height is questionable. Ground effects near the source are already included in measured data. Uneven surfaces due to ditches for drainage are not covered by the model. Ground effects near the receiver can be excluded by an appropriate minimum height. Consequently, it has been decided that German guidelines for prediction of rail traffic noise, presently under revision, shall not account for frequency-dependent ground effects.

In contrast, U.S. requirements for road traffic computations consist of a more detailed computer model based on acoustical theory—the U.S. Federal Highway Administration (FHWA) Traffic Noise Model (TNM).[11] Version 2.5 of this model has been compared with more than 100 h of measured sound levels (L_{eq}) at 17 highway sites, most with flat terrain, with and without noise barriers. Measurements ranged to 400 m from the roadway during daytime hours under varying wind conditions. Under these conditions, the average difference between results of computations and measurements was within approximately 1 dB.[12]

Barriers

Any large and dense object that causes an acoustical shadow zone in the path of a sound ray is considered as a sound barrier. Such objects include walls, roofs, buildings, or the ground itself if it forms an edge or a hill. Sound penetrates into such shadow zones by diffraction and thereby suffers barrier attenuation D_z. The amount of attenuation is primarily determined by the Fresnel number

$$N = \frac{2z}{\lambda} \tag{5.16}$$

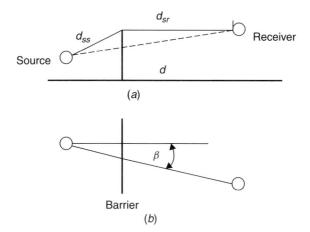

FIGURE 5.6 Barrier between source and receiver: (*a*) cross sectional view; (*b*) plan view.

where z is the increase in path length (the extension of a hypothetical rubber band between source and receiver) caused by one or more of the barrier edges and $\lambda = c/f$ is the wavelength of sound at frequency f. Note that z decreases with increasing angle of sound incidence β (see Fig. 5.6).

For $z = 0$ there is already a small barrier attenuation of about 5 dB plus an interference pattern at the edge of the shadow zone. Details of this interference pattern require a substantial computational effort that is not needed for engineering purposes. Therefore, preferred are simple approximations that are derived from extrapolations of the diffraction model and the ray-tracing model. When both models provide the same result for the line of sight over a hill or over undulating ground, the calculation of barrier insertion loss should be based on a limiting ($z = 0$) ground attenuation of about 5 dB independent of frequency, as described by Eq. (5.7). This consideration leads to the barrier insertion loss:

$$A_{\text{bar}} = \begin{cases} D_z - A_{\text{gr}} > 0 & \text{for diffraction over top edge of barrier} \\ D_z > 0 & \text{for diffraction around vertical edge of barrier} \end{cases} \quad (5.17)$$

where D_z is defined in Eq. (5.18).

Inclined barrier edges are considered to be top edges whenever the angle of inclination against the ground is less than 45°. Otherwise they are considered to be vertical edges.

Note in Eq. (5.17) that an intervening barrier causes the loss of all ground effects. At a height of 4 m or more over porous ground, this approximation is relatively accurate. Close to hard ground, however, it is not correct. Pressure doubling due to hard ground, which results in negative values of A_{gr}, persists even with intervening noise barriers and cannot result in larger values of A_{bar} than for porous ground.

The calculation of barrier attenuation is based on Huygens' model for wave fields, which takes all insonified points in a plane perpendicular to the ray from source to receiver as new sound sources. Points on a radius around the ray radiate with the same phase, varying between positive and negative values with increasing radius. Intervals between "phase zeros" in the hypothetical plane are called Fresnel zones. In free space, the contributions from neighboring Fresnel zones cancel each other, except for the center zone with the Fresnel radius $\sqrt{\lambda d_{so}}$, where d_{so} is the distance from the source to the plane. This center zone is the only uncanceled portion of the secondary wave front of Huygens' model and therefore is responsible for the entire sound field at the receiver. When this center zone is blocked by a barrier, the direct ray disappears.

In addition, a barrier partially blocks outer Fresnel zones and thereby inhibits the complete cancellation of their contributions. In the limit of many blocked Fresnel zones both above and on either side of the direct ray from the source to the receiver by a sufficiently high and long barrier, the intensity of the uncanceled portion decays with $1/N$; that is, the barrier attenuation increases with increasing path length difference over the top edge and with frequency.

Refined calculations of barrier attenuation include sound reflection at the barrier surface. When either the source or the receiver is close to a reflecting barrier surface, then the image source behind the barrier contributes almost as much to the sound field as does the actual source. This effect reduces the barrier attenuation but is often negligibly small.

Ground reflections can be important. When either the source or the receiver is high above the ground, ground-reflected contributions are relatively small due to relatively large path length differences z. But such high sources and receivers are generally not the case. Consequently, barrier attenuation over ground is computed 3 dB lower than the attenuation by a diffracting half-plane.

For outdoor sound propagation, the coherence of sound at different points in the sound field is reduced by wind and temperature fluctuations. This sets a limit for interference, depending upon the Fresnel radius and Fresnel number. Furthermore, one has to account for the curvature of rays due to gradients of wind and temperature. In the downwind direction, this curvature reduces the effective barrier height and the diffraction angle and, consequently, reduces the barrier attenuation—depending on the distance between the source, barrier, and receiver.

Theoretical and heuristic models have been developed to account for many or all of the effects just described. For prediction purposes, they suffer from the need for extremely detailed or unavailable input. In contrast, ISO 9613-2 requires less input—perhaps the upper limit of detail that can typically be obtained. In that standard, barrier attenuation is approximated by

$$D_z = 10 \lg(3 + 2NC_2C_3 K_{met}) \text{ dB} < 20 \text{ dB for } C_3 = 1 (<25 \text{ dB for } C_3 > 1) \tag{5.18}$$

where $C_2 = 20$ and includes the effects of (1) sound rays from the source via the barrier edge to the receiver, (2) sound rays from the image source in the ground, and (3) sound rays to the image receiver in the ground (if ground reflections are taken into account separately in special cases, then $C_2 = 40$); $C_3 = 1$ for diffraction at a single edge but can increase to 3 for diffraction at two or more edges in series, according to

$$C_3 = \frac{1 + (5\,\lambda/e)^2}{1/3 + (5\,\lambda/e)^2} \tag{5.19}$$

with e as the path length from the first to the last diffracting edge; and K_{met} is a meteorological correction factor which accounts for down-wind ray curvature and is calculated from

$$K_{met} = \exp\left(-\frac{1}{2000\text{m}}\sqrt{\frac{d_{ss}d_{sr}d}{2z}}\right) \tag{5.20}$$

where d_{ss} is the path length from the source to the first diffracting edge, d_{sr} is the path length from the last diffracting edge to the receiver, and d is the path length without the barrier.

All of Eqs. (5.17)–(5.20) are conventions that are based on first-order approximations of theoretical relations, plus field experience, for sound propagation in a refracting atmosphere with curvature parameter $a > 0$. They can be applied for barriers of infinite or finite length on the ground, with one to three perpendicular diffracting edges. The barrier length must be large enough to meet the requirements for reflectors (see below). For example, this is true for typical railway barriers of 2 m height above the railhead. At a distance of 25 m and a height of 3.5 m above the railhead, the A-weighted sound pressure level of freight cars is reduced by about 11 dB.

Calculations of ISO 9613-2 do not include the case of barriers aloft, like shelter roofs at gas stations. If diffraction at a single edge is dominant for the received sound, the calculation should still be applicable, however. Contributions from more than one edge may be assumed as incoherent for sufficiently large barriers. Minimum dimensions must meet the requirements for reflectors.

Reflectors and Reverberation

Any large object—a wall, roof, building, or road sign, for example—that has plane surfaces in the path of a sound ray and that blocks an area with Fresnel radius at least $\sqrt{\lambda d_{so}}$ is considered to be a reflector. More specifically according to ISO 9613-2, the projection of the minimum dimension of the object in the direction of sound incidence shall meet the requirement

$$l_{min} \cos\beta > \sqrt{\frac{2\lambda}{1/d_{so} + 1/d_{or}}} \tag{5.21}$$

where β is the angle of sound incidence on the reflecting plane and d_{so} and d_{or} are the path lengths from the reflection point to the source and to the receiver, respectively. Planes approximate nonplanar surfaces. Specular reflections are considered exclusively in terms of an image source (see Fig. 5.7).

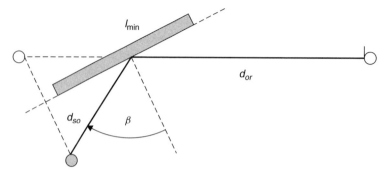

FIGURE 5.7 Specular reflection from a plane with minimum dimension l_{min}.

A reflection coefficient* $R < 1$ of the reflector reduces the sound power level of the image source by $-10 \lg(R)$ dB. Diffuse reflections from rough building facades, which result in scattering of sound toward the sky, are generally taken into account with a reflection loss of 1 or 2 dB. The directivity of a source needs to be considered in the direction of a receiver image.

Atmospheric Absorption

The attenuation due to atmospheric absorption

$$A_{atm} = \alpha d \tag{5.22}$$

is theoretically well understood and fully described in ISO 9613-1.[13] For prediction purposes, one has to select the appropriate conditions of relative humidity, temperature, and ambient pressure. When low temperature and high humidity are assumed, the attenuation is relatively low around 500 Hz. In contrast, in the frequency bands at and above 4000 Hz, attenuation is high enough to reduce contributions to the A-weighted overall level to negligible values at distances of a few hundred meters. On the other hand, A_{atm} itself is mostly negligible in the frequency bands at and below 125 Hz. Table 5.2 contains selected values.

TABLE 5.2 Atmospheric Attenuation Coefficient α for Octave Bands of Noise at Selected Nominal Midband Frequencies[6]

Temperature, °C	Relative Humidity, %	Atmospheric Attenuation Coefficient α, dB/km							
		63 Hz	125 Hz	250 Hz	500 Hz	1000 Hz	2000 Hz	4000 Hz	8000 Hz
10	70	0.1	0.4	1.0	1.9	3.7	9.7	33	117
20	70	0.1	0.3	1.1	2.8	5.0	9.0	23	77
15	50	0.1	0.5	1.2	2.2	4.2	10.8	36	129
15	80	0.1	0.3	1.1	2.4	4.1	8.3	24	83

*ISO 9613-2 uses ρ instead of R.

Effects of Ground Cover and Trees

The consideration of ground cover and trees is often excluded from engineering prediction of excess attenuation in the propagation path of sound. The reason is twofold. The effect is small and unreliable. The situation is not only subject to seasonal variations but may change with different land use.

However, much money can be wasted building roadway noise barriers longer than truly needed if tree attenuation is ignored. This happens when barrier lengths are needlessly extended to reduce sound coming from far up and down the roadway—needlessly because trees adequately reduce that flanking sound by themselves. Note that sound arriving perpendicularly from the roadway might typically pass through only 50 m of trees, whereas sound arriving from far up and down the roadway passes through 500–1000 m of trees—for significant attenuation.

5.5 INTERACTION OF ELEMENTS AND CONTRIBUTIONS FROM VARIOUS PATHS

Standard Regulations for Consideration of Interactions

The attenuations discussed for individual elements in the propagation path of a sound ray are added to yield the total attenuation:

$$A = A_{\text{div}} + A_{\text{gr}} + A_{\text{bar}} + A_{\text{atm}} \tag{5.23}$$

where A_{div} = attenuation due to spherical spreading, predicted according to Eq. (5.5)

A_{gr} = attenuation due to ground effects, predicted according to Eqs. (5.7)–(5.10) and (5.14)–(5.15)

A_{bar} = attenuation due to diffraction at barrier edge, predicted according to Eqs. (5.16)–(5.20)

A_{atm} = attenuation due to atmospheric absorption, predicted according to Eq. (5.22)

The only interaction taken into account is that of barriers and ground, as described by Eq. (5.17). Reflections are assumed to result in incoherent contributions at the receiver.

Interaction of Barriers and Ground at Various Distances

Barriers on both sides of a roadway or railway provide at least two contributions to the total sound field at a receiver outside of the barriers—one by diffraction over the barrier at the receiver side and the second by reflection from the barrier on the other side—which may also suffer diffraction over the barrier on the receiver side (see Fig. 5.8). The second contribution is lower due to higher geometrical attenuation A_{div} but possibly higher due to lower barrier attenuation

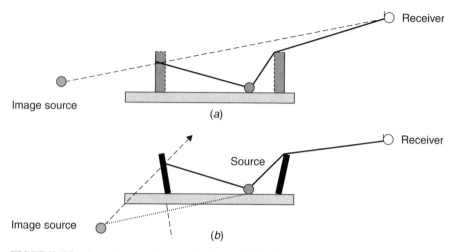

FIGURE 5.8 Sound rays from traffic lane shielded on both sides by (a) absorptive barriers and (b) inclined barriers.

A_{bar}. It generally needs to be reduced by absorptive or inclined barrier surfaces, which reduce the reflection toward the receiver.

Frequently discussed is the question of whether ground attenuation should be completely neglected for sound diffracted over the top of a barrier, as in Eq. (5.17). The reasoning to account for some ground attenuation in the barrier shadow zone is based on the assumption of an equivalent source at the diffracting edge. However, this assumption is not consistent with diffraction theory. There is no ray going out from the edge suffering attenuation just from divergence, but the diffracted field penetrates deeper and deeper into the shadow zone until it reaches the receiver. Since the ground effect on the receiver side is mostly determined by interference with a ground reflection close to the receiver, one has to consider two differently diffracted contributions, which are differently attenuated. Thus, the chance for considerable interference is reduced, particularly at high receiver positions. Of course, such an argument is valid only for meteorological conditions that provide no shadow formation. For greater accuracy at receiver positions near the ground, this interference is not neglected in the United States and Canada.

Barriers Close to Trees, with Gaps and Slots

The attenuation performance of barriers can be reduced by various influences. Past concern centered on trees or bushes that exceed the height of a nearby barrier. Experiments have shown that high-frequency components of noise are scattered by leaves into the barrier shadow zone.[14] However, this effect seems to be small compared to the better acceptability of "green" barriers.

Calculations with Eq. (5.18) assume that diffracted sound dominates in the barrier shadow zone. For this to be true, sound passing through the barrier must be negligible (down 5 dB or more) compared to the diffracted sound, which holds for a barrier relatively free of holes in its face and also free of a continuous gap between it and the ground. More sound energy gets through such holes/gaps than geometrically incident on the openings. Also required is a sufficiently large transmission loss through the surface material of the barrier. A mass per unit surface area (excluding framing) of 20 kg/m^2 is usually sufficient to provide a transmission loss of 25 dB or more at 500 Hz. Such weight is easily achievable with thickness and materials needed for mechanical stability of the barrier.

Procedures Applied by Computer Software

Engineering computer software for noise mapping is based on geometrical data provided by geographical information systems. The ground is often described by plane triangles; roads, railways, and barriers by sections of straight lines; and individual sources by points. Houses are modeled by rectangles. The handling of numerous data for larger areas requires a high degree of sophistication.

A sound ray may experience several reflections and may be diffracted on the path from source to receiver. Because of computer speed, current computer software can deal with many reflections with reasonable computation times. However, current software is not necessarily precise, due to approximate modeling of reflectors and reflection losses. Normally only three reflections are taken into account. Former consideration of reverberation in street canyons is excluded from current computations.

An essential part of the software is its validation by test cases[15]. This task does not aim for physical correctness but for consistency with the referenced calculation schemes. For this purpose, the program must be run in the same mode for test cases as for real applications—for example, the setting of software switches for accelerated computation.

5.6 ENGINEERING APPROACHES TO ACCOUNT FOR METEOROLOGY

Correction Terms C_{met} and C_0

The transfer function of Eq. (5.2), as specified by the attenuation in Eq. (5.23), leads from source properties to the equivalent continuous A-weighted sound pressure level $L_{AT}(\text{DW})$ for meteorological conditions that are favorable to sound propagation. This may be appropriate to meet specific restrictive requirements. A slightly better balance between interests of industry and neighborhood is the long-term A-weighted sound pressure level $L_{AT}(\text{LT})$, where the time interval T is several months or a year. Such a period will normally include a variety of sound propagation conditions, both favorable and unfavorable.

According to ISO 9613-2, a value $L_{AT}(LT)$ may be obtained from the value of $L_{AT}(DW)$ by subtraction of a meteorological correction term,

$$C_{met} = \begin{cases} 0 & \text{if } d \leq 10(h_S + h_R) \\ C_0\left(1 - 10\dfrac{h_S + h_R}{d}\right) & \text{if } d > 10(h_S + h_R) \end{cases} \quad (5.24)$$

where C_0 (in decibels) depends on local meteorological statistics for wind speed and directions plus temperature gradients. Similar to Eq. (5.7), this subtracted term depends on the ratio of the effective height of the sound ray and distance, so that Fig. 5.4 gives a qualitative impression, except for distance. The correction term C_{met} is applied at somewhat larger distances. Experience indicates that values of C_0 are limited in practice between 0 and approximately 5 dB, with values in excess of 2 dB being exceptional.

Local Weather Statistics

Among the various procedures proposed for the calculation of C_0 from weather statistics is the following:

$$C_0(\alpha, g) = -10 \lg \left(\sum_{i=0}^{I-1} Q \frac{W_i(\theta_i)}{2} [1 + g - (1-g)\cos(\theta_{rec} - \theta_i)] + 1 - Q \right) \text{ dB} \quad (5.25)$$

where θ_{rec} = angle between North and the source-receiver line
 g = parameter between 0.01 and 0.1 that accounts for attenuation in upwind direction
 θ_i = angle between North and ith wind direction
 W_i = probability for ith direction of wind
 I = number of wind directions
 $1 - Q$ = probability for no wind (calm)

Values of $g = 0.1$ and $g = 0.01$ correspond to a maximum upwind attenuation of 10 and 20 dB, respectively. Crosswind attenuation is less than 3 dB. Such values are effective for single point sources and wind always from the same direction. For extended sources (e.g., traffic lines or larger industrial areas), the appropriate averaging procedure typically results in values slightly smaller than 2 dB.

5.7 UNCERTAINTIES

General

The uncertainty of sound pressure levels calculated for a receiver position results from the uncertainty ΔL_W of the source level, determined, among other matters, by uncertain assumptions about the operating conditions of a sound source, the number of vehicles on a road, the roughness of rail heads, and the uncertainty

ΔA of the propagation losses. In the following, the uncertainty ΔA is discussed exclusively. However, note that uncertainties about source levels cannot always be neglected.

Elements

Uncertainty about spreading losses is generally small. Except for air traffic, distances between source and receiver can be determined very precisely. Effects of road traffic distribution on several parallel lanes are generally small if the most inner and outer lanes are modeled separately.

Uncertainty is significantly affected by effects of ground and meteorology. Very pronounced uncertainty has been obtained by Schomer[16] from long-term measurements with a loudspeaker centered at a height of 0.6 m above grassland and receivers at 1.2 m at distances up to 800 m. The standard deviation describing high portions of received octave-band levels is shown in Fig. 5.9.

At low frequencies up to 125 Hz and also at high frequencies above 2000 Hz, the standard deviation increases approximately continuously with distance and frequency, as one would expect. Anomalies can be seen in the intervening frequency bands. These anomalies can be attributed to variable ground interference. The low measurement height of 1.2 m may exaggerate the effect, though this receiver height is important to computations in the United States and Canada.

For special barrier-top design, a considerable uncertainty about barrier attenuation is found in the comparison of standard calculations with results calculated or measured in laboratories. However, in field tests such designs have hardly ever shown level differences of more than 1 dB at a distance of more than 25 m from a roadside or railway barrier. Source or receiver positions close to a barrier,

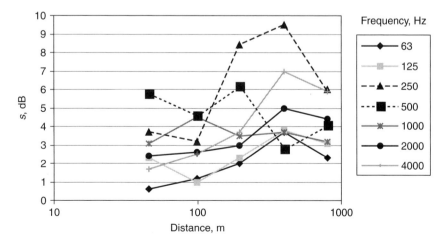

FIGURE 5.9 Standard deviation s of octave-band sound pressure levels for downwind condition at various distances from a loudspeaker centered at a height of 0.6 m and receiver height 1.2 m above grassland.

where reflections from the barrier surface or deep shadow zones play a role, are rarely met in practice, and therefore the laboratory effects are not observed in the field.

There is substantial uncertainty about the validity of the factor C_3 in Eq. (5.19). In many laboratory cases, multiple diffraction has shown an increase in barrier attenuation by more than 5 dB. But these cases are often related to barrier attenuations in excess of 20 dB. In practice, they depend significantly upon specific propagation paths from source to receiver.

Significant uncertainty exists for the meteorological correction factor of Eq. (5.20). It is matched to numerous data obtained for downwind situations at distances up to 1000 m over porous ground. Estimates for this uncertainty are about 2 dB, independent of distance, since the magnitude of barrier attenuation decreases with distance.

Uncertainty due to the roughness and absorption of reflecting surfaces is approximately 1 dB. In addition, limiting the number of reflections to three is sometimes not sufficient to let even a single computed sound ray enter through a passageway into a courtyard. It is obvious from such considerations that ray theory and specular reflections are limited to relatively open areas. Otherwise, standard deviations of more than 5 dB may occur. It is this consideration that partially argues for the U.S. and Canadian retention of line segment sources instead of point source approximations.

Uncertainty about atmospheric absorption may be calculated from the uncertainties of relative humidity, temperature, and ambient pressure. At low frequencies, that uncertainty is very small. At high frequencies, it is of little concern for engineering purposes. At frequencies around 500 Hz, the standard deviation is usually small compared to that of ground effects.

Overall A-Weighted Sound Pressure Level

The uncertainty for sound attenuation along the path of a ray in any frequency band increases with the square root of the sum of variances due to geometry, ground effects, barriers, reflectors, and atmospheric absorption. When several rays contribute to the sound pressure level at the receiver, the variances must be weighted according to their energy and the sum is divided by the total energy. This summation of uncorrelated partial uncertainties results in a reduction of the total uncertainty.

The uncertainty for the A-weighted overall sound pressure level may be calculated from the variances in octave bands, weighted according to their energy. When a number of n frequency bands equally contribute to the overall level, the uncertainty is reduced by a factor $1/\sqrt{n}$.

Approximately consistent with estimated accuracies in ISO 9613-2 is the standard deviation reported by Schomer[16] for the A-weighted overall sound pressure level received from a source of pink noise.[14] At distances from 100 to 800 m and an average height above grassland of 0.9 m, that standard deviation was about 3 dB. At an average height above 5 m, ISO 9613-2 estimates an accuracy of 1 dB. Advanced procedures for calculation of outdoor noise levels aim for an accuracy with a standard deviation of 1 dB up to a distance of 1000 m over flat terrain and of 2.5 dB over hilly terrain.[3]

REFERENCES

1. Lord Rayleigh, *Theory of Sound*, 2nd ed., Dover, New York, 1877, reissued 1945, Vol. 2, p. 128 ff.
2. W. A. Kuperman, "Propagation of Sound in the Ocean," in M. J. Crocker (Ed.), *Encyclopedia of Acoustics*, Wiley, New York, 1997, Chapter 36.
3. F. de Roo and I. M. Noordhoek, "Harmonoise WP2—Reference Sound Propagation Model," Proc. DAGA 03, Aachen, Germany, pp. 354–355
4. K. Attenborough et al., "Benchmark Cases for Outdoor Sound Propagation," *J. Acoust. Soc. Am.*, **97**(1), 173–191 (1995).
5. R. Matuschek and V. Mellert, "Vergleich von technischen Prognoseprogrammen für die Schallimmission mit physikalischen Berechnungen der Schallausbreitung im Freien" ("Comparison of Technical Prediction Programs for Sound Immission with Physical Calculations of Sound Propagation Outdoors"), Proc. DAGA 03, Aachen, Germany, pp. 428–429.
6. ISO 9613-2, *"Acoustics—Attenuation of Sound During Propagation Outdoors—Part 2: General Method of Calculation,"* International Organization for Standardization, Geneva, Switzerland, 1996.
7. H. Gidamy, C. T. Blaney, C. Chiu, J. E. Coulter, M. Delint, L. G. Kende, A. D. Lightstone, J. D. Quirt, and V. Schroter, "ORNAMENT: Ontario Road Noise Analysis Method for Environment and Transportation," Environment Ontario, Noise Assessment and Systems Support Unit, Advisory Committee on Road Traffic Noise, 1988.
8. L. C. Sutherland and G. A. Daigle, "Atmospheric Sound Propagation," in M. J. Crocker (Ed.), *Encyclopedia of Acoustics*, Wiley, New York, 1997, Chapter 32.
9. J. Kragh, "A New Meteo-Window for Measuring Environmental Noise from Industry," Technical Report LI 359/93, Danish Acoustical Institute, Lyngby, Denmark, 1993.
10. A. L'Espérance, P. Herzog, G. A. Daigle, and J. R. Nicolas, "Heuristic Model for Outdoor Sound Propagation Based on an Extension of the Geometrical Ray Theory in the Case of a Linear Sound Speed Profile," *Appl. Acoust.* **37**, 111–139, (1992).
11. G. S. Anderson, C. S. Y. Lee, G. G. Fleming, and C. W. Menge, *FHWA Traffic Noise Model, Version 1.0: User's Guide*, Report FHWA-PD-96-009 and DOT-VNTSC-FHWA-98-1, U.S. Department of Transportation, Federal Highway Administration, Washington, DC, January 1998.
12. J. L. Rochat and G. G. Fleming, *Validation of FHWA's Traffic Noise Model (TNM): Phase 1, Addendum*, Addendum to Report No. FHWA-EP-02-031 and DOT-VNTSC-FHWA-02-01 U.S. Department of Transportation, Federal Highway Administration, Washington, DC (in publication).
13. ISO 9613-1, *"Acoustics—Attenuation of Sound During Propagation Outdoors—Part 1: Calculation of the Absorption of Sound by the Atmosphere*, International Organization for Standardization, Geneva, Switzerland, 1993.
14. T. van Renterghem and D. Botteldooren, "Effect of a Row of Trees Behind Noise Barriers in Wind," *Acta Acustica* united with *Acustica*, **88**, 869–878 (2002).

15. DIN 45687, *"Acoustics—Software Products for the Calculation of the Sound Propagation Outdoors—Quality Requirements and Test Conditions" (draft)*, Deutsches Institut für Normung, Berlin, Germany, 2004.
16. P. D. Schomer, "A Statistical Description of Ground-to-Ground Sound Propagation," *Noise Control Eng. J.*, **51**(2), 69–89 (2003).

■■■■■ **CHAPTER 6**

Sound in Small Enclosures

DONALD J. NEFSKE AND SHUNG H. SUNG

Vehicle Development Research Laboratory
General Motors Research & Development Center
Warren, Michigan

6.1 INTRODUCTION

Only in an anechoic room may sound waves travel outwardly in any direction without encountering reflecting surfaces. In practice, one must deal with every shape and size of enclosure containing an infinite variety of sound-diffusing, sound-reflecting, and sound-absorbing objects and surfaces. However, in small rooms of fairly regular shape, with smooth walls, the sound is not statistically diffuse and will depend on the acoustical modal response in the room. Typical examples where this is the case are the passenger compartments of transportation vehicles, ductwork and small rooms in buildings, enclosures used to enhance the response of audio equipment, and enclosures designed for sound isolation. Because of the need to understand the modal nature of sound in such enclosures, this chapter will describe the governing equations, the modal theory, and its application for determining and controlling the interior acoustical response. While the modal theory for rectangular enclosures is described in most standard textbooks on acoustics,[1-5] the extended approach for irregular geometries and the use of advanced numerical techniques are more amenable to practical application and are also described here.

6.2 SOUND IN A VERY SMALL ENCLOSURE

Before considering the more general case, it will be instructive to consider the sound pressure response in a very small enclosure. Frequently, in noise control problems, a noise source is enclosed in a very small box to prevent it from radiating noise to the exterior, as conceptually represented in Fig. 6.1a. When the noise source has a frequency low enough so that the wavelength of the

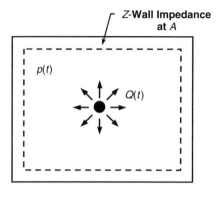

(a) Interior Noise Source

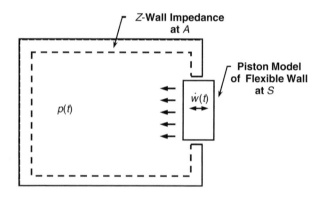

(b) Vibrating Flexible Wall

FIGURE 6.1 Sound in small enclosure with impedance boundaries generated by (a) interior noise source and (b) piston model of flexible-wall vibration. Symbols: $p(t)$, sound pressure; $Q(t)$, volume velocity; Z, wall impedance; A, impedance wall area; $\dot{w}(t)$, piston velocity; S, piston area.

sound is long compared to the largest distance of the box, the sound pressure produced by the source will be uniform throughout the entire cavity. A uniform sound pressure field will also be created in an enclosure when the walls of the enclosure are forced to vibrate at low frequency by the pressures of an external sound source, as conceptually represented by the vibrating piston in Fig. 6.1b. Frequently, the walls of the enclosure will be damped by applying panel-damping treatment or absorption material over the walls.

The uniform sound pressure $p(t)$ in such an enclosure will satisfy an extended form of Eq. (2.9) as $(1/P)dp/dt + (\gamma/V)d\tau/dt = (\gamma/V)Q(t)$, where $Q(t)$ is the *volume velocity* of the sound source (m³/s), V is the enclosure volume, P is the ambient air pressure, and $\tau(t)$ is the volume variation. At the wall area A (m²)

SOUND IN A VERY SMALL ENCLOSURE

in Fig. 6.1a or b, the specific acoustical impedance is $Z = p/u$ and the particle velocity is $u(t)$, so $d\tau/dt = Au = Ap/Z$. Since $c^2 = \gamma P/\rho$ by Eq. (2.12), the uniform sound pressure in the enclosure satisfies

$$\frac{dp}{dt} + 2\delta p = \left(\frac{\rho c^2}{V}\right) Q(t) \tag{6.1}$$

where $\delta = \rho c^2 A/(2VZ)$. For steady-state excitation, $Q(t) = \hat{Q}\cos(\omega t + \phi)$, which is the volume velocity of the source operating at the forcing frequency $f = \omega/2\pi$ (hertz) with the volume velocity amplitude $\hat{Q}$ and phase ϕ. The steady-state sound pressure in the enclosure is then $p(t) = p_0 \cos(\omega t + \theta_0)$, with the sound pressure amplitude and phase given by

$$p_0 = \frac{\rho c^2 \hat{Q}}{V[(\omega + 2\delta_i)^2 + (2\delta_r)^2]^{1/2}} \quad \text{N/m}^2$$

$$\theta_0 = \phi - \tan^{-1}\frac{\omega + 2\delta_i}{2\delta_r} \quad \text{rad} \tag{6.2}$$

Here $\delta = \delta_r + i\delta_i$ is a complex damping factor ($i = \sqrt{-1}$) that accounts for the acoustical impedance $Z = R + iX$ of the wall area A, where

$$\delta_r = \frac{cA}{2V}\text{Re}\frac{\rho c}{Z} \qquad \delta_i = \frac{cA}{2V}\text{Im}\frac{\rho c}{Z} \quad \text{s}^{-1} \tag{6.3}$$

For the interior noise source in Fig 6.1a, $Q(t)$ is taken as positive for outward volume flow, while for the flexible-wall vibration in Fig. 6.1b, $Q(t) = -S\dot{w}(t)$ is an equivalent volume velocity with outward piston velocity $\dot{w}(t)$ taken as positive.

From Eq. (6.2), we see that the amplitude p_0 of the sound pressure in the enclosure depends, not only on the noise source amplitude $\hat{Q}$ and forcing frequency $f = \omega/2\pi$, but also on the enclosure volume V and total wall impedance in δ. For a rigid-wall enclosure, $|Z| \to \infty$, so that $\delta_r, \delta_i = 0$, and

$$p_0 = \frac{\rho c^2 \hat{Q}}{\omega V} \quad \text{N/m}^2$$

$$\theta_0 = \phi - \tfrac{1}{2}\pi \quad \text{rad} \tag{6.4}$$

In this case, the sound pressure depends only on the amplitude and frequency of the volume velocity of the source and on the enclosure volume. The sound pressure lags the volume velocity by exactly 90°, indicating that the source does not radiate any acoustical power.

Example 6.1. The piston of Fig. 6.1b has an area of 1 cm² and is driven harmonically with a peak-to-peak displacement of 4 mm at a frequency of 100 Hz.

148 SOUND IN SMALL ENCLOSURES

The volume of the cavity is 0.0125 m³ and it has rigid walls. What is the sound pressure level in the cavity?

Solution For harmonic displacement excitation $w = \hat{w} \sin \omega t$, and we obtain $\dot{w} = (\omega \hat{w}) \cos \omega t = \hat{\dot{w}} \cos \omega t$, so that $\hat{\dot{w}} = \omega \hat{w}$. Therefore, $\hat{Q} = S\hat{\dot{w}} = S\omega\hat{w}$ and from Eq. (6.4),

$$p_0 = \frac{\rho c^2 S \hat{w}}{V} = \rho c^2 \frac{\Delta \hat{V}}{V} \quad \text{N/m}^2 \tag{6.5}$$

where $\Delta \hat{V} = S\hat{w}$ is the volume change of the enclosure due to the piston displacement. Inserting the appropriate numerical values, we have $\Delta \hat{V} = 10^{-4} \times 2 \times 10^{-3} = 2 \times 10^{-7}$ m³, so that

$$p_0 = 1.21 \times 343^2 \times \frac{2 \times 10^{-7}}{1.25 \times 10^{-2}} = 2.28 \text{ N/m}^2$$

The sound pressure level is then

$$L_p = 20 \log_{10} \frac{p_{\text{rms}}}{p_{\text{ref}}} = 20 \log_{10} \frac{0.707 \times 2.28}{2 \times 10^{-5}} = 98 \text{ dB}$$

Example 6.2. A noise source is enclosed in a very small box, as in Fig. 6.1a, that has flexible but very stiff walls. Determine the formula for the interior sound pressure.

Solution For a mass–spring–dashpot model of the box walls, the acoustical impedance is $Z = A^{-1}[C + i\omega(M - K/\omega^2)]$, where A is the wall surface area and C, M, K are overall values of wall damping, mass, and stiffness. For very stiff walls, $Z \approx -iK/A\omega$ so that, from Eq. (6.3),

$$\delta_r = 0 \qquad \delta_i = \frac{\omega \rho c^2 A^2/V}{2 K} = \frac{\omega}{2} \frac{K_{\text{air}}}{K} \quad \text{s}^{-1}$$

where $K_{\text{air}} = \rho c^2 A^2/V$ (N/m) is the *stiffness* of the air and K (N/m) is the wall stiffness. Substituting δ_r and δ_i into Eq. (6.2) then gives

$$p_0 = \frac{1}{1 + K_{\text{air}}/K} \frac{\rho c^2 \hat{Q}}{\omega V} \quad \text{N/m}^2 \tag{6.6}$$

$$\theta_0 = \phi - \tfrac{1}{2}\pi \quad \text{rad}$$

Note that the sound pressure in an enclosure with compliant walls is lower than the sound pressure in an equivalent rigid-wall enclosure given by Eq. (6.4), and the addition of damping to the walls can be shown to produce a similar effect.

Example 6.3. A *Helmholtz resonator* is a rigid-wall enclosure with a small aperture of cross-sectional area S that connects the enclosure to a column of air of length L that oscillates as the piston in Fig. 6.1b. Determine the natural frequency.

Solution The mass of the column of air is $M = \rho S L$ and the force on it is pS. Hence we must have $M\ddot{w} = pS$, where $\dot{w}$ is the velocity of the air column. Since $\hat{\ddot{w}} = \omega \hat{\dot{w}}$, we obtain $\hat{Q} = S\hat{\dot{w}} = S^2 p_0/\omega M$. Substituting this for $\hat{Q}$ in Eq. (6.4) gives $(1 - \rho c^2 S^2/\omega^2 M V) p_0 = 0$ or $(1 - K_{\text{air}}/\omega^2 M) p_0 = 0$, where $K_{\text{air}} = \rho c^2 S^2/V$ is the air stiffness. The natural frequency is then $\omega_0 = \sqrt{K_{\text{air}}/M} = c\sqrt{S/LV}$.

6.3 GOVERNING EQUATIONS FOR ACOUSTICAL MODAL RESPONSE

For larger enclosures or for higher frequencies, the sound pressure field in the enclosure is no longer uniform but depends on the acoustical modal response in the enclosure. More importantly, the sound pressure can be amplified considerably near discrete frequencies corresponding to the acoustical cavity resonances. This modal nature of sound in an enclosure results from the superposition of sound waves that propagate according to the well-known *acoustical wave equation* [Eq. (2.13)],

$$\nabla^2 p - \frac{\ddot{p}}{c^2} = 0 \quad \text{N/m}^4 \tag{6.7}$$

where $\ddot{p}$ denotes the second partial derivative with respect to time t.

Noise sources interior to an enclosed cavity can be included as forcing terms in the wave equation. For the example of a monopole source (e.g., a loudspeaker in a cabinet) as in Fig. 6.1a, the time-varying mass flow rate is $\dot{m}(x, y, z, t) = \rho Q(x, y, z, t)$ (kg/s), so that

$$\nabla^2 p - \frac{\ddot{p}}{c^2} = -\frac{\rho \dot{Q}}{V} \quad \text{N/m}^4 \tag{6.8}$$

where $\dot{Q} = \partial Q/\partial t$. Other interior sources can be represented as combinations of monopole sources or else can be included directly in the wave equation in a similar manner. For simple harmonic motion, $p(x, y, z, t) = \text{Re}[\hat{p}(x, y, z) \exp(i\omega t)]$ and $Q(x, y, z, t) = \text{Re}[\hat{Q}(x, y, z) \exp(i\omega t)]$, and we obtain the inhomogeneous *Helmholtz equation* for the steady-state sound pressure response,

$$\nabla^2 \hat{p} + \left(\frac{\omega}{c}\right)^2 \hat{p} = -\frac{i\omega \rho \hat{Q}}{V} \quad \text{N/m}^4 \tag{6.9}$$

where $f = \omega/2\pi$ (Hz) is the forcing frequency of the vibration and $\lambda = c/f$ (m) is the wavelength of the sound produced by the source.

The boundary conditions for p determine the reflection, absorption, and transmission of the sound waves at the enclosure's surfaces and are derived from fluid mechanical considerations.[6] For small-amplitude motions, a momentum balance at the boundary requires that the air particle velocity u normal to the boundary surface be related to p through

$$\frac{1}{\rho}\frac{\partial p}{\partial n} = -\dot{u} \quad \text{m/s}^2 \qquad (6.10)$$

where $\partial/\partial n$ is the outward surface-normal derivative. For an impervious wall surface, u is the normal-velocity component of the air into the pores of the surface. Table 6.1 lists the boundary conditions for different wall surfaces and air interfaces that are characterized by their acoustical impedance Z and surface-normal vibration velocity $\dot{w}$. In what follows, we will investigate the steady-state, random, and transient sound pressure response in enclosures with flexible, absorbent boundaries where Z and $\dot{w}$ are defined as in Table 6.1.

6.4 NATURAL FREQUENCIES AND MODE SHAPES

Acoustical resonances that result in high sound pressure occur at discrete natural frequencies in an enclosed cavity. The acoustical resonances are found from the free-vibration solution of the wave equation by substituting $p_n = p_{n0}\Psi_n(x, y, z)\exp(i\omega_n t)$, which gives

$$\nabla^2 \Psi_n + \left(\frac{\omega_n}{c}\right)^2 \Psi_n = 0 \quad \text{m}^{-2} \qquad (6.11)$$

with free-vibration boundary conditions for the walls. The nondimensional pressure distributions $\Psi_n(x, y, z)$ for $n = 0, 1, 2, \ldots$ are the *mode shapes*, with $f_n = \omega_n/2\pi$ (Hz) being the corresponding *natural frequencies*. For absorbent boundaries, the mode shapes and natural frequencies are complex, and the acoustical modes are damped. However, for rigid boundaries ($|Z| \to \infty$) or for fully reflective open boundaries ($|Z| = 0$), one obtains real, undamped modes, or *standing waves*, that depend only on the geometrical shape of the cavity.

Table 6.2 gives formulas for the natural frequencies and mode shapes of the undamped acoustical modes of a few regular-shape enclosures with rigid walls as well as those of a tube with fully reflective and open boundaries. (A more complete list can be found in ref. 7.) One can show that the acoustical modes are orthogonal over the cavity volume such that

$$\int_V \Psi_m \Psi_n \, dV = \begin{cases} 0 & m \neq 0 \\ V_n & m = n \end{cases} \quad \text{m}^3 \qquad (6.12)$$

and constitute *normal modes* of the enclosure. In Table 6.2, the number of indices used to identify each mode is based on the dimensionality of the enclosure, whereas we have used a single index to identify a mode, so the equivalence for a three-dimensional enclosure would be $f_n \equiv f_{ijk}$ and $\Psi_n \equiv \Psi_{ijk}$.

TABLE 6.1 Acoustical Boundary Conditions

Type	Boundary Condition	Air Particle Velocity
1. Rigid wall	$\dfrac{\partial p}{\partial n} = 0$	$u = 0$
2. Flexible wall	$\dfrac{1}{\rho}\dfrac{\partial p}{\partial n} = -\ddot{w}$	$u = \dot{w}$
3. Absorber on rigid wall	$\dfrac{1}{\rho}\dfrac{\partial p}{\partial n} = -\dfrac{1}{Z_a}\dfrac{\partial p}{\partial t}$	$u = \dfrac{p}{Z_a}$
4. Absorber on flexible wall	$\dfrac{1}{\rho}\dfrac{\partial p}{\partial n} = -\dfrac{1}{Z_a}\dfrac{\partial p}{\partial t} - \ddot{w}$ $= -\left(\dfrac{1}{Z_a} + \dfrac{1}{Z_w}\right)\dfrac{\partial p}{\partial t}$	$u = \dfrac{p}{Z_a} + \dot{w}$ $= \dfrac{p}{Z_a} + \dfrac{p}{Z_w}$
5. Open pressure release	$p = 0$	Determine from analysis
6. Open plane wave	$\dfrac{1}{\rho}\dfrac{\partial p}{\partial n} = -\dfrac{1}{Z_{\text{air}}}\dfrac{\partial p}{\partial t}$	$u = \dfrac{p}{Z_{\text{air}}}$

TABLE 6.2 Acoustical Modes and Natural Frequencies[a]

Description	Figure	Natural Frequency, f_{ijk} (Hz)	Mode Shape, Ψ_{ijk}
1. Slender tube both ends closed		$\dfrac{ic}{2L}$ $D \ll \lambda$ where $\lambda = c/f$	$\cos \dfrac{i\pi x}{L} \quad i = 0, 1, 2, \ldots$
2. Slender tube one end closed one end open		$\dfrac{ic}{4L}$ $D \ll \lambda$ where $\lambda = c/f$	$\cos \dfrac{i\pi x}{L} \quad i = 1, 3, 5, \ldots$
3. Slender tube both ends open		$\dfrac{ic}{2L}$ $D \ll \lambda$ where $\lambda = c/f$	$\sin \dfrac{i\pi x}{L} \quad i = 1, 2, 3, \ldots$
4. Closed rectangular volume		$\dfrac{c}{2}\left(\dfrac{i^2}{L_x^2} + \dfrac{j^2}{L_y^2} + \dfrac{k^2}{L_z^2}\right)^{1/2}$	$\cos \dfrac{i\pi x}{L_x} \cos \dfrac{j\pi y}{L_y} \cos \dfrac{k\pi z}{L_z}$ $i, j, k = 0, 1, 2, \ldots$
5. Closed cylindrical volume		$\dfrac{c}{2\pi}\left(\dfrac{\lambda_{jk}^2}{R^2} + \dfrac{i^2\pi^2}{L^2}\right)^{1/2}$ λ_{jk} from Table 6.2a below	$J_j\left(\lambda_{jk}\dfrac{r}{R}\right)\cos\dfrac{i\pi x}{L}\begin{cases}\sin j\theta \\ \text{or} \\ \cos j\theta\end{cases}$ $i, j, k = 0, 1, 2, \ldots$

Modes Symmetric about Center

6. Closed spherical volume

$$\frac{\lambda_i c}{2\pi R} \qquad \frac{R}{\lambda_i r}\sin\frac{\lambda_i r}{R}$$
$$i = 0, 1, 2, \ldots$$

λ_i from Table 6.2b below

7. Arbitrary closed volume

L—Maximum linear dimension
fundamental natural frequency
(approximate): $\dfrac{c}{2L}$

Finite-element analysis
Boundary element analysis

Table 6.2a

λ_{jk}				j			
k	0	1	2	3	4	5	6
0	0	1.8412	3.0542	4.2012	5.3176	6.4156	7.5013
1	3.8317	5.3314	6.7061	8.0152	9.2824	10.5199	11.7349
2	7.0156	8.5363	9.9695	11.3459	12.6819	13.9872	15.2682
3	10.1730	11.7060	13.1704	14.5859	15.9641	17.3128	18.6374

$\lambda_{j=0,k} = \pi(k + 1/4)$ for $k \geq 3$ $(J'_j(\lambda_{jk}) = 0)$

Table 6.2b

i	0	1	2	3	4
λ_i	0	4.4934	7.7253	10.9041	14.0662

$\lambda_i = \pi(i + 1/2)$ for $i \geq 4$ $(\tan\lambda_i = \lambda_i)$

[a]From Ref. 6.

154   SOUND IN SMALL ENCLOSURES

It is noteworthy that the first natural frequency of a cavity that is completely enclosed by rigid walls is zero, $f_0 = 0$, and is a *uniform pressure mode* (sometimes called a *Helmholtz mode*) with $\Psi_0 = 1$ and

$$V_0 = \int_V \Psi_0^2 \, dV = V \quad \text{m}^3 \tag{6.13}$$

where V is the volume of the enclosure. Also, the fundamental frequency of the first spatially varying mode in an enclosed cavity is approximately $f_1 = c/2L$ (Hz), where L is the maximum linear dimension of the cavity.

For a cavity with an open boundary, the $n = 0$ mode is absent because a (nonzero) uniform pressure cannot be sustained. Also, the natural frequencies and mode shapes in Table 6.2 for a tube with an open boundary are approximate because the boundary condition ($p \approx 0$) does not fully model the exact physics of the acoustical behavior at an open boundary. The accuracy of the formulas in Table 6.2 for an open tube increases with increasing slenderness of the tube. In general, the dimension of the open boundary must be small compared with the acoustical wavelength so that the sound is fully reflected from the open boundary.

Example 6.4.   A closed rectangular cavity has dimensions $0.41 \times 0.51 \times 0.61$ m. From Table 6.2, frame 4, the formulas for the normal-mode frequencies and mode shapes are

$$f_{ijk} = \frac{c}{2} \sqrt{\left(\frac{i}{L_x}\right)^2 + \left(\frac{j}{L_y}\right)^2 + \left(\frac{k}{L_z}\right)^2} \quad \text{Hz}$$

$$\Psi_{ijk} = \cos \frac{i\pi x}{L_x} \cos \frac{i\pi y}{L_y} \cos \frac{i\pi z}{L_z} \tag{6.14}$$

Let $L_x = 0.61$ m, $L_y = 0.51$ m, and $L_z = 0.41$ m. (a) Find the natural frequencies of the $i = 2, j = 0, k = 0$; the $i = 1, j = 1, k = 0$; and the $i = 2, j = 1, k = 0$ normal modes of vibration. (b) Plot the sound pressure distribution for these three normal modes of vibration. The speed of sound, adjusted for temperature, is 347.3 m/s.

Solution

(a) From Eq. (6.13) we have

$$f_{2,0,0} = \frac{347.3}{2} \sqrt{\left(\frac{2}{0.61}\right)^2} = 569.3 \text{ Hz}$$

$$f_{1,1,0} = \frac{347.3}{2} \sqrt{\left(\frac{1}{0.61}\right)^2 + \left(\frac{1}{0.51}\right)^2} = 443.8 \text{ Hz}$$

$$f_{2,1,0} = \frac{347.3}{2} \sqrt{\left(\frac{2}{0.61}\right)^2 + \left(\frac{1}{0.51}\right)^2} = 663.4 \text{ Hz}$$

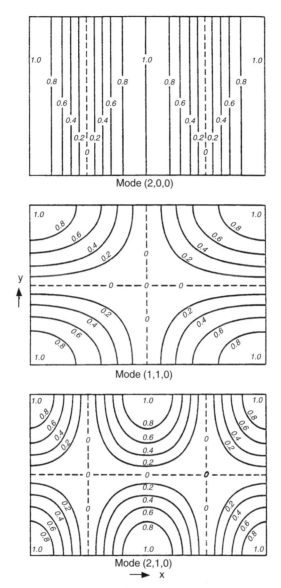

FIGURE 6.2 Sound pressure contour plots of a section through a rectangular enclosure for three different modes of vibration. The numbers on the plots indicate the relative sound pressure amplitude.

(b) The pressure distributions $|\Psi_{ijk}|$ of the modes are shown in Fig. 6.2, where the normalization is such that the maximum pressure at the corners is $|\Psi_{ijk}| = 1$. The minimum sound pressure occurs at the nodal surfaces for which $|\Psi_{ijk}| = 0$. Since $k = 0$ for all three modes, the pressure distributions are uniform in the z direction.

6.5 NUMERICAL METHODS FOR ACOUSTICAL ANALYSIS

For simple geometries and boundary conditions, the acoustical modes can be expressed analytically as in Table 6.2, but for more complicated geometries and boundary conditions numerical approaches such as the finite-element method or the boundary-element method are required.[8] The finite-element method is based upon the representation of the interior of the enclosure by a finite number of interconnected *volume* elements. The boundary-element method is based upon the representation of only the boundary surface of the enclosure by a finite number of interconnected *planar* elements. While both the finite-element method and the boundary-element method were originally developed for structural analysis, they are applicable to numerous nonstructural problems, and the acoustical cavity problem falls into this category. The major advantage of the boundary-element method is the reduction in modeling effort that is required because only the surface of the enclosure needs to be meshed. On the other hand, the finite-element method generally results in banded, frequency-independent, real matrices that require significantly less computational time when solving for the acoustical response.

To illustrate a numerical acoustical modal analysis of a complex-shaped enclosure, Fig. 6.3 shows the application of the acoustical finite-element method to an automobile passenger compartment. The acoustical finite-element model in Fig. 6.3a was developed from linear hexahedral and pentahedral elements, where only a half-model is required because of symmetry.[9] The model can be solved to obtain the acoustical modes by implementing a structural finite-element code and using the *structural–acoustical analogy*[10] outlined in Table 6.3. By specifying the structural material properties in terms of the acoustical material properties as shown in Table 6.3, the structural equations of motion reduce to the acoustical wave equation and the structural boundary conditions reduce to the required acoustical boundary conditions. The accuracy of the modal solution will be proportional to $(n/N)^2$ when using linear elements, where n is the mode number and N is the number of elements in a particular direction. Generally, accuracy to within 10% in frequency can be expected for the first four modes in a particular direction when there are about 10 linear elements in that direction.

The first acoustical mode predicted by the passenger compartment finite-element model in Fig. 6.3a is shown in Fig. 6.3b. It is a longitudinal mode with the nodal surface passing vertically through the passenger compartment. The nodal surface on which $|\Psi_n| = 0$ and the antinodal surfaces on which $|\Psi_n|$ is a local maximum are of practical importance because they indicate low- and high-noise regions, respectively, which result from excitation of the mode. For the example in Fig. 6.3b, excitation of the mode may go unnoticed by a front-seat occupant because of the proximity of the nodal surface, but it may be heard by a rear-seat occupant. The situation is just the opposite for the second mode illustrated in Fig. 6.3c. Because of the irregular geometrical shape of the compartment, these modes could not be predicted by simple formulas like those presented in Table 6.2. However, the numerical approach provides a method to predict the

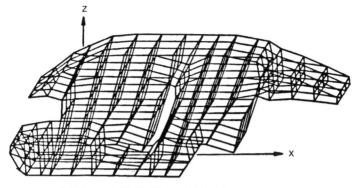

(a) Acoustic Finite Element Model

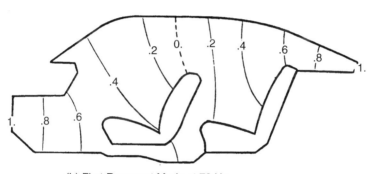

(b) First Resonant Mode at 73 Hz

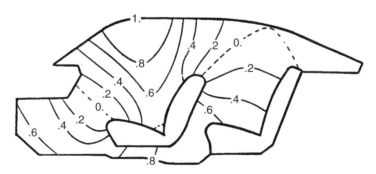

(c) Second Resonant Mode at 130 Hz

FIGURE 6.3 Acoustical finite-element analysis of automobile passenger compartment to determine the acoustical mode shapes and natural frequencies. The mode shapes become increasingly more complex as the natural frequency increases and as the complexity of the enclosure geometry increases.

TABLE 6.3 Structural–Acoustical Analogy

Description	Structure	Analogy	Acoustical Cavity
Finite element	(diagram with nodes w_x, w_y, w_z along axes x, y, z) $$w_k = \sum_{l=1}^{N} (W_k)_l N_l^k \quad k = x, y, z$$	$w_x = p$ $N_l^z = N_l$	(diagram with nodes (p) along axes x, y, z) $$p = \sum_{l=1}^{N} p_l N_l$$
Constitutive equations	$$\begin{pmatrix} \sigma_{xx} \\ \tau_{xy} \\ \tau_{xz} \end{pmatrix} = \begin{pmatrix} G_{11} & G_{14} & G_{16} \\ G_{14} & G_{44} & G_{46} \\ G_{16} & G_{46} & G_{66} \end{pmatrix} \begin{pmatrix} \varepsilon_{xx} \\ \gamma_{xy} \\ \gamma_{xz} \end{pmatrix}$$	$\sigma_{xx} = u_x$ $\tau_{xy} = u_y$ $\tau_{xz}a = u_z$ $\dfrac{\partial w_x}{\partial x} = \dfrac{\partial p}{\partial x}$ $\dfrac{\partial w_x}{\partial y} = \dfrac{\partial p}{\partial y}$ $\dfrac{\partial w_x}{\partial z} = \dfrac{\partial p}{\partial z}$ Material properties $G_{11} = 1/\rho$ $G_{44} = 1/\rho$ $G_{66} = 1/\rho$ $G_{14} = 0$ $G_{16} = 0$ $G_{46} = 0$ $\rho_s = 1/\rho c^2$	$$\begin{pmatrix} u_x \\ u_y \\ u_z \end{pmatrix} = \begin{pmatrix} -1/\rho & 0 & 0 \\ 0 & -1/\rho & 0 \\ 0 & 0 & -1/\rho \end{pmatrix} \begin{pmatrix} \partial p/\partial x \\ \partial p/\partial y \\ \partial p/\partial z \end{pmatrix}$$
Equilibrium equations	$$\dfrac{\partial \sigma_{xx}}{\partial x} + \dfrac{\partial \tau_{xy}}{\partial y} + \dfrac{\partial \tau_{xz}}{\partial z} = -\rho_s \ddot{w}_x$$ $([K] - \omega^2 [M])\{w\} = \{0\}$		$$\dfrac{\partial u_x}{\partial x} + \dfrac{\partial u_y}{\partial y} + \dfrac{\partial u_z}{\partial z} = -\dfrac{1}{\rho c^2} \ddot{p}$$ $([K] - \omega^2 [M])\{p\} = \{0\}$
Finite-element equations	$$K_{ij} = \int_V \{\nabla N_i^x\}^{\mathrm{T}} [G] \{\nabla N_i^x\} dV$$ $$M_{ij} = \int_V \rho_s N_i N_j \, dV$$		$$K_{ij} = \int_V \dfrac{1}{\rho} \{\nabla N_i\}^{\mathrm{T}} \{\nabla N_j\} dV$$ $$M_{ij} = \int_V \dfrac{1}{\rho c^2} N_i N_j \, dV$$

acoustical modes, which can then be used to predict the forced sound pressure response by the methods of the following sections.

6.6 FORCED SOUND PRESSURE RESPONSE IN ENCLOSURE

Often the sound pressure response in an enclosure for known sound sources is of interest. The enclosure may be complicated in shape, and it may have walls that are both flexible and absorbent as well as multiple interior surfaces. If the undamped modes $\Psi_n(x, y, z)$ and natural frequencies $f_n = \omega_n/2\pi$ for $n = 0, 1, 2, \ldots$ are known, the forced acoustical response in such an enclosure can be directly expressed using the *modal analysis* technique as the *normal-mode expansion*,[11,12]

$$p(x, y, z, t) = \sum_n P_n(t)\Psi_n(x, y, z) \quad \text{N/m}^2 \tag{6.15}$$

where the P_n are time-varying coefficients that must be determined to satisfy the acoustical wave equation, the boundary conditions, and the initial conditions.

For flexible and absorbent boundaries described by the boundary conditions in Table 6.1, the modal coefficients $P_n(t)$ are determined from

$$\ddot{P}_n + 2\delta_n \dot{P}_n + \omega_n^2 P_n = \frac{\rho c^2}{V_n} F_n(t) \quad \text{N/m}^2 \cdot \text{s}^{-2} \tag{6.16}$$

where V_n is given by Eq. (6.12) and $F_n(t)$ is the *modal force*,

$$F_n(t) = \int_V \frac{\dot{Q}}{V} \Psi_n \, dV - \int_S \ddot{w} \Psi_n \, dS \quad \text{m}^3/\text{s}^2 \tag{6.17}$$

Q is the volume velocity of interior noise sources, and $\dot{w}$ is the vibration velocity of the wall panels at the boundary surface S. The integrations are carried out over the enclosure volume V and the wall panel surface S, respectively. In Eq. (6.16), δ_n is the *modal damping constant* or *modal damping decrement*, which as defined here is complex and given by

$$\delta_n = \delta_n^r + i\delta_n^i = \frac{\rho c^2}{2V_n} \int_A \frac{\Psi_n^2}{Z} \, dA \quad \text{s}^{-1} \tag{6.18}$$

where $Z = R + iX$ is the complex wall impedance and the integration is carried out over the impedance wall area A. Table 6.1 determines the particular interpretation of $\dot{w}$ and Z in Eqs. (6.17) and (6.18) for different boundaries. The undamped modes Ψ_n are assumed to satisfy the rigid-wall boundary condition $(\partial \Psi_n/\partial n = 0)$ on both the wall panel surface S and the impedance boundary A.

From Eq. (6.18), the real and imaginary parts of δ_n can be expressed as [cf. Eq. (6.3)]

$$\delta_n^r = \frac{cA_n}{2V_n} \operatorname{Re} \frac{\rho c}{Z_n} \qquad \delta_n^i = \frac{cA_n}{2V_n} \operatorname{Im} \frac{\rho c}{Z_n} \quad \text{s}^{-1} \tag{6.19}$$

where Z_n is an average impedance for the nth mode and

$$\frac{\rho c}{Z_n} = \frac{1}{A_n} \int_A \frac{\rho c}{Z} \Psi_n^2 \, dA \quad \text{where} \quad A_n = \int_A \Psi_n^2 \, dA \quad (6.20)$$

When the wall impedance Z does not vary over the wall surface area A, Eq. (6.20) results in $\rho c/Z_n \equiv \rho c/Z$, as expected. Recalling the real and imaginary interpretation of the acoustical impedance in Eq. (6.19), we observe that δ_n^r relates to the wall resistivity ($\sim$ damping) and δ_n^i relates to the wall reactance ($\sim$ flexibility). In the above formulation, we have assumed that "light damping" exists such that $|\delta_n^r| \ll \omega_n$ and $|\delta_n^i| \ll \omega_n$. For more "heavily damped" enclosures that often occur in practice, the modal equations (6.16) are coupled through the boundary impedance and the wall flexibility, and they must be solved simultaneously. A complete discussion of the procedure is given in references 12 and 13.

Example 6.5. The acoustical modes Ψ_n and natural frequencies ω_n are given by the finite-element analysis in Table 6.3. Determine the normal-mode expansion for the forced sound pressure response.

Solution From the finite-element analysis, the acoustical modes are given as the $M \times N$ matrix

$$\{\Psi_1 \Psi_2 \cdots \Psi_n \cdots \Psi_N\} = \begin{bmatrix} \psi_{11} & \psi_{12} & \cdots & \psi_{1n} & \cdots & \psi_{1N} \\ \psi_{21} & \psi_{22} & \cdots & \psi_{2n} & \cdots & \psi_{2N} \\ \vdots & \vdots & \ddots & \vdots & \ddots & \vdots \\ \psi_{m1} & \psi_{m2} & \cdots & \psi_{mn} & \cdots & \psi_{mN} \\ \vdots & \vdots & \ddots & \vdots & \ddots & \vdots \\ \psi_{M1} & \psi_{M2} & \cdots & \psi_{Mn} & \cdots & \psi_{MN} \end{bmatrix} \quad (6.21)$$

where each row m corresponds to the response at a particular grid point of the finite-element model and each column n corresponds to the response of a particular mode. In matrix form, the normal-mode expansion in Eq. (6.15) for the forced sound pressure response then becomes

$$\begin{bmatrix} p_1(t) \\ p_2(t) \\ \vdots \\ p_m(t) \\ \vdots \\ p_M(t) \end{bmatrix} = \begin{bmatrix} \psi_{11} & \psi_{12} & \cdots & \psi_{1n} & \cdots & \psi_{1N} \\ \psi_{21} & \psi_{22} & \cdots & \psi_{2n} & \cdots & \psi_{2N} \\ \vdots & \vdots & \ddots & \vdots & \ddots & \vdots \\ \psi_{m1} & \psi_{m2} & \cdots & \psi_{mn} & \cdots & \psi_{mN} \\ \vdots & \vdots & \ddots & \vdots & \ddots & \vdots \\ \psi_{M1} & \psi_{M2} & \cdots & \psi_{Mn} & \cdots & \psi_{MN} \end{bmatrix} \begin{bmatrix} P_1(t) \\ P_2(t) \\ \vdots \\ P_n(t) \\ \vdots \\ P_N(t) \end{bmatrix} \quad (6.22)$$

Each modal coefficient $P_n(t)$ is determined from Eq. (6.16) with ω_n known and $V_n = \{\Psi_n\}^T[M]\{\Psi_n\}$, where superscript T indicates the transpose and $[M]$ is developed in the finite-element analysis as in Table 6.3. The modal forcing in Eq. (6.16) is

$$\{F_n(t)\} = [\Psi]^T (\{\dot{Q}\} - [S]\{\ddot{w}\}) \tag{6.23}$$

where $\{\dot{Q}\}$ and $\{\ddot{w}\}$ are vectors of the grid point volume acceleration and wall panel acceleration and $[S]$ is a matrix of surface areas associated with each wall panel grid. Similarly, the damping constant in Eq. (6.16) is

$$\{\delta_n\} = \frac{\rho c^2}{2V_n} [\Psi]^T \left[\frac{A}{Z}\right] [\Psi] \tag{6.24}$$

where $[A/Z]$ is a matrix of grid point impedance values weighted by the surface area A_m associated with each grid point m. Most structural finite-element computer codes, when adapted for acoustical normal-mode analysis according to Table 6.3, also have capabilities for modal frequency response and modal transient response to give the forced sound pressure response in the form of Eq. (6.22).

6.7 STEADY-STATE SOUND PRESSURE RESPONSE

The steady-state sound pressure in an enclosure is often of interest where the noise emanates from a point source, as in Fig. 6.1a. For a point source located at (x_0, y_0, z_0) and having a steady-state volume velocity $\hat{Q}\cos(\omega t + \phi)$, the mass flow rate from the source can be mathematically represented using the Dirac delta function as $\dot{m}(x, y, z, t) = \rho \hat{Q}\delta(x - x_0)\delta(y - y_0)\delta(z - z_0)\cos(\omega t + \phi)$. At a particular forcing frequency $f = \omega/2\pi$, the sound pressure response in the enclosure can then be obtained from Eqs. (6.15)–(6.17) by using the *frequency response function* technique,

$$p(x, y, z, t) = \sum_n p_n(x, y, z, \omega) \cos(\omega t + \theta_n) \quad \text{N/m}^2 \tag{6.25}$$

where

$$p_n(x, y, z, \omega) = \frac{\rho c^2 \omega \hat{Q} \Psi_n(x, y, z) \Psi_n(x_0, y_0, z_0)}{V_n[(\omega^2 - \omega_n^2 + 2\delta_n^i \omega)^2 + (2\delta_n^r \omega)^2]^{1/2}} \quad \text{N/m}^2 \tag{6.26}$$

and

$$\theta_n(\omega) = \phi - \tan^{-1} \frac{\omega^2 - \omega_n^2 + 2\delta_n^i \omega}{2\delta_n^r \omega} \quad \text{rad} \tag{6.27}$$

The amplitude of the modal sound pressure is $|p_n|$. When $\hat{Q} = 1$, $|p_n|$ is the amplitude of the modal frequency response function.

Equation (6.25) shows that the steady-state sound pressure at one point in the enclosure can be considered as the superposition of numerous components of the same frequency ω but with different amplitudes $|p_n|$ and phase angles θ_n. If the source location (x_0, y_0, z_0) or the observer location (x, y, z) in Eq. (6.26) is on the nodal surface of a particular mode n, then the minimum participation of that mode will be observed in the response. It is also noteworthy that the solution is symmetric in the coordinates (x_0, y_0, z_0) of the sound source and the point of observation (x, y, z). If we put the sound source at (x, y, z), we observe at point (x_0, y_0, z_0) the same sound pressure as we did at (x, y, z) when the source was at (x_0, y_0, z_0). This is the famous reciprocity theorem (see Chapter 10) that can sometimes be applied with advantage to measurements in room acoustics.

When only the uniform-pressure mode ($\omega_0 = 0$) is excited, Eqs. (6.25)–(6.27) reduce to $p = p_0 \cos(\omega t + \theta_0)$, where

$$p_0(\omega) = \frac{\rho c^2 \hat{Q}}{V[(\omega + 2\delta_0^i)^2 + (2\delta_0^r)^2]^{1/2}} \quad \text{N/m}^2$$

$$\theta_0(\omega) = \phi - \tan^{-1} \frac{\omega + 2\delta_0^i}{2\delta_0^r} \quad \text{rad}$$

(6.28)

which are equivalent to Eqs. (6.2). The uniform-pressure mode applies to the case of sound sources operating at very low frequencies, well below the first ($n = 1$) resonance of the enclosure. This requirement is generally met if $L < \lambda/10$, where L is the maximum linear dimension of the cavity and $\lambda = c/f$ is the wavelength of sound generated by the source operating at the forcing frequency $f = \omega/2\pi$.

Example 6.6. For the rectangular cavity of Example 6.4, determine the steady-state sound pressure response in the cavity for loudspeaker excitation when (a) the enclosure walls are rigid and there is no damping, (b) one wall is covered uniformly with absorption material of known impedance Z, and (c) the damping is expressed in terms of the *critical damping ratio* ζ_n. Assume the loudspeaker acts as a simple monopole source.

Solution

(a) For an undamped ($\delta_n^r = 0$) and rigid-wall ($\delta_n^i = 0$) enclosure, we obtain, from Eqs. (6.25)–(6.27),

$$p(x, y, z, t) = \rho c^2 \hat{Q} \sum_{i=0}^{I} \sum_{j=0}^{J} \sum_{k=0}^{K} \frac{\omega \Psi_{ijk}(x, y, z) \Psi_{ijk}(x_0, y_0, z_0)}{V_{ijk}|\omega^2 - \omega_{ijk}^2|}$$

$$\times \cos\left(\omega t + \phi - \frac{\pi}{2}\right) \quad \text{N/m}^2 \quad (6.29)$$

where Ψ_{ijk} and $f_{ijk} = \omega_{ijk}/2\pi$ are given in Eq. (6.14), V_{ijk} is determined from Eq. (6.12), and I, J, K are the number of modes we include to obtain a converged solution. For a closed rectangular enclosure,

$$\frac{V_{ijk}}{V} = \varepsilon_i \varepsilon_j \varepsilon_k \quad \text{where} \quad \varepsilon_n = \begin{cases} 1 & \text{for } n = 0 \\ \frac{1}{2} & \text{for } n \geq 1 \end{cases} \quad (6.30)$$

with $V = L_x L_y L_z$. Note from Eq. (6.14) that for every mode of vibration the sound pressure is a maximum at the corners of the rectangular enclosure. Also, for every mode of vibration for which one of the indexes i, j, or k is *odd*, the sound pressure is zero at the *center* of the enclosure; hence at the geometrical center of the enclosure only one-eighth of the modes of vibration produce a finite sound pressure. Extending this further, at the center of any one wall, the modes for which two of the indexes (i, j, k) are odd will have zero pressure, so that only one-fourth of them will participate. Finally, at the center of one edge of the enclosure, the modes for which one index is odd will have zero pressure, so that only one-half of them participate there.

(b) Assume the uniform absorption material is on the $z = 0$ wall. Then, from Eqs. (6.19) and (6.20),

$$\begin{aligned} \delta_{ijk}^r &= \frac{cA_{ijk}}{2V_{ijk}} \operatorname{Re} \frac{\rho c}{Z_{ijk}} = \frac{c}{2L_z \varepsilon_k} \operatorname{Re} \frac{\rho c}{Z} \quad \text{s}^{-1} \\ \delta_{ijk}^i &= \frac{cA_{ijk}}{2V_{ijk}} \operatorname{Im} \frac{\rho c}{Z_{ijk}} = \frac{c}{2L_z \varepsilon_k} \operatorname{Im} \frac{\rho c}{Z} \quad \text{s}^{-1} \end{aligned} \quad (6.31)$$

where $\rho c/Z_{ijk} = \rho c/Z$ and $A_{ijk} = \varepsilon_i \varepsilon_j A_z$ with $A_z = L_x L_y$. Substituting Eq. (6.31) into Eqs. (6.25)–(6.27), one can evaluate the series solution for the sound pressure when Z is known.

(c) From the theory of vibration (Chapter 13), the critical damping ratio ζ_n is related to the damping constant through $\delta_n = \zeta_n \omega_n$. In complex form, $\zeta_n = \zeta_n^r + i\zeta_n^i$, so that $\delta_n^r = \zeta_n^r \omega_n$ and $\delta_n^i = \zeta_n^i \omega_n$, which can be substituted into Eqs. (6.25)–(6.27) to evaluate the sound pressure for given ζ_n.

Figure 6.4 shows the predicted versus measured sound pressure level in a rectangular enclosure for volume–velocity (loudspeaker) excitation. The response was predicted by using Eqs. (6.25)–(6.27) as developed in Example 6.6. Figure 6.4a is the response in the undamped, rigid-wall enclosure, where the resonance peaks resulting from the excitation of the rigid-wall cavity modes are identified. Theoretically, the response at these resonances in an undamped enclosure should be infinite. In practice, however, damping is present even in an enclosure with very rigid walls since viscous and thermal losses occur in the air and at the boundaries. Modal damping provides a convenient method of accounting for these losses, and it is easily included in the solution as described in Example 6.6(c). Since the losses are primarily resistive, a real value of modal damping is used. Figure 6.4b shows the sound pressure response when sound-absorptive material of known impedance covers the bottom wall of the enclosure. The predicted response is obtained using the complex "damping" formulas given

164 SOUND IN SMALL ENCLOSURES

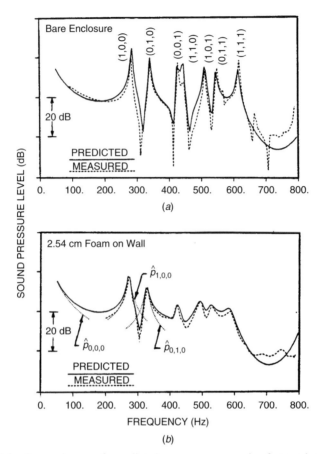

FIGURE 6.4 Comparisons of predicted versus measured of sound-pressure-level curves for a constant volume–velocity source in a rectangular enclosure of dimensions $0.61 \times 0.51 \times 0.41$ m. The microphone was at $(0.51 \text{ m}, 0.10 \text{ m}, 0.30 \text{ m})$ and the source was at $(0.0 \text{ m}, 0.10 \text{ m}, 0.10 \text{ m})$. (a) Bare enclosure with 0.5% modal damping. (b) Acoustical foam 2.54 cm thick with measured impedance $\rho c/Z = (0.12 + 0.47i)(f/800)$ for $0 \leq f < 800$ Hz covering the 0.61×0.51-m wall.

in Eq. (6.31) and by using the modes based on the reduced height of the enclosure due to the absorption material thickness. The resonant responses of the $|p_{0,0,0}|$, $|p_{1,0,0}|$, and $|p_{0,1,0}|$ terms in the series in Eq. (6.25) are also shown. The attenuation and frequency shift of the resonant peaks are evident by comparing Figs. 6.4a and 6.4b, and these are due to the resistive and reactive nature of the absorption material.

6.8 ENCLOSURE DRIVEN AT RESONANCE

If a point source of volume velocity Q is driven by a steady-state generator of frequency ω and if that frequency equals a normal-mode frequency ω_n, then the modal sound pressure from Eq. (6.26) is

$$p_n(x, y, z, \omega_n) = \frac{\rho c^2 \hat{Q} \Psi_n(x, y, z) \Psi_n(x_0, y_0, z_0)}{V_n[(2\delta_n^i)^2 + (2\delta_n^r)^2]^{1/2}} \quad \text{N/m}^2 \quad (6.32)$$

From this we conclude that $|p_n| \to \infty$ when both $\delta_n^r = 0$ and $\delta_n^i = 0$, so that ω_n is the *resonance frequency* when we have an enclosure with undamped and rigid walls.

For an enclosure with reactive walls ($\delta_n^i \neq 0$, $\delta_n^r \equiv 0$), the frequency dependence in the denominator of Eq. (6.26) can be approximated by

$$\omega^2 - \omega_n^2 + 2\delta_n^i \omega \approx \omega^2 - \omega_n^2 \left(1 - \frac{2\delta_n^i}{\omega_n}\right) = \omega^2 - \omega_{n0}^2 \quad (6.33)$$

where

$$\omega_{n0} = \omega_n \sqrt{1 - \frac{2\delta_n^i}{\omega_n}} \quad \text{rad/s} \quad (6.34)$$

Since now $|p_n| \to \infty$ when $\omega \to \omega_{n0}$, we see that the new resonance frequency of an enclosure with undamped, reactive walls is ω_{n0}. The resonance frequency *shifts* from the rigid-wall resonance value ω_n due to the reactance δ_n^i of the walls, and ω_{n0} can be either higher or lower than the rigid-wall frequency ω_n, depending on whether $\delta_n^i < 0$ or $\delta_n^i > 0$, respectively. From Eq. (6.27), the phase angle between the modal pressure and the volume velocity source at resonance (i.e., when $\omega = \omega_{n0}$) is either 0° or 180°, and this provides a means of experimentally identifying the resonance frequency ω_{n0} and the enclosure reactance δ_n^i.

The enclosure dissipation δ_n^r controls the amplitude of the resonant peak when $\omega = \omega_{n0}$ since

$$p_n(x, y, z, \omega_{n0}) = \frac{\rho c^2 \hat{Q} \Psi_n(x, y, z) \Psi_n(x_0, y_0, z_0)}{2\delta_n^r V_n} \quad \text{N/m}^2 \quad (6.35)$$

The dissipation also shifts the resonance frequency an additional amount. The actual resonance frequency ω_{res} at which $|p_n|$ becomes a maximum can be shown to be

$$\omega_{\text{res}} = \sqrt{\omega_{n0}^2 - 2(\delta_n^r)^2} = \sqrt{\omega_n^2 - 2\delta_n^i \omega_n - 2(\delta_n^r)^2} \quad \text{rad/s} \quad (6.36)$$

The dissipative effect (δ_n^r) is always to reduce the resonance frequency, but this effect is of second order and generally less important than the reactive effect (δ_n^i), which is of first order, unless δ_n^r is large. In general, therefore, to first order, the resonance frequency of an enclosure is $\omega_{\text{res}} = \omega_{n0}$ with ω_{n0} given by Eq. (6.34). The *half-power bandwidth* of the resonant peak is the width of the resonance curve at 3 dB below the peak power (6 dB below the peak pressure) and is

$$\Delta\omega_{\text{res}} = 2\delta_n^r \quad \text{rad/s} \quad (6.37)$$

166 SOUND IN SMALL ENCLOSURES

This provides a means of experimentally determining the enclosure dissipation δ_n^r.

Example 6.7. Determine the resonance frequency of the first mode in Fig. 6.4b. The measured wall impedance is $\rho c/Z = (0.12 + 0.47i)(f/800)$.

Solution For the (1, 0, 0) mode, we have $f_{1,0,0} = 284.8$ Hz from Eq. (6.14). For this frequency, $\rho c/Z = (0.12 + 0.47i)(284.8/800) = 0.043 + 0.167i$. Noting that the thickness of the absorption material is 2.54 cm, we have $L_z = 0.41 - 0.0254 = 0.3846$ m, and from Eq. (6.31),

$$\delta_{1,0,0}^r = \frac{347.3}{2 \times 0.3846 \times 1} \times 0.043 = 19.6$$

$$\delta_{1,0,0}^i = \frac{347.3}{2 \times 0.3846 \times 1} \times 0.167 = 76.2$$

Substituting these into Eq. (6.36) gives the resonance frequency

$$f_{\text{res}} = \frac{\omega_{\text{res}}}{2\pi} = f_n \sqrt{1 - \frac{2\delta_n^i}{\omega_n} - 2\left(\frac{\delta_n^r}{\omega_n}\right)^2}$$

$$= 284.8(1 - 0.085 - 0.00024)^{1/2} = 272.4 \text{ Hz}$$

Actually, the calculated frequency should be slightly greater than this because the impedance is frequency dependent and should be evaluated at f_{res}. An iterative calculation gives $f_{\text{res}} = 273$ Hz, while the measured resonance frequency in Fig. 6.4b is 274 Hz.

6.9 FLEXIBLE-WALL EFFECT ON SOUND PRESSURE

A flexible wall may exhibit structural resonances and affect the sound pressure in two ways: (a) by acting as a noise source when exterior structural or pressure loads excite vibrations of the wall and (b) by acting as a boundary impedance that can alter the cavity sound pressure. Both of these effects may occur simultaneously in a room with flexible walls, in which case a coupled *structural–acoustical analysis* is required to predict the sound pressure response. We shall first consider each of the two effects of a flexible wall separately and then consider the coupled structural–acoustical response.

Flexible Wall as Noise Source

When the vibration velocity $\dot{w}(x, y, z, t)$ of a flexible wall is known, the boundary condition in Table 6.1, frame 2 or 4, can be applied to Eqs. (6.16) and (6.17). The forcing is through the equivalent modal volume–velocity of the wall vibration,

$$Q_n = \hat{Q}_n \cos(\omega t + \phi_n) = -\int_S \dot{w} \Psi_n \, dS \quad \text{m}^3/\text{s} \tag{6.38}$$

where the vibration velocity of the wall $\dot{w} = \hat{\dot{w}} \cos(\omega t + \phi_w)$ is assumed to have a known amplitude $\hat{\dot{w}}(x, y, z)$ and phase $\phi_w(x, y, z)$. The steady-state sound pressure response is then obtained as in Eqs. (6.26) and (6.27),

$$p_n(x, y, z, \omega) = \frac{\rho c^2 \omega \hat{Q}_n \Psi_n(x, y, z)}{V_n[(\omega^2 - \omega_n^2 + 2\delta_n^i \omega)^2 + (2\delta_n^r \omega)^2]^{1/2}} \quad \text{N/m}^2 \quad (6.39)$$

$$\theta_n(\omega) = \phi - \tan^{-1} \frac{\omega^2 - \omega_n^2 + 2\delta_n^i \omega}{2\delta_n^r \omega} \quad \text{rad} \quad (6.40)$$

where δ_n^r, δ_n^i relate to the layer of absorption material of acoustical impedance Z_a covering the flexible wall.

Figure 6.5a shows the calculation of the forced acoustical response in the automobile passenger compartment using the above method with measured panel acceleration amplitude and phase data to represent the wall vibration and with measured wall impedance data to represent the absorption materials. The solution is expressed in matrix form as in Eq. (6.22) with the acoustical modes obtained from a finite-element analysis (Fig. 6.3). Figure 6.5a shows that a large spatial variation in sound pressure level occurs in the passenger compartment because of the modal nature of the acoustical response. Figure 6.5b shows the sound pressure computed separately for the vibration of each individual wall panel. The individual sound pressures can be combined, by considering their magnitude and phase, to obtain the resultant sound pressure, which is shown in Fig. 6.5b. This provides a method to identify the major noise sources and the extent to which they must be controlled to yield a specified noise reduction. For example, Fig. 6.5b illustrates the large amplitude of the sound pressure due to the back-window vibration. Also note that the elimination of the roof vibration would increase the noise at the driver's ear.

Flexible Wall as Reactive Impedance

Wall panel vibration can generally be expressed in the normal-mode expansion form

$$w = \sum_m W_m(t) \Phi_m(x, y, z) \quad \text{m} \quad (6.41)$$

where Φ_m are the structural vibration mode shapes and W_m are the modal amplitudes. If we consider the coupling between the wall panel and the cavity, for the mth structural mode and the nth acoustical mode, the modal acoustical impedance of the wall can be expressed as

$$Z_{mn} = (S_{mn})^{-1} \left[C_m + i\omega \left(M_m - \frac{K_m}{\omega^2} \right) \right] \quad \text{N/m}^3 \cdot \text{s}^{-1} \quad (6.42)$$

168 SOUND IN SMALL ENCLOSURES

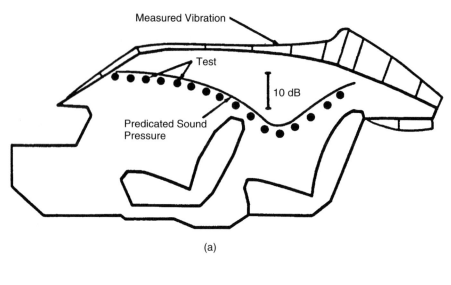

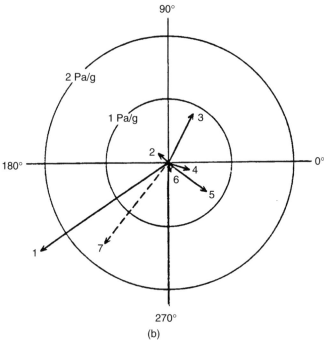

FIGURE 6.5 (*a*) Comparison of measured versus predicted sound-pressure-level spatial variation in automobile passenger compartment for 40-Hz structural excitation. (*b*) Polar amplitude–phase diagram indicating panel contributions to resultant sound pressure at driver's ear (1, back window; 2, rear floor; 3, roof; 4, windshield; 5, rear shelf; 6, front floor; 7, resultant).

where M_m, C_m, and K_m are the modal mass, damping, and stiffness of the mth wall mode, and

$$S_{mn} = \int_S \Phi_m \Psi_n \, dS \quad \text{m}^2 \tag{6.43}$$

where S_{mn} is the *structural–acoustical coupling*. [Note that when $S_{mn} = 0$, there is no coupling between the structural mode m and the acoustical mode n, and Eq. (6.42) does not apply. In practice, when S_{mn} is sufficiently small, the coupling can be neglected to simplify the solution.]

For an undamped enclosure, the structural–acoustical coupling modifies Eq. (6.26) to

$$p_n(x, y, z, \omega) = \frac{\rho c^2 \omega \hat{Q} \Psi_n(x, y, z) \Psi_n(x_0, y_0, z_0)}{V_n |\omega^2 - \omega_{n0}^2|} \left| \frac{\omega^2 - \Omega_m^2}{\omega^2 - \Omega_{m0}^2} \right| \quad \text{N/m}^2 \tag{6.44}$$

where $\Omega_m = \sqrt{K_m/M_m}$ (rad/s) is the natural frequency of the mth wall mode and ω_{n0}^2 and Ω_{m0}^2 are solutions of

$$(\omega^2 - \omega_n^2)(\omega^2 - \Omega_m^2) - D^2 \omega^2 = 0 \tag{6.45}$$

where $D = \sqrt{K_{\text{air}}/M_m}$ with $K_{\text{air}} = \rho c^2 S_{mn}^2 / V_n$ (N/m) being the *stiffness* of the nth acoustical mode relative to the mth structural mode.

Equation (6.44) shows that the flexible wall introduces an additional resonance at $\omega = \Omega_{m0}$, which will appear as a resonance peak in the sound pressure response and corresponds to the wall resonance. It can be shown from Eq. (6.45) that, with wall modes at frequencies below the rigid-wall cavity resonance ($\Omega_m < \omega_n$, the case of *mass-controlled* boundaries), the cavity resonance is raised ($\omega_{n0} > \omega_n$) and the wall resonance is lowered ($\Omega_{m0} < \Omega_m$). With wall modes at frequencies above the rigid-wall cavity resonance ($\Omega_m > \omega_n$, the case of *stiffness-controlled* boundaries), the cavity resonance is lowered ($\omega_{n0} < \omega_n$) and the wall resonance is raised ($\Omega_{m0} > \Omega_m$). In addition, the wall acts as a vibration absorber when $\omega = \Omega_m$, so that the modal sound pressure $p_n = 0$. However, the total sound pressure p may not be zero because of the participation of other acoustical modes. A Helmholtz resonator used as a vibration absorber as described in Chapter 8 has impedance Z_R and performs in a similar manner.

Coupled Structural–Acoustical Response

Above we have considered the coupling of one acoustical mode n with one structural mode m, which is the case when a single coupling coefficient S_{mn} dominates for the mode pair (m, n). In practice, however, each acoustical mode n may couple with several structural modes, and the general case can be treated by substituting Eq. (6.41) into the second integral of Eq. (6.17) to obtain, from Eq. (6.16),

$$\ddot{P}_n + 2\delta_n \dot{P}_n + \omega_n^2 P_n + \frac{\rho c^2}{V_n} \sum_m S_{mn} \ddot{W}_m = \frac{\rho c^2}{V_n} F_n(t) \quad \text{N/m}^2 \cdot \text{s}^2 \qquad (6.46)$$

where the summation is carried out over all coupled structural modes m. Similarly, each structural mode m may in general couple with several acoustical modes, and the corresponding equation for the structural modal amplitude $W_m(t)$ in Eq. (6.41) is

$$\ddot{W}_m + 2\Delta_m \dot{W}_m + \Omega_m^2 W_m - \frac{1}{M_m} \sum_n S_{mn} P_n = \frac{R_m(t)}{M_m} \quad \text{m/s}^2 \qquad (6.47)$$

where the summation is carried out over all coupled acoustical modes n. In this latter equation for the structure, Δ_m is the modal damping constant, Ω_m is the natural frequency, M_m is the modal mass, and $R_m(t)$ is the modal force. The coupling between the acoustical and structural systems makes it necessary to solve Eqs. (6.46) and (6.47) simultaneously for all coupled P_n and W_m. A complete derivation of the equations and discussion of the procedure can be found in reference 12, and its finite-element implementation is described in references 14 and 15.

To illustrate a typical coupled structural–acoustical response, Fig. 6.6 shows an application to a small metallic box used to isolate instruments from an external noise field (cf. Fig. 6.1b). The analysis of such enclosures to predict their noise attenuation is discussed in detail in Chapter 11. The finite-element method can be used to model the box wall panels and the acoustical cavity (Figs. 6.6b, c) in order to compute the uncoupled structural and acoustical modes. Equations (6.46) and (6.47) are then solved for the coupled frequency response, where the exterior noise field in Fig. 6.6a is specified as an oscillating external pressure $\hat{p}_E \cos \omega t$ applied uniformly to the wall panels. The noise attenuation inside the enclosure is characterized by its *insertion loss*, defined in Chapter 11 as $20 \log_{10}(\hat{p}_I/\hat{p}_E)$, where $\hat{p}_I$ is the interior sound pressure. The predicted versus measured insertion loss is shown in Fig. 6.6d. The structural–acoustical coupling effect is particularly evident in the rms panel vibration shown in Fig. 6.6e, which compares the coupled versus the uncoupled (i.e., in vacuo) panel response and illustrates the significant effect that the acoustical cavity modes have on the structural vibration response.

6.10 RANDOM SOUND PRESSURE RESPONSE

For steady-state, deterministic sound source excitation, the sound pressure response can be obtained by the frequency response function technique of Eqs. (6.25)–(6.27). However, for sound source excitation that is nondeterministic, it is generally necessary to utilize a random-analysis technique

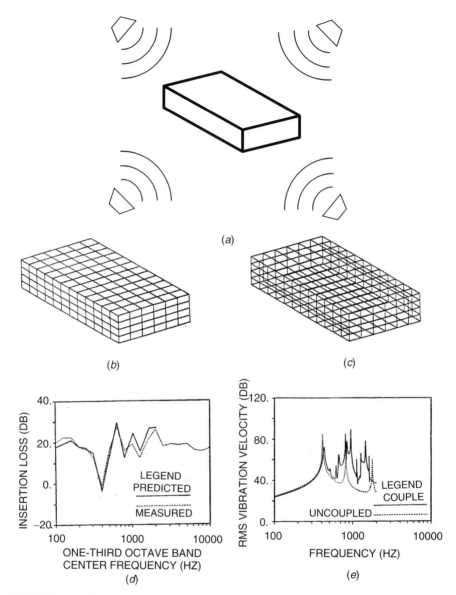

FIGURE 6.6 Structural–acoustical analysis to predict insertion loss of a 30 × 15 × 5-cm unlined aluminum box with 0.16-cm-thick walls: (*a*) box in uniform noise environment; (*b*) structural finite-element model of box wall panels; (*c*) acoustical finite-element model of box cavity; (*d*) predicted versus measured insertion loss at box center; (*e*) coupled versus uncoupled surface-averaged vibration of 30 × 15-cm wall panel. [Measured data in (*d*) from Chapter 11].

to obtain the sound pressure response. For example, irregular pressure fluctuations generated by turbulence can result in complicated sound sources that are more readily measured or defined as random excitations than as deterministic excitations. Also, interactions with irregular surfaces such as roads, runways, and pathways are more readily represented as random loads than as deterministic loads. Under certain conditions (stationarity) the average frequency content of these excitations may be represented by a spectral density function (Chapter 3). Then by using the spectral density representation of the input and the frequency response function technique to determine the sound pressure output, conventional random-analysis theory may be utilized to obtain the sound pressure spectral density response.

From random-analysis theory, nondeterministic stationary processes can be characterized by their auto- (power) spectral density function and their cross-spectral density function.[16] The power spectral density (PSD) function of a physical variable such as sound pressure $p(t)$ is defined as a real-magnitude function $S_p(\omega)$, where

$$S_p(\omega) = \lim_{T\to\infty} \frac{2}{T} \left| \int_0^T e^{-i\omega t} p(t)\, dt \right|^2 \quad (\text{N/m}^2)^2/\text{Hz} \qquad (6.48)$$

Similarly, the PSD function $S_Q(\omega)$ for a volume–velocity point source $Q(t)$ is

$$S_Q(\omega) = \lim_{T\to\infty} \frac{2}{T} \left| \int_0^T e^{-i\omega t} Q(t)\, dt \right|^2 \quad (\text{m}^3/\text{s})^2/\text{Hz} \qquad (6.49)$$

The PSD functions of the sound pressure response and sound source excitation are related via the frequency response function as

$$S_p(x, y, z, \omega) = \sum_n p_n^2(x, y, z, \omega) S_Q(\omega) \quad (\text{N/m}^2)^2/\text{Hz} \qquad (6.50)$$

where $S_p(x, y, z, \omega)$ is the sound pressure PSD response and $p_n(x, y, z, \omega)$ is the modal frequency response function in Eq. (6.26) for unit sound source excitation $\hat{Q} = 1$. Equation (6.50) is the random-response equivalent of Eq. (6.25) for the steady-state response. However, for random response, only the amplitude of the modal frequency response function is required and not the phase information in $\theta_n(\omega)$ in Eq. (6.27).

Another useful result from random-analysis theory is that, if several sound sources $Q_1(t), Q_2(t), \ldots, Q_a(t), \ldots$ are statistically independent, then the cross-correlation between any pair of sources is zero, and the total sound pressure PSD response is equal to the sum of the PSD responses due to the individual sources. That is,

$$S_p(x, y, z, \omega) = \sum_a S_{p_a}(x, y, z, \omega)$$

$$= \sum_a \sum_n p_{an}^2(x, y, z, \omega) S_{Q_a}(\omega) \quad (\text{N/m}^2)^2/\text{Hz} \qquad (6.51)$$

Finally, when two sound sources $Q_a(t)$, $Q_b(t)$ are statistically correlated, the degree of correlation is related by a cross-spectral density function $S_{Q_aQ_b}(\omega)$ given by

$$S_{Q_aQ_b}(\omega) = \lim_{T\to\infty} \frac{2}{T}\left[\left(\int_0^T e^{-i\omega t}Q_a(t)\,dt\right)^*\left(\int_0^T e^{-i\omega t}Q_b(t)\,dt\right)\right]$$
$$(\text{m}^3/\text{s})^2/\text{Hz} \qquad (6.52)$$

where the asterisk denotes complex conjugate. Unlike the PSD function, the cross-spectral density function is generally a complex number that includes both amplitude and phase information. For correlated sound sources, the spectral density of the sound pressure response is then a complex number given by

$$S_p(x,y,z,\omega) = \sum_a\sum_b\sum_m\sum_n (p_{am}(x,y,z,\omega)e^{i\theta_m(\omega)})(p_{bn}(x,y,z,\omega)e^{i\theta_n(\omega)})^*$$
$$\times S_{Q_aQ_b}(\omega) \quad (\text{N/m}^2)^2/\text{Hz} \qquad (6.53)$$

where the phase relationships $\theta_m(\omega)$, $\theta_n(\omega)$ are from Eq. (6.27) with $\phi = 0$. In the case where the sound sources are uncorrelated, Eq. (6.53) reduces to the previous real-magnitude sound pressure PSD functions in Eqs. (6.50) and (6.51).

Example 6.8. Determine the sound pressure PSD response in the very small enclosure in Fig. 6.1a with rigid walls when (a) there is a single sound source Q_a in the enclosure with PSD amplitude $S_{Q_a} = S(\omega)$; (b) there are two uncorrelated interior sound sources Q_a, Q_b in the enclosure with PSD amplitudes $S_{Q_a} = S(\omega)$, $S_{Q_b} = S(\omega)$; and (c) there are two interior sound sources with $S_{Q_a} = S(\omega)$, $S_{Q_b} = S(\omega)$ but perfectly correlated so that $S_{Q_aQ_b} = S(\omega)$.

Solution

(a) For a rigid-wall enclosure with a single sound source, substituting Eq. (6.4) into Eq. (6.50) gives

$$S_p(x,y,z,\omega) = p_0^2(x,y,z,\omega)S_{Q_a} = \left(\frac{\rho c^2}{\omega V}\right)^2 S_{Q_a} = \left(\frac{\rho c^2}{\omega V}\right)^2 S(\omega)$$

(b) For two uncorrelated sound sources in the enclosure, Eq. (6.51) gives

$$S_p(x,y,z,\omega) = \left(\frac{\rho c^2}{\omega V}\right)^2 (S_{Q_a} + S_{Q_b}) = 2\left(\frac{\rho c^2}{\omega V}\right)^2 S(\omega)$$

(c) For two correlated sound sources in the enclosure, Eq. (6.54) gives

$$S_p(x,y,z,\omega) = \left(\frac{\rho c^2}{\omega V}\right)^2 (S_{Q_a} + S_{Q_b} + 2S_{Q_aQ_b}) = 4\left(\frac{\rho c^2}{\omega V}\right)^2 S(\omega)$$

Thus, for uncorrelated sound sources, doubling the number of the sound sources in the enclosure *doubles* the sound pressure PSD response (3 dB increase), as in (b). On the other hand, for perfectly correlated sound sources, doubling the number of the sound sources *quadruples* the sound pressure PSD response (6 dB increase), as in (c).

The random-analysis formulas in Eqs. (6.50)–(6.53) are also applicable to structural analysis, by replacing the acoustical frequency response functions and sound source excitations with the corresponding structural frequency response functions and structural excitations.[16] Also, the coupled structural–acoustical analysis provided by Eqs. (6.46) and (6.47) can be applied to predict the sound pressure PSD response for random excitations.[17] As an example, Fig. 6.7 illustrates the predicted interior sound pressure PSD response in an automobile passenger compartment for the vehicle traveling at constant speed V over a randomly rough road. In this example, random road excitation occurs at the tire patch (Fig. 6.7a) and the sound pressure response in the passenger compartment results from the transmitted vibration to the vehicle body panels. Equations (6.46) and (6.47) can be used to obtain the sound pressure frequency response functions by employing the vehicle finite-element models in Figs. 6.7b,c. Equation (6.51) is then used to predict the sound pressure PSD response by applying the road profile PSD function in Fig. 6.7d as excitation at each tire patch at the vehicle speed V. Figure 6.7e shows the predicted sound pressure PSD response in the passenger compartment versus the 95% confidence band based on the measured responses in nominally identical vehicles. The relative participation of the sound sources and body panels to the sound pressure PSD response can also be identified by the methods in Chapter 3 and reference 17.

6.11 TRANSIENT SOUND PRESSURE RESPONSE

When the source of sound in a room is turned off, the sound dies out, or decays, at a rate that depends on the dissipation, or damping, in the room. The acoustical response that exists in the room can be found from the transient solution of Eq. (6.16), which is

$$P_n(t) = \frac{\rho c^2}{V\omega_{nD}} \int_0^t F_n(\tau) e^{-\delta_n^r(t-\tau)} \sin[\omega_{nD}(t-\tau)]\, d\tau \quad \text{N/m}^2 \qquad (6.54)$$

where $\omega_{nD} = \sqrt{\omega_n^2 - (\delta_n^r)^2 + (\delta_n^i)^2}$ is the "damped" modal frequency of the acoustical response in the enclosure. Equation (6.54) is the particular solution that satisfies zero initial conditions. For general initial conditions, the following free-vibration response must be added to the above equation:

$$P_n(t) = e^{-\delta_n^r t}\left[\frac{\dot{P}_n(0) + P_n(0)\delta_n^r}{\omega_{nD}} \sin \omega_{nD}t + P_n(0)\cos \omega_{nD}t\right] \quad \text{N/m}^2 \quad (6.55)$$

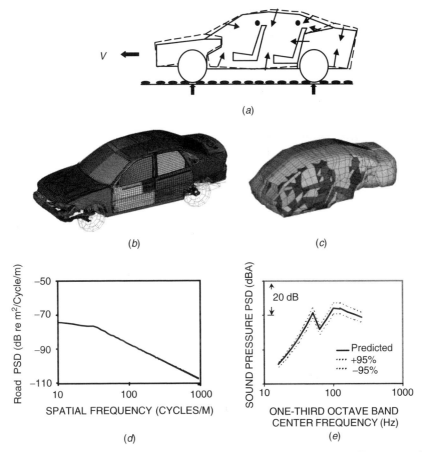

FIGURE 6.7 Structural–acoustical analysis to predict sound pressure PSD response in vehicle traveling at constant speed V on randomly rough road: (*a*) interior road noise generation; (*b*) structural finite-element model of vehicle; (*c*) acoustical finite-element model of passenger-trunk compartment; (*d*) road profile power spectral density; (*e*) predicted A-weighted sound pressure PSD response at front seat occupant ear location versus 95% confidence interval.

Equation (6.55) can be used to determine the decay of the modes after the sound source is turned off at $t = 0$, when the initial conditions $P_n(0)$ and $\dot{P}_n(0)$ are known. For light damping, each mode of vibration behaves independently of the others, and the total process of sound decay is the summation in Eq. (6.15) of the sound pressures associated with all of the individual modes of vibration that fall within the frequency band of interest. The long-time sound decay may be different than the short-time sound decay because of the different damping factors and initial conditions for the modes.

The *reverberation time T* is defined as the time in seconds required for the level of sound to drop by 60 dB or for the pressure to drop to $\frac{1}{1000}$ of its initial value (see Chapter 7). One can similarly define a *modal* reverberation time T_n as that for which the sound pressure decays in that mode by 60 dB or $\frac{1}{1000}$ of its initial value. Since the reverberation time is that associated with the decaying part $(-\delta_n^r t)$ of the solution in Eq. (6.55),

$$T_n = \frac{6.91}{\delta_n^r} = \frac{13.82 V_n}{c A_n \operatorname{Re}(\rho c/Z_n)} \quad \text{s} \qquad (6.56)$$

where we have substituted for δ_n^r from Eq. (6.19). With $c = 343$ m/s (20°C) and expressed in terms of the random-incidence absorption coefficient, $\alpha_n = 8 \operatorname{Re}(\rho c/Z_n)$, with V in cubic meters and A in square meters, one obtains the modal reverberation time $(n > 0)$

$$T_n = 0.322 \frac{V_n}{\alpha_n A_n} \quad \text{s} \qquad (6.57)$$

which can be evaluated for a given Z by using the formulas in Eqs. (6.20) and (6.12). The uniform-pressure mode $(n = 0)$ must be treated separately. The general transient solution of Eq. (6.16) when $n = 0$ reduces to

$$P_0(t) = P_0(0) e^{-2\delta_0^r t} + \frac{\rho c^2}{V} \int_0^t \int_0^\tau F_0(\sigma)\, d\sigma \, e^{-2\delta_0^r(t-\tau)}\, d\tau \quad \text{N/m}^2 \qquad (6.58)$$

and the reverberation time for the decay of a uniform sound pressure is $T_0 = 6.91/2\delta_0^r = 6.91 V/cA \operatorname{Re}(\rho c/Z)$ so that, for V in cubic meters and A in square meters,

$$T_0 = 0.161 \frac{V}{\alpha A} \quad \text{s} \qquad (6.59)$$

In this case, the formula for the reverberation time of a uniform sound pressure is identical to the formula of Sabine for the reverberation time of a diffuse sound field.

If the pressure–time history is dominated by a single mode, then the reverberation time is equal to the appropriate modal reverberation time. If the sound source is turned off at $t = 0$, the magnitude of the sound pressure associated with a particular mode at a response location (x, y, z) is given by Eq. (6.55). The decay of the response from its maximum amplitude $[\dot{P}_n(0) = 0]$ can then be written as

$$p_n(x, y, z, t) = P_n(0) \Psi_n(x, y, z) e^{-\delta_n^r t} \cos(\omega_n t + \theta_n) \quad \text{N/m}^2 \qquad (6.60)$$

where $\theta_n = \tan^{-1}(\delta_n^r/\omega_{nD})$ is the modal phase. If we take the rms time average of $\cos(\omega_n t + \theta_n)$ and the rms spatial average of $\Psi_n(x, y, z)$ and we designate the resultant as $\overline{p}_n(t)$, then Eq. (6.60) indicates that on a plot of $\log \overline{p}_n$ versus

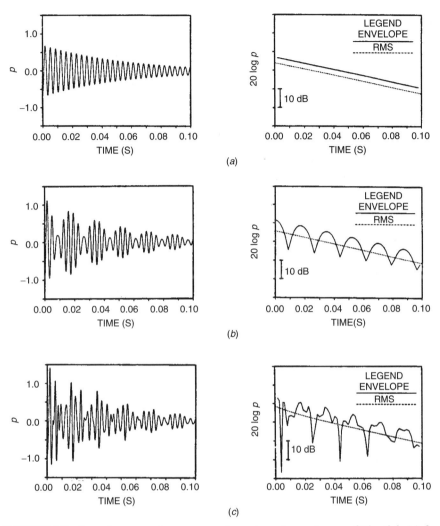

FIGURE 6.8 Sound pressure decay curves for rectangular enclosure of Fig. 6.6: (*a*) for the (1, 0, 0) mode of vibration; (*b*) for the (1, 0, 0) and (0, 1, 0) modes together; (*c*) for all modes up to 800 Hz. The graphs on the left show the course of the instantaneous sound pressure at the microphone location, and those on the right show the curve of the envelope of the left graphs and the rms pressure of Eq. (6.61) plotted on a log *p*-versus-*t* coordinate system.

time, as in Fig. 6.8*a*, both the envelope of the sound pressure and the rms sound pressure decay linearly with time and have the constant reverberation time T_n.

When many normal modes of vibration (each with its own amplitude, phase, resonance frequency, and damping constant) decay simultaneously, the total rms (spatial and temporal) sound pressure is obtained from Eq. (6.15) as

$$\overline{p}(t) = \sqrt{\overline{p}_I^2 + \overline{p}_{I+1}^2 + \cdots + \overline{p}_{I+N}^2} \quad \text{N/m}^2 \qquad (6.61)$$

where I is the first mode in the frequency band considered and $I + N$ is the last mode in the decay. In this case, the decay envelope is generally not linear as above, even when the modal reverberation times do not vary greatly from one mode to another so that all modes in the frequency band have a similar damping constant. The decay envelope is irregular, as in Figs. 6.8b,c, because the modes of vibration have different frequencies and beat with each other during the decay. However, the rms decay from Eq. (6.61) will be nearly linear if the modal reverberation times are similar (Fig. 6.8b) or if the long-time decay is governed by a least-damped mode, which may occur when the modal reverberation times or initial conditions differ (Fig. 6.8c).

This brief description of sound decay in small rooms is applicable to rooms of any size and shape, and it can be extended to include the effects of structural–acoustical coupling. The fact that sound fields in large enclosures involve too many modes for the calculations to be practical does not mean that there are not distinct normal modes of vibration, each with its own natural frequency and damping constant. Alternate and more practical methods for handling large enclosures involving large numbers of modes are discussed in the next chapter.

REFERENCES

1. P. M. Morse and K. U. Ingard, *Theoretical Acoustics*, Princeton University Press, Princeton, NJ, 1986.
2. L. L. Beranek (Ed.), *Noise and Vibration Control*, rev. ed., Institute of Noise Control Engineers, Poughkeepsie, NY, 1988.
3. H. Kutruff, *Room Acoustics*, Applied Science, London, 1979.
4. L. E. Kinsler, A. R. Frey, A. B. Coppins, and J. V. Saunders, *Fundamentals of Acoustics*, 4th ed., Wiley, New York, 1999.
5. L. L. Beranek and I. L. Ver, *Noise and Vibration Control Engineering*, Wiley, New York, 1992.
6. E. H. Dowell, C. F. Chao, and D. A. Bliss, "Absorption Material Mounted on a Moving Wall-Fluid/Wall Boundary Condition," *J. Acoust. Soc. Am.*, **70**(1), 244–245 (1981).
7. R. D. Blevins, *Formulas for Natural Frequency and Mode Shape*, Krieger, Malabar, FL, 1995.
8. M. J. Crocker (Ed.), *Handbook of Noise and Vibration Control*, Chapter 6: "Numerical Acoustical Modeling (Finite Element Modeling)" and Chapter 7: "Numerical Acoustical Modeling (Boundary Element Modeling)," Wiley, New York, 2006.
9. M. M. Kamal and J. A. Wolf, Jr. (Eds.), *Modern Automotive Structural Analysis*, Van Nostrand Reinhold, New York, 1982.
10. G. C. Everstine, "Structural Analogies for Scalar Field Problems," *Int. J. Numer. Methods Eng.*, **17**, 471–476 (1981).

11. P. M. Morse and R. H. Bolt, "Sound Waves in Rooms," *Rev. Modern Phys.*, **16**(2), 69–150 (1944).
12. E. H. Dowell, G. F. Gorman III, and D. A. Smith, "Acoustoelasticity: General Theory, Acoustic Natural Modes and Forced Response to Sinusoidal Excitation, Including Comparisons with Experiment," *J. Sound Vib.*, **52**(4), 519–542 (1977).
13. E. H. Dowell, "Reverberation Time, Absorption, and Impedance," *J. Acoust. Soc. Am.*, **64**(1), 181–191 (1978).
14. D. J. Nefske, J. A. Wolf, Jr., and L. J. Howell, "Structural-Acoustic Finite Element Analysis of the Automobile Passenger Compartment: A Review of Current Practice," *J. Sound Vib.*, **80**(2), 247–266 (1982).
15. S. H. Sung and D. J. Nefske, "A Coupled Structural-Acoustic Finite Element Model for Vehicle Interior Noise Analysis," *ASME J. Vib. Acoust. Stress Reliabil. Design*, **106**, 314–318 (1984).
16. D. E. Newland, *An Introduction to Random Vibration, Spectral and Wavelet Analysis*, Wiley, New York, 1994.
17. S. H. Sung and D. J. Nefske, "Component Mode Synthesis of a Vehicle Structural-Acoustic System Model," *Am. Inst. Aero. Astro. J.*, **24**(6), 1021–1026 (1986).

CHAPTER 7

Sound in Rooms

MURRAY HODGSON
University of British Columbia
Vancouver, Canada

JOHN BRADLEY
National Research Council of Canada
Ottawa, Canada

Acoustical problems experienced by room users include difficult verbal communication in classrooms and conference rooms, impaired student learning and teacher voice problems in classrooms, hearing loss in industrial workshops, and inadequate speech privacy in open-plan offices. Room sound fields comprise both "signals" (useful sounds, such as speech) and detrimental "noise". To modify room sound fields and improve acoustical conditions, the acoustician must understand the relationship between the room user and activity, the sound sources, the room and its contents, and the characteristics of room sound fields. Models for predicting room sound fields are of primary importance. They allow the acoustical conditions to be optimized during design and permit sound control measures to be evaluated for cost effectiveness in new or existing rooms. In this chapter, we take an energy-based approach, ignoring phase, interference, and modal effects discussed in Chapter 6. This is justified, except perhaps at low frequencies, since we are interested in rooms which are large compared with the sound wavelength, which are of complex shape and may not be empty. Further, we are interested in wide-band (total A-weighted, octave- or third-octave band) results. In this chapter, the sound field is quantified by the time-averaged, mean-square sound pressure p^2 in (N/m^2)^2 or the associated sound pressure level L_p in decibels ($L_p = 10 \log_{10}[p^2/p_0^2]$, $p_0 = 2 \times 10^{-5}$ N/m^2). In the case of multiple sources or sound reflections, the total energy is the sum of the individual energy contributions: $p_{\text{tot}}^2 = \sum p_i^2$, $L_{p,\text{tot}} = 10 \log_{10}\left(\sum 10^{L_{pi}/10}\right)$.

Sound sources radiate energy over some frequency range and, at each frequency, with a certain intensity. The rate of emission of energy is described by the sound power W in watts or the associated sound power level L_w in decibels ($L_w = 10 \log_{10}(W/W_0)$, $W_0 = 10^{-12}$ W), usually measured in octave- or third-octave frequency bands. As discussed in Section 7.8, the presence of surfaces near a source effectively increases its sound power output. Depending on its physical characteristics and frequency, a source may radiate uniformly in all directions (omnidirectional source) or more or less in certain directions (directional source). Source directivity is quantified by the directivity factor Q, which is the ratio of the mean-square sound pressure radiated in the receiver direction to the average of that radiated in all directions (see also Chapter 4). Small sources with three similar major dimensions can often be modeled as compact sources. Most real sources (large machines, conveyors, building walls) are more complex than a compact source, being extended in space in one or more dimensions. Readers wishing to do further reading on more fundamental aspects of room acoustics should consult reference 1. Throughout this chapter, references are provided for the benefit of readers who wish to do more in-depth reading on the topic under discussion.

7.1 ROOM SOUND FIELDS AND CONTROLLING FACTORS

Propagation of Sound

Sound waves propagate away from a source, being reflected (or scattered) by room boundaries, barriers, and furnishings. The resulting sound field at the receiver is composed of two parts. Sound which propagates directly from the source to the receiver is the direct sound. Its amplitude decreases with increasing distance from the source, as in a free field (discussed in Chapters 1 and 2). Sound which reaches the receiver after reflection from room surfaces or furnishings is the reverberant sound. Its amplitude decreases with increasing source distance at a rate which depends on the source geometry and on the acoustical properties of the room and its contents, as will be discussed below. To examine sound propagation in rooms, independent of the source sound power, to which it is directly related, we define the sound propagation function $SPF(r) = L_p(r) - L_w$ in decibels as the variable characterizing the effect of the room alone on the sound field. Sound pressure levels may then be obtained using $L_p(r) = SPF(r) + L_w$. Thus, from Eqs. (1.27)–(1.31), for a compact source with $Q = 1$ in a free field, $SPF(r) = -20 \log_{10} r - 11$ dB. In the case of multiple sources, the individual source contributions are obtained from the source sound power levels and the room sound propagation function values at the relevant source/receiver distances. The total sound pressure level at the receiver position is obtained by summing the individual levels on a power (p^2) basis.

Sound Decay

Shortly after a steady-state sound source begins to radiate, an equilibrium (or steady state) is established between the rate of energy absorption in the room and the rate of energy emission by the source. If the source then ceases to radiate, the energy in the room decreases with time at a rate determined by the rate of energy absorption. This is the room sound decay. It is usually characterized by the reverberation time T_{60}, which is the time in seconds required for the sound pressure level to decrease by 60 dB. Calculations of T_{60} are based on the average rate of decay over some part of the decay curve—usually the −5- to −35-dB part. Also of interest, since it quantifies perceived reverberance, is the early-decay time (EDT), based on the average rate of decay over the first 10 dB.

Air Absorption

Energy is absorbed continuously as sound propagates in air. This process follows an exponential law, $E(r) = E_0 e^{-2mr}$, where the constant $2m$ in the exponent is called the energy air absorption exponent, expressed in nepers per meter (1 Np = 8.69 dB), with $e = 2.7173$. Air absorption depends on air temperature, relative humidity, ambient pressure, and sound frequency. Table 7.1 presents some typical values calculated using reference 2.

Surface Absorption and Reflection

The ability of a surface to absorb incident sound energy is characterized by the energy absorption coefficient α. This is the fraction of the energy striking the surface that is not reflected, as discussed in more detail in Chapters 8 and 11. It is usually measured in octave- or third-octave bands. If E_i is the energy incident on a surface with absorption coefficient α, the reflected energy is $E_r = E_i(1 - \alpha)$. Ignoring transmission, acoustical energy not absorbed by a surface that it strikes is reflected. If the surface is flat, hard, and homogeneous, reflection is specular; that is, the angle of reflection is equal to the incident angle. In practice, because of finite absorber size, surface roughness, as well as physical or impedance discontinuities, the energy is scattered or diffused (reflected into a range of angles). Such diffuse reflections usually result in a more rapid decay of sound and lower reverberation times in the room.

TABLE 7.1 Energy Air Absorption Exponents ($2m$ in Terms of 10^{-3} Np/m) Predicted Using Reference 2 Assuming Ambient Pressure of 101.3 kPa

Temperature, °C	Relative Humidity, %	Air Absorption Exponents at Selected Frequencies						
		125 Hz	250 Hz	500 Hz	1000 Hz	2000 Hz	4000 Hz	8000 Hz
10	25	0.1	0.2	0.6	1.7	5.8	18.7	43.0
	50	0.1	0.2	0.5	1.0	2.8	9.8	33.6
	75	0.1	0.2	0.5	0.9	2.0	6.5	23.6
20	25	0.1	0.3	0.6	1.2	3.5	12.4	41.6
	50	0.1	0.3	0.7	1.2	2.3	6.5	22.4
	75	0.1	0.2	0.6	1.3	2.2	5.0	15.8
30	25	0.1	0.4	0.9	1.5	3.0	8.3	28.8
	50	0.1	0.2	0.8	1.7	3.0	5.8	16.4
	75	0.0	0.2	0.6	1.7	3.3	5.8	13.4

Furnishings

Many rooms are not empty, containing furnishings (the various obstacles, such as desks in a classroom or machines in a workshop, also called fittings). Often furnishings occupy the lower region of a room and there are fewer objects at higher elevations. Of course, the density and horizontal distribution of room furnishings may vary considerably. Acoustical energy propagating in furnished regions is scattered as well as partially absorbed, significantly modifying the sound field. In particular, it increases the rate of sound decay, reducing reverberation times. It also causes a redistribution of steady-state sound energy toward sources, due to back scattering, resulting in higher sound pressure levels near sources and a higher rate of decrease of levels with distance from sources.

7.2 DIFFUSE-FIELD THEORY

By far the best known theoretical models for predicting room sound fields are based on diffuse-field theory. Diffuse-field theory is widely applied because of its simplicity. Often forgotten is the fact that it may be of limited applicability because of its restrictive assumption of a diffuse field. The sound field in a room is diffuse if it has the following attributes:

1. At any position in the room, energy is incident from all directions with equal intensities (and random phase, though we are ignoring phase in this chapter).
2. The reverberant sound does not vary with receiver position.

These conditions are approximated only in specially designed acoustical test rooms called reverberation chambers, discussed in Section 7.8.

Average Diffuse-Field Surface Absorption Coefficient

Diffuse-field theory uses the average rate of random-incidence surface sound absorption averaged over all of the room surfaces. Thus, we define the average diffuse-field surface absorption coefficient as

$$\overline{\alpha}_d = \frac{\sum S_i \overline{\alpha}_{di}}{\sum S_i} \qquad (7.1)$$

where S_i and α_d are the surface area and diffuse-field absorption coefficient of the ith surface. If α_d is to be useful, it is necessary that no part of the room be strongly absorbing, since in this case a diffuse sound field cannot exist. Absorbing objects such as seats, tables, and people must be included when calculating α_d, despite the fact that such objects have ill-defined surface areas. In such cases it is common practice to assign an absorption A_i in square meters to each object, where $A_i = \alpha_d S_i$, with S_i the surface area. The absorptions of all objects are summed, and the total absorption is included in the calculation of α_d, with no modification made to the total area. In other words, the total area $\sum S_i$ in Eq. (7.1) is taken to be that of the room boundaries, excluding objects and people. In the case of closely spaced absorbers, such as auditorium seats or suspended ceiling baffles, caution is necessary, since the total absorption may depend on the total area covered and may not be the sum of the absorptions of the individual objects. Also, the absorption coefficients of small patches of absorbing material (e.g., as measured in a reverberation room) are higher than those of a large area of the material (e.g., covering all of a room surface), due to the effects of diffraction at the material's edges. In the former case, the absorption coefficients of highly absorptive materials can exceed 1.0 (see also Section 7.7).

Sabine and Eyring Approaches

Several diffuse-field approaches exist; two are considered here. Both are expressed by the following equations, where Eq. (7.3) is applicable only for a compact source:

$$T_{60} = \frac{55.3V}{cA} \quad \text{(sound decay)} \tag{7.2}$$

$$L_p(r) = L_w + 10 \log_{10}\left(\frac{Q}{4\pi r^2} + \frac{4}{R}\right) \quad \text{(steady-state conditions)} \tag{7.3}$$

where T_{60} = reverberation time, s
L_w = source sound power level, dB re 10^{-12} W
L_p = sound pressure level, dB re 2×10^{-5} N/m²
c = speed of sound in air, ≈344 m/s at standard temperature and pressure (STP), m/s
R = room constant, $= A/(1 - \alpha_d)$, m²
A = total room absorption, $= -S \ln(1 - \alpha_d) + 4mV$ (Eyring)
     $\approx \alpha_d S + 4mV$ (Sabine), m²
r = source/receiver distance, m
Q = directivity factor
$2m$ = energy air absorption coefficient, Np/m
V = room volume, m³
α_d = average diffuse-field surface absorption coefficient
S = total room (surface and barrier) surface area, m²

The Eyring approach is applicable to rooms with arbitrary average surface sound absorption coefficient. It should be used, for example, to determine absorption coefficients of room surfaces from measured reverberation times. If, however, the average surface absorption coefficient is sufficiently low (say, $\alpha_d < 0.30$), it is accurate to use the Sabine approach; that is, Eqs. (7.2) and (7.3) with $-\ln(1 - \alpha_d)$ replaced by α_d. Thus, the Sabine approach can be used when determining the sound absorption coefficients of materials in reverberation rooms. Of course, the approach applied in a particular sound field prediction must be consistent with that used to obtain the sound absorption coefficient data used in the prediction. Refer to reference 1 for further discussion of the relative merits of the Sabine, Eyring, and other diffuse-field theories.

Diffuse-field theory predicts the following characteristics of the sound field:

1. *Sound Decay/Reverberation Time.* After a sound source ceases to radiate, the mean-square pressure at the receiver decays exponentially with time. The corresponding sound pressure level decreases linearly with time. To a first approximation, the reverberation time is directly proportional to V/S and inversely proportional to α_d.
2. *Steady State.* The total field is the sum of direct and reverberant components described, respectively, by the first and second terms in parentheses of Eq. (7.3). The direct field, which dominates near the source, is independent of the room's properties. Its sound pressure level decreases at 6 dB/dd (dd = distance doubling). The reverberant field, which dominates far from the source, does not vary with source/receiver distance. Its sound pressure level is, to a first approximation (when both α_d and m are small), inversely proportional to α_d and S. The sound propagation function is as shown in Fig. 7.1.

Note that diffuse-field theory accounts for only some of the relevant room acoustical parameters and for some of these in an approximate manner. In particular, room geometry and source directivity are modeled only approximately. Neither the distribution of the surface absorption nor the presence of barriers or furnishings is modeled. This and the related fact that the theory is based on restrictive hypotheses seriously limit its applicability. To give one concrete example, the constant

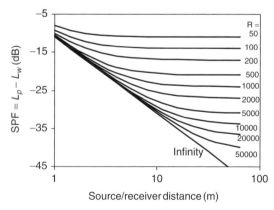

FIGURE 7.1 Sound propagation function curves predicted by diffuse-field theory for various values of the room constant R in m².

sound pressure level at sufficiently large distances from sources that is predicted by diffuse-field theory is only (approximately) found in practice in small rooms with low surface absorption (e.g., reverberation rooms, discussed in Section 7.8). Noise levels due to a single source in most rooms—especially large or absorptive ones—decrease monotonically with distance from the source (see, e.g., Fig. 7.4b below). Reference 3 discusses the applicability of diffuse-field theory more fully.

Diffuse-field theory, when applicable, can be used to estimate the reductions of reverberation time and noise level that occur when absorptive material is added to a room. Let A_b and A_a be the total room absorptions before and after treatment; $A_a = A_b + \Delta A$, where ΔA is the increase in room absorption. If A_b and ΔA are known, values of A_b and A_a can be used directly in Eqs. (7.2) and (7.3). Alternatively, if $T_{60,b}$ before treatment is known, $T_{60,a}$ after treatment can be determined from Eq. (7.2) as follows: since $T_{60,b} = 55.3V/(cA_b)$, then $A_b = 55.3V/(cT_{60,b})$. Now, $T_{60,a} = 55.3V/(cA_a) = 55.3V/[c(A_b + \Delta A)]$, and thus

$$T_{60,a} = \frac{55.3V}{c\left\{[55.3V/(cT_{60,b})] + \Delta A\right\}} \tag{7.4}$$

The steady-state sound pressure level $L_{p,a}$ after treatment can be calculated from that before treatment, $L_{p,b}$, as follows:

$$L_{p,a}(r) = L_{p,b}(r) + 10\ \log_{10}\left[\frac{Q/(4\pi r^2) + 4/R_a}{Q/(4\pi r^2) + 4/R_b}\right] \tag{7.5}$$

At positions near sources, where the direct field dominates, the reduction in sound pressure level is small. At large distances from all sources, where the reverberant field dominates, the reduction approaches that of the reverberant sound pressure level alone. This, to a first approximation (when both α_d and m are small), is given by $10\ \log_{10}(A_a/A_b)$. Doubling the room absorption reduces the sound pressure level of the reverberant field by 3 dB. Similarly, using Eq. (7.2), the reverberation time after treatment is $T_{60,a} = T_{60,b}(A_b/A_a)$. Doubling the sound absorption halves the reverberation time.

As an example, consider a room with dimensions 10 m × 5 m × 2.4 m (volume $V = 120$ m^3, surface area $S = 172$ m^2) and all concrete surfaces with $\alpha_b = 0.05$. Thus, ignoring air absorption and using the Sabine approach, we have $A_b = \alpha_b S = 0.05 \times 172 = 8.6$ m^2 and $T_{60,b} = 55.3V/(cA_b) = 55.3 \times 120/(344 \times 8.6) = 2.2$ s. A suspended acoustical ceiling with $\alpha_c = 0.9$ and surface area of 50 m^2 is installed to improve the acoustical environment. After treatment, $A_a = 8.6 + (0.9 - 0.05) \times 50 = 51.1$ m^2, so $T_{60,a} = 2.2 \times 8.6/51.1 = 0.4$ s.

7.3 OTHER PREDICTION APPROACHES

When diffuse-field theory cannot be applied, alternative approaches can be used to predict the steady-state sound pressure level, sound decay, and reverberation times in a room. These models apply only to compact sources. Extended sources must be approximated by an array of compact sources and the prediction model applied to each source. Two main approaches are available to accomplish this objective: computer algorithms (the method of image sources, ray or beam tracing, radiosity, hybrid models, etc.) and empirical models. The different approaches make assumptions which define and limit their applicabilities and accuracies. Furthermore, they take the various room acoustical parameters into account to a greater or lesser extent. To take advantage of the strengths of the different approaches and avoid their weaknesses, hybrid models combining several approaches into one model have been developed. Empirical prediction models are usually limited in application to specific building types.

Method of Image Sources

The method of image sources is based on the assumption that reflections from surfaces, assumed specular, can be replaced by the direct sound contributions of image sources. Figure 7.2 shows the simple example of a partial two-dimensional array of image rooms and sources. The simplest implementation of the method of image sources applies to empty, rectangular–parallelepiped rooms with no barriers.[4] In this case, the image sources corresponding to the infinite number of reflections are located on a three-dimensional grid. The total mean-square pressure is the sum of the contributions of all of the image sources, allowing for spherical divergence and energy losses due to absorption by the air and at each surface encountered by the ray from the image source. For decaying sound, the sound pressure at time t after the cessation of sound generation is calculated by summing over all image sources located at distances greater than ct from the receiver, where c is the speed of sound. The method of image sources can be extended to empty rooms of arbitrary shape bounded by planes; however, calculation times are greatly increased and become impracticable. The method of image sources can also account for the presence of furnishings, under the assumption that they are isotropically distributed in random fashion throughout

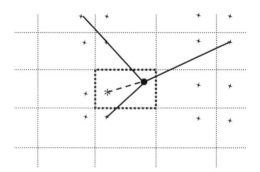

FIGURE 7.2 Image space of an empty two-dimensional room, showing the source (✽) and receiver (●), the image sources (★), the image surface planes (·········), the direct sound (- - - -), and the propagation paths of sound from three of the image sources (———).

the room.[5] In this case, the furnishings, as well as the sources, are mirrored in the room surfaces. The steady-state and decaying sound fields are formed as before. However, the contribution of each image source is modified due to scattering and absorption by the furnishings; absorption at surfaces is also increased due to scattering.

Ray and Beam Tracing

Ray-tracing techniques[6] can be used to predict sound fields in rooms of arbitrary shape with arbitrary surface absorption distributions, different surface reflection properties, and variable furnishing density. Source directivity can also be modeled. A computer program simulates the emission of a large number of rays (or beams) from each source in either random or deterministic fashion. Each ray is followed as it propagates in the room, being reflected and scattered by surfaces, barriers, and furnishings until it reaches a receiver position. The energy of a ray is attenuated according to spherical divergence as well as surface, furnishing, and air absorption. In principle, any surface reflection law (specular, diffuse, etc.) can be modeled. Ondet and Barbry[7] proposed an algorithm for accounting for quasi-arbitrary furnishing distributions in a ray-tracing model. Figure 7.3, taken from reference 8, illustrates the potential of ray-tracing techniques. It shows the contour map of the predicted reductions of total A-weighted noise levels over the floor of a furnished workshop containing eight machinery sources, which result from the introduction of an acoustical barrier around a noise-sensitive assembly bench and a partial ceiling absorption treatment suspended over the region of the sources. Of course, the program must be run for each frequency band (i.e.,

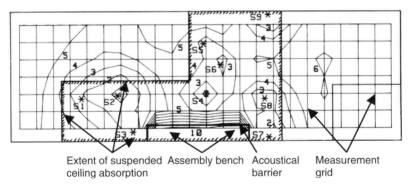

FIGURE 7.3 Floor layout of a furnished workshop (dimensions $46.0 \times 15.0 \times 7.2$ m) showing a contour map of the predicted reductions of total A-weighted noise levels which result from the introduction of an acoustical barrier around a noise-sensitive assembly bench and of an absorptive treatment suspended below the ceiling over the region of the eight machinery sources (the shaded, L-shaped, enclosed area). The map was produced from noise levels predicted before and after treatment at positions on an imaginary 2×2-m grid, 1.5 m above the floor, as shown. Also shown are the noise sources (e.g., S4*) and the contour lines and their values (e.g., 5).

octave- or third-octave bands) to account for the frequency-dependent nature of the sound absorption coefficient of the room surfaces and that of the barrier insertion loss. This kind of detailed analysis is not possible using diffuse-field theory. It would only be possible to estimate noise reductions from the increase in absorption. These reductions could then be corrected as a function of the proximity of the receiver to the acoustical treatment and of the estimated insertion loss of the barrier.

Empirical Models

Empirical models are developed from experimental data. They usually focus on a particular category of room—models for offices, classrooms, and industrial workshops are discussed below. Measurements are made, in a number of rooms of some category, of the acoustical quantities to be predicted. Empirical algorithms or equations which predict the experimental results from appropriate predictor parameters describing the rooms are developed. This can, for example, involve multivariable regression modeling, whereby statistical modeling software is used to find the minimum set of mutually statistically independent predictors, each of which is highly correlated with the data to be predicted, that most accurately predicts the measured data. For further information on multivariable regression modeling, consult reference 9.

Auralization

Auralization is the process of rendering audible sounds "played" in a virtual room for which the acoustical response has been predicted. It allows subjective evaluation of the room. If done accurately, auralization allows full three-dimensional sound perception, with source externalization (i.e., the sound is perceived to be outside the head), spatial localization, and spatiousness. Auralization involves predicting the combined response of a virtual room and a virtual listener in it. The listener response is quantified by head-related transfer functions (HRTFs) quantifying how the acoustical response of the human external auditory system varies with frequency and angle. The room response is predicted using models, such as ray tracing, which predict the individual sound path contributions arriving at the receiver; since impulse responses involving phase effects are required, the room's phase response must also be considered. The result is the binaural impulse response (comprising the impulse responses at the left and right eardrums) of the room–listener combination. This is convolved with a sound signal for replay to a real listener. Readers interested in further details of this technique should consult reference 10.

7.4 DOMESTIC ROOMS AND CLOSED OFFICES

Schultz[11] investigated the prediction of sound pressure levels in domestic rooms and closed offices. He measured the variation of sound pressure level with distance from a source in a variety of small rooms, observing that levels never become constant at larger distances as predicted by diffuse-field theory. In fact, he found that the curves always had a slope of about −3 dB/dd, though the absolute levels of the curves varied considerably. He further found that existing models developed for predicting sound propagation in, for example, workshops and corridors could not predict his experimental results. Schultz therefore proposed the following empirical formula:

$$L_p(r) = L_w - 10 \log_{10} r - 5 \log_{10} V - 3 \log_{10} f + 12 \text{ dB} \quad (7.6)$$

where r is the source–receiver distance in meters, V is the room volume in cubic meters, and f is the frequency in hertz. Note that this formula does not explicitly contain a room absorption term. Such behavior contrasts markedly with that predicted by diffuse-field theories and is an excellent example of their limitations.

7.5 CLASSROOMS

Relevance and Characteristics

Classrooms are acoustically critical spaces in which verbal communication is crucial for teaching and learning. Acoustically, conference rooms, gymnasiums, and other rooms for speech have much in common with classrooms. Many classrooms are built with little or no attention paid to the acoustical design. Studies have shown that nonoptimal acoustical conditions in classrooms result in impaired verbal communication between teachers and students, impaired student language development and learning, and teacher voice and other problems.[12,13] The problems are particularly acute for "acoustically challenged" listeners, including young and hearing-impaired people and those using a second language.

Classrooms vary from small seminar rooms for a few occupants to school classrooms for several tens of children to larger university lecture rooms and auditoria accommodating hundreds of listeners. Smaller classrooms are usually of rectangular shape. Larger lecture rooms can have, for example, fan plan shape, inclined seating, and nonflat ceiling profiles. In smaller classrooms, talkers and listeners can be anywhere in the classroom, and source–receiver distances can vary from less than a meter to several meters. In lecture rooms, the talker is usually at one end of the room, with the listeners spread out in front; source–receiver distances can vary from several meters to several tens of meters. For hygiene reasons, classrooms may have hard, nonabsorptive surfaces, though carpets and wall and ceiling absorption are not uncommon. Lecture and conference rooms can have nonabsorptive or padded, sound-absorptive seating. Of course, the occupants themselves contribute significant absorption to the classroom. This and the fact that classroom occupancy can vary considerably must be considered in the acoustical design. A detailed discussion of the physical and acoustical characteristics of (university) classrooms is found in reference 14. Of particular interest is the rate at which speech and noise levels decrease with distance from their sources, which is small in small, nonabsorptive classrooms. It increases with classroom size and absorption and can be large in large, absorptive (e.g., occupied) classrooms. One consequence of this is that speech levels in large classrooms may decrease by 10 or more decibels from the front to the back of the seating area.

Speech Intelligibility

In classrooms, as in other rooms for speech, quality and ease of verbal communication are prime concerns (see also Section 7.7). The quality can be quantified in human terms by speech intelligibility, the percentage of words correctly identified by the listener. Here, in general, we consider the effect of the room on the accurate transmission of speech signals from a talker to a listener and not, for example, the effects of individual talker and listener characteristics. Verbal communication is considered to be affected by two main factors—the classroom reverberation and the speech-signal-to-background-noise-level difference (often called the signal-to-noise ratio). The effect of reverberation can be quantified by the early-to-late energy ratio C_{50} (which is highly correlated with early decay time and reverberation time in many cases). This frequency-dependent ratio is usually measured in octave bands. The signal-to-noise level difference depends on the levels of speech and the levels of noise at the listener position. Total A-weighted or octave- or third-octave-band equivalent continuous levels are often considered. If the voice levels remain constant, the speech intelligibility decreases with increased reverberation and increases with increased signal-to-noise level difference. A number of physical metrics exist which combine these two quantities into a single speech intelligibility measure; these include the useful-to-detrimental energy ratio U_{50}, articulation index AI (discussed in Section 7.7), speech transmission index STI, and speech intelligibility index SII.

Numerous studies have found that many classrooms have nonoptimal acoustical conditions, with excessive noise levels and reverberation. A review of published data and of the many complex issues associated with speech intelligibility in classrooms is contained in reference 15. The classroom speech sources are the teachers' and the students' voices. Classroom noise sources include mechanical services (e.g., ventilation outlets), classroom equipment (projectors, computers), and the teachers' or students' voices when another person is generating the signal to be heard. Noise breaking into the classroom from outside can be significant when the classroom is located near transportation (such as highways and airports) or in cases when children are active in nearby corridors or play areas. Finally, classroom activity itself generates significant noise, including impact noise from furniture, toys, and so on. Data on the typical sound pressure levels generated by human talkers and classroom noise sources are available.[16] It is likely that a number of complex factors, including room acoustics, affect what these levels are at listener positions in a given classroom at a given time.

It is generally considered that, for excellent speech intelligibility, background noise levels should not exceed about 40 dBA for acoustically unchallenged listeners and 30 dBA for acoustically challenged listeners. The question of what is the optimal reverberation for speech intelligibility is a complex one. One body of opinion considers that reverberation should always be minimized. However, this is based on the results of speech intelligibility tests that were done under conditions that did not account for the acoustical behavior of classrooms. Investigations involving the useful-to-detrimental energy ratio speech intelligibility metric and realistic room acoustical modeling show that some reverberation can be beneficial (since it increases reverberant speech levels). The results indicate that the reverberation should be optimized as a function of the classroom volume and background noise. Optimal reverberation times vary from low values in small, quiet classrooms to over 1 s in large, noisy ones. For a full discussion of this issue, refer to reference 17. In any case, reverberation should be minimized in classrooms with high signal-to-noise level differences (e.g., resulting from voice amplification). Moreover, it is likely that less reverberation can be tolerated by acoustically challenged listeners. It has been shown that speech intelligibility is not very sensitive to variations in reverberation, and it has been suggested that a classroom reverberation time of about 0.5 s is usually appropriate.[18]

Standards and guidelines exist to help design professionals achieve high acoustical quality in classrooms and other educational spaces. They specify the acoustical conditions to be achieved in new and renovated rooms. They discuss reverberation times and background noise levels to be achieved in the unoccupied room. They also provide advice on how to achieve the design targets. Two recent examples of such standards are ANSI S12.60—2002 in the United States[19] and Building Bulletin 93 in the United Kingdom.[20]

Predicting Classroom Acoustics

Predicting acoustical quality in a classroom involves predicting room reverberation and steady-state sound pressure levels generated by the speech and noise sources, including student activity. Prediction should take into account the absorption contributed by the room occupants and furnishings (e.g., absorptive seating). This involves an appropriate prediction model and accurate input data. In the case of small, untreated classrooms, it is likely that diffuse-field theory is reasonably accurate. For treated and larger classrooms, comprehensive techniques such as ray tracing can be used. Empirical models are also available and inherently incorporate realistic data. Based on measurements of reverberation times and steady-state levels generated by a speech source in a wide variety of typical university classrooms and on published information, empirical models for predicting octave-band reverberation times and total A-weighted speech levels have been developed.[21] Approximate values of $T_{60,u}$ in the unoccupied classroom can be determined using Eq. (7.2), with the average surface absorption coefficient calculated as the sum of contributions (presented in Table 7.2) to the room average value, which are associated with the sound-absorbing features present in the classroom. An alternative empirical formula for predicting the 1-kHz octave-band $T_{60,u}$ is

$$T_{60,u} = 0.874 + 0.0021(LW) + 0.303\text{refl} + 0.412\text{basic}$$
$$- 0.384\text{absorb} - 0.804\text{upseat} \quad (7.7)$$

TABLE 7.2 Typical Octave-Band Contributions of Classroom Sound-Absorptive Features to Room-Average Surface Absorption Coefficient

Surface Feature	125 Hz	250 Hz	500 Hz	1000 Hz	2000 Hz	4000 Hz	8000 Hz
Basic	0.12	0.10	0.10	0.09	0.10	0.09	0.10
Carpeted floor	0.00	0.00	0.02	0.04	0.07	0.09	0.17
Wall/ceiling absorption	0.01	0.04	0.08	0.10	0.10	0.10	0.10
Upholstered seats	0.20	0.20	0.16	0.12	0.09	0.07	0.05

TABLE 7.3 Octave-Band Values of Average Absorption per Person

Band, Hz	125	250	500	1000	2000	4000	8000
A_p, m²	0.25	0.45	0.67	0.81	0.82	0.83	1.14

where the parameters are defined following Eq. (7.10). The absorption of the classroom occupants can be included using the data in Table 7.3, and the number of occupants can be used to estimate the associated increase in classroom absorption. For an average talker speaking between a normal and a raised voice (total A-weighted sound power level ≈74 dBA), total A-weighted speech levels in the unoccupied classroom can be predicted from

$$\text{SLA}_u(r) = I_u + s_u \frac{\log_{10}(r)}{\log_{10}(2)} \quad (7.8)$$

with

$$I_u = 65.8 - 0.0105(LW) + 1.52\text{fwdist} - 1.41\text{absorb} - 4.32\text{upseat} \quad (7.9)$$

and

$$s_u = -1.21 - 0.088\ L + 1.14\text{basic} \quad (7.10)$$

where $T_{60,u}$ = 1-kHz octave-band reverberation time, s
SLA_u = total A-weighted speech level, dBA
I_u = intercept of total A-weighted sound propagation curve, dBA
s_u = slope of total A-weighted sound propagation curve, dBA/dd
r = source–receiver distance, m
L = average classroom length, m
W = average classroom width, m
fwdist = distance from source to nearest wall, m
refl = 1 if classroom contains beneficial reflectors, = 0 if not

absorb = factor quantifying extent of sound-absorptive wall or ceiling treatment

absorb = 1 corresponds to full-coverage wall or ceiling treatment; in other cases, values should be scaled proportionately

upseat = 0 for nonabsorptive seating; = 1 for padded, sound-absorptive seating

basic = 1 if all classroom surfaces and seats are not sound absorptive; = 0 otherwise

As an example, consider a large classroom with average dimensions of $L = 20$ m long, $W = 10$ m wide, and 6 m high (volume $V = 1200$ m^3, surface area $S = 760$ m^2). The distance from the typical lecturing position to the nearest (front) wall is fwdist = 2.5 m. The classroom has a profiled ceiling directing sound to the back (i.e., beneficial reflectors, refl = 1). Consider the cases where N_p is 50 or 200 occupants. Before treatment, the classroom has no sound-absorptive surfaces (absorb = 0, basic = 1) and nonabsorptive seating (upseat = 0). Thus, the average surface absorption coefficient is that corresponding to the "basic" configuration in Table 7.2 ($\alpha = 0.09$). Equation (7.2) gives the corresponding $T_{60,u}$'s. Assuming air absorption exponents for a temperature of 20°C and a relative humidity of 50% ($m = 0.0012$ Np/m; see Table 7.1) and using the Eyring approach, the total absorption in the unoccupied room at 1 kHz is $A_u = -S \ln(1 - \alpha) + 4mV = -760 \ln(1 - 0.09) + 4 \times 0.0012 \times 1200 = 77.4$ m^2. From Eq. (7.2) the 1-kHz $T_{60,u} = 55.3V/(cA_u) = 55.3 \times 1200/(344 \times 77.4) = 2.49$ s. With 50 occupants, the total room absorption $A_o = A_u + N_p A_p = 77.4 + 50 \times 0.81 = 117.9$ m^2, and $T_{60,o} = 55.3V/(cA_o) = 55.3 \times 1200/(344 \times 117.9) = 1.64$ s. Similarly, with 200 occupants $T_{60,o} = 0.81$ s. Here, T_{60} is excessive with low occupancy and decreases sharply with the number of occupants. To control the reverberation, the classroom is treated with full-coverage wall absorption (absorb = 1) and padded, sound-absorptive seating (upseat = 1); thus, basic = 0. The average surface absorption coefficients are now the sum of the values for basic, wall/ceiling absorption, and upholstered seats given in Table 7.2 (e.g., $0.09 + 0.10 + 0.12 = 0.31$ at 1 kHz). The 1-kHz T_{60}'s in the classroom when unoccupied and with 50 or 200 occupants are 0.67, 0.63, and 0.52 s. The treatment significantly reduces T_{60}—in the unoccupied classroom by as much as 200 occupants do—and decreases the variation of T_{60} with the number of occupants.

As for speech levels, consider listeners seated at the front and back of the classroom at distances of 3 and 12 m from the talker. Before treatment, from Eq. (7.9), $I_{u,b} = 65.8 - 0.0105LW + 1.52\text{fwdist} - 1.41\text{absorb} - 4.32\text{upseat} = 65.8 - 0.0105 \times 20 \times 10 + 1.52 \times 2.5 - 1.41 \times 0 - 4.32 \times 0 = 67.5$ dBA and, from Eq. (7.10), $s_{u,b} = -1.21 - 0.088L + 1.14\text{basic} = -1.21 - 0.088 \times 20 + 1.14 \times 1 = -1.83$ dBA/dd. Using Eq. (7.8), the speech level at any position is $\text{SLA}_{u,b}(r) = 67.5 - 1.83 \log_{10}(r)/\log_{10}(2)$; thus, at 3 and 12 m speech levels are 64.6 and 61.0 dBA. Levels in the occupied classroom can be calculated using Eq. (7.5). Assuming $Q = 2$ to the front of a talker, at the front and back of the classroom speech levels are 63.1 and 59.2 dB with 50 occupants and 60.9 and 56.2 dBA with 200 occupants. After treatment, $I_{u,a} = 65.8 - 0.0105 \times 20 \times 10 + 1.52 \times 2.5 - 1.41 \times 0 - 4.32 \times 1 = 63.2$ dBA and $s_{u,a} = -1.21 - 0.088 \times 20 + 1.14 \times 0 = -3.0$ dBA/dd, giving speech levels of 57.1 and 51.1 dBA unoccupied at the front and back seats; occupants decrease levels by no more than 1 dB. Acoustical treatment causes speech levels at the front and back of the unoccupied classroom to decrease by 7.5 and 9.9 dBA for 50 occupants and with 200 occupants by 4.5 and 6.1 dB. These results must be interpreted with care, as they do not account for certain relevant nonacoustical factors. For example, they assume that the talker's vocal output is invariant, as would be the case for a loudspeaker. In fact, research suggests that a talker's vocal output varies with the prevailing classroom acoustical conditions; a preliminary empirical model for predicting this phenomenon is presented in reference 16.

Noise levels generated by classroom noise sources (e.g., projectors or ventilation outlets) can be estimated from their output power levels and receiver distances using the above empirical model after adjusting for the difference between the actual and the assumed output power levels. Student activity noise should also be considered; a preliminary empirical model for predicting this is presented in reference 16. Then signal-to-noise level differences can be determined. Classroom early-decay times and signal-to-noise level differences can then be used to determine values of speech intelligibility metrics.

Controlling Classroom Sound

Controlling and optimizing the acoustical conditions in a classroom or other rooms for speech involve three fundamental considerations:

1. *Promoting High Speech Levels.* Avoid excessive classroom cubic volume due, for example, to high and vaulted ceilings. Use room geometries which direct sound to the back of the room. In large lecture rooms, this can include angled reflectors around teaching areas and profiled ceilings. Keep at least the central part of the ceiling sound reflective to promote the reflection of speech sounds to the back of the classroom. Use approximately square floor plans to avoid long and wide rooms. Amplification by a speech reinforcement system or a sound field enhancement system may be an option; their design is beyond the scope of this chapter; see reference 22. One important issue to consider at the classroom design stage is that the optimal acoustical conditions for unaided speech may not be the same as with a speech reinforcement system.

2. *Controlling Background Noise.* Avoid open-plan design. Control the noise and vibration of mechanical services (see Chapter 16). Locating ventilation outlets toward the front of lecture rooms where speech levels are highest helps optimize signal-to-noise level differences throughout the room. Avoid high terminal velocities of supply air terminal devices and place air-volume control devices at distances of 0.5 m or more upstream to minimize noise generated by turbulent flow. Choose quiet equipment for use in the classroom. Impact noise due to student activity can be reduced by the use of carpets and cushioning materials

(split tennis balls on desk, table, and chair legs are widely used). The partitions bounding the classroom must provide adequate sound isolation (see Chapter 10); in critical cases, this might require the use of nonopenable windows.

3. *Optimizing Reverberation.* Apply appropriate sound-absorptive materials to the room surfaces. Avoid applying sound absorption to the central part of the ceiling, which provides useful reflections between talkers and listeners. Using sound-absorptive seating allows the ceiling to be left reflective and reduces the sensitivity of the classroom's acoustical conditions to the number of occupants.

7.6 INDUSTRIAL WORKSHOPS

Relevance and Characteristics

Industrial workshops often have serious acoustical problems due to excessive noise and reverberation. Most jurisdictions have regulations aimed at limiting the risk of hearing damage by limiting the noise exposure of industrial workers. Excessive reverberation can also lead to poor verbal communication and a reduced ability to identify warning signals, and thus danger, as well as to stress and fatigue.

Industrial buildings come in every shape and size. However, many are rectangular in floorplan (with widely varying dimension ratios and floorplan sizes), with flat or nonflat (e.g., pitched or sawtooth) roofs. The floors of most workshops are made of concrete. The walls are often of brick or blockwork, sometimes of metal cladding. Workshop roofs are usually of suspended-panel construction, consisting of metal or other panels supported by metal trusswork or portal frames. A common modern construction is the steel deck, consisting of, for example, profiled metal as the internal surface, a vapor barrier, several centimeters of thermal insulation, tar paper, and gravel ballast as the external surface, again supported by metal trusswork. Acoustical steel decks exist; the inner metal layer is perforated, and its profiles are filled with sound-absorptive material, providing high sound absorption over a broad range of frequencies. Acoustical decks can support loads almost as great as normal decks, cost only about 10% more, and should always be considered at the design stage. The average surface absorption coefficients of untreated industrial buildings are 0.08–0.16 in the 125- and 250-Hz octave bands, varying with construction, and 0.06–0.08 in the 500–4000-Hz octave bands.

Figures 7.4a and b show the reverberation times and 1-kHz sound propagation function curves measured in an empty, untreated workshop with average dimensions of 45 m × 42.5 m and height of 4 m and a double-panel roof.[23] It is often the case that the sound propagation function—$L_p(r) - L_w$ as defined in Eq. (7.3)—at most source–receiver distances and the reverberation time are found to be highest at midfrequencies. The values of both quantities decrease at low and high frequencies due to increased panel and air absorption.

The effects of workshop furnishings are also illustrated in Figs. 7.4a and b, showing the reverberation times and sound propagation function curves after first 25 and then an additional 25 printing machines were introduced. These

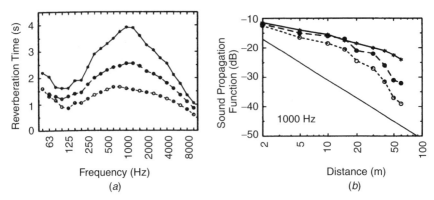

FIGURE 7.4 Measured (a) third-octave-band reverberation times and (b) 1-kHz sound propagation curves in an industrial workshop when empty (×———×) and containing 25 (●- - -●) and 50 (○·····○) metal machines; (———) free-field sound.

metal machines had average dimensions of 3 m × 3 m and height of 2 m. Introduction of the furnishings significantly decreased reverberation times and sound propagation function values. According to diffuse-field theory, the decreases of reverberation time for this particular workshop correspond to the introduction of as much as 900 m² of acoustical absorption.[23] In fact, they are related to increased diffusion, not increased absorption. The percentage changes of reverberation time and the magnitudes of the changes of the sound propagation function with increasing furnishing density vary little with frequency. These results are supported by similar measurements in other workshops. They again illustrate the limitations of diffuse-field theory, which cannot accurately predict the effects of complex room geometry or furnishings.

Workshop Noise Prediction

Models developed for predicting noise levels and reverberation times in industrial workshops fall into two categories—*comprehensive* models, based on approaches such as ray tracing and the method of image sources, and *simplified*, based on simple theoretical or empirical approaches. Comprehensive models have been evaluated experimentally, concluding that the ray-tracing model of Ondet and Barbry[7] was most inherently accurate.[24] Ray tracing can account for complex room shape, barriers, and nonisotropic absorption and furnishing distributions. Simplified models have also been evaluated experimentally, concluding that a model developed by Kuttruff was inherently accurate in furnished workshops.[26]

Kuttruff[26] developed a simplified model for long and wide furnished workshops based on radiosity theory; the numerous furnishings on the floor and the ceiling were modeled as diffusely reflecting surfaces. According to this model, in a workshop of height h and with average absorption coefficient α, the sound pressure level $L_p(r)$ generated by an omnidirectional compact source of sound power level L_w at a distance r is given by

$$L_p(r) = L_w + 10 \log_{10}\left\{\frac{1}{4\pi r^2} + \frac{1-\alpha}{\pi h^2}\left[\left(1+\frac{r^2}{h^2}\right)^{-1.5}\right.\right.$$
$$\left.\left.+\frac{\beta(1-\alpha)}{\alpha}\left(\beta^2+\frac{r^2}{h^2}\right)^{-1.5}\right]\right\} \quad (7.11)$$

where $\beta \approx 1.5\alpha^{-0.306}$. Despite the assumption of infinite length and width, the model performed well in comparison with measurements made in workshops of a wide range of lengths and widths.[25] One disadvantage of the model is that it does not include a parameter that can be varied to account for variations in the number of furnishings. However, as can be seen from Fig. 7.4b, the variations are small at the smaller source–receiver distances that dominate noise levels in many practical cases. An alternative is empirical models for predicting workshop noise levels and reverberation times.[27]

When dealing with multiple sources in workshops, the effect of noise control measures may be determined to a first approximation by considering what happens at a distance corresponding to the average source–receiver distance. Changes in total noise levels due to noise control measures are approximated by changes in the sound propagation function at this distance. For machine operator positions, the average source–receiver distance is small—typically 1–2 m. For a room with a more or less square floorplan and with sound sources uniformly distributed over the floor, the average source–receiver distance is about one-half the average horizontal dimension.

As an example, let us use the Kuttruff model to calculate the 1-kHz noise level in the workshop illustrated in Fig. 7.3 (but without the acoustical screen and partial suspended ceiling shown) before and after introduction of full-coverage suspended ceiling absorption. Consider a position midway along the assembly bench and 1.5 m above the floor. The workshop dimensions are length 48.0 m, width 16.0 m, and height 7.2 m. Assume that the average midfrequency absorption coefficients of the workshop surfaces are 0.07 (untreated) and 0.4 (treated). The workshop contains nine noise sources located as shown in Fig. 7.3, with 1-kHz sound power levels and source–receiver distances as shown in Table 7.4. Table 7.4 shows the individual noise-level contributions of the sources before and after treatment. Total levels are determined by energy-based addition of the individual source contributions. The result is 82.4 dB before treatment and 78.7 dB after treatment. Introduction of the ceiling treatment is predicted to reduce the 1-kHz noise level at the assembly bench by 3.7 dB (confirming that the treatment illustrated in Fig. 7.3 is highly cost-effective).

Accurate prediction relies on an inherently accurate model and accurate input data. Workshop surface absorption coefficients were discussed above. Many workshop prediction models also include a parameter describing the workshop furnishing density. It is not known how to determine this quantity accurately. In theory, it can be estimated as the total surface area of the furnishings (or, more practically, of imaginary boxes that would fit around the individual objects)

TABLE 7.4 Details of Calculation of 1-kHz Noise Levels in Workshop before and after Acoustical Treatment Using Kuttruff Simplified Prediction Model[a]

Source Number	L_w, dB	Distance, r (m)	Before $L_{p,a}(r)$, dB	After $L_{p,b}(r)$, dB
1	94	13.5	70.7	65.3
2	101	9.8	79.1	74.8
3	92	7.6	71.1	67.5
4	94	3.9	75.6	73.5
5	86	9.5	64.2	60.0
6	90	7.6	69.1	65.5
7	91	8.7	69.6	65.6
8	89	8.5	67.6	63.8
9	93	15.5	69.0	63.2
Total noise level			82.4	78.7

[a] Reference 26.

divided by 4 times the volume of the furnished region.[24] However, there is some evidence that this procedure underestimates the correct value.[8] Research has shown that average workshop furnishing densities may attain 0.2 m^{-1} or higher, those of a workshop's furnished regions 0.5 m^{-1} or higher.[28] In any case, no simplified model will ever predict noise levels in workshops of complex shape or furnishing distributions or that contain, for example, barriers with high accuracy. In such cases, ray tracing may be the only viable option. Finally, a problem shared by all prediction models is the accurate estimation of the sound power levels of the noise sources. This problem is discussed in Chapter 4.

Workshop Noise Control

To achieve the reduction of workshop noise levels (as required by modern occupational noise regulations) in a cost-effective manner, the application of noise control principles should be incorporated into new designs and renovation in the following order of priority:

1. *Control at Source.* Reduce the sound power outputs of the equipment by design or by retrofit acoustical treatment.

2. *Control of Direct Field.* Isolate receiver positions from noisy sources by increasing the distance between them, by the use of source enclosures (see Chapter 12), and by surrounding the receiver by a cabin or screen. Barriers and screens are often not practical or cost-effective alternatives in workshops. Often separation can be achieved by appropriate planning of the workshop layout, taking full advantage of the building geometry, natural barriers such as stockpiles, and furnishings at the design stage.

3. *Control of Reverberant Field.* Apply sound-absorptive materials to the room surfaces. Such treatments should be accorded lower priority since they tend to be

expensive and not very effective near noise sources, such as at operator positions. In practice, reductions of 0–6 dBA are possible. The best surfaces to treat are those closest to noise sources and/or receiver positions. In low-height industrial workshops, this usually implies the ceiling. Consider an acoustical steel deck at the design stage or acoustical treatments consisting of sound-absorptive materials applied to or suspended from the ceiling—for example, acoustical baffles (rectangular pieces of absorptive material) hung in appropriate patterns at appropriate densities. A particularly cost-effective treatment consists of suspending sound absorbers directly above noise sources. Difficulties may arise due to interference with overhead cranes and lighting and sprinkler systems and with respect to fire regulations and hygiene requirements. The treatment of walls may be warranted in more regularly shaped enclosures and where noise sources are located close to walls when it is important to absorb the strong wall reflections. Note that in workshops of any shape surface absorption, even when it has little effect on the total noise levels, may significantly reduce reverberation, reducing the perceived "noisiness" of the work environment and improving verbal communication.

Readers interested in the low-noise design of industrial workshops should be aware of relevant ISO standards available to help in the task. ISO 11688[29,30] provides advice on the design of low-noise machinery and equipment. Part 1 discusses planning, and Part 2 discusses the principles of low-noise design. ISO 14257[31] defines criteria for evaluating the acoustical quality of workshops with respect to noise control and describes how to perform the measurements to evaluate an existing workshop.

7.7 OPEN-PLAN OFFICES

Open-plan offices are one of the most common modern work environments. In these large spaces, many seated workers are separated by low barriers that provide partial visual and acoustical separation between workstations. The barriers may be free-standing but today are usually in the form of integrated furniture or cubicles. The ceiling and the carpeted floor form two large, extended sound-absorbing planes; their horizontal dimensions are much greater than the height of the ceiling. The space between these surfaces is filled with office furniture and barriers that are usually sound absorbing. Diffuse-field theory certainly does not apply in such a space; the sound pressure level decreases continuously with distance from sources (an example is shown in reference 3). The principal problem in an open-plan office is not propagation to large distances but the provision of privacy between neighboring workstations. The important sound paths are, therefore, the short-range ones. Sound from one work position reflects from extended surfaces (ceiling, walls, and windows) and diffracts over or around the edges of barriers, as illustrated in Fig. 7.5. These sound paths must be controlled to provide acoustical privacy for workers.

Open-plan offices can provide a reasonable degree of acoustical privacy if they are carefully designed as a complete system and if adjacent work functions are

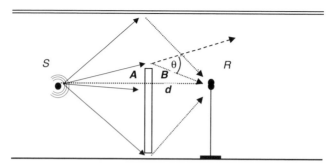

FIGURE 7.5 Possible sound paths between workstations in an open-plan office cross section. Reflection from the ceiling is reduced by using highly absorptive ceiling materials, transmission through the barrier by selecting a sufficiently heavy panel with adequate sound transmission loss, and diffraction over the barrier by increasing the barrier height.

compatible and not too close together. If any aspect of the system is neglected, then adequate acoustical privacy will not be achieved. Obtaining adequate acoustical privacy is first a question of attenuating unwanted speech sounds from adjacent work spaces, because speech sounds are much more distracting than most other types of noises. It is usually also necessary to include masking sound to render intruding speech sounds less intelligible while not being a source of annoyance. Special protection should be provided against noisy devices such as printers and copiers—for example, by locating them in a shielded area. Conference rooms with full-height partitions, providing good sound insulation, should be available for activities which require low background noise levels and particularly high privacy.

Open-Plan Office Barriers

Barriers (also called screens, partial-height partitions, workstation panels, or office dividers) provide sound attenuation and visual privacy between workstations for seated persons. They are the basic component of systems furniture (usually forming cubicles) that combines the functions of a barrier and supporting amenities such as storage compartments, lighting, power, communications, and work surfaces into a single unit. A barrier should attenuate sound that passes through it so that the transmitted sound is negligible.

Sound diffracts over the top of the barrier to reach the next workstation. For an infinitely wide barrier between a source and a receiver position, the insertion loss IL relative to the level at the receiver in the absence of the barrier can be approximated by[32]

$$\text{IL} = 13.9 + 7 \log_{10} N + 1.4 (\log_{10} N)^2 \quad \text{dB} \qquad \text{for } N \geq 0.001 \qquad (7.12)$$

where the Fresnel number $N = 2f(A + B - d)/c$, d is the straight-line distance from the source to the receiver in meters, A is the distance from the source to the top of barrier in meters, B is the distance from the top of the barrier to the receiver in meters, and c is the sound speed (Fig. 7.5 shows A, B, and d). The greater the angle that sound has to bend to reach the receiver position on the other side of the barrier (θ in Fig. 7.5), the greater is the insertion loss of the barrier. Thus, higher barriers are more effective than lower ones, and barriers placed close to the talker or listener are more effective than those equidistant from both.

The effects of diffraction around and transmission through a barrier can be combined to determine the total insertion loss of a barrier. Sound transmitted through the screen should be negligible relative to the sound diffracted around it, especially at those frequencies important for speech intelligibility. Specifying that, at 1000 Hz, the normal-incidence sound transmission loss should be 6 dB greater than the theoretical insertion loss due to diffraction is a satisfactory criterion. This criterion leads to the requirement that the minimum mass per unit area of the barrier, ρ_s in kilograms per square meter, should be $\rho_s \geq 2.7(A + B - d)$. For an isolated screen, the total effect of the barrier can be calculated by applying Eq. (7.12) to each edge in turn and then summing the acoustical energies.

Maximum barrier dimensions are usually limited by physical convenience, possible interference with airflow, and preservation of the open look. Recommended minimum dimensions are height 1.7 m and width 1.8 m. Of course, a complete cubicle is equivalent to a long barrier, for which there is negligible propagation around the ends. If the attenuation of a high barrier is desired but visual openness must be maintained, a plate of glass or transparent plastic panel can be fitted to the top of a low barrier to increase its "acoustical height." The gap between the bottom edge of the barrier and the floor should be small; otherwise sound can reflect under the barrier to the opposite side. This is not critical, because sound following this path tends to be diffused or absorbed by furniture and carpets; a gap of up to 100 mm can be left when the floor is carpeted.

Measures for Rating Acoustical Privacy

Acoustical privacy is usually referred to as speech privacy because it is speech sounds that are usually the most disturbing (see also Section 7.5). Speech privacy is essentially the opposite of speech intelligibility. That is, the lower the intelligibility of the speech, the greater the speech privacy. Speech privacy (and speech intelligibility) is related to the level of the intruding speech sounds relative to the less meaningful ambient noise. Hence, speech privacy is related to measures involving the signal-to-noise level difference, where speech is the signal and the general ambient noise is the noise component.

Articulation index (AI) is a frequency-weighted signal-to-noise level difference measure which indicates the expected speech intelligibility in particular conditions. The signal-to-noise level differences in each frequency band are weighted according to their relative importance to the intelligibility of speech. These weighted signal-to-noise level differences are summed to obtain the AI value, between 0 and 1. An AI of 1 is intended to indicate conditions in which near-perfect intelligibility is expected. An AI close to 0 is expected to indicate conditions of near-perfect speech privacy. The AI has been widely used as a measure of speech privacy; AI ≤ 0.15 is considered to provide *normal* or *acceptable* speech privacy in open-plan office situations. The AI has now been replaced by the SII. It is similar to the AI but has slightly larger values, so the criterion for acceptable speech privacy becomes SII ≤ 0.20. Both have been described in ANSI standards.[33,34]

Figure 7.6 illustrates the suitability of this criterion. In this figure, median speech intelligibility scores are plotted versus SII for situations simulating conditions in open-plan offices. It is seen that SII = 0.2 represents the point below which privacy increases rapidly with decreasing SII. One can think of it as the point at which improvements to the acoustical design start to improve speech privacy. In practice, it is also a practically achievable goal if all aspects of the acoustical design are carefully considered.

Acoustical Design of Complete Workstations

Sound propagation between workstations in an open-plan office with modular workstations (cubicles) involves many different sound paths, not just propagation over a simple screen as described above. There are reflections from the ceiling and the other panels of the workstation as well as from nearby walls. These consist of not only simple first-order reflections but also paths with multiple reflections, such as those involving the floor and the ceiling or the vertical surfaces of the workstation. The problem of sound propagation between workstations has been addressed using the method of image sources (see Section 7.3) and implemented as a computer algorithm. Diffraction over the screen was modeled using Maekawa's results described in Eq. (7.12). The method of image sources assumes that all reflections are specular, so the angle of incidence equals the angle of reflection. However, the model includes an empirical correction to the

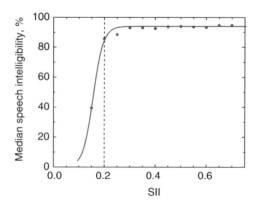

FIGURE 7.6 Variation of median speech intelligibility score with SII for situations simulating a range of conditions in open-plan offices.[35]

absorption coefficients of ceiling materials (obtained from standard reverberation room tests) to account for the limited range of angles of incidence for sound propagation between workstations, compared to that in a diffuse reverberant test room. The model has been validated with respect to a wide range of measurement conditions and is used here to illustrate the effects of various design parameters. See references 36–39 for details of the model.

Speech and Noise Levels

Two factors that have very large effects on speech privacy are the speech levels of the talkers in the open offices and the general ambient noise levels. These cannot be ignored when striving for acceptable speech privacy. Fortunately, people do not usually talk with *normal* vocal effort in open-plan offices. Measurements of talkers in open-plan offices have found mean voice levels close to those described as *casual* vocal effort. Using normal voice levels to calculate expected speech privacy greatly exaggerates the lack of speech privacy. The intermediate office speech level (IOSL) shown in Fig. 7.7 is recommended as the speech source level for calculating speech privacy in open-plan offices. It was obtained by averaging the mean speech levels measured in open-plan offices and the casual voice data and then adding 3 dB, corresponding to one standard deviation above the average talker level. The IOSL spectrum is therefore representative of louder talkers in open-office situations and corresponds to a speech source level at a distance of 1 m in a free field of 53.1 dBA. References 40–42 provide more details of speech levels.

One important aspect of achieving speech privacy is to encourage the use of lower voice levels as a form of open-plan office etiquette. Extended discussions should not be held in open-office areas but should be moved to closed meeting rooms.

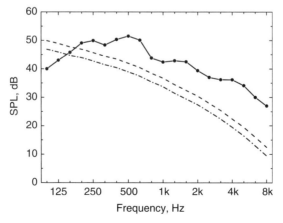

FIGURE 7.7 The IOSL spectrum and noise-masking spectra for open-office design calculations: (——•——) IOSL; (-·····-··) optimum masking; (- - - - -) maximum masking.

It is difficult to obtain acceptable speech privacy if the general ambient noise level is low. On the other hand, if ambient noise levels are high, they can become annoying and cause people to talk louder. There is a narrow range of ambient noise levels that are expected to be acceptable and that can also partially mask intruding speech sounds from adjacent workstations. (The masking of speech by noise is a well-researched field, and ANSI S3.5[34] includes many references on this topic.) For this reason, electronic sound-masking systems are often included in successful open-office designs. They can be designed to provide nearly ideal noise levels to mask speech sounds and enhance privacy without being unduly disturbing. The masking noise should be adjusted to sound like natural ventilation system noise and to be evenly distributed throughout the office. Figure 7.7 includes an example of a noise-masking spectrum that was judged to be at an optimal level, corresponding to 45 dBA. It has been found that the maximum acceptable level for masking sound is approximately 48 dBA. Hence, the maximum masking spectrum in Fig. 7.7 represents an example of such a maximum noise-masking spectrum. These two noise spectra are indicative of the narrow range of masking noises that provides *acceptable* speech privacy. Spatial variations of masking noise should be less than 3 dBA.

Conventional sound-masking systems have included those with centrally located electronics and others with distributed units. Distributed systems, with many small independent units, have the advantage of avoiding correlated sources, which can lead to annoying spatial variations in the masking sound. Manufacturers of both types claim various practical advantages. Although propagation into the ceiling void may aid the homogeneity of the masking sound in the office below, the masking sound will be modified by propagation through the ceiling tiles, and transmission through lighting fixtures can lead to localized areas of higher sound levels below. More recently, sound-masking systems with loudspeakers mounted in the ceiling tiles and on the panels of workstations have been introduced to provide better control over the resulting masking sound. The installation of noise-masking systems is best left to experienced professionals.

Important Design Parameters

In the design of an open-plan office, it is first assumed that the transmission loss of the workstation panels is sufficient to attenuate, to an insignificant level, the sound propagating directly through the panels. The requirement for surface density given by Eq. (7.12) usually leads to a minimum value corresponding to STC $\geq$ 20 (STC is the sound transmission class, obtained from a standard sound transmission loss test[43]; see also Chapter 10). After this, the two most important parameters are the sound absorption of the ceiling and the height of the separating barriers or panels. If these are not adequate, it will not be possible to achieve acceptable speech privacy.

The shaded area of Fig. 7.8 indicates combinations of ceiling absorption and panel height that can provide SII $\leq$ 0.20 if the other details considered below are also acceptable. Ceiling absorption is described in terms of the sound absorption

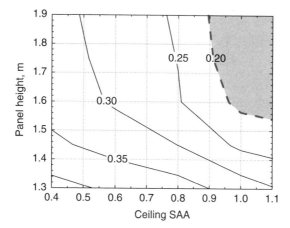

FIGURE 7.8 Shaded area shows combinations of workstation panel height and ceiling absorption that lead to *acceptable* speech privacy corresponding to SII ≤ 0.20.[44] Contours in unshaded part indicate degrees of unacceptable speech privacy. SAA is the average of the absorption coefficients in the third-octave bands from 250 Hz to 2.5 Hz.

average (SAA), which is the average of the absorption coefficients in the third-octave frequency bands from 250 Hz to 2.5 kHz. It replaces the older noise reduction coefficient (NRC) and has very similar values. Values of SAA greater than 1.0 are included in Fig. 7.8 because such values can result from the standard test procedure, due to the effects of edge diffraction (as mentioned in Section 7.2). It is seen that the sound absorption of the ceiling should correspond to SAA $\geq$ 0.90 and the separating panels should be ≥ 1.7 m high. Since the results in Fig. 7.8 were calculated with other details set to nearly ideal values, combinations of ceiling absorption and screen height outside of the shaded area make it impossible to achieve acceptable speech privacy.

A third key design parameter is the size of the workstation plan. The results in Fig. 7.8 are for a workstation with dimensions of 3 m × 3 m. If the workstation length and width each decrease by 1 m, then the resulting SII increases by 0.05. That is, for a 2 m × 2 m workstation plan, meeting the SII ≤ 0.20 criterion would require very high screens and the most absorptive ceiling materials.

Other Design Parameters

The sound-absorptive properties of the workstation panels and the floor have smaller effects on speech privacy. The height of the ceiling and the presence of lighting fixtures in the ceiling can also influence speech privacy.

The results shown in Fig. 7.8 were calculated for workstation panels with SAA = 0.9. If this were reduced to 0.75, the corresponding SII values would decrease by 0.01. If the panel absorption were decreased to 0.6, SII values would decrease by 0.02. These seem to be small degradations in the overall performance, but one must always remember that most designs will at best be on the edge of just providing acceptable speech privacy; even these small improvements may help to meet the SII ≤ 0.20 criterion. Of course, if the panels were not highly absorptive (i.e., SAA < 0.60), there would be a more significant increase in the resulting SII values. For example, if the workstation panels were non-absorptive in an otherwise ideal workstation design, SII would increase from 0.2 to 0.3.

Varying floor absorption and varying ceiling height generally have very small effects on the resulting SII values. In most cases, varying these parameters changes SII by no more than about 0.01. On the other hand, lighting fixtures in the ceiling can degrade the speech privacy of the open-office design. The effect of a lighting fixture depends on the type of lighting unit and its location. Lights with a flat plastic or glass surface produce the strongest unwanted reflections and are most troublesome when located over the separating panel between two workstations. They will change SII values most when installed in a highly absorptive ceiling. Evaluation of the effects of several types and locations of lighting fixtures has shown that, for a ceiling with SAA = 0.90, SII values could increase by up to 0.08. Open-grill lighting fixtures have smaller effects but still reduce the effectiveness of a highly absorptive ceiling.

Practical Issues and Other Problems

Sound propagating in the horizontal plane can bypass barriers by reflection from vertical surfaces and hence reduce the sound attenuation between work positions. To minimize these effects, surfaces such as walls, office barriers, square columns, backs of cabinets, and systems furniture, as well as bookcases, should be covered with sound-absorptive material having SAA values of 0.7 or higher. A thickness of 2.5 cm or more of glass fiber with a porous fabric cover will satisfy this requirement. The application of carpets directly to hard surfaces is not an effective solution, since typical carpets have low sound absorption coefficients. Round columns with diameters of 0.5 m or less may be left uncovered. A simple way to avoid wall reflections is to avoid gaps between the wall and barriers. To prevent reflections from their surfaces, office barriers should be covered with sound-absorptive material on both faces.

Where workstations are adjacent to windows, there are often large gaps between the workstation panels and the window. To provide adequate acoustical privacy between these workstations, this gap must be filled with additions to the workstation panels. This problem cannot easily be solved using drapes, which would need to be heavy and closed; furthermore, most slatted blinds do not reduce reflections. Using the area next to the windows as a corridor is another solution to this problem.

Standard procedures exist for the evaluation of the degree of speech privacy in an existing office, in a mock-up of a proposed office,[45] or for the evaluation of open-plan noise-masking systems.[46] It is very important to evaluate the speech and noise in the office to completely determine the existing degree of speech privacy.

7.8 REVERBERATION ROOMS

Reverberation rooms are designed and equipped to give a close approximation to a diffuse sound field. Measurements in these rooms are used to characterize the sound-absorptive properties of materials, the sound power of sources, and the sound transmission through building elements, among other things, under standard diffuse-field conditions according to various standards.[43,47–52] A typical reverberation room has a volume of about 200 m^3; some are constructed with nonparallel walls. The walls and surfaces of the room are made highly reflective so that reverberation times are high and the region dominated by the direct field of a source is as small as possible.

At low frequencies, the frequency response to wide-band noise shows peaks corresponding to individual room modes. As frequency increases, the spacing between modes becomes less, the modes begin to overlap, and the individual modes are less obvious. At some transition frequency the room response for bands of noise becomes approximately constant, the properties of the sound field become more uniform, and the room response may be described in statistical terms. This transition point is usually defined by the Schroeder frequency[53] $f_S = 2000(T_{60}/V)^{1/2}$ hertz. For a $V = 250$ m^3 room with $T_{60} = 5$ s, $f_S = 282$ Hz.

To make the response of the room more uniform at low frequencies, it is usually advisable to add low-frequency sound-absorbing elements. Even in rooms of approximately 200 m^3 volume, correctly chosen dimensions, and the recommended amount of sound absorption, the spatial variations of pressure and the sound decay rate are often too large to satisfy the precision requirements of standards. It is therefore common to add fixed panels suspended at random positions and orientations throughout the room to perturb the room modes and to create more diffuse conditions. In many cases rotating diffusers are also used for this purpose.

Fixed diffusers are most important in rectangularly shaped rooms used for sound absorption measurements. They may not be necessary in rooms that are sufficiently nonrectangular in shape. They help to create a more diffuse sound field during the measurement of sound decays by ensuring that, throughout the sound decay, a portion of the decaying sound energy is redirected toward the sound-absorbing sample. When commissioning a new reverberation room for sound absorption measurements, it is necessary to systematically increase the number of diffusers until measured absorption coefficients just reach maximum values due to the increasingly diffuse conditions in the chamber. Of course, adding too many diffusers can limit the number of valid positions for microphones, which must always be located more than a half wavelength (at the lowest frequency of interest) from reflecting surfaces such as walls and diffusing panels.

Rotating diffusers are particularly useful in reverberation rooms used for measuring the sound power levels of devices with strong tonal components. These types of sound sources lead to large spatial variations in sound levels that can be effectively reduced using a rotating diffuser. By continuously changing the geometry of the room, they shift the modal patterns and hence average out some of the spatial variations of the sound levels. Making the rotating diffuser in the form of panels of revolution can reduce the required drive power and aerodynamic noise.

In addition to such measures, it is still necessary to sample the room volume in order to measure mean-squared sound pressures and decay rates accurately. To provide statistically independent samples of the sound field, microphone positions must be more than one-half wavelength apart (at the lowest frequency of interest) as well as be located more than one-half wavelength away from reflecting objects to avoid nonrepresentative measures of the room average sound level.

It has been shown that close to one, two, or three infinite reflecting planes the sound pressure and energy increase. The mean-squared pressure p^2 can be expressed as

$$p^2 = 1 + \sum_{n=1}^{N_{im}} \frac{\sin(kr_n)}{kr_n} \quad (N/m^2)^2 \qquad (7.13)$$

where $k = 2\pi/\lambda (\lambda = \text{wavelength})$
p^2 = mean-square pressure normalized to unity in absence of images or far from reflecting surfaces
r_n = distances from images of measurement point to measurement point
N_{im} = number of images: 1 for measurement near a plane surface, 3 near an edge, and 7 near a corner

Closer than $\lambda/2$ to highly reflecting surfaces, sound pressure increases significantly because of these positive interference effects. To account for the increase in sound energy close to surfaces, an adjustment term $(1 + S\lambda/8V)$ is included in the calculation of the sound power W from the space-averaged mean-square sound pressure. Thus, the relationship used to determine the sound power W of a source from the space-averaged mean-square sound pressure p^2 that it creates in the room is[54]

$$W = \frac{55.3 p^2 V}{4\rho c^2 T_{60}} (1 + S\lambda/8V) \quad N \cdot m/s \qquad (7.14)$$

where ρ = density of air, kg/m^2
c = speed of sound in air, m/s
V = room volume, m^3
T_{60} = reverberation time, s

It is assumed that sampling of the sound field in the room is confined to the central regions away from room surfaces. The Sabine formula, Eq. (7.2), relating reverberation time to room absorption is assumed to hold and is the basis for determinations of sound absorption in reverberation rooms.[47,48]

The planning and qualification of reverberation rooms are complex and best done by experienced professional acousticians.

7.9 ANECHOIC AND HEMI-ANECHOIC CHAMBERS

An anechoic chamber is a room with all of its interior surfaces highly absorptive such that a source in the room radiates in essentially free-field conditions. The resulting sound field has only a direct component. Hemi-anechoic chambers have a hard floor with all other interior surfaces highly sound absorptive. They are used for measuring the sound power level and radiation pattern of equipment (such as road vehicles, appliances, etc.) that are operated over a hard surface. The acoustical performance of an anechoic or hemi-anechoic room can be evaluated by determining how closely conditions in the room approximate those in a free field. This is usually done by measuring the variation of sound levels with distance from an approximately omnidirectional source; it should ideally be −6 dB per doubling of distance; see Eq. (7.3) and Fig. 7.1.

Tests are made in an anechoic room when it is necessary to accurately measure the unperturbed sound radiated by a source—for example, when measuring its radiation pattern (directivity) or sound power. The surfaces of anechoic chambers are made highly absorptive by lining them with deep, sound-absorptive materials. The lining typically consists of wedges of mineral wool or glass fiber.[55] All anechoic chambers are more anechoic at high than at low frequencies. The lowest frequency at which an anechoic chamber can be used depends primarily on the chamber volume and the depth of the wedges. This cutoff frequency, at which the wedges absorb 99% of the incident sound energy, is usually achievable with a wedge depth (including any air space between the base of the wedge and the hard wall) of approximately one-quarter wavelength. A large chamber with 1-m-deep wedges may be effective down to 80–100 Hz. To provide a walking surface to facilitate setting up experiments in an anechoic chamber, an open-metal-floor grid that is removed after the setup or a permanent wire-mesh floor can be provided. The perimeter frame of such a wire-mesh floor and to a lesser extent the wire mesh itself may degrade the high-frequency performance of the chamber.

REFERENCES

1. L. Cremer, H. A. Müller, and T. J. Schultz, *Principles and Applications of Room Acoustics*, Applied Science, New York, 1982; H. Kuttruff, *Room Acoustics*, 4th ed., Spon, London, 2000.
2. ANSI S1.26, "Method for the Calculation of the Absorption of Sound by the Atmosphere," Acoustical Society of America, Melville, New York.
3. M. R. Hodgson, "When Is Diffuse-Field Theory Applicable," *Appl. Acoust*, **49**(3), 197–207 (1996).
4. J. B. Allen and D. A. Berkeley, "Image Method for Efficiently Simulating Small-Room Acoustics," *J. Acoust. Soc. Am.*, **65**(4), 943–950 (1979).
5. E. A. Lindqvist, "Noise Attenuation in Factories," *Appl. Acoust.*, **16**, 183–214 (1983).
6. A. Krokstad, S. Strom, and S. Sorsdal, "Calculating the Acoustical Room Response by the Use of a Ray-Tracing Technique," *J. Sound Vib.*, **8**(1), 118–125 (1968).

7. A. M. Ondet and J. L. Barbry, "Modeling of Sound Propagation in Fitted Workshops Using Ray Tracing," *J. Acoust. Soc. Am.*, **85**(2), 787–796 (1989).
8. M. R. Hodgson, "Case History: Factory Noise Prediction Using Ray Tracing—Experimental Validation and the Effectiveness of Noise Control Measures," *Noise Control Eng. J.*, **33**(3), 97–104 (1989).
9. J. Neter, W. Wasserman, and G. A. Whitmore, *Applied Statistics*, 4th ed., Allyn and Bacon, Boston, 1993.
10. M. Kleiner, B.-I. Dalenbäck, and P. Svensson, "Auralization—An Overview," *J. Audio Eng. Soc.*, **41**(11), 861–875 (1993).
11. T. J. Schultz, "Improved Relationship between Sound-Power Level and Sound-Pressure Level in Domestic and Office Spaces," Report No. 5290, Bolt Beranek and Newman, Cambridge, MA, 1983.
12. B. M. Shield and J. E. Dockrell, "The Effects of Noise on Children at School: A Review," *J. Building Acoust.*, **10**(2), 97–116 (2003).
13. P. B. Nelson, S. D. Soli, and A. Seltz, *Acoustical Barriers to Learning*, Acoustical Society of America, Melville, NY, 2002.
14. M. R. Hodgson, "Experimental Investigation of the Acoustical Characteristics of University Classrooms," *J. Acoust. Soc. Am.*, **106**(4), 1810–1819 (1999).
15. M. Picard and J. S. Bradley, "Revisiting Speech Interference in Classrooms," *Audiology*, **40**(5), 221–244 (2001).
16. M. R. Hodgson, R. Rempel, and S. Kennedy, "Measurement and Prediction of Typical Speech and Background-Noise Levels in University Classrooms during Lectures," *J. Acoust. Soc. Am.*, **105**(1), 226–233 (1999).
17. M. R. Hodgson and E.-M. Nosal, "Effect of Noise and Occupancy on Optimum Reverberation Times for Speech Intelligibility in Classrooms," *J. Acoust. Soc. Am.*, **111**(2), 931–939 (2002).
18. S. R. Bistafa and J. S. Bradley, "Reverberation Time and Maximum Background-Noise Level for Classrooms from a Comparative Study of Speech Intelligibility Metrics," *J. Acoust. Soc. Am.*, **107**(2), 861–875 (2000).
19. ANSI S12.60, "Acoustical Performance Criteria, Design Requirements and Guidelines for Schools," Acoustical Society of America, Melville, New York.
20. Building Bulletin 93, *Acoustic Design of Schools*, Department of Education and Skills, UK, 2004.
21. M. R. Hodgson, "Empirical Prediction of Speech Levels and Reverberation in Classrooms," *J. Building Acoust.*, **8**(1), 1–14 (2001).
22. D. Davis and C. Davis, *Sound System Engineering*, 2nd ed., Focal Press, Burlington, MA, 1997.
23. M. R. Hodgson, "Measurement of the Influence of Fittings and Roof Pitch on the Sound Field in Panel-Roof Factories," *Appl. Acoust.*, **16**, 369–391 (1983).
24. M. R. Hodgson, "On the Accuracy of Models for Predicting Sound Propagation in Fitted Rooms," *J. Acoust. Soc. Am.*, **88**(2), 871–878 (1989).
25. M. R. Hodgson, "Experimental Evaluation of Simplified Methods for Predicting Sound Propagation in Industrial Workrooms," *J. Acoust. Soc. Am.*, **103**(4), 1933–1939 (1998).
26. H. Kuttruff, "Sound Propagation in Working Environments," *Proc. 5th FASE Symp.*, 17–32 (1985).

27. N. Heerema and M. R. Hodgson, "Empirical Models for Predicting Noise Levels, Reverberation Times and Fitting Densities in Industrial Workrooms," *Appl. Acoust.*, **57**(1), 51–60 (1999).
28. M. R. Hodgson, "Effective Fitting Densities and Absorption Coefficients of Industrial Workrooms," *Acustica*, **85**, 108–112 (1999).
29. ISO/TR 11688-1, "Recommended Practice for the Design of Low-Noise Machinery and Equipment—Part 1: Planning," International Organization for Standardization, Geneva, Switzerland, 1995.
30. ISO/TR 11688-2, "Recommended Practice for the Design of Low-Noise Machinery and Equipment—Part 2: Introduction to the Physics of Low-Noise Design," International Organization for Standardization, Geneva, Switzerland, 1998.
31. ISO 14257, "Measurement and Parametric Description of Spatial Sound Distribution Curves in Workrooms for Evaluation of Their Acoustical Performance," International Organization for Standardization, Geneva, Switzerland, 2001.
32. Z. Maekawa, "Noise Reduction by Screens," *Appl. Acoust.*, **1**(3), 157–173 (1968).
33. ANSI S3.5, "American National Standard Methods for the Calculation of the Articulation Index," Acoustical Society of America, Melville, New York, 1969.
34. ANSI S3.5, "Methods for Calculation of the Speech Intelligibility Index," Acoustical Society of America, Melville, New York, 1997.
35. J. S. Bradley and B. N Gover, "Describing Levels of Speech Privacy in Open-Plan Offices," Report IRC-RR-138, Institute for Research in Construction, National Research Council, Ottawa, 2003.
36. C. Wang and J. S. Bradley, "Sound Propagation between Two Adjacent Rectangular Workstations in an Open-Plan Office, I: Mathematical Modeling," *Appl. Acoust.*, **63**(12), 1335–1352 (2002).
37. C. Wang and J. S. Bradley, "Sound Propagation between Two Adjacent Rectangular Workstations in an Open-Plan Office, II: Effects of Office Variables," *Appl. Acoust.*, **63**(12), 1353–1374 (2002).
38. C. Wang and J. S. Bradley, "Prediction of the Speech Intelligibility Index behind a Single Screen in an Open-Plan Office," *Appl. Acoust.*, **63**(8), 867–883 (2002).
39. J. S. Bradley and C. Wang, "Measurements of Sound Propagation between Mock-Up Workstations," Report IRC-RR 145, Institute for Research in Construction, National Research Council, Ottawa, 2001.
40. A. C. C. Warnock and W. Chu, "Voice and Background Noise Levels Measured in Open Offices," Report IR-837, Institute for Research in Construction, National Research Council, Ottawa, 2002.
41. W. O. Olsen, "Average Speech Levels and Spectra in Various Speaking/Listening Conditions: A Summary of the Pearson, Bennett, and Fidell (1977) Report," *J. Audiol.*, **7**, 1–5 (October 1998).
42. J. S. Bradley, "The Acoustical Design of Conventional Open-Plan Offices," *Can. Acoust.*, **27**(3), 23–30 (2003).
43. ASTM E90, "Standard Test Method for Laboratory Measurement of Airborne Sound Transmission Loss of Building Partitions," American Society for Testing and Materials, Conshohocken, PA, 2004.
44. J. S. Bradley, "A Renewed Look at Open Office Acoustical Design," Paper N1034, Proceedings Inter Noise 2003, Seogwipo, Korea, August 25–28, 2003.

45. ASTM E1130, "Standard Test Method for Objective Measurement of Speech Privacy in Open Offices Using the Articulation Index," American Society for Testing and Materials, Conshohocken, PA, 2002.
46. ASTM E1041, "Standard Guide for Measurement of Masking Sound in Open Offices," American Society for Testing and Materials, Conshohocken, PA, 1985.
47. ASTM C423, "Standard Test Method for Sound Absorption and Sound Absorption Coefficients by the Reverberation Room Method," American Society for Testing and Materials, Conshohocken, PA, 2002.
48. ISO 354, "Measurement of Sound Absorption in a Reverberation Room," International Organization for Standardization, Geneva, Switzerland, 2003.
49. ANSI S1.31, "Precision Methods for the Determination of Sound-Power Levels of Broad-Band Noise Sources in Reverberation Rooms," and ANSI S1.32, "Precision Methods for the Determination of Sound-Power Levels of Discrete-Frequency and Narrow-Band Noise Sources in Reverberation Rooms," Acoustical Society of America, Melville, New York.
50. ISO 3740, 3741, 3742, "Determination of Sound-Power Levels of Noise Sources," International Organization for Standardization, Geneva, Switzerland, 2000.
51. ISO 140/III, "Laboratory Measurements of Airborne Sound Insulation of Building Elements," International Organization for Standardization, Geneva, Switzerland, 1995.
52. ASTM E492, "Standard Method of Laboratory Measurement of Impact Sound Transmission Through Floor-Ceiling Assemblies Using the Tapping Machine," American Society for Testing and Materials, Conshohocken, PA, 1992.
53. M. R. Schroeder, "Frequency-Correlation Functions of Frequency Responses in Rooms," *J. Acoust. Soc. Am.*, **34**(12), 1819–1823 (1962).
54. R. V. Waterhouse, "Output of a Sound Source in a Reverberation Chamber and in Other Reflecting Environments," *J. Acoust. Soc. Am.*, **30**(1), 4–13 (1958).
55. L. L. Beranek and H. P. Sleeper, Jr., "Design and Construction of Anechoic Sound Chambers," *J. Acoust. Soc. Am.*, **18**, 140–150 (1947).

CHAPTER 8

Sound-Absorbing Materials and Sound Absorbers*

KEITH ATTENBOROUGH

University of Hull
United Kingdom

ISTVÁN L. VÉR

Consultant in Acoustics, Noise, and Vibration Control
Stow, Massachusetts

8.1 INTRODUCTION

One of the most frequent problems faced by noise control engineers is how to design sound absorbers that provide the desirable sound absorption coefficient as a function of frequency in a manner that minimizes the size and cost, does not introduce any environmental hazards, and stands up to hostile environments such as high temperatures, high-speed turbulent flow, or contamination. The designer of sound absorbers must know how to choose the proper sound-absorbing material, the geometry of the absorber, and the protective facing. The theory of sound-absorbing materials and sound absorbers has progressed considerably during the last 10 years. Much of this progress is documented in K. U. Ingard's *Notes on Sound Absorption Technology*[1] and in F. P. Mechel's *Schall Absorber* (*Sound Absorbers*)[2] and the separately sold computer program on CD-ROM.[3] Ingard's book is a paperback edition of modest length and price and comes with a CD-ROM that allows the reader to make easy numerical predictions from almost all of the equations used in the book. The underlying physical processes are explained with great clarity and the mathematical treatment is kept simple. Derivations and difficult explanation are presented in appendices. The books cover all aspects of sound absorption in great detail. To derive full benefit, it helps if the reader has a reasonably strong mathematical background. However, even those with no such background will find the figures, which give the

*The authors acknowledge the contributions of Dr. Fridolin P. Mechel to the several parts of this chapter retained from the 1992 edition.

difficult mathematical results in graphical form, useful and practical. Mechel's book consists of three hard-covered volumes totaling 2866 pages. The associated computer program is available in MAC and Window versions but is expensive. The classical book of H. Kutruff, *Room Acoustics*,[4] deals predominantly with the room acoustics aspects of sound absorption and should be studied by acousticians designing buildings for the performing arts. Beranek's recently updated classic book *Acoustic Measurements*[5] describes the methods and experimental hardware for measuring all acoustical descriptors of sound-absorbing materials and sound absorbers. The reader with serious interest in sound absorption technology is advised to study all of these texts.

How Sound Is Absorbed

Sound is the organized superposition of particle motion on the random thermal motion of the molecules. The speed of the organized particle motion in air is typically six orders of magnitude smaller than that of the thermal motion. All sound absorbers facilitate the conversion of the energy carried by the organized particle motion into random motion. All forces, other than those that compress and accelerate the fluid, caused by the oscillatory particle flow in the presence of solid material result in loss of acoustical energy. The most important contribution to the conversion is associated with the drag forces caused by friction between the interface of a rigid or flexible wall or the skeleton of the porous or fibrous sound absorber material and the fluid in the thin acoustical boundary layer. For porous and fibrous sound-absorbing materials the acoustical flow speed is low and the flow remains laminar, and the drag force is proportional to the acoustical particle velocity. At high velocities, which occur at the mouth of resonators, the flow separates, turbulence is created, and the friction force becomes proportional to the square of the velocity. Other loss mechanisms, discussed in more detail elsewhere,[1,2] include the isothermal compression of air at low frequencies, direct conversion of acoustical energy into heat owing to a time lag between compression, and heat flow that takes place in closed cell foams. In the case of plate and foil (very thin plate) absorbers the acoustical energy is converted into heat in the vibrating flexible plate and radiated as sound from the rear of the plate or is transmitted in the form of vibration energy into connected structures.

A sound absorber can absorb only that part of the incident sound energy not reflected at its surface. Consequently, it is important to keep the reflection at the surface as low as possible. Essentially, the part of the incident acoustical energy that enters the absorber should be dissipated before it returns to the surface after traversing the absorber and reflecting from a rigid backing. Otherwise, the absorber gives back acoustical energy to the fluid on the receiver side that is in addition to the initial reflection. This requires a sufficient thickness. The challenge in sound absorber design is to keep the absorber thickness to a minimum.

Sound Absorption Coefficients

The acoustical performance of flat sound absorbers is characterized by their sound absorption coefficient α, defined as the ratio of the sound power, W_{nr}, that is not reflected (i.e., dissipated in the absorber, transmitted through the absorber into a room to its rear, or conducted, in the form of vibration energy, to a connected structure) and the sound power incident on the face of the absorber, W_{inc}:

$$\alpha \equiv \frac{W_{\text{nr}}}{W_{\text{inc}}} \quad (8.1)$$

For convenience in analyses, the absorption coefficient is defined in terms of sound pressure reflection factor R of the absorber interface, namely

$$\alpha = 1 - |R|^2 \quad (8.2)$$

The vertical lines bracketing R indicate the absolute value. The reflection factor R is usually a function of the angle of sound incidence, the frequency, the material, and the geometry of the absorber. The absorber is characterized by its wall impedance (otherwise known as surface impedance) Z_w, defined as

$$Z_w = \frac{p}{v_n} \quad \text{N} \cdot \text{s/m}^3 \quad (8.3)$$

where p is the sound pressure and v_n is the normal component of the particle velocity, both evaluated at the interface. A substantial part of this chapter concerns the prediction of the wall impedance offered by a large variety of sound absorbers. The sound absorption coefficient as defined in Eqs. (8.1) and (8.2) is further differentiated according to the angular composition of the incident sound field, such as *normal incidence, oblique incidence, and random incidence*, and whether the absorber is *locally reacting* (sound cannot propagate in it parallel to the interface) or *non–locally reacting* (where sound can propagate in it parallel to the interface). At normal incidence there is no difference between locally and non–locally reacting materials.

Measurement of Normal-Incidence Sound Absorption Coefficient α_0**.** The normal incidence sound absorption coefficient α_0 can be measured in an impedance tube according to the American Society for Testing and Materials (ASTM) standard C384-98 as $\alpha_0 = 4\zeta/(\zeta + 1)^2$, where $\zeta = p_{\max}/p_{\min}$ is the ratio of the maximum and minimum standing-wave sound pressure pattern in the tube upstream of the sample. Normal-incidence sound absorption coefficients, measured in an impedance tube, never exceed unity.

Measurement of Random-Incidence Sound Absorption Coefficient α_R**(rev) in Reverberation Room.** The random-incidence sound absorption coefficient $\alpha_R(\text{rev})$, which is also referred to as the *Sabine absorption coefficient*, can be measured directly in a reverberation room according to ASTM C423-02. It is defined as $\alpha_R(\text{rev}) \equiv (55.3V/S)[(1/T_S) - (1/T_0)]$, where V is the volume of the reverberation chamber in m^3, $S = 6.7$ m^2 is the standardized surface area of the test sample, and T_S and T_0 are the reverberation times in seconds (see

Chapter 7) measured with and without the sample, respectively. It is important to distinguish between the two random-incidence sound absorption coefficients $\alpha_R(\text{rev})$ and α_R. The coefficient $\alpha_R(\text{rev})$, measured in the reverberation room, can yield values which exceed unity (measured values up to 1.2, as can be seen in Fig. 8.10). Such values obviously violate the theoretical definition of α given in Eq. (8.36) implying that the panel absorbs more energy than is incident on it. This fact that measured values of $\alpha_R(\text{rev}) > 1$ results from the finite size of the test sample, which means that there is diffraction at the edges of the sample. In design calculations, it is customary to replace all values of $\alpha_R(\text{rev}) > 1$ with unity. The random-incidence sound absorption coefficient α_R is computed from the wall impedance and never yields absorption coefficients that exceed unity. Based on the analyses of $\alpha_R(\text{rev})$ measured according ASTM C423-02 for a large variety of thicknesses and flow resistivity of Owens Corning series 700 fiberglass boards, Godfrey[6] has proposed an empirical prediction scheme to relate $\alpha_R(\text{rev})$ of a standard size (6.7 m^2) sample to α_R predicted theoretically from the computed wall impedance of the samples as

$$\alpha'_R(\text{rev}) = \left(\frac{21.3}{f^{0.5}} + 0.73\right)\alpha_R$$

where $\alpha'_R(\text{rev})$ is the empirically predicted Sabine absorption coefficient of a standard size sample and f is the frequency in hertz. This formula provides reasonably accurate predictions for low flow resistivities (in the range of 9000 N · m/s^4 = 0.56 $\rho_0 c_0$/in. to 17,000 N · m/s^4 = 1.1 $\rho_0 c_0$/in.) and small layer thicknesses from 1 to 3 in. (2.5–7.5 cm). For flow resistivities greater than 32,000 N · m/s^4 = 2 $\rho_0 c_0$/in. and layer thicknesses ≥ 76 mm = 3 in., the analytical method[7] yields the best prediction.

8.2 SOUND ABSORPTION BY NON–SOUND ABSORBERS

Although inefficiently, all rigid and flexible structures absorb sound. If efficient sound absorbers are present in the room, the contribution of these marginal sound absorbers may be neglected. However, if they represent the only sound-absorbing mechanism (such as is in the case of reverberation rooms without low-frequency absorbers), they, and at high frequencies the air absorption, account for the total sound absorption and set an upper limit for the maximum achievable reverberation time and for the buildup of reverberant sound pressure in these special rooms.

Sound Absorption of Rigid Nonporous Wall

The general assumption that a rigid, nonporous wall, however massive, gives rise to total reflection of the incident sound is not valid. In the immediate vicinity of the interface there are two phenomena that cause small but finite dissipation of sound energy. The first is that the wall-parallel component of the particle velocity of the incident sound results in shear forces in the acoustical boundary layer. The acoustical boundary layer is identical to that which would develop at the interface between a stationary fluid and a plane, rigid wall oscillating with the frequency and velocity amplitude of the wall-parallel component of the sound-induced particle motion. The second component of the unavoidable dissipation results from the large thermal capacity of the wall, which makes it impossible to fully recover the heat energy in the rarefaction phase that was built up in the compression phase. According to reference 4, for randomly incident sound, the combined effect of these two dissipation processes results in a lower limit of the sound absorption coefficient of

$$\alpha_{\min} = 1.8 \times 10^{-4} (f)^{1/2} \tag{8.4}$$

where f is the frequency in hertz. For frequencies of 1 and 10 kHz, Eq. (8.4) yields values of α_{mim} of 0.006 and 0.018, respectively.

Sound Absorption of Flexible Nonporous Wall

Building partitions such as walls, windows, and doors absorb low-frequency sound. The sound power lost, W^{Loss}, when the incident sound interacts with a nonporous, flexible, homogeneous, and isotropic single partition that has another room or the outdoors on the receiver side is made up of three components:

$$W^{\text{Loss}} = W^{\text{ForcedTrans}} + W^{\text{ResTrans}} + W^{\text{ResDiss}} \quad \text{N} \cdot \text{m/s} \tag{8.5}$$

where $W^{\text{ForcedTrans}}$ is the sound power transmitted by the forced bending waves (i.e., mass law portion radiated from the receiver side by the forced bending waves), W^{ResTrans} is the sound power transmitted by the free resonant bending waves, and W^{ResDiss} is the sound power dissipated in the partition by the free resonant bending waves. The first two terms in Eq. (8.5) can be expressed in the form of sound transmission loss, TL, of the partition and the third term as a function of the space–time mean-square value of the sound-induced resonant vibration velocity $\langle v^2 \rangle$ of the plate, yielding

$$W^{\text{Loss}} = W^{\text{inc}} 10^{-\text{TL}/10} + \langle v^2 \rangle \rho_S \omega \eta S \quad \text{N} \cdot \text{m/s} \tag{8.6}$$

where η is the composite loss factor of the plate accounting for energy loss through dissipation in the plate as well as for that lost through structural coupling to neighboring structures at the plate boundaries and S is the surface area of the plate in square meters. The sound absorption coefficient is defined as

$$\alpha = \frac{W^{\text{Loss}}}{W^{\text{inc}}} \tag{8.7}$$

where W^{inc} is the sound power incident on the source side of the partition. As shown in Chapter 10, for random incidence, where the sound energy has equal probability of angle of incidence, the incident sound power is

$$W^{inc} = S\left(\frac{\langle p^2 \rangle}{4\rho_0 c_0}\right) \quad \text{N·m/s} \tag{8.8}$$

where S is the surface area of the partition (one side) in square meters, $\langle p^2 \rangle$ (N²/m⁴) is the mean-square space–time average sound pressure in the source room, ρ_0 is the density of air in kilograms per cubic meter, and c_0 is the speed of sound of the gas at design temperature in meters per second. Using the results of the analyses in Chapter 10, for the power dissipated in the partition by the free bending waves, one obtains the following approximate formula for the random-incidence sound absorption coefficient α^{rand} of a single, flexible, homogeneous, isotropic, nonporous partition:

$$\alpha^{rand} \approx 10^{(-TL_{rand}/10)} + \frac{2\pi\sqrt{12}\rho_0 c_0^3 \sigma_{rad}}{\rho_M h^2 c_L \omega^2} \tag{8.9}$$

where TL_{rand} = random-incidence sound transmission loss of panel (measured or predicted according to Chapter 10)
ω = angular frequency, $= 2\pi f$
f = frequency, Hz
σ_{rad} = radiation efficiency of free bending waves (see Chapter 10)
ρ_M = density of plate material, kg/m³
h = plate thickness, m
c_L = speed sound in plate material, in m/s

The second term on the right side of Eq. (8.9) is strictly valid only if the ratio of the power loss through dissipation and sound radiation by the free bending waves exceeds unity, $(h\rho_M \omega \eta_C / \rho_0 c_0 \sigma_{rad}) > 1$, which is almost always the case because σ_{rad} strongly decreases with decreasing frequency below the coincidence frequency of the plate. In Eq. (8.9) the first term represents the loss of energy by the transmission of sound into the receiver space as defined in Chapter 10. The second term represents the combined contribution of the dissipation of the free resonant bending waves in the plate and energy loss in the form of vibration energy transmission to neighboring structures at the boundaries. It is interesting to note that this second term does not depend of the composite loss factor η_C of the plate (the TL of the plate of course depends on η_C in the frequency range at and above the critical frequency where the free bending waves control the sound radiation). The physical reason for this is that in the case of a small loss factor the plate vibrates more vigorously and the combination of the increased vibration and decreased loss factor results in the same dissipation as happens in the case of a larger loss factor and less vigorous vibration response.

While the first term in Eq. (8.9) is valid for any type of partition (including, e.g., double walls and inhomogeneous and non isotropic plates), the second term is valid only for single, homogeneous, isotropic, platelike partitions. In case of double walls or double windows the first term will exhibit a peak at the double-wall resonance frequency where the TL has a sharp minimum. In the case of double partitions made of two isotropic, homogeneous plates with air space in between, the second term in Eq. (8.9) can be crudely approximated by entering the parameters of the source-side plate into the second term.

8.3 SOUND ABSORBERS USING THIN LAYERS

For the purposes of this chapter, we call those constructions "sound absorbers" which have been designed on purpose to yield high sound absorption in a narrow or wide frequency band. The rest of this chapter deals with the design of such absorbers. We start with those absorber configurations which can be described by a few directly measurable parameters, such as a thin, flow-resistive layer in front of an air space backed by a rigid wall that can be fully characterized by its flow resistance and mass per unit surface area, and will proceed to configurations utilizing thick porous layers and to tuned absorbers.

Thin, Flow-Resistive Layer in Front of Rigid Wall

A sound absorber that consists of a thin, flow-resistive material in front of a rigid wall with an air space in between, such as shown in Fig. 8.1, is the configuration for which it is the easiest to predict acoustical performance. Accordingly, it is logical to treat it first. The flow-resistive layer might be rigid or flexible.

The sketch on the left in Fig. 8.1 depicts a configuration where the air space between the rigid porous layer and the rigid wall is partitioned in a honeycomb pattern to prohibit propagation of sound parallel to the plane of the porous layer. This absorber configuration is called *locally reacting* because the sound field in the absorber depends only on the sound pressure on the interface at the location of the particular honeycomb cell. The sound in the absorber can propagate only perpendicular to the plane of the interface. The sketch on the right in Fig. 8.1 has no partitioning of the air space and the sound can propagate within the absorber parallel to the plane of the interface. The sound pressure at any particular location in the air space depends on the sound pressure on all the locations on the sound-exposed face of the absorber. This configuration is called a *non–locally reacting* sound absorber.

Rigid Porous Layer

A thin, rigid, flow-resistive layer can be fully characterized by a single parameter, namely its flow resistance R_f or its normalized impedance $z' = R_f/(\rho_0 c_0)$, where ρ_0 is the density and c_0 is the speed of sound of the fluid. The normalized impedance of the partitioned air space is $z'' = j \cot(kt)$, where $k = 2\pi/\lambda$ is the wavenumber, λ is the wavelength of the sound, and t is the depth of the air space.

222 SOUND-ABSORBING MATERIALS AND SOUND ABSORBERS

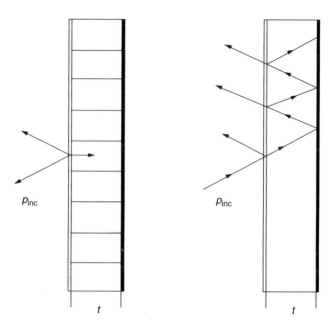

FIGURE 8.1 Thin porous sound-absorbing layer in front of an air space backed by a rigid wall. Left: Partitioned, locally reacting air space. Right: Not partitioned, non–locally reacting air space.

The normalized wall impedance for normal incidence, $\theta = 0$, is then given by

$$z_{w0} = z' + jz'' = \frac{R_f}{\rho c} + j \cot(kt) \tag{8.10}$$

The impedance of the air space is zero when $\cot(kt) = 0$, which is the case at those frequencies where the depth of the air space is an odd number of quarter wavelengths. The maximum value of the normal-incidence sound absorption coefficient is

$$\alpha_{0,\max} = \frac{4z'}{(1+z')^2} \tag{8.11}$$

indicating that total absorption occurs for $z' = 1$. According to reference 4, the normal-incidence sound absorption coefficient is given by

$$\alpha_0(f) = \left\{ \left[\left(\frac{R_f}{\rho_0 c_0}\right)^{0.5} + \left(\frac{\rho_0 c_0}{R_f}\right)^{0.5} \right]^2 + \frac{\rho_0 c_0}{R_f} \cot^2\left(\frac{2\pi f t}{c_0}\right) \right\}^{-1} \tag{8.12}$$

Ingard[1] has computed the sound absorption coefficient of a rigid, flow-resistive layer in front of a locally reacting (partitioned) and non–locally reacting (unpartitioned) air space backed by a rigid wall for both normal and random incidence, as shown in Fig. 8.2.

Figure 8.2 shows that for a locally reacting air space the maximum normal-incidence sound absorption coefficient occurs at those frequencies where the depth of the air space corresponds to an odd multiple of quarter wavelengths $[t = (n+1)\lambda/4$, where $n = 0, 1, 2, \ldots$]. The physical reason for this is that the sound pressure in the partitioned air space just behind the porous layer is zero at these frequencies (the sound that is fully reflected from the hard wall is 180° out of phase with the incident sound). Consequently, the sound pressure gradient across the porous layer, Δp, reaches its maximum value that is numerically equal to the sound pressure on the incident side of the layer, namely $\Delta p = p_{\text{inc}}(1 + R)$, where p_{inc} is the amplitude of the incident sound and R is the reflection factor:

$$R = \frac{R_f - \rho_0 c_0}{R_f + \rho_0 c_0}$$

This pressure gradient produces a particle velocity $v = \Delta p/R_f$ through the porous layer and the sound power dissipated per unit surface area of the absorber,

$$W^{\text{diss}} = v \Delta p = \frac{(p_{\text{inc}})^2 (1 + R)^2}{R_f}$$

In the case of $R_f = \rho_0 c_0$, $R = 0$ and $W^{\text{diss}} = (p_{\text{inc}})^2/(\rho_0 c_0) = W^{\text{inc}}$, indicating that all of the incident sound energy is dissipated in the porous layer. For values of $R_f \neq \rho_0 c_0$, the sound absorption coefficient still is a maximum but less than unity. The more R_f differs from $\rho_0 c_0$, the smaller is the sound absorption coefficient.

The maxima of the random-incidence sound absorption coefficient for a partitioned air space occur at the same frequencies as those at normal incidence. However, there is no value of R_f that would yield total absorption.

Another significant feature to be noted from Fig. 8.2 is that for a partitioned, locally reacting air space curves for both the normal and random-incidence sound absorption coefficients versus frequency have notches (zero absorption) at frequencies that correspond to an even multiple of half wavelengths. This occurs because at the corresponding frequencies the sound reflected from the hard wall combines at the rear of the porous layer with the incident sound in such a manner that there is no sound pressure gradient across the porous layer and consequently no sound is absorbed at these frequencies. This does not affect the random-incidence sound absorption coefficient obtained for the not partitioned (non–locally reacting) air space, as shown in the curves marked c. The advantage of not having notches in curves of the random-incidence sound absorption coefficient versus frequency must be "paid for" by having substantially lower sound absorption at low frequencies. The most important observation from Fig. 8.2 is that it is not possible to achieve a high sound absorption coefficient with a rigid porous layer in front of an air space backed by a rigid wall unless the thickness

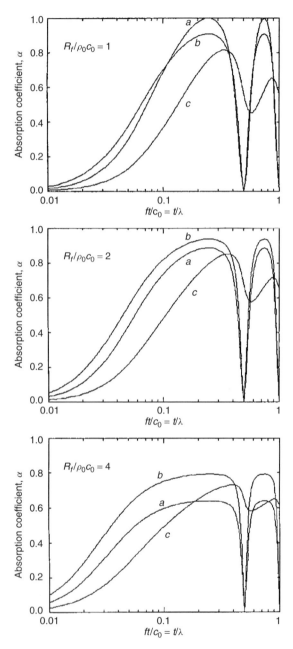

FIGURE 8.2 Sound absorption coefficient of a single, rigid, flow-resistive layer in front of an air space of thickness t backed by a hard wall as a function of the normalized frequency in the form of air space thickness–wavelength ratio $t/\lambda = tf/c_0$ with the normalized flow resistance of the layer, $R_f/\rho_0 c_0$, as the parameter: a, normal incidence; b, random-incidence, locally reacting air space; c, random-incidence, non–locally reacting air space. (After Ref. 1.)

of the air space exceeds one-eighth of the wavelength. For example, an air space thickness of 0.4 m (17 in.) or more is necessary to achieve a high degree of sound absorption at 100 Hz.

As we will discuss below, substantial low-frequency sound absorption can be achieved with an air space thickness of less than one-eighth of the wavelength if the porous layer is not rigid.

Limp Porous Layer

It is little known that sound absorbers using a limp, flow-resistive layer can provide high sound absorption at much lower frequencies than those with a rigid layer. However, caution should be exercised in designing such absorbers for applications where the sound pressure is very high. The limp, flow-resistive layer (e.g., heavy glass fiber cloth or stainless steel wire mesh) should be able to withstand the stresses owing to the high-amplitude sound-induced motion without fatigue. If the porous sound-absorbing layer is limp instead of being rigid, the mass per unit area of the layer and the stiffness per unit area of the air space between the layer and the hard wall result in a resonant system. Near the resonance frequency the limp porous layer will exhibit large-amplitude motion. Approximately, the resonance frequency f_{res} is given by[1]

$$f_{res} \approx \frac{1}{2\pi} \left(\frac{\kappa \rho_0 (c_0)^2}{t m''} \right)^{1/2} \quad \text{Hz} \quad (8.13)$$

where $\kappa = 1.4$ is the adiabatic compression coefficient, t is the thickness of the air space in meters, and m'' is the mass per unit area of the limp porous layer in kilograms per square meter. The analytical model to predict the sound absorption coefficient of sound absorbers utilizing a limp porous screen is documented in reference 1. The computer program on the CD included in reference 1 allows prediction of the normal- and random-incidence sound absorption coefficient of such absorbers with locally reacting and non–locally reacting air space. The data in Figs. 8.3–8.5 were computed this way and show the sound absorption coefficient as a function of the normalized frequency, $f_n = t/\lambda = tf/c_0$, with the normalized flow resistance of the limp porous layer ($R_f/\rho_0 c_0$) as a parameter for various ratios of the mass of the limp layer and the mass of air in the air space behind, MR $= m''/t\rho_0$.

Figure 8.3 shows that the first maximum of the normal-incidence sound absorption coefficient occurs close to the mass–spring resonance frequency given by Eq. (8.13). The higher is the mass ratio MR $= m''/\rho_0 t$, the lower is the frequency of the peak sound absorption. For MR $= 16$ the frequency of the peak occurs at a frequency that corresponds to a normalized frequency $f_n = 0.03$. This is more than a factor of 8 lower than the peak at $f_n = 0.25$ one would obtain with a rigid porous layer. Also note that the first peak of the normal-incidence sound absorption coefficient is unity (100% absorption) if the mass ratio equals the normalized flow resistance (i.e., when MR $= R_f/\rho_0 c_0$). To achieve a relatively high

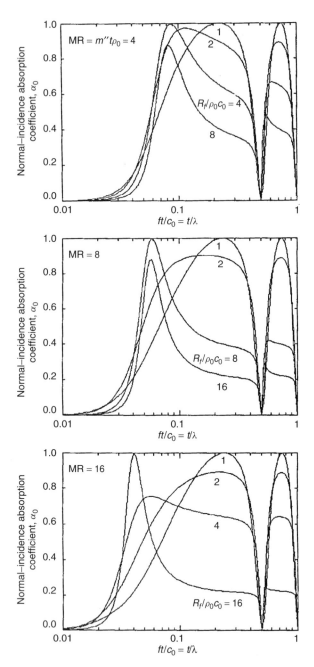

FIGURE 8.3 Normal-incidence absorption coefficient of a single, limp, flow-resistive layer in front of an air space of thickness t backed by a rigid wall as a function of the normalized frequency in the form of the thickness–wavelength ratio $t/\lambda = tf/c_0$ with the normalized flow resistance of the layer, $R_f/\rho_0 c_0$, as the parameter for mass ratios MR. (After Ref. 1.)

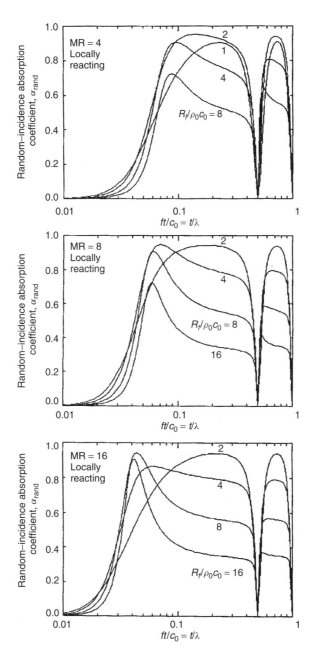

FIGURE 8.4 Random-incidence absorption coefficient of a single, limp, flow-resistive layer in front of a partitioned (locally reacting) air space of thickness t backed by a rigid wall as a function of the normalized frequency in the form of the thickness–wavelength ratio $t/\lambda = tf/c_0$ with the normalized flow resistance of the layer, $R_f/\rho_0 c_0$, as the parameter for mass ratios MR. (After Ref. 1.)

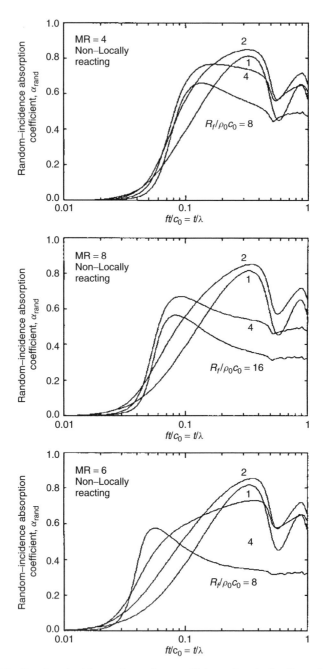

FIGURE 8.5 Random-incidence absorption coefficient of a single, limp, flow-resistive layer in front of a not-partitioned (non–locally reacting) air space of thickness t backed by a rigid wall as a function of the normalized frequency in the form of the thickness–wavelength ratio $t/\lambda = tf/c_0$ with the normalized flow resistance of the layer, $R_f/\rho_0 c_0$, as the parameter for mass ratios MR. (After Ref. 1.)

low-frequency absorption in a wide frequency band, a normalized flow resistance $R_f/\rho_0 c_0$ of between 2 and 3 seems to be a good choice.

Figure 8.4 shows the random-incidence sound absorption coefficient as a function of the normalized frequency computed for a locally reacting (partitioned) air space. Sufficient distance must be provided between the limp porous layer and the honeycomb configuration used to partition the air space to prevent the limp layer from hitting the honeycomb layer at the largest expected displacement amplitude. Curves of the absorption coefficient versus the normalized frequency have similar shape as those in Fig. 8.3 for normal incidence, except that no combination of the design parameters results in total absorption. To obtain substantial low-frequency sound absorption, the mass ratio MR ≈ 2 seems to be a good choice.

Figure 8.5 shows the random-incidence sound absorption coefficients for a non-locally reacting (unpartitioned) air space. Note the very different behavior from that observed for the partitioned air space in Figs. 8.2 an 8.3. The first, and undesirable, difference is that no combinations of parameters yield random-incidence sound absorption coefficients much above $\alpha_{\text{rand}} = 0.8$. The second, and desirable, attribute of the nonpartitioned air space is that there are no notches in the frequency response where the absorption coefficient would go to zero, as was the case for the partitioned air space.

Multiple Limp Porous Layers

As documented in reference 1, very high sound absorption at low frequencies can be achieved over nearly four octaves by placing a large number (up to 16) of limp, porous layers of low-flow resistance in front of a rigid wall. The computer program supplied with reference 1 allows the prediction of the normal- and random-incidence sound absorption coefficients of such absorbers and the optimization of performance by trial and error varying the flow resistance, mass per unit area, and number and position of the layers.

8.4 POROUS BULK SOUND-ABSORBING MATERIALS AND ABSORBERS

Porous bulk sound-absorbing materials are utilized in almost all areas of noise control engineering. This section deals with the following aspects:

1. description of the key physical attributes and parameters that cause a porous material to absorb sound,
2. description of the acoustical performance of porous sound absorbers used to perform specific noise control functions,
3. compilation of acoustical parameters that allow the quantitative design of sound-absorbing configurations on the bases of material and geometric parameters, and

4. experimental methods for measuring the acoustical parameters of porous sound-absorbing materials and the acoustical performance of porous sound absorbers.

Porous materials used for sound absorption may be fibrous, cellular, or granular. Fibrous materials may be in the form of mats, boards, or preformed elements manufactured of glass, mineral, or organic fibers (natural or man made) and include felts and felted textiles. Applications of fibrous materials in silencers require some form of protective covering. The most common covering consists of perforated metal with a fiberglass cloth behind. Sometimes metal screens are inserted between the perforated metal and fiberglass cloth. Cellular materials include polymer foams of varying degrees of rigidity and porous metals. For special applications increasing use is being made of porous metals (e.g., aluminum foams) as sound absorbers. During manufacture, before solidification, metal foams have closed cells, that is, without interconnections. At this stage such materials are poor sound absorbers. However, as liquid-metal foams solidify, thermal stresses occur. The solidified foams usually have cracked cell walls, which significantly increases sound absorption. In addition, by slightly rolling thin sheets of foam from, say, a thickness of 10 to 9 mm, further mechanical cracking occurs, and the interconnections between adjacent cells widen. This increases sound absorption even more. The result is an absorption coefficient versus frequency that has a maximum between 1 and 5 kHz and a peak value of up to 95%. By placing an air gap between metal foam and a rigid wall, one can shift the frequency curve to lower frequencies. In comparison with other materials, glass wool, for example, which gives high absorption over a wide frequency range, metal foams are not very good sound absorbers. However, the high weight-specific stiffness, good crash-energy absorption ability, and fire resistance of porous metals make them suitable for sound absorption panels in the aircraft and automotive industries.

Granular materials can be regarded as an alternative to fibrous and foam absorbers in many indoor and outdoor applications.[8] Sound-absorbing granular materials combine good mechanical strength and very low manufacturing costs. Granular materials may be unconsolidated (loose) or consolidated through use of some form of binder on the particles as in wood-chip panels, porous concrete, and pervious road surfaces. As well as man-made granular materials, there are many naturally occurring granular materials, including sands, gravel, soils, and snow. The acoustical properties of such materials are important for outdoor sound propagation.

A common feature of porous sound-absorbing materials is that the pores are interconnected and have typical dimensions below 1 mm, that is, much less than the wavelengths of the sounds of interest in noise control.

Porous materials must be considered to be elastic if they are in the form of flexible sheets. However, in many cases, the stiffness of the solid frame of the material is much greater than that of air so that the material may be treated as if it were rigid over a wide range of frequencies. This means that such materials can be treated as lossy, homogeneous media. This chapter will be confined to porous materials that may be treated as rigid framed.

How Rigid Porous Materials Absorb Sound

When excited by an incident sound wave, the air molecules in the pores of a porous material are made to oscillate. The proximity of the surrounding solid means that the oscillations result in frictional losses. An important factor that determines the relative contribution of frictional losses is the size (width) of the pores relative to the thickness of the viscous boundary layer. At low frequencies the viscous boundary layer thickness might be comparable with the pore width and the viscous loss is high. At high frequencies the viscous boundary layer thickness may be significantly less than the pore size and the viscous loss is small. At such high frequencies the oscillating flow is "pluglike." However, the presence of the solid causes changes in the flow direction and expansions and contractions of the flow through irregular pores, resulting in loss of momentum in the direction of wave propagation. This mechanism is relatively important at higher frequencies. In the larger pores and at lower frequencies heat conduction also plays a part in energy loss. During the passage of a sound wave, the air in the pores undergoes periodic compression and decompression and an accompanying change of temperature. If the solid part of the material is relatively heat conducting, then the large surface-to-volume ratio means that during each half-period of oscillation there is heat exchange and the compressions are essentially isothermal. At high frequencies the compression process is adiabatic. In the frequency range between isothermal and adiabatic compression the heat exchange process results in further loss of sound energy. In a fibrous material this loss is especially high if the sound propagates parallel to the plane of the fibers and may account for up to 40% of sound attenuation (energy lost per meter of propagation). The losses from forced mechanical oscillations of the skeleton of a porous material are generally so low that it is reasonable to neglect them.

Physical Characteristics of Porous Materials and Their Measurement

Porosity. The gravimetric measurement of porosity requires the weighing of a known volume of dry material. Since in inexpensive fibrous materials, such as mineral wool, the large droplets (or shot with diameter over 100 μm) may constitute up to 30% of the weight, their contribution to the acoustical porosity should be disregarded. Shot can be separated from fiber by a centrifuge process. The dry weight can be used together with the sample volume to calculate the bulk density ρ_B. Subsequently an assumed solid density is used to calculate the porosity h from

$$h = 1 - \frac{\rho_A}{\rho_B} \tag{8.14}$$

For glass fiber and mineral wool products the density of the fiber material is $\rho_B = 2450$ kg/m^3. For silica sand, the density of the mineral grains is 2650 kg/m^3. A gravimetric method which may be used with some consolidated granular materials is to saturate the sample with water and deduce the porosity from the relative weights of the saturated and unsaturated samples. Mercury has been used as the pore-filling fluid in some applications (e.g., soils), but for many materials the introduction of liquids affects the pores. Following a proposal of Cremer and Hubert,[9] Champoux et al.[10] have developed a *dry* method of porosity determination which is based on the measurements of the change in pressure within a sample container subject to a small known change in volume. The lid of the container is a plunger, which is driven by a precise micrometer. The pressure inside the chamber is monitored by a sensitive pressure transducer and an air reservoir connected to the container through a valve serves to isolate the system from fluctuations in atmospheric pressure. The system has been estimated to deliver values of porosity accurate to within 2%. An important feature of the method from the point of view of acoustical properties is that it measures the porosity of connected air-filled pores. However, the gravimetric methods do not differentiate between sealed pores and connected pores. Recently a new method for determining porosity based on a simple measurement of displaced air volume has been proposed.[11] This has the advantage that it does not require compensation for temperature, as does the method of Champoux et al.[10]

An alternative method which may be used with some consolidated materials is to saturate the sample with water and deduce the porosity from the relative weights of the saturated and unsaturated samples.

An acoustical (ultrasonic) impulse method for measuring porosity using the impulse reflected at the first interface of a slab of air-saturated porous material has been proposed and has been shown to give good results for plastic foams[12-14] and random bead packing.[15]

The porosity of a stack of spherical particles depends on the form of the packing, ranging from 0.26 for the densest packing (face-centered-cubic) to 0.426 for simple-cubic packing.[16] A random packing of spheres has a porosity of 0.356.[17] An approximate value that may be assumed for the porosity of a granular material is 0.4.

Typical ranges of values of porosity are listed in Table 8.1.

Tortuosity. The tortuosity of a porous solid is a measure of the irregularity of the fluid-filled paths through the solid matrix. At very high frequencies, it is responsible for the difference between the speed of sound in air and the speed of sound through a rigid porous material. Tortuosity is related to the formation factor used to describe the electrical conductivity of a porous solid saturated with a conducting fluid. Indeed, tortuosity can be measured using an electrical conduction technique in which the electrical resistivity of such a saturated porous sample is compared to the resistivity of the saturating fluid alone. Thus

$$T = \frac{F}{h} \qquad (8.15)$$

TABLE 8.1 Measured and Calculated Values of Porosity and Tortuosity

Material	Porosity, h	Method of Determination	Tortuosity, T	Method of Determination
Lead shot, 3.8 mm particle size	0.385	Measured by weighing	1.6	Estimated as $\alpha = 1/\phi^{0.5}$ (fits acoustic data)
			1.799	Cell model predictions[a]
Gravel, 10.5 mm grain size	0.45	Measured by weighing	1.55	Deduced by fitting surface admittance data
Gravel, 5–10 mm grain size	0.4	Measured by weighing	1.46	Deduced by fitting surface admittance data
Glass beads, 0.68 mm[b]	0.375	Unspecified	1.742	Measured[c]
			1.833	Cell model predictions[a]
Coustone[b]	0.4	Unspecified	1.664	Measured[c]
Foam YB10[b]	0.61	Unspecified	1.918	Measured[c]
Porous concrete	0.312	Measured	1.8	Deduced by fitting surface admittance data
Clay granulate, Laterlite, 1–3 mm grain size	0.52		1.25	Deduced by fitting surface admittance data (assuming porosity of 0.52)
Olivine sand	0.444	Measured	1.626	Cell model predictions[a]
Aluminum foam	0.93	Measured by weighing	1.1	Ultrasonic measurements (laser-generated pulses)
			1.07	Deduced by fitting surface admittance data
Polyurethane foam (Recticel Wetter, Belgium)				
Sample w1[d]	0.98	Measured	1.06	Measured
Sample w2[d]	0.97	Measured	1.12	Measured

[a] From Ref. 20.
[b] From Ref. 26.
[c] From Ref. 11.
[d] From Ref. 27.

where F is the formation factor defined by

$$F = \frac{\sigma_S}{\sigma_f} \tag{8.16}$$

where σ_f and σ_S are the electrical conductivities of the fluid and the fluid-saturated sample, respectively. These in turn are defined by

$$\sigma = \frac{GL}{A} \tag{8.17}$$

where L is the length of the sample, A is the area of the end of the sample, and G is the ratio of the resulting current to the voltage applied across the sample.

To measure the formation factor, first a cylindrical sample of the material is saturated with a conducting fluid (brine solution is convenient). Saturation is achieved by drawing the fluid through the sample after forming a vacuum above it. Agitation of the sample is also required if the pore sizes are small. A voltage is applied across the saturated sample placed between two similarly shaped electrodes at a known separation. The conductivity of the fluid is measured at similar voltages within a separate fluid-tight unit. The use of separate current and voltage probes assures a good contact between the end of the sample and the electrodes, eliminates problems associated with voltage drop at the current electrodes, and allows the simultaneous measurement of the electrical resistivities of the fluid and the saturated porous material.

The tortuosity of a random stacking of glass spheres is given by $1/\sqrt{h}$.[18] This has been verified for a range of porosities from 0.33 to 0.38 and is a special case of the relationship

$$T = h^{-n'} \tag{8.18}$$

where n' depends on grain shape and has the value 0.5 for spheres.[19]

An alternative derivation for stacked identical spheres[20] gives

$$T = 1 + \frac{1-h}{2h} \tag{8.19}$$

The equivalent formula for the tortuosity of a system of identical parallel fibers is

$$T = \frac{1}{h} \tag{8.20}$$

This means that the tortuosity of a typical fibrous material used in noise control is a little larger than 1 since the porosity is close to (but never greater than) 1. In a rigid porous material, tortuosity is one of the properties that contribute to the "structure factor" used in classical descriptions and is related to the "added mass" that is used in the theory of propagation of sound in porous and elastic materials.[21] The structure factor, introduced in classical texts, is intended to include frequency-dependent thermal effects (complex bulk modulus) and frequency-dependent effects due to fiber motion. This means it can *only* be determined by acoustical means. For characterizing the acoustical properties of a bulk porous material, tortuosity is preferred over the structure factor since it has clear physical meaning and can be deduced by nonacoustical means in many cases.

At the highest frequencies, the speed of sound inside the rigid porous material is equal to the speed of sound in air divided by the square root of tortuosity. Since it is primarily responsible for the acoustical properties of porous materials at high audible or ultrasonic frequencies, tortuosity can be deduced from ultrasonic measurements.[22]

Some representative values of tortuosity are listed in Table 8.1.

Flow Resistance and Flow Resistivity. The most important parameter in determining the acoustical behavior of thin porous materials is the airflow resistance R_f. For bulk-, blanket-, or board-type porous materials the flow resistivity (specific flow resistance per unit thickness) $R_1 = R_f/\Delta x$, where Δx is the thickness of the layer, is the key acoustical parameter. The flow resistivity is a measure of the resistance per unit thickness inside the material experienced when a steady flow of air moves through the test sample. Flow resistance R_f represents the ratio of the applied pressure gradient to the induced volume flow rate and has units of pressure divided by velocity. If a material has a high flow resistivity (high flow resistance per unit thickness), it means that it is difficult for air to flow through the surface. In disciplines other than noise control engineering (e.g., geophysics), it is more common to refer to air permeability (k), which is related to the inverse of flow resistivity ($k = \eta/R_1$, where η is the dynamic coefficient of air viscosity). Since flow resistivity is related to the inverse of permeability, high flow resistivity implies low permeability. Typically, low permeability results from very low surface porosity.

Flow resistance is measured as

$$R_f = \frac{\Delta p}{v} = \frac{TS \, \Delta p}{V} \quad \text{N} \cdot \text{s} \cdot \text{m}^{-3} \text{ or } \text{Pa} \cdot \text{s} \cdot \text{m}^{-1} \text{ (mks rayls)} \tag{8.21}$$

From this the flow resistivity is obtained as

$$R_1 = \frac{R_f}{\Delta x} \quad \text{N} \cdot \text{s} \cdot \text{m}^{-4} \text{ or } \text{Pa} \cdot \text{s} \cdot \text{m}^{-2} \text{ (mks rayls/m)} \tag{8.22}$$

where Δp is the static pressure differential across a homogeneous layer of thickness Δx, v is the velocity of the steady flow through the material, V is the volume of air passing through the test sample during the time period T, and S is face area (one side) of the sample.

Since R_f generally depends on the velocity v, it is customary to measure it at a number of different flow rates and extrapolate measured R_f versus v to R_f ($v = 0.05$ cm/s) because below this particle velocity the flow resistance of most fibrous materials does not depend any more on the velocity. The fact that, in general, flow resistivity depends on particle velocity becomes important when considering the behavior of porous absorbers at high sound levels.

The measurement of the flow resistance and flow resistivity of porous building materials has been standardized on a compressed-air apparatus.[23] A similar apparatus may be used to determine the flow resistivity of soils or granular materials (Fig. 8.6). In this measurement the pressure gradient across the sample in a fixed sample holder is monitored together with various (low) flow rates. Compressed air is passed through a series of regulating valves and a very narrow opening into chamber E. This creates an area of low pressure immediately in front of the three tubes connected to the rest of the system. Air is drawn from the environment through the sample as a result of the pressure differential. The rate of airflow through the system is controlled by three flowmeters, giving a total measurement

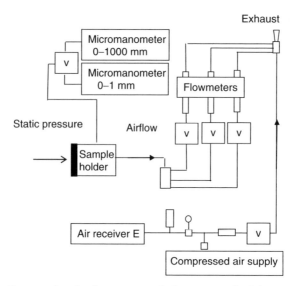

FIGURE 8.6 Concept sketch of a compressed-air apparatus for laboratory measurement of flow resistance.

range between 8.7 and 0.1 L/min. Typically the flow rate must be kept below 3 L/min to avoid structural damage to the sample.

A comparative method[24] makes use of a calibrated known resistance (a laminar-flow element) placed in series with the test sample. Variable-capacitance pressure transducers are used to measure pressure differences across both the test sample and the calibrated resistance. For steady, nonpulsating flow, the ratio of flow resistances equals the ratio of measured pressure differences. The airflow may also be controlled electronically. A unique method of measuring flow resistance by Ingard[1] does not require any flow-moving device, flow-rate meter, or static pressure differential sensor, just a stop watch. In this test setup the airflow (after the piston reaches its terminal velocity) is forced trough the sample, located at the open bottom end of the tube, at a constant rate driven by the static pressure differential $\Delta p = Mg/S$, determined by the weight, Mg, of a tightly sliding piston in the sample holder tube and by the cross-sectional area S of the piston. The factor $g = 9.81$ m/s² is the acceleration due to gravity. The flow resistance of the sample is then inversely proportional to the time T_L required for the piston to travel a specific distance L. The flow resistance of the sample is determined as $R_f = C[Mg \cos(\Phi)/LA^2] ST_L$, where A is the free surface area of the sample and Φ is the angle of tube axis with the vertical. For vertical tube orientation $\Phi = 0$. The small leakage flow between the piston and the tube wall, which is of consequence only if the sample has a very high flow resistance, is accounted for by the correction factor

$$C = \frac{1 + T_{L0}/T_L}{1 - T_L/T_{\text{Closed}}}$$

where T_{L0} and T_{Closed} are respectively the piston travel times with the sample removed and when the end of the tube is air tightly closed. The variation of the flow rate through the sample (needed to extrapolate the flow resistance to 5×10^{-4} m/s) can be accomplished by either changing the tube axis angle Φ or increasing the mass of the piston.

Empirical Prediction of Flow Resistivity. According to private communications with Richard Godfrey[6] of Owens Corning Corporation, the flow resistivity of fibrous sound-absorbing materials can be approximately predicted by a slightly modified version of Eq. 10.4 of the 1971 edition of this book. The modified equation is

$$R_1 \approx \frac{3450}{d^2} \left[\left(\frac{\text{SpGrGlass}}{\text{SpGrFiber}} \right) \rho_{\text{bulk}} \right]^{1.53} \quad \text{N} \cdot \text{m/s}^4 \quad (8.23)$$

where d is the average fiber diameter in micrometers (10^{-6} m), SpGrGlass and SpGrFiber are the specific gravities of glass and that of the actual fiber material, and ρ_{bulk} is the bulk density (with shot contribution deducted) of the fibrous material in kilograms per cubic meter. Note that 1 μm $= 4 \times 10^{-5}$ in. and that $\rho_{\text{bulk}} \approx \rho_M h$, where ρ_M is the density of the fiber material and h is the porosity. It has been found that flow resistivity values measured according to the American Society of Mechanical Engineers (ASME) Standard E 522 for the entire glass fiber-based product line of different bulk densities and fiber diameters are successfully predictable by Eq. (8.23) with a correlation coefficient of 0.93. Equation (8.23) also can be used for predicting the flow resistivity of polymer-based fibrous sound-absorbing materials. For the same bulk density, the polymer fibers have approximately 2.5 times as much surface area and flow resistivity as glass fibers.

Theoretical Prediction of Flow Resistivity. For almost all granular or fibrous sound-absorbing materials, the flow resistivity cannot be predicted analytically on the basis of the geometry of the skeleton. It must be measured. For a few idealized absorber materials, such as materials made of identical spheres or parallel identical fibers, it is possible to make such a prediction. Since the prediction formulas identify important parameters that might be used to scale flow resistivity data measured for a specific material composition to other compositions in cases where direct measurement of the flow resistivity is not feasible, it is instructive to consider at least one idealized case where theoretical prediction is possible.

For stacked identical spheres with radius r it can be shown that[20]

$$R_1 = \frac{\eta}{k} = \frac{9\eta(1-h)}{2r^2 h^2} \frac{5(1-\Theta)}{5 - 9\Theta^{1/3} + 5\Theta - \Theta^2} \quad (8.24)$$

where h is the porosity and

$$\Theta = \frac{3}{\sqrt{2\pi}}(1-h) \cong 0.675(1-h) \quad (8.25)$$

The equivalent formula for the flow resistivity of a system of identical parallel fibers of radius r is[25]

$$R_1 = \frac{16\eta h(1-h)}{r^2 h^2 \left[(1-8\Gamma) + 8\Gamma(1-h) - (1-h)^2 - 2\ln(1-h)\right]} \tag{8.26}$$

where $\Gamma \approx 0.577$. This formula overestimates the flow resistivity of fibrous materials as a consequence of shot content, random fiber orientation, and distribution of fiber diameters in a real material.

Analytical Characterization of Porous Granular or Fibrous Sound-Absorbing Materials. For certain idealized microstructures such as parallel, identical cylindrical or slitlike pores, it is possible to make straightforward analytical predictions of the acoustical properties.[1,19,21] Although analytical characterization of porous granular or fibrous sound-absorbing materials with arbitrary microstructures is possible, it is a relatively complex procedure.[27–31] Moreover, at present, many of the required parameters are not routinely available to practicing noise control engineers. Consequently, the analysis is not reviewed in this chapter. For a comprehensive introduction to recent analytical treatments, the reader should consult reference 21. Semiempirical formulas for the acoustical properties of fibrous materials have been derived based on such a sophisticated analytical treatment.[32] It should be noted, also, that empirical formulas have been derived that predict the acoustical properties of granular media from knowledge only of porosity, grain density, and mean grain size.[33]

Acoustical Properties of Porous Materials. The acoustical properties of porous sound-absorbing materials are characterized by their complex propagation constant k_a and complex characteristic impedance Z_{Ca}, which are defined by

$$p(x,t) = \hat{p}e^{-jk_a x}e^{j\omega t} \quad \text{N/m}^2 \tag{8.27}$$

$$\langle p(x,t)\rangle_y = Z_{Ca}\langle v_x(x,t)\rangle_y \quad \text{N/m}^2 \tag{8.28}$$

where $\hat{p}$ represents the amplitude at $x = 0$ and $\langle \cdots \rangle$ represents an average perpendicular to the propagation direction (x) over an area that is small compared with the wavelength but large compared with the size of pores. The propagation constant

$$k_a = \beta - j\alpha$$

includes the attenuation constant α, which may be obtained by using a probe tube microphone to measure the decrease of the sound pressure level (in nepers per meter) of a plane sound wave propagating in a very thick layer of material. The phase constant β is obtained by measuring the change of phase with distance. It is equal to the angular frequency divided by the frequency-dependent sound speed within the material. The complex characteristic impedance $Z_{Ca} = R_{Ca} - jX_{Ca}$ is obtained by measuring the surface impedance of a thick layer (sufficiently thick that reflection from the end is not detectable) of the absorber material placed in an impedance tube.

An alternative to use of a thick sample with a practically "infinite length" is to use a sample of known finite length and to measure the transfer function between two microphones located in an impedance tube with the loudspeaker source at one end and the sample at the other (closed) end.[34] A broadband input signal such as white noise or a sine sweep can be used. By means of a frequency analysis, output signals from the microphones are used to calculate the transfer function, which is converted to the surface impedance of a sample. The characteristic impedance and propagation constant of a sample can be obtained as long as two distinct sets of surface impedance are measured. This can be realized by using either two samples of different thicknesses[35] or a single sample backed with two different lengths of air cavities.[36] For the former, the two-thickness method, it is convenient if the length of the second sample is double that of the first. In the two-cavity method, the difference in the lengths of the air cavities needs to be tuned to fit the frequency range of interest together with the tube diameter and the microphone spacing.

Empirical Predictions from Regression Analyses of Measured Data

There are two families of parameters that determine the sound absorption coefficient. The characterization would be simplest if both k and Z_C could be expressed in terms of a single parameter. Figure 8.7a shows the measured[37] normal-incidence sound absorption coefficient α_0 of different Rockwool materials of practically infinite thickness (between 0.5 and 1 m, depending on the bulk density of the absorber material) as a function of frequency with bulk density of the material as parameter. Figure 8.7a clearly indicates that the bulk density is not the parameter that would collapse the measured data points into a single curve.

In 1970, Delany and Bazley[38] used many measurements on fibrous materials to deduce semiempirical formulas based on the dimensioned parameter (f/R_1). Equations (8.29a) and (8.29b) are valid only if the frequency f is entered in hertz (s^{-1}) and the flow resistivity R_1 in mks rayls per meter ($N \cdot s \cdot m^{-4}$). Although they have been used widely and successfully, Delany and Bazley's formulas give rise to unphysical results at low frequencies. In particular, the real part of the surface impedance of a rigid-backed layer goes negative. Miki's model[39] represents a modification of these formulas based on Delany–Bazley's data and an electrical analogy for the acoustical properties of porous materials. Miki's formulas are as follows:

$$k_{an} = \frac{k}{k_0} = 1 + 0.109 \left(\frac{\rho_0 f}{R_1}\right)^{-0.618} - j0.160 \left(\frac{\rho_0 f}{R_1}\right)^{-0.618} \quad (8.29a)$$

$$Z_{Cn} = \frac{Z_C}{Z_0} = 1 + 0.070 \left(\frac{\rho_0 f}{R_1}\right)^{-0.632} - j0.107 \left(\frac{\rho_0 f}{R_1}\right)^{-0.632} \quad (8.29b)$$

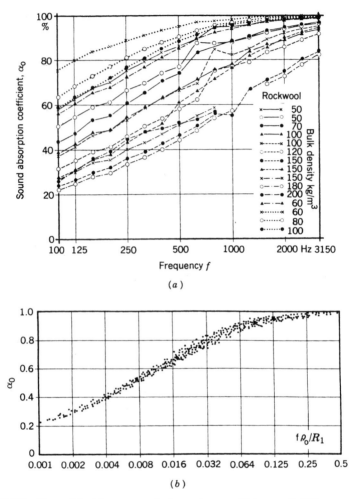

FIGURE 8.7 Normal-incidence sound absorption coefficients of different rock wool materials of practically infinite thickness (0.5–1 m) measured in an impedance tube plotted as a function of (a) frequency with bulk density ρ_A as parameter and (b) nondimensional frequency parameter $\rho_0 f/R_1 = \rho_0 c_0/(R_1 \lambda)$.

Figure 8.7b shows the same data points as Fig. 8.7a but now as a function of the *dimensionless* variable $E = \rho_0 f/R_1$ on the horizontal scale, where ρ_0 is the density of air, f is the frequency, and R_1 is the flow resistivity of the bulk material at the density at which α_0 was measured. Clearly $E = \rho_0 f/R_1 = \rho_0 c_0/\lambda R_1$ is the single parameter that collapses all measured data. According to Fig. 8.7b, even for a layer of practically infinite thickness, very high sound absorption is obtainable only in the frequency range where $(\lambda/4)R_1 < \rho_0 c_0$. To fulfill this requirement at low frequencies, where λ is large, requires the use of a sound-absorbing material of low flow resistivity. The normalized, dimensionless frequency variable

$E = \rho_0 f/R_1$ is useful in describing not only the sound-absorbing capability of the semi-infinite layer of fibrous material but also the propagation constant and characteristic impedance of the bulk material.

It is useful to present the propagation constant and characteristic impedance in a dimensionless manner as

$$k_{an} = \frac{k}{k_0} = \frac{a - jb}{k_0} = a_n - jb_n \tag{8.30a}$$

$$Z_{Cn} = \frac{Z_C}{Z_0} = R - jX \tag{8.30b}$$

where $k_0 = \omega/c_0$ is the wavenumber in air and $Z_0 = \rho_0 c_0$ is the characteristic impedance of the gas filling the voids between the fibers for plane waves.

Figures 8.8 and 8.9 show plots of the real and imaginary parts of the normalized propagation constant k_{an} and that of the normalized characteristic impedance Z_{Cn} for a large variety of mineral wool sound-absorbing materials plotted as a function of the normalized frequency parameter, indicating that indeed the normalized frequency parameter $E = \rho_0 f/R_1$ is a universal descriptor of fibrous porous sound-absorbing materials. The data presented in Figs. 8.8 and 8.9 and similar curves for glass fiber materials were obtained by careful measurements of the acoustical and material characteristics (a, b, R, X, and R_1) of over 70 different types of materials.[38]

The solid lines in Figs. 8.8 and 8.9 result from the regression analyses of the data and have the form

$$k_{an} = \frac{k}{k_0} = (1 + a'' E^{-\alpha''}) - ja' E^{-\alpha'} \tag{8.31a}$$

$$Z_{Cn} = \frac{Z_C}{Z_0} = (1 + b' E^{-\beta'}) - jb'' E^{-\beta''} \tag{8.31b}$$

The regression parameters a', a'', b', b'', α', α'', β', and β'' in Eqs. (8.31) are compiled in Table 8.2 There are different regression parameters for the normalized frequency regions below and above $E = 0.025$. It was found that measured fibrous materials could be divided into two categories: (1) mineral wool and basalt wool and (2) glass fiber.

Polyester Fiber Materials

Polyester fiber materials are used increasingly to replace glass and mineral fiber materials in situations where there is concern to keep the air completely free of fibers suspected to have an adverse influence on health. An example material consists of a mix of two kinds of fibers: (1) polyethylenterephtalate and (2) a core of polyethylenterephtalate and a lining of copolyester. The raw mix is treated at 150°C to melt the external lining of the "bicomponent" fibers and hence form a skeleton of thermally bound fibers. The fiber diameters are between

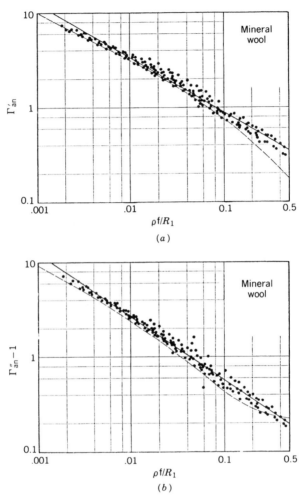

FIGURE 8.8 Measured normalized propagation constant $k_{an} = k/k_0 = a_n + ib_n$ for mineral wool as a function of normalized frequency parameter $E = \rho_0 f/R_1$: (———) regression line: Eq. (8.31a) and Table 8.2; (- - - -) prediction for cylindrical pores.[19,21,27]

TABLE 8.2 Regression Coefficients for Predicting Propagation Constant and Characteristic Impedance of Fibrous Sound-Absorbing Materials

Material	E Region	b'	β'	b''	β''	a''	α''	a'	α'
Mineral and	$E \leq 0.025$	0.081	0.699	0.191	0.556	0.136	0.641	0.322	0.502
basalt wool	$E > 0.025$	0.0563	0.725	0.127	0.655	0.103	0.716	0.179	0.663
Glass fiber	$E \leq 0.025$	0.0668	0.707	0.196	0.549	0.135	0.646	0.396	0.458
	$E > 0.025$	0.0235	0.887	0.0875	0.770	0.102	0.705	0.179	0.674

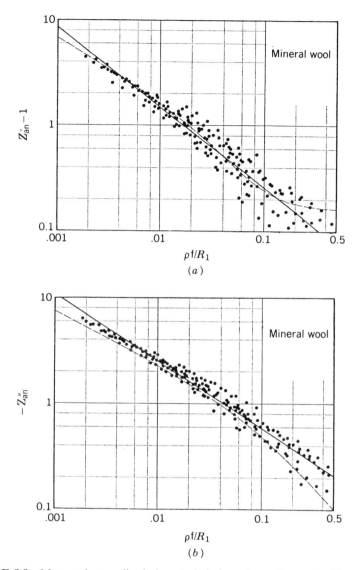

FIGURE 8.9 Measured normalized characteristic impedance $Z_{Cn} = Z_C/Z_0 = R + iX$ for mineral wool as a function of normalized frequency parameter $E = \rho_0 f/R_1$: (―――) regression line; Eq. (8.29b); (- - - -) prediction for cylindrical pores.

17.9 and 47.8 μm (mean 33 μm) and have a mean length of 55 mm. Based on measurements of 38 samples, the flow resistivity has been found to obey the relationship[40]

$$R_1 D^2 = 26\rho_A^{1.404} \qquad (8.32)$$

where D is the mean diameter in micrometers and ρ_A the bulk density in kg/m³.

TABLE 8.3 Regression Coefficients for Predicting Propagation Constant and Characteristic Impedance of Polyester Fiber Materials

Material	b'	β'	b''	β''	a'	α'	a''	α''
Polyester fiber	0.159	0.571	0.121	0.530	0.078	0.623	0.074	0.660

Regression coefficients for predicting acoustical properties of polyester fiber materials are given in Table 8.3.[38]

Plastic Foams

Since concerns about fire hazard and release of toxic combustion products can be overcome largely by suitable treatments, plastic foams are used increasingly for noise control applications. A large variety of plastic foams are available with several different types of physical structure. Polyurethane foams, based on polyester or polyether polyols, are used most commonly. These foams can be fully or partially reticulated, that is, with all membranes or with varying proportions of membranes between cells removed. Formulas (8.31) with similar regression coefficients to those for fibrous materials are applicable to plastic foams.[41,42] However, unlike with fibrous materials, it has not been found necessary to distinguish between E-regions. The relevant regression coefficients are listed in Table 8.4.

The first row of Table 8.4 also shows the comparable values obtained from the empirical formulas of Delany and Bazley (ref. 38). Although these formulas

TABLE 8.4 Regression Coefficients for Predicting Propagation Constant and Characteristic Impedance of Plastic Foams

Material	b'	β'	b''	β''	a'	α'	a''	α''
Mineral and glass fiber	0.0571	0.754	0.087	0.732	0.0978	0.700	0.189	0.595
Fully reticulated polyurethane foam, $60 \leq R_1 \leq 6229$ (Cummings/Beadle)	0.0953	0.491	0.0986	0.665	0.174	0.372	0.167	0.636
Mixed plastic foams,[42] $2900 \leq R_1 \leq 24300$	0.209	0.548	0.105	0.607	0.188	0.554	0.163	0.592
Fully reticulated polyurethane foam, $380 \leq R_1 \leq 3200$[a]	0.114	0.369	0.0985	0.758	0.136	0.491	0.168	0.795
Partly reticulated polyurethane foam, $R_1 = 10,100$[b]	0.279	0.385	0.0881	0.799	0.267	0.461	0.158	0.700

[a] From ref. 43.
[b] From ref. 44.

have been superseded by those due to Mechel (ref. 37) and Miki (ref. 39), they serve to indicate that the coefficient values for plastic foams are significantly different and that the Delany and Bazley relationships would not generally give satisfactory predictions for the bulk properties of plastic foams.

Effects of Temperature

In sound absorbers or silencers designed to operate at high temperatures, it is important to know the acoustical characteristics of the fibrous porous sound-absorbing materials at design temperature. Fortunately, it is not necessary to measure the propagation constant k and characteristic impedance Z_C at design temperature T because the values of these acoustical characteristics measured at room temperature, T_0, can be scaled. The dimensionless frequency variable $E = \rho_0 f / R_1$ must be evaluated at design temperature. Consider that the influence of temperature on ρ and η is

$$\rho(T) = \rho(T_0)\frac{T_0}{T} \tag{8.33}$$

$$\eta(T) = \eta(T_0)\left(\frac{T}{T_0}\right)^{0.5} \tag{8.34}$$

where T and T_0 are the design temperature and room temperature, respectively, measured on an absolute scale (i.e., Kelvin and Rankine; K = °C + 273). Considering that $\rho = \rho_0(T/T_0)^{-1}$ and $c = c_0(T/T_0)^{1/2}$, the acoustical characteristics at design temperature T (in Kelvin) are determined as follows:

$$b(T) = b[E(T)]\frac{\omega}{c_0}\left(\frac{T}{T_0}\right)^{-1/2} \tag{8.35a}$$

$$Z_C(T) = Z_{Cn}[E(T)](\rho c_0)\left(\frac{T}{T_0}\right)^{-1/2} \tag{8.35b}$$

where $b[E(T)]$ and $Z_{Cn}[E(T)]$ are the normalized attenuation constant and normalized characteristic impedance according to Eqs. (8.30) computed for $E = E(T)$ determined from Eqs. (8.34) and (8.35). When fibrous sound-absorbing materials are first used in high-temperature applications, the material undergoes changes. For mats and boards the binder burns off. This decreases the bulk density to a negligible amount. The burning off of the binder does not appreciably change the flow resistivity. After the first high-temperature exposure the glass or mineral fibers thicken. Their virgin diameter d_V increases to the burned-off diameter $d_{\mathrm{BOF}} = d_V + 0.5$. This increase is approximately 0.5 μm. This increase in fiber diameter and corresponding decrease in flow resistivity can be accounted for by correcting flow resistivity measured at room temperature by multiplying it by the factor $(1 + 0.5/d_V)^{-2}$ when calculating $E_{\mathrm{BOF}}(T_0) = f\rho_0/R_1(1 + 0.5/d_V)^{-2}$.

Example 8.1. Compute the real and imaginary parts of the normalized propagation constant and characteristic impedance of a very (practically infinite) thick layer of glass fiber at a frequency of 100 Hz at 20°C and for 500°C and determine α_N using the regression parameter values in Table 8.2. The flow resistivity of the fibrous material at 20°C is $R_1(T_0) = 16{,}000 \text{ N} \cdot \text{s/m}^4$ and $\rho_0 = 1.2 \text{ kg/m}^3$.

Solution

$$T_0 = 273 + 20 = 293 \text{ K} \qquad T = 273 + 500 = 773 \text{ K}$$

$$E(T_0) = \frac{\rho_0 f}{R_1} = 1.2 \times \frac{100}{16{,}000} = 0.0075, \quad \text{which is } < 0.025$$

$$E(T) = E(T_0) \left(\frac{T}{T_0}\right)^{-1.65} = 0.0075 \left(\frac{773}{293}\right)^{-1.65} = 1.5 \times 10^{-3},$$

which is < 0.025

The regression parameters from Table 8.2 are

$a' = 0.396$	$a'' = 0.135$	$\alpha' = 0.458$	$\alpha'' = 0.646$
$b' = 0.0668$	$b'' = 0.196$	$\beta' = 0.707$	$\beta'' = 0.549$

According to Eqs. (8.35), we obtain the following:

Parameter	At 20°C	At 500°C
$b = a' E^{-\alpha'}$	3.72	7.78
$a = 1 + a'' E^{-\alpha''}$ a	4.18	10.0
$R = 1 + b' E^{-\beta'}$ R	3.12	7.62
$X = b'' E^{-\beta''}$	2.88	6.9

The normal-incidence sound absorption coefficient according to Eq. (8.36) is

$$\alpha_N = 4 \frac{Z'_{an}}{Z'^2_{an} + 2 Z'_{an} + 1 + Z''^2_{an}}$$

yielding

$$\alpha_N = \begin{cases} 0.49 & \text{for } 20°\text{C} \\ 0.25 & \text{for } 500°\text{C} \end{cases}$$

8.5 SOUND ABSORPTION BY LARGE FLAT ABSORBERS

When considering sound absorption, the portion of a sound wave incident on the absorber that is absorbed is a quantity of interest. This is easiest to define when the surface of the absorber is flat and sufficiently large that sound waves scattered at the edges of the absorber can be neglected. For a plane incident sound wave, it is possible to assign a sound energy absorption coefficient α for each point on the absorber surface given by

$$\alpha = \frac{\text{absorbed energy}}{\text{incident energy}} = 1 - |R^2| \tag{8.36}$$

where R is the reflection factor, which is defined as the ratio of the reflected and incident sound pressures at the interface. A high sound absorption coefficient ($\alpha \to 1$) requires that $|R| \to 0$. Note that $|R| = 0.1$ corresponds to $\alpha = 0.99$. In this chapter we shall deal only with infinitely large, flat, homogeneous absorbers. Edge effects manifest themselves in increased sound absorption with increasing perimeter–surface area ratio of the absorber.[45] Edge effects are involved also when numerical values greater than 1 are obtained for random-incidence absorption coefficients measured in reverberation rooms.

Plane Sound Waves at Normal Incidence

For plane waves of sound incident perpendicularly to an absorber it is sufficient to know the complex normal specific *surface impedance* $Z_1 = Z_1' - jZ_1''$ of the absorber, which is the ratio of sound pressure to the *normal* component of the particle velocity at the interface (see Chapter 1). The reflection factor and absorption coefficient are related by

$$R = \frac{Z_1 - Z_0}{Z_1 + Z_0} \qquad \alpha = \frac{4Z_1' Z_0}{(Z' + Z_0)^2 + Z_1''^2} \tag{8.37}$$

where $Z_0 = \rho_0 c_0$ is the characteristic impedance of air for plane waves and ρ_0 is the density and c_0 the speed of sound in air.

For sound absorbers consisting of highly porous layers with perforated or cloth facings of smaller porosity, a modified air-side surface impedance must be used in Eq. (8.37). This is obtained by taking the air-side volume velocity averaged over the face area of the absorber. If an absorber with surface impedance Z_i is covered by a perforated facing with porosity h, then $Z_1 = Z_i/h$.

If $|R| \to 0$, Eq. (8.37) indicates that $Z_1 \to Z_0$. This means that the ideal absorber should have a surface impedance similar to that of unbounded air. For a thick (effectively semi-infinite) layer, this requires that the characteristic impedance of porous sound-absorbing materials, Z_C, should be only slightly above that for air, Z_0, and so it is imperative to keep the porosity high. For fibrous sound-absorbing materials, the porosity should be in the range of 0.95–0.99 and for perforated facings the porosity should be above 0.25.

This is illustrated in Fig. 8.10, which shows random-incidence absorption coefficients $\sigma_R(\text{rev})$ for various configurations with a glass fiber fabric or perforated plate covering measured in a reverberation room. Note the relatively

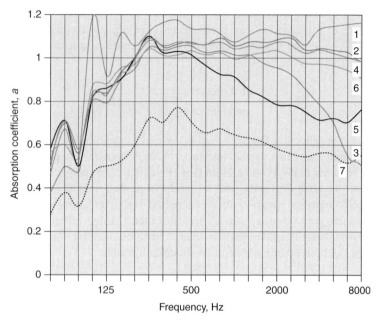

FIGURE 8.10 Random-incidence sound absorption coefficients of 200-mm-thick exhaust silencer pillows (basalt fiber core, density 125 kg/m^3) as a function of frequency measured in a reverberation room: (1) high-porosity fabric covering, equivalent flow resistivity 12 kPa · s · m^{-2}; (2) perforated sheet (32% open area) over high-porosity fabric covering; (3) perforated sheet (17% open area) over high-porosity covering; (4) high-porosity fabric covering and fiberglass facing over basalt fiber core; (5) low-porosity fabric covering, equivalent flow resistivity 43 kN · m · s^{-4}; (6) perforated sheet (32% open area) and fiberglass facing over basalt fiber core; (7) perforated sheet (17% open area), low-porosity fabric covering and fiberglass facing. (Courtesy of John Bowman, Lancaster Glass Fiber, Ltd.)

poor results for the configurations of low-percentage open-area perforated-plate (17%), low-porosity, high-flow-resistivity fabric covering (43,000 mks rayls/m = 100 $\rho_0 c_0$ per meter = 2.4 $\rho_0 c_0$ per inch). Further note that the measured values of α_R(rev) exceed unity. The reasons for this behavior have been discussed in the introductory part of this chapter.

Normal Incidence on a Porous Layer in Front of a Rigid Wall

A rigid wall represents a nearly [see Eq. (8.4)] perfect acoustical reflector. It is an acoustically hard surface. When a porous layer is placed on such an acoustically hard backing, the surface impedance is controlled by the combination of the incident and (multiple) reflected sound waves in the layer, yielding

$$Z_1 = Z_{Ca} \coth(jk_a d) = -jZ_{Ca} \cot(k_a d) \quad \text{N} \cdot \text{s/m}^3 \tag{8.38a}$$

where d is the layer thickness and Z_{Ca} is the characteristic impedance.

Equation (8.38a) may be expressed as

$$Z_1 = Z_{Ca} \frac{\sinh(bd)\cosh(bd) - j\sin(ad)\cos(ad)}{\cosh^2(bd) - \cos^2(ad)} \quad \text{N} \cdot \text{s/m}^3 \quad (8.38b)$$

The impedance given by Eqs. (8.38) exhibits the following behavior:

(a) When $d \ll \frac{1}{4}\lambda_a$ [$\lambda_a = 2\pi/a$ is the wavelength in the absorbing material, and a is defined in Eq. (8.30a)], which is the case when the layer is thin or the frequency is low, the magnitude of $\coth(jk_{an}d)$ is always large and the lack of impedance matching between Z_1 and Z_0 leads to a small sound absorption coefficient. This is the reason there is no "sound-absorbing paint" and that rugs absorb sound only modestly.

(b) When d is sufficiently large compared with the wavelength inside the material and the attenuation constant is not too small, the acoustical behavior of the layer approximates that of an "infinitely thick" layer. In this case $\coth(jk_{an}d) \to 1$ and the surface impedance is the same as the characteristic impedance.

(c) When $\coth(jk_{an}d)$ is minimum, the surface impedance of the hard-backed layer is minimum and the absorption coefficient takes its maximum value. Consideration of the expanded form in Eq. (8.38b) indicates that, for $bd > 0$, which is the only situation of interest, the real part is never zero. So the normalized surface impedance is minimum approximately when the imaginary part is zero. In fact, this condition gives a slight underestimate of the frequency of maximum absorption, whereas $\cos(ad) = 0$, corresponding to $d = \frac{1}{4}\lambda_a$, gives an overestimate.

In general the absorption coefficient is maximum at a frequency that is somewhat less than the frequency at which the layer is a quarter-wavelength thick. Doubling the layer thickness halves the frequency of maximum absorption.

For practical purposes a layer of thickness d can be considered "infinitely thick" if $bd > 2$. For such conditions $Z_1 = Z_{Ca}$. The higher the flow resistivity of the porous material, the more Z_1 exceeds Z_0. Even in the extreme case of practically infinite layer thickness (which at low frequencies and low flow resistivity becomes very large), the sound absorption coefficient will be small. As frequency increases, the general tendency is $Z_{Ca} \to Z_0/h$, and for highly porous absorbers where $h \approx 1$, the good impedance matching results in a high absorption coefficient.

Sound-Absorbing Layer Separated from a Rigid Wall by an Air Space. The sound-absorbing configuration depicted in Fig. 8.11 is frequently used in practical applications (e.g., hung acoustical ceilings). The impedance of a layer of air of thickness t in front of a rigid wall is

$$Z_2 = -jZ_0 \cot(k_0 t) \quad \text{N} \cdot \text{s/m}^3 \quad (8.39)$$

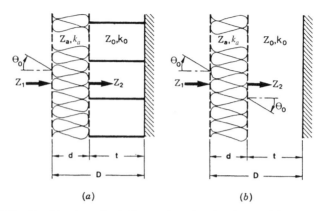

FIGURE 8.11 Combination of a bulk-reacting absorber layer with (*a*) an air gap such that sound can only travel perpendicular to the hard wall and (*b*) an air gap such that sound can travel parallel to the hard wall.

where $k_0 = \omega/c_0 = 2\pi\lambda_0/c_0$ is the acoustical wavenumber for air. The internal reflection coefficient at the back of the layer is $R_B = (Z_2 - Z_{Ca})/(Z_2 + Z_{Ca})$ and the surface impedance at the front of the layer is

$$Z_1 = Z_C \left(\frac{Z_2 + jZ_{Ca} \tan(k_a d)}{Z_{Ca} + jZ_2 \tan(k_a d)} \right) \quad \text{N} \cdot \text{s/m}^3 \quad (8.40a)$$

where Z_C is the normalized characteristic impedance of the porous layer, k_a is the (not-normalized) propagation constant in the porous layer, and d is the layer thickness. Note that Z_2 is zero when $t = \frac{1}{4}\lambda_0$; that is, the impedance Z_s of the absorbing layer backed by zero impedance is given by $Z_s = jZ_C \tan(k_a d) = Z_C \tanh(jk_a d)$. If the impedance of the absorbing layer backed by a hard wall [i.e., with $Z_2 \to \infty$ in Eq. (8.40a) or from Eq. (8.38a)] is denoted by Z_h, then $Z_{Ca} = \sqrt{Z_h Z_s}$. So Eq. (8.40a) can be written in the form

$$Z_1 = \frac{Z_h(Z_2 + Z_s)}{Z_2 + Z_s} = Z_h \frac{1 + Z_s/Z_2}{1 + Z_s/Z_2} \quad \text{N} \cdot \text{s/m}^3 \quad (8.40b)$$

The reflection factor and absorption coefficient *a* are computed by using Z_1 in Eq. (8.37). For small air space thickness such that $t/\lambda_0 < \frac{1}{8}$, $|Z_h| \gg |Z_w|$ and the following approximation is valid:

$$Z_1 \approx \frac{Z_h}{1 + Z_h/Z_2} \quad \text{N} \cdot \text{s/m}^3 \quad (8.41)$$

When the thickness of the absorbing layer is not too small, that is, $|Z_h| \ll |Z_2|$, Eq. (8.41) results in $Z_1 \approx Z_h$. Consequently at low frequencies a very thin air space between the absorbing layer and the rigid wall is ineffective.

As long as the layer thickness is small compared with the wavelength ($d < \frac{1}{8}\lambda_a$), the air space behind the absorbing layer results in a reduction of the surface impedance and at low frequencies the magnitude of the surface impedance shifts toward the characteristic impedance of air Z_0. This results in a decrease of the reflection factor R and a corresponding increase of the normal-incidence sound absorption coefficient. The largest improvement is observed when the thickness of the air space is a quarter of a wavelength, $t = \frac{1}{4}\lambda_0$.

When the air space thickness corresponds to a multiple of $\frac{1}{2}\lambda_0$, $\sin(k_0 t) \to 0$ and Z_2 is very large [see Eq. (8.39)]. At such frequencies, the air space becomes totally ineffective.

Oblique Sound Incidence

For oblique sound incidence one must distinguish between locally reacting and bulk-reacting absorbers. Locally reacting absorbers are those where sound propagation parallel to the absorber surface is prohibited, as, for example, in partitioned porous layers, porous layers backed by a partitioned air space as depicted on the left in Fig. 8.1, in Helmholtz resonators with partitioned volumes, and in small plate absorbers. Local reaction is a good approximation to absorber behavior also when the flow resistivity is relatively high, causing the transmitted wave to bend toward the normal to the surface. This is true, for example, at high densities or in granular media consisting of small particles. In bulk-reacting absorbers, such as a low-flow-resistivity porous layer, possibly with an unpartitioned air space behind it (Fig. 8.11*b*), sound propagation in the direction parallel to the absorber surface is possible in the sound-absorbing layer or in the air space behind it. The term "locally reacting" results from the fact that the particle velocity at the interface depends only on the *local* sound pressure. For bulk-reacting absorbers, the particle velocity at the interface depends not only on the local sound pressure but also on the particular distribution of the sound pressure in the entire absorber volume. Certain materials are inherently anisotropic; that is, the properties vary with direction through the material. This is true, for example, of fibrous materials where the fibers lie in planes parallel to the surface. Strictly, in such materials the propagation constant and characteristic impedance are a function of angle. However, this complication will not be considered further here.

Oblique-Incidence Sound on Locally Reacting Absorber. Locally reacting absorbers are characterized by a surface impedance Z_1 that is independent of the angle of incidence θ_0. The reflection factor $R(\theta_0)$ for locally reacting absorbers is determined by how well the surface impedance Z_1 matches the field impedance of $Z_0/\cos\theta_0$ and is given by

$$R(\theta_0) = \frac{Z_1 \cos\theta_0 - Z_0}{Z_1 \cos\theta_0 + Z_0} \tag{8.42}$$

Equation (8.42) indicates that best impedance match, and correspondingly the lowest reflection factor for a given incidence angle θ_0, is achieved by $|Z_1| > Z_0$

and that such "overmatched" surface impedances will yield absorption maxima at a specific angle of incidence.

Oblique-Incidence Sound on Bulk-Reacting Absorber. Mineral-fiber and open-pore foams are the most frequently used bulk-reacting sound-absorbing materials. As long as the materials may be treated as isotropic, the acoustical behavior of these materials can be fully characterized by their propagation constant k_a and characteristic acoustical impedance Z_a. Methods for measuring and predicting these key acoustical parameters are given in Section 8.4 under Physical Characteristics of Porous Materials and their Measurement.

Semi-Infinite Layer. Figure 8.12 shows the incident, reflected, and transmitted waves at the interface. The combination of these waves must satisfy the following boundary conditions: (1) equal normal components of the impedances and (2) equality of the wavenumber components parallel to the interface. The second of these requirements yields the refraction law

$$\frac{\sin \theta_1}{\sin \theta_0} = \frac{k_0}{k_a} \tag{8.43}$$

where θ_1 is the complex propagation angle of the sound inside the absorber. The complex reflection factor is given by

$$R(\theta_0) = \frac{1 - (Z_0/Z_1) \cos \theta_0}{1 + (Z_0/Z_1) \cos \theta_0} \tag{8.44a}$$

and

$$Z_1 = \frac{Z_{Ca}}{[1 - (k_0/k_a)^2 \sin \theta_0^2]^{1/2}} \quad \text{N} \cdot \text{s/m}^3 \tag{8.44b}$$

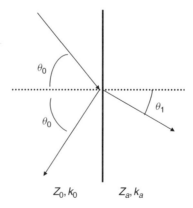

FIGURE 8.12 Reflection and transmission of an oblique-incidence sound wave by a semi-infinite, bulk-reacting absorber.

Finite-Layer Thickness. For a finite layer of bulk-absorbing material of thickness d in front of a hard wall, the reflection factor is

$$R(\theta_0) = \frac{1 - (Z_0/Z_{1d})/\cos\theta_0}{1 + (Z_0/Z_{1d})\cos\theta_0} \tag{8.45}$$

where $Z_{1d} = (Z_{Ca}/\cos\theta_1)\coth(jk_a d \cos\theta_1)$ and θ_1 is defined in Eq. (8.43),

Multiple Layers. Absorbers may consist of a porous sound-absorbing layer with an air space behind, as illustrated in Fig. 8.11, or a number of porous layers of different thicknesses and different acoustical characteristics (Fig. 8.13). Analytical expressions describing the absorber functions and design charts to predict the sound absorption coefficients for those multiple-layer absorbers are provided in reference 46.

Random Incidence. In a diffuse sound field where the intensity $I(\theta) = I$ is independent of the incident angle θ, the sound power incident on a small surface area dS of an absorber is given as

$$dW_{\text{inc}} = I\,dS \int_0^{2\pi} d\phi \int_0^{\pi/2} \cos\theta \sin\theta\,d\theta = \pi I\,dS \tag{8.46}$$

and the absorbed portion as

$$dW_a = I\,ds \int_0^{2\pi} d\phi \int_0^{\pi/2} \alpha(\theta) \cos\theta \sin\theta\,d\theta$$

$$= 2\pi I\,dS \int_0^{\pi/2} \alpha(\theta)\cos\theta\sin\theta\,d\theta \tag{8.47}$$

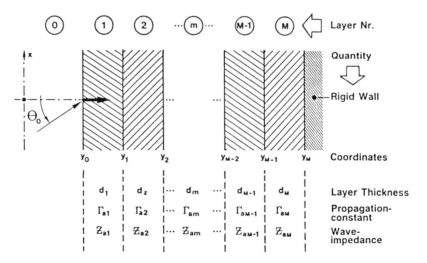

FIGURE 8.13 Multiple-layer absorber.

So the random-incidence sound absorption coefficient is given by

$$\alpha_R = \frac{dW_a}{dW_{\text{inc}}} = 2 \int_0^{\pi/2} \alpha(\theta) \cos\theta \sin\theta \, d\theta \qquad (8.48)$$

Random-Incidence Sound Absorption Coefficient of Locally Reacting Absorbers. When the constituent materials of a multiple-layer absorber are locally reacting, the combination of Eqs. (8.2), (8.42), and (8.48) will yield the random-incidence sound absorption coefficient. The surface impedance Z_1 that is required for the computation of the random-incidence absorption coefficient can be measured in an impedance tube or computed, and the true random-incidence sound absorption coefficient α_R can be determined as[4]

$$\alpha_R = \frac{8}{|z|} \cos(\beta) \left\{ 1 + \frac{\cos 2\beta}{\sin \beta} \arctan\left(\frac{|z| \sin \beta}{1 + |z| \cos \beta}\right) \right. \\ \left. - \frac{\cos \beta}{|z|} \ln(1 + 2|z| \cos \beta + |z|^2) \right\} \qquad (8.49)$$

where $z = Z/\rho_0 c_0$ is the normalized wall impedance and
$\beta = \arctan[\text{Im}(z)/\text{Re}(z)]$.

The maximum value of the true $\alpha_R(\max) = 0.955$ that occurs at a normalized wall impedance of $|z| = 1.6$ and $\beta \approx \pm 30°$ indicates that no locally reacting sound absorber can achieve total absorption or even the 99% absorption that is required as a minimum for the walls of anechoic chambers. The random-incidence sound absorption coefficient measured in a reverberation room, $\alpha_R(\text{rev})$, as described in Chapter 7, can yield physically impossible values above unity (sometimes in excess of 1.2). To avoid confusion, the computed physically correct random-incidence coefficient represented by α_R and that measured with the reverberation room method should have a different symbol, namely $\alpha_R(\text{rev})$.

Random-Incidence Sound Absorption Coefficient of Bulk-Reacting Absorbers. For bulk-reacting absorbers the integration in Eq. (8.48) can only be evaluated numerically. It is more appropriate to use the geometric and acoustical characteristics of the porous liner than its impedance to generate design charts for general use.

8.6 DESIGN CHARTS FOR FIBROUS SOUND-ABSORBING LAYERS

For design use it is useful to have charts where the sound absorption coefficient is plotted as a function of the first-order parameters of the absorber, such as thickness d, flow resistivity R_1, and frequency f. This section contains such design charts. Design charts for a multiple-layered absorber, such as shown in Fig. 8.13, are given in reference 46.

Monolayer Absorbers

As shown in Fig. 8.7b, the key acoustical parameters k_a and Z_{Ca} of fibrous sound-absorbing materials depend only on a single material parameter, namely the flow resistivity R_1 (specific flow resistance per unit thickness). Consequently, the absorption coefficient can be computed and plotted as a function of two dimensionless variables:

$$F = \frac{d}{\lambda} = \frac{fd}{c_0} \quad \text{and} \quad R = \frac{R_1 d}{Z_0} \tag{8.50}$$

Then $k_0 d = 2\pi F$ and $E = \rho_0 f / R_1 = F/R$. At oblique incidence, the angle θ is the third input variable. In the design charts contour lines of constant values of α are plotted on a log-log chart of R versus F. A variation of frequency f produces a horizontal path, an increase of the flow resistivity R_1 causes an upward move, and an increase in layer thickness d leads to a diagonally upward shift to the right. Frequency curves for an absorber layer may be derived by a horizontal intersection, with a starting point determined by the individual parameter values.

Figure 8.14 shows lines of constant absorption of a monolayer absorber for *normal incidence* (the absorber may be bulk or locally reacting). The maximum absorption is reached at about $R = 1.2$ for $F = 0.25$, that is, at a thickness d equal to a quarter of a free-field wavelength and at the flow resistance of the layer of $1.2 Z_0$. Higher resonances at odd multiples of a quarter wavelength are visible for a small flow resistance. The "summit line" of the first maximum, which goes through the points of relative maxima of a at the first resonance, is inclined so

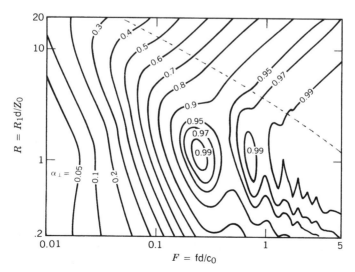

FIGURE 8.14 Lines of constant absorption at normal-incidence sound; absorption coefficient for fibrous absorbers of thickness d and flow resistivity R_1; no air space.

that the first relative maximum will occur below $F = 0.25$ for flow resistances such that $R > 1$, and the first relative maximum of the frequency curve will occur at somewhat higher frequencies than $F = 0.25$ for flow resistances such that $R < 1$. The higher orders of resonances are relatively stationary in frequency as R varies.

At *oblique incidence* the sound absorption for small angles, below about $\theta = 30°$, is only modified slightly. The modifications in $\alpha(\theta)$ are pronounced for larger angles with locally reacting absorbers. The curves on the left in Figs. 8.15 a–c were computed for a locally reacting absorber for θ values of $30°$, $45°$, and $60°$. The resonance structure of the contour plots becomes very distinct at low flow resistances and high frequencies.

Curves on the right in Figs. 8.15a–c show the analogous contour plots of $\alpha(\theta)$ for the same values of θ but now for bulk-reacting (isotropic) absorber layers. Compared to those for a local-reacting absorber, the contours are relatively continuous. The maxima of absorption are shifted downward toward smaller R values for larger angles.

The *random-incidence* absorption coefficients α_R for locally reacting and bulk-reacting absorbers are plotted on the left and right, respectively, in Fig. 8.16. For locally reacting absorbers the absolute maximum of $\alpha_R = 0.95$ is attained for a matched flow resistance $R = 1$ at $F = 0.367$, that is, at a thickness $d = 0.367\lambda_0$, which is larger than a free-field quarter wavelength $\frac{1}{4}\lambda_0$. With homogeneous isotropic absorbers, the absorption coefficient α_R becomes larger than with locally reacting absorbers. Only a weak resonance maximum (belonging to the second resonance) is observed with homogeneous absorbers. The contour lines of absorption are smooth and steady.

Key Information Contained in Design Charts. Many of the general conclusions that were known qualitatively can be answered now in a quantitative manner by studying Figs. 8.14–8.16.

Determining Thickness of Absorber. What is the thickness d_∞ from which an absorber layer starts to behave acoustically as infinitely thick and yields no further increase in absorption? This thickness starts where the contour lines become $45°$ diagonally straight lines. The exact position of the curve for thickness d_∞ will depend on the criterion chosen and on the tolerance allowed. If, at this limit, the final value of α is supposed to be reached with a deviation between 1 and 3%, then the limit curve for normal incidence (in Fig. 8.14) is defined by

$$F = 7.45 R^{-1.67} \tag{8.51}$$

The layer becomes practically infinite when the sound attenuation during propagation of the sound wave through the layer from the surface to the rigid rear wall is $8.68\Gamma'_a d_\infty = 24$ dB, as first derived in reference 2. This is a much higher attenuation than usually assumed as sufficient (about 6–10 dB) for neglecting the influence of the reflection from the rigid wall on the sound absorption at the front side of the absorber layer.

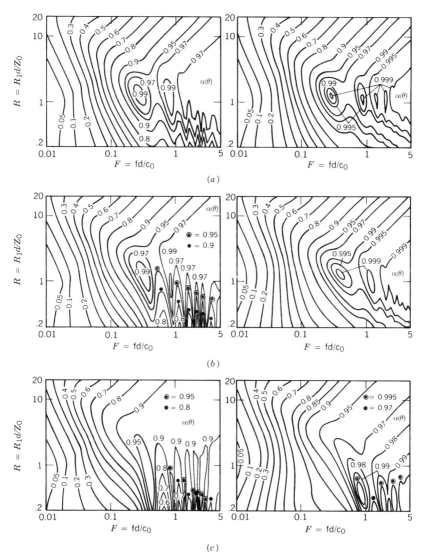

FIGURE 8.15 Lines of constant absorption $\alpha(\theta)$ for fibrous absorbers of thickness d and flow resistivity R_1 for discrete angles of incidences θ; no air space. Left: locally reacting. Right: bulk reacting. (a) $\theta = 30°$; (b) $\theta = 45°$; (c) $\theta = 60°$.

The relationships for "practically infinite" layer thickness d_∞ for which $\Delta L(d_\infty) = 24$ dB can be formulated as

$$f d_\infty^{2.67} R_1^{1.67} = 59 \times 10^6 \quad \text{for } \theta = 0° \tag{8.52a}$$

$$f d_\infty^{2.56} R_1^{1.56} = 34.3 \times 10^6 \quad \text{for } \theta = 45°,$$

locally reacting and random incidence (8.52b)

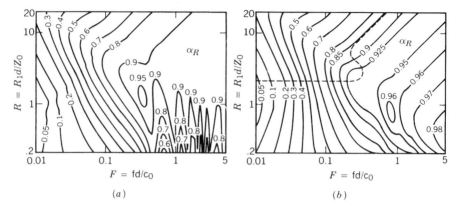

FIGURE 8.16 Lines of constant random-incidence absorption coefficient α_R for fibrous absorbers of thickness d; no air space: (a) locally reacting liner; (b) bulk-reacting liner.

$$fd_\infty^{2.4} R_1^{1.4} = 7.7 \times 10^6 \quad \text{for } \theta = 45°, \text{ bulk reacting} \quad (8.52c)$$

$$fd_\infty^{2.2} R_1^{1.2} = 1.1 \times 10^6 \quad \text{for random-incidence bulk reacting} \quad (8.52d)$$

where f = frequency, Hz
d_∞ = layer thickness, m
R_1 = flow resistivity, N · s/m^4

Optimal Choice of Flow Resistance. Optimal values for the normalized specific flow resistance of the layer, $R = R_1 d/Z_0$, plotted on the vertical scale of Figs. 8.14–8.16, depend on what the designer wants to accomplish.

If the aim is to maximize the sound absorption coefficient, then the optimal choice of normalized specific flow resistance R is in the range of 1–2 for both locally reacting and bulk-reacting absorbers and for normal, oblique, or random angles of sound incidence with the exception of bulk-reacting absorbers for $\theta > 45°$ incidence where $R = 0.7$ yields the best results.

The absorption coefficient depends strongly on flow resistivity R_1, where the curves in Figs. 8.14–8.16 have nearly horizontal contours. The design charts reveal that the typical orientation of the curves of constant α is nearly vertical, indicating only a slight dependence on R_1. This is fortunate because our ability to determine R_1 and our accurate control over its value in the manufacturing process are limited.

Bulk-Reacting versus Locally Reacting Absorbers. A comparison of two graphs in Fig. 8.16 indicates that the random-incidence sound absorption coefficient of locally reacting and bulk-reacting absorbers of the same thickness is practically identical (within 10%) in the left upper quadrant, where $R > 2$ and $F = fd/c_0 < 0.25$, above the dotted line in the graph on the right in Fig. 8.16.

A comparison of $\theta = 45°$ curves in Fig. 8.15b with the random-incidence curves presented in Fig. 8.16 indicates that the $\theta = 45°$ curves for both locally reacting and bulk-reacting absorbers agree with the random-incidence curves within an error of 5%, except in the range where $\alpha > 0.9$, which is of little practical interest. This close agreement between $\alpha(45°)$ and α_R indicates that it is permissible to compute $\alpha(45°)$ instead of the much more difficult α_R.

Two-Layer Absorbers

The two-layer absorbers shown in Fig. 8.11 have a layer of fibrous sound-absorbing material of thickness d backed by an air space of thickness t, with total thickness $D = d + t$. Figure 8.11a shows a partitioned (e.g., in a honeycomb pattern) air space resulting in a locally reacting absorber, and Fig. 8.11b shows an unpartitioned air space resulting in a bulk-reacting absorber. The sound absorber types shown in Fig. 8.11 are usually employed when it is impractical to fill the entire absorber thickness with porous material because no fibrous material of sufficiently low flow resistivity is available to keep the normalized flow resistance $R = R_1 d/Z_0 < 2$. Based on the analyses described in reference 5, design charts similar to those presented in a preceding section for monolayer absorbers were computed.

Figure 8.17 shows contour maps of the random-incidence sound absorption coefficient α_R in a diffuse sound field for a bulk absorber layer of thickness d in front of a locally reacting air gap of thickness t (see Fig. 8.11a) for three fractions, $d/D = 0.25, 0.5, 0.75$, absorber thickness d, and total thickness $D = d + t$. Because the two-layer design curves in Fig. 8.17 are not plotted as functions of the total layer thickness D, they are not directly comparable with the monolayer design curves presented in Fig. 8.16. Some differences may be noticed by comparing these figures with Fig. 8.16, which applies to the bulk monolayer absorber.

First, the variable $F = fd/c_0$ for maximum absorption in the first resonance maximum is shifted toward smaller values with decreasing d/D, which results in higher absorption at low frequencies for less absorber material thickness. Generally, however, the reduction in weight of the material is not as large. The horizontal shift of the absorption maximum toward lower F values, when realized by a reduction in d, makes a larger flow resistivity necessary on the ordinate in order to hold R on a constant value. The increase in R_1 is achieved mostly by an increase of the material bulk density ρ_A. For most fibrous absorber materials R_1 is proportional to about $\rho_A^{1.5}$. Hence, only a small net reduction in absorber material is possible by the addition of an air gap.

For thick absorber layers, that is, for large d, the air gap has no effect, as expected. The limit for the onset of an "infinite" thickness is the same as with the monolayer absorber.

The resonance structure of the layered absorber with small d/D resembles more that of the locally reacting monolayer absorber (see Fig. 8.16a). The character of the bulk-reacting monolayer absorber plot (see Fig. 8.16b), which must be the asymptotic limit for increasing d/D, becomes dominant for $d/D \geq 0.5$.

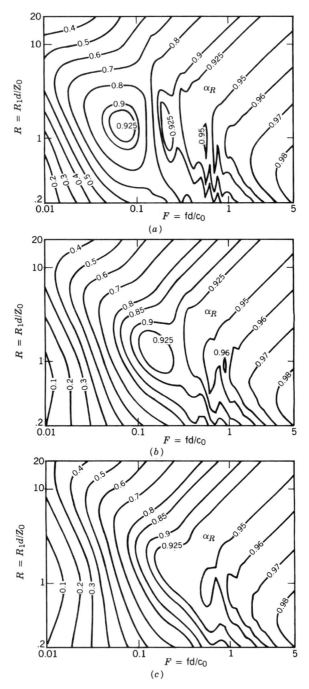

FIGURE 8.17 Lines of constant random-incidence absorption coefficient α_R for a bulk-reacting absorber layer of thickness d in front of a locally reacting air gap of thickness t for thickness ratios $d/D = d/(d+t)$ of (a) 0.25, (b) 0.5, and (c) 0.75.

The sound absorption coefficient of a porous layer of thickness d with an air gap behind it is generally higher than that of a monolayer with equal d. However, the absorption of the layered absorber is smaller than that of a monolayer absorber with equal total thickness $d' = D$. In the range $R > 1$, the absorption of a layered absorber with $d/D = 0.75$ corresponds to that of a monolayer absorber with thickness $d' = 1.25d$ and the absorption of a layered absorber with $d/D = 0.5$ corresponds to that of a monolayer having a thickness $d' = 1.67d$. The quite different character of the plot for $d/D = 0.25$ excludes a similar equivalence.

Finally, the lines of constant α_R of the layered absorber have strong deflections toward the left for values of α_R between about 0.7 and 0.9. As a consequence, the optimum flow resistance $R_1 d$ (which is defined by the leftmost point of a curve) has a more distinct meaning than with monolayer absorbers.

Design graphs for layered absorbers with bulk-reacting absorber layers in front of a bulk-reacting air gap (see Fig. 8.7b) are given in reference 45.

Multilayer Absorbers

Multilayer absorbers, such as shown in Fig. 8.13, have been treated in reference 4. Best results are obtained if the flow resistivity of the layers, R_1, increases from the interface toward the rigid wall. These results obtained with such multilayer absorbers are somewhat better than those obtained with a monolayer absorber of equal thickness. However, the improvements seldom justify the added cost and complexity. Of course, Fig. 8.13 represents an idealization of practical constructions (see Fig. 8.18) which combine multiple porous layers with perforations and slots.

It should be noted, however, that even the most elaborate multiple-layer absorber cannot match the random-incidence sound absorption performance of anechoic wedges.

Thin Porous Surface Layers

Thin porous surface layers such as mineral wool felt sprayed on plastic, steel wool, mineral wool, or glass fiber cloth; wire mesh cloth; and thin perforated metal are frequently used to provide mechanical protection. They also reduce the

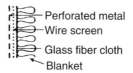

FIGURE 8.18 Multilayered absorber with perforated porous layer structure.

loss of fibers. The acoustical properties of such thin layers are characterized by their flow resistance $R_s = \Delta p/v$ and their mass per unit area ρ_s. The surface impedance is

$$Z_s = \begin{cases} R_s & \text{(fixed)} \quad (8.53a) \\ \dfrac{(j\omega\rho_s)R_s}{j\omega\rho_s + R_s} & \text{(free)} \quad (8.53b) \end{cases}$$

which must be added to the wall impedance of the absorber. If the porous surface layer is not free to move, Eq. (8.53a) is used. For surface layers that are free to move due to the pressure differential produced by their flow resistance, Eq. (8.53b) is used.

Tables 8.5 and 8.6 provide design information on wire mesh cloth and glass fiber cloth, respectively. The flow resistance values listed in these tables represent the linear part of the flow resistance, which is appropriate for design use only if the particle velocity is low. At high sound pressure levels (above 140 dB), the nonlinear behavior of the flow resistance must be determined by measuring the flow resistance as a function of the face velocity.

Sintered porous metals have been developed for silencers in jet engine inlets. If backed by a honeycomb-partitioned air space, they provide good sound absorption and remain linear up to high sound pressure levels and for high-Mach-number grazing flow. Their additional advantage is that they do not require any surface protection. Table 8.7 gives the flow resistance R_s and mass per unit area ρ_s for some commercially available porous metal sheets.

Steel wool mats typically have a thickness from 10 to 45 mm, mass per unit area ρ_s from 1 to 3.8 kg/m², and a specific flow resistance R_s from 100 to 500 N · s/m³. Recently, such steel wool mats needled on mineral wool of specified thickness and density have become available on special order.

Perforated metal facings, when they rest directly on a porous sound-absorbing layer, can be accounted for by a series impedance Z_s given by

$$Z_s = \frac{j\omega\rho_0}{\varepsilon}\left\{l + \Delta l\left[H(1 - |v_h|) - j\frac{\Gamma_a}{k_0}\frac{Z_a}{Z_0}\right]\right\} \quad (8.54)$$

TABLE 8.5 Mechanical Characteristics and Flow Resistance R_s of Wire Mesh Cloths

Wires/cm	Wires/in.	Wire Diameter		Mass per Unit Area		Flow Resistance R_s	
		μm (10^{-6} m)	mils (10^{-3} in.)	kg/m²	lb/ft²	N · s/m³	$\rho_0 c_0$
12	30	330	13.0	1.6	0.32	5.7	0.014
20	50	220	8.7	1.2	0.25	5.9	0.014
40	100	115	4.5	0.63	0.13	9.0	0.022
47	120	90	3.6	0.48	0.1	13.5	0.033
80	200	57	2.25	0.31	0.63	24.6	0.06

DESIGN CHARTS FOR FIBROUS SOUND-ABSORBING LAYERS

TABLE 8.6 Mechanical Characteristics and Flow Resistance R_s of Glass Fiber Cloth

Manufacturer[b]	Cloth Number	Surface Density[a] oz/yd²	g/m²	Construction, Ends × Picks	Flow Resistance, mks rayls (N · s/m³)
1, 2, 3	120	3.16	96	60 × 58	300
1, 2, 3	126	5.37	164	34 × 32	45
1, 2, 3	138	6.70	204	64 × 60	2200
1, 2, 3	181	8.90	272	57 × 54	380
3	1044	19.2	585	14 × 14	36
2	1544	17.7	535	14 × 14	19
3	3862	12.3	375	20 × 38	350
1	1658	1.87	57	24 × 24	10
1	1562	1.94	59	30 × 16	<5
1	1500	9.60	293	16 × 14	13
1	1582	14.5	442	60 × 56	400
1	1584	24.6	750	42 × 36	200
1	1589	12.0	366	13 × 12	11

[a] Averaged over a large sample.
[b] Code numbers for manufacturers are as follows: (1) Burlington Glass Fabrics Company; (2) J. P. Schwebel and Company; (3) United Merchants Industrial Fabrics.

TABLE 8.7 Specific (Unit-Area) Flow Resistance R_s, Thickness, and Mass per Unit Area of Sintered Porous Metals Manufactured by the Brunswick Corporation[a]

Specific Flow Resistance Air, 70°F		NLF[b] 500/20	Designation	Thickness		Mass per Unit Area	
$\rho_0 c_0$	N · s/m³			mm	in.	kg/m²	lb/ft²
0.25	100	3.6	FM 125	1.0	0.04	3.9	0.79
		5.0	FM 127	0.76	0.03	3.3	0.67
		2.6	FM 185	0.5	0.02	2.0	0.4
		2.0	347-10-20-AC3A-A	0.5	0.02	1.32	0.27
		2.0	347-10-30-AC3A-A	0.76	0.03	1.1	0.23
		2.0	FM 802	0.5	0.02	1.3	0.27
0.88	350	4.7	FM 134	0.89	0.035	3.8	0.77
1.25	500	1.8	FM 122	0.76	0.03	1.4	0.28
		3.6	FM 126	0.66	0.026	3.7	0.76
		3.3	FM 190	0.41	0.016	2.0	0.4
		2.0	347-50-30-AC3A-A	0.76	0.03	1.4	0.29

[a] Except FM 802, which is made of Hastelloy X, all materials are type 347 stainless steel. (Courtesy of Brunswick Corporation.)
[b] Nonlinearity factor, calculated as the ratio of flow resistances obtained at flow velocities of 500 and 20 cm/s, respectively.

where ε is the fractional open area of the perforated facing, l is the plate thickness, Δl is the end correction length of the perforations given in Section 8.5, H is a step function that is unity for $|v_h| \leq 1$ m/s and zero for $|v_h| > 1$ m/s, and v_h is the particle velocity in the holes of the perforations. For thin, unrestrained perforated surface protection plates, it is necessary to take the parallel combination of Z_s and $j\omega\rho_s$ according to Eq. (8.53b).

8.7 RESONANCE ABSORBERS

In building acoustics the most frequently used type of resonant absorber is the Helmholtz resonator, consisting of the mass of an air volume in a cross-sectional area restriction (such as holes or slits in a covering plate) and the compliance of the air volume behind the covering plate. In the narrow frequency range, centered at their resonance frequency, the resonators "soak in" the acoustical energy from a large room volume in the vicinity of the neck. This high energy concentration results in high local sound pressures at the neck (and in the internal volume) which are much higher than that one would register with the resonator opening closed. On the room side, this near-field sound pressure decays exponentially with increasing distance. In architectural application of resonance absorbers care should be exercised not to locate them too near to any member of the audience to avoid distortion of the auditory experience. The open bottles embedded in the vertical portion of the seats in ancient Greek amphitheaters indicate that this local amplification of the sound field in the vicinity of acoustical resonators was known more than 2000 years ago. What was not known, which misguided their application, was that the amplification occurs only in a narrow frequency band and cannot locally amplify such broadband signals as speech. Such a Helmholtz resonator (named after the nineteenth-century German physicist Ludwig Helmholtz, who was first to utilize its narrow-band tuning to analyze the spectral composition of complex sounds) is shown in Fig. 8.19. In the following treatment it is assumed that all dimensions of a single resonator are small compared with the acoustical wavelength (except in the case of two-dimensional resonators with slits) and that the skeleton of the resonator is rigid.

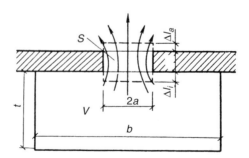

FIGURE 8.19 Key geometric parameters of a Helmholtz resonator.

Acoustical Impedance of Resonators

The specific acoustical impedance of the resonator opening Z_R is the sum of the impedance of the enclosed air volume Z_v and that of the air volume that oscillates in and around the resonator mouth Z_m, namely,

$$Z_R = Z_v + Z_m = (Z'_v + jZ''_v) + (Z'_m + jZ''_m) \quad \text{N} \cdot \text{s/m}^3 \tag{8.55}$$

The volume impedance $Z_v = jZ''_v$ is purely imaginary and predominantly of spring character while the mouth impedance has a real part Z'_m and an imaginary part Z''_m and is predominantly of mass character.

The impedance of a rectangular resonator volume is

$$Z_v = jZ''_v = -j\rho_0 c_0 \cot(k_0 t) \frac{S_a}{S_b} \quad \text{N} \cdot \text{s/m}^3 \tag{8.56a}$$

where S_b is the surface area of the resonator cover plate, $S_a = \pi a^2$ is the area of the resonator mouth, t is the depth of the resonator cavity, and $V = S_b t$ is the resonator volume. If the resonator dimensions are small compared with the wavelength ($k_0 t \ll 1$), Eq. (8.56a) yields

$$Z_v = jZ''_v = -j\frac{\rho_0 c_0^2}{\omega} \frac{S_a}{V} \quad \text{N} \cdot \text{s/m}^3 \tag{8.56b}$$

Thus, when all the dimensions are smaller than the wavelength of interest, the shape of the air cavity does not matter.

The impedance of the resonator mouth Z_m consists of components related to the oscillations of air internal to the mouth, within the mouth, and external to the mouth of the resonator plus the impedance of the screen (Z_s) that may be placed across the resonator mouth to provide resistance. For circular resonator openings of radius a and orifice plate thickness l, the resulting total impedance is given by

$$Z'_m = \rho_0 \left[\sqrt{8\nu\omega}\left(1 + \frac{l}{2a}\right) + \frac{(2\omega a)^2}{16 c_0} \right] + Z_s \quad \text{N} \cdot \text{s/m}^3 \tag{8.57a}$$

$$Z''_m = \omega \rho_0 \left[l + \left(\frac{8}{3\pi}\right) 2a \right] + \left(l + \frac{1}{2a}\right)\sqrt{8\nu\omega} \quad \text{N} \cdot \text{s/m}^3 \tag{8.57b}$$

where ν is the kinematic viscosity ($\nu = 1.5 \times 10^{-6}$ m^2/s for air at room temperature). The quantity $0.5(8\nu/\omega)^{1/2}$ is the viscous boundary layer thickness.[1]

In general, the first term in Eq. (8.57a) applies only for smooth orifice edges. For sharp orifice edges it can be many times higher. Also note that the second term in Eq. (8.57b) is usually small compared with the first term and can be neglected.

The results of Eqs. (8.57) for circular orifices and for a single resonator can be generalized for other orifice shapes and for situations where the orifice is located on a large wall or at a two- or three-dimensional corner. This generalized form is

$$Z'_m = \frac{P(l + P/2\pi)}{4S_a}\rho_0\sqrt{8v\omega} + \rho_0 c_0\frac{k_0^2 S_a/\Omega}{1 + k_0^2 S_a/\Omega} + Z_s \quad \text{N} \cdot \text{s/m}^3 \quad (8.58a)$$

$$Z''_m = j\omega\rho_0\left[l + 2\Delta l + \frac{P(l + P/2\pi)}{4S_a}\sqrt{8v/\omega}\right] \quad \text{N} \cdot \text{s/m}^3 \quad (8.58b)$$

where P is the perimeter of the orifice, S_a is its surface area, and Ω is the spatial angle the resonator "looks into":

$$\Omega = \begin{cases} 4\pi & \text{for resonator away from all walls} \\ 2\pi & \text{flush mounted on wall far from corners} \\ \pi & \text{flush mounted on wall at two-dimensional corners} \\ \tfrac{1}{2}\pi & \text{flush mounted on wall at three-dimensional corners} \end{cases}$$

The quantity $\Delta l = 16a/3\pi$ represents the combined internal and external end corrections.

Resonance Frequency

The resonance frequency f_0 of the Helmholtz resonator occurs where Z''_m and Z''_v are equal in magnitude. Combination of Eqs. (8.56b) and (8.57b) yields

$$f_0 = \frac{c_0}{2\pi}\sqrt{\frac{S_a}{V\langle\cdots\rangle}} \simeq \frac{c_0}{2\pi}\sqrt{\frac{S_a}{V(l + 16a/3\pi)}} \quad (8.59)$$

where $\langle\cdots\rangle = l + \Delta l$ represents the quantity in square brackets in Eq. (8.58b).

The viscous terms in (8.57b) are ignored in (8.59). However, this is usual when calculating a resonance frequency because the viscous contribution is small. For resonators where the area of the mouth, S_a, is not much smaller than the area of the top plate of the resonator, the proximity of the side walls of the cavity influences the flow in the resonator mouth. In this case it is necessary to determine the combined end correction from $\Delta l = 8a/3\pi + \Delta l_{\text{int}}$.[47]

Absorption Cross Section of Individual Resonators

For individual resonators (or groups of resonators where the distance between the individual resonators is large enough that interaction is negligible), the absorption coefficient α is meaningless. In this case, the sound-absorbing performance must be characterized by the absorption cross section A of the individual resonator. The absorption cross section is defined as that surface area (perpendicular to the direction of sound incidence) through which, in the undisturbed sound wave (resonator not present), the same sound power would flow through as the sound power absorbed by the resonator.

The power dissipated in the mouth of a resonator is

$$W_m = 0.5|v_a|^2 S_a R_T = \frac{0.5 \times 2^n |p_{\text{inc}}|^2 S_a R_T}{|Z_R|^2} \quad \text{W} \tag{8.60}$$

Accordingly, the absorption cross section is

$$A = \frac{2^n \rho_0 c_0 S_a R_T}{|Z_R|^2} \quad \text{m}^2 \tag{8.61}$$

where $n = 0$ for resonators placed in free space, $n = 1$ if the resonators are flush mounted in a wall, $n = 2$ if the resonators are at the junction of two planes, and $n = 3$ if the resonators are in a corner; $R_T = Z'_m + Z_{\text{rad}}$ is the total resistance.

At the resonance frequency f_0, where $Z''_v + Z''_m = 0$, the absorption cross section reaches its maximum value A_0:

$$A = 2^n \frac{S_a \rho_0 c_0}{R_{\text{rad}}} \left[\frac{R_T/R_{\text{rad}}}{|1 + R_T/R_{\text{rad}}|^2} \right] \quad \text{m}^2 \tag{8.62a}$$

For a flush-mounted ($n = 1$) circular resonator area $S_a = \pi a^2$, the radiation resistance is $R_{\text{rad}} = 2(\pi a)^2 \rho_0 c_0 / \lambda_0^2$, and Eq. (8.62a) yields

$$A = \frac{1}{\pi} \lambda_0^2 \left[\frac{R_T/R_{\text{rad}}}{|1 + R_T/R_{\text{rad}}|^2} \right] \quad \text{m}^2 \tag{8.62b}$$

where λ_0 is the wavelength at the resonance frequency of the resonator. The maximum value of the absorption cross section is obtained by matching the *internal* resistance to the radiation resistance at the resonance frequency, yielding ($R_T = R_{\text{rad}}$)

$$A_0^{\max} = \frac{\lambda_0^2}{4\pi} \quad \text{m}^2 \tag{8.63}$$

According to Eq. (8.63), a matched resonator tuned to 100 Hz can achieve absorption cross section $A_0^{\max} = 0.92 \text{ m}^2$.

According to Eq. (8.61), the frequency dependence of A can be presented in generalized form as

$$\frac{A(f)}{A_0} = \frac{1}{1 + Q^2 \phi^2} \tag{8.64}$$

where the Q of the resonator and the normalized frequency parameter ϕ are given as

$$Q = \frac{Z''_m(\omega_0)}{Z'_m(\omega_0)} \quad \phi = \frac{f}{f_0} - \frac{f_0}{f}$$

Figure 8.20 shows the normalized absorption cross section A/A_0 as a function of the normalized frequency ϕ with Q as the parameter, indicating that the

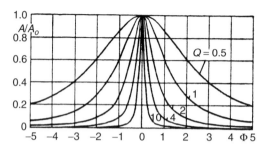

FIGURE 8.20 Normalized absorption cross section A/A_0 as function of normalized frequency ϕ with Q as parameter.

normalized bandwidth of the absorption cross section f_0/Q corresponds to the relative bandwidth $\Delta f/f_0 = 1/Q$.

Consequently, the choice of Q determines the relative shape of the absorption curve and the factor in the square brackets in Eq. (8.62b) determines the height of the curve at resonance. To obtain optimal absorption characterized by high absorption at resonance A_0 and wide bandwidth requires resistance matching $R_T = R_{\text{rad}}$ and a relatively low value of Q.

Nonlinearity and Grazing Flow

The oscillating air mass (that includes the end-correction term) is a measure for the reversible kinetic energy of the acoustical resonator. Reversibility and spatial coherence diminish with increasing turbulence due to jet flow through the orifice caused by high-amplitude sound and by the grazing flow of velocity U_∞. Both reduce the air mass that participates in the oscillating motion and consequently increase resonance frequency and increase the losses of the resonator.

Nonlinearity and grazing flow affect only the mouth impedance Z_m of the resonator orifice. Table 8.8 is a compilation of the nonlinear effects owing to high sound pressure level and grazing flow for a perforated plate. The round holes are regularly distributed. The porosity of the perforated plate is $\varepsilon = \pi a^2/b^2$, where a is the hole radius and b the hole spacing. The amplitude nonlinearity is characterized by the particle-velocity-based Mach number $M_0 = v/c_0$, which is the ratio of the particle velocity in the resonator orifice and the speed of sound in air. The grazing flow nonlinearity is characterized by the flow-velocity-based Mach number $M_\infty = U_\infty/c_0$, where U_∞ is the velocity of the grazing flow far from the wall. The formulas presented in Table 8.8 are analytical results adjusted to yield agreement with measured data.[48] The impedances in Table 8.8 are values averaged over the plate surface.

Internal Resistance of Resonators

The most difficult part of resonator design is the prediction of the internal resistance. In the earlier section Acoustical Impedance of Resonators (page 265) formulas were given to predict the friction loss resistance [first term in Eq. (8.57a)], which is valid only for rounded orifice perimeter. Additional resistance due to sharp edges cannot be predicted analytically. Table 8.8 contains formulas to predict resonator resistance due to nonlinear effects owing to high sound pressure level and grazing flow.

Should it be desirable to obtain higher loss resistance than provided by friction and nonlinear effects (e.g., obtaining larger absorption bandwidth), porous materials such as a screen, felt, or layer of fibrous sound-absorbing material must be placed in (or behind) the resonator orifice. Figure 8.21 shows some of the more frequently used ways to increase resonator resistance. In Fig. 8.21, R_1 is the flow resistivity of the porous material, d is its thickness, t_1 is its distance from the back side of the cover plate, t is the depth of the resonator cavity, $2a$ is the diameter of the holes, Δl is the end correction, and ε is the porosity of the cover plate.

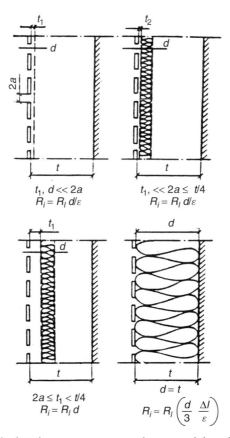

FIGURE 8.21 Methods to increase resonator resistance and the achieved specific acoustical resistance R_i (R_1 = flow resistivity of porous material, d is its thickness, and $\Delta l = 16a/3\pi$).

TABLE 8.8 Nonlinearity of Orifice Impedance $Z_m = Z'_m + jZ''_m$ of Perforated-Plate Resonator Covers Caused by High Sound Pressure Level L_p and Grazing Flow U_∞ [a]

Grazing Flow	Low SPL, $L_p < L_{0l}$	Medium SPL, $L_{0l} \leq L_p \leq L_{0h}$	High SPL, $L_p > L_{0h}$
No flow or very low flow, $M_\infty \leq 0.025$, $U_\infty \leq 8$ m/s	$L_{0l} = 107 + 27 \log[4(1-\epsilon^2)\omega\rho_0 v(1+l/2a)^2]$ dB $Z'_m = R_0 \quad Z''_m = X_0(\delta)$ $R_0 = \dfrac{\rho_0}{\epsilon}\sqrt{8v\omega}\left(1+\dfrac{l}{2a}\right) + (\rho_0/8\epsilon c_0)(2a\omega)^2$ $X_0 = \dfrac{\omega\rho_0}{\epsilon}\left[\sqrt{\dfrac{8v}{\omega}}\left(1+\dfrac{l}{2a}\right) + l + \delta\right]$ $\delta = \delta_0 = 0.85(2a)\phi_0(\epsilon)$ $\phi_0(\epsilon) = 1 - 1.47\sqrt{\epsilon} + 0.47\sqrt{\epsilon^3}$	$L_{0h} = L_{0l} + 30$ dB $Z'_m = \sqrt{R_h^2 - R_0^2} \quad Z''_m = X_0(\delta)$ $\delta = \delta_0\phi_1(M_0)$ $M_0 = \dfrac{10^{(-2.25 + 0.025L_p)}}{\sqrt{0.5\rho_0 c_0^2(1+\epsilon^2)}}$	$L_{0h} = L_{0l} + 30$ dB $Z'_m = R_h \quad Z''_m = X_0(\delta)$ $R_h = \dfrac{1}{\epsilon}\sqrt{2\rho_0(1-\epsilon^2)} \times 10^{(-2.25 + 0.018 L_p)}$ $\delta = \delta_0\phi_1(M_0)$ $\phi_1(M_0) = \dfrac{1 + 5\times 10^3 M_0^2}{1 + 10^4 M_0^2}$
With flow, $M_\infty \geq 0.025$, $U_\infty \geq 10$ m/s	$L_p \leq L_{ml} \quad L_{ml} = 175 + 40 \log M_\infty$ dB $Z'_m = R_m \quad Z''_m = X_0(\delta)$ $R_m = 0.6\rho_0 c_0 \dfrac{1-\epsilon^2}{\epsilon}(M_\infty - 0.025) -$ $40R_0(M_\infty - 0.05); M_\infty \leq 0.05$ $R_m = 0.3\rho_0 c_0 \dfrac{1-\epsilon^2}{\epsilon}M_\infty$ $\delta = \delta_0\phi_2(M_\infty)$ $\phi_2(M_\infty) = 1/(1 + 305 M_\infty^3)$	$L_{ml} \leq L_p \leq L_{mh}$ $Z'_m = \sqrt{R_m^2 + R_h^2} \quad Z''_m = X_0(\delta)$ $M_\infty > 0.05$ $\delta = \delta_0\phi_2(M_\infty)$	$L_p > L_{mh} \quad L_{mh} = L_{ml} + 18$ dB $Z'_m = R_h \quad Z''_m = X_0(\delta)$ $\delta = \delta_0\phi_1(M_\infty)$

[a] L_p is octave-band SPL in dB re 20 µPa at the peak of the spectrum.

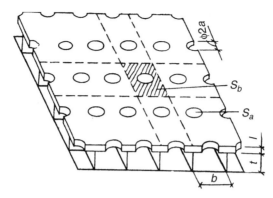

FIGURE 8.22 Resonators arranged on a raster to form a surface array.

Placing of flow-resistive materials in or behind the resonator opening reduces nonlinearity and sensitivity to grazing flow.

Spatial Average Impedance of Resonator Arrays

Locally reacting absorbers such as resonators are frequently used in the form of a surface array, shown in Fig. 8.22, where the individual resonators are arranged periodically. If the dimension b or $\sqrt{S_b}$ in Fig. 8.22 is much smaller than the acoustical wavelength, it is not necessary to take into account the interaction of the individual elements (resonators) with the sound field. To characterize the absorber, it is sufficient in this case to compute the spatial average wall impedance Z_1:

$$Z_1 = \frac{\langle p \rangle}{\langle v \rangle} = \frac{S_b}{S_a} \frac{p_a}{v_a} = \frac{S_b}{S_a} Z_R = \frac{Z_R}{\varepsilon} \quad \text{N} \cdot \text{s/m}^3 \qquad (8.65)$$

where Z_R is the impedance of the resonator as given in Eqs. (8.55) "measured" in the orifice of the resonator and $\varepsilon = S_a/S_b$ is the surface porosity. The effective wall impedance Z_1 is then used in Eq. (8.37) to determine the absorption coefficient. The impedances Z_m in Table 8.8 are effective impedances averaged over the plate surface (resonator mass impedance divided by porosity).

8.8 PLATE AND FOIL ABSORBERS

Plate absorbers are used in absorbing low-frequency tonal noise in situations where Helmholtz resonators are not feasible. The resonator consists of the mass of a thin plate and usually the stiffness of the air space between the plate and the elastic mounting at the perimeter of the plate. Foils are a special case of a thin limp plate.

Limp Thin Plate or Foil Absorber

The resonance frequency of a foil absorber is

$$f_0 = \frac{\sqrt{\rho_0 c_0^2 / m'' t}}{2\pi} \tag{8.66}$$

where t is the depth of the partitioned air space in meters and m'' is the mass per unit area of the limp foil in kilograms per square meter. If the air space depth is not small ($t > \frac{1}{8}\lambda$), there will be many resonance frequencies ω_n ($n = 0, 1, 2, \ldots$) given by

$$\frac{\omega_n t}{c_0} \tan \frac{\omega_n t}{c_0} = \frac{\rho_0 t}{m''} \tag{8.67}$$

For an unpartitioned air space behind the foil the resonance frequency depends on the angle of sound incidence θ_0. In this case t must be replaced in Eqs. (8.66) and (8.67) by $t \cos \theta_0$. If the air space is filled (without obstructing the movement of the foil) with a porous sound-absorbing material, then $(t/c_0)\Gamma_a''$ is used instead of t/c_0 in Eq. (8.67).

Foil-Wrapped Porous Absorber

Sound absorbers consisting of a protective layer of thin foil, a porous layer, and an air space behind are frequently used where the porous material must be protected from dust, dirt, and water. The effect of the thin foil can be taken into account by a series impedance $Z_s = j\omega m''$ that is added to the surface impedance Z_1 [see Eq. (8.37)]. It is essential that the foil is not stretched so that its inertia and not its membrane stress controls its response. Inserting a large-mesh (≥ 1 cm) wire cloth between the porous sound-absorbing material and the protective foil is a practical way to assure this.

Elastically Supported Stiff Plate

A form of resonant plate absorber that can be easily treated analytically is the elastically supported stiff plate. The elastic support may be localized by discrete points to occur along the perimeter in the form of a resilient gasket strip. The effective stiffness of such a resonator is

$$s'' = \frac{s_e}{S_p} + \frac{\rho_0 c_0^2}{t} \quad \text{N/m}^3$$

and the resonance frequency is

$$f_0 = \frac{1}{2\pi}\sqrt{\frac{s''}{m''}} = \frac{1}{2\pi}\sqrt{\frac{s_e/S_p + \rho_0 c_0^2/t}{m''}} \quad \text{Hz} \tag{8.68}$$

where $s_e = S_p \Delta p/\Delta x$ is the dynamic stiffness of resilient plate mounting and S_p is the surface area of the plate. Note that the elastic mounting increases the resonance frequency (compared with that obtainable with a limp foil of the same m''). This can be compensated for by an appropriate increase of m''. However, an increase of m'' increases the Q of the resonator ($Q = \sqrt{s''/m''}/R$) resulting in a narrower bandwidth.

Elastic Foil Absorber

Elastic foil absorbers consist of small air volumes enclosed in thin (200–400-μm) foil coffers. The coffers have typical dimensions of a few centimeters. A typical

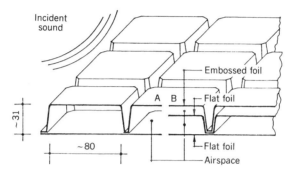

FIGURE 8.23 Construction of a foil absorber made of cold-drawn PVC foil.

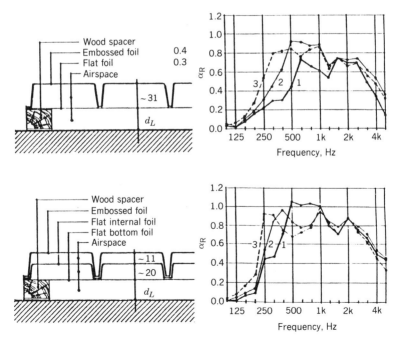

FIGURE 8.24 Random-incidence sound absorption coefficient α_R of absorbers made of cold-drawn PVC foil: (1) $d_L = 0$; (2) $d_L = 25$ mm; (3) $d_L = 50$ mm.

foil absorber is shown in Fig. 8.23. The wavelength of the incident sound is large compared with the typical dimensions of the coffers. Consequently, the incident sound periodically compresses the air in the individual coffers by exiting all of the volume-displacing modes of vibration of the various coffer walls. Because of the small thickness of the coffer walls, these resonances fall into the frequency range of 200–3150 Hz, which is of primary interest in building applications.[49] The foil material can be plastic (PVC) or metal. It is beneficial to emboss the foil because this leads to a more even distribution of the resonance frequencies. Figure 8.24 shows the measured random-incidence sound absorption coefficient obtained with two different foil absorber configurations, indicating that significant broadband sound absorption is achievable.

The key advantage of these foil absorbers is that they are nonporous, do not support bacterial growth, are lightweight, and can be made light transparent. Their main application is in breweries, packaging plants, hospitals, and computer chip manufacturing areas, where high emphasis is placed on hygiene and dust cannot be tolerated. They also lend themselves as sound absorbers in high-moisture environments such as swimming pools and as muffler baffles in cooling towers. Their light transparency is a distinct advantage in industrial halls. Considerable experience has been gained in such applications in Europe.[49]

REFERENCES

1. K. U. Ingard, *Notes on Sound Absorption Technology*, Version 94-02, Noise Control Foundation, Poughkeepsie, NY, October 1994.
2. F. P. Mechel, *Schall Absorber [Sound Absorbers]*, Vol. 1: *Aeussere Schallfelder, Wechselwirkungen [Outer Sound Fields, Interactions]*; Vol. 2: *Innere Schallfelder, Structuren [Internal Sound Fields, Structures]*; Vol. 3: *Anwendungen [Application]*, S. Hirzel Verlag, Stuttgart, 1989.
3. F. P. Mechel, *MAPS-Mechels Acoustic Program System*, CD-ROM for Windows, S. Hirzel, Stuttgart, Germany, 2001.
4. H. Kutruff, *Room Acoustics*, Applied Science, London, 1973.
5. L. L. Beranek, *Acoustic Measurements*, Acoustical Society of America, Melville, New York, 1988.
6. R. D. Godfrey, Owens Corning Fiberglass Company, private communication, June 2004.
7. T. Northwood, *J. Acoust. Soc. Am.*, **31**, 596 (1959); **35**, 1173 (1963).
8. M. J. Swift, "The Physical Properties of Porous Recycled Materials," Ph. D. Thesis, University of Bradford, United Kingdom, 2000.
9. L. Cremer and M. Hubert, *Vorlesungen Ueber Technische Akustik [Lectures on Technical Acoustics]*, 3rd ed., Springer, Berlin, 1985.
10. Y. Champoux, M. R. Stinson, and G. A. Daigle, "Air-Based System for the Measurement of Porosity," *J. Acoust. Soc. Am.*, **89**, 910–916 (1991).

11. P. Leclaire, O. Umnova, K. V. Horoshenkov, and L. Maillet, "Porosity Measurement by Comparison of Air Volumes," *Rev. Sci. Instrum.*, **74**, 1366–1370 (2003).
12. Z. E. A. Fellah, S. Berger, W. Lauriks, C. Depollier, C. AristBgui, and J.-Y. Chapelon, "Measuring the Porosity and the Tortuosity of Porous Materials via Reflected Waves at Oblique Incidence," *J. Acoust. Soc. Am.*, **113**, 2424–2433 (2003).
13. Z. E. A. Fellah, S. Berger, W. Lauriks, C. Depollier, and M. Fellah, "Measuring the Porosity of Porous Materials Having a Rigid Frame via Reflected Waves: A Time Domain Analysis with Fractional Derivatives," *J. Appl. Phys.*, **93**, 296–303 (2003).
14. Z. E. A. Fellah, S. Berger, W. Lauriks, C. Depollier, and J. Y. Chapelon, "Inverse Problem in Air-Saturated Porous Media via Reflected Waves," *Rev. Sci. Int.*, **74**, 2871–2879 (2003).
15. Z. E. A. Fellah, S. Berger, W. Lauriks, C. Depollier, W. Lauriks, P. Trompette, and J. Y. Chapelon, "Ultrasonic Measurement of the Porosity and Tortuosity of Air-Saturated Random Packings of Beads," *J. Appl. Phys.* to appear.
16. A. S. Sangani and A. Acrivos, "Slow Flow through Periodic Array of Spheres," *Int. J. Multiphase Flow*, **8**, 343–360 (1982).
17. W. S. Jodrey and E. M. Tory, "Computer-Simulation of Close Random Packing of Equal Spheres," *Phys. Rev. A*, **32**, 2347–2351 (1985).
18. N. Sen, C. Scala, and M. H. Cohen, "A Self-Similar Model for Sedimentary Rocks with Application to the Dielectric Constant of Fused Glass Beads," *Geophysics*, **46**, 781–795 (1981).
19. K. Attenborough, "Acoustical Characteristics of Rigid Absorbents and Granular Media," *J. Acoust. Soc. Am.*, **83**(3), 785–799 (March 1983).
20. O. Umnova, K. Attenborough, and K. M. Li, "Cell Model Calculations of Dynamic Drag Parameters in Packings of Spheres," *J. Acoust. Soc. Am.*, **107**(3), 3113–3119 (2000).
21. J. F. Allard, *Propagation of Sound in Porous Media*, Elsevier Applied Science, London, 1993.
22. J. F. Allard, B. Castagnede, M. Henry, and W. Lauriks, "Evaluation of Tortuosity in Acoustic Porous Materials Saturated by Air," *Rev. Sci. Instrum.*, **65**, 754–755 (1994).
23. ASTM C 522-87, *"Standard Test Method for Airflow Resistance of Acoustical Materials,"* American Society for Testing and Materials, West Conshohocken, PA, pp. 169–173.
24. M. R. Stinson and G. A. Daigle, "Electronic System for the Measurement of Flow Resistance," *J. Acoust. Soc. Am.*, **83**, 2422–2428 (1988).
25. O. Umnova, University of Salford, United Kingdom, September, 2004. private communication.
26. P. Leclaire, M. J. Swift, and K. V. Horoshenkov. "Determining the Specific Area of Porous Acoustic Materials from Water Extraction Data," *J. Appl. Phys.*, **84**, 6886–6890 (1998).
27. K. Attenborough, "Models for the Acoustical Properties of Air-Saturated Granular Media," *Acta Acustica*, **1**, 213–226 (1993).
28. K. V. Horoshenkov, K. Attenborough, and S. N. Chandler-Wilde, "Pade Approximants for the Acoustical Properties of Rigid Frame Porous Media with Pore Size Distribution," *J. Acoust. Soc. Am.*, **104**, 1198–1209 (1998).

29. K. V. Horoshenkov and M. J. Swift, "The Acoustic Properties of Granular Materials with Pore Size Distribution Close to Log-normal," *J. Acoust. Soc. Am.*, **110**, 2371–2378 (2001).
30. D. L. Johnson, J. Koplik, and R. Dashen, "Theory of Dynamic Permeability and Tortuosity in Fluid-Saturated Porous Media," *J. Fluid Mech.*, **176**, 379–402 (1987).
31. O. Umnova, K. Attenborough, and K. M. Li, "A Cell Model for the Acoustical Properties of Packings of Spheres," *Acustica* combined *with Acta Acustica*, **87**, 26–235 (2001).
32. J. F. Allard and Y. Champoux, "New Empirical Relations for Sound Propagation in Rigid Frame Fibrous Materials," *J. Acoust. Soc. Am.*, **91**, 3346–3353 (1992).
33. N. N. Voronina and K. V. Horoshenkov, "A New Empirical Model for the Acoustic Properties of Loose Granular Media," *Appl. Acoust.*, **64**(4), 415–432 (2003).
34. J. Y. Chung and D. A. Blaser, "Transfer Function Method of Measuring In-Duct Acoustic Properties. I. Theory, II Experiment," *J. Acoust. Soc. Am.*, **68**(3), 907–921 (1980).
35. C. D. Smith, and T. L. Parrott, "Comparison of Three Methods for Measuring Acoustic Properties of Bulk Materials," *J. Acoust. Soc. Am.*, **74**(5), 1577–1582 (1983).
36. H. Utsuno et al., "Transfer Function Method for Measuring Characteristic Impedance and Propagation Constant of Porous Materials," *J. Acoust. Soc. Am.*, **86**(2), 637–643 (1989).
37. P. Mechel, *Akustische Kennwerte von Faserabsorbem [Acoustic Parameters of Fibrous Sound Absorbing Materials]*, Vol. I, *Bericht* BS 85/83; Vol. II, *Bericht* BS 75/82, Fraunhofer Inst. Bauphysik, Stuttgart, Germany.
38. M. E. Delany and E. N. Bazley, "Acoustical Properties of Fibrous Absorbent Materials," *Appl. Acoust.*, **3**, 105–116 (1970).
39. Y. Miki, "Acoustical Properties of Porous Materials—Modifications of Delany-Bazley Models," *J. Acoust. Soc. Jpn. (E)*, **11**, 19–24 (1990).
40. M. Garai and F. Pompoli, "A Simple Empirical Model of Polyester Fibre Materials for Acoustical Applications," *Appl. Acoust.* (in press).
41. A. Cummings and S. P. Beadle, "Acoustic Properties of Reticulated Plastic Foams," *J. Sound Vib.*, **175**(1), 115–133 (1993).
42. Q. Wu, "Empirical Relations between Acoustical Properties and Flow Resistivity of Porous Plastic Open-Cell Elastic Foams," *Appl. Acoust.*, **25**, 141–148 (1988).
43. L. P. Dunn and W. A. Davern "Calculation of Acoustic Impedance of Multilayer Absorbers," *Appl. Acoust.*, **19**, 321–334 (1986).
44. R. J. Astley and A. Cummings, "A Finite Element Scheme for Attenuation in Ducts Lined with Porous Material: Comparison with Experiment," *J. Sound Vib.*, **116**, 239–263 (1987).
45. J. Royar, "Untersuchungen zum Akustishen Absorber-Kanten-Etfekt an einem zwei-dimensionalen Modell" ["Investigation of the Acoustic Edge-Effect Utilizing a Two-Dimensional Model"], Ph.D. Thesis, Faculty of Mathematics and Nature Sciences, University of Saarbrueken, Germany, 1974.
46. F. P. Mechel, "Design Charts for Sound Absorber Layers," *J. Acoust. Soc. Am.*, **83**(3), 1002–1013 (1988).

47. F. Mechel, "Sound Absorbers and Absorber Functions," in G. L. Osipova and E. J. Judina, (Eds.), *Reduction of Noise in Buildings and Inhabited Regions* (in Russian), Strojnizdat, Moscow, 1987.
48. J. L. B. Coelho, "Acoustic Characteristics of Perforate Liners in Expansion Chambers," Ph.D. Thesis, Institute of Sound Vibration, Southampton, 1983.
49. F. Mechel and N. Kiesewetter, "Schallabsorberaus Kunstotf-Folie" ["Plastic-Foil Sound Absorber"], *Acustica*, **47**, 83–88 (1981).

CHAPTER 9
Passive Silencers

M. L. MUNJAL

Indian Institute of Science
Bangalore, India

ANTHONY G. GALAITSIS

BBN Systems and Technologies Corporation
Cambridge, Massachusetts

ISTVÁN L. VÉR

Consultant in Acoustics, Noise, and Vibration Control
Stow, Massachusetts

9.1 INTRODUCTION

Chapter Organization

This chapter is organized in seven sections. The present section reviews some of the significant milestones in silencer design, citing relevant references. Section 9.2 reviews the acoustical criteria commonly used to determine silencer performance. Section 9.3 discusses the decomposition of a silencer into basic acoustical transmission elements, presents transmission matrix expressions for some elements used in the Section 9.4 examples, and provides references for additional elements not explicitly covered in this chapter. Section 9.4 discusses the predicted performance of several silencer designs and provides selected examples of designs aimed at achieving specific acoustical goals. Section 9.5 provides a more qualitative discussion of perforated-element mufflers along with a few muffler design examples. Section 9.6 considers dissipative silencers most widely used to attenuate broadband noise in ducts with gas flow with a minimum of pressure drop across the silencer. Section 9.7 discusses designs that combine reactive silencer low-frequency tonal performance with passive silencer high-frequency broadband performance. A reader solely interested in designing a silencer for a specific application may proceed directly to Sections 9.3–9.7.

Background Information

The noise generated by air/gas handling/consuming equipment, such as fans, blowers, and internal combustion engines, is controlled through the use of two types of devices: (1) passive silencers and lined ducts whose performance is a function of the geometric and sound-absorbing properties of their components and (2) active noise control silencers whose noise cancellation features are controlled by various electromechanical feed-forward and feedback techniques. This chapter focuses on passive silencers while active noise control is discussed in Chapter 17 and lined ducts are covered in Chapter 16.

In the remainder of this chapter, the term *silencer* is often used generically to refer to any type of passive noise control device. Specific names are introduced primarily in cases where the use of "silencer" may lead to ambiguities.

Over the last four decades silencers have been the focus of many research programs that improved the understanding of the basic phenomena and resulted in more accurate design methods. The groundwork for the behavior of lined ducts with no flow was laid even earlier by Morse[1] and Cremer,[2] and the first systematic evaluation of mufflers with no flow was conducted by Davis et al.[3] Those works were followed by numerous theoretical and experimental investigations which addressed additional important issues such as uniform flow, temperature gradient, and the behavior of new silencer components that are common in modern applications. Reviews of the major accomplishments in this area have been presented periodically in the form of individual articles in professional journals, chapters in engineering handbooks,[4-7] and, more recently, books by Munjal[8,9] solely devoted to mufflers.

The analytical work on this subject benefited substantially from the early adaptation of the transfer matrix approach to silencer modeling. This method, which was promoted for the description of mechanical systems early on,[10] was used along with electrical analogs to describe the behavior of basic silencer components[11] for plane-wave sound propagation in the absence of flow. Subsequent investigations began addressing the effects of flow on the response of elements consisting of area discontinuities[12,13] and generating transfer matrix models for silencer elements. The predicted performance of various reactive silencers using transfer matrices with convective and dissipative effects in the presence of flow agreed well with measured data.[14]

Concurrently, systematic studies were initiated on the behavior of perforated plates (perforates) to take advantage of their dissipative properties[15-17] for general applications in the transportation industry. Such perforates are used in automotive applications as part of the two- and three-pipe elements featuring one and two perforated pipes, respectively, contained within a larger diameter rigid-wall cylindrical cavity. Such configurations were investigated[18-23] by combining an orifice model with the transfer matrices of the axially segmented perforated tube or tubes and the unpartitioned cavity.

At higher frequencies or for large mufflers, three-dimensional effects become significant. The approaches used in three-dimensional silencer analysis include the finite-element, boundary-element, and acoustical-wave finite-element methods. Each has some inherent advantages at higher frequencies where higher order modes will start propagating, but on the average they are much more complex and time consuming to implement. Therefore, the plane-wave transfer matrix method is presently the most widely used approach, particularly for synthesis of an initial configuration of a passive silencer.[24-26]

Experimental work conducted in parallel with the analysis validated the developed models and pointed out areas requiring further investigation. Methods for the direct measurement of transfer matrices were developed[27-31] to validate/modify the models of existing silencer elements. Substantial attention was also paid to the source impedance, which was expected to influence the insertion loss of a silencer and the net radiated acoustical power of the system. The source impedance of a six-cylinder engine was measured with an impedance tube using pure tones,[32,33] and the impedance of electroacoustical drivers and multicylinder engines was subsequently measured[34-37] using the much quicker two-microphone methods.[38-40] More recent indirect methods characterized the source impedance through measurements using two, three, or four different acoustical loads.[41-46]

The research performed and the knowledge developed in the field of passive silencers are very extensive and cannot be adequately covered in a single chapter. Accordingly, this chapter reviews the basic background information, summarizes the methodologies for silencer design, demonstrates these approaches through selected examples, and discusses additional topics and references useful to general applications. A reader wishing to acquire more information on internal combustion engine silencers is urged to review references 47–50 for a summary of a general automotive silencer development approach and of typical predicted and measured results, references 51 and 52 for condensed qualitative reviews of the accomplishments in this field, and references 8 and 9 for an extensive bibliography on and a detailed discussion of most silencer-related topics.

9.2 SILENCER PERFORMANCE METRICS

The majority of noise sources have an intake and an exhaust and, generally, they require both intake and exhaust silencers. The two cases are characterized by different flow direction, back pressure, average gas temperature, and average sound pressure levels, but the corresponding hardware is developed using the same principles and design methods. The following discussion is tailored to exhaust silencers but it is equally applicable to intake silencers.

Silencer Selection Factors

The use of a silencer is prompted by the need to reduce the radiated noise of a source but, in most applications, the final selection is based on trade-offs between the predicted acoustical performance, mechanical performance, volume/weight, and cost of the resulting system.

During development, the acoustical performance (insertion loss) of a candidate silencer is determined from the free-field sound pressure levels measured at the same relative locations with respect to the noise outlet of the unsilenced and silenced sources. If the noise outlet is inserted into a reverberant space, then the difference can be measured anywhere within the reverberant field, and the difference sound pressure level represents the power-based insertion loss.

282 PASSIVE SILENCERS

The impact of the silencer on the mechanical performance of the source is determined from the change in the silencer back pressure. For a continuous-flow source, such as a fan or a gas turbine, the impact is determined from the increase in the average back pressure; by contrast, for an intermittent-flow source, such as a reciprocating engine, the impact is a function of the increase in the exhaust manifold pressure when an exhaust valve is open.

Most silencers are subject to volume/weight constraints, which also influence the silencer design process. In addition, the initial purchase/installation cost and the periodic maintenance cost are other important factors that influence the silencer selection process.

Since noise is the root cause of silencer design, the remainder of this chapter focuses on the prediction of a silencer's acoustical performance and also on the estimation of back pressure. The additional performance criteria associated with hardware volume, weight, and cost are trade-off parameters to be determined separately for each application.

Factors Influencing Acoustical Performance

Reactive and Dissipative Silencers. The net change ΔW in the acoustical power radiated from a source can be expressed as

$$\Delta W = W_1 - W_2 = W_1 - (W_1' - W_d) = (W_1 - W_1') + W_d \quad \text{N} \cdot \text{m/s} \quad (9.1)$$

where W_1 and W_2 correspond to the unsilenced and silenced sources (Fig. 9.1), $W_1 - W_1'$ is the net change in the acoustical power output of the source resulting from changes in the silencer's reflection coefficient, and W_d is attributable to the dissipative properties of the silencer.

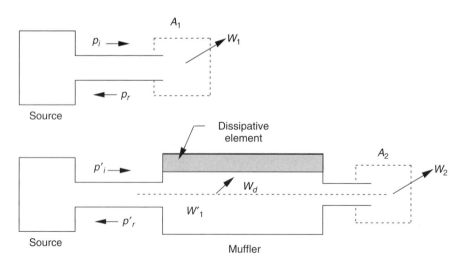

FIGURE 9.1 Radiated noise reduction mechanisms of a silencer.

The reflection coefficient is manipulated primarily by introducing cross-sectional discontinuities in the piping system, and the dissipated part W_d depends largely on the dissipative properties of the silencer elements. In practice, all silencer elements reflect and dissipate acoustical energy to some extent. However, traditionally, they have been categorized into reactive and dissipative silencers when their insertion loss is dominated by reflective and dissipative mechanisms, respectively. This convention has also been adopted in this chapter, but the reader should be aware that most common applications involve various degrees of overlap between the two extremes.

Flanking Paths and Secondary Source Mechanisms. The acoustical power radiated from a silenced source, such as an engine, includes airborne noise (W_{EX}) from the exhaust pipe outlet, silencer shell noise (W_{SH}) radiated from the vibrating walls of the silencer and exhaust line pipes, and a contribution (W_{AD}) from additional sources, such as source shell and source intake, located upstream of the silencer.

The shell noise results from the excitation of silencer components by the vibrating engine, by the intense internal acoustical pressure fields, and by aerodynamic forces. These contributions can be reduced substantially through various methods, but they cannot be totally eliminated and, eventually, they set the limit for the achievable insertion loss for high-performance silencers.

The acoustical power contributions W_{SH} and W_{AD} are excluded from further consideration in the remainder of the chapter; that is, the discussion is focused on the impact of silencers on the airborne noise contributions W_{EX}. However, one should be aware of their existence and significance, particularly in experimental applications where limited resources may lead to the measurement of W_{EX} in the presence of W_{SH} and W_{AD}.

Silencer System Modeling

Silencer Component Representation. The basic components of a typical silencer system comprise the noise source, silencer, connecting pipes, and surrounding medium (Fig. 9.2a). The pipe connecting the source to the silencer is treated as part of the source. Figure 9.2b shows the corresponding electrical analog of the acoustical system, which is widely used to facilitate the representation and handling of acoustical transmission lines. This analog model, which uses acoustical pressure p and mass velocity $\rho_0 S u$ in kg/s, instead of voltage and current, represents the noise source with a source pressure p_s and an internal impedance Z_s, the silencer and pipe segment with four-pole elements (T_{ij} and D_{ij}), and the surrounding medium with a termination (or radiation load) impedance Z_T. Analytical expressions for the transfer matrices of pipes and selected silencer elements are presented in references 8 and 9. Some of them are given later in Section 9.3.

Let p_i, u_i, $i = 1, 2, 3$, designate the acoustical pressure and particle velocity at the interfaces of acoustical components of the source–silencer–load system illustrated in Fig. 9.2b. These quantities may be obtained by solving the system

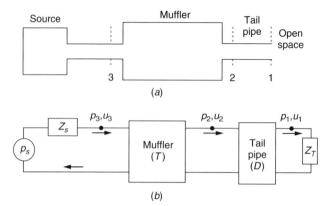

FIGURE 9.2 (*a*) Acoustical and (*b*) electrical analog components of a silencer.

of equations describing the response of the components depicted in Fig. 9.2*b*, namely,

$$p_s = p_3 + \rho_0 Z_s S_3 u_3 \tag{9.2}$$

$$\begin{bmatrix} p_3 \\ \rho_0 S_3 u_3 \end{bmatrix} = \begin{bmatrix} T_{11} & T_{12} \\ T_{21} & T_{22} \end{bmatrix} \begin{bmatrix} p_2 \\ \rho_0 S_2 u_2 \end{bmatrix} \tag{9.3}$$

$$\begin{bmatrix} p_2 \\ \rho_0 S_2 u_2 \end{bmatrix} = \begin{bmatrix} D_{11} & D_{12} \\ D_{21} & D_{22} \end{bmatrix} \begin{bmatrix} p_1 \\ \rho_0 S_1 u_1 \end{bmatrix} \tag{9.4}$$

$$p_1 = Z_T \rho_0 S_1 u_1 \tag{9.5}$$

where ρ_0 is the mean gas density and S_i is the duct cross-sectional area at the ith location. The calculated quantities p_i, $\rho_0 S_i u_i$, $i = 1, 2, \ldots$, can then be combined with the selected performance criterion to determine the effectiveness of the silencer.

Acoustical Performance Criteria. The most frequently used performance criteria are the insertion loss (IL), noise reduction (NR), and transmission loss (TL). All three use a sound-pressure-level difference as a performance indicator; therefore, they require no explicit knowledge of the source strength p_s, of Eq. (9.2). The number of required system parameters (source impedance Z_s, transfer matrix elements D_{ij}, and termination impedance Z_T*) other than the silencer matrix T_{ij} can be further reduced by the specific choice of the performance criterion.

The IL is defined as the change in the radiated sound pressure level resulting from the insertion of the muffler, that is, the replacement of a pipe segment of length l_1 (Fig. 9.3*a*) located downstream from the source by the silencer and a new tail-pipe segment of length l_2 (Fig. 9.3*b*). Insertion loss is expressed as

$$\text{IL} = L_b - L_a = 20 \log \left| \frac{p_b}{p_a} \right| \tag{9.6}$$

*The impedance Z in this chapter is defined as $p_i/(\rho_0 S_i u_i)$.

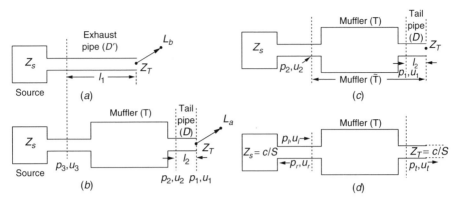

FIGURE 9.3 Quantities use for the determination of (*a,b*) insertion loss, (*c*) noise reduction, and (*d*) transmission loss.

where p_b, p_a and L_b, L_a are the measured sound pressures and sound pressure levels at the same relative location (distance and orientation) with respect to the exhaust outlet (Fig. 9.3*a*) before and after the installation of the muffler and "log" refers to a logarithm with respect to the base 10. Mathematically,

$$\text{IL} = 20 \, \log \left| \frac{\tilde{T}_{11} Z_T + \tilde{T}_{12} + \tilde{T}_{21} Z_s Z_T + \tilde{T}_{22} Z_s}{D_{11} Z_T + D_{12} + D_{21} Z_s Z_T + D_{22} Z_s} \right| \qquad (9.7)$$

where $\tilde{T}_{ij}$ is the combined transfer matrix element of the silencer and its tail pipe ($\tilde{T} = TD$), and D'_{ij} is the transfer matrix element of the replaced exhaust pipe segment. If the silencer is simply added to the end of the source ($l_1 = l_2 = 0$ in Figs. 9.3*a* and *b*), then both D' and D are identity matrices (see Transfer Matrices for Reactive Elements in Section 9.3) and Eq. (9.7) is reduced to

$$\text{IL}(l_1 = l_2 = 0) = 20 \, \log \left| \frac{T_{11} Z_r + T_{12} + T_{21} Z_r Z_s + T_{22} Z_s}{Z_T + Z_s} \right| \qquad (9.8)$$

The noise reduction is given by

$$\text{NR} = L_2 - L_1 = 20 \, \log \left| \frac{p_2}{p_1} \right| \qquad (9.9)$$

that is, it is the difference between the sound pressure levels measured at two points 1 and 2 (Fig. 9.3*c*) of the silenced system located upstream and downstream of the silencer, respectively. If D_{ij} and $\tilde{T}_{ij}$ represent the transfer matrix elements for the silencer portions downstream of points 1 and 2, respectively, the noise reduction is given by

$$\text{NR} = 20 \, \log \left| \frac{\tilde{T}_{11} Z_T + \tilde{T}_{12}}{D_{11} Z_T + D_{12}} \right| \qquad (9.10)$$

The TL is the acoustical power-level difference between the incident and transmitted waves of an anechoically terminated silencer (see Fig. 9.3d). In terms of the corresponding transfer matrix, the transmission loss is given by

$$\text{TL} = 20 \, \log \left| \frac{T_{11} + (S/c)T_{12} + (c/S)T_{21} + T_{22}}{2} \right| \quad (9.11)$$

where c is the speed of sound at design temperature and S is the cross-sectional area of the reference pipe.

This expression assumes that (a) the area of cross section of the exhaust pipe upstream of the silencer is the same as that of the tailpipe, that is, S; (b) the silencer is simply added at the end of the source ($l_1 = l_2 = 0$); and (c) the tail pipe end (or radiation end) is anechoic ($Z_T = c/S$).

The IL is the most appropriate indicator of a silencer's performance because it is the level difference of the acoustical power radiated from the unsilenced and silenced systems. It is easy to quantify from pre- and postsilencing data, but it is hard to predict because it depends on Z_s and Z_T, which vary from one application to another. By contrast, the TL is easy to predict but is only an approximation of the silencer's actual performance because it does not account for the source impedance and it models all silencer outlets with anechoic terminations. Equations (9.7) and (9.11) show that the TL and IL of a silencer become identical when the noise source and silencer termination are anechoic, that is, $Z_s = Z_T = c/S$. Similarly, it can readily be shown[8] that for a constant-pressure source ($Z_s = 0$), the NR and IL become identical, provided the cross-sectional area of the exhaust pipe upstream of the silencer is the same as that of the tailpipe.

The ultimate selection of an evaluation criterion is based on trade-offs between the desired accuracy in the predictions and the amount of available resources. For example, Z_s, required for IL predictions, can be determined experimentally through various methods,[32–46] but the procedure is too costly for the majority of applications. As a result, the silencer design is, generally, based on predicted TL, with a clear understanding of the associated approximations, while the final evaluation of the hardware during field tests is based on the measured IL.

9.3 REACTIVE SILENCER COMPONENTS AND MODELS

Reactive silencers consist typically of several pipe segments that interconnect a number of larger diameter chambers. These silencers reduce the radiated acoustical power primarily through impedance mismatch, that is, through the use of acoustical impedance discontinuities to reflect sound back toward the source. In essence, the more pronounced the discontinuities, the higher the amount of reflected power. Acoustical impedance discontinuities are commonly achieved through (a) sudden cross-sectional changes (i.e., expansions or contractions), (b) wall property changes (i.e., transition from a rigid-wall pipe to an equal-diameter absorbing wall pipe), or (c) any combination thereof.

Silencer Representation by Basic Silencer Elements

Every silencer can be divided into a number of segments or elements each represented by a transfer matrix. The transfer matrices can then be combined to obtain the system matrix, which may then be substituted into Eqs. (9.7), (9.10), or (9.11) to predict the corresponding acoustical performance for the silencer system.

The procedure is illustrated by considering the silencer of Fig. 9.4, which is divided into the basic elements, labeled 1–9, indicated by the dashed lines. Elements 1, 3, 5, 7, and 9 are simple pipes of constant cross section. Element 2 is a simple area expansion, element 4 is an area contraction with an extended outlet pipe, element 6 is an area expansion with an extended inlet pipe, and element 8 is a simple area contraction. The nine elements are characterized by the transfer matrices $T^{(1)}$ through $T^{(9)}$; therefore, the system matrix $T^{(S)}$ for the entire silencer is obtained from

$$T^{(S)} = T^{(1)} \cdot T^{(2)} \cdot \cdots \cdot T^{(9)} \qquad (9.12)$$

through matrix multiplication. The matrices for each of the above elements may be derived from the formulas presented later in this section.

A few of the most common reactive muffler elements are discussed in this section and some are used in illustrative design examples. Additional elements that were the subject of previous investigations include gradually varying area ducts (such as horns), uniform-area compliant-wall ducts (such as hoses), Quincke tubes, inline cavities, inline bellows, catalytic converter elements, branch sub systems, side inlets/outlets, cavity-backed resistive-wall elements, and perforated-duct elements with and without mean flow through the perforations. Transfer matrices for each of these elements have been derived over the last three decades by different researchers,[53–70] and explicit expressions for their four-pole parameters are given in detail in references 8 and 9.

The variety of silencer elements and the multitude of elements per silencer result in numerous silencer configurations; therefore, a comprehensive study of all possible silencer applications is beyond the scope of this work. For this

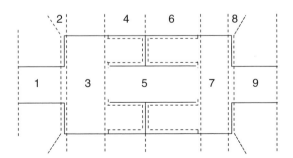

FIGURE 9.4 Decomposition of a silencer into basic elements.

reason, the remainder of this chapter uses selected silencer configurations to demonstrate

(a) the typical process for modeling a basic silencer element with a (four-pole parameter) transmission matrix,
(b) the process for predicting the acoustical performance of a given silencer configuration from the properties (transmission matrices) of its constituent elements, and
(c) the process for selecting a silencer design to achieve approximately a specified acoustical performance.

Furthermore, Section 9.5 provides additional information and sketches for perforated-tube elements and analyzes the performance of selected perforated-tube muffler configurations. The reader could also apply the demonstrated design process to any other silencer configurations by obtaining and utilizing the relevant transmission matrix information from the corresponding references 8, 9, 53–69.

Transfer Matrices for Reactive Elements

The transfer matrix of a silencer element is a function of the element geometry, state variables of the medium, mean flow velocity, and properties of duct liners, if any. The results presented below correspond to the linear sound propagation of a plane wave in the presence of superimposed flow. In certain cases, the matrix may also be influenced by nonlinear effects, higher order modes, and temperature gradients; these latter effects, which can be included in special cases, are discussed qualitatively later in this section, but they are excluded from the analytical procedure described below. The following is a list of variables and parameters that appear in most transfer matrix relations of reactive elements:

p_i = acoustical pressure at ith location of element
u_i = particle velocity at ith location of element
ρ_0 = mean density of gas, kg/m^3
c = sound speed m/s, = $331\sqrt{\theta/273}$
θ = absolute temperature, K = °C + 273
S_i = cross section of element at ith location, m^2
$Y_i = c/S_i$
A, B = amplitudes of right- and left-bound fields
$k_c = k_0/(1 - M^2)$ assuming negligible frictional energy loss along straight pipe segments
$k_0 = \omega/c = 2\pi f/c$
$\omega = 2\pi f$
f = frequency, Hz
$M = V/c$
V = mean flow velocity through S
T_{ij} = (ij)th element of transmission matrix or transfer matrix

Symbols without subscripts, such as V, c, and M, describe quantities associated with the reference duct.

Pipe with Uniform Cross Section. The acoustical pressure and mass velocity fields p, $\rho_0 S u$ in a pipe with uniform cross section S and a mean flow V_0 are given by

$$p(x) = (Ae^{-jk_c x} + Be^{jk_c x})e^{j(Mk_c x + \omega t)} \tag{9.13a}$$

$$\rho_0 S u(x) = (Ae^{-jk_c x} - Be^{jk_c x})\frac{e^{j(Mk_c x + \omega t)}}{Y} \tag{9.13b}$$

where S is the cross section of the pipe. Equations (9.13a) and (9.13b) can be evaluated at $x = 0$ and $x = l$ to obtain the corresponding fields p_2, $\rho_0 S u_2$ and p_1, $\rho_0 S u_1$, respectively. Upon elimination of the constants A and B, one obtains[8,13]

$$\begin{bmatrix} p_2 \\ \rho_0 S u_2 \end{bmatrix} = \begin{bmatrix} T_{11} & T_{12} \\ T_{21} & T_{22} \end{bmatrix} \begin{bmatrix} p_1 \\ \rho_0 S u_1 \end{bmatrix} \tag{9.14}$$

where the transmission matrix T_{pipe} is given by

$$T_{\text{pipe}} = \begin{bmatrix} T_{11} & T_{12} \\ T_{21} & T_{22} \end{bmatrix}_{\text{pipe}} = e^{-jMk_c l} \begin{bmatrix} \cos k_c l & jY_0 \sin k_c l \\ \frac{j}{Y} \sin k_c l & \cos k_c l \end{bmatrix} \tag{9.15}$$

In the transfer matrix of Eq. (9.15) the acoustical energy dissipation that may result from friction between the gas and the rigid wall as well as from turbulence is neglected. These effects, which would result in a slightly different matrix,[8] may be noticeable in very long exhaust systems but they are negligible for most silencer applications.

Cross-Sectional Discontinuities. The transition elements used to model most cross-sectional discontinuities are shown in Figs. 9.5b,c,e,f and in the first column of Table 9.1. Using decreasing element-subscript values with distance from the noise source, the cross-sectional areas upstream, at, and downstream of the transition (S_3, S_2, and S_1) are related through[8]

$$C_1 S_1 + C_2 S_2 + S_3 = 0 \tag{9.16}$$

where the constants C_1 and C_2 (Table 9.1) are selected so as to satisfy the compatibility of the cross-sectional areas across the transition.

For each configuration, Table 9.1 also shows the pressure loss coefficient K, which accounts for the conversion of some mean-flow energy and acoustical field energy into heat at the discontinuities. As indicated, $K \leq 0.5$ for area contractions, while $K \to (S_1/S_3)^2$ for area expansions at large values of S_1/S_3.

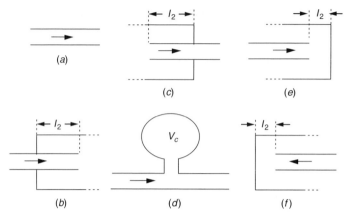

FIGURE 9.5 Basic silencer element: (*a*) plain pipe; (*b,c*) extended tube inlet/outlet; (*d*) resonator; (*e,f*) reversal expansion/contraction.

TABLE 9.1 Parameter Values of Transition Elements

Element Type	C_1	C_2	K
$S_3 \to S_1$, S_2, L_2	-1	-1	$\dfrac{1 - S_1/S_3}{2}$
$S_3 \to S_1$, S_2, L_2	-1	1	$\left(\dfrac{S_1}{S_3} - 1\right)^2$
S_1, S_2, S_3, L_2	1	-1	$\left(\dfrac{S_1}{S_3}\right)^2$
S_3, S_1, S_2, L_2	1	-1	0.5

Transfer matrices for cross-sectional discontinuities (csd) in the presence of mean flow that include terms proportional up to the fourth power (M^4) of the Mach number are presented in reference 8. However, in most silencer design applications, $M \ll 1$; therefore, terms of the form $1 + M^n, n \geq 2$, are set to unity. This simplification reduces the matrix T_{csd}, which relates the upstream and downstream acoustical fields p_3, $\rho S u_3$ and p_1, $\rho S u_1$ through

$$\begin{bmatrix} p_3 \\ \rho_0 S_3 u_3 \end{bmatrix}_{\text{upstream}} = T_{\text{csd}} \begin{bmatrix} p_1 \\ \rho_0 S_1 u_1 \end{bmatrix}_{\text{downstream}} \tag{9.17}$$

to

$$[T_{\text{csd}}] = \begin{bmatrix} 1 & kM_1 Y_1 \\ \dfrac{C_2 S_2}{C_1 S_2 Z_2 + S_2 M_3 Y_3} & \dfrac{C_2 S_2 Z_2 - M_1 Y_1 (C_1 S_1 + S_3 K)}{C_2 S_2 Z_2 + S_3 M_3 Y_3} \end{bmatrix} \tag{9.18}$$

where $Z_2 = -j(c/S_2) \cot k_0 l_2$

l_2 = length of extended inlet/outlet pipe, m

(9.19)

Letting length l_2 tend to zero (as in the case of sudden expansion and contraction shown in Fig. 9.5) yields the transfer matrix

$$\begin{bmatrix} 1 & KM_1 Y_1 \\ 0 & 1 \end{bmatrix} \tag{9.20}$$

Resonators. A resonator is a cavity-backed opening in the sidewall of a pipe Fig. (9.5d). The opening may consist of a single hole on the pipe wall (Fig. 9.6a) or a closely distributed group of holes (Fig. 9.6b). The volume behind this opening can comprise a throat of length l_t and cross section S_n terminated by a straight pipe of cross-sectional area S_c and depth l_c (Fig. 9.6c), a concentric cylinder extending a length l_u and l_d upstream and downstream of the opening (Fig. 9.6d), an odd-shaped chamber of total volume V_c (Fig. 9.6e), or an extended-tube (quarter-wave) resonator of length l_2 (Figs. 9.5b,c,e,f).

The resonator opening in the wall of the main duct is assumed to be well localized; that is, the axial dimension of the perforated section is much smaller than the wavelength and typically smaller than a duct diameter. This requirement ensures that the duct–cavity interaction is in phase over the entire surface of the connecting opening. Transmission matrices for multihole openings with substantial axial dimensions requiring modeling by perforated-tube elements may be found in references 8 and 9.

The transfer matrix for a resonator for a stationary medium is given by[9]

$$T_r = \begin{bmatrix} 1 & 0 \\ \dfrac{1}{Z_r} & 1 \end{bmatrix} \tag{9.21}$$

where $Z_r = Z_t + Z_c$, Z_t is the impedance of the throat connecting the pipe to the cavity, and Z_c is the impedance of the cavity. Here, Z_c is independent of the flow in the main duct and is given by one of the following expressions:

$$Z_c = \begin{cases} Z_{tt} = -j\dfrac{c}{S_c} \cot k_0 l_c \\ Z_{cc} = -j\dfrac{c}{S_c} \dfrac{1}{\tan k_0 l_u + \tan k_0 l_d} \\ z_{gv} = -j\dfrac{c}{k_0 V_c} \end{cases} \quad (\text{m} \cdot \text{s})^{-1} \tag{9.22}$$

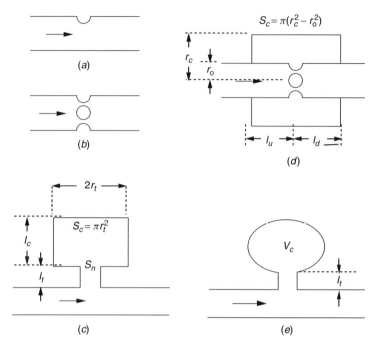

FIGURE 9.6 Resonator components and configurations.

where the subscripts tt, cc, and gv refer to the transverse-tube (quarter-wave resonator), concentric-cylinder, and general-volume cavities. These quantities and the lengths l_u, l_d, and l_c, the cross section S_c, and the volume V_c are illustrated in Fig. 9.6. It should be noted that at low frequencies ($k_0 l_c \ll 1$ and $k_0 l_u$, $k_0 l_d \ll 1$) both Z_{tt} and Z_{cc} are reduced to the expression of Z_{gv}.

The impedance Z_t of the throat connecting the duct to the cavity changes dramatically with grazing flow; therefore, it is characterized by two sets of values. For $M = 0$, this quantity is given by

$$Z_t^{[M=0]} = \frac{1}{n_h}\left(\frac{ck_0^2}{\pi} + j\frac{ck_0(l_t + 1.7r_0)}{S_0}\right) \quad (\text{m} \cdot \text{s})^{-1}, \quad (9.23)$$

where l_t is the length of the connecting throat, r_o the orifice radius, S_o the area of a single orifice, and n_h the total number of perforated holes. In the absence of mean flow, this expression has yielded good agreement between predicted and measured results for single-hole and well-localized multihole resonators.

The presence of grazing flow ($M \neq 0$ in the duct) has a strong effect on the impedance Z_t of the resonator throat. Measurements conducted using single- and multihole throats[15,57] have led to the empirical expression

$$Z_t^{[M \neq 0]} = \frac{c}{\sigma S_0}[7.3 \times 10^{-3}(1 + 72M) + j2.2$$

$$\times 10^{-5}(1 + 51 l_t)(1 + 408 r_0)f] \quad (\text{m} \cdot \text{s})^{-1} \quad (9.24)$$

where the parameters l_t and r_o in Eq. (9.23) are in meters and σ is the porosity. The expression of Eq. (9.24) is more appropriate for predictions in the presence of grazing flow.

Transmission Matrices for Other Elements. The mass and momentum conservation laws used to derive the transmission matrices for a pipe segment, area discontinuities, and resonators can be similarly deployed to obtain similar expressions for other elements. The reader may find additional information on transmission matrices in references 8, 9, and 53–69.

9.4 PERFORMANCE PREDICTION AND DESIGN EXAMPLES FOR EXPANSION CHAMBER MUFFLERS

Simple Expansion Chamber Muffler (SECM)

This silencer consists of an exhaust pipe, an expansion chamber, and a tail pipe of different transverse dimensions as shown in Fig. 9.7. The TL for this simple configuration may be obtained by substituting into Eq. (9.11) the transmission matrices for these elements obtained from Eqs. (9.15) and (9.18).

In a simplified case, where the flow is or is assumed to be insignificant, the product of the three matrices is reduced to the expression

$$\text{TL} = 10 \log \left[1 + 0.25 \left(N - \frac{1}{N} \right)^2 \sin^2 kL \right] \quad \text{dB} \quad (9.25)$$

where k is the wavenumber, L is the chamber length, and N is the area ratio given by $N = S_2/S_1$, where S_2 is the area of cross section of the chamber and S_1 that of the tail pipe or exhaust pipe (assumed to be equal).

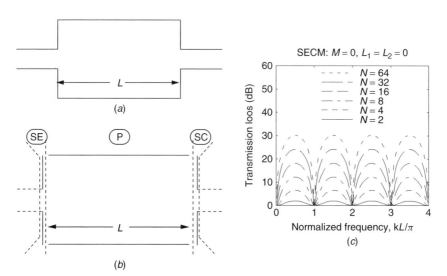

FIGURE 9.7 Simple expansion chamber muffler: (*a*) typical cross section; (*b*) muffler elements; (*c*) predicted transmission loss for different area ratios.

The predicted TL is presented in Fig. 9.7c for a few values of the area ratio N and displayed versus the dimensionless quantity $q = kL/\pi$. The troughs of the TL curve occur at $kL/\pi = n$, while the peaks occur at $kL/\pi = (2n-1)/2$, where $k = \omega/c$ is the wave number and n is an integer.

Example 9.1. The predicted curves of Fig. 9.7c can be used to develop a starting design of a SECM with a given performance goal. For example, assume that we need a 10-dB reduction for a 180-Hz tone generated by the exhaust of a small engine venting through a pipe with a diameter $d = 2$ in. (5.1 cm) at a temperature of 50°C and at a negligible flow speed.

The typical design process aligns the 180-Hz tone with one of the predicted acoustical performance peaks. Generally, alignment with a peak on a low-area-ratio (S_2/S_1) curve will minimize the muffler diameter and alignment with a peak at a low value of kL/π will minimize the length of the resulting muffler. In the present example, the design may proceed along the following steps:

- Step 1: The 180-Hz tone is aligned with the predicted 12-dB peak of the $S_2/S_1 = 8$ curve at $kL/\pi = 0.5$ (Fig. 9.7c).
- Step 2: At 50°C (323 K), the sound speed $c = 331$ radical $\sqrt{323/273} = 360$ m/s.
- Step 3: The corresponding wavenumber $k = 2\pi f/c = 2\pi \times 180/360 = 3.14$ rad/m.
- Step 4: The required muffler length L (given by $kL/\pi = 0.5$) $= 0.5\pi/k = 0.5\pi/3.14 = 0.5$.
- Step 5: The required muffler diameter $D = d\sqrt{S_2/S_1} = 5.1\sqrt{8} = 14.4$ cm.

One important implication of antilog addition and subtraction is that raising the trough levels of the IL (or, for that matter, the TL) by an amount DL has a more favorable impact than raising the peaks by the same amount DL. Therefore, it is desirable to try to raise the troughs (particularly, the first few at the lower end frequency range) through design adjustments, even at the expense of some reduction in the peaks of the revised muffler configuration.

A major disadvantage of the simple expansion chamber silencer is that in certain applications time-varying tones and their harmonics may align simultaneously with the periodic troughs and cause a severe deterioration in acoustical performance. This problem may be resolved to varying degrees by using extended-tube (inlet and/or outlet) elements.

Extended-Outlet Mufflers

An extended-outlet muffler (Fig. 9.8a) represents the first incremental step toward improving the performance of a single-chamber muffler. This configuration is decomposed into the three basic elements illustrated in Fig. 9.8b:

- a sudden expansion element (SE) at the inlet, flush with the left wall of the chamber;

PERFORMANCE PREDICTION AND DESIGN EXAMPLES FOR EXPANSION CHAMBER MUFFLERS 295

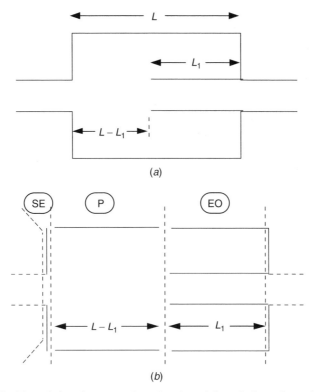

FIGURE 9.8 Extended-outlet expansion chamber: (a) typical configuration; (b) constituent muffler elements.

- a pipe element (P); and
- an extended-outlet element (EO) at the right end of the chamber.

The extended-outlet muffler design introduces a new length, L_1, which increases by 1 the number of dimensionless parameters (L/d, S_2/S_1, and L_1/L) that can be used to optimize the muffler's TL.

Transmission matrices for the silencer elements shown in Fig. 9.8b are obtained from Eq. (9.15) (for the pipe element P), Eq. (9.18) (extended-outlet element EO), and Eq. (9.20) (sudden expansion element SE).

It should be noted that at certain frequencies the impedance Z_2, introduced by the transmission matrix of Eq. (9.19) for the extended-outlet element, would approach zero, and the branch element would generate a pressure release condition. Under these conditions, the incident wave would appear to interact with a closed-end cavity, and no acoustical power would be transmitted downstream. For rigid end plates, this condition would occur when $\cot(kL_1) = 0$, corresponding to $kL_1/\pi = (2n - 1)/2$. Accordingly, a dominant peak in the source spectrum can be reduced significantly by proper selection of the tube length extended into the chamber.

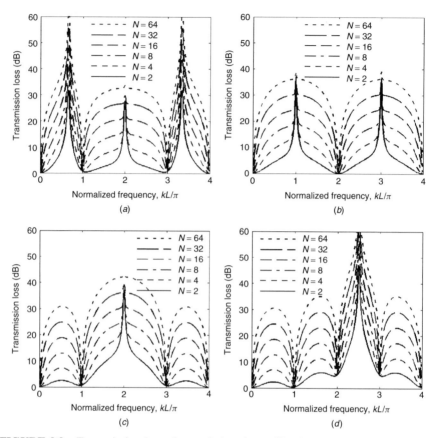

FIGURE 9.9 Transmission loss of extended-outlet muffler for different outlet pipe lengths: $M = 0$; L_1/L equals (a) 0.75, (b) 0.5, (c) 0.25, and (d) 0.2.

Figures 9.9a–9.9d illustrate the variation of the TL for an extended-outlet expansion chamber for four values of the dimensionless parameter L_1/L. As expected, the maximum peak of TL shifts toward higher frequencies as L_1/L decreases. Furthermore, a suitable choice of L_1/L, resulting in $kL_1/\pi = (2n - 1)/2$, n being an integer, relocates some performance peaks so as to eliminate some of the troughs that would otherwise occur at $kL = n\pi$ (Fig. 9.7c) for the simple expansion chamber muffler. This behavior is exploited in the following two design examples.

Example 9.2. Consider the application requiring a 15-dB reduction of a steady 120-Hz tone generated by the exhaust of a small engine venting through a pipe with a diameter $d = 1.5$ in. (3.8 cm) at a temperature of 100°C and at a negligible flow speed.

Since the tone is steady, that is, its does not vary appreciably versus time, we can consider aligning it with one of the sharp peaks in the predicted TL. This suggests selection of the Fig. 9.9a configuration, characterized by $L_1/L = 0.75$, which has a family of relatively sharp peaks at $kL/\pi = 0.75$.

In the present example, the design may proceed along the following steps:

- Step 1: The 120-Hz tone is aligned with the predicted >12-dB peak of the $S_2/S_1 = 8$ curve at $kL/\pi = 0.75$ (Fig. 9.9a).
- Step 2: At 100°C (373 K), the sound speed $c = 331\sqrt{373/273} = 387$ m/s.
- Step 3: The corresponding wavenumber $k = 2\pi f/c = 2\pi \times 120/387 = 1.95$ rad/m.
- Step 4: The required muffler length (given by $kL/\pi = 0.75$) $L = 0.75\pi/1.95 = 1.2$ m.
- Step 5: The required muffler diameter $D = d\sqrt{S_2/S_1} = 3.8\sqrt{8} = 10.8$ cm.

Example 9.3. Consider the application requiring a 10-dB reduction of a time-varying 150-Hz tone generated by the exhaust of a small engine venting through a pipe with diameter $d = 2.5$ in. (6.3 cm) at a temperature of 80°C and at a negligible flow speed. Furthermore, assume that the a variable load in the engine causes its frequency to vary ±15%, that is, operating anywhere in the 127 < f < 172-Hz range.

Since the frequency of the tone is unsteady, we can consider aligning it with one of the "broad" peaks in the predicted TL. This suggests selection of the Fig. 9.9b configuration, characterized by $L_1/L = 0.5$, which has a family of relatively "broad hill peaks" around the spikes occurring at $kL/\pi = 1$.

All curves with $S_2/S_1 \geq 4$ appear to have TL > 10 dB within 15% of $kL/\pi = 1$. For a conservative result, we select the curve with $S_2/S_1 = 8$ and complete the muffler design using the following steps:

- Step 1: The 150-Hz tone is aligned with the predicted >10-dB peak of the $S_2/S_1 = 8$ curve at $kL/\pi = 1$ (Fig. 9.9b).
- Step 2: At 80°C (353 K), the sound speed $c = 331\sqrt{353/273} = 376$ m/s.
- Step 3: The corresponding wavenumber $k = 2\pi f/c = 2\pi \times 150/376 = 2.5$ rad/m.
- Step 4: The required muffler length (given by $kL/\pi = 1$) $L = 1.0\pi/2.5 = 1.25$ m.
- Step 5: The required muffler diameter $D = d\sqrt{S_2/S_1} = 6.3\sqrt{8} = 18$ cm.

The TL for extended-inlet silencers can be derived in a similar manner. For negligible values of flow, their acoustical performance is similar to that of the extended-outlet silencers; in other words, for negligible inflow, either side of the muffler can be used as the inlet with no appreciable change in the TL.

Double-Tuned Expansion Chamber (DTEC)

The "filling up" of the troughs achieved through the extended-outlet muffler can be further enhanced through the simultaneous deployment of an extended-inlet pipe of length L_2 (double tuning). An extended inlet introduces an additional parameter, L_2/L, which used in conjunction with the extended-outlet parameter L_1/L can fill up additional troughs and further improve the achievable TL over a broader frequency range. For optimum results, L_1 and L_2 are selected so as to neutralize different sets of troughs to the extent possible.

The new double-tuned expansion chamber (DTEC) silencer design (Fig. 9.10a) is decomposed into three basic elements (Fig. 9.10b):

- an extended-inlet element (EI) at the left end of the chamber,
- a pipe element (P), and
- an extended-outlet element (EO) at the right end of the chamber

Thus, the introduction of the new length, L_2, for the extended inlet enhances the optimization of the DTEC's transmission loss through an increased number of dimensionless parameters, which now include d/L, S_2/S_1, L_1/L, and L_2/L.

Transmission matrices for the Fig. 9.10b silencer elements are obtained from Eq. (9.15) (for the pipe element P) and from Eq. (9.18) (extended-inlet and extended-outlet elements EI and EO).

Of the infinite number of $(L_2/L, L_1/L)$ combinations, a selection of $L_1 = L/2$ and $L_2 = L/4$ leads to a particularly favorable TL. Specifically, since the extended-inlet element includes a term proportional to $\cot(kL_2)$ [see Eqs. 9.18 and 9.19], it produces stop bands at $kL_2 = (2n-1)\pi/2$ or, equivalently, at $kL = 2(2n-1)\pi$, filling up the troughs at $kL/\pi = 2, 6, 10, \ldots$. Similarly, since the EO element includes a term proportional to $\cot(kL_1)$, it produces stop bands at

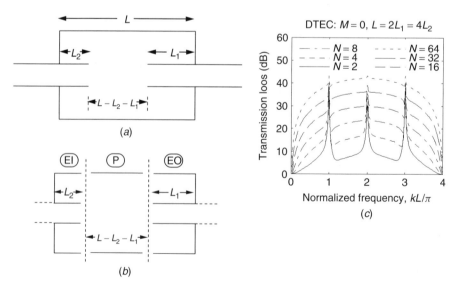

FIGURE 9.10 Double-tuned expansion chamber: (a) typical configuration; (b) constituent elements; (c) predicted TL for $L_2 = L/4$, $L_1 = L/2$.

$kL_1 = (2n-1)\pi/2$ or, equivalently, at $kL = (2n-1)\pi$, filling up the troughs at $kL/\pi = 1, 3, 5, \ldots$. In other words, this design features troughs only at $kL/\pi = 0, 4, 8, 12, \ldots$.

The corresponding predicted TL for a few area ratio values is shown in Fig. 9.10c. Its most prominent feature is the lack of troughs at locations other than at $kL/\pi = 4n$, n being an integer. Accordingly, this design offers a better solution for broadband performance than any of the previous configurations. High reduction limited to a narrow band (like Fig. 9.9a) can be achieved by adjusting the inlet/outlet lengths away from their $(L_1/L, L_2/L) = (\frac{1}{2}, \frac{1}{4})$ values.

Example 9.4. Consider the application requiring a 10-dB reduction of 200–600-Hz broadband noise generated by a noise source venting through a pipe with diameter $d = 2$ in. (5.1 cm) at a temperature of 120°C and at a negligible flow speed. Because of the broadband requirements, a DTEC muffler design is selected through the following steps:

- Step 1: The 400-Hz center of the subject frequency band is aligned with the center, $kL/\pi = 2$ of the broad TL peaks of Fig. 9.10c; this places the upper and lower frequency bounds (200 and 600 Hz) at kL/π values of 1 and 3, respectively.
- Step 2: The $S_2/S_1 = 4$ curve (Fig. 9.10c) is selected because it provides > 10 dB TL in the $1 < kL/\pi < 3$ range.
- Step 3: The required muffler diameter $D = d\sqrt{S_2/S_1} = 5.1\sqrt{4} = 10.1$ cm.
- Step 4: At 120°C (393 K), the sound speed $c = 331\sqrt{393/273} = 397$ m/s.
- Step 5: Given that $kL/\pi = 1$ corresponds to 200 Hz, the required muffler length $L = k = \pi c/(2\pi f) = 397/(2 \times 200) = 0.99$ m, $L_1 = L/2 = 0.495$ m and $L_2 = L/4 = 0.247$ m.

General Design Guidelines for Expansion Chamber Mufflers

Transmission loss curves with higher levels and fewer or less pronounced troughs can be obtained with more complex muffler configurations. Specifically, one can cascade two or more DTECs to further optimize performance through an increased number of system parameters. However, an increasing system complexity is not without limits, since muffler size and weight are important design parameters in practical applications. For example, increasing the number of elements reduces the average length/diameter of individual elements, which reduces low-frequency performance, and increases the number of partitions, which increases weight. These competing factors improve certain features while degrading others and lead to design trade-offs that are specific to each application.

A close examination of a number of designs (not detailed here) featuring one to three DTECs leads to the following observations or design considerations:

- The TL of a SECM features nulls (troughs) at $kL/\pi = 1, 2, 3, 4, 5$.

300 PASSIVE SILENCERS

- If an SECM is augmented with an extended-inlet (or outlet, for that matter) pipe equal in length to half the chamber length, then the troughs corresponding to $kL/\pi = 1, 3, 5, 7, 9, \ldots$ are eliminated.
- If an SECM is augmented with an extended-inlet (or outlet, for that matter) pipe equal in length to quarter the chamber length, then the troughs corresponding to $kL/\pi = 2, 6, 10, 14, \ldots$ are eliminated.
- If an SECM is augmented with extended-inlet and extended-outlet pipes equal in length to half and quarter the chamber length, respectively (DTEC), then the TL retains troughs only at $kL/\pi = 4, 8, 12, 16, \ldots$.
- The TL of two identical cascaded DTECs occupying the same envelope as a single DTEC features less pronounced troughs and generally higher levels, except at low frequencies ($kL/\pi < 0.5$), where some degradation is observed.
- The TL of two unequal cascaded DTECs can be further improved over the TL of two identical cascaded DTECs occupying the same envelope.

Additional TL improvements can be achieved through the use of three or more cascaded DTECs occupying the same total envelope as one DTEC. However, such improvements lead to further degradation of low-frequency performance; therefore, commercial application mufflers generally include no more than two or three cascaded DTEC chambers. Nevertheless, a designer who is forced to use a larger number of chambers by the needs of a specific application may recover low-frequency performance through the use of other types of elements, as discussed in the next section.

9.5 REVIEW OF PERFORATED-ELEMENT MUFFLERS

Perforated-element silencers have long been known to be acoustically more efficient than the corresponding simple tubular-element silencers. However, a systematic aeroacoustical analysis of perforated elements began only in late 1970s, when Sullivan introduced his segmentation model.[18-20] This was followed by the distributed-parameter model of Munjal et al., which produced experimentally verified four-pole parameters for perforated elements, namely, concentric-tube resonators (Fig. 9.11a), plug chambers (Fig. 9.11b), and three-duct cross-flow (Fig. 9.11c) and reverse-flow chambers.[8,9,21-23].

This section conducts a parametric study on some perforated-element mufflers to identify representative trends and to develop basic design guidelines.[25] Again, TL has been selected as an appropriate performance index. Explicit expressions for the impedance and the transfer matrix parameters of perforated muffler elements in terms of the geometric and operating variables and formulas for TL and IL are given in Chapter 3 of reference 8.

Range of Variables

Perforated-tube mufflers share several parameters with the expansion chamber mufflers discussed earlier, but they also include additional parameters accounting

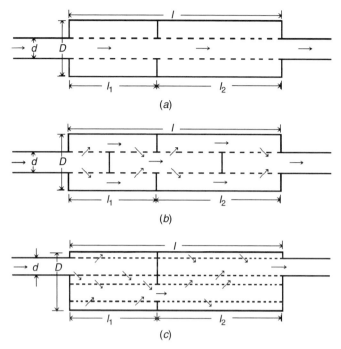

FIGURE 9.11 Schematic of the two-chamber configurations of three types of perforated-tube mufflers: (*a*) Concentric-tube resonator; (*b*) plug muffler; (*c*) three-duct cross-flow muffler.

for the additional features of the perforated tubes. Thus, the physical parameters influencing the performance of the Fig. 9.11 perforated mufflers include the following:

M = mean-flow Mach number in exhaust pipe
l = total length of muffler shell
D = internal diameter of muffler shell
d = internal diameter of exhaust pipe and tail pipe
d_h = diameter of perforated holes (Fig. 9.12)
C = center-to-center distance between consecutive holes (Fig. 9.12)
N_c = number of chambers within fixed-length muffler shell
t_w = wall thickness of perforated tube

Another parameter, often used instead of explicit values of d_h or C, is the porosity (or open-area ratio) of the perforated-tube wall, which is defined by

$$\sigma = \frac{\pi d_h^2}{4C^2} \qquad (9.26)$$

Extensive simulations conducted for representative configurations encountered in most practical applications showed that the performance of perforated-tube

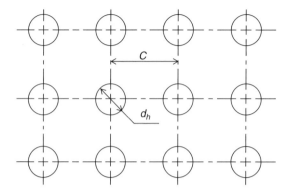

FIGURE 9.12 Parameters d_h and C of perforated-tube walls.

mufflers is relatively insensitive (within ±1 dB) to the hole diameter d_h and wall thickness t_w. For this reason, all of the following predictions have been conducted at the typically encountered values of $d_h = 4$ mm and $t_w = 1$ mm.

Furthermore, to reduce the simulations to a manageable number, the pipe diameter was fixed to a commonly encountered value of $d = 30$ mm for all cases considered in this section. This constraint does not influence the corresponding predicted TL curves, because they are invariant with respect to the specific value of d when plotted against the nondimensional frequency parameter kR ($R = D/2$ is the shell radius and $k = \omega/c$).

Despite the simplification resulting from the fixed values of d, d_h, and t_w, the number of perforated-tube muffler configurations is still too large for a comprehensive study. Under the circumstances, the influence of each parameter on muffler performance is demonstrated by selecting a reference (default) configuration and then by changing individual parameters (one at a time) about the default values. Table 9.2 shows the range of parameter values investigated including the default values (last column).

The TL was calculated as a function of the dimensionless frequency kR to extend the validity of the predictions over all geometrically similar hardware, that is, mufflers differing only by a length scale. The nondimensional frequency parameter kR was varied in steps of 0.025 from 0.025 to 3, which is well

TABLE 9.2 Parameter Values Used in Perforated Tube Muffler Performance Predictions

Parameter Name	Description	Range	Default
M	Mach number	0.05, 0.1, 0.15, 0.2	0.15
D/d	Expansion ratio	2, 3, 4	3
C/d_h	Center-to-center distance	3, 4, 5, 6	4
N_c	Number of chambers	1, 2, 3	2
L_2/L	Chamber size	$1, \frac{1}{2}, \frac{1}{3}$	$\frac{1}{2}$
L/d	Muffler length–pipe diameter	15, 20, 25	20

within the plane-wave cutoff limit of 3.83 for the axisymmetric mufflers of Figs. 9.11a,b. By comparison, the cutoff limit of kR for the asymmetric configuration of Fig. 9.11c is 1.84.

Acoustical Performance

Figures 9.13–9.17 show the computed TL for the Fig. 9.11 mufflers as a function of the normalized frequency kR. The three plots (a, b, and c) in each figure display the TL of a concentric-tube resonator muffler, plug muffler, and three-duct cross-flow muffler, respectively, and the different traces within each plot correspond to different combinations of muffler parameter values listed in Table 9.2.

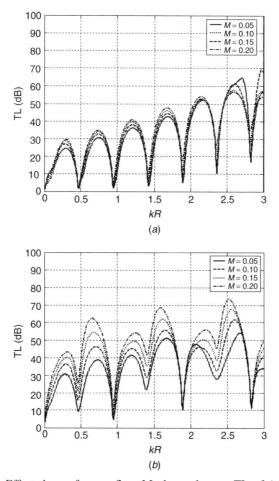

FIGURE 9.13 Effect three of mean flow Mach number on TL of (a) CTR, (b) plug muffler, and (c) duct muffler: (—) $M = 0.05$; (- - -) $M = 0.1$; (...) $M = 0.15$; (-----) $M = 0.2$.

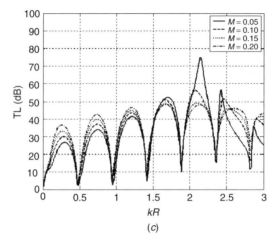

FIGURE 9.13 (*continued*)

Figure 9.13 shows the effect of the mean-flow Mach number on the acoustical performance of three perforated-element muffler designs. These predictions show a general increase in TL when the mean-flow Mach number increases. The effect is minimum for the concentric-tube resonator (CTR) and maximum for the plug muffler; this is not surprising given that (because of their intrinsic geometric features) the former and latter configurations present the minimum and maximum "blockages," respectively, to the mean flow through the muffler.

Clearly, the choice of the Mach number is not in the designer's control, as it is determined by the engine displacement, speed, and exhaust pipe diameter. Nevertheless, the designer may combine the information in Fig. 9.13 with backpressure information (discussed in the next section) in the selection of a muffler design appropriate for the specific application.

The effect of the expansion ratio, or the diameter ratio D/d, is illustrated in Fig. 9.14. The TL of all three perforated-element mufflers improves considerably with higher diameter ratio.

In most practical applications, the shell diameter D is directly related to muffler volume, weight, and cost and influences the choice of the expansion ratio D/d and the type of muffler. The alternate troughs in the TL curves are higher for a plug muffler (Fig. 9.14*b*) than for the other two designs; therefore, when space is constrained, the plug muffler typically offers a higher acoustical performance. Unfortunately, it also leads to a typically higher back pressure, as discussed later. As a result, the final choice is usually based on a trade-off between acoustical and mechanical performance.

The effect of the center-to-center spacing between consecutive holes (which determines the porosity) on the TL of the mufflers is illustrated in Fig. 9.15. As can be seen, the performance of all perforated mufflers tends to that of the simple expansion chamber muffler for sufficiently high values of porosity ($\sigma > 0.1$) but is considerably better for lower values of porosity, particularly for the plug

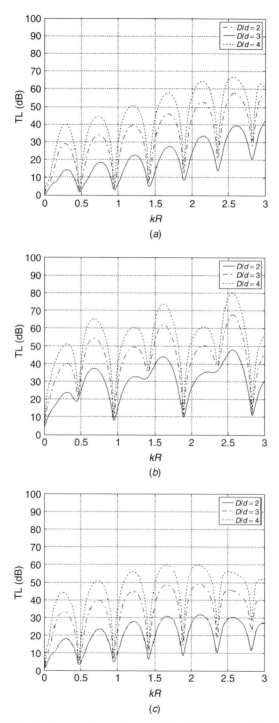

FIGURE 9.14 Effect of expansion ratio on TL of (*a*) CTR, (*b*) plug muffler, and (*c*) three-duct muffler: (—) $D/d = 2$; (- - -) $D/d = 3$; (...) $D/d = 4$.

muffler and the three duct muffler configurations. On the other hand, a reduced porosity raises back pressure, except for the CTR, where the back pressure is nearly independent of porosity, as demonstrated later in this section. Consequently, the porosity provides another parameter that can be used to trade off between acoustical and mechanical performance.

Another, and perhaps more meaningful, parameter often used instead of the porosity is the open area ratio x, defined as

$$x = \frac{\text{total area of perforations}}{\text{cross-sectional area of pipe}} \tag{9.27}$$

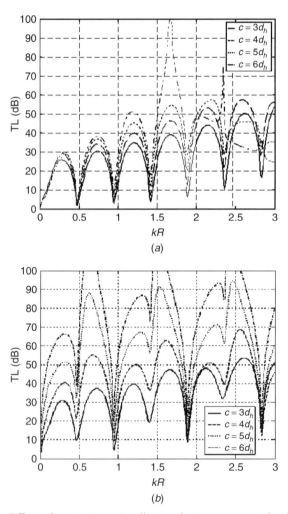

FIGURE 9.15 Effect of center-to-center distance between consecutive holes on TL of (a) CTR, (b) plug muffler, and (c) three-duct muffler: (——) $C = 3d_h$; (- - -) $C = 4d_h$; (...) $C = 5d_h$; (-----) $C = 6d_h$.

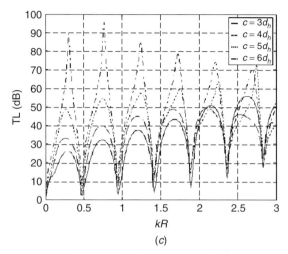

FIGURE 9.15 (*continued*)

The denominator refers to the inlet/outlet pipe cross section that accommodates the entire mean flow. For example, the CTR (Fig. 9.15a) and three-duct muffler (Fig. 9.15c), may be described in terms of either the porosity associated with the indicated $C = 3d_h, 4d_h, 5d_h, 6d_h$ or in terms of the corresponding open-area ratio values of 3.49, 1.78, 1.25, and 0.87, respectively. The corresponding values of x for the plug muffler (Fig 9.15b) are half as much.

The plot in Fig. 9.16 shows the effect of multiple equal partitioning of a muffler of overall length l on the TL. The normalized frequency spacing between the consecutive troughs is proportional to the number of partitions and, as in the case of the simple expansion chamber mufflers, the troughs occur at frequencies given by

$$\sin k l_{1,2} = 0 \quad k l_{1,2} = n\pi \quad n = 1, 2, 3 \ldots \quad (9.28)$$

In general, the partitioning improves the TL of all three mufflers, except at the low-frequency end of the CTR curves, where the TL decreases. An increased number of partitions N_c would be particularly desirable for a CTR, where an increase in partitions would not increase the back pressure. However, the back pressure will be proportional to the cube of the number of chambers for the other two types of mufflers. Therefore, increased partitioning must be closely considered with an increased porosity to optimize acoustical performance while maintaining the back pressure within acceptable levels.

Figure 9.17 illustrates the effect of unequal partitioning of a muffler of overall length l on the TL by comparing the results of an equally and an unequally partitioned muffler for each of the three designs. One can observe that unequal partitioning (a) has no significant impact on the envelope of the TL peaks, (b) influences significantly the location of individual peaks and troughs, and (c) raises noticeably the level of each alternate trough. On average, the additional

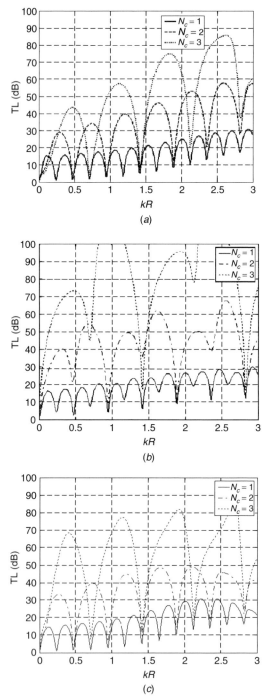

FIGURE 9.16 Effect of number of chambers within the same overall length on TL of (a) CTR, (b) plug muffler, and (c) three-duct muffler: (—) $N_c = 1$; (- - -) $N_c = 2$; (...) $N_c = 3$.

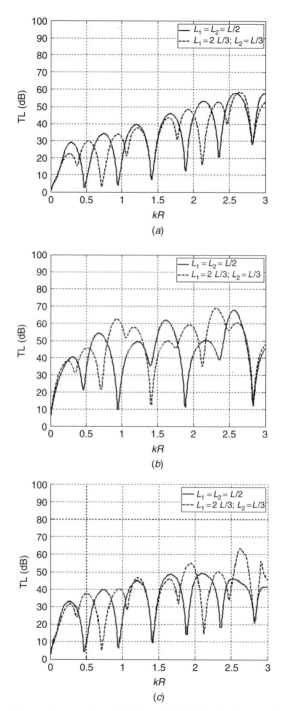

FIGURE 9.17 Effect of unequal partitioning of (*a*) CTR, (*b*) plug muffler, and (*c*) three-duct muffler: (——) $l_1 = l_2 = \frac{1}{2}$; (- - -) $l_1 = 2l/3$; $l_2 = l_3$.

Back Pressure

The static pressure drop resulting from energy dissipation in sheared flow regions of the muffler elements results in a substantial back pressure on the piston of a reciprocating engine. This generally has an adverse effect on the volumetric efficiency, power, and specific fuel consumption of a multicylinder engine.

A systematic experimental study was undertaken[25] to derive empirical expressions for pressure drop or head loss across all the three types of perforated-element chambers of Fig. 9.11. The results were expressed as a coefficient that normalized the pressure drop with respect to the incoming dynamic head in the pipe, namely,

$$y \equiv \frac{\Delta p}{H} \qquad H = \tfrac{1}{2}\rho U^2 \qquad (9.29)$$

The parameter y was dependent on the open-area ratio x of the perforate defined by Eq. (9.27), and its value was determined to vary as follows:

Concentric-tube resonator:

$$y_{ctr} = 0.06x \qquad 0.6 < x \qquad (9.30)$$

Plug muffler:

$$y_{pm} = 5.6e^{-0.23x} + 67.3e^{-3.05x} \qquad 0.25 < x < 1.4 \qquad (9.31)$$

Three-duct cross-flow chamber:

$$y_{cfc} = 4.2e^{-0.06x} + 16.7e^{-2.03x} \qquad 0.4 < x < 5.8 \qquad (9.32)$$

During the measurements of the parameter y,[25] the desired variation in the open-area ratio x was achieved by varying the length l_2 (see Fig. 9.11). The measured y [Eqs. (9.30)–(9.32)] was found to be independent of the mean-flow velocity U (for Mach number $M \leq 0.2$) and very weakly dependent on the expansion ratio D/d and lengths l_1 and l_3 for values of x in the range indicated by Eqs. (9.30)–(9.32). However, the expressions for y in these equations are least-squares fits and therefore are not limited to the indicated range; they cover the entire practical range ($0.2 < x < 6.0$).

The back-pressure coefficient y_{ec} for the expansion chamber mufflers of Section 9.3 is analogously given by[8]

$$y_{ec} = (1-n)^2 + \tfrac{1}{2}(1-n) \qquad n = \left(\frac{d}{D}\right)^2 \qquad (9.33)$$

Equations (9.30)–(9.32) show that the static pressure drop across a CTR, which involves no net flow through the perforated-tube walls, is much smaller than that across the corresponding plug muffler and the three-duct cross-flow chamber. Similarly, Equations (9.30) and (9.33) indicate that the pressure drop across a CTR is less than that across a simple expansion chamber of identical shell size and pipe diameter.

9.6 DISSIPATIVE SILENCERS

Dissipative silencers are the most widely used devices to attenuate the noise in ducts through which gas flows and in which the broadband sound attenuation must be achieved with a minimum of pressure drop across the silencer. They are frequently used in the intake and exhaust ducts of gas turbines, air conditioning and ventilation ducts connected to small and large industrial fans, cooling-tower installations, and the ventilation and access openings of acoustical enclosures, and they have an allowed pressure drop that typically ranges from 125 to 1500 Pa (0.5–6 in. H$_2$O). Unlike reactive silencers, which mostly reflect the incident sound wave toward the source, dissipative silencers attenuate sound by converting the acoustical energy propagating in the passages into heat caused by friction in the voids between the oscillating gas particles and the fibrous or porous sound-absorbing materials, as described in detail in Chapter 8.

The theories of dissipative silencers were developed long ago,[1,2,71–74] and they are highly complex. This chapter provides design information in a form that can be readily used by engineers not thoroughly trained in acoustics. Though there are a large variety of geometries used, the most common configurations include parallel-baffle silencers, round silencers, and lined ducts.

Lined Ducts

The sound attenuation of lined and unlined ducts and lined and unlined bends are treated in Chapter 16. The geometry of frequently used dissipative silencers is shown schematically in Fig. 9.18.

Figure 9.19 shows some of the baffle constructions that have been used in typical applications. Only the acoustically significant features are shown. Protective treatments such as perforated facing, fiberglass cloth, and porous screens are omitted. If the same concepts depicted in Figs. 9.19*a–i* are applied as a duct lining, one-half of the baffles depicted in Fig. 9.19 constitutes the lining.

Figure 9.19 illustrates the cross sections of frequently used silencer baffle configurations.

The full-depth porous baffle depicted in Fig. 9.19*a* is the most frequently used in parallel-baffle silencers. The other baffle configurations are used in special-purpose silencers custom deigned to yield high attenuation in specific

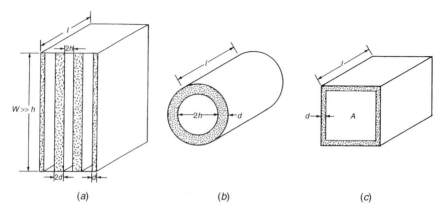

FIGURE 9.18 Geometry of frequently used silencer types: (*a*) parallel baffle; (*b*) round; (*c*) lined ducts.

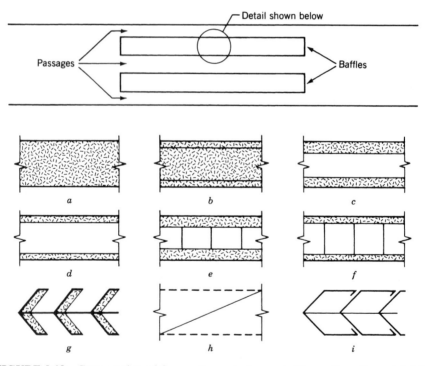

FIGURE 9.19 Cross section of frequently used silencer baffle configurations: (*a*) full-depth porous; (*b*) porous center layer with thin resistive facing on both sides; (*c*) thick porous surface layer with unpartitioned center air space; (*d*) thin porous surface layers with unpartitioned air space; (*e*) thick porous layer with partitioned center air space; (*f*) thin porous surface layers with partitioned center air space; (*g*) tuned cavity (Christmas tree), porous material in cavities protected from flow; (*h*) small percentage open-area perforated surface plates exposed to grazing flow, partitioned center air space; (*i*) Helmholtz resonators.

frequency ranges or to work in hostile environments of contaminated flow or high temperatures.

Key Performance Parameters

The key design parameters of silencers are acoustical insertion loss (IL), pressure drop (Δp), flow-generated noise, size cost, and life expectancy. The challenge of silencer design is to obtain the needed IL without exceeding the allowable pressure drop and size for a minimum of cost. These are frequently opposing requirements, and the optimal design represents a balance compromise between them.

Insertion Loss. The IL of a silencer is defined as

$$\text{IL} = 10 \log \frac{W_0}{W_M} \quad (9.34)$$

where W_0 and W_M represent the sound power in the duct without and with the silencer, respectively. Provided that structure-borne flanking along the muffler casing and sound radiation from the casing is kept low, the sound power in the duct with the silencer in place is given by

$$W_M = W_0 \times 10^{-(\Delta L_l + \Delta L_\text{ENT} + \Delta L_\text{EX})/10} + W_\text{SG} \quad (9.35)$$

where W is in watts (N·m/s); ΔL_l represents the attenuation of the silencer of length l; ΔL_ENT and ΔL_EX are the entrance and exit losses, respectively; and W_SG is the sound power generated by the flow exiting the silencer.

Combining Eqs. (9.34) and (9.35) yields

$$\text{IL} = -10 \log \left(\frac{W_\text{SG}}{W_0} + 10^{-(\Delta L_l + \Delta L_\text{ENT} + \Delta L_\text{EX})/10} \right) \quad (9.36)$$

In the extreme case of very high silencer attenuation, the second term on the right-hand side of Eq. (9.36) becomes comparable to the first, and the achievable IL is affected by the self-generated noise of the silencer, and in this case Eq. (9.36) is nonlinear. When the flow velocity in the silencer passages is sufficiently low, flow noise is negligible. In this case, Eq. (9.36) is linear and simplifies to

$$\text{IL} \cong \Delta L_l + \Delta L_\text{ENT} + \Delta L_\text{EX} \quad \text{dB} \quad (9.37)$$

Entrance Loss ΔL_ENT. In most dissipative silencers, the entrance losses ΔL_ENT are small if the incident sound energy is in the form of a plane wave normally incident on the silencer entrance. This is always the case for straight ducts at low frequencies. The small entrance loss should be considered as a safety factor.

However, if the cross dimensions of the duct are much larger than the wavelength, the incident sound field usually contains a very large number of higher order modes. The conversion of the semidiffuse sound field in the entrance duct into a plane-wave field in the narrow silencer passages typically results in an entrance loss of 3–6 dB. The engineer can also assign any entrance loss between 0 dB at low frequencies and up to 8 dB at high frequencies on the basis of prior experience with similar situations or on the basis of scale model measurements. If no such information is available, Fig. 9.20 can be used to estimate the entrance loss.

Exit Loss ΔL_{EX}. Most exit losses ΔL_{EX} are generated when the silencer is located at the open end of a duct and the typical cross dimensions of the opening are small compared with the wavelength. In this case, the exit loss is predominantly determined by the end reflection. Exit losses for silencers inserted in ducts are usually small and can either be neglected or considered as part of the safety margin.

It should be noted that the relative importance of the entrance and exit losses diminishes as the silencer length increases because both quantities are independent of this length. Figure 9.21 shows qualitatively a typical sound pressure level versus distance recorded by a microphone traveling through the silencer and indicates the three components of IL.

Silencer Attenuation ΔL_l. The silencer attenuation ΔL_l is proportional to its length (tail and nose of baffle not included) and to the lined perimeter of the passage, P, and inversely proportional to the cross-sectional area of the passage,

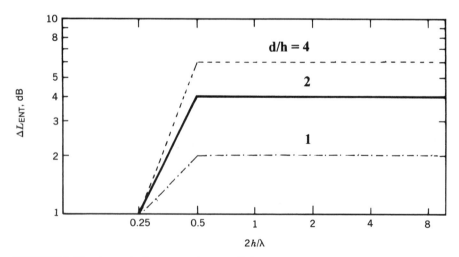

FIGURE 9.20 Acoustical entrance loss coefficient, ΔL_{ENT}, of silencers in a large duct with a semireverberant sound field in the entrance duct: $2h$ = silencer passage cross dimension.

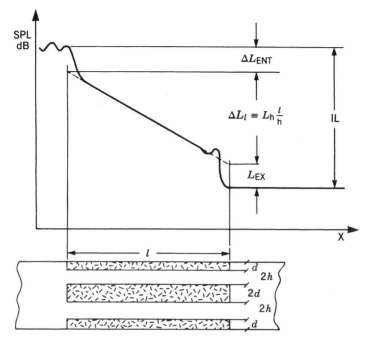

FIGURE 9.21 Typical SPL-versus-distance curve obtained when microphone traverses through a silencer.

A. It can be expressed as

$$\Delta L_l = \left(\frac{P}{A}\right) l L_h \tag{9.38}$$

where L_h is the parameter that depends in a complex manner on the geometry of the passage and the baffle, acoustical characteristics of the porous sound-absorbing material filling the baffles, frequency, and temperature. The quantity L_h, which also depends on the velocity of the flow in the passage, is usually referred to as the attenuation per channel height. The major part of this section is devoted to the determination of this important normalized sound attenuation parameter.

Pressure Drop Δp. The total pressure drop Δp_T across a muffler is made up of entrance, exit, and friction losses:

$$\Delta p_T = 1/2 \rho v_P^2 \left[K_{\text{ENT}} + K_{\text{EX}} + \left(\frac{P}{A}\right) l K_F \right] = \Delta p_{\text{ENT}} + \Delta p_{\text{EX}} + \Delta p_F \quad \text{N/m}^2 \tag{9.39}$$

where ρ is the density of the gas and v_P its face velocity in the passage of the silencer. The constants K_{ENT} and K_{EX} are the entrance and exit head loss coefficients, which depend only on the geometry of the baffle/passage configuration.

The third term on the right-hand side of Eq. (9.39) represents the friction losses, given by

$$\Delta p_F = \frac{P}{A} l (K_F \tfrac{1}{2} \rho v_P^2) \quad \text{N/m}^2 \qquad (9.40)$$

Comparing Eq. (9.38) with Eq. (9.40), one notes that baffle configurations that tend to yield high acoustical attenuation ΔL_l also tend to yield high frictional pressure losses Δp_F. Both ΔL_l and Δp_F are proportional to $(P/A)l$, indicating that high silencer attenuation and low frictional pressure drop are contradictory requirements. This finding emphasizes the need to optimize L_h by beneficial choice of the acoustical parameters of the baffles before one resorts to obtaining increased sound attenuation by increasing the factor $(P/A)l$.

Parallel-Baffle Silencers

The parallel-baffle-type silencer shown in Fig. 9.18a is the most frequently used because of its good acoustical performance and low cost. The attenuation of such a silencer is proportional to the perimeter–area ratio P/A, the length l, and L_h. Therefore, it is maximized by maximizing $(P/A)L_h$. The largest perimeter–area ratio obtained for narrow passages is $1/h$. Allowing for entrance losses, Eq. (9.38) yields the following simple formula for silencer attenuation:

$$\Delta L_l = L_h \frac{l}{h} + \Delta L_{\text{ENT}} \quad \text{dB} \qquad (9.41)$$

where ΔL_{ENT} can be approximated from Fig. 9.20. The following discussion shows how to obtain L_h from the geometric and acoustical parameters of the silencer.

Prediction of Attenuation. The sound energy traveling in the passages of a parallel-baffle silencer, depicted in Fig. 9.19, is attenuated effectively in a wide bandwidth if (1) the sound enters the porous sound-absorbing material in the baffles and (2) a substantial part of the energy of the sound wave entering the baffle is dissipated before it can reenter the passage. Formulas for wall impedance that yield maximum attenuation in a narrow-frequency band are given in reference 2.

Requirement 1 is fulfilled if the passage height is small compared with wavelength (i.e., $2h < \lambda$) and the porous sound-absorbing material is open enough and has sufficiently low flow resistivity so that the sound wave enters the baffle rather than being reflected at the interface. This requires a "fluffy" material of low flow resistivity. Requirement 2 is fulfilled by a porous material of moderate flow resistivity. The requirements of easy sound penetration and high dissipation are contradictory unless the baffle is very thick and is packed with porous sound-absorbing material of low flow resistivity. Consequently, the choice of baffle thickness and flow resistivity of the porous sound-absorbing material is always a compromise.

Generally, the silencer geometry is controlled by the shape of the attenuation-versus-frequency curve we aim to achieve. To provide reasonable attenuation at the low end of the frequency spectrum, the baffle thickness $2d$ must be on the order of one-eighth of the wavelength. To provide reasonable attenuation at the high end of the frequency spectrum, the passage height $2h$ must be not much larger than the wavelength. To allow reasonable penetration of the sound and yield the needed dissipation, the total flow resistance $R_1 d$ of a baffle of thickness $2d$ must be 2–6 times the characteristic impedance of the gas in the silencer passages at design temperature.

Quantitative Considerations. The normalized attenuation constant L_h is obtained by solving the coupled wave equation in the passage and in the porous material of the baffle and requiring that (1) the coupled wave, which propagates axially in both the passage and baffle, has a common propagation constant Γ_c and (2) both particle velocity and the sound pressure are continuous at the passage baffle interface.

The coupled wave equation[1,71-74] can be solved by numerical iteration methods to yield the common propagation constant Γ_c. The normalized attenuation L_h is then obtained from

$$L_h = 8.68 h \ \mathrm{Re}\{\Gamma_c\} \quad \mathrm{dB/m} \tag{9.42}$$

where Γ_c depends on the characteristic impedance ρc of the gas in the passage, the characteristic impedance Z_a and the propagation constant Γ_c of the porous material in the baffle, and the geometry.

The characteristic impedance Z_a and the propagation constant Γ_c of the porous sound-absorbing materials (which are complex quantities and vary with frequency) are generally not available. As discussed in Chapter 8, one can approximate these important parameters with reasonable accuracy if the flow resistivity of the porous sound-absorbing material R_1 is known. For fibrous sound-absorbing materials the characteristic impedance Z_a and propagation constant Γ_a in the bulk porous material can be estimated using the empirical formulas presented in Chapter 8.

The most accurate characterization of the porous materials is achieved by measuring the characteristic impedance and propagation constant on a sufficiently large number of samples as a function of frequency at room temperature and by scaling these data to design temperature. Note that this attenuation, which is computed on the basis of the acoustical parameters of the porous material predicted from flow resistivity by employing the formulas given in Chapter 8, compares very well with experimental data if the porous sound-absorbing material is homogeneous.

Normalized Graphs to Predict Acoustical Performance of Parallel-Baffle Silencers. The normalized attenuation L_h has been computed for various percentages of open area of the silencer cross section (i.e., for various h/d) and for various values of the normalized flow resistance $R = R_1 d/\rho c$ of isotropic porous sound-absorbing material in the baffles. It is presented in Figs. 9.22a–c

for open-area ratios of 66, 50, and 33%, respectively. The effect of nonisotropic material is covered in reference 72. The vertical axis in Fig 9.22 represents the normalized attenuation L_h in decibels on a logarithmic scale. The lower horizontal scale represents the normalized frequency parameter $\eta = 2h/c$. It is valid for all temperatures and gases provided that the speed of sound c is taken at the actual temperature. *The upper horizontal scale, which is valid only for air at room temperature, represents the product of the half-passage height h in centimeters and the frequency f in kilohertz.*

Figures 9.22a–c show that the attenuation starts to decrease rapidly above the frequency where the passage height $2h$ becomes large compared to the wavelength (i.e., $\eta > 1$). The attenuation in this frequency region can be increased by up to 10 dB by utilizing a two-stage silencer with staggered-baffle arrangement. Note that the bandwidth of appreciable attenuation increases with decreasing percentage of open area of the silencer cross section.

Figure 9.22 shows that in the range of $R = R_1 d/\rho c$ from 1 to 5, the attenuation does not depend strongly on the specific choice of the flow resistance; this coincidence is welcome because the present lack of adequate knowledge of and control over the material characteristics represents the weakest link in the prediction process. Note that if the normalized flow resistance becomes too large

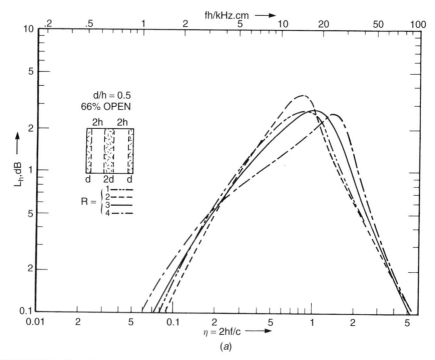

FIGURE 9.22 Normalized attenuation-versus-frequency curves for parallel-baffle silencers with normalized baffle flow resistance $R = R_1 d/\rho c$ as parameter: (*a*) 66% open; (*b*) 50% open; (*c*) 33% open.

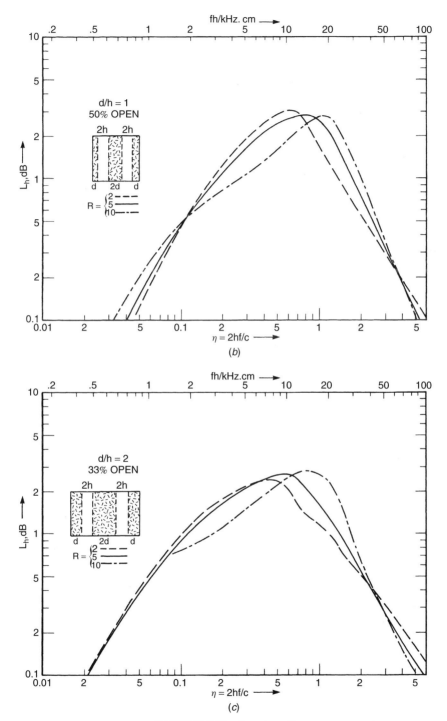

FIGURE 9.22 (*continued*)

($R \geq 10$), a substantial decrease of attenuation occurs in the frequency region from $\eta = 0.2$ to $\eta = 1$ accompanied by a modest increase of attenuation at very low and at high frequencies.

The normalized attenuation-versus-frequency curves presented in Figs. 9.22a–c correspond to zero flow; corrections to account for flow are presented in a later section.

The use of the design information presented in Figs. 9.22a–c is illustrated by a few examples.

Example 9.5. Predict the attenuation-versus-frequency curve of a parallel-baffle silencer consisting of 200-mm- (8-in.-) thick parallel baffles 1 m (40 in.) long spaced 400 mm (16 in.) off center when the flow resistance of the baffle $R_1 d = 5\rho c$ and the duct carries a very low velocity air at 20°C; $h = 0.1$ m; $d = 0.1$ m; $L = 1$ m; $c = 340$ m/s; $\rho = 1.2$ kg/m³; $R = R_1 d/\rho c = 5$.

Solution

1. Determine frequency f^*, which corresponds to $\eta = 1$:

$$f^* = \frac{c}{2h} = \frac{340 \text{ m/s}}{0.2 \text{ m}} = 1700 \text{ Hz}$$

2. Determine $1/h = 1$ m/0.1 m $= 10$.
3. Determine the applicable normalized attenuation-versus-frequency curve for $d/h = 1$ and $R = 5$; the solid curve in Fig. 9.22b is applicable.
4. Mark the frequency $f^* = 1700$ Hz on the horizontal scale of a sheet of transparent graph paper that has the same horizontal and vertical scales as Fig. 9.22b and align it with $\eta = 1$ in Fig 9.22b.
5. Shift the transparent graph paper vertically until the mark $L_h = 1$ in Fig. 9.22b corresponds to 10 ($1/h = 10$) on the transparent overlay.
6. Copy the solid curve in Fig. 9.22b that corresponds to $R = 5$ on the overlay.
7. The copied curve then corresponds to the attenuation-versus-frequency curve of the silencer according to $\Delta L_l = L_h(1/h)$; the above procedure is sketched in Fig. 9.23.

Example 9.6. Design a parallel-baffle silencer that yields the attenuation listed below:

f, Hz	100	200	500	1000	2000	4000
ΔL, dB	4	9	19	26	10	5

Design Steps

1. Find the graph in Figs. 9.22a–c that best matches the shape of the desired attenuation-versus-frequency curve plotted in Fig. 9.23. Overlay the transparent paper of Fig. 9.23 on the curve that yields the best match and shift the transparent overlay horizontally and vertically until all of the

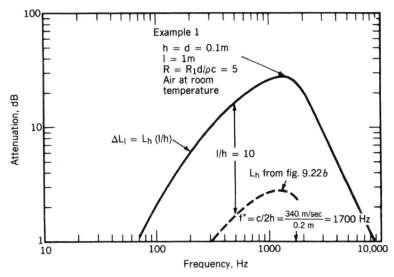

FIGURE 9.23 Attenuation prediction for parallel-baffle silencer, Example 9.5.

desired attenuation-versus-frequency points fall below the chosen normalized attenuation-versus-frequency curve. In this case, the solid curve in Fig. 9.22c gives the best match.

2. On the overlay, mark the frequency f^* that corresponds to $\eta = 1$ on the horizontal scale of the appropriate design curve under the overlay, and on the vertical scale, mark the attenuation ΔL^*, which corresponds to $L_h = 1$ on the design curve below the overlay, as shown in Fig. 9.24. In this case, these will be $f^* = 2000$ Hz and $\Delta L^* = 10$ dB.
3. From the design curve below the overlay, note the values of the parameters d/h and R that correspond to the curve that provides the best match. In this case, these are $d/h = 2$ and $R = R_1 d/\rho c = 5$.
4. On the basis of the information obtained in steps 3 and 4, one obtains the geometric and acoustical parameters of the silencer that will yield the specified attenuation as follows:

- Passage height $2h$:

$$\eta = 1 = \frac{2hf^*}{c} \quad \text{yields} \quad 2h = \frac{340 \text{ m/s}}{2000 \text{ s}^{-1}} = 0.17 \text{ m}$$

- Baffle thickness $2d$:

$$2d = 2(2h) = 2 \times 0.17 \text{ m} = 0.34 \text{ m}$$

- Silencer length:

$$\Delta L^* = 10 = \frac{l}{h} \quad \text{yields} \quad l = \Delta L^* h = 10 \times \frac{0.17}{2} \text{ m} = 0.85 \text{ m}$$

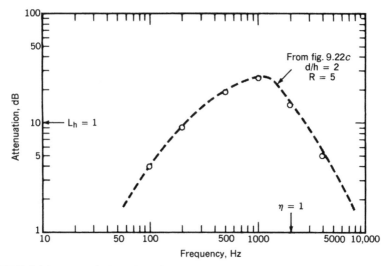

FIGURE 9.24 Acoustical design of a parallel-baffle silencer, Example 9.6: (○) design requirements; (- - - - - -) curve of matching attenuation versus frequency from Fig. 9.22c.

- Flow resistance per unit thickness of porous material:

$$\frac{R_1 d}{\rho c} = 5 \quad \text{yields} \quad R_1 = \frac{5\rho c}{d} = \frac{5}{0.17 \text{ m}} \rho c = 29.4 \rho c / \text{m}$$

or

$$R_1 = 0.7 \rho c / \text{in.} \quad \text{or} \quad R_1 = 1.2 \times 10^4 \text{ N} \cdot \text{s/m}^4$$

Cross-Sectional Area. The cross-sectional area of the muffler is determined by the maximum allowable pressure drop and self-generated noise as discussed later.

Effect of Temperature. The design temperature affects the acoustical and aerodynamic performance of the silencers because the following key parameters depend on the temperature: (1) speed of sound, (2) density of the gas, and (3) viscosity. The effects of temperature are taken into account as follows:

$$c(T) = c_0 \sqrt{\frac{273 + T}{293}} \quad \text{m/s} \tag{9.43}$$

$$\rho(T) = \rho_0 \frac{293}{273 + T} \quad \text{kg/m}^3 \tag{9.44}$$

$$R(T) = R_0 \left(\frac{273 + T}{293}\right)^{1/2} \tag{9.45}$$

where T is the design temperature in degrees Celsius, c_0 is the speed of sound, ρ_0 is the density of the gas (usually air) at 20°C, and

$$R_0 = \frac{R_1(20°C)d}{\rho_0 c_0} \tag{9.46}$$

where $R_1(20°C)$ is the flow resistivity of the porous bulk material at 20°C. This material parameter is usually provided by the manufacturer or is measured.

Example 9.7. To illustrate how to account for the effect of temperature, let us predict the attenuation provided by the silencer of Example 9.5 at $T = 260°C$ (500°F).

Solution The effect of temperature must be accounted for in the flow resistivity and in the speed of sound according to Eqs. (9.45) and (9.43), yielding $c(T) = 457$ m/s and $R(T) = 10$. From now on, the solution proceeds according to the same steps followed in Example 9.5.

1. $f^* = c(T)/2h = (457 \text{ m/s})/0.2 \text{ m} = 2285 \text{ Hz}; l/h = (1 \text{ m}/0.1 \text{ m}) = 10$.
2. The applicable normalized attenuation curve that corresponds to $d/h = 1$ and $R = 10$ is the short/long-dashed curve in Fig. 9.22b.

This results in the attenuation-versus-frequency curve shown as the solid line in Fig. 9.25. The dashed curve in Fig. 9.25 is the attenuation of the same silencer at room temperature (20°C), as determined in Example 9.5.

Comparing the solid curve obtained for 260°C with the dashed curve obtained for 20°C, one notes a shift in the attenuation-versus-frequency curve toward higher frequencies with increasing temperature. This shift is mainly due to the increase of propagation speed of sound with increasing temperature. In addition, one also observes distortion in the shape of the attenuation-versus-frequency curve. This is caused by the increase in the flow resistivity of the porous material,

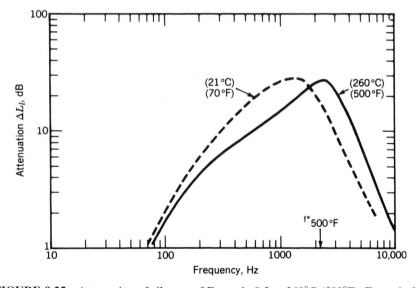

FIGURE 9.25 Attenuation of silencer of Example 9.5 at 260°C (500°F), Example 9.7.

Baffle Thickness Considerations. A particular percentage of open area of a silencer can be accomplished either by a small number of thick baffles or by a large number of thin baffles. Figure 9.26 shows the attenuation-versus-frequency curves computed for a 2-m- (6.5-ft-) long silencer of 50% open area with baffle thickness $2d = 2h$ of 152 mm (5 in.), 203 mm (8 in.), 254 mm (10 in.), and 305 mm (12 in.). The baffles are filled with a fibrous sound-absorbing material that has a flow resistivity $R_1 = 51,500$ N · s/m^4 at 500°C (1ρc/in. at 20°C). The outstanding feature of the data presented in Fig. 9.26 is that in the midfrequency region, the sound attenuation decreases with increasing baffle thickness. At 500 Hz, the 152-mm- (6-in.-) thick baffles yield 25 dB attenuation while the 305-mm- (12-in.) thick baffles yield only 11 dB. This is because the sound does not fully penetrate into the fibrous sound-absorbing material in the thick baffles. Consequently, the material and the space in the center of the thick baffles are wasted.

Figure 9.27 shows the predicted sound attenuation of a silencer of the same geometry as that in Fig. 9.26, but the baffles in this case are filled with fibrous sound-absorbing material of low flow resistivity such that the total normalized flow resistance of each baffle was $R = R_1 d/\rho c = 2$, which allows full penetration of the sound into even the thickest baffle. Consequently, the sound attenuation at

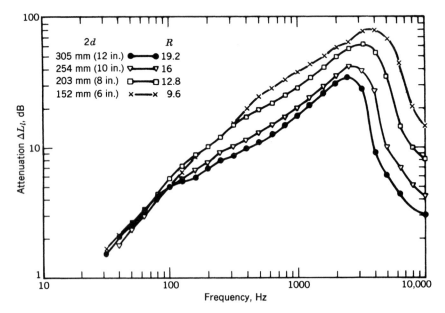

FIGURE 9.26 Computed attenuation-versus-frequency curves for a 2-m- (6.6-ft-) long parallel-baffle silencer of 50% open area with baffle thickness $2d$ as parameter.

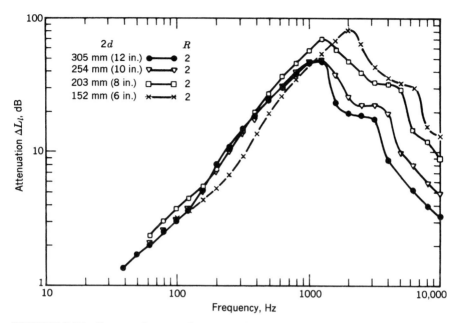

FIGURE 9.27 Computed attenuation-versus-frequency curves of the same silencer as Fig. 9.26 but fibrous fill is chosen to yield $R = R_1 d/\rho c = 2$ for all baffle thicknesses.

low and midfrequencies depends only slightly on the baffle thickness. Comparing Figs. 9.26 and 9.27 reveals that using a few thick baffles (which is more economical than using many thinner baffles) results in decreased sound attenuation at midfrequencies unless the baffles are filled with a sound-absorbing material of low enough resistivity so that the normalized baffle flow resistance at design temperature, $R = R_1 d/\rho c$, is much less than 10. Fibrous sound-absorbing materials that fulfill this requirement for thick baffles used in elevated temperatures may not be readily available. Consequently, silencer baffles of traditional design in high-temperature applications should be kept at a thickness that seldom exceeds $2d = 200$ mm (8 in.).

Note that the attenuation-versus-frequency curve of parallel-baffle silencers increases monotonically up to a frequency $f = c/2d$, where the passage width corresponds to a wavelength, and decreases sharply above it.

Round Silencers

Round silencers, as depicted in Fig. 9.18*b*, are used in connection with round ducts. The curvature of the duct casing results in a high form stiffness that yields high sound transmission loss of the silencer wall at low frequencies. The acoustical performance of round silencers with respect to normalized attenuation L_h is very similar to that of parallel-baffle silencers. The diameter of the round passage, $2h$, and the thickness of the homogeneous isotropic lining, d, resemble the passage width $2h$ and the half-baffle thickness d of a parallel-baffle silencer. The silencer attenuation $\Delta L^r(l)$ of a round muffler of length l is obtained from

$$\Delta L^r(l) = L_h^r \frac{l}{h} \qquad (9.47)$$

The sound attenuation in round ducts has been studied by Scott[71] and Mechel.[74] Their work forms the theoretical foundations for computing their performance.

Figure 9.28 shows curves of normalized attenuation L_h^r versus normalized frequency $\eta = 2hf/c$ for round silencers with homogeneous isotropic lining for thickness–passage radius ratios d/h of 0.5, 1, and 2, respectively, with the normalized lining flow resistance $R = R_1 d/\rho c$ as parameter. The lower horizontal scale is valid for all temperatures, while the upper scale is valid only for air at room temperature. Figure 9.28 shows that the normalized attenuation increases monotonically with frequency until the wavelength corresponds to the diameter of the passage ($\eta = 2hf/c = 2h/\lambda = 1$). Above this frequency the attenuation decreases rapidly with increasing frequency, as was the case for parallel-baffle silencers. Note that the maximum normalized attenuation is about $L_{h,\max}^r \approx 6$ dB

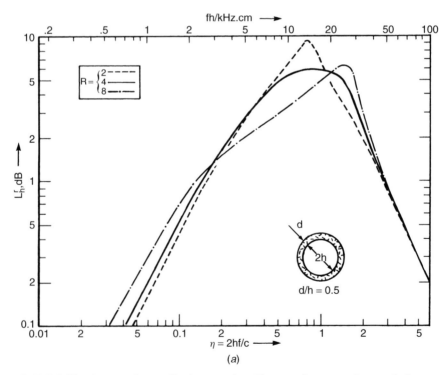

FIGURE 9.28 Curves of normalized attenuation L_h^r versus frequency for round silencers with normalized lining flow resistance $R = R_1 d/\rho c$ as parameter: (a) $d/h = 0.5$; (b) $d/h = 1$; (c) $d/h = 2$.

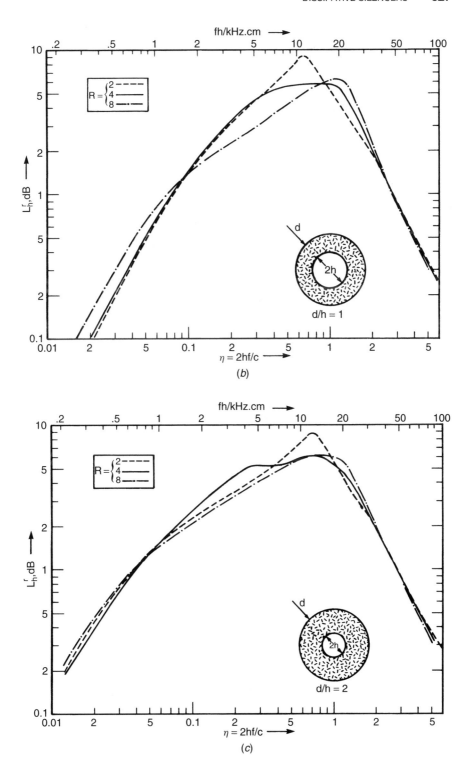

FIGURE 9.28 (*continued*)

for round silencers compared to $L^r_{h,\text{max}} \approx 3$ dB for parallel-baffle silencers. This is because a round passage has twice as high perimeter–area ratio $(2/h)$ than a narrow passage of a parallel-baffle silencer $(1/h)$. As expected, the attenuation bandwidth increases toward low frequencies with increasing lining thickness d. The curves presented in Fig. 9.28 are used in the same way as the corresponding curves for parallel-baffle silencers by applying the design steps listed earlier in the Examples 9.5 and 9.7. Incidentally, similar curves and trends have been predicted by Mechel.[75]

Pod Silencers

Round silencers have a generic disadvantage of providing poor high-frequency performance when the passage diameter is large compared with the wavelength. This disadvantage can be overcome by inserting a center body or pod into the passage[76]. With the center body in place, the round silencer has a narrow annular passage, just like a parallel-baffle silencer, and the attenuation will continue to increase monotonically until the wavelength of the sound becomes equal to the width of the narrow annular passage, as illustrated in the example of Fig. 9.29. Both the rigid and absorbing center bodies result in a modest increase in attenuation at low and midfrequencies, which is mostly due to reduction of the passage cross-sectional area. The attenuation of this silencer without center body

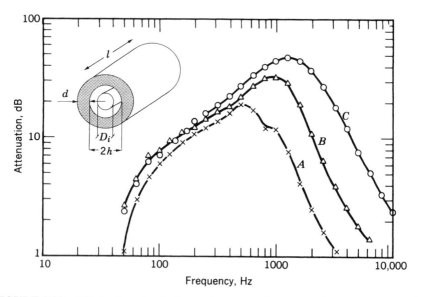

FIGURE 9.29 Effect of center body on the attenuation-versus-frequency curve of a round silencer, passage diameter $2h = 0.6$ m (24 in.), center body diameter $D_i = 0.3$ m (12 in.), lining thickness $d = 0.2$ m (8 in.), length $l = 1.2$ m (48 in.), $T = 20°C$ (68°F), $R_1 = 16,000$ N · s/m^4 ($1\rho c$/in.): A, no center body; B, rigid center body; C, absorbing center body.

decreases sharply above 560 Hz, the frequency where the wavelength equals the diameter of the passage. With the rigid center body the attenuation continues to increase up to 1130 Hz and with the porous center body up to 2260 Hz.

If center bodies are impractical, the high-frequency attenuation of round silencers can also be increased by inserting parallel baffles into the round passage. Similarly, the high-frequency attenuation of silencers with rectangular cross section can be increased by inserting parallel baffles (oriented perpendicular to the plane of the thick, low-frequency liner, or side branches on the sidewalls) into the rectangular passage. As described in detail by Kurze and Ver[77], the inserted parallel baffles also increase the low-frequency attenuation by a beneficial interaction with the low-frequency part of the silencer located at the walls. The mechanism of this interaction is that the reduction of the wave speed owing to the structure factor of the parallel baffles increases the attenuation performance of the low-frequency liner on the sidewalls. Similarly, the structure factor of the low-frequency liner on the sidewalls increases the attenuation of the parallel baffles. Consequently, the beneficial interaction results in increased sound attenuation at both low and high frequencies. This type of insert silencer also has the benefit of requiring substantially less total silencer length than the traditional series combination of a low- and high-frequency silencer section.

Effect of Flow on Silencer Attenuation

The flow affects the attenuation of sound in silencers in three ways:

1. It changes slightly the effective propagation speed of the sound.
2. By creating a velocity gradient near the passage boundaries, it refracts the sound propagating in the passage toward the lining if the propagation is in the flow direction (exhaust silencers) and "focuses" the sound toward the middle of the passage if the propagation is opposite to the flow direction (intake silencers).
3. It increases the effective flow resistance of the baffle.

Figure 9.30a shows the effect of flow on the attenuation performance of an exhaust silencer when the flow direction coincides with the direction of sound propagation. Note that the attenuation is decreased at low frequencies and is increased slightly at high frequencies. In most cases, a Mach number $M > 0.1$ is not permissible because of material deterioration or self-noise.

Figure 9.30b shows what happens when the sound propagates against the flow. In this case the attenuation at low frequencies increases because it takes the sound a longer time to traverse through the *muffler* passage. The high-frequency attenuation decreases because the velocity gradients in the passage "channel" the sound toward the center of the passage. The effect of flow on attenuation can be approximated by appropriate shift of the attenuation-versus-frequency curve obtained without flow ($M = 0$).

The attenuation in the presence of flow is obtained by shifting the no-flow ($M = 0$) attenuation curve as illustrated in Fig. 9.31a when the direction of sound

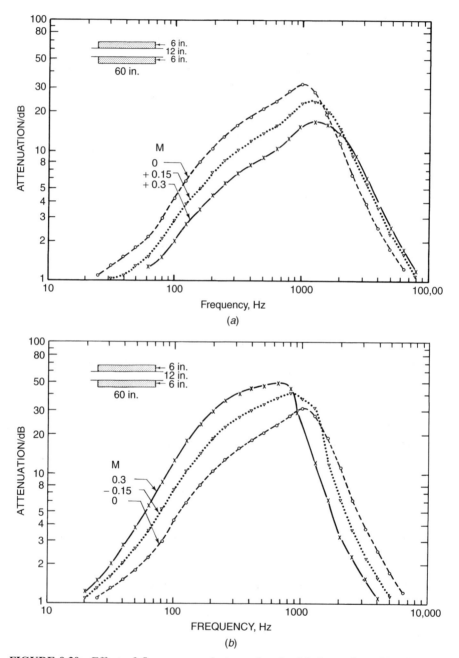

FIGURE 9.30 Effect of flow on sound attenuation for Mach numbers $M = 0$, 0.15, 0.3: sound propagation (a) in flow direction and (b) against flow direction.

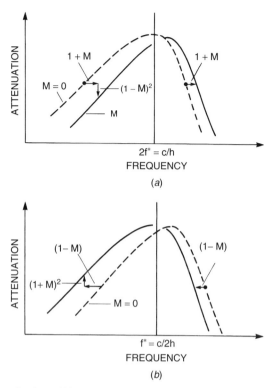

FIGURE 9.31 Rules for shifting the no-flow ($M = 0$) attenuation-versus-frequency curve to account for flow effects: sound propagation (*a*) in flow direction and (*b*) against flow direction.

propagation and the gas flow are the same and in Fig. 9.31*b* when they are opposite.

Figure 9.32 shows the attenuation-versus-frequency curve of a silencer with $M = 0.15$ flow in the direction of sound propagation. The values represented by the crosses are those directly computed for $M = +0.15$ (dotted line in Fig. 9.30*a*) and those represented by the open circles were obtained by shifting the $M = 0$ curve in Fig. 9.30*a* according to the guidelines discussed in conjunction with Fig. 9.31 The agreement between the curve obtained by shifting and by computations (which takes into account flow effects in the wave equation) is good.

The empirical flow correction procedure illustrated in Fig. 9.31 is based on experience with parallel-baffle silencers. We have no experience at present to gauge its applicability and accuracy for other silencer geometries.

Flow-Generated Noise

At present there is no universally accepted method for predicting the flow-generated noise of silencers. Information provided by silencer manufacturers shall be used wherever available. The empirical predictive scheme presented here

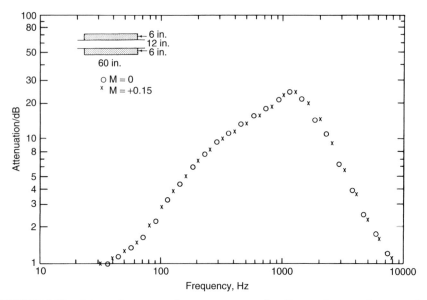

FIGURE 9.32 Attenuation-versus-frequency curve of a silencer for sound propagation in flow direction, $M = +0.15$: ($\times$) computed; ($\circ$) obtained by shifting the $M = 0$ curve according to Fig. 9.31a.

is based on a broad range of experimental data on flow-generated noise of duct silencers and is reproduced from ISO 14163:1998(E).[78]

An estimate for the octave-band sound power level of regenerated sound can be obtained from Eq. (9.48):

$$L_{W,\text{oct}} = B + \left\{ 10 \lg \frac{PcSn}{W_0} + 60 \lg \text{Ma} + 10 \lg \left[1 + \left(\frac{c}{2fH} \right)^2 \right] - 10 \lg \left[1 + \left(\frac{f\delta}{v} \right)^2 \right] \right\} \quad (9.48)$$

where B = a value depending on type of silencer and frequency, dB
v = flow velocity in narrowest cross section of silencer, m/s
c = speed of sound in medium, m/s
Ma = Mach number (Ma = v/c)
P = static pressure in duct, Pa
S = area of narrowest cross section of passage, m^2
n = number of passages
f = octave-band center frequency, Hz
H = maximum dimension of duct perpendicular to baffles, m
δ = length scale characterizing high-frequency spectral content of regenerated noise, m
$W_0 = 1 \text{ W} = 1 \text{ N} \cdot \text{m/s}$

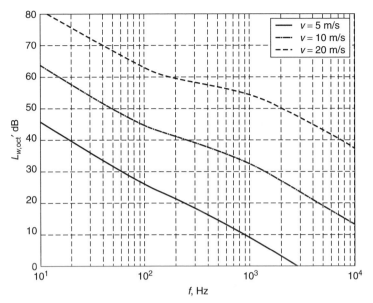

FIGURE 9.33 Octave-band sound power level $L_{W,\text{oct}}$ of regenerated sound versus frequency f for air under ambient conditions in a ducted silencer with narrowest cross section $S = 0.5 \text{ m}^2$, maximum transverse dimension $H = 1$ m, and different flow velocities v.

The sound power level of regenerated sound will vary with the temperature T approximately with $-25 \lg(T/T_0)$ decibels. For smooth-walled dissipative splitter silencers used in heating, ventilation, and air-conditioning equipment, an approximation is given by $B = 58$ and $\delta = 0.02$ m. For this case, a graph of Eq. (9.48) is shown in Fig. 9.33, and the A-weighted sound power level of a duct cross section of 1 m² is then calculated from

$$L_{WA} = \left(-23 + 67 \lg\left[\frac{v}{v_0}\right]\right) \quad \text{dB} \tag{9.49}$$

where $v_0 = 1$ m/s. For other types of silencers, particularly for resonator silencers, B may be larger in certain frequency bands. However, no general information can be given on the values of B and δ.

Prediction of Silencer Pressure Drop

The maximum permissible pressure drop at design flow rates and at design temperatures, together with the flow-generated noise, determines the cross-sectional dimensions of a silencer. It is important that the silencer designer makes good use of all the available pressure drop, though with an adequate factor of safety. The maximum permissible pressure drop $\Delta p_{\max}$ must be carefully allocated among the pressure drop of transition ducts (Δp_{trans}) and the total pressure drop of the silencer, Δp_{tot}:

$$\Delta p_{\max} \geq F_s(\Delta p_{\text{trans}} + \Delta p_{\text{tot}}) \quad \text{N/m}^2 \tag{9.50}$$

The specific choice of the safety factor ($F_s > 1$) is influenced by the degree of inhomogeneity of the inflow and by guaranty obligations regarding pressure drop performance of the silencer system. The silencer system usually includes the transition ducts both upstream and downstream of the silencer.

Detailed information on how to predict the pressure drop of inlet and exit transitions has been compiled by Idel'chik.[79] The silencer pressure losses are expressed as a product of the dynamic head in the muffler passage, $0.5\rho v_p^2$, and a loss coefficient. The terms K_{ENT}, K_F, and K_{EX} represent the entrance, friction, and exit loss coefficients. They are predicted according to the formulas given in Table 9.3

The required silencer face area is determined by an iteration process. First, Eq. (9.39) is solved for the passage velocity v_p. The initial face area A'_F is obtained as

$$A'_F = \frac{Q}{v_p}\left(\frac{100}{\text{POA}}\right) \quad \text{m}^2 \tag{9.51}$$

TABLE 9.3 Pressure Loss Coefficient for Parallel Baffle Silencers

Geometry	Loss Coefficient
Square-edge nose	$K_{\text{ENT}} = \dfrac{0.5}{1 + h/d}$
Rounded nose	$K_{\text{ENT}} \simeq \dfrac{0.05}{1 + h/d}$
Typical perforated metal facing	$K_F \simeq 0.0125$ ℓ = baffle length, tail and nose not included
Square tail	$K_{\text{EX}} = \left(\dfrac{1}{1 + h/d}\right)^2$
Rounded tail	$K_{\text{EX}} = 0.7\left(\dfrac{1}{1 + h/d}\right)^2$
Faired tail, 7.5°	$K_{\text{EX}} = 0.6\left(\dfrac{1}{1 + h/d}\right)^2$

where Q is the volume flow rate through the silencer (in cubic meters per second) v_p is the passage velocity (in meters per second), and POA is the percentage of open area of the silencer cross section (e.g., for parallel-baffle mufflers POA = 50 means that $h/d = 1$; the passage height is the same as the baffle thickness). Based on this initial value of $A_F = A'_F$, the pressure loss of the inlet and exit transitions is calculated and added to the muffler pressure drop. The total $F_s(\Delta p_{\text{trans}} + \Delta p_T)$ is compared with maximum permissible pressure drop Δp_{max} according to Eq. (9.50). If $F_s(\Delta p_{\text{trans}} + \Delta p_T) > \Delta p_{\text{max}}$, the face area of the silencer must be increased. This results in a decrease of Δp_T and a relatively smaller increase in Δp_{trans}, and the iteration is continued until the inequality, expressed in Eq. (9.50), is satisfied.

Economic Considerations

For large silencers such as used in electric power plants, the product of the pressure drop and volume flow rate, $Q\Delta p$, represents a substantial power that is lost (converted to heat). The cost to produce this power during the entire design life of the installation usually far exceeds the purchase cost of the silencer. Therefore, it is important to specify a silencer pressure drop that yields the lowest total cost. The total cost includes the present worth of revenue requirements to purchase and install the silencer (which decreases with increasing pressure drop allowance) and the present worth of revenue requirements of the energy cost caused by operating the silencer (which increases with increasing pressure drop allowance). The optimal pressure drop is that which yields the lowest total cost. Information on how to predict the optimal silencer pressure drop is given in reference 80.

9.7 COMBINATION MUFFLERS

It is clear from Sections 9.1–9.3 that the TL of reactive mufflers is generally characterized by several troughs that limit the overall value of TL notwithstanding the peaks, owing to the arithmetic of antilog summation. The acoustically lined ducts and parallel-baffle mufflers (or splitter silencers) do not suffer from this characteristic; they are characterized by a wide-band TL curve. However, they have rather poor performance (attenuation) at low frequencies where the reactive mufflers have a relative edge. One way out of these limitations is to combine reflective or reactive elements and dissipative elements into a combination muffler (or hybrid muffler). An acoustically lined plenum chamber is one obvious example of a combination muffler. Figures 9.34–9.37 illustrate the concept and present some design guidelines in the process.[81]

Figure 9.34a is an unlined simple axisymmetric expansion chamber. Figures 9.34b,c show the chamber with lining (with resistivity of 5000 Pa · s/m^2) of 18.2 and 36.4 mm thickness, respectively. The plots in Fig. 9.34 compare the computed values of their axial transmission loss TL$_a$ (assuming rigid shell). Clearly, the role of lining is to raise and even out the troughs, particularly at the medium and high frequencies.

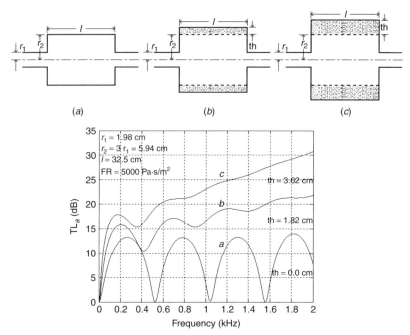

FIGURE 9.34 Use of acoustical lining in combination with a simple expansion chamber or plenum; th = lining thickness in centimeters.

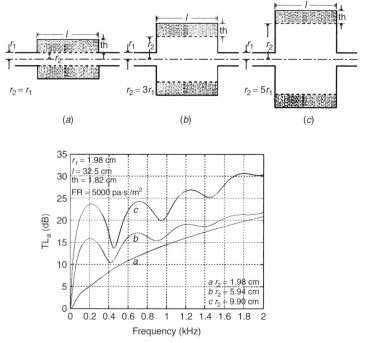

FIGURE 9.35 Use of sudden area discontinuities in combination with lined duct.

COMBINATION MUFFLERS

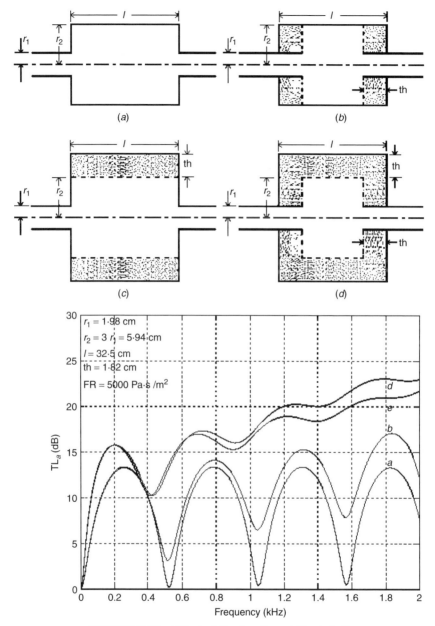

FIGURE 9.36 Effect of lined walls and lined shell.

Figure 9.35 shows the effect of building a sudden expansion and contraction into an acoustically lined circular duct, keeping the radial thickness of the lining constant. As can be noticed, the area discontinuities (sudden impedance mismatch) increase TL, particularly at the low frequencies (≤ 400 Hz).

338 PASSIVE SILENCERS

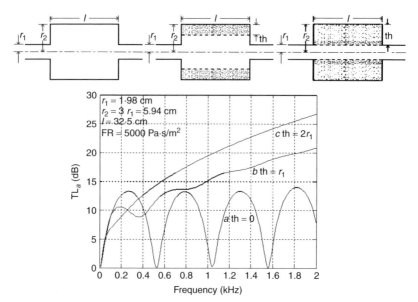

FIGURE 9.37 Reactive elements versus dissipative elements within the same overall shell radius.

A plenum chamber has lining not only on the shell but also on the sidewalls/plates. Figure 9.36 shows (a) a simple expansion chamber (plenum), (b) one with lined sidewalls, (c) one with lined shell only, and (d) one with lined shell as well as lined sidewalls (or end plates). It may be noticed that while the sidewall lining has considerable effect (in raising the troughs), the effect of shell lining is much more pronounced. In fact, a comparison of curves b and c of Fig. 9.36 shows that the relative effect of sidewall lining is only marginal.

In all the three foregoing figures, provision of lining has increased the overall radius of the muffler shell. Sometimes, it may not be feasible or advisable to increase the shell radius. Then, the internal lining of the shell would decrease the ratio of sudden expansion and sudden contraction. Figure 9.37 shows the effect of this compromise (between reaction and dissipation), keeping the overall shell radius constant. It may be observed that the performance of lined ducts is relatively poor at very low frequencies, when a simple unlined chamber has an edge. The two can supplement each other in a combination muffler (configuration b in Fig. 9.37).

REFERENCES

1. P. M. Morse, "Transmission of Sound in Pipes," *J. Acoust. Soc. Am.*, **11**, 205–210 (1999).
2. L. Cremer, "Theory of Attenuation of Airborne Sound in a Rectangular Duct with Sound-Absorbing Walls and the Maximum Achievable Attenuation" (in German), *Acustica*, **3**, 249–263 (1953).

3. D. D. Davis, Jr., M. Stokes, D. Moore, and L. Stevens, "Theoretical and Experimental Investigation of Mufflers with Comments on Engine Exhaust Design," NACA Report 1192, 1954.
4. D. Davis, "Acoustical Filters and Mufflers," in C. Harris (Ed.), *Handbook of Noise Control*, McGraw-Hill, New York, 1957, Chapter 21.
5. N. Doelling, "Dissipative Mufflers," in L. Beranek (Ed.), *Noise Reduction*, McGraw-Hill, New York, 1960, Chapter 17.
6. T. Embleton, "Mufflers," in L. Beranek (Ed.), *Noise and Vibration Control*, McGraw-Hill, New York, 1971, Chapter 12.
7. L. Ericsson, "Silencers," in D. Baxa (Ed.), *Noise Control in Internal Combustion Engines*, Wiley, New York, 1982, Chapter 5.
8. M. L. Munjal, *Acoustics of Ducts and Mufflers*, Wiley, New York, 1987.
9. M. L. Munjal, "Muffler Acoustics," in F. P. Mechel (ed.), *Formulas of Acoustics*, Springer-Verlag, Berlin, 2002, Chapter K.
10. C. L. Molloy, "Use of Four Pole Parameters in Vibration Calculations," *J. Acoust. Soc. Am.*, **29**, 842–853 (1957).
11. M. Fukuda and J. Okuda, "A Study on Characteristics of Cavity-Type Mufflers," *Bull. Jpn. Soc. Mechan. Eng.*, **13**(55), 96–104 (1970).
12. R. J. Alfredson and P. O. A. L. Davies, "Performance of Exhaust Silencer Components," *J. Sound Vib.*, **15**(2), 175–196 (1971).
13. M. L. Munjal, "Velocity Ratio Sum Transfer Matrix Method for the Evaluation of a Muffler with Mean Flow," *J. Sound Vib.*, **39**(1), 105–119 (1975).
14. P. T. Thawani and A. G. Doige, "Effect of Mean Flow and Damping on the Performance of Reactive Mufflers," *Can. Acoust.*, **11**(1), 29–47 (1983).
15. D. Ronneberger, "The Acoustical Impedance of Holes in the Wall of Flow Ducts," *J. Sound Vib.*, **24**, 133–150 (1972).
16. T. Melling, "The Acoustic Impedance of Perforates at Medium and High Sound Pressure Levels," *J. Sound Vib.*, **29**(1), 1–65 (1973).
17. P. D. Dean, "An In Situ Method of Wall Acoustic Impedance Measurement in Flow Ducts," *J. Sound Vib.*, **34**(1), 97–130 (1974).
18. J. W. Sullivan and M. Crocker, "Analysis of Concentric-Tube Resonators Having Unpartitioned Cavities," *J. Acoust. Soc. Am.*, **64**(1), 207–215 (1978).
19. J. W. Sullivan, "A Method for Modeling Perforated Tube Mufflers Components. I. Theory." *J. Acoust. Soc. Am.*, **66**(3), 772–778 (1979).
20. J. W. Sullivan, "A Method for Modeling Perforated Tube Mufflers Components. II. Applications." *J. Acoust. Soc. Am.*, **66**(3), 779–788 (1979).
21. K. Jayaraman and K. Yam, "Decoupling Approach to Modeling Perforated Tube Muffler Components," *J. Acoust. Soc. Am.*, **69**(2), 390–396 (1981).
22. P. T. Thawani and K. Jayaraman, "Modeling and Applications of Straight-Through Resonators," *J. Acoust. Soc. Am.*, **73**(4), 1387–1389 (1983).
23. M. L. Munjal, K. N. Rao, and A. D. Sahasrabudhe, "Aeroacoustic Analysis of Perforated Muffler Components," *J. Sound Vib.*, **114**(2), 173–188 (1987).
24. M. L. Munjal, M. V. Narasimhan, and A. V. Sreenath, "A Rational Approach to the Synthesis of One-Dimensional Acoustic Filters," *J. Sound Vib.*, **29**(3), 148–151 (1973).

25. M. L. Munjal, S. Krishnan, and M. M. Reddy, "Flow-Acoustic Performance of the Perforated Elements with Application to Design," *Noise Control Eng. J.*, **40**(1), 159–167 (1993).
26. M. L. Munjal, "Analysis and Design of Pod Silencers," *J. Sound Vib.*, **262**(3), 497–507 (2003).
27. C. W. S. To and A. G. Doige, "A Transient Testing Technique for the Determination of Matrix Parameters of Acoustic Systems, I: Theory and Principles," *J. Sound Vib.*, **62**(2), 207–222 (1979).
28. C. W. S. To and A. G. Doige, "A Transient Testing Technique for the Determination of Matrix Parameters of Acoustic Systems, II: Experimental Procedures and Results," *J. Sound Vib.*, **62**(2), 223–233 (1979).
29. C. W. S. To and A. G. Dogie, "The Application of a Transient Testing Method to the Determination of Acoustic Properties of Unknown Systems," *J. Sound Vib.*, **71**(4), 545–554 (1980).
30. T. Y. Lung and A. G. Doige, "A Time-Averaging Transient Testing Method for Acoustic Properties of Piping Systems and Mufflers with Flow," *J. Acoust. Soc. Am.*, **73**(3), 867–876 (1983).
31. M. L. Munjal and A. G. Doige, "Theory of a Two Source Location Method for Direct Experimental Evaluation of the Four-Pole Parameters of an Aeroacoustic Element," *J. Sound Vib.*, **141**(2), 323–334 (1990).
32. A. Galaitsis and E. K. Bender, "Measurement of Acoustic Impedance of an Internal Combustion Engine," *J. Acoust. Soc. Am.*, **58**(Suppl. 1) (1975).
33. M. L. Kathuriya and M. L. Munjal, "Experimental Evaluation of the Aeroacoustic Characteristics of a Source of Pulsating Gas Flow," *J. Acoust. Soc. Am.*, **65**(1), 240–248 (1979).
34. M. G. Prasad and M. J. Crocker, "Insertion Loss Studies on Models of Automotive Exhaust Systems," *J. Acoust. Soc. Am.*, **70**(5), 1339–1344 (1981).
35. M. G. Prasad and M. J. Crocker, "Studies of Acoustical Performance of a Multi-Cylinder Engine Exhaust Muffler System," *J. Sound Vib.*, **90**(4), 491–508 (1983).
36. M. G. Prasad and M. J. Crocker, "Acoustical Source Characterization Studies on a Multicylinder Engine Exhaust System," *J. Sound Vib.*, **90**(4), 479–490 (1983).
37. D. F. Ross and M. J. Crocker, "Measurement of the Acoustic Source Impedance of an Internal Combustion Engine," *J. Acoust. Soc. Am.*, **74**(1), 18–27 (1983).
38. A. F. Seybert and D. F. Ross, "Experimental Determination of Acoustic Properties Using a Two-Microphone Random Excitation Technique," *J. Acoust. Soc. Am.*, **61**, 1362–1370 (1977).
39. J. Y. Chung and D. A. Blaser, "Transfer Function Method of Measuring In-Duct Acoustic Properties II. Experiment," *J. Acoust. Soc. Am.*, **68**, 914–921 (1980).
40. M. L. Munjal and A. G. Doige, "The Two-microphone Method Incorporating the Effects of Mean Flow and Acoustic Damping," *J. Sound Vib.*, **137**(1), 135–138 (1990).
41. H. S. Alves and A. G. Doige, "A Three-Load Method for Noise Source Characterization in Ducts," in J. Jichy and S. Hay, (Eds.), *Proceedings of NOISE-CON*, National Conference on Noise Control Engineering, State College, PA, Noise Control Engineering, New York, 1987, pp. 329–339.
42. M. G. Prasad, "A Four Load Method for Evaluation of Acoustical Source Impedance in a Duct," *J. Sound Vib.*, **114**(2), 347–356 (1987).

43. H. Boden, "Measurement of the Source Impedance of Time Invariant Sources by the Two-Microphone Method and the Two-Load Method," Department of Technical Acoustics Report TRITA-TAK-8501, Royal Institute of Technology, Stockholm, 1985.
44. H. Boden, "Error Analysis for the Two-Load Method Used to Measure the Source Characteristics of Fluid Machines," *J. Sound Vib.*, **126**(1), 173–177 (1988).
45. V. H. Gupta and M. L. Munjal, "On Numerical Prediction of the Acoustic Source Characteristics of an Engine Exhaust System," *J. Acoust. Soc. Am.*, **92**(5), 2716–2725 (1992).
46. L. Desmonds, J. Hardy, and Y. Auregan, "Determination of the Acoustical Source Characteristics of an Internal Combustion Engine by Using Several Calibrated Loads," *J. Sound Vib.*, **179**(5), 869–878 (1995).
47. P. T. Thawani and R. A. Noreen, "Computer-Aided Analysis of Exhaust Mufflers," ASME paper 82-WA/NCA-10, American Society of Mechanical Engineers, New York, 1982.
48. L. J. Eriksson, P. T. Thawani, and R. H. Hoops, "Acoustical Design and Evaluation of Silencers," *J. Sound Vib.*, **17**(7), 20–27 (1983).
49. L. J. Eriksson and P. T. Thawani, "Theory and Practice in Exhaust System Design," SAE Technical paper 850989, Society of Automotive Engineers, Warrendale, PA, 1985.
50. M. L. Munjal, "Analysis and Design of Mufflers—An Overview of Research at the Indian Institute of Science," *J. Sound Vib.*, **211**(3), 245–433 (1998).
51. A. Jones, "Modeling the Exhaust Noise Radiated from Reciprocating Internal Combustion Engines—A Literature Review," *Noise Control Eng. J.*, **23**(1), 12–31 (1984).
52. M. L. Munjal, "State of the Art of the Acoustics of Active and Passive Mufflers," *Shock Vib. Dig.*, **22**(2), 3–12 (1990).
53. V. Easwaran and M. L. Munjal, "Plane Wave Analysis of Conical and Experimental Pipes with Incompressible flow," *J. Sound Vib.*, **152**, 73–93 (1992).
54. E. Dokumaci, "On Transmission of Sound in a Non-uniform Duct Carrying a Subsonic Compressible Flow," *J. Sound Vib.*, **210**, 391–401 (1998).
55. M. L. Munjal and P. T. Thawani, "Acoustic Performance of Hoses—A Parametric Study," *Noise Control Eng. J.*, **44**(6), 274–280 (1996).
56. K. S. Peat, "A Numerical Decoupling Analysis of Perforated Pipe Silencer Element," *J. Sound Vib.*, **123**, 199–212 (1988).
57. K. N. Rao and M. L. Munjal, "Experimental Evaluation of Impedance of Perforates with Grazing Flow," *J. Sound Vib.*, **108**, 283–295 (1986).
58. M. L. Munjal, B. K. Behera, and P. T. Thawani, "Transfer Matrix Model for the Reverse-Flow, Three-Duct, Open End Perforated Element Muffler," *Appl. Acoust.*, **54**, 229–238 (1998).
59. M. L. Munjal, B. K. Behera, and P. T. Thawani, "An Analytical Model of the Reverse Flow, Open End, Extended Perforated Element Muffler," *Int. J. Acoust. Vib.*, **2**, 59–62 (1997).
60. M. L. Munjal, "Analysis of a Flush-Tube Three-Pass Perforated Element Muffler by Means of Transfer Matrices," *Int. J. Acoust. Vib.*, **2**, 63–68 (1997).
61. M. L. Munjal, "Analysis of Extended-Tube Three-Pass Perforated Element Muffler by Means of Transfer Matrices," in C. H. Hansen (Ed.), *Proceedings of ICSV-5*,

International Conference on Sound and Vibration, Adelaide, Australia, International Institute of Acoustics and Vibration, Auburn, AL, 1997.
62. A. Selamet, V. Easwaran, J. M. Novak, and R. A. Kach, "Wave Attenuation in Catalytical Converters: Reactive versus Dissipative Effects," *J. Acoust. Soc. Am.*, **103**, 935–943 (1998).
63. M. L. Munjal and P. T. Thawani, "Effect of Protective Layer on the Performance of Absorptive Ducts," *Noise Control Eng. J.*, **45**, 14–18 (1997).
64. M. E. Delany and B. N. Bazley, "Acoustical Characteristics of Fibrous Absorbent Materials," *Appl. Acoust.*, **3**, 106–116 (1970).
65. F. P. Mechel, "Extension to Low Frequencies of the Formulae of Delany and Bazley for Absorbing Materials" (in German), *Acustica*, **35**, 210–213 (1976).
66. V. Singhal and M. L. Munjal, "Prediction of the Acoustic Performance of Flexible Bellows Incorporating the Convective Effect of Incompressible Mean Flow," *Int. J. Acoust. Vib.*, **4**, 181–188 (1999).
67. A Selamet, N. S. Dickey, and J. M. Novak, "The Herchel–Quincke Tube: A Theoretical, Computational, and Experimental Investigation," *J. Acoust Soc. Am.*, **96**, 3177–3199 (1994).
68. M. L. Munjal and B. Venkatesham, "Analysis and Design of an Annular Airgap Lined Duct for Hot Exhaust Systems," in M. L. Munjal (Ed.), *IUTAM Symposium on Designing for Quietness*, Kluwer, Dordrecht, 2002.
69. M. O. Wu, "Micro-Perforated Panels for Duct Silencing," *Noise Control Eng. J.*, **45**, 69–77 (1997).
70. P. O. A. L. Davies, "Plane Acoustic Wave Propagation in Hot Gas Flows," *J. Sound Vib.*, **122**(2), 389–392 (1998).
71. R. A. Scott, "The Propagation of Sound between Walls of Porous Material," *Proc. Phys. Soc. Lond.*, **58**, 338–368 (1946).
72. U. J. Kurze and I. L. Ver, "Sound Attenuation in Ducts Lined with Non-Isotropic Material," *J. Sound Vib.*, **24**, 177–187 (1972).
73. F. P. Mechel, "Explicit Formulas of Sound Attenuation in Lined Rectangular Ducts" (in German), *Acustica*, **34**, 289–305 (1976).
74. F. P. Mechel, "Evaluation of Circular Silencers" (in German), *Acustica*, **35**, 179–189 (1996).
75. F. P. Mechel, *Schallabsorber* (Sound Absorbers), Vol. III, *Anwendungen* (Applications), S. Hirzel, Stuttgart 1998.
76. M. L. Munjal, "Analysis and Design of Pod Silencers," *J. Sound Vib.*, **262**(3), 497–507 (2003).
77. U. J. Kurze and I. L. Ver, "Sound Silencing Method and Apparatus," U.S. Patent No. 3,738,448, June 12, 1973.
78. ISO 14163-1998(E), "Acoustics—Guidelines for Noise Control by Silencers," International Organization for Standardization, Geneva, Switzerland, 1998.
79. I. D. Idel'chik, *Handbook of Hydrodynamic Resistance, Coefficients of Local Resistance and of Friction*, translation from Russian by the Israel Program for Scientific Translations, Jerusalem, 1966 (U. S. Development of Commerce, Nat. Tech. Inf. Service AECTR-6630).

80. I. L. Ver and E. J. Wood, *Induced Draft Fan Noise Control*, Vol. 1: *Design Guide*, Vol. 2: *Technical Report*, Research Report EP 82-15, Empire State Electric Energy Research Corporation, New York January 1984.
81. S. N. Panigrahi and M. L. Munjal, "Combination Mufflers—A Parametric Study," *Noise Control Eng. J.*, Submitted for publication.

CHAPTER 10

Sound Generation

ISTVÁN L. VÉR
Consultant in Acoustics, Noise, and Vibration Control
Stow, Massachusetts

Acoustical engineers can be grouped into two categories: those who try to *minimize* the efficiency of sound radiation and are called *noise control engineers* and those who try to *maximize* it and are called *audio engineers*. This book addresses the first group exclusively.

The main duty of noise control engineers is to know how to reduce, in a cost-effective manner, the noise produced by equipment at the location of an observer. The noise reduction can be achieved (1) *at the source*; (2) *along the propagation path* by building extensive barriers between the source and receiver, installing silencers, and so on, as discussed in Chapters 5 and 9; and (3) *at the receiver* by enclosing it, as discussed in Chapter 12, or by creating a limited "zone of silence" around it by active noise control, as described in Chapter 18.

The most effective, and by far the least expensive, noise control can be achieved at the source by reducing its sound radiation efficiency. A reduction of the efficiency of noise generation of the source results in a commensurate *reduction of the noise at all observer locations*. In comparison, the erection of barriers, enclosing the receiver, and creating a localized "zone of silence" at the receiver are *effective only in limited spatial regions*. Putting silencers on the noisy equipment is cumbersome and usually increases the size and weight and reduces the mechanical efficiency of the equipment. The last two measures are often referred to as the *brute-force* methods.

The reduction of the efficiency of noise radiation at the source can be achieved passively by changing the shape of the body if its vibration velocity is given and by changing both the shape and mass of the body if the vibratory force acting on it is given. Reducing the efficiency of sound radiation by active means is generally accomplished by placing a secondary sound source of the same volume displacement magnitude but opposite phase in the immediate vicinity of the body, as discussed in detail in Chapter 18, or by a combination of these passive and active measures.

346 SOUND GENERATION

If constructional changes are possible, always the passive measures should be implemented first, even if the needed additional reduction will be achieved by active means. Even if they do not achieve all the needed noise reduction, passive noise control measures reduce the demand on the power-handling capability of the noise-canceling loudspeakers. In addition, passive noise control measures remain effective in case the active control fails. The most desirable, but also the most difficult to achieve, passive noise control measures are those which require constructional changes of the equipment. Not infrequently, the same constructional changes that reduce the efficiency of sound radiation also may improve the mechanical efficiency of the equipment. The primary purpose of every machine is to perform a specific, usually mechanical, function. The noise is an unwanted by-product. The reduction of the efficiency of sound radiation at the source must always be accomplished without interfering with the primary function of the equipment. As we will show later, the reduction in the efficiency of sound radiation of any noise source is always achieved by reducing the force or forces that act on the surrounding fluid.

10.1 BASICS OF SOUND RADIATION

This chapter deals with the sound radiation of small pulsating and oscillating rigid bodies. Throughout this chapter the adjective *small* means that the dimensions of the body are small *compared with the acoustical wavelength*. Sound radiation of large bodies is treated in Chapter 11. An exhaustive treatment of the sound radiation of all types of bodies is given in reference 1.

Historical Overview

Technical acoustics is a branch of applied physics. All acoustical phenomena can be derived from Newton's basic laws. During the early part of the last century the science of technical acoustics has been developed mainly by physicists with a strong background in electromagnetism or by electronics engineers. The reason for this is that only they had the mathematical background to deal with dynamic phenomena in general and, most importantly, with the wave equation. In those days, the education of mechanical engineers was almost exclusively in static deformation of structures and *slow* phenomena in hydrodynamics. Physicists often describe the acoustical phenomena by mathematical formalisms that facilitate the mathematical analyses. Unfortunately, these formalisms, such as *monopoles, dipoles, quadrupoles, "equivalent electric circuits," and "short circuiting,"* were often not familiar to mechanical engineers. This built artificial barriers for the mechanical engineers which have been overcome only in the last few decades due to the dramatic increase in the education of mechanical engineers in dynamic phenomena.

This chapter attempts to explain the important phenomena of sound generation in a form, the author believes, that might have been done if the theory of sound generation had been developed by mechanical engineers. As we will show in the remainder of the chapter, the sound radiation phenomena can be fully described without knowing anything about monopoles, dipoles, quadrupoles, and equivalent electric circuits.

The added advantage of characterizing the sound sources by the force and the moment they exert on the fluid rather than assigning abstract descriptors such as *equivalent point sound source, dipole strength*, and *quadrupole strength* means that such characterization gives a better understanding of the physical phenomena involved and identifies the actual physical parameters that control all types of sound generation processes.

Qualitative Description of Sound Radiation of Small Rigid Bodies

Small rigid bodies oscillating in an unbounded fluid experience three types of reaction forces; (1) inertia force needed to accelerate the fluid to move out of the way of the body, (2) force to compress the fluid, (3) friction forces owing to the viscosity of the fluid.

The inertia force, which is by far the largest of them, is 90° out of phase with the velocity of the body. The compressive force is in phase with the velocity and is much smaller than the inertia force. However, it has to be considered because it determines the sound radiation. The friction force is usually small enough compared with the two other components that it can be neglected by assuming that the fluid has no viscosity.

Sound is generated by phenomena that cause a localized compression of the fluid (gases or liquids). This chapter deals with the most common sources of sound generated by pulsation and oscillatory motion of small rigid bodies and by pointlike forces and moments acting on the fluid.

Anyone who has operated a bicycle pump can attest to it that it takes considerable force to maintain a repetitive pumping motion of the piston when the valve is closed. The compression of the air in the closed cavity generates a large reaction force that opposes the motion of the piston. The opposing force depends only on the magnitude of the stroke and does not depend on the rapidity of the pumping.

It takes considerably less force to perform the same repetitive pumping motion when the valve is open and the hose is not connected to the tire. The force that resists the stroke in this case is the inertia force needed to push the air mass (stroke times surface area of the piston times the density of the air) through the open valve. The inertia force is proportional to the acceleration (i.e., the rapidity of the pumping motion) of this mass. The reaction force is very small if the rate of repetition is low and increases with increasing repetition rate. At sufficiently high repetition rate it becomes harder to accelerate the air mass that must be pushed through the open valve during increasingly shorter time periods than to compress the air in the pump's cavity. In this case the force needed to operate the leaky pump approaches that needed to operate the pump with the valve closed.

As we will see, similar behavior is observed in the sound radiation of small rigid bodies where the surrounding fluid is compressed in earnest only when the force needed to move the fluid in the vicinity of the pulsating or oscillating body out of the way becomes the same order of magnitude as that required to compress it.

Pulsating Small Rigid Bodies. If a small body, exemplified by a pulsating sphere, operates at low frequencies where the wavelength of the sound is large compared with the acoustical wavelength, the pressure has two components. One is 90° out of phase with the velocity of the pulsating surface and a much smaller component is in phase with it. The former component is owing to the inertia of the surrounding fluid that is pushed into a larger spherical area during the outward motion of the pulsating surface and pulled back during the inward movement without much compression. The latter component of the pressure, which is in phase with the surface velocity, is owing to the compression of the surrounding fluid.

The force needed to overcome the inertia of this back-and-forth sloshing fluid volume increases with increasing frequency and so does the in-phase component of the surface pressure. Above a frequency where the wavelength becomes smaller than six times the radius of the sphere, it becomes easier to compress the fluid than to accelerate the fluid volume and the pulsating sphere becomes an efficient radiator of sound. Both the in-phase and 90° out-of-phase components of the pressure are evenly distributed over the surface of the sphere and the pulsating sphere radiates sound omnidirectionally.

At low frequencies not only the pulsating sphere but also all pulsating small bodies have an omnidirectional sound-radiating pattern and both the far-field sound pressure and radiated sound power depend only on the net volume flux and the frequency irrespective of how the vibration pattern is distributed over their surface. Because of these unique properties, pulsating small bodies are referred to as monopoles.

Oscillating Small Rigid Bodies. In the case of an oscillating rigid body, exemplified by an oscillating sphere, the fluid pushed aside by the forward-moving half of the sphere is "sucked in" by the void created by the receding half of the sphere. The fluid moves back and forth around the sphere between the two poles. The velocity of the fluid is highest near the surface of the sphere and decreases exponentially with increasing radial distance. This exponentially decreasing velocity field is referred to as the near field. The kinetic energy flux of the near field depends on the geometry of the body, the direction of the oscillatory motion, and the density of the fluid surrounding the body. The quantity *added mass/fluid density* is a property of the body and its motion. It is referred to in the acoustical literature as the *added volume*. In the hydrodynamic literature the *added mass coefficient* is defined as the ratio of the added volume, V_{ad}, and volume of the body, V_b.

Push–pull action between the poles (located in the direction of the oscillatory motion) makes it easy to move the fluid back and forth and the surface of the oscillating small rigid sphere, for the same surface velocity, experiences much smaller reaction pressure than the pulsating sphere and radiates substantially less sound. The magnitude of the reaction pressure on the surface of the oscillating sphere has its maxima at the poles where the radial velocity has its maxima and zero at the equator where the radial velocity is zero. Consequently, the sound radiation pattern is maximum in the direction of the oscillatory motion and zero in the direction perpendicular to it. This behavior is not restricted to the oscillating sphere but holds for small oscillating rigid bodies of any shape. In analogy to the similar radiation patter of the magnetic field caused by an oscillating charge, oscillating small rigid bodies are referred to as dipole sound sources.

The net oscillatory force exerted on the fluid is obtained by integrating the pressure distribution over the surface of the sphere. The force points in the direction of the oscillatory motion. As we will show later, with respect to sound radiation, a small, oscillating rigid body is equivalent to a pointlike force (i.e., pressure exerted by a small surface) acting on the fluid. The radiated sound power is obtained as the product of the body's velocity and the component of the net reaction force that is in phase with the velocity.

The distribution of the reaction pressure on the surface is known only for a few bodies. Consequently, the net oscillatory force they exert on the fluid (and consequently the radiated sound field) cannot be obtained directly. As we will show later, the force exerted on the fluid by such bodies—and consequently also their sound radiation pattern—can be predicted on the bases of the oscillatory velocity and geometry of the body and its added mass in the direction of the oscillatory motion. The latter can be found frequently in the hydrodynamic literature or can be determined experimentally by measuring the resonance frequency of the body supported on a spring; first in a vacuum then immersed in a fluid (preferably a liquid).

Moment Excitation: Lateral Quadrupole. Two closely spaced identical small rigid bodies oscillating in the same direction but 180° out of phase, or the equivalent two closely spaced parallel pointlike forces acting on the fluid 180° out of phase, are referred to as a *lateral quadrupole*. The forces constituting the moment are frequently due to shear strain. Both of these descriptions are helpful in understanding the sound radiation properties.

If we consider the representation of the two closely spaced small rigid bodies oscillating in opposite directions, it becomes apparent that the fluid pushed aside by the pole of one of the bodies does not need to be moved around to the opposite pole but is easily moved into the void created by the receding pole of the second body. The reaction pressure on the surface of each of the bodies, and consequently the sound field they produce, is substantially smaller than would result when only one of the bodies would oscillate.

If we consider the model of two parallel closely spaced pointlike forces, we realize that the lateral quadrupole represents excitation of the fluid by a moment.

The source strength is characterized by the moment, regardless of whether the moment is constituted by large forces spaced very close to each other or by smaller forces spaced at a larger distance apart, provided that the distance is still much smaller than the wavelength.

The sound radiation pattern of a lateral quadrupole is zero in two planes. The first of this is perpendicular to the plane defined by the two parallel forces and cuts the plane of the forces halfway between them. All points located in this plane have an equal distance from the two bodies moving in opposite directions and their contribution to the sound field cancels each other.[2] The second plane of zero sound pressure is perpendicular to the direction of the motion of the oscillating bodies and cuts through their center. The sound pressure is zero at all points on this plane because the two oscillating bodies constituting the lateral dipole have zero radial velocity in these directions.

The sound radiation pattern has maxima in two planes. Both of these planes cut the zero planes at 45° angles. In two dimensions the radiation pattern resembles an old-fashioned, four-bladed propeller where the four maxima are represented by the four blades and the minima by the void between the blades.

If the two opposing forces are of different magnitude, the sound radiation pattern is the superposition of the quadrupole radiation pattern associated with a moment constituted by two opposing forces having the same magnitude as the smaller force and a dipole pattern associated with a single pointlike force that equals the difference between the two forces.

Because the maxima of the dipole pattern "fill" the zero planes of the quadrupole pattern, the resulting radiation pattern has no zero value in any direction.

Sound Excitation by Two Opposing Forces Aligned Along a Line: Longitudinal Quadrupole. If two equal-magnitude but opposing dynamic forces are placed on a line a small distance apart (or the equivalent of two identical small rigid bodies are oscillating 180° out of phase), the sound radiation pattern resembles that of a single pointlike force. However, the two maximum lobes, pointing in the direction of the forces, are much narrower and the radiated sound power is much smaller than if only one of the forces acted on the fluid. In analogy to the directivity pattern of the magnetic field generated by two closely spaced charges oscillating in the opposite direction, this type of sound source is referred to as a longitudinal quadrupole. This type of radiation pattern occurs when a small rigid body oscillates in water near to and perpendicular to the water–air interface. In this case the mirror image represents the identical second body that oscillates 180° out of phase.

If the two opposing forces are not equal, then the resulting radiation pattern is the superposition of the pattern of a longitudinal quadrupole and a dipole. Because both radiation patterns have zero value in a plane that is perpendicular to the forces and cuts through the center of the force pair, the resulting radiation pattern has zero value in this plane. The contribution of the dipole component widens the two maximum lobes.

BASICS OF SOUND RADIATION   351

Radiated Sound Power. The sound power radiated by all of the above-discussed sound sources can also be determined by integrating the intensity $p_{\text{rms}}^2(r, \vartheta, \theta)/\rho_0 c_0$ over a large radius ($kr \gg 1$) sphere centered at the acoustical center of the source, where $p_{\text{rms}}^2(r)$ is the mean-square value of the sound pressure at distance r in the direction designated by the angles ϑ and θ.

Acoustical Parameters of Sound Radiation

This section contains a brief introductory discussion of the acoustical parameters of fluids, including the speed of sound, density, and plane-wave acoustical impedance.

Propagation Speed of Sound in Fluids. The molecules of fluids and gases in equilibrium undertake a random motion with equal probability of direction, c_t, and the propagating speed in a stationary fluid, c_0, given by[2]

$$c_0 = c_t\sqrt{\gamma} = \sqrt{\frac{\gamma RT}{M}} \quad \text{m/s} \quad (10.1)$$

where $\gamma = c_p/c_v$ is the specific heat ratio with $\gamma \approx 1.4$ for air and $\gamma \approx 1.3$ for steam, R is the universal gas constant ($R = 8.9$ J/K), T is the absolute temperature in degrees Kelvin (K), (0 K $= -273°$C), and M is the molar mass equal to 0.029 kg/mole for air and 0.018 kg/mole for steam. For air Eq. (10.1) yields

$$c_0 = 20.02\sqrt{T(\text{K})} = 342.6\sqrt{\frac{T(\text{K})}{293}} = 342.6\sqrt{\frac{T(°\text{C}) + 273}{293}} \quad \text{m/s} \quad (10.2)$$

For air at $20°$C $= 68°$F $= 293$ K the propagating speed is $c_0 = 342.6$ m/s, at which a sound wave travels in air. The individual molecules collide with each other in a random fashion exchanging momentum with each other. If they impinge on a solid impervious wall, they rebound back from it. Sound is generated when dynamic forces acting on the fluid superimpose a very small but organized dynamic velocity on the very large random velocity of the molecules. The word "organized" means that in the case of periodic motion the superimposed velocity has a specific direction and a specific frequency and in the case of random motion it has a direction and its frequency composition can be described either by its power spectrum or autocorrelation function, as described in Chapter 3. In a stationary fluid the kinetic energy of the molecules acquired through the excitation causes compression and the compression then accelerates the molecules. In a stationary fluid the kinetic energy and potential energy are in balance in any small volume of the fluid. If the sound source acts on a moving media, the kinetic energy depends on the velocity and the direction of the fluid movement. In this chapter we will deal with stationary media only. Sound radiation in moving media is covered in Chapter 15.

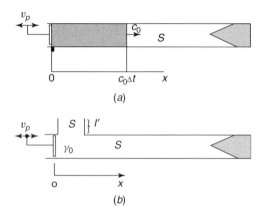

FIGURE 10.1 Oscillating plane rigid piston radiating sound into a small-diameter rigid pipe that is either infinitely long or anechoically terminated: (a) Hypothetical experiment to determine the plane wave impedance of fluids, No side branch; (b) open side branch of the same cross section area S as the face area of the piston and effective length l'.

The superimposed organized particle velocity of sound we experience is usually in the range 10^{-8} m/s (corresponding to a sound pressure level of 14 dB re 20 μPa) to 10^{-4} m/s (corresponding to 94 dB re 20 μPa).

Plane-Wave Impedance. The plane-wave impedance can be obtained by performing a hypothetical experiment similar to that suggested by Ingard,[2] as shown in Fig. 10.1a.

Hypothetical Experiment 1. Assume, as sketched in Fig. 10.1a, that a plane, rigid piston of surface area S is located on the left end of a small-diameter (much smaller than the wavelength), infinitely long or anechoically terminated tube with rigid walls filled with a fluid of density ρ_0 and speed of sound c_0. If the piston moves to the right with a constant speed of v_p, the face of the piston experiences a reaction pressure. To determine the plane-wave impedance defined as $Z_0 = p(x=0)/v_p$, we need to find that particular pressure $p(x=0)$ at the piston–fluid interface (at position $x=0$) that results from the piston velocity v_p. At the time Δt, the shaded volume of the fluid on the left, $S_0 c_0 \Delta t$, has acquired the particle velocity v_0 and the rest of the fluid on the right is motionless. The momentum M acquired by the fluid during the time Δt is then $M = v_0 \rho S_0 c_0 \Delta t$. According to Newton's law the force on the interface is

$$F(x=0) = Sp(x=0) = \frac{d}{dt} M \, \Delta t = v_p \rho_0 S c_0 \quad \text{N} \qquad (10.3)$$

yielding the plane wave impedance Z_0,

$$Z_0 \equiv \frac{p(x=0)}{v_p} = \rho_0 c_0 \quad \text{N} \cdot \text{s/m}^3 \qquad (10.4)$$

Performing the same thought experiment on a solid bar yields the plane-wave impedance of a solid as

$$Z_0 \approx \rho_M c_L \quad \text{N} \cdot \text{s/m}^3 \tag{10.5}$$

where c_L = propagation speed of longitudinal waves in bar, $(E/\rho)^{1/2}$ m/s
ρ_M = density of solid material, kg/m³
E = Young's modulus, N/m²

If the piston oscillates with a velocity of the form $v_p e^{j\omega t}$, starting at $t = 0$, the sound pressure in the tube filled with a fluid (air) is

$$p(x, t) = v_p \rho_0 c_0 e^{-jkx} e^{j\omega t} \quad \text{N/m}^2 \tag{10.6}$$

and the sound power radiated by the piston, W_{rad}, is

$$W_{\text{rad}} = \tfrac{1}{2} \hat{v}_p \; \text{Re}[\hat{p}_p(x = 0)] \; S\rho_0 c_0 \quad \text{N} \cdot \text{m/s} \tag{10.7}$$

where in Eqs. (10.6) and (10.7) $\omega = 2\pi f$, f is the frequency in hertz, $k = \omega/c_0$ is the wavenumber, and the hat above the symbol signifies the peak value. Observing Eqs. (10.6) and (10.7), note that the higher is the product of density and speed of sound, the larger are the sound pressure and the radiated power that a given excitation velocity v will generate.

For example, when plane waves are generated in a column of air, water, and steel, an enforced velocity v = 1 mm/s = 10^{-3} m/s will produce the sound pressure p, sound pressure level SPL (in dB re 2×10^{-5} N/m²) listed below. Assuming that the cross-sectional area is 10^{-4} m², we also can predict the radiated sound power W_{rad} in watts:

Material	ρ (kg/m³)	c (m/s)	p (N/m²)	SPL (dB)	W_{rad} (W)
Air inside rigid tube	1.21	344	0.42	86	4.2×10^{-8}
Water inside rigid tube	998	1481	1478	157	1.5×10^{-4}
Steel bar in air	7700	5050	3.9×10^4	186	3.9×10^{-3}

In plane waves all of the volume displacement produced by the vibratory excitation at the interface is converted into compression and the resulting sound pressure at the interface is in phase with the excitation velocity. In this case, the sound radiation is that which is the maximum achievable. In other situations, where the volume displaced by the vibratory excitation is mainly used for "pushing aside" rather compressing the fluid, the resulting sound pressure at the interface is much smaller. That is, the pressure has only a small component that is in phase with the vibration velocity of the excitation, which is the reason for the inefficient low-frequency sound radiation of small bodies. For such small sound sources, maintaining the same velocity with increasing frequency means that it is increasingly more difficult to push aside the fluid. When it becomes much easier to compress the fluid than to push it aside, the pressure produced at the interface reaches its maximum and is nearly in phase with the excitation velocity and the efficiency of sound radiation approaches that of a plane wave. The remaining very small but finite phase difference is essential because it determines the radius of curvature of the wave front.

Piston Generating Sound in Rigid Tube with Open Side Branch. To illustrate the effect in a quantitative manner that the efficiency of sound generation is decreased if it is easier to push the fluid aside than to compress it, consider the situation shown in Fig. 10.1b. It differs from the situation depicted in Fig. 10.1a by the addition of an open side branch to the pipe in the immediate vicinity of the piston. Let assume that the piston, the main pipe, and the side-branch pipe all have the same cross-sectional area S and that the effective length of the open-ended side branch is l'. The mass of the air in the side branch is $\rho_0 S l'$. The force needed to push this mass aside, $j\omega v_p \rho_0 S l'$, is smaller than the force $v_p \rho_0 c_0 S$ needed to compress the fluid, and the presence of the open side branch will reduce the ability of the piston to exert force on the fluid.

The sound pressure that the oscillating piston with peak velocity amplitude $\hat{v}_p$ produces is given by

$$\hat{p}(x=0) = \hat{v}_p \frac{1}{1/(j\omega\rho_0 l') + 1/(\rho_0 c_0)} \quad \text{N/m}^2 \qquad (10.8)$$

Separating the real and imaginary parts in Eq. (10.7) yields

$$\hat{p}(x=0) = \hat{v}_p \frac{(\omega\rho_0 l')^2 \rho_0 c_0 + (j\omega\rho_0 l')(\rho_0 c_0)^2}{(\rho_0 c_0)^2 + (\omega\rho_0 l')^2} \quad \text{N/m}^2 \qquad (10.9)$$

At low frequencies where $\omega\rho_0 l' \ll \rho_0 c_0$, Eq. (10.8) yields

$$\hat{p}(x=0) \approx \hat{v}_p (j\omega\rho_0 l') \quad \text{N/m}^2 \qquad (10.10)$$

At high frequencies where $\omega\rho_0 l' \gg \rho_0 c_0$, Eq. (10.8) yields

$$\hat{p} \approx \hat{v}_p \rho_0 c_0 \quad \text{N/m}^2 \qquad (10.11)$$

Inspecting Eqs. (10.9) and (10.10), note that the sound pressure produced by the piston is small, is nearly 90° out of phase of the velocity of the piston, and increases linearly the frequency ω. Using the vocabulary of electric engineers, the low impedance of the mass in the open side branch, $j\omega\rho_0 l'$, "shunts" $\rho_0 c_0$, the plane-wave impedance of the fluid in the tube.

More specifically at low frequencies most of the volume displaced by the oscillating piston flows into the open side branch, where it encounters little "resistance," and only a very small portion flows into the main tube. Consequently, the magnitude of the reaction force exerted on the piston owing to the presence of both the side branch and the main pipe is only a minute amount larger than the small force needed to keep the fluid mass in the side branch oscillating. The same small force is exerted on the fluid in the main pipe. This force is substantially smaller than it would be if the entrance of the side branch were closed. Accordingly, the sound pressure in the main tube is also substantially smaller.

The sound power that the piston is radiating into the main tube is the product of the velocity of the piston and the small component of the reaction force that is in phase with the velocity and is given by

$$W_{\text{rad}} = \tfrac{1}{2}\hat{v}_p \, \text{Re}[\hat{p}(x=0)] \, S = \tfrac{1}{2}\hat{v}_p^2 S \frac{\rho_0 c_0 (\omega \rho_0 l')^2}{(\rho_0 c_0)^2 + (\omega \rho_0 l')^2} \quad \text{N·m/s} \quad (10.12)$$

The hat above the variables indicates peak amplitude. If the rms value of the piston' velocity were used, the factor $\tfrac{1}{2}$ would be omitted in Eq. (10.12).

Inspecting Eq. (10.12), note that at low frequencies where $\omega \rho_0 l' \ll \rho_0 c_0$, the radiated sound power approaches zero with decreasing frequency. At high frequencies where $\omega \rho_0 l' \gg \rho_0 c_0$, the radiated sounds power becomes independent of frequency and approaches that obtained in Eq. (10.7) for the pipe without an open side branch.

Pipe with Open Side Branch Representing Sound Radiation of Pulsating Sphere. Following the qualitative suggestion by Cremer and Hubert,[3] we are assigning the specific values for S and l' in Eqs. (10.9) and (10.12), namely, $S = 4\pi a^2$, $\hat{v}_p = \hat{v}_s$, and $l' = a$, and considering that the wavenumber $k = \omega/c_0$ and $\hat{v}_p S = \hat{v}_s 4\pi a^2 = \hat{q}$ is the peak volume velocity of our sound source, Eq. (10.9) yields the well-known[3] formula for the peak value of the sound pressure on the surface of the sphere:

$$\hat{p}(a) = \hat{v}_s \frac{j \omega \rho_0 a}{1 + jka} \quad \text{N/m}^2 \quad (10.13)$$

Also note that the volume of fluid in the side branch is $V_{\text{ad}} = 4\pi a^3$ and the added mass of the fluid is $M_{\text{ad}} = 4\pi a^3 \rho_0$. For a pulsating sphere the volume V_{ad} is three times the volume of the sphere and is called the *added volume* in the acoustical literature and in M_{ad} is called the *added mass*. The added mass plays a central role in the sound radiation of small pulsating and oscillating rigid bodies. The added mass is a measure of the ability of these bodies to exert a dynamic force on the surrounding fluid.

Substituting the same values into Eq. (10.12) yields the well-known formula[4] for the sound power radiated by the a pulsating sphere of radius a and peak surface velocity $\hat{v}_s$:

$$W_{\text{rad}} = \hat{v}_s^2 4\pi a^2 \left(\frac{\rho_0 c_0}{2}\right) \frac{(ka)^2}{1 + (ka)^2} = \hat{q}^2 \left(\frac{\rho_0 c_0}{2}\right) \frac{(ka)^2}{1 + (ka)^2} \quad \text{N/m}^2 \quad (10.14)$$

Sound Radiation by Volume-Displacing Sound Sources

To describe the sound radiation of small volume-displacing sound sources, we take the unusual approach of starting with the case where the source operates in the most restrained environment and then gradually weakening the restraints to arrive at the unrestrained case where the sound source operates in an undisturbed infinite fluid.

On the top of Figure 10.2 is shown the starting point where the volume-displacing sound source is represented by a piston that vibrates with an axial velocity v_p in the mouth of a rigid linear horn. The horn represents the restraint. The linear horn is a rigid hollow cone with the tip cut off. The cutoff end becomes the mouthpiece of the horn. The diameter of the cross-sectional area of the linear horn increases linearly with increasing distance r from the tip and its cross-sectional area increases with r^2. The cross-sectional area of the horn at its mouth, $l\theta$, is chosen to fit the mouth of a human speaker. The symbol θ is the solid angle (θ equals π for a quarter space, 2π for a half space, and 4π for a full space, as illustrated in Fig. 10.3).

The cross-sectional area at the large end of the horn is chosen to be $S_{End} \geq \lambda^2 \geq c_0^2/f_{int}^2$, where f_{int} is the lowest frequency of interest (usually 500 Hz for voice communication) so that practically no sound is reflected from the far end of the horn. If this condition is fulfilled, the use of the linear horn is equivalent to increasing the size of the mouth of the human speaker by the ratio of the end and mouthpiece cross-sectional areas of the horn, as illustrated by the lower sketch in Fig. 10.11.

The cross-sectional area of the horn can be expressed as

$$S(r) = \theta r^2 \quad m^2 \tag{10.15}$$

According to Cremer and Hubert,[3] the sound pressure inside the horn as a function of distance from the tip can be expressed as

$$\tilde{p}(r) = \frac{\tilde{v}_p \, (l/r)}{1/(\rho_0 c_0) + 1/(j\omega \rho_0 l)} \quad N/m^2 \tag{10.16}$$

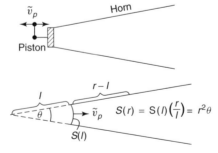

FIGURE 10.2 Vibrating piston sound source driving a linear acoustical horn (see text for explanation of symbols).

where v_p is the vibration velocity of the piston at the mouth of the horn and l is the radial length from the cutoff tip of the cone to the mouthpiece end, as illustrated in Fig. 10.2. Observing Eq. (10.16), note the following:

1. The term $j\omega\rho_0 l$ represents the added mass that is the consequence of the expansion of the cross section, which makes it possible to push the fluid off axis. In the extreme case when $\theta \to 0$ the horn turns into a rigid pipe and the added mass becomes infinite. This means that the near field in the pipe extends to infinity.
2. The sound propagates in the linear horn at all frequencies (which is not the case in an exponential horn, where no sound propagates below its cutoff frequency).
3. At low frequencies when $kl = (\omega/c_0)l \ll 1$ the sound pressure increases linearly with frequency as $p(r) \cong j\omega\rho_0(l^2/r)\tilde{v}_p$.
4. Above the frequency $\omega \geq c_0/l$ the linear horn works at full efficiency and the sound pressure as a function of distance becomes $p(r) \cong v_p(l/r)\rho_0 c_0$.

Considering that the volume velocity at the mouth of the horn is $q = v_p \theta l^2$, the radiated sound power takes the form

$$W_{\text{rad}} = q^2 \rho_0 c_0 \left(\frac{k^2}{\theta}\right) \frac{1}{1+(kl)^2} = q^2 \frac{\rho_0 \omega^2}{c_0 \theta}\left[\frac{1}{1+\pi(S/\lambda^2)(4\pi/\theta)}\right] \quad \text{N} \cdot \text{m/s} \tag{10.17}$$

Figure 10.3 shows what happens if we gradually weaken the restraint by increasing the solid angle θ from zero to 4π. In the extreme case, when the solid angle

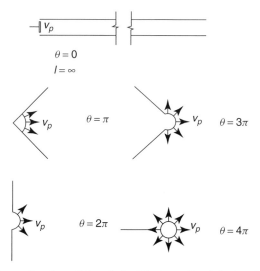

FIGURE 10.3 Increasing the solid angle θ of the horn; Top left: $\theta = 0$, pipe. Middle left: $\theta = \pi$, quarter space. Bottom left: $\theta = 2\pi$, half space. Middle right: $\theta = 3\pi$, three-quarter space. Bottom right: pulsating sphere in unbounded space.

358 SOUND GENERATION

of the horn θ is zero and tip length l approaches infinity, the horn becomes a cylindrical pipe, as shown in the top sketch. In this extreme case the sound pressure in the pipe is

$$p(r, t) = (q/S) \, e^{-jkr} \, e^{j\omega t} \quad \text{N/m}^2 \qquad (10.18)$$

indicating that theoretically the amplitude of the sound pressure would not decrease with distance. A practical use of this attribute is shown on the left in Fig. 10.11.

Consider now the effect of increasing the solid angle of the horn from $\theta = 0$ by an infinitesimally small $\Delta\theta$. Consequence of this is a phenomenological change in the behavior of the sound field, namely the sound pressure approaches zero at an infinitely large axial distance while for the pipe it would remain just as high as on the surface of the piston. Such quantum jumps are not allowed in Newtonian physics. The change in the behavior of the sound field is rapid but continuous. In the constant cross-sectional area of the pipe the near field extends infinitely. In the linear horn with infinitesimally small solid angle the near field extends nearly infinitely. Since in practice there are only finite-length pipes and horns, a microphone near the end of the anechoically terminated finite-length horn with an infinitesimally small solid angle would practically sense the same sound pressure as it would in a pipe.

Let us now explore the behavior of our linear horn given in Eq. (10.17) at solid angles of practical interest. Equation (10.17) yields

$$W_{\text{rad}} = q^2 (\rho_0 c_0) \left(\frac{k^2}{4\pi} \right) n \quad \text{N} \cdot \text{m/sec} \qquad (10.19)$$

where $n = 1$ when the volume source is located in an infinite, undisturbed fluid;
$n = 2$ when the volume source is flash mounted in an infinite, hard, flat baffle and radiates into a half space;
$n = 4$ when the sound source is located in a two-dimensional corner and radiates into a quarter space; and
$n = 8$ when located in a three-dimensional corner and radiating into an eighth space.

If there is a net volume displacement (as far as the far-field sound pressure and the radiated sound power are concerned), the specific distribution of the velocity over the vibrating surface is unimportant. This is because the non–volume-displacing vibration patterns (dipole, quadrupole, etc.) radiate low-frequency sound much less efficiently than volume-displacing patterns.

Figure 10.4 shows volume sources with different distributions of the vibration velocity but having the same net volume velocity. The solid and dashed lines represent the location of the vibrating surface at two time instances. The solid line shows the maximum and the dashed line the minimum volume displacement positions, respectively.

BASICS OF SOUND RADIATION

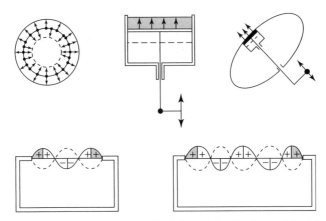

FIGURE 10.4 Small volume sources with different distributions of the vibration velocity but the same net volume velocity.

Limits of Maximum Achievable Volume Velocity of Pulsating Sphere.

The maximum peak volume velocity $\hat{q}_{max}$ achievable with a small pulsating sphere of volume $V_{sphere} = 4\pi a^3/3$ is limited by (1) the impossibility of having a larger volume displacement than the volume of the sphere, (2) the desire of not having supersonic surface velocity, and (3) the desire of keeping nonlinear distortion within acceptable limits. These requirements are accounted for in Eq. (10.20) below.

$$\frac{\hat{q}_{max}}{V_{sphere}} \leq \frac{n}{a/(3c_0) + 1/(12\pi\omega)} \quad \text{s}^{-1} \qquad (10.20)$$

where the first summand in the denominator accounts for the first limit, the second for the second limit, and the factor $n \ll 1$ in the numerator for the third limit.

Combining Eqs. (10.19 and 10.20) and solving for $\hat{W}_{rad}$ yield the maximum peak acoustical power a pulsating sphere of radius a can produce:

$$\hat{W}_{rad}(max) \leq 2\pi \left(\frac{\rho_0 c_0^3 a^4 k^2 n^2}{\{1 + [3/(12\pi)][1/(ka)]\}^2} \right)$$

$$= 2\pi \left(\frac{\rho_0 \omega^2 c_0 a^4 n^2}{1 + [3/(12\pi)][c_0/(\omega a)]} \right) \quad \text{N} \cdot \text{m/s} \qquad (10.21)$$

If we use the pulsating sphere as a loudspeaker to reproduce voice and music, it would be appropriate to choose n in the range of 0.001–0.01 to keep distortion at peak events within tolerable limits.

A dodecahedron-shaped enclosure with a loudspeaker flash mounted on each of the 12 faces of the enclosure is often used to simulate an omnidirectional volume sound source. The top photograph in Fig. 10.5 shows such a special reciprocity transducer[5-7] (loudspeaker) that the author has developed for the NASA

360 SOUND GENERATION

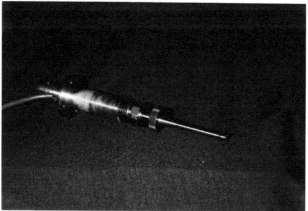

FIGURE 10.5 Photographs of a 4-in. edge-length dodecahedron loudspeaker to simulate a pulsating sphere as an omnidirectional sound source with volume velocity calibration and compensation for acoustical loading. Top: View of the loudspeaker. Bottom: Removed internal $\frac{1}{4}$-in. microphone used for volume velocity calibration and for correcting the acoustical loading when the source is placed in a confined environment.

Langley Research Center. The lower photograph in Fig. 10.5 shows the internal microphone removed from the center of the speaker. The internal microphone is used for volume velocity calibration at low frequencies, for measuring and compensating for the effect of acoustical loading (when the loudspeaker is placed in a small, confined environment such as inside an automobile or in the cockpit of an airplane or placed near a boundary) on the volume velocity, and for measuring nonlinear distortion at low frequencies. The volume velocity calibration of the transducer at mid- and high frequencies was accomplished experimentally by calibrating the transducer as a microphone and utilizing the principle of reciprocity to obtain the volume velocity as a function of the applied voltage to the loudspeakers.

Example 10.1. determine the minimum radius $a_{\min}$ of a pulsating sphere that should radiate 1 W peak acoustical power at 100 Hz for a distortion parameter $n = 0.005$ and $n = 0.01$.

Solution solving Eq. (10.21) for the radius a yields

$$a_{\min} = \begin{cases} 0.088 \text{ m} \cong 3.5 \text{ in.} & \text{for } n = 0.005 \\ 0.075 \text{ m} \cong 3 \text{ in.} & \text{for } n = 0.01 \end{cases}$$

The investigations above indicates that to radiate a practically meaningful amount of sound power at low frequencies requires a sizable sound source. Consequently, a *point sound source* is a practical impossibility. Furthermore, if the point sound source does not exist, then there are no physically meaningful acoustic dipoles and quadrupoles.

Consequently, it wound be physically more meaningful to substitute the following in the engineering acoustical vocabulary:

small-volume sound source instead of *monopole* or *point source*,
force (acting on the fluid) instead of *dipole*, and
moment (acting on the fluid) instead of *quadrupole*.

Sound Radiation by Non-Volume-Displacing Sound Sources

Parts (*a*) and (*b*) in Figure 10.6 show the response of an unbounded fluid to rigid, non-volume-displacing sound sources and parts (*c*) and (*d*) that of volume-displacing sound sources. All of the sources are small compared with the acoustical wavelength. The upper row depicts the body and shows the variables that are responsible for the sound generation. The lower row shows, in a schematic manner, how the fluid particles are pushed aside, creating a near field. The symbol $\tilde{v}_b$ represents the dynamic velocity of the body and $\tilde{v}_n$ the particle velocity in the near field. As we will show latter, the kinetic energy flux of this near field, $\omega(0.5\rho_0\tilde{v}_n)$, integrated over the entire fluid volume outside the body represents the added mass, which is a measure for the vibrating body's ability to exert a dynamic force on the fluid ("grab onto it").

In Fig. 10.6 (*a*) and (*b*) are examples of non-volume-displacing sound sources and (*c*) and (*d*) of volume-displacing sound sources.

In Fig. 10.6*a* the upper sketch depicts a small rigid body that exhibits random motion in one dimension with a velocity $\tilde{v}_b$ and the lower sketch depicts the direction and path of the particle velocity in the near field at a given moment in time. Note that the fluid "pushed aside" by the forward face of the body is

362 SOUND GENERATION

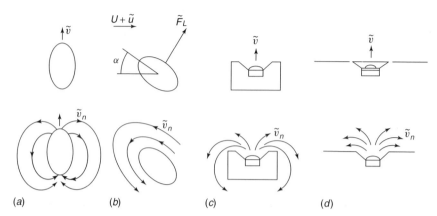

FIGURE 10.6 Schematic representation of various non-volume-displacing and volume-displacing sound sources and the associated particle velocity in the near field. The symbol $\tilde{v}$ is the vibration velocity of the body and $\tilde{v}_n$ is the particle velocity in the near field. (a) Non-volume-displacing sound source; small, rigid body of arbitrary shape moving with a velocity $\tilde{v}$ in an unbounded fluid. (b) Stationary, small, rigid body exposed to a flow field of stationary velocity U with a fluctuating component $\tilde{u}$. The angle of attack is α and the resulting dynamic force is $\tilde{F}_L = C_L \left(\frac{1}{2}\rho_0 \tilde{u}^2\right)$ and C_L is the lift coefficient of the body. (c) Volume-displacing small sound source typified by a loudspeaker mounted on the face of an airtight, rigid enclosure. (d) Volume-displacing small sound source flash mounted in an infinitely large, plane, rigid baffle.

"sucked in" by the void created by the receding backward face of the body. This "push–pull" situation results in a very small force. Borrowing from the vocabulary of electronic engineers, this situation is referred to as an "acoustical short circuit."

In Fig. 10.6b the upper sketch depicts a small stationary body exposed to a flow field of average velocity U and a fluctuating component $\tilde{u}$. The fluctuating component of the flow produces a fluctuating lift force $\tilde{F}_L$ and a fluctuating drag force $\tilde{F}_D$ (the drag force is not shown to preserve clarity).

In Fig. 10.6c the upper sketch shows a loudspeaker mounted on a face of a rigid-walled enclosure. In this case the rigid enclosure prevents the "pulling" by the receding back face of the speaker. Moving aside the fluid by "push" alone, without the help of the "pull," requires more force. More force acting on the fluid translates into higher sound radiation. This is the reason that a small oscillating surface radiates sound more efficiently if its back side is enclosed than the same surface would radiate if the back side is not enclosed.

Sound sources with an enclosed back side displace the same volume in the back side as on the front side. However, the volume displaced by the back side is prevented from "communicating" with the surrounding fluid by the rigid enclosure. If a dynamic force of constant amplitude drives the vibration velocity (which is the case for our loudspeaker shown in the upper sketch in Fig. 10.6c), the obtainable vibration velocity is limited either by the volume stiffness of the air in the enclosure volume or by the mass of the moving part of loudspeaker.

If the enclosure is small compared with the acoustical wavelength, the sound pressure that is produced at the surface of the active part has time to spread around the outside wall of the enclosure during each vibration cycle. As a direct consequence of this, all small volume-displacing sound sources, independent of their geometry, have an omnidirectional sound radiation pattern in the far field.

In Fig. 10.6d the upper sketch shows a vibrating surface (in the form of a loudspeaker membrane) flush mounted in a flat infinite rigid baffle. This is a situation that basically resembles the configuration in Fig. 10.6c, except that the sound radiation must take place in a hemisphere (half space). Consequently, for the same volume displacement the mean-square sound pressure p_{rms}^2 at a given radial distance r will be twice as large as an omnidirectionally radiating source would produce.

Response of Unbounded Fluid to Excitation by Oscillating Small Rigid Body. The dynamic "near-point" force acting on the fluid is always in the form of a dynamic pressure over a small area. By *small*, we mean here and in the rest of this chapter that the body is small compared with the acoustic wavelength ($\lambda = c_0/f$).

The quintessential question is, "How does the organized motion that a small rigid body imparts on the fluid spread out, that is, diffuse in various directions with increasing distance through the molecular collision process.

Diffusion of Sound Energy in Response to Excitation of Fluid by Oscillating Small Rigid Sphere. This section deals with the prediction of the sound radiation pattern and the radiated sound power of small, rigid, oscillating bodies for one-dimensional, harmonic, and random motion.

Harmonic Excitation. The sound field that the harmonic, one-dimensional, oscillatory velocity $v(\omega) = \hat{v}_b \, e^{j\omega t}$ of a small rigid sphere generates is well known and for $r \geq a$ is given by[4]

$$\hat{p}(r, \varphi, \omega) = \hat{v}_b(\omega) \cos(\varphi) \left(\frac{j\omega \rho_0 a^3}{2 - k^2 a^2 + j2ka} \right) \left(\frac{1 + jkr}{r^2} \right) e^{-jk(r-a)} \, e^{j\omega t} \; \text{N/m}^2 \tag{10.22}$$

where $\hat{p}(r, \varphi, \omega)$ = peak sound pressure (N/m²) at distance r (m) in direction ϕ
$\hat{v}_b$ = peak velocity of body in $\phi = 0$ direction, m/s
ω = radian frequency, $2\pi f$ Hz
f = frequency of oscillation, Hz
ρ_0 = density fluid, kg/m³
a = radius of sphere, m
k = wavenumber, ω/c_0, m⁻¹
c_0 = propagation speed of sound, m/s

It is instructive to evaluate Eq. (10.22) at the two extreme cases, namely on the surface of the sphere ($r = a$) and in the geometric far field where $kr \gg 1$.

On the surface of the oscillating sphere Eq. (10.22) yields the reaction pressure of the fluid

$$\hat{p}(a, \varphi, \omega) = \hat{v}_b(\omega) \cos(\varphi)\, e^{j\omega t} \frac{k^4 a^4 + jka(2 + k^2 a^2)}{4 + k^4 a^4} \quad \text{N/m}^2 \quad (10.23)$$

The radiation impedance of the sphere, defined as the ratio of the surface pressure and the radial velocity $\hat{v}_b \cos(\varphi)$, is

$$Z_{\text{rad}} = \begin{cases} \rho_0 c_0 \left\{ \dfrac{k^4 a^4 + jka(2 + k^2 a^2)}{4 + k^4 a^4} \right\} \quad \text{Ns/m}^3 & (10.24a) \\[1em] \rho_0 c_0 \left(\dfrac{k^4 a^4}{4} + \dfrac{jka}{2} \right) \quad \text{for } ka \ll 1 & (10.24b) \\[1em] \rho_0 c_0 \quad \text{for } ka \gg 1 & (10.24c) \end{cases}$$

The dynamic force exerted on the fluid in the direction of the oscillation, F_{ax}, is obtained by integrating the axial component of the elementary radial force, $dF_x = \cos(\varphi)\, dF_{\text{rad}} = p(a, \varphi) \cos(\varphi)\, dS$ over the surface of the sphere, yielding

$$\hat{F}_{\text{ax}} = \begin{cases} \dfrac{2}{3} \hat{v}_0 \rho_0 c_0 (2\pi a^2) \dfrac{k^4 a^4 + jka(2 + k^2 a^2)}{4 + k^4 a^4} \quad \text{N} & (10.25a) \\[1em] j\omega \rho_0 \hat{v}_0 \dfrac{2\pi a^2}{3} = j\omega \hat{v}_0 \rho_0 V_{\text{ad}} \quad \text{for } ka \ll 1 & (10.25b) \\[1em] \dfrac{4\pi a^2}{3} \rho_0 c_0 \hat{v}_0 \quad \text{for } ka \gg 1 & (10.25c) \end{cases}$$

where V_{ad} is the added volume and $\rho_0 V_{\text{ad}}$ is the added mass (as we will define latter on). Observing Eq. (10.24b) note that at low frequencies $\hat{F}_{\text{ax}}$ is the force needed to produce the same acceleration as that of the sphere ($j\omega \hat{v}_0$) on a concentrated mass that is equal the added mass.

The radiated sound power, W_{rad}, is obtained as a product of the oscillatory velocity and the component of the axial force that is in-phase with it, yielding

$$W_{\text{rad}} = \begin{cases} \dfrac{1}{2} |\hat{v}_0| \text{Re}\{F_{\text{ax}}\} = \dfrac{1}{2} |\hat{v}_0|^2 \rho_0 c_0 (4\pi a^2) \dfrac{k^4 a^4}{4 + k^4 a^4} \quad \text{Nm/s} & (10.26a) \\[1em] \dfrac{1}{2} |\hat{v}_0|^2 \rho_0 c_0 (\pi a^2/3) k^4 a^4 \quad \text{for } ka \ll 1 & (10.26b) \\[1em] \dfrac{1}{2} |\hat{v}_0|^2 \rho_0 c_0 (4\pi a^2/3) \quad \text{for } ka \gg 1 & (10.26c) \end{cases}$$

Equation (10.25) indicates that at low frequencies ($ka \ll 1$) the radiated sound power is small and increases with the fourth power of the frequency and at high frequencies ($ka \gg 1$) reaches a frequency-independent value.

In the far field ($kr \gg 1$) Eq. (10.22) takes the form

$$\hat{p}(r, \varphi, \omega) = \begin{cases} -\dfrac{\hat{v}_0}{r} \cos(\varphi) \rho_0 c_0 k^3 a^3 \dfrac{(2 - k^2 a^2) + j2ka}{4 + k^4 a^4} & \text{N/m}^2 \quad (10.27a) \\[6pt] -\dfrac{\hat{v}_0(\omega) \cos(\varphi)}{4\pi c_0 r} \omega^2 \rho_0 2\pi a^3 & \text{for } ka \ll 1 \quad (10.27b) \\[6pt] -\hat{v}_b \rho_0 c_0 \left(\dfrac{a}{r}\right) \cos(\varphi) & \text{for } ka \gg 1 \quad (10.27c) \end{cases}$$

Sound Radiation of Oscillating Small Rigid Nonspherical Bodies. Realizing that $2\pi a^3/2 = 4\pi a^3/3 + 2\pi a^3/3$ and that $V_b = 4\pi a^3/3$ is the volume of the sphere and $V_{\text{ad}} = 2\pi a^3/3$ is the added mass of an oscillating rigid sphere, we can write Eq. (10.24) in the form

$$\hat{p}(r, \varphi, \omega) \approx -\frac{\omega^2 \rho_0}{4\pi rc} \hat{v}_b(\omega) \cos(\varphi) [V_b + V_{\text{ad}}(\square, \Psi)] e^{j(\omega t - kr)} \quad \text{N/m}^2 \quad (10.28)$$

where the symbols $\square$ and Ψ in the argument of V_{ad} signify that the added mass depends on the geometry of the body and the specific direction of its oscillatory motion.

While Eq. (10.27b) is valid only for a small oscillating rigid sphere, Eq. (10.28) is exact for an oscillating sphere and is a good engineering approximation for small rigid bodies of any shape. It is instructive to rewrite Eq. (10.28) in the form

$$\hat{p}(r, \varphi, \omega) \simeq -\frac{\omega^2 \rho_0}{4\pi rc_0} \hat{v}_b(\omega) \cos(\varphi) (S_p l_v) e^{j(\omega t - kr)} \quad \text{N/m}^2 \quad (10.29)$$

where S_p is the projected area of the body on a plane perpendicular to the oscillatory motion and

$$l_v \equiv \frac{V_b + V_{\text{ad}}(\square, \Psi)}{S_p} \quad \text{m} \quad (10.30)$$

is a length defined by the author in Eq. (10.30), where the angle Ψ corresponds to the direction of the oscillation of the body. The importance of the length l_v is that it defines the frequency range to $ka \ll 1$, where the body can be considered small and where Eq. (10.29) is valid.

The $\cos(\varphi)$ dependence of the directivity pattern in the vicinity of the oscillating body is strictly valid only for spheres. However, in the far field of any oscillating small rigid body (i.e., $kr \gg 1$ and $kl_v \ll 1$) the same $\cos(\varphi)$ dependence applies provided that the fluid displaced by the forward-moving face of the body is free to move to the void left by the opposite face. A detailed mathematical proof of this is given by Koopmann and Fahnline.[8]

Prediction of l_v. The length l_v of various three-dimensional and two-dimensional bodies are listed in Table 10.1. The l_v values in Table 10.1 where computed utilizing the formulas for V_{ad} given by Newman[9]

For the spheroid with major axis $2a$ and minor axes $2b$ and aspect ratio b/a, the Ver-lengths l_{vx}, l_{vy}, and l_{vz} are obtained by finding $m''_{11} = V_{ad}(x)/V_b$ and $m''_{22} = V_{ad}(y)/V_b = V_{ad}(z)/V_b$ in Fig. 10.7 at the desired b/a ratio and performing the operation indicated in Table 10.1. The increase in moment of inertia (in addition to that measured in a vacuum) when the ellipsoid body rotates around any of the minor axes is obtained by finding the numerical value of m'_{55} that belongs to the appropriate b/a ratio and performing the operation indicated in Table 10.1.

Predicting Radiated Sound Power. The sound power radiated by small oscillating rigid bodies is obtained by integrating the far-field intensity $I = p^2(r, \varphi, \omega)/\rho_0 c_0$ over the surface of a large sphere of radius r centered at the oscillating small rigid body, namely

$$W_{rad} \simeq \int_0^\pi \frac{|\hat{p}|^2(r, \varphi, \omega)}{\rho_0 c_0} 2\pi r^2 \sin(\varphi)\, d\varphi \quad \text{N} \cdot \text{m/s} \tag{10.31}$$

Combining Eqs. (10.29) and (10.31) yields

$$W_{rad} \simeq \frac{k^4 \rho_0 c_0 S_p^2 (l_v)^2}{12\pi} v_{b,rms}^2 \quad \text{N} \cdot \text{m/s} \tag{10.32}$$

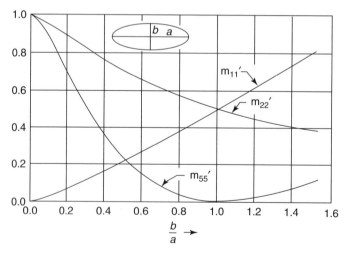

FIGURE 10.7 Added-mass coefficient V_{ad}/V_b for a spheroid, of length $2a$ and maximum diameter $2b$. The added-mass coefficient m'_{11} corresponds to longitudinal acceleration, m'_{22} corresponds to lateral acceleration in the equatorial plane, and m'_{55} denotes the added moment of inertia coefficient for rotation about an axes in the equatorial plane representing the ratio of the added moment of inertia and the moment of inertia of the displaced volume of the fluid. After Ref. 9. Reproduced by permission of the MIT Press.

TABLE 10.1 Added Volume, V_{ad}, Ver-length, I_v, and Added Moment of Inertia M of Oscillating and Rotating Rigid Bodies

Name and Sketch	V_{ad} and I_v	M
Sphere of Radius a 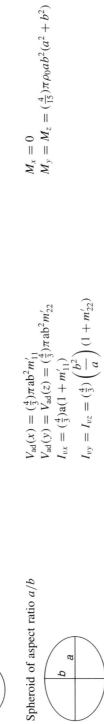	$V_{ad}(x) = V_{ad}(y) = V_{ad}(z) = 2\pi a^3/3$ $I_{vx} = I_{vy} = I_{vz} = 2a$	$M_x = M_y = M_z = 0$
Spheroid of aspect ratio a/b	$V_{ad}(x) = (\tfrac{4}{3})\pi ab^2 m'_{11}$ $V_{ad}(y) = V_{ad}(z) = (\tfrac{4}{3})\pi ab^2 m'_{22}$ $I_{vx} = (\tfrac{4}{3})a(1 + m'_{11})$ $I_{vy} = I_{vz} = (\tfrac{4}{3})\left(\dfrac{b^2}{a}\right)(1 + m'_{22})$	$M_x = 0$ $M_y = M_z = (\tfrac{4}{15})\pi\rho_0 ab^2(a^2 + b^2)$

(continued overleaf)

TABLE 10.1 (*continued*)

Name and Sketch	V_{ad} and I_v	M
Long thin plate strip $L \gg 2a; kL \ll 1$	$V_{ad}(x) = V_{ad}(z) = 0$ $V_{ad}(y) = \pi a^2 L$ $I_{vx} = I_{vz} = 0$ $I_{vy} = (\pi/2)a$	$M_x = (\tfrac{1}{8})\pi\rho_0 a^4 L$
Long round finned rod $L \gg 2a; kL \ll 1$	$V_{ad}(x) = \text{NPI}^a$ $V_{ad}(y) = \pi a^2 L$ $V_{ad}(z) = \pi[a^2 + (b^2 - a^2)^2/b^2]L$ $I_{vx} = \text{NPI}^a$ $I_{vy} = \pi a$ $I_{vz} = \pi[a^2/b + (b^2 - a^2)^2/2b^3]$	M_x = see footnote b

Long rectangular plates crossing perpendicularly $L \gg 2a; kL \ll 1$

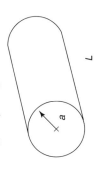

$V_{ad}(x) = \text{NPI}^a$
$V_{ad}(y) = V_{ad}(z) = \pi a^2 L$
$I_{vx} = \text{NPI}^a$
$I_{vy} = I_{vz} = (\pi/2)a$

$M_x = (2/\pi)a^4 L$

Long cylindrical rod of radius a
$L \gg 2a; kL \ll 1$

$V_{ad}(x) = \text{NPI}^a$
$V_{ad}(y) = V_{ad}(z) = \pi a^2 L$
$I_{vx} = \text{NPI}^a$
$I_{vy} = I_{vz} = \pi a$

$M_x = 0$

(continued overleaf)

TABLE 10.1 (*continued*)

Name and Sketch	V_{ad} and I_v	M
Long rod with elliptical cross section $L \gg 2a; kL \ll 1$	$V_{ad}(x) = \text{NPI}^a$ $V_{ad}(y) = \pi a^2 L$ $V_{ad}(z) = \pi b^2 L$ $I_{vx} = \text{NPI}^a$ $I_{vy} = I_{vz} = (\pi/2)(a+b)$	$M_x = \left(\frac{1}{8}\right)\pi \rho_0 a^4 L$
Long rod of square cross section $L \gg 2a; kL \ll 1$	$V_{ad}(x) = \text{NPI}^a$ $V_{ad}(y) = V_{ad}(z) = 4.754 a^2 L$ $I_{vx} = \text{NPI}^a$ $I_{vy} = I_{vz} = 2.38 a$	$M_x = 0.725 a^4 L$

[a] NPI = not of practical interest

[b] $M_x = \{\pi^{-1} \csc^4(\alpha)[2\alpha^2 - \alpha \sin(4\alpha) + (1/2)\sin^2(2\alpha)] - \pi/2\}$, $\sin(\alpha) = 2ab/(a^2 + b_2)$, and $\pi/2 < \alpha < \pi$

For a oscillating, small, rigid spherical body of radius a the area projected on the plane perpendicular to the direction of the oscillatory motion $S_p = \pi a^2$ and $l_v = 2a$ and Eq. (10.32) yield

$$W_{\text{rad}}(\text{Sphere}) = \frac{\rho_0 c_0 (4\pi a^2)(ka)^4}{24} |v_b|^2 \quad \text{N} \cdot \text{m/s} \tag{10.33}$$

This is the well-known[3,4] formula for the sound power radiated by a small ($ka \ll 1$) sphere of radius a oscillating with an rms velocity amplitude of $v_{b,\text{rms}}$.

Random Excitation. Let us investigate the case where a small rigid sphere exhibits random motion in one dimension (e.g., in the x direction). As described in Chapter 3, the random motion is characterized by the power spectrum of its velocity in the x direction, $S_{vv}(x)$.

The *power spectrum*, as defined in Chapter 3, represents the quantity obtained by filtering the random time signal in a 1-Hz bandwidth multiplying by itself and time averaging the product. In this chapter we use the symbol S with a double subscript to represent a power spectrum or cross spectrum. The double-letter subscript of S identifies which of the two filtered signal are multiplied with the other before time averaging. For example the symbol $S_{vv}(\omega)$ represents the autospectrum of the random velocity signal filtered in a 1-Hz bandwidth centered at the angular frequency $\omega = 2\pi f$, where f is the frequency in hertz. The symbol $S_{xy}(\omega)$ represents a cross spectrum of two random signals $x(t)$ and $y(t)$ obtained by filtering both signals in a 1-Hz bandwidth, multiplying them, and time averaging the product. The power spectrum of a signal resulting from the superposition of two signals $z(t) = x(t) + y(t)$ is computed as $S_{zz}(\omega) = S_{xx}(\omega) + S_{yy}(\omega) + 2S_{xy}(\omega)$. If the two random processes $x(t)$ and $y(t)$ are not correlated, $S_{xy}(\omega) = 0$. If the two random processes of the same power spectral density are fully correlated (i.e., they have the same cause), the cross spectrum is identical to the autospectrum, $S_{xy}(\omega) = S_{xx}(\omega) = S_{yy}(\omega)$ and $S_{zz}(\omega) = 4S_{xx}(\omega) = 4S_{yy}(\omega)$.

If the velocity of the vibrating body is exclusively in the x direction and is a random function of time, with a power spectrum of $S_{v_x v_x}(\omega)$ the power spectrum of the sound at distance r in the direction ϕ from the positive x axis, $S_{pp}(r, \varphi, \omega)$, is

$$S_{pp}(\omega, r, \varphi) = S_{v_x v_x}(\omega) |h_1(\omega, \rho_0, k, a, r, \varphi)|^2 \quad \text{N}^2/\text{m}^4 \tag{10.34}$$

where

$$h_1(\omega, \rho_0, k, l_v, r, \varphi) = \frac{\omega^2 \rho_0}{4\pi r c_0} \cos(\varphi)(S_p l_v) e^{jkr} \tag{10.35}$$

Hypothetical Experiment 2: Force Acting on Fluid. To find the dynamic force acting on the fluid F_f, let us perform the thought experiment shown in Fig. 10.8. Let a dynamic force $\mathbf{F}_b$ act on a small, rigid, hollow body of arbitrary shape at its center of gravity and determine the oscillatory velocity $\mathbf{v}_b$ of the

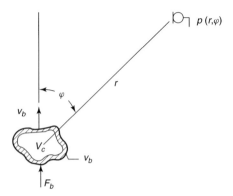

FIGURE 10.8 Hypothetical experiment to determining the force F_f that acts on the fluid and the sound field it produces by letting the wall thickness of an oscillating, hollow rigid body approach zero.

body responding to the force as

$$\mathbf{F}_b = \mathbf{v}_b(j\omega m_b + Z_{\text{rad}}) \quad \text{N} \tag{10.36}$$

where m_b is the mass of the body and $Z_{\text{rad}} = F_f/v_b$ is the radiation impedance:

$$\mathbf{F}_b = j\omega \mathbf{v}_b[\rho_M(V_b - V_c) + \rho_0 V_c + \rho_0 V_{\text{ad}}(\Box, \Psi)] \quad \text{N} \tag{10.37}$$

where V_b is the volume of the oscillating, hollow, rigid body and V_c is the volume of the inside cavity, filled with the same fluid that surrounds the body. Both volumes are in cubic meters. The symbol ρ_M represents the density of the solid part of the body in kilograms per cubic meter. The symbol V_{ad}, in cubic meters, is the added volume, or in the hydrodynamic vocabulary the added mass coefficient. The symbols $\Box$ and Ψ in the argument of V_{ad} is to remind us that it is a function of the body's geometry and the particular angle of its oscillatory motion to one of the designated axes of the body.

Let us now investigate the very important extreme case of Eq. (10.37) where the wall thickness of the hollow body approaches zero (and still remains rigid) by setting $V_c \to V_b$. In this case the entire applied dynamic excitation force $F_b = F_f$ is acting directly on the fluid, yielding

$$\mathbf{F}_f = v_b j\omega \rho_0 V_b + \mathbf{v}_b j\omega \rho_0 V_{\text{ad}}(\Box, \Psi) \quad \text{N} \tag{10.38}$$

The first term in Eq. (10.38) represents the flux of the momentum of the fluid volume displaced by the body, $\rho_0 V_b[\partial(v_b)/\partial t]$, and the second term, $\rho_0 V_{\text{ad}}(\Box, \Psi)[\partial(v_b)/\partial t]$, the flux of the momentum of the near field obtained by integrating the component of the particle velocity that is out phase with the velocity of the body over the entire volume outside of the body. Fortunately, we do not need to perform this integration to obtain $V_{\text{ad}}(\Box, \Psi)$ because

(a) hydrodynamists have computed it for many body geometries of interest (see Table 10.1 and Fig. 10.7),
(b) for most bodies of practical interest it can be well approximated by spheroids of the appropriate aspect ratio (see Fig. 10.7), and
(c) it can be determined experimentally.

Equation (10.38) is of fundamental importance. It signifies the following:

1. The force acting on the fluid acts in the direction of the oscillatory motion.
2. Two-dimensional bodies, which have no volume ($V_b = 0$) can exert force on the fluid provided their velocity has a component perpendicular to their plane and, consequently, $V_{ad} \neq 0$.
3. No force can be exerted on the fluid without a finite added mass (i.e., $V_{ad} \neq 0$).

The far-field sound pressure as a function of distance r, direction φ, and frequency ω produced by a pointlike force F_f is obtained by solving Eq. (10.38) for v_b and inserting this value into Eq. (10.25), yielding

$$p(r, \varphi, \omega) = -j \frac{k F_f \cos(\varphi)}{4\pi r} e^{j(\omega t - kr)} = -j \frac{F_f \cos(\varphi)}{2r\lambda} e^{j(\omega t - kr)} \quad \text{N/m}^2 \quad (10.39)$$

The radiated sound power W_{rad} is obtained by integrating the intensity $I = p^2(r, \vartheta, \omega)/\rho_0 c_0$ over the surface of a sphere of radius $r \gg (1/k)$ centered on the excitation point. Carrying out the integration yields

$$W_{rad} = \frac{k^2 F_{f,rms}^2}{12\pi \rho_0 c_0} = \frac{\omega^2 F_{f,rms}^2}{12\pi \rho_0 c_0^3} \quad \text{N} \cdot \text{m/s} \quad (10.40)$$

If the excitation is random, the excitation and response parameters have to be replaced by their respective power spectra.

Equation (10.39) describes the diffusion law of acoustics. It gives the full answer to the question "How does this organized motion—that is, in the direction of the force at the application point—diffuse with increasing distance in other directions through the molecular collision process?"

Equation (10.39) signifies that:

1. The sound pressure in any particular direction is proportional the projection of the force vector in that direction, $F_f \cos(\phi)$, indicating that the strength of the diffusion varies as $\cos(\phi)$. The radiated sound pressure has its maximum in the direction of the force and it is zero perpendicular to it. This is consistent with the fact that the oscillatory motion of the body exerts on the fluid the maximum force in the motion direction and no force at all in the direction perpendicular to it. The directivity pattern $\cos(\phi)$ in three dimensions is represented by two spheres of unit diameter aligned in the

374 SOUND GENERATION

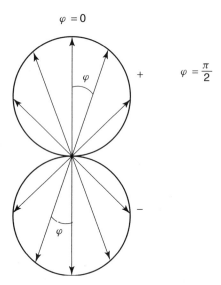

FIGURE 10.9 Far-field sound radiation pattern of a small rigid body oscillating in the $\phi = 0$ direction or an oscillatory force acting on the fluid in the $\phi = 0$ direction.

direction of the oscillatory motion or the direction of the force and centered on the line of the oscillating body or on the point where the force acts on the fluid, as shown (in two dimensions) in Fig. 10.9.
2. The sound pressure in any direction is inversely proportional to the source–receiver distance r.
3. Independent of what causes the dynamic force F_f which acts on the fluid, the sound pressure fully determines the sound radiation pattern and sound power output of small rigid bodies. The force can be, for example, the reaction force to the oscillatory movement of the body or that caused by fluctuating lift or drag forces generated on small stationary bodies by the fluctuation of the inflow velocity or by vortex shedding at the trailing edge of airfoils. Fluctuating inflow is caused by incoming turbulence of a scale that is large compared with the size of the body. If the scale of the convected pressure fluctuations (turbulence) is large compared with the size of the rigid body, the force would be identical to that obtained by moving the body in a stationary fluid in the direction and with the velocity of the fluctuating component of the unsteady inflow and predicting lift and drag forces.

However, a rigid stationary body exposed to fluctuating aerodynamic forces radiates sound into a moving media. Consequently, the radiated sound pressure and sound power is different from that an oscillating rigid body, that exerts the same dynamic force on the fluid, would produce in a stationary media. The effect of moving media on sound radiation is treated in Chapter 15.

4. The sound pressure is independent of the density of the surrounding fluid. This is because the acceleration of the fluid caused by the force is inversely proportional to the density and the sound pressure is proportional to the product of the density and acceleration.

If the force applied to the F_f is of random nature, it is characterized by its power spectral density $S_{FF}(\omega)$ and the sound pressure it produces by its power spectral density $S_{pp}(r, \varphi, \omega)$, as defined in Chapter 3. In this case Eq. (10.39) takes the form

$$S_{pp}(\omega, r, \varphi) = S_{FF}(\omega,) \left| \frac{\omega \cos(\varphi)}{4\pi r c_0} \right|^2 \quad (\text{N/m}^2)^2 \qquad (10.41)$$

and the power spectrum of the radiated sound power takes the form

$$S_w(\omega) = S_{FF}^2(\omega) \left| \frac{\omega^2}{12\pi \rho_0^2 c_0^3} \right|^2 \quad (\text{N} \cdot \text{m/s})^2 \qquad (10.42)$$

Response of Bounded Fluid to Point Force Excitation. It is instructive to investigate briefly what effects nearby boundaries have on the sound radiation pattern and on the sound power radiated by a point force. We will consider

1. the most severe restricting boundaries, surrounding the oscillating rigid body by an infinitely long (or finite-length but at both ends anechoically terminated) small-diameter rigid pipe;
2. rigid plane boundaries; and
3. yielding plane boundaries.

Rigid Pipe Surrounding Point Force. Figure 10.10 shows the situation where a point force of arbitrary orientation is acting in a fluid inside of a pipe of diameter d and cross-sectional area $S = d^2\pi/4 = r^2\pi$ that is anechoically terminated at both ends. Let the axis of the pipe coincide with the x axis of the Cartesian coordinate system and let the point force, which acts at $x = 0$, have components F_x, F_y, and F_z. [If we place a small rigid oscillating body with known velocity v_b into the tube, we must first determine the oscillatory force F_f according to Eq. (10.38).] Let restrict our analysis to the frequency range where the acoustical wavelength is large compared with the internal diameter of the pipe (i.e., $\lambda \gg d$) when only plain sound waves can propagate in the pipe and assume that

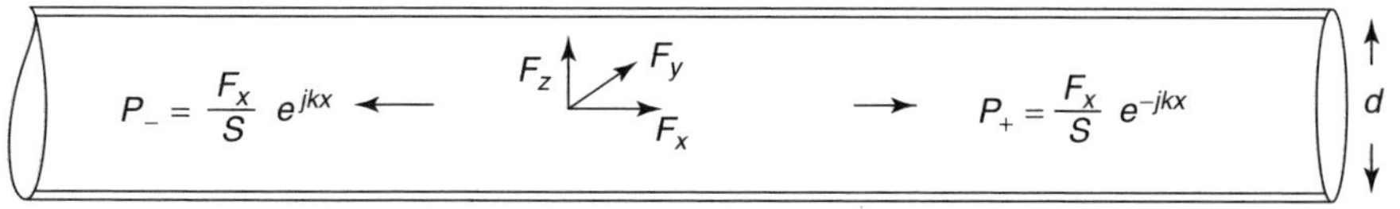

FIGURE 10.10 Force exerted on the fluid inside of a small-diameter, infinitely long, or on both ends anechoically terminated rigid pipe.

the near field of the body giving rise to the force is not influenced by the presence of the pipe (i.e., $kl_v \ll d$). This is not an unreasonable assumption because we can always choose a reasonably small size body and compensate for it by appropriately increasing its oscillatory velocity (as long as $v_b \ll c_0$) to produce the same force on the fluid. With these minor restrictions, the sound pressure at a distance $x \gg d$ the sound pressure $p(x)$ and the power spectrum S_{pp} for any $x \gg d$ will be

$$p(x) = -p(-x) = \left(\frac{F_x}{S}\right) e^{-kx} e^{j\omega t} \quad \text{N/m}^2 \quad (10.43)$$

$$S_{pp}(f) = \left(\frac{S_{FF}(f)}{S}\right) \quad \text{N/m}^2 \quad (10.44)$$

Note that Eq. (10.40) signifies that the amplitude of the sound propagating inside the rigid pipe does not diminish with increasing distance, as was the case when the point force was acting on an unbounded fluid [see Eq. (10.39)]. Before electroacoustical communication was invented, sea captains communicated with distant crew in the belly of a ship through pipes, as shown in Fig. 10.11a, and on the deck by using an acoustical horn, as depicted in Fig. 10.11b.

Equal sound power is radiated toward the right and left and the total sound power is

$$W'_{rad} = \frac{2[(F_x)^2/S]}{\rho_0 c_0} \quad \text{N/m}^2 \quad (10.45)$$

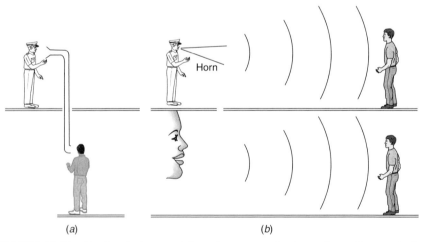

(a)   (b)

FIGURE 10.11 Passive devices used for voice communication to large distances before the invention of electroacoustical devices. (a) Captain gives orders to crew member located in a lower deck. (b) Top: Captain gives orders to distant crew member on deck using an acoustical horn. Bottom: Using a horn is equivalent to enlarging the mouth of the speaker to the size of the downstream cross-sectional area of the horn.

$$S'_{ww}(f) = \frac{S^2_{F_x F_x}(f)}{|S(\rho_0 c_0)|^2} \quad \text{N/m}^2 \tag{10.46}$$

where $S'_{ww}(f)$ is the power spectrum density of the radiated sound power. The prime notation signifies that the parameter is obtained when the point force is acting inside a small-diameter ($kr \ll 1$) rigid pipe.

The ratio W'_{rad}/W_{rad} is a measure of the effect that enclosing the force with a rigid pipe has on the radiated sound power. It is obtained by dividing Eq. (10.45) by Eq. (10.39), yielding

$$\frac{W'_{rad}}{W_{rad}} = \left(\frac{6\pi}{S}\right)\left(\frac{1}{k^2}\right)\left\{\frac{1}{1 + (F_y/F_x)^2 + (F_z/F_x)^2}\right\} \tag{10.47}$$

Equation (10.47) indicates that no sound power is radiated if $F_x = 0$, independent of how large F_y and F_z might be. The effect of placing the force inside the pipe reaches its maximum if the force is aligned with the axis of the tube (i.e., $F_y = F_z = 0$) and the value inside the curly brackets is unity. In this case Eq. (10.47) simplifies to

$$\frac{W'_{rad}}{W_{rad}} = \left(\frac{3}{2\pi}\right)\frac{c_0^2}{sf^2} \tag{10.48}$$

Example 10.2. Calculate the increase in radiated sound power and the sound power level when an axially oriented force located in a rigid pipe of $a = 2.5$ cm $= 0.025$ m internal radius at a frequency $f = 100$ Hz at a temperature of $20°C = 68°F$.

Solution Utilizing Eq. (10.2), we find $c_0 = 342.6$ m/s and with the above values Eq. (10.48) yields

$$\frac{W'_{rad}}{W_{rad}} = \frac{(3/2\pi)(342.6)^2}{\pi(0.025)^2(100)^2} = 2854$$

indicating that the same point force oriented axially in this rigid tube radiates 2854 times as much sound power than it does in a unbounded fluid. This corresponds to a $10\log_{10}(2854) = 34.6$ dB increase of the sound power level.

10.2 EFFECT OF NEARBY PLANE BOUNDARIES ON SOUND RADIATION

In this section we only consider entirely rigid and fully yielding plane boundaries. Other boundaries of practical interest, such as lining the rigid boundary with a porous sound-absorbing material of finite thickness, is beyond the scope of this chapter. For guidance in this situation see Chapter 6. The infinitely rigid plane boundary is approached by an infinitely large, nonporous wall and the fully yielding plane boundary by the undisturbed horizontal air–water interface. With regard to their effect on the sound power radiated by a point force (or by an oscillating small rigid body), the plane boundaries do not need to be infinitely large. For practical purposes it is sufficient that they extend, in each direction from the position of the force, many acoustical wavelengths.

A water tank with practically all boundaries pressure releasing (pressure reflection coefficient $R \simeq -1$) is realized by lining the retaining sidewalls and the bottom of the pool with a thin impervious foil with an elastic layer behind. These pools are the equivalent of reverberation chambers with the important exception that in this case the sound pressure reflection coefficient $R \approx -1$ instead of $R \approx +1$ in the case of the (air-filled) reverberation chamber. In both cases the walls have theoretically zero sound absorption coefficient, $\alpha \cong [1 - |R^2|]$ (see Chapter 8). As we will show latter, the vicinity of a rigid boundary increases the sound power output and that of a yielding boundary decreases the sound power output provided that the sound source is less than a half wavelength from the plain boundary.

Effect of Nearby Plane Rigid Boundaries. Rigid plane boundaries are places where the particle velocity is zero and the sound pressure doubles. If the sound source is a small, oscillating, rigid body or a dynamic force acting on the fluid, the directivity pattern of the image source must have the same strength but a mirrored directivity pattern of the actual sound source to produce zero particle velocity in the direction perpendicular to the plane of the boundary and, at the same time, double the sound pressure at all locations in the plane of the rigid boundary, as illustrated in Fig. 10.12.

The sound field in the receiver side of the rigid plane boundary can be constructed as the superposition of the sound fields that would result from the real sound source and its mirror image if both were placed in an infinite fluid. Then, as the sound source approaches the plane rigid interface, the sound field on the source side approaches that of a single source with doubled sources strength.

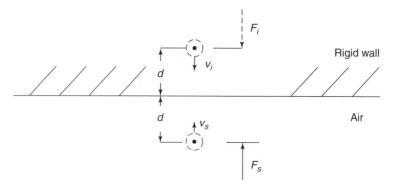

FIGURE 10.12 Representation of a plane rigid boundary, such as at air–hard wall interface, by mirroring the original sound source and superimposing the sound field of the real and in–phase image source in the source-side half space.

If it approaches a rigid two-dimensional corner (i.e., the intersection of two perpendicular plane rigid surfaces), there will be three collapsing mirror images and the sound field in the source-side quarter space approaches that of a single source with 4 times larger source strength. Finally, if the sound source approaches a three-dimensional corner (the intersection of three perpendicular rigid plane surfaces), there will be seven collapsing mirror images and the sound field in the source-side one-eighth space approaches that of a single source with 8 times the source strength.

Effect of Nearby Plane Pressure-Release Boundaries. The pressure-release boundary is well approximated by the plane water–air interface shown schematically in Fig. 10.13 for the force acting perpendicular to the interface and Fig. 10.14 when it acts parallel to the interface. For an oblique-incidence force, the force vector can be decomposed into a component that is perpendicular to the interface and into components that are parallel to the interface. Consequently, we will investigate the perpendicular and parallel cases only. If the sound source is a small, oscillating, rigid body or a dynamic force acting on the fluid, the directivity pattern of the image source must have the same strength but opposite phase and a mirrored directivity pattern of the actual sound source to produce zero sound pressure on the entire interface and, at the same time, double the sound particle velocity at all locations in the plane of the pressure release boundary.

Excitation Is Perpendicular to Pressure-Release Boundary. For the case shown in Fig. 10.13, where the distance to the interface is small compared to the wavelength ($kd \ll 1$) and the body exerting the force is small ($kl_v \ll 1$) and the direction of motion and the force exerted on the fluid are perpendicular to the plane pressure-release interface, the superposition of the sound pressures produced by the real source and that produced by its negative mirror image yields

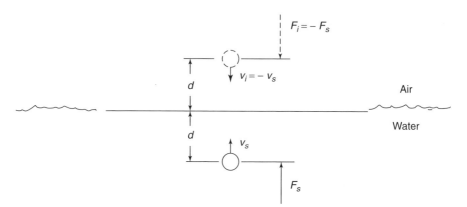

FIGURE 10.13 Representation of a plane pressure-release boundary, such as the water–air interface, by mirroring the original sound source and the 180° out-of-phase mirror image superimposing the sound field of the real and image sources in the source-side half space. Motion is perpendicular to the interface.

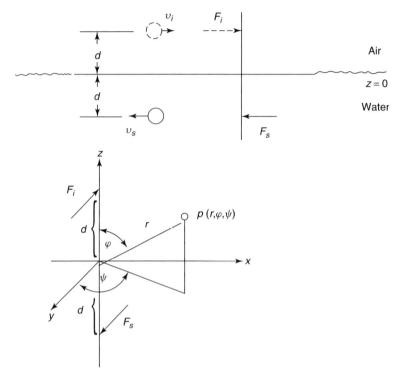

FIGURE 10.14 Representation of a plane pressure-release boundary, such as the water–air interface, by mirroring the original sound source 180° out of phase and superimposing the sound field of the original sound source and the image in the source-side half space. Motion is parallel to the interface.

the sound pressure, intensity, and radiated sound power in the liquid obtained by substituting $F = \rho_0 \omega Q h$ in the respective formulas given by Bies and Hansen[10] for the sound radiation of longitudinal quadrupoles constituted from two dipoles of strength Qh, where Q is the volume velocity of the monopoles that make up the dipole and $2h$ is the distance between them:

$$p(r, \varphi, \omega) = \left(\frac{1}{\pi r}\right)\left(\frac{\omega^2}{c_0^2}\right) F d \cos^2(\varphi) = \left(\frac{1}{\pi r}\right) k^2 F d \cos^2(\varphi) \quad \text{N/m}^2 \quad (10.49)$$

$$I(r, \varphi, \omega) = \left(\frac{F^2 d^2 k^4 \cos^4(\varphi)}{(\pi r)^2 \rho_0 c_0}\right) = \frac{F^2 d^2 \omega^4 \cos^4(\varphi)}{(\pi r)^2 \rho_0 c_0^5} \quad \text{N/m} \cdot \text{s} \quad (10.50)$$

$$W_{\text{rad}} = \left(\frac{4}{5\pi}\right)\frac{F^2 d^2 k^4}{\rho_0 c_0} = \left(\frac{4}{5\pi}\right) F^2 d^2 \omega^4 \cos^4(\varphi) \quad \text{N} \cdot \text{m/s} \quad (10.51)$$

where $p(r, \varphi, \omega)$ is the sound pressure at distance r at an angle φ from the direction of the motion or from the force at the angular frequency $\omega = 2\pi f$, $I(r, \varphi, \omega)$ is the sound intensity in the radial direction from the excitation point,

W_{rad} is the radiated sound power, F is the rms value of the force, and d is the distance between the excitation point and the plain pressure-release boundary (i.e., the depth of the excitation point in the water).

Note that Eqs. (10.49)–(10.51) are valid for both cases:

(a) A single excitation source is at a distance d from the plane pressure-release boundary.
(b) Two identical small bodies at a distance $2d$ oscillate with the same amplitude but opposite phase in an infinite fluid (for practical purposes far enough from the interface).

Observing Eq. (10.49), note that (1) the radiation pattern is much more directional than that obtained when the body radiates into an infinite fluid [$\cos^2(\varphi)$ vs. $\cos(\varphi)$]; (2) here again the sound pressure is independent of the density of the fluid; and (3) comparing Eq. (10.49) to Eq. (10.39) reveals that the vicinity of the pressure-release boundary substantially reduces the efficiency of the sound radiation at low frequencies. Those familiar with quadrupoles will realize that Eqs. (10.49)–(10.51) describe a longitudinal quadrupole constituted either by two opposing forces of equal magnitude acting on the fluid a short distance $2d$ apart or by two equal-size small rigid bodies oscillating in opposite directions in a small distance $2d$ apart.

Excitation Is Parallel to Pressure-Release Boundary. Figure 10.14 shows the situation where the excitation is parallel to the plane pressure-release boundary. For this case the sound field descriptors, obtained in a similar manner as for the longitudinal quadrupole above, are

$$p(r, \varphi, \omega) = \left(\frac{1}{\pi c_0^2 r}\right) \omega^2 Fd \cos(\varphi) \sin^2(\psi) \quad \text{N/m}^2 \qquad (10.52)$$

$$I(r, \varphi, \psi, \omega) = \left(\frac{1}{\rho_0 c_0^5}\right)\left(\frac{1}{\pi r}\right)^2 \omega^4 (Fd)^2 \cos^2(\varphi) \sin^4(\psi) \quad \text{N/m} \cdot \text{s} \qquad (10.53)$$

$$W_{\text{rad}} = \left(\frac{4}{15\pi}\right)\left[\frac{(Fd)^2 k^4}{\rho_0 c_0}\right] = \left(\frac{4}{15\pi}\right)\left[\frac{(Fd)^2 \omega^4}{\rho_0 c_0^5}\right] \quad \text{N} \cdot \text{m/s} \qquad (10.54)$$

Those familiar with quadrupoles will realize that Eqs. (10.52)–(10.54) describe a lateral quadrupole.

Sound Radiation of Oscillating Moment in Infinite Liquid. An oscillating moment consists of a pair of oscillating forces which (a) act in the same plane, (b) have the same magnitude but opposite direction and phase, and (c) are at a short distance apart.

Observing Fig. 10.14, we realize that the force pair has all the above attributes. Consequently, if we substitute M in place of Fd in Eqs. (10.52)–(10.54), the

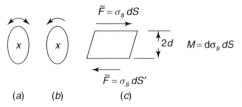

FIGURE 10.15 Different ways of exerting a moment on a fluid: (*a*) rotary oscillation around axes of nonrevolution; (*b*) rotation around axes of nonrevolution; (*c*) by shear stresses.

equations describe the sound radiation of an oscillating moment, yielding

$$p(r, \varphi, \omega) = \left(\frac{1}{\pi c_0^2 r}\right) \omega^2 M \cos(\varphi) \sin^2(\psi) \quad \text{N/m}^2 \tag{10.55}$$

$$I(r, \varphi, \psi, \omega) = \left(\frac{1}{\rho_0 c_0^5}\right)\left(\frac{1}{\pi r}\right)^2 \omega^4 M^2 \cos^2(\varphi) \sin^4(\psi) \quad \text{N/m} \cdot \text{s} \tag{10.56}$$

$$W_{\text{rad}} = \left(\frac{4}{15\pi}\right)\left[\frac{M^2 k^4}{\rho_0 c_0}\right] = \left(\frac{4}{15\pi}\right)\left[\frac{M^2 \omega^4}{\rho_0 c_0^5}\right] \quad \text{N} \cdot \text{m/s} \tag{10.57}$$

The moment can be any combination of F and d provided that $kd \ll 1$.

As shown in Fig. 10.15, the moment M can be the result of the oscillatory rotation of solid bodies around axes of nonrotational symmetry, shear stresses acting on the fluid (as it is the main source of noise of high-speed jets), or any other acceleration motion that results in a increase of the rotary inertia of the body above that which would exist if the oscillatory rotation would be taking place in a vacuum. For a large number of body shapes this additional moment of rotary inertia M that acts on the fluid is given in Table 10.1 under heading M_x and for ellipsoids of different aspect ratios in Fig. 10.7.

Figure 10.15 shows various ways for the moment excitation of a fluid. Figure 10.15a depicts moment excitation by an oscillatory rotation of a body around a nonsymmetric axis, and Fig. 10.15b shows excitation by steady rotation along the same axes. Figure 10.15c shows moment excitation by shear forces, where $\tilde{F} = \tilde{\sigma} \, dS$ acts on a small fluid volume $dV = 2(dS)d$. The symbol $\tilde{\sigma}$ is the fluctuating shear stress in newtons per square meter. This is the type of moment excitation found in the shear layer of high-speed jets and is treated in Chapter 15.

10.3 MEASURES TO REDUCE SOUND RADIATION

In this section we will discuss some of the measures noise control engineers and machine designers might employ to minimize the sound radiation of vibrating bodies. Most of these measures have been already discussed in a qualitative manner in Section 10.1. In this section we will show certain specific measures in the form of conceptual sketches and give references where the reader can find more detailed design information.

Fish as Example for Minimized Added Mass. Through evolutionary development, fish have developed body shapes that minimize the reaction force on them when they accelerate their motion in the forward direction. If not in motion, most fish have their pectoral fins deployed perpendicular to their body. If they sense danger and want to dart away as fast as possible, the pectoral fins move rapidly backward (just like the oars of a paddleboat). For moving their pectoral fin in this direction they have a large added mass and can produce a large impulsive force on the water, which propels them forward. As the pectoral fins come near their body, the fins rotate 90° to align their body so that they contribute only a negligible amount to the friction coefficient. After this first impulsive burst, the pectoral fins are not deployed again until the fish want to use them to rapidly stop forward motion. With pectoral fins aligned the forward motion is propelled by the movement of the tail. At rest they deploy the pectoral fins again to be ready for the next dart.

To survive, predator fish must develop a body that minimizes the added mass for swimming in the forward direction so that they can generate a minimum of waterborne sound when they increase their speed (accelerate) when closing in on their prey. Otherwise the prey would sense the sound generated by their acceleration early enough to dart into a safe hiding space. In choosing body shapes that, for a given oscillatory velocity, radiate the least amount of sound in the direction of the body's motion, noise control engineers are well advised to choose the shape of the vibrating body to resemble the body of a fish.

As shown in Fig. 10.16, the most promising body shape would resemble that obtained by cutting two identical-size fish perpendicularly at the location of their largest cross-sectional area and fitting the two head-parts together so that the two mouths point in the opposite direction. For motion in a direction different

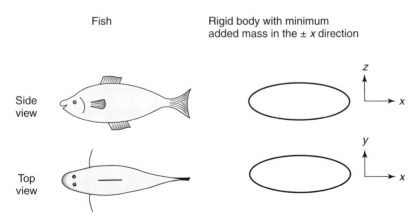

FIGURE 10.16 Learning from fish how to design bodies of minimum sound radiation owing to minimum added mass when oscillating in a specific direction.

from forward, especially in the direction perpendicular to it, the added mass of the fish is much larger, indicating the importance of the directional and shape dependence of the added mass of vibrating bodies of a geometry that is different from a sphere. The sphere has the distinction of having the same, large added mass for accelerating in any direction of motion. This large added mass is the reason that fish do no have a spherelike body (except some slow- swimming variety that inflate themselves to deceive their would-be predator). If fish would rotate instead of swim (mostly along straight paths), all of them would have a near-sphere-like body. On the other hand, if we wish to reduce the noise generated by the spinning of a rigid body, the sphere is the best choice. Surrounding a rotating nonspherical body with a spherical shell that rotates with the body will substantially reduce the noise radiation.

Reduction of Sound Radiation at Specific Frequencies.

There is a class of noise sources, such as power transformers and turbine-generators, that radiate exclusively or predominantly tonal noise. Figures 10.17 and 10.18 show how the tonal noise produced by such equipment can be controlled by lining their radiating surface with Helmholtz resonator arrays tuned to the frequency (or frequencies) of their tonal noise output.

As shown in the lower sketch in Fig. 10.18, the noise radiation at the tuning frequency of the resonators is achieved by the low impedance of the resonator mouth in the vicinity of its resonance frequency. The fluid displaced by the solid part of the face of the resonator finds it easier to enter the resonator volume than to undergo compression. The smaller the compression, the lower is the radiated sound. Information on how to design Helmholtz resonators can be found in Chapter 8.

In the case of power transformers these tonal frequencies are 120, 240, 360, and 480 Hz in the United States and 100, 200, 300, and 400 Hz in Europe. Noise

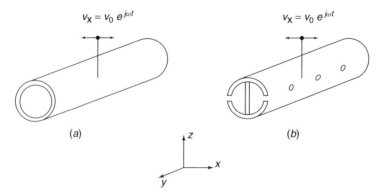

FIGURE 10.17 Frequency-selective reduction of sound generation owing to incorporation of Helmholtz resonators into the body: stiffened structural pipe beam vibrating perpendicular to its axes or rotating at constant rate as a spoke of a wheel.

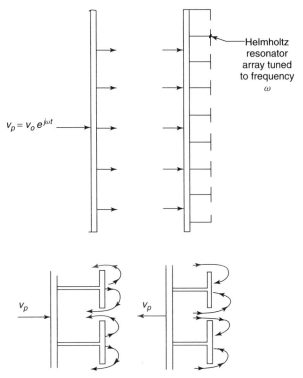

FIGURE 10.18 Frequency-selective reduction of sound radiation of a large flat surface, such as the outside wall of transformer tank, by lining it with a Helmholtz resonator array.

control measures for such noise sources and their advantages, disadvantages, and cost are compiled in reference 11.

Vacuum Bubbles. Figure 10.19 shows the concept of reducing the sound radiation of a transformer tank by lining the oil-side surface of the transformer tank with vacuum bubbles and how the presence of the vacuum bubbles, owing to their very large compliance, relieves the dynamic pressure on the tank wall and, consequently, reduces the sound radiation. As shown schematically in Fig. 10.20a, vacuum bubbles are twin curved thin shells with a small volume in between. When the cavity is evacuated, the shell nearly flattens out. In this evacuated final state they yield a compliance (inverse of the dynamic volume stiffness) that can be up to 50 times larger than the compliance of the air volume they replace and up to 1000 times that of the compliance of oil volume they replace. A unique feature of the vacuum bubbles is that there is vacuum in the cavity and there is no limit for the dynamic compliance, as there would be if the cavity were filled with a fluid. Below their resonance frequency, which is determined by the surface mass of the shell and its dynamic volume compliance, their volume compliance is independent of frequency. They work well even if the frequency approaches

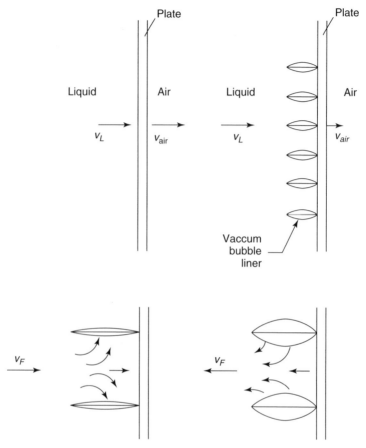

FIGURE 10.19 Reduction of the sound radiation from a transformer tank by lining the oil-side face of the tank wall with an array of vacuum bubbles of extremely high volume compliance. The vacuum bubbles can be oriented either perpendicular to the wall (as shown here) or parallel to it. Lower sketches indicate how the vacuum bubbles, which are much more compliant than either the oil or the tank wall, "soaks up" the momentum of the periodically approaching and receding oil. The velocity of the oil is due to the vibration of the transformer core driven by magnetostrictive forces.

zero. Information on how to design a vacuum bubbles for specific applications is given in references 12–14.

Figure 10.20 shows the geometry of a vacuum bubble (also called Scillator by its inventor Dr. Bschorr[12]) before and after evacuation of the cavity. As the name vacuum bubble implies, it is useful only in the evacuated state.

Vacuum bubbles are very sensitive to changes in static pressure (barometric pressure in air and depth-dependent pressure in liquids). Consequently, their most promising application would be in space stations, where the static air pressure can be controlled accurately.

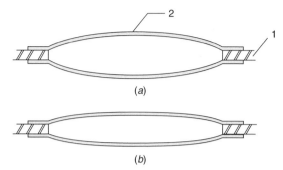

FIGURE 10.20 Conceptual sketch of a vacuum bubble. (*a*) Cross-sectional shape before evacuation of the cavity: 1, double-curved thin shell in the form of a top segment of a large-radius hollow sphere; 2, flat-plate strip base with punched holes. (*b*) Shape after evacuation of the cavity.

REFERENCES

1. M. C. Junger and D. Feit, *Sound Structures and Their Interaction*, 2nd ed., MIT Press, Cambridge, MA, 1986.
2. U. K. Ingard, *Notes on Sound Absorption Technology*, Version 94-02, Noise Control Foundation, Arlington Branch, Poughkeepsie, NY, pp B-5 (out of print). A CD version can be purchased from the author at www.ingard.com.
3. L. Cremer and M. Hubert, *Vorlesungen ueber Technische Akustik*, [Lectures about Technical Acoustics], 3rd ed., Springer, Berlin, 1985.
4. M. Heckel and H. A Mueller, *Taschenbuch der Technischen Akustik*, 2nd ed., Springer, Berlin, 1994.
5. I. L. Ver, "Reciprocity as a Prediction and Diagnostic Tool in Reducing the Transmission of Structureborne and Airborne Noise into an Aircraft Fuselage," BBN Report No. 4985, NASA Contract No. NASI-16521, January 3, 1982, Bolt Beranek and Newman Inc., Cambridge MA.
6. I. L. Ver "Using Reciprocity and Superposition as a Diagnostic and DesignTool in Noise Control," *Acustica, Acta Acoustica*, **82**, 62–69 (1996).
7. I. L. Ver and R. W. Oliphant, "Acoustic Reciprocity for Source-Path-Receiver Analyses," *Noise Vibration*, March 1996, pp. 14–17.
8. G. H. Koopmann and J. B. Fahnline, *Designing Quiet Structures; A Sound Power Minimization Approach*, Academic, New York, 1997.
9. J. N. Newman, *Marine Hydrodynamics*, MIT Press, Cambridge, MA, 1980, p. 144.
10. D. A. Bies and C. H. Hansen, *Engineering Noise Control Theory and Practice*, Unwin Hyman, Boston, 1988.
11. I. L. Ver, C. L. Moore, et al., "Power Transformer Noise Abatement," BBN Report No. 4863, October 1981. Submitted to the Empire State Electric Energy Research Corporation (ESEERCO).
12. O. Bschorr and E. Laudien, "Silatoren zur Daempfung und Daemmung von Schall" [Vacuum Bubbles for Absorption and Reduction of Sound], *Automobil-Industrie*, **2**, 159–166 (1988).

13. I. L. Ver, "Potential Use of Vacuum Bubbles in Noise Control," BBN Report No. 6938, NASA CR-181829, December 1988, Bolt Beranek and Newman, Inc., Cambridge MA.
14. I. L. Ver, "Noise Reduction from a Flexible Transformer Tank Liner," Phase I, Feasibility Study, BBN Report No. 6240, February 1987, Bolt Beranek and Newman, Inc., Cambridge MA.

CHAPTER 11

Interaction of Sound Waves with Solid Structures

ISTVÁN L. VÉR

Consultant in Acoustics, Noise, and Vibration Control
Stow, Massachusetts

The response of structures to dynamic forces or dynamic pressures is the subject of structural dynamics and acoustics. Structural dynamics is concerned predominantly with dynamic stresses severe enough to endanger structural integrity. Structural acoustics deals with low-level dynamic processes in structures resulting from excitation by forces, moments, and pressure fields. The primary interest in airborne and structure-borne noise problems is the prediction of

1. power input into the structure,
2. the response of the excited structure,
3. propagation of structure-borne sound to connecting structures, and
4. sound radiated by the vibrating structure.

The subject matter of this chapter is specific to noise control problems and thus is restricted to the audible frequency range and to air as the surrounding medium, though many of the concepts are directly applicable to liquid media or to higher or lower frequencies.

A typical noise control problem is illustrated in Fig. 11.1. A resiliently supported floor slab in the room to the left (source room) is excited by the periodic impacts of a tapping machine, while the microphone in the receiving room to the right registers the resulting noise. A part of the vibrational energy is dissipated in the floating slab, a part is radiated directly as sound into the source room, and the remainder is transmitted through the resilient layer into the building structure.

The radiated sound energy builds up a reverberant sound field in the source room that in turn excites the walls. The vibrations in the wall separating the two rooms, identified as path 1 in Fig. 11.1, radiate sound directly into the receiver room. The vibrations in the other partitions of the source room travel in the form of structure-borne sound to the six partitions of the receiving room and radiate

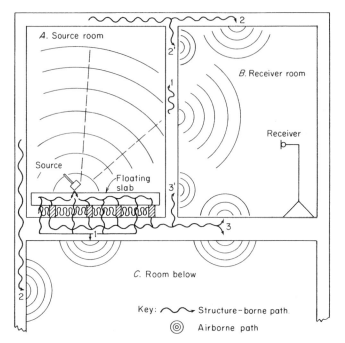

FIGURE 11.1 Sound transmission paths between an impact source in room A and a receiver in room B. Also shown are the paths to room C below.

sound into the receiver room. The structure-borne paths are identified by waves, 2, 2' and 3, 3'.

To reduce the noise level in the receiving room by a desired amount, the acoustical engineer must estimate the sound power transmitted from the source room to the receiver room through each path and then design appropriate measures to (1) reduce power input to the floating slab, (2) increase power dissipation in the slab, (3) increase vibration isolation, (4) reduce the structure-borne noise along its propagation path between source and receiver room, (5) reduce the sound radiated by the partitions of the source room, and (6) reduce the reverberation buildup of the sound in the receiver room by increasing sound absorption.

The aim of this chapter is to provide the information needed to solve such typical noise control engineering problems.

11.1 TYPES OF WAVE MOTION IN SOLIDS

Wave motion in solids can store energy in shear as well as in compression. The types of waves possible in solids include compressional waves, flexural waves, shear waves, torsional waves, and Rayleigh waves. Compressional waves are of practical importance in gases and liquids, which can store energy only in compression. The different types of waves in solids result from different ways of stressing. For a wave to propagate in solids, liquids, and gases, the medium must be capable of storing energy alternatively in kinetic and potential form. Kinetic energy is stored in any part of a medium that has mass and is in motion, while potential energy is stored in parts that have undergone elastic deformation.

Solid materials are characterized by the following parameters:

Density	ρ_M	kg/m^3
Young's modulus	E	N/m^2
Poisson's ratio	ν	
Loss factor	η	

The shear modulus G is related to Young's modulus as

$$G = \frac{E}{2(1+\nu)} \quad \text{N/m}^2 \tag{11.1}$$

The sketches on the left of Table 11.1 illustrate the deformation pattern typical for compressional, shear, torsional, and bending waves in bars and plates. Acoustically important parameters of solid materials are compiled in Table 11.2. Note that all tables in this chapter are placed at the end of the text. Thickness and weight per unit surface area of steel plates are listed in Table 11.10. Percentage open area of perforated metal is given in Table 11.11.

In dealing with structure-borne sound, we must distinguish between velocity and propagation speed. The term *velocity* refers to vibration velocity of the structure (i.e., the time derivative of the local displacement), which is linearly proportional to the excitation and will be designated by the letter v. As illustrated on the left of Table 11.1, the displacement and velocity are in the direction of wave propagation for longitudinal waves and perpendicular to it for shear and bending waves.

The term *speed* refers to the propagation speed of structure-borne sound, which is a characteristic property of the structure for each type of wave motion and is independent of the strength of excitation provided that the deformations are small enough to avoid nonlinearities. Propagation speeds will be designated by the letter c. We must distinguish between *phase speed* and *group speed* (or energy speed).

The phase speed c is defined in terms of (1) the wavelength λ, which is the distance in the propagation direction for which the phase of a sinusoidal wave changes 360°, and (2) the frequency f of the sinusoid. Consequently, $c = \lambda f$. Formulas to calculate phase speed of the various wave types are given on the right side of Table 11.1. Note that the phase speed for longitudinal, shear, and torsional waves is independent of frequency. Consequently, they are referred to as nondispersive waves. This means that the time history of an impulse, such as caused by a hammer blow striking axially on one end of a semi-infinite bar, as illustrated in Fig. 11.2a, will have the same shape regardless of where the

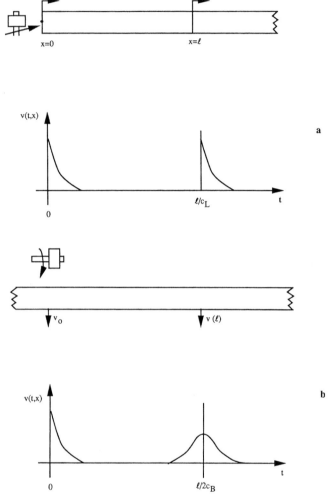

FIGURE 11.2 Time history of beam motion for impulsive excitation: (*a*) axial impact-generating nondispersive compressional waves; (*b*) normal impact-generating dispersive bending waves.

axial motion is sensed. Because the pulses sensed at different locations are time-delayed versions of each other, the propagation speed of the longitudinal wave can be determined experimentally as the ratio of the axial distance and the difference of arrival time. Since all frequency components travel at the same speed, the phase speed (which is defined only for a steady-state sinusoidal excitation) is identical with the speed of energy transport.

The phase speed of the bending wave can be easily measured by exciting the infinite beam with a steady-state sinusoidal force acting normal to the beam axis and measuring the gradient of the phase $d\phi/dx$ along the beam. The phase velocity c_B is defined as

$$c_B = \frac{2\pi f}{|d\phi/dx|} \quad \text{m/s} \tag{11.2}$$

However, the situation is much more complicated in the case of a complex waveform such as illustrated in Fig. 11.2b, where an infinite beam is impacted normally and a bending-wave pulse that contains a wide band of frequencies is created. Because of their higher phase speed, high-frequency components speed ahead of the low-frequency component. Consequently, the width of the pulse increases as it travels along the beam. The speed of energy transport for the bending-wave pulse is not obvious. In case of light fluid loading (when the energy carried in the surrounding fluid is negligible), it is meaningful to define the energy transport speed as the ratio of transmitted power W and the energy density E'' (per unit length of beam). For a pulse whose spectrum peaks at frequency f the energy speed c_{BG} is twice the phase speed[1]:

$$c_{BG}(f) = 2c_B(f) \quad \text{m/s} \tag{11.3}$$

Accordingly, the power transmitted by a plane bending wave in the direction of propagation per unit area of a structure normal to that direction is

$$W_S = c_{BG} E'' = 2c_B \rho_M v^2 \quad \text{W/m}^2 \tag{11.4}$$

where v is the rms value of the bending-wave velocity and ρ_M is the density of the material. According to Eq. (11.4), the bending-wave power in an infinite beam of cross section S is given by

$$W = 2c_B S \rho_M v^2 \quad \text{W} \tag{11.5}$$

and power transmitted by a plate of thickness h across a line length l aligned parallel to the wave front is

$$W = 2c_B h l \rho_M v^2 \quad \text{W} \tag{11.6}$$

11.2 MECHANICAL IMPEDANCE AND POWER INPUT

Except for the initial transient, the velocity at the excitation point is zero if a structure is subjected to a static force. Accordingly, a static force does not transmit power into a fixed structure. For dynamic excitation by a point force or moment [the moment $M = F l_m$ is defined as a pair of forces (F) of equal magnitude but opposite direction acting simultaneously at a short distance ($\frac{1}{2} l_M$) on both sides of the excitation point], there is always a finite dynamic velocity or dynamic rotation at the excitation point. If the velocity at the excitation point has a component that is in phase with the force, or the angular velocity has a component that is in phase with the moment, then power is fed continuously into the structure.

Mechanical Impedance

The mechanical impedance is a measure of how a structure "resists" outside forces or moments. A thorough understanding of mechanical impedances is a necessary requirement for understanding and for solving most noise control problems. The concept of mechanical impedance is best introduced by considering a harmonic force $F = \hat{F}e^{j\omega t}$ acting on an ideal lumped-parameter system such as a rigid unrestrained mass, a massless spring, or a dashpot. The velocity response is also a harmonic function of the form $v = \hat{v}e^{j\omega t}$. The force impedance Z_F is defined as

$$Z_F = \frac{\hat{F}}{\hat{v}} \quad \text{N} \cdot \text{s/m} \tag{11.7}$$

and the moment impedance Z_M as

$$Z_M = \frac{\hat{M}}{\hat{\dot{\theta}}} \quad \text{N} \cdot \text{s/m} \tag{11.8}$$

where M is the exciting moment (torque) and $\dot{\theta}$ is the angular velocity at the excitation point.

The point force impedance for these basic lumped elements are listed below:

Element	Z_F
Mass	$j\omega m$
Spring	$-js/\omega$
Dashpot	r

where m is the mass of a rigid body, s is the stiffness of the spring, and r is the resistance of the dashpot. Note that the direction of the force must go through the center of gravity of the mass so that the velocity response is free of rotation and is in the direction of the force. For the rigid mass and the massless spring the force impedance is imaginary, indicating that the force and velocity are in quadrature and the power input is zero.

Formulas for computing the point force and moment impedances of infinite structures are given in references 2–12 and in Tables 11.3 and 11.4, respectively. Both force and moment impedances are idealized concepts, defined for point forces and point moments. Forces or moments can be considered applied at a point, provided that the surfaces over which they act are smaller than one-sixth of the structural wavelength. Naturally, the excited surface area must be large enough so that the force does not cause plastic deformation of the structure. There is proportionality between force and moment impedances in the form of

$$Z_M = l_M^2 Z_F \quad \text{N} \cdot \text{m/s} \tag{11.9}$$

where l_M is the effective length given in Table 11.5.

Power Input

For the dashpot the force impedance is real. The velocity is in phase with the force and the power that is fed continuously into the dashpot and dissipated in it is given by $\frac{1}{2}|\hat{F}|^2/r$. More generally, for a system with a complex force impedance Z_F, the input power is

$$W_{\text{in}} = \frac{\omega}{2\pi} \int_0^{2\pi/\omega} F(t)v(t) \, dt = \tfrac{1}{2}|\hat{F}| \, |\hat{v}| \cos\phi = \tfrac{1}{2}|\hat{F}|^2 \, \text{Re}\left\{\frac{1}{Z_F}\right\} \quad \text{W} \tag{11.10}$$

where $\hat{F}$ and $\hat{v}$ are the peak amplitude of the exciting force and the velocity, ϕ is the relative phase between $F(t)$ and $v(t)$, and Re{} refers to the real part of the bracketed quantity. Equation (11.10) is valid for any situation, not only for the dashpot.

Rigid masses, massless springs, and dashpots are useful abstractions that represent the dynamic properties of structural elements. At sufficiently low frequencies, the housing of a resiliently mounted pump behaves like a rigid mass, while the resilient rubber mount that supports it behaves like a spring. This lumped-parameter characterization of mechanical elements becomes invalid when, with increasing frequency, the structural wavelength becomes smaller than the largest dimensions of the structural element.

To deal with finite structures that are large compared with the structural wavelength, we need another useful abstraction. Regarding their input impedance—which governs power input—it is useful to consider such structures as infinite. The strategy for predicting the vibration response and sound radiation for these structures is to estimate the power input as if they were infinite and then predict the average vibration that such input power will cause in the actual finite structure. Because most structures are made of plates and beams, our strategy requires us to characterize the impedance of infinite plates and beams. The abstraction of infinite size is invoked so that no structural waves reflected from the boundaries come back to the excitation point. In this respect, structures can be considered infinite if no reflected waves that are coherent with the input come back to the excitation point because the boundaries are completely absorbing or because there is sufficient damping so that the response attributable to the reflected waves is small compared to that attributable to the local excitation.

In respect to point force excitation, infinite structures are characterized by their point force impedance $\tilde{Z}_{F\infty}$ or point input admittance (also called mobility) $\tilde{Y}_{F\infty}$, defined as

$$\tilde{Z}_{F\infty} = \frac{\tilde{F}_0}{\tilde{v}_0} \quad \text{N} \cdot \text{s/m} \tag{11.11}$$

$$\tilde{Y}_{F\infty} = \frac{1}{\tilde{Z}_{F\infty}} = \frac{\tilde{v}_0}{\tilde{F}_0} \quad \text{m/N} \cdot \text{s} \tag{11.12}$$

where $\tilde{F}_0$ is the applied point force and $\tilde{v}_0$ is the velocity response at the excitation point. The tilde above the symbols (which we will drop in further considerations) signifies that F_0 and v_0 are complex scalar quantities characterized by a magnitude and phase. Since $\tilde{v}_0$ usually has a component that is in phase with $\tilde{F}_0$ and one that is out of phase with $\tilde{F}_0$, the parameters $\tilde{Z}_F$ and $\tilde{Y}_F$ usually have both real and imaginary parts.

The power inputs to infinite structures excited by a point force or by a point velocity source (i.e., a vibration source of high internal impedance) are, respectively,

$$W_{\text{in}} = |\tfrac{1}{2}\hat{F}_0^2| \operatorname{Re}\left\{\frac{1}{Z_{F\infty}}\right\} \quad \text{W} \tag{11.13a}$$

$$W'_{\text{in}} = G_{FF} \operatorname{Re}\left\{\frac{1}{Z_{F\infty}}\right\} \quad \text{W/Hz} \tag{11.13b}$$

$$W_{\text{in}} = (\tfrac{1}{2}\hat{v}_0^2) \operatorname{Re}\{Z_{F\infty}\} \quad \text{W} \tag{11.14a}$$

$$W'_{\text{in}} = G_{VV} \operatorname{Re}\{Z_{F\infty}\} \quad \text{W/Hz} \tag{11.14b}$$

where $\hat{F}_0$ and $\hat{v}_0$ are peak amplitudes of the exciting sinusoidal force and velocity and G_{FF} and G_{VV} are the force and velocity spectral densities when the excitation is of broadband random nature. For excitation by a moment of peak amplitude $\hat{M}$ or by an enforced angular velocity of peak amplitude $\hat{\theta}$, the corresponding expressions are

$$W_{\text{in}} = |\tfrac{1}{2}\hat{M}_0^2| \operatorname{Re}\left\{\frac{1}{Z_M}\right\} \quad \text{W} \tag{11.15a}$$

$$\frac{W'_{\text{in}}}{\text{Hz}} = G_{MM} \operatorname{Re}\left\{\frac{1}{Z_M}\right\} \quad \text{W/Hz} \tag{11.15b}$$

$$W_{\text{in}} = |\tfrac{1}{2}\hat{\theta}| \operatorname{Re}\{Z_M\} \quad \text{W} \tag{11.16a}$$

$$W'_{\text{in}} = G_{\theta\theta} \operatorname{Re}\{Z_M\} \quad \text{W/Hz} \tag{11.16b}$$

Table 11.3 lists the point force impedance $Z_{F\infty}$ and Table 11.4 the moment impedance $Z_{M\infty}$ for semi-infinite and infinite structures. Table 11.5 lists the effective length l_M that connects the point force and moment impedances according to Eq. (11.9). Information regarding power input to infinite structures for force, velocity, moment, and angular velocity excitation is compiled in Table 11.6. The most complete collection of impedance formulas, which includes beams, box beams, orthotropic and sandwich plates with elastic or honeycomb core, grills, homogeneous and rib-stiffened cylinders, spherical shells, plate edges, and plate intersections and transfer impedances of various kinds of vibration isolators, is presented in reference 12.

For approximate calculations or for cases where no formulas are given in Tables 11.3–11.6, the power input to infinite structures can be estimated according to Heckl[10,11] as

$$W_{in} = \frac{\frac{1}{2}\hat{F}_0^2}{Z_{eq}} \quad W \tag{11.17}$$

$$W_{in} = (\tfrac{1}{2}\hat{v}_0^2)Z_{eq} \quad W \tag{11.18}$$

where Eq. (11.17) is used for localized force excitation (i.e., low-impedance vibration source) and Eq. (11.18) for localized velocity excitation (i.e., high-impedance vibration source) and Z_{eq} is estimated as

$$Z_{eq} = \omega \rho_M S \epsilon [\alpha \lambda] \quad N \cdot s/m \quad (beam) \tag{11.19a}$$

$$Z_{eq} = \omega \rho_M h \epsilon [\pi (\alpha \lambda)^2] \quad N \cdot s/m \quad (plate) \tag{11.19b}$$

$$Z_{eq} = \omega \rho_M \epsilon [\tfrac{4}{3}\pi (\alpha \lambda)^3] \quad N \cdot s/m \quad (half\ space) \tag{11.19c}$$

where ρ_M is the density of the material, S is the cross-sectional area of the beam, h is the plate thickness, λ is the wavelength of the motion excited most strongly (bending, shear, torsion, or compression), and ϵ is 1 when the structure is excited in the "middle" and 0.5 when it is excited at the end or at the edge (semi-infinite structure) and α is a number in the range of 0.16–0.6. If not known, it is customary to use $\alpha = 0.3$.

The equivalent impedance Z_{eq} in Eq. (11.19) has an instructive and easy-to-remember interpretation, namely that the magnitude of the point input impedance is roughly equal to the impedance of a lumped mass that lies within a sphere of radius $r \simeq \tfrac{1}{3}\lambda$ centered at the excitation point. This portion of the structure is shown in the third column of Table 11.3. This principle can be extended to composite structures[10] such as beam-stiffened plates where the combination of Eqs. (11.19a) and (11.19b) yields

$$Z_{eq} \cong \omega \rho_B S_B(\tfrac{1}{3}\lambda_{BB}) + \omega \rho_p h(\tfrac{1}{3}\lambda_{BP}) \quad N \cdot s/m \tag{11.20}$$

Another useful rule of thumb is that, regarding power input, excitation by a moment $\hat{M}$ and by an enforced angular velocity $\hat{\theta}$ can be represented by an equivalent point force $\hat{F}_{eq}$ or equivalent velocity $\hat{v}_{eq}$,[11]

$$\hat{F}_{eq} \cong \frac{\hat{M}}{0.2\lambda_B} \quad \text{N} \tag{11.21}$$

$$\hat{v}_{eq} \cong \hat{\theta}(0.2\lambda_B) \quad \text{m/s} \tag{11.22}$$

If the excitation force extends over an area that is large compared with the bending wavelength, the power transmitted into the structure becomes substantially less than it would be for a concentrated force. For extended velocity source the power transmitted becomes significantly larger than it would be for a concentrated velocity excitation.

According to reference 5, the real part of the point force impedance, Re$\{Z_{F\infty}\}$, can be predicted according to Bode's theorem as

$$\text{Re}\{Z_{F\infty}\} \cong |Z_{eq}| \cos(\tfrac{1}{2}\epsilon\pi) \quad \text{N} \cdot \text{s/m} \tag{11.23}$$

where ϵ is defined here as an exponent of the Z_{eq}-versus-frequency curve (i.e., $Z_{eq} \sim \omega^{\epsilon}$), where $\epsilon = 0$ for homogeneous isotropic thick plates, $\epsilon = 0.5$ for beams in bending, and $\epsilon = \tfrac{2}{3}$ for cylinders below the ring frequency.

Parameters Influencing Power Input. When the excitation point is near a structural junction, part of the incident vibration wave is reflected. This reflected wave influences the velocity at the excitation point and thereby also the driving point impedance and consequently the power input. Fig. 11.3a shows the effect of the vicinity of a T-junction comprised of identical, anechoically terminated beams on the point force impedance $Z_2(x_0)$ as a function of the distance x_0 between the junction and the excitation point.[13] As expected, the impedance increases rapidly with decreasing distance when $x_0 < \tfrac{1}{3}\lambda_B$. The junction has a considerable effect even for large x_0, causing approximately a 2:1 fluctuation in the magnitude of the impedance. Figure 11.3b shows the measured distribution of the beam vibration velocity $|v(x)|$ along each of the three beam branches, indicating that the observed strong standing-wave pattern is limited to that part of beam 2 that lies between the junction and the excitation point.

The impedance formulas presented in Tables 11.3–11.6 are strictly valid only when the surrounding fluid medium has no appreciable effect on the response of the structure. This is generally true for air as the surrounding medium but not for water, which is 800 times more dense than air. The effect of fluid loading at low frequencies is equivalent[14] to adding a virtual mass per unit area ρ'_s,

$$\rho'_s = \frac{\rho_0 c_B(f)}{\omega} \cong \rho_0[\tfrac{1}{6}\lambda_B(f)] \quad \text{kg/m}^2 \tag{11.24}$$

to the mass per unit area of the plate. In Eq. (11.24), ρ_0 is the density of the fluid and $c_B(f)$ and $\lambda_B(f)$ are the bending-wave speed and the wavelength of the free bending waves in the unloaded plate (see Table 11.1). Because ρ'_s is inversely proportional to $\omega^{1/2}$, fluid loading results in greater reduction of the response

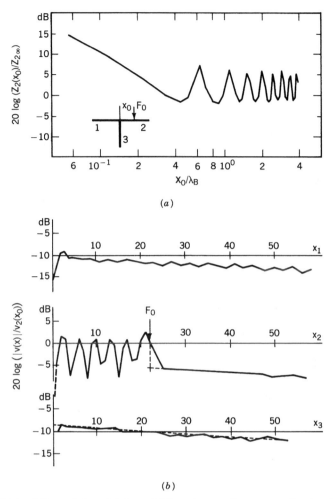

FIGURE 11.3 Effect of the vicinity of T-junction on the measured point force impedance and velocity response[13]: (a) normalized force impedance $Z_1(x_0)/Z_{1\infty}$ as a function of the normalized distance x_0/λ_B and x_i is the distance from the junction along beam i; (b) distribution of normalized velocity, $v(x)/v_0$, along the three beam branches as numbered in (b).

at the excitation point at low than at high frequencies. Fluid-loading effects are treated in references 14–17.

11.3 POWER BALANCE AND RESPONSE OF FINITE STRUCTURES

For finite structures, the response at the excitation point, and accordingly also the input admittance, $Y = 1/Z$, depends on the contribution of the waves reflected from the boundaries or discontinuities. Consequently, Y varies with the location of the excitation point and with frequency. However, both the space-averaged and frequency-averaged input admittances of the finite structure $\langle Y(x,f)\rangle_x$ and $\langle Y(x,f)\rangle_f$ equal the point input admittance of the equivalent infinite structure,[18] namely,

$$\langle Y(x,f)\rangle_x = \langle Y(x,f)\rangle_f = Y_\infty \quad \text{m/N}\cdot\text{s} \tag{11.25}$$

Figures 11.4a and 11.4b show the typical variation of the real and imaginary parts of the point input admittance of a finite plate with location x and frequency f, respectively. This particular behavior has considerable practical importance:

1. The power introduced into a finite structure by a point force of random-noise character can be well approximated by the power that the same force would introduce into an equivalent infinite structure.
2. The power introduced into a finite structure by a large number of randomly spaced point forces can be approximated by the power the same forces would introduce into an equivalent infinite structure.

Resonant Modes and Modal Density

The peaks in Fig. 11.4a correspond to resonance frequencies of the finite structure, where waves reflected from the boundaries travel in closed paths such that they arrive at their starting point in phase. The spatial deformation pattern that corresponds to such a particular closed path is referred to as the *mode shape* and the frequency where it occurs as the *eigenfrequency* or *natural frequency* of the finite structure. The importance of such resonances lies in the high transverse velocity caused by the in-phase superposition of the multiple reflections that may result in increased sound radiation or fatigue.

Exact calculation of the natural frequencies is possible only for a few highly idealized structures. Fortunately, the modal density $n(f)$, which is defined as the average number of natural frequencies in a 1-Hz bandwidth, depends not too strongly on the boundary conditions. Accordingly, one can make a reliable statistical prediction of the modal density using the formulas obtained for the equivalent idealized system.

For example, the modal density of a thin, flat, homogeneous, isotropic plate of not too high aspect ratio two octaves above the first plate resonance is already well approximated by the modal density of a rectangular plate of equal surface area, which is given by[1]

$$n(f) \approx \frac{\sqrt{12}S}{2c_L h} \quad \text{s} \tag{11.26}$$

where S is the area (one side) and h is the thickness. Note that $n(f)$ is independent of frequency and that it is large for large and thin plates.

The modal density of rooms of volume V one octave above the first room resonance is usually well approximated by the first term of the modal density of

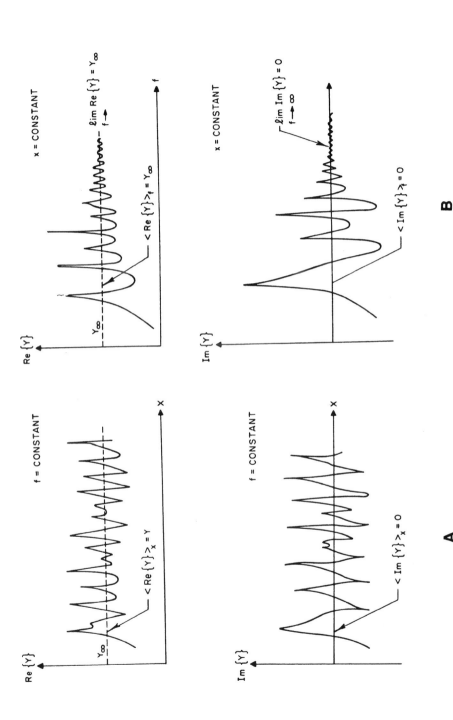

FIGURE 11.4 Typical fluctuations of the real part, Re{Y}, and imaginary part, Im{Y}, of the point input admittance of a finite plate as a function of (*a*) location x, $\langle \cdots \rangle_x$ is spatial average at a fixed frequency; (*b*) frequency f, $\langle \cdots \rangle_f$ is frequency average at a fixed location.

a rectangular room with hard walls,

$$n(f) = \frac{4\pi V}{c_0^3} f^2 + \frac{\pi S}{c_0^2} + \frac{L}{8c_0} \quad \text{s} \qquad (11.27)$$

where S is the total wall surface area and L is the longest dimension of the rectangular room. The first resonance frequency, modal densities, and mode shapes of a number of finite structures and the point input impedance (the inverse of the admittance) of the equivalent infinite structure are compiled in Table 11.7. More detailed information on mode shapes, natural frequencies, and modal densities are given in reference 19. The following important relationship exists between the modal density and the real part of the point force admittance of the equivalent infinite system[1]:

$$\text{Re}\{Y_\infty\} = \frac{n(f)}{iM} \quad \text{m/N} \cdot \text{s} \quad i = \begin{cases} 4 \text{ for thin plate} \\ 2 \text{ for thin cylinder} \\ 1 \text{ for thin sphere} \end{cases} \qquad (11.28)$$

where M is the total mass of the finite system.

Power Balance

The power balance given in Eq. (11.29),

$$W_{in} = W_d + W_{tr} + W_{rad} \quad \text{W} \qquad (11.29)$$

states that, in steady state, the power introduced, W_{in}, equals the power dissipated in the structure, W_d, the power transmitted to connected structures, W_{tr}, and the power radiated as sound into the surrounding fluid, W_{rad}. If the excitation is by a point force of peak amplitude $\hat{F}_0$ and the structure is a homogeneous isotropic thin plate of thickness h, area S (one side), density ρ_M, with longitudinal wave speed c_L, immersed in a fluid of density ρ_0, and speed of sound c_0, Eq. (11.29) takes the form

$$\tfrac{1}{2}\hat{F}^2 \, \text{Re}\{Y_F\} = \langle v^2 \rangle S[\rho_s \omega \eta_d + \rho_s \omega \eta_{tr} + 2\rho_0 c_0 \sigma] \quad \text{W} \qquad (11.30)$$

where $\rho_s = \rho_M h$ is the mass per unit area of the plate, $\omega = 2\pi f$ is the radian frequency, η_d and η_{tr} are the dissipative and transmissive loss factors, and σ is the sound radiation efficiency of the plate (see Section 11.6). Combining dissipative and transmissive losses into a single composite loss factor $\eta_c = \eta_d + \eta_{tr}$ and assuming $\text{Re}\{Y_F\} = Y_{F\infty} = 1/Z_{F\infty} = 1/(2.3\rho_s C_L t)$, as given in Table 11.3, and solving for the space–time averaged mean-square plate velocity $\langle v^2 \rangle$ yield

$$\langle v^2 \rangle = \frac{\hat{F}_0^2}{4.6\rho_s^2 c_L t \omega \eta_c S(1 + 2\rho_0 c_0 \sigma / \rho_s \omega \eta_c)} \quad \text{m}^2/\text{s}^2 \qquad (11.31)$$

Example 11.1. Predict the space–time averaged velocity response $(\langle v^2 \rangle)^{1/2}$ of a 2-mm-thick, 1×2-m steel plate to a point force $\hat{F}_0 = 10$ N peak amplitude at a frequency of 10 kHz when $\sigma = 1$ and $\eta_c = 0.01$. Compare $\langle v^2 \rangle$ with the mean-square velocity at the excitation point v_0^2. The input parameters to be used in connection with Eq. (11.31) and Table 11.2 are

$$\hat{F}_0 = 10 \text{ N} \qquad \rho_s = \rho_M t = 7700 \text{ kg/m}^3 \times 2 \times 10^{-3} \text{ m} = 15.4 \text{ kg/m}^2$$
$$c_L = 5050 \text{ m/s} \qquad \omega = 2\pi f = 2\pi \times 10^4 \text{ rad/s} \qquad \eta_c = 0.01$$
$$S = 2 \text{ m}^2 \qquad \rho_0 = 1.2 \text{ kg/m}^3 \qquad c_0 = 340 \text{ m/s} \qquad \sigma = 1$$

yielding

$$\langle v^2 \rangle = \frac{10^2}{4.6 \times (15.4)^2 \times 5.05 \times 10^3 \times 2 \times 10^{-3} \times 2\pi \times 10^4 \times 10^{-2}}$$
$$\times \frac{1}{2(1 + 2 \times 1.2 \times 340/15.4 \times 2\pi \times 10^4 \times 2 \times 10^{-2})}$$
$$= 6.7 \times 10^{-6} \text{ m/s}^2$$
$$\sqrt{\langle v^2 \rangle} = 2.6 \times 10^{-3} \text{ m/s}$$
$$v_0^2 = \frac{1}{2}\hat{F}_0^2 Y_\infty = \frac{\hat{F}^2}{4.6 \rho_s c_L t} = \frac{10^2}{4.6 \times 15.4 \times 5.03 \times 10^3 \times 2 \times 10^{-3}}$$
$$= 0.14 \text{ m/s}$$
$$\frac{v_0^2}{\langle v^2 \rangle} = \frac{(0.14)^2}{6.7} \times 10^6 = 2925 \qquad \sqrt{\frac{\langle v^2 \rangle}{v_0^2}} = 54$$

11.4 REFLECTION AND TRANSMISSION OF SOUND AT PLANE INTERFACES

When a plane sound wave traveling in a homogeneous medium encounters a plane interface with another medium, it may be (1) totally reflected, (2) partially reflected, or (3) totally transmitted depending upon the angle of incidence, the propagation speed of sound, and the density of the materials on both sides of the interface. Interfaces of practical importance are between air and water, between air and solid materials, between air and porous materials (such as ground or sound-absorbing materials), and between layers of fibrous sound-absorbing material such as treated in Chapter 8.

The simplest case is when the wave front of the incident plane wave is parallel to the plane of the interface (i.e., the wave propagates normal to the interface), as shown schematically in Fig. 11.5. In this simple case the transmitted wave p_{tran} retains the propagation direction of the incident wave p_{inc} and the following

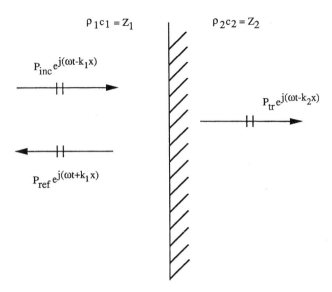

FIGURE 11.5 Reflection and transmission of a plane sound wave normally incident on the plane interface of an infinitely thick medium: p_{inc}, p_{ref}, p_{tr}, peak amplitudes of incident, reflected, and transmitted pressure; ρ, c, density and speed of sound of respective media.

relationships apply:

$$\frac{p_{tran}}{p_{inc}} = \frac{4\mathrm{R_e}\{Z_2\}/\mathrm{R_e}\{Z_1\}}{|Z_2/Z_1 + 1|^2} = \frac{2Z_2}{Z_2 + Z_1} \qquad (11.32)$$

$$\frac{p_{ref}}{p_{inc}} = 1 - \frac{W_{tran}}{W_{ref}} = \frac{Z_2 - Z_1}{Z_2 + Z_1} \qquad (11.33)$$

and for the power transmission and reflections

$$\frac{W_{tran}}{W_{inc}} = 1 - \left|\frac{Z_2 - Z_1}{Z_2 + Z_1}\right|^2 \qquad (11.34)$$

$$\frac{W_{ref}}{W_{inc}} = \left|\frac{Z_2 - Z_1}{Z_2 + Z_1}\right|^2 \qquad (11.35)$$

where $Z_i = \rho_i c_i$ is the characteristic impedance of media i for plane waves and ρ_i and c_i are the density and sound speed of the medium. For an air–steel interface and for a water–steel interface where $Z_2/Z_1 = 3.9 \times 10^7/4.1 \times 10^2$ and $Z_2/Z_1 = 3.9 \times 10^7/1.5 \times 10^6$, respectively, Eqs. (11.32)–(11.35) yield

Interface	p_{tran}/p_{inc}	p_{ref}/p_{inc}	W_{tran}/W_{inc}	W_{ref}/W_{inc}
Air–steel	1.99998	0.99998	0.00004	0.99996
Water–steel	1.927	0.927	0.141	0.859

indicating that plane waves normally incident from air onto bulk solid materials transmit only an extremely small portion of the incident energy while those incident from liquids are able to transmit an important fraction of the incident sound energy.

If the plane wave arrives at the plane interface at an oblique angle ϕ_1 ($\phi_1 = 0$ is normal incidence), the angle of the reflected wave $\phi_r = \phi_i$ but the angle of the transmitted wave ϕ_2 depends on the ratio of the propagation speeds in the two materials according to Snell's law:

$$\frac{c_1}{c_2} = \frac{\sin \phi_1}{\sin \phi_2} \tag{11.36}$$

If the plane interface is between two semi-infinite fluids or between a fluid and a porous sound-absorbing material, only compressional waves are generated. However, at interfaces between a fluid and a solid the energy transmitted into the semi-infinite solid contains both compressional and shear waves. As illustrated in Fig. 11.6, the transmitted wave breaks toward the normal of the interface if $c_2 < c_1$ and away from the normal if $c_2 > c_1$.

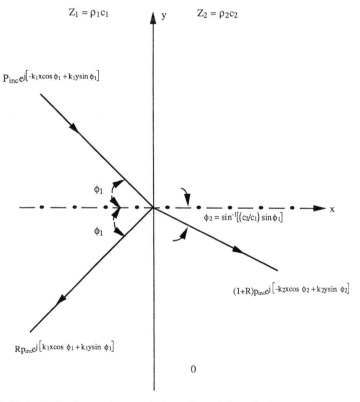

FIGURE 11.6 Reflection and transmission of an oblique-incidence plane sound wave at the plane interface of two semi-infinitely thick medium, $k_1 = \omega/c_1$, $k_2 = \omega/c_2$.

If $c_2 < c_1$, there is always a transmitted wave even for grazing incidence ($\phi_1 = 90°$), and for lossless media the power transmission reaches 100% at an oblique-incidence angle ϕ_1 when

$$\left(\frac{Z_1}{Z_2}\right)^2 = \frac{1 - (c_2/c_1)^2 \sin^2 \phi_1}{\cos^2 \phi_1} \tag{11.37}$$

If $c_2 > c_1$ sound transmission occurs only in a limited incidence angle range $0 < \phi_1 < \phi_{1L} = \sin^{-1}(c_1/c_2)$. For angles $\phi_1 > \phi_{1L}$ there is a total reflection and the sound wave penetrates into the second medium only in the form of a near field that exponentially decays with distance from the interface. The pressure reflection coefficient for oblique-incidence sound is given by

$$R(\phi_1) = \frac{p_{\text{ref}}(\phi_1)}{p_{\text{inc}}(\phi_1)} = \begin{cases} \dfrac{Z_2/\sqrt{1 - [(c_2/c_1)\sin\phi_1]^2} - Z_1/\cos\phi_1}{Z_2/\sqrt{1 - [(c_2/c_1)\sin\phi_1]^2} + Z_1/\cos\phi_1} \\ \quad \text{for } c_2 < c_1 \text{ or } c_2 > c_1 \text{ and } \phi_1 < \phi_{1L} \\ 1 \text{ for } c_2 > c_1 \text{ and } \phi_1 > \phi_{1L} \end{cases} \tag{11.38}$$

The limiting angle for plane-wave sound transmission from air to steel is $\phi_{LC} = \sin^{-1}(c_0/c_s) = 3.8°$ for compressional waves and $\phi_{LS} = \sin^{-1}(c_0/c_s) = 4.5°$ for shear waves. For sound transmission from water to steel, the corresponding limiting angles are $\phi_{LC} = 13°$ and $\phi_{LS} = 15°$, respectively. For angles larger than these, there is total reflection and only exponentially decaying near fields exist in the solid.

11.5 POWER TRANSMISSION BETWEEN STRUCTURAL ELEMENTS

In the preceding sections, the power lost to a connected structure was considered only as an additional mechanism that increases the loss factor of the excited structure. In many practical problems, however, the power transmitted to a neighboring structure is the prime reason for a noise reduction program.

The power balance equation states that the power introduced into the directly excited structure is either dissipated in it or is transmitted to neighboring structures. Accordingly, if in a noise reduction problem the power is to be *confined* to the excited structure, the power *dissipated* in the structure must greatly exceed the power *transmitted* to the neighboring structures. This requires a *high loss factor* for the excited structure and a construction that *minimizes the power transmission* to neighboring structures. Methods to achieve high damping are the subject of Chapter 14.

Reduction of Power Transmission through a Change in Cross-Sectional Area

The simplest construction that causes a partial reflection of an incident compression or bending wave is a sudden change in cross-sectional area, as shown schematically in Fig. 11.7.

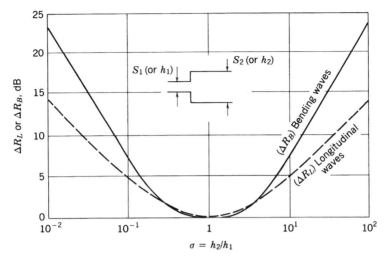

FIGURE 11.7 Attenuation at discontinuity in cross section as a function of thickness ratio. (After reference 1.)

Reflection Loss of Compression Waves. The *reflection loss* ΔR_L, defined as the logarithmic ratio of the incident to the transmitted power (for both sections of the same material), is calculated as[1]

$$\Delta R_L = 20 \log[\tfrac{1}{2}(\sigma^{1/2} + \sigma^{-1/2})] \quad \text{dB} \tag{11.39}$$

where $\sigma = S_2/S_1$ is ratio of the cross-sectional areas (see Fig. 11.7).

The reflection loss as a function of cross-sectional area ratio is plotted as the dashed line in Fig. 11.7. Since Eq. (11.39) is symmetrical for S_1 and S_2, the reflection loss is independent of the direction of the incident wave. This equation is also valid for plates where $\sigma = h_2/h_1$ is the ratio of the thicknesses. Note that a 1 : 10 change in cross-sectional area yields only 4.8 dB reflection loss. To achieve 10 dB reflection loss, a 1 : 40 change in cross-sectional area would be necessary!

Reflection Loss of Bending Waves. The reflection loss for bending waves of perpendicular incidence at low frequencies is independent of frequency and is given by[1]

$$\Delta R_B = 20 \log \frac{\tfrac{1}{2}(\sigma^2 + \sigma^{-2}) + (\sigma^{1/2} + \sigma^{-1/2}) + 1}{(\sigma^{5/4} + \sigma^{-5/4}) + (\sigma^{3/4} + \sigma^{-3/4})} \quad \text{dB} \tag{11.40}$$

The equation is also plotted in Fig. 11.7 (solid line).

We conclude from Fig. 11.7 that a change in cross-sectional area is not a practical way to achieve high reflection loss in load-bearing structures.

Reflection Loss of Free Bending Waves at an L-Junction

Structural elements that necessitate a change in the direction of a bending wave play an important role in structures. We consider here normally incident bending waves at a junction between two plates (or beams) at right angles: For low frequencies both the transmitted and reflected energy is predominantly in the form of bending waves. In this frequency range the reflection loss (logarithmic ratio of incident to transmitted power) for plates and beams of the same materials is given by[1]

$$\Delta R_{BB} = 20 \log \left[\frac{\sigma^{5/4} + \sigma^{-5/4}}{\sqrt{2}} \right] \quad \text{dB} \qquad (11.41)$$

This equation is plotted in Fig. 11.8. Because ΔR_{BB} is symmetric in σ, the reflection factor does not depend upon whether the original bending wave is incident from the thicker or from the thinner beam or plate. Note that the lowest reflection loss of 3 dB occurs for equal thicknesses ($\sigma = 1$). If the two plates or beams constituting the junction are of different material, replace the ratio $\sigma = h_2/h_1$ by

$$\sigma = \left(\frac{B_2 c_{B1}}{B_1 c_{B2}} \right)^{2/5} \qquad (11.42)$$

where B and c_B are the bending stiffness and propagation speed of free bending waves, respectively (see Table 11.1). At higher frequencies the incident bending wave also excites longitudinal waves in the second structure.[1]

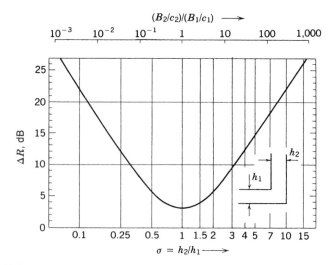

FIGURE 11.8 Attenuation of bending waves at corners (in absence of longitudinal wave interactions) as a function of thickness ratio. (After reference 1.)

Reflection Loss of Bending Waves through Cross Junctions and T-Junctions

Other structures that may provide a substantial reflection of an incident bending wave are the *cross junction* of walls shown schematically in Fig. 11.9 and the *T-junction* in Fig. 11.10. If a bending wave of perpendicular incidence reaches

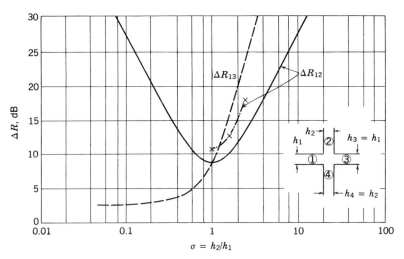

FIGURE 11.9 Attenuation of bending waves at plate intersections (in absence of longitudinal wave interactions) as a function of thickness ratio. (After reference 1.) (—·—) ΔR_{12} for random incidence computed by Kihlman.[20]

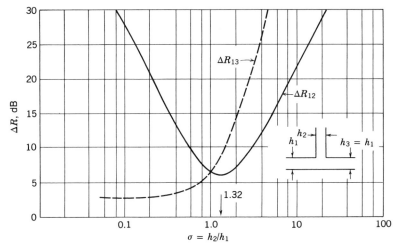

FIGURE 11.10 Attenuation of bending waves at plate intersections (in absence of longitudinal wave interactions) as a function of thickness ratio. (After reference 1.)

the cross junction from plate 1, it is partially reflected and partially transmitted to the other plates. The transmitted power splits into a number of different wave types, namely, bending waves in plate 3 and longitudinal and bending waves in plates 2 and 4. Because of the symmetry in the geometry, plates 2 and 4 will have the same excitation.[1]

The reflection loss (defined as the logarithmic ratio of the power in the incident to that in the transmitted bending wave) is given as a function of the ratio of the plate thickness for plates or beams of the same material and is shown in Figs. 11.9 and 11.10. When the plates are made of different material, the ratio $\sigma = h_2/h_1$ is given by Eq. (11.42). The amplitudes of the bending waves transmitted without a change in direction are restrained by the perpendicular plate, and the reflection loss in this direction, ΔR_{13}, increases monotonically with increasing thickness of the restraining plate, h_2. Since this plate effectively stops the vertical motion of the horizontal plate at the junction, even for very thin vertical walls, ΔR_{13} remains level at 3 dB, indicating that only the power carried by the bending moment can pass the junction. For those bending waves that change direction at the junction, the reflection loss becomes a minimum ($\Delta R_{12} = 9$ dB) at a thickness ratio $\sigma = h_2/h_1 = 1$ for the cross junction; corresponding numbers for the T-junction are 6.5 dB at a thickness ratio $\sigma = h_2/h_1 = 1.32$. The reflection loss then increases symmetrically for increasing or decreasing thickness ratio (h_2/h_1).

The transmission of free bending waves at cross junctions for random incidence has been computed and the reflection loss for a number of combinations of dense and lightweight concrete plates determined.[20] The results for ΔR_{12} are plotted in Fig. 11.9 (as x's), which indicate that ΔR_{12} for random incidence is somewhat higher than that for normal incidence. It was also found that for random incidence ΔR_{12} is independent of frequency but ΔR_{13} decreases with increasing frequency.

Power Transmission from a Beam to a Plate

The structural parts of a modern building frequently include columns and structural floor slabs. Consequently, the power transmission from a beam to a plate, which models this situation, is of practical interest. Let us first consider the reflection loss for longitudinal and bending waves incident *from* the beam onto an infinite homogeneous plate.

Reflection Loss for Longitudinal Waves.
When a longitudinal wave in the beam reaches the plate, its energy is partly reflected back up the beam and is partly transmitted to the plate in the form of a bending wave. The reflection loss (equal to the logarithmic ratio of the incident to the transmitted power) is given by[21]

$$\Delta R_L = -10 \log \left(1 - \left|\frac{Y_b - Y_p}{Y_b + Y_p}\right|^2\right) \quad \text{dB} \qquad (11.43)$$

where $Y_b = 1/Z_b$ = admittance of semi-infinite beam for longitudinal waves, m/N · s
Y_p = point admittance of infinite plate, $1/Z_p$, m/N · s

Both Y_p and Y_b are real and frequency independent and can be found for infinite beams and plates from Table 11.3 by taking the reciprocal of the impedances, that is, $Y_b = 1/\rho_b c_{Lb} S_b$ and $Y_p = 1/2.3\rho_p c_{Lp} h^2$.

Complete Power Transmission between Beam and Plate for Longitudinal Waves. Inspecting Eq. (11.43), we note that the reflection factor is zero (all the incident energy is transmitted to the plate) when $Y_b = Y_p$. Equating the values for Y_p and Y_p given above yields the requirements for complete power transmission from the beam to the plate:

$$S_b = 2.3 \frac{\rho_p c_{Lp} h^2}{\rho_b c_{Lb}} \quad m^2 \tag{11.44}$$

If the column and the slab are of the same material so that $\rho_p c_{Lp} = \rho_b c_{Lb}$, Eq. (11.44) simplifies to

$$S_b = 2.3 h^2 \quad m^2 \tag{11.45}$$

This equation says that for perfect power transfer from a beam of square cross section to a large plate, the cross dimension of the beam must be 1.52 times the thickness of the plate and for a beam of circular cross section the radius must be 0.86 times the plate thickness. Actually, this is well within the range of slab thicknesses–column cross section ratios commonly found in architectural structures. The reflection factor for different geometries of steel beam and plate connections has been measured.[21] The results for a substantial mismatch (*a*) and for a near matching (*b*) are plotted in Fig. 11.11.

Reflection Loss for Bending Waves. When a beam carries a free bending wave, a part of the energy carried by the wave is transmitted to the plate by the effective bending moment and excites a radially spreading free bending wave in the plate. A part of the incident energy is reflected from the junction. Here the reflection loss is determined by the respective moment impedances[1,21] of the plate and the beam:

$$\Delta R_b = -10 \log \frac{Y_b^M - Y_p^M}{Y_b^M + Y_p^M} \quad dB \tag{11.46}$$

where the moment admittances Y_b^M and Y_p^M are given by[1,20]

$$Y_b^M = \frac{2}{1+j} \frac{k_b^2}{\rho_b S_b c_{B_b}} \quad m/N \cdot s \tag{11.47}$$

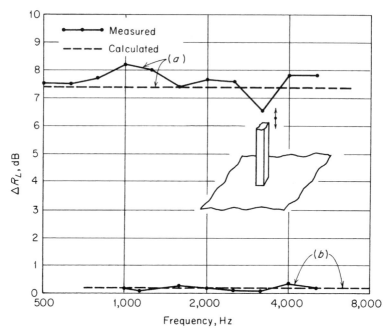

FIGURE 11.11 Reflection loss for longitudinal waves, ΔR_L, for a steel beam plate system: (a) plate thickness 2 mm, beam cross section 10×20 mm; (b) plate thickness 4 mm, beam cross section 5×10 mm. (After reference 21.)

and

$$Y_b^M = \frac{\omega}{16 B_p} \left(1 + j\frac{4}{\pi} \ln \frac{1.1}{k_p a}\right) \quad \text{m/N} \cdot \text{s} \tag{11.48}$$

where k = wavenumber, m^{-1}
 B_p = bending stiffness per unit width of plate, N · m
 a = effective distance of pair of point forces making up moment on plate, m; for rectangular and circular beam cross section $a_r = \frac{1}{3}d$ and $a_c = 0.59r$
 d = side dimension of rectangular beam cross section (in direction of bending), m
 r = radius of circular beam cross section, m

The reflection loss obtained[21] for the bending-wave excitation of a steel rod of 1×2 cm cross-sectional area attached to a 0.2-cm-thick semi-infinite steel plate for two perpendicular directions of bending of the rod is plotted in Fig. 11.12.

Complete Power Transmission for Bending Waves. Since the moment impedances of the plate and beam are both frequency dependent, complete power transmission ($\Delta R_B = 0$) can occur only at a single frequency. The criteria for

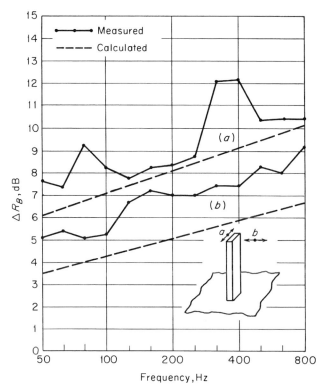

FIGURE 11.12 Reflection loss for bending waves, ΔR_B, for a steel beam plate system for two different directions of bending of the beam (a and b). Plate thickness 2 mm, beam cross section 10×20 mm. (After reference 21.)

perfect power transmission is achieved when both the real and imaginary parts of the moment impedances of the beam and plate are equal, which requires that

$$\lambda_b = 0.39 \frac{B_b}{B_p} \quad \text{m} \tag{11.49}$$

$$\lambda_p = 2.6a \quad \text{m} \tag{11.50}$$

where B_p = bending stiffness of beam, $N \cdot m^2$
λ_b = bending wavelength in beam, m
λ_p = bending wavelength in plate, m

Reduction of Power Transmission between Plates Separated by Thin Resilient Layer

In architectural structures it is customary to provide a so-called vibration break by inserting a thin layer of resilient material between structural elements. The

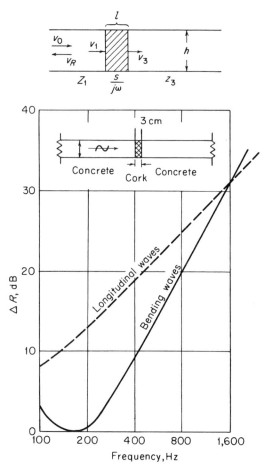

FIGURE 11.13 Attenuation due to an elastic interlayer of 3-cm-thick cork between 10-cm-thick slabs of concrete as a function of frequency. (After reference 1.)

geometry of such a vibration break is shown in Fig. 11.13. Often this construction serves also as an expansion joint.

Reflection Loss for Compression Waves. The reflection loss is given by[1]

$$\Delta R_L = 10 \log \left[1 + \left(\frac{\omega Z_1}{2 s_i} \right) \right] \quad \text{dB} \tag{11.51}$$

where Z_1 = impedance of solid structure for compressional waves, $N \cdot s/m$
s_i = stiffness of resilient layer in compression, N/m

Above a certain frequency ($\omega = 2s_i/Z_1$) the ΔR_L-versus-frequency curve increases with a 20-dB/decade slope with increasing frequency. Below this frequency the resilient layer transmits the incident wave almost entirely. As an example,[1] the ΔR_L-versus-frequency curve for a 3-cm-thick layer of cork inserted between two 10-cm-thick concrete slabs (or columns) is shown in Fig. 11.13.

To achieve a high reflection loss, the resilient layer must be as soft as is permitted by the load-bearing requirements ($s_i/\omega \ll Z_1$). However, the stiffness of the layer cannot be reduced indefinitely by increasing the thickness. For frequencies where the thickness of the layer is comparable with the wavelength of compressional waves in the resilient material, the layer can no longer be considered as a simple spring characterized by its stiffness alone. The reflection loss in this frequency region is given by[1]

$$\Delta R_L = 10\log\left[\cos^2 k_i l + \frac{1}{4}\left(\frac{Z}{Z_i} + \frac{Z_i}{Z}\right)^2 \sin^2 k_i l\right] \quad \text{dB} \tag{11.52}$$

where k_i = wavenumber for compressional waves in resilient material, m^{-1}
Z = impedance of equivalent infinite structure for compression waves,
    = $Z_1 = Z_3$, N · s/m
Z_i = impedance of equivalent infinite (length) resilient material for compression waves, N · s/m
l = length of resilient layer, m

The impedances Z, Z_i and the wavenumber k_i are assumed real; thus Eq. (11.52) does not account for wave damping in the resilient material. As expected, $\Delta R_L = 0$ for $Z_i = Z$. The maximum of ΔR_L is reached when $k_i l = (2n+1)(\frac{1}{2}\pi)$; here the reflection loss becomes

$$\Delta R_{L,\max} = 20\log\frac{Z_i^2 + Z^2}{2ZZ_i} \quad \text{for } l = \tfrac{1}{2}(2n+1)\lambda_i \tag{11.53}$$

For $k_i l = n\pi$ ($l = \tfrac{1}{2}n\lambda_i$), the denominator of Eq. (11.51) becomes unity independent of the magnitude Z, yielding

$$\Delta R_{L,\min} = 0 \quad \text{for } l = \tfrac{1}{2}n\lambda_i$$

Finally for $k_i l \ll 1$ and $Z_i \ll Z$, Eq. (11.52) simplifies to Eq. (11.51).

Reflection Loss for Bending Waves. The geometry in Fig. 11.13 suggests that for bending waves of perpendicular incidence the moments and forces acting on both sides of the junction must be equal. However, the transverse velocity and angular velocity on both sides of the junction are different because of shear and compressional deformation of the resilient layer. The elastic layer behaves quite differently for bending waves than it does for compressional waves.[1] The most striking difference is the *complete transmission* of the incident bending wave at a certain frequency and a *complete reflection* of it at another, higher frequency. Unfortunately, it turns out that the frequency of complete transmission for architectural structures of interest usually occurs in the audio frequency range. As an example, Fig. 11.13 shows the reflection factor for bending waves as a function of frequency for a layer of 3-cm-thick cork between 10-cm-thick concrete slabs. The transmission is complete at a frequency of 170 Hz and then decreases with increasing frequency. The reflection loss for bending waves can be approximated by[1]

$$\Delta R_B \cong 10 \log \left[1 + \frac{1}{4} \left(1 - \frac{E}{E_i} \frac{2\pi^3 l h^2}{\lambda_B^3} \right)^2 \right] \quad \text{dB} \qquad (11.54)$$

where E = Young's modulus of structural material
E_i = Young's modulus of elastic material
l = length of elastic layer, m
h = thickness of structure, m
λ_B = wavelength of bending waves in structure, m

The bending wavelength in the structure for which the elastic layer provides a complete transmission of bending waves is given by

$$\lambda_{B,\text{trans}} = \pi \left(\frac{E}{E_i} h^2 l \right)^{1/3} \quad \text{m} \qquad (11.55)$$

If one wishes to reduce the frequency of complete transmission, the ratio E/E_i and the length of the resilient layer must be large. However, the length of the resilient material should always be small compared with the wavelength of bending waves in the elastic layer to avoid resonances.

A complete reflection of the incident bending wave occurs when the bending wavelength in the plate equals π times the plate thickness, $\lambda_{Bs} = c_b/f_s = \pi h$. Consequently, the frequency where *complete reflection* occurs is *independent* of the dynamic properties and the length of the elastic layer and is given by the thickness and the dynamic properties of the plate or beam.

11.6 SOUND RADIATION

The sound radiation of small rigid bodies is treated in Chapter 10. This section deals exclusively with the sound radiation of thin flexible plates excited to vibration by point forces or by sound fields. Sound radiation of small rigid bodies is treated in Chapter 10.

Vibration of rigid and elastic structures forces the surrounding fluid or gas particles at the interface to oscillate with the same velocity as the vibrating structure and thus causes sound. The sound waves propagate in the form of compressional waves that travel with the speed of sound in the surrounding medium.

Infinite Rigid Piston

Conceptually the simplest sound-radiating structure is an infinite plane rigid piston. The motion of the piston forces the fluid particles to move along parallel lines that are perpendicular to the plane of the piston. There is no divergence that could lead to inertial reaction forces, such as those that can occur along the edges of a finite piston where the fluid can move to the side. Consequently, the reaction force per unit area (i.e., the sound pressure) is fully attributable to compression effects. This is the same situation as if the piston would be placed in a tube of rigid walls, as discussed already in Chapter 10. If the piston vibrates with velocity $\hat{v}\cos\omega t$, it generates a plane sound wave traveling perpendicular to the plane of the piston. The sound pressure as a function of distance is

$$p(x,t) = \hat{v}\rho_0 c_0 \cos(\omega t - k_0 x) \quad \text{N/m}^2 \tag{11.56}$$

and the radiated sound power per unit area is

$$W'_{\text{rad}} = 0.5\hat{v}^2 \rho_0 c_0 = \langle v^2 \rangle_t \rho_0 c_0 \quad \text{W/m}^2 \tag{11.57}$$

where ρ_0 and c_0 are the density and speed of sound of the medium, $\omega = 2\pi f$ is the radian frequency, $k_0 = \omega/c_0 = 2\pi/\lambda_0$ is the wavenumber, $\lambda_0 = c_0/f$ is the wavelength of the radiated sound, and $\langle v^2 \rangle_t$ is the time-averaged mean-square velocity (i.e., $v = v_{\text{rms}}$).

Infinite Thin Plate in Bending

If a plane bending wave of velocity amplitude $\hat{v} = \sqrt{2}v$ and bending wave speed c_B travels on a thin plate in the positive x direction, the sound pressure as a function of x and perpendicular distance z is given by[1]

$$\hat{p}(x,y) = \frac{j\hat{v}\rho_0 c_0 e^{j\omega t}}{\sqrt{(k_B/k_0)^2 - 1}} e^{-jk_B z} \exp(-z\sqrt{k_B^2 - k_0^2}) \quad \text{N/m}^2 \tag{11.58}$$

where $k_B = 2\pi f/c_B = 2\pi/\lambda_B$ and $k_0 = 2\pi f/c_0 = 2\pi/\lambda_0$ are the bending wavenumber in the plate and the wavenumber in the air, respectively. Inspecting Eq. (11.58), one finds that for $c_B \langle c_0$ $(k_B/k_0 \rangle 1)$ the sound pressure is 90° out of phase with the velocity at the interface so that no sound power is radiated by the plate. The sound pressure constitutes a near field that decays exponentially with increasing z. For $c_B > c_0$; $k_B/k_0 < 1$, Eq. (11.58) has the form

$$\hat{p}(x,y) = \frac{\hat{v}\rho_0 c_0 e^{-j\omega t}}{\sqrt{1-(k_B/k_0)^2}} e^{-jk_B x} \exp\left(-jk_0 z \sqrt{1-\left(\frac{k_B}{k_0}\right)^2}\right) \quad \text{N/m}^2 \tag{11.59}$$

where the pressure and velocity are in phase at the interface ($z = 0$) and the sound power radiated by a unit area of the plate is

$$W'_{\text{rad}} = \begin{cases} \dfrac{0.5\hat{v}^2 \rho_0 c_0}{\sqrt{1 - (\lambda_0/\lambda_B)^2}} & \text{W/m}^2 \quad \text{for } \lambda_B > \lambda_0 \\ 0 & \text{for } \lambda_B < \lambda_0 \end{cases} \quad (11.60)$$

which, with increasing frequency ($\lambda_0/\lambda_B \ll 1$), approaches that of an infinite rigid piston as given in Eq. (11.57).

Radiation Efficiency

It is customary to define the radiation efficiency of a vibrating body as

$$\sigma_{\text{rad}} = \frac{W_{\text{rad}}}{\langle v_n^2 \rangle \rho_0 c_0 S} \quad (11.61)$$

where $\langle v_n^2 \rangle$ is the normal component of the space–time average mean-square vibration velocity of the radiating surface of area S and W_{rad} is the radiated sound power. With this definition, Eqs. (11.57) and (11.60) yield for the piston

$$\sigma_{\text{rad}} = \begin{cases} 1 & \text{for infinite rigid piston} \\ 0 & \text{for } \lambda_B < \lambda_0 \text{ for infinite plate in bending} \\ [1 - (\lambda_0/\lambda_B)]^{-1/2} & \text{for } \lambda_B > \lambda_0 \text{ for infinite plate in bending} \\ (k_0 a)^2/[1 + (k_0 a)^2] & \text{for pulsating sphere} \end{cases} \quad (11.62)$$

It is important to know that the radiation efficiency depends not only on the size and shape of the radiating body as compared with the wavelength but also on the manner the body is vibrating. If a sphere vibrates back and forth instead of pulsating, then the net volume displacement is zero and the radiation efficiency is[22]

$$\sigma_{\text{rad}} = \frac{(k_0 a)^4}{4 + (k_0 a)^4} \quad (11.63)$$

Comparing Eqs. (11.62) and (11.63) reveals that at low frequencies a rigid body vibrating in translation radiates much less (approximately $\frac{1}{4}(k_0 a)^2$ times) sound power than a pulsating body of the same surface area. Radiation efficiencies of some typical structural elements are given in Table 11.8 (see references 22 and 23 for a more extensive collection). Radiation efficiencies for oscillating three-dimensional bodies of near-unity aspect ratio (such as a sphere or cube) are plotted in Fig. 11.14 and those of pipes of circular rods oscillating as rigid bodies in Fig. 11.15. Note that with increasing frequency, when the distance between the neighboring, out-of-phase moving parts of the vibrating body becomes larger than the wavelength of the radiated sound (i.e., $2\pi a > \lambda_0$ for oscillating bodies and $\lambda_B > \lambda_0$ for plates in bending), it becomes more difficult to push the air

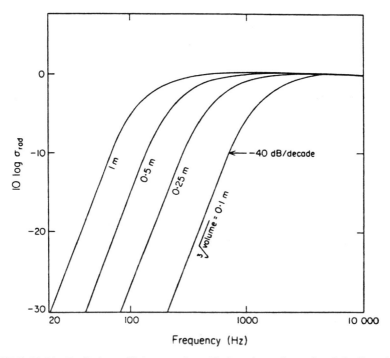

FIGURE 11.14 Radiation efficiency of oscillating three-dimensional bodies. (After reference 22.)

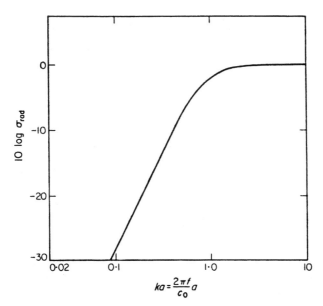

FIGURE 11.15 Radiation efficiency of oscillating rigid pipes and rods of radius a. (After reference 22.)

rapidly enough aside then to compress it, and the radiation efficiency approaches unity.

Point-Excited Infinite Thin Plate

When a very large homogeneous isotropic thin plate is excited by a point force of amplitude $\hat{F}_0$ or by an enforced local velocity of amplitude $\hat{v}_0$, free bending waves with a propagation speed of $c_B(f) \sim \sqrt{f}$ propagate radially from the excitation point. At low frequencies, where $c_B < c_0$, the free bending waves radiate no sound. Far from the excitation point, where the wave front approaches a straight line, the sound radiation that does occur is due to the in-phase vibration of the plate in the vicinity of the excitation point. The sound intensity radiation pattern has a $\cos^2 \phi$ dependence, where ϕ is the angle to the normal of the plate. The mechanical power input W_{in}, the radiated sound power W_{rad}, and the acoustical–mechanical conversion efficiency $\eta_{am} = W_{rad}/W_{in}$ are listed below:

Point Force Excitation   Point Velocity Excitation

$$W_{in} = \tfrac{1}{2}\hat{F}_0^2 \frac{1}{2.3\rho_s c_L h} \qquad (\tfrac{1}{2}\hat{v}_0^2) 2.3 \rho_s c_L h \qquad (11.64)$$

$$W_{rad}(f < f_c) = \hat{F}^2 \frac{\rho_0}{4\pi \rho_s^2 c_0} \qquad \hat{v}^2 \frac{c_L^2 h^2 \rho_0}{2.38 c_0} \qquad (11.65)$$

$$\eta_{am}(f < f_c) = 0.37 \frac{\rho_0}{\rho_M} \frac{c_L}{c_0} \qquad 0.37 \frac{\rho_0}{\rho_M} \frac{c_L}{c_0} \qquad (11.66)$$

where $\rho_s = \rho_M h$ is the mass per unit of the plate, ρ_M is the density and c_L the speed of longitudinal waves in the plate material, h is the plate thickness, and ρ_0 and c_0 are the density and speed of sound in the surrounding fluid.

Equations (11.64)–(11.66) contain the following, quite surprising, information:

1. W_{in}, W_{rad}, and η_{am} are independent of frequency and plate loss factor.
2. For point force excitation, the radiated sound power depends only on the mass per unit area of the plate ($W_{rad} \sim 1/\rho_s^2$) and not on stiffness.
3. For point velocity excitation, the radiated sound power depends only on stiffness ($W_{rad} \sim c_L^2 h^2$) and not on the plate material density.
4. The acoustical–mechanical conversion efficiency is the same for point force or point velocity excitation, and it is independent of plate thickness h and is a material constant [$\eta_{am} \sim (c_L/\rho_M)(\rho_0/c_0)$].

For noise control engineers, observations 2 and 3, embodied in Eq. (11.65), are very important. To minimize sound radiation from highly damped, point-excited, thin, platelike structures, the plate should have large mass per unit area for force excitation (e.g., by a low-impedance vibration source) and low stiffness (low E/ρ_M) for velocity excitation (by a high-impedance vibration source).

Equation (11.66) supplies the rationale for violin makers, who want to convert as large a portion of the mechanical power of the vibrating string as possible into radiated sound, to use wood ($\eta_{am} = 0.024$) and not steel ($\eta_{am} = 0.0023$) or lead ($\eta_{am} = 0.0004$) for the body of the violin.

Note also that the radiated sound power given in Eq. (11.65) usually represents the minimum achievable for a finite-size plate since it accounts only for the sound radiated from the vicinity of the excitation point.

Point-Excited Finite Plates

For finite-size plates, the sound power radiated has two components. The first component is radiated from the vicinity of the excitation point given in Eq. (11.65). The second component is radiated by the free bending waves as they interact with plate edges and discontinuities. The contribution of these two components to total radiated noise is represented by the first and second terms in Eqs. (11.67a) and (11.67b), where the first equation is valid for point force and the second for point velocity excitation:

$$W_{rad}^F \cong \hat{F}^2 \left[\frac{\rho_0}{4\pi \rho_s^2 c_0} + \frac{\rho_0 c_0 \sigma_{rad}}{4.6 \rho_s^2 c_L h \omega \eta_c} \right] \quad W \quad (11.67a)$$

$$W_{rad}^v \cong \hat{v}^2 \left[\frac{\rho_0 c_L^2 h^2}{2.38 c_0} + 1.15 \left(\frac{c_L h}{\omega \eta_c} \right) \rho_0 c_0 \sigma_{rad} \right] \quad W \quad (11.67b)$$

where η_c is the composite loss factor and σ_{rad} is the radiation efficiency of the plate for free bending waves. The second term in Eqs. (11.67) has been derived by the power balance (see Section 11.3) of the finite plate, assuming that the mechanical power input to the finite plate can be well approximated by the power input to the equivalent infinite plate.

Equations (11.67a) and (11.67b) can be used to assess the useful upper limit of the composite loss factor (η_c^{max}) which, if exceeded, results only in added expense but no meaningful reduction of the radiated noise. This is done by equating the first and second terms in the square brackets and solving for η_c.

11.7 SOUND EXCITATION AND SOUND TRANSMISSION

The sound transmission process has the following three components: (1) description of the sound field at the source side, (2) prediction of the vibration response, and (3) sound radiation from the receiver side of the partition into the receiver room. Item 1 is treated in Chapter 7. Items 2 and 3 will be treated in this section. Regarding the appropriate way of analysis, partitions can be classified as either small or large compared with the acoustical wavelength.

Sound Transmission of Small Partitions

In the following analyses we define a partition small if its dimensions are small compared with the acoustical wavelength. Consequently, if the frequency is sufficiently low, even partitions of large size are considered "acoustically small."

The subject of sound transmission of acoustically small plates has been mostly ignored or treated only qualitatively in books on noise control engineering. The reason for this is that the traditional definition of sound transmission loss, which is based on the ratio of the incident and transmitted sound power, does not make sense in this low-frequency region where the sound power radiated into the receiver room is independent of the angle of incidence of the sound striking the source side of the partition. In this low-frequency region the partition will radiate the same sound power for grazing incidence (sound wave on the source side propagates parallel to the partition) than it does at normal incidence (where it propagates perpendicular to the plane of the partition).

It is instructive to start the investigation of sound transmission with a small, single, homogeneous, isotropic plate that is flash mounted in the test section between two reverberation rooms, as shown schematically in Fig. 11.16.

Let our investigation start at very low frequencies and proceed by gradually increasing the frequency of the incident sound. We will investigate the following cases:

1. The size of the panel, L, is small compared with the acoustical wavelength λ and the frequency of the incident sound wave, f, is small compared with the frequency of the first mechanical resonance of the plate, f_{M1}:

$$L \ll \lambda \quad \text{and} \quad f \ll f_{M1}$$

2. The size of the panel, L, is small compared with the acoustical wavelength λ and the frequency of the incident sound wave, f, matches the frequency of the first mechanical resonance of the plate, f_{M1}.

$$L \ll \lambda \quad \text{and} \quad f = f_{M1}$$

3. The size of the panel, L, is small compared with the acoustical wavelength λ and the frequency of the incident sound wave, f, is much larger than the frequency of the first mechanical resonance of the plate, f_{M1}.

$$L \ll \lambda \quad \text{and} \quad f \gg f_{M1}$$

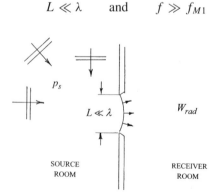

FIGURE 11.16 Sound Transmission of a small homogeneous isotropic partition.

Case 1 ($L \ll \lambda$ and $f \ll f_{M1}$). If the size of the partition is much smaller than the acoustical wavelength, the sound pressure, even at grazing incidence (i.e. the sound wave propagates parallel to the plate), is nearly constant across the source side of the plate. At any other angle of incidence, the sound pressure is even more evenly distributed across the plate than at grazing. At normal incidence, where the sound wave propagates perpendicular to the plane of the plate, the sound pressure is absolutely evenly distributed over the surface of the plate independent of the wavelength (for normal incidence the trace wavelength is infinitely large at all frequencies). In the case 1 frequency range the sound transmission of the plate is controlled by its volume compliance C_v defined as

$$C_v \equiv \frac{\Delta V}{p} \quad \text{m}^5/\text{N} \tag{11.68}$$

where C_v = volume compliance of plate, m^5/N
$p \approx$ sound pressure on source side of plate, N/m^2

As long as the dimensions of the plate are small compared with the acoustical wavelength, $p_s = 2p_{\text{inc}}$, where p_{inc} is the amplitude of the incident sound wave.

Considering now that the volume velocity $q = d\,\Delta V/dt = j\omega C_v p_s$ and using Eq. (10.49) from Chapter 10 the formula for the radiated sound power yields

$$W_{\text{rad}} = q^2(\rho_0 c_0)\left(\frac{k^2}{2\pi}\right) = \frac{\omega^4 C_v^2(\rho_0/c_0)}{2\pi} p_s^2 \quad \text{N}\cdot\text{m/s} \tag{11.69}$$

where ρ_0 = density of gas, kg/m^3
c_0 = speed of sound in gas, m/s.

Considering that our small plate radiates omnidirectionally into a half space, the sound pressure as a function of distance is

$$p(r) = \frac{p_s}{2\pi r}(\omega^2 C_v \rho_0) \quad \text{N/m}^2 \tag{11.70}$$

where r is the radial distance in meters from the center of the panel.

Formulas to predict the volume compliance C_v and resonance frequency f_{M1} for rectangular plates with simply supported and clamped edges are given in Chapter 12. Panels found in practical applications have a C_v that is larger than predicted for clamped edges and smaller than that predicted for simply supported edges. Also the resonance frequency of practical panels falls between those resonance frequencies predicted for clamped and simply supported edges.

Case 2 ($L \ll \lambda$ and $f \approx f_{M1}$). At the first resonance frequency of the panel, f_{M1}, the mass part of the panel impedance (which increases with increasing frequency) becomes equal in magnitude but opposite in phase to the stiffness part of the panel impedance, which is decreasing with increasing frequency. At and in the vicinity of the resonance frequency even a small sound pressure on the source side can cause a large volume velocity and the panel motion is limited only by energy dissipation in the panel and by its radiation resistance. In the vicinity of f_{M1} the sound power radiated by the receiver side of the panel reaches a maximum. The dissipative losses in the panel cannot be predicted analytically. Consequently, the sound transmission cannot be predicted with satisfactory accuracy. In designing enclosures for equipment with a high-intensity tonal component at low frequencies the panels or subpanels should be chosen that their first resonance frequency should be substantially higher than the frequency of the low-frequency tonal component of the enclosed equipment. The first resonance frequency of platelike elements can be increased by curving them in two perpendicular directions to gain form stiffness.

Case 3 ($L \ll \lambda$ *and* $f \gg f_{M1}$). Above the first mechanical resonance frequency the velocity of the plate is controlled by the mass per unit area $\rho_s = \rho_M h$
ρ_M = density of plate material, kg/m³
h = plate thickness, m

The volume velocity of the plate in this frequency region can be approximated by

$$q \approx p_s \frac{S}{j\omega \rho_s} \quad \text{m}^3/\text{s} \tag{11.71}$$

the radiated sound power by

$$W_{\text{rad}} \approx p_s^2 \frac{\omega^2 S^2 (\rho_0 c_0)}{2\pi} \quad \text{N} \cdot \text{m/s} \tag{11.72}$$

and the sound pressure at a distance r by

$$p(r) \approx p_s \frac{\omega^2 \rho_0^2 S}{2\pi r} \quad \text{N/m}^2 \tag{11.73}$$

where S is surface area (one side) of the plate in square meters.

More accurate prediction of the transmitted sound power can be obtained by determining the response of each of the volume-displacing modes of the plate responding to the spatially uniform sound pressure on the source side of the plate and determining the net volume velocity as the sum total of the volume velocities of all modes, as done in Chapter 6.

If the spatial distribution of the plate vibration $v(x, y)$ is known, the sound pressure in the receiver-side half space $p(x, y, z, \omega)$ can be computed as[1]

$$p(x, y, z, \omega) = \frac{j\omega \rho_0}{2\pi} \int v(x, y, \omega) \frac{e^{-jkr}}{r} dS \quad \text{N/m}^2 \tag{11.74}$$

where r is the distance between the small surface element of the plate and the observation point. Equation (11.74) is valid only for a flat plate flash mounted in an infinite, flat, hard wall.

In the low-frequency region, where the plate dimensions are small compared with the acoustical wavelength ($\sqrt{S} \ll \lambda$), the transmitted sound power depends only on the sound pressure on the source side of the plate p_s and does not depend on the angle of incidence or the sound power incident on the source side of the plate, W_{inc}. Consequently, the traditional definition of the sound transmission loss given in Eqs. (11.79) and (11.80) makes no sense in this frequency region.

Sound Transmission of Large Partitions

This section deals with the sound transmission loss of acoustically large partitions, where it is meaningful to characterize the excitation by the sound power incident on the source side of the partition and define the transmission coefficient τ and the sound transmission loss R as given in Eqs. (11.75) and (11.76), respectively.

The most common problem that noise control engineers have to deal with is the transmission of sound through solid partitions such as windows, walls, and floor slabs. The problem may be either prediction or design. The prediction problem is typically this: Given a noise source, a propagation path up to the partition, and the size and construction of the partition and the room acoustics parameters of the receiver room, predict the noise level in the receiver room. The design problem is typically stated as: Given a source, a propagation path, the room acoustics parameters of the receiver room, and a noise criterion (in the form of octave-band sound pressure levels), determine the construction of those partitions that would assure that the noise criteria are met with a sufficient margin of safety.

The transmission coefficient τ and sound transmission loss R, which characterize sound transmission through partitions, are defined as

$$\tau(\phi, \omega) = \frac{W_{trans}(\phi, \omega)}{W_{inc}(\phi, \omega)} \tag{11.75}$$

$$R(\phi, \omega) = 10 \log \frac{1}{\tau(\phi, \omega)} = 10 \log \frac{W_{inc}(\phi, \omega)}{W_{trans}(\phi, \omega)} \quad \text{dB} \tag{11.76}$$

where $W_{inc}(\phi, \omega)$ is the sound power incident at angle ϕ at frequency $\omega = 2\pi f$ on the source side and $W_{trans}(\phi, \omega)$ is the power transmitted (radiated by the receiver side).

Though knowledge of the sound transmission loss of a window or curtain wall for a particular angle of incidence may be desirable sometimes, it is a more common problem to characterize the transmission of sound between two adjacent rooms where the sound is incident on the separating partition from all angles with approximately equal probability. In such "random incident" sound fields, the sound intensity (energy incident on a unit area) I_{random} is related to the space-averaged mean-square sound pressure in the source room $\langle p^2 \rangle$ as

$$I_{random} = \frac{\langle p^2 \rangle}{4\rho_0 c_0} \quad \text{W/m}^2 \tag{11.77}$$

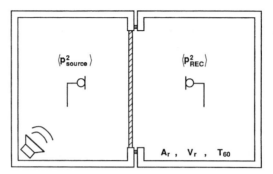

FIGURE 11.17 Experimental setup for laboratory measurements of the sound transmission loss of partitions (see text for definition of symbols).

Laboratory procedures ASTM E90-02[24] and ISO 140-1:1997,[25] adopted for measuring the sound transmission loss of partitions, are based on such a diffuse sound field obtained by utilizing large reverberation rooms (see Chapter 7) as source and receiver rooms. The measurement procedure depicted in Fig. 11.17 has three steps. The first step is to measure the sound power incident on the source-side face of the test partition of area S_w:

$$W_{inc} = I_{inc} S_w = \frac{\langle p^2_{source} \rangle S_w}{4\rho_0 c_0} \quad W \tag{11.78}$$

by measuring the space–time average mean-square sound pressure level $\langle p^2_{source} \rangle$ in the source room by spatially sampling the sound field. The second step is to measure the transmitted sound power W_{trans} from the power balance of the receiver room,

$$W_{trans} = \frac{\langle p^2_{rec} \rangle A_r}{4\rho_0 c_0} \quad W \tag{11.79}$$

yielding the laboratory sound transmission loss

$$R_{lab} = 10 \log \frac{W_{inc}}{W_{trans}} = \langle L_p \rangle_s - \langle L_p \rangle_R + 10 \log \frac{S_w}{A_r} \quad dB \tag{11.80}$$

where A_r is the total absorption in the receiving room. The third step is to determine A_r from the known volume and the measured reverberation time T_{60} of the receiver room, as described in Chapter 7.

Once R_{lab} has been measured, it can be used for predicting the mean-square sound pressure in a particular receiver room acoustically characterized by its total absorption A_r through a partition of surface area S_w for an incident sound field of intensity I_{inc} as

$$\langle p^2_r \rangle = \frac{\tau I_{inc} S_w (4\rho_0 c_0)}{A_r} = \frac{\tau S_w < p^2_{source}}{A_r} \quad N^2/m^4 \tag{11.81a}$$

$$\langle \text{SPL}_R \rangle = \langle \text{SPL}_s \rangle - R_{\text{lab}} + 10 \log \frac{S_w}{A_r} \quad \text{dB} \tag{11.81b}$$

provided that the partition is not much smaller than tested and that edge conditions are not much different. Curves of measured random-incidence sound transmission loss versus frequency for standard windows, doors, and walls are available from manufacturers and should be used in design and prediction work. Measured sound transmission losses of some selected partitions are given by Bies and Hansen in their Table 8.1 (see the Bibliography).

The purpose of the discussion that follows is to identify the physical processes and the key parameters that control sound transmission through partitions and to provide analytical methods that further the development of an informed judgment needed for working with data obtained in laboratory measurements. Most importantly, however, this will focus on predicting sound transmission loss for situations that are different from those employed in the standardized laboratory measurements (i.e., for near-grazing incidence) and for the task of designing nonstandard partitions for unique applications.

Excitation of structures by an incident sound wave is significantly different from excitation by localized forces, moments, enforced velocities, or angular velocities. As discussed in the previous sections, the response of thin, platelike structures to localized excitation results in radially spreading free bending waves. The propagation speed of these waves, $c_B(f)$, is as unique a characteristic of the plate as is the period of a pendulum. The thinner the plate and the lower the frequency, the lower is the propagation speed.

If the structure is excited by an incident sound wave, forcing occurs simultaneously over the entire exposed surface of the plate. The incident sound enforces its spatial pattern on the plate, causing it to instantaneously conform to its trace. The trace "runs" along the plate with a speed that approaches the speed of sound in the source-side media when the sound runs nearly parallel to the plate (grazing incidence) and approaches infinity as the sound incidence approaches normal. This "sound-forced" response of the thin plate to sound excitation is referred to as the "forced wave." In contrast to the free bending waves, the speed of the forced bending wave is independent of frequency, plate thickness, and mass per unit area (though the amplitude of the response depends on them). Because of their supersonic speed, forced waves radiate sound very efficiently at all frequencies (e.g., their radiation efficiency $\sigma_F \geq 1$), except for panels that are small compared with the acoustical wavelength.

Transmission of Normal-Incidence Plane Sound Waves through an Infinite Plate

It is advisable to introduce the complex process of sound transmission by considering first the least complicated case when a plane sound wave is normally incident on a uniform homogeneous, isotropic, flat plate of thickness h, as shown in Fig. 11.18. Because the pressure exerted on the plate is in phase over the

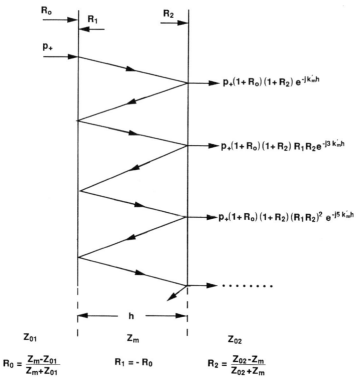

FIGURE 11.18 Transmission of a normally incident sound wave through a flat, homogeneous, isotropic plate. The transmitted pressure p_T is the sum of all the infinite components on the right. (See text for definition of symbols.)

entire surface of the plate, only compressional waves are excited. The sound wave of pressure amplitude p_+ propagates in a gas of characteristic plane-wave impedance $Z_{01} = \rho_{01} c_{01}$. It encounters the solid plate of characteristic impedance $Z_M = \rho c_L$ and on the transmitting side (right) will radiate sound into another gas of characteristic impedance $Z_{02} = \rho_{02} c_{02}$. The multiple reflection and transmission phenomena at the interfaces are governed by Eqs. (11.32) and (11.33). The amplitude of the transmitted sound pressure p_T in the receiver-side media is obtained by the summation of the transmitted components, as illustrated in Fig. 11.18, yielding[26]

$$p_T = p_+ \frac{(1+R_0)(1+R_2)e^{-jk_m h}}{1 - R_1 R_2 e^{-j2k_m h}} \quad \text{N/m}^2 \quad (11.82)$$

where R_0, R_1, and R_2 are the reflection factors at the interfaces and directions indicated in Fig. 11.18 and h is the plate thickness. Propagation losses can be taken into account by a complex wavenumber $k'_m = k_m(1 + \frac{1}{2}j\eta)$, where $k_m = \omega/c_L$ is the wavenumber of compressional waves in the plate and η is the loss factor. The normal-incidence sound transmission loss is defined as

$$R_N = 10 \log \frac{W_{\text{inc}}}{W_{\text{rad}}} = 10 \log \frac{p_+^2}{p_T^2} \quad \text{dB} \tag{11.83a}$$

Assuming that the same gas is on both sides of the plate, Eq. (11.82) yields

$$R_N = 10 \log \left[\cos^2 k'_m h + 0.25 \left(\frac{Z_0}{Z_m} + \frac{Z_m}{Z_0} \right)^2 \sin^2 k'_m h \right] \quad \text{dB} \tag{11.83b}$$

which for $|k'_m h| \ll 1$ yields the simple expression

$$R_N \cong 10 \log \left[1 + \left(\frac{\rho_s \omega}{2 \rho_0 c_0} \right)^2 \right] \quad \text{dB} \tag{11.83c}$$

known as the normal-incidence mass law. In Eq. (11.83c), $\rho_s = \rho h$ is the mass per unit area of the plate and $\rho_0 c_0 = Z_0$ is the characteristic impedance of the gas, assumed the same on both sides.

Figure 11.19 shows the computed normal-incidence sound transmission loss of a 0.6-m- (2-ft-) thick dense concrete wall. At low frequencies where the plate thickness is less than one-sixth of the compressional wavelength ($f < c_L/6h$) the sound transmission loss follows the normal-incidence mass law of Eq. (11.83c), increasing by approximately 6 dB for each doubling of the frequency or the mass per unit area. At high frequencies compressional wave resonances in the plate

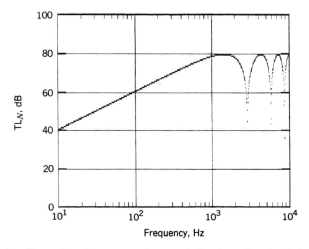

FIGURE 11.19 Normal-incidence sound transmission loss R_0 of a 0.6-m-thick dense concrete wall computed according to Eq. (11.83b) assuming $\eta = 0$.

occur and the R_N-versus-frequency curve exhibit strong minima at $f_n = nc_L/2h$, yielding

$$R_N^{\min} = 20\log\left(1 + \frac{\pi}{4}\frac{Z_m}{Z_0}\eta n\right) \quad \text{dB} \tag{11.84a}$$

and the maximal achievable sound transmission loss can be approximated as

$$R_N^{\max} \cong 20\log\frac{Z_m}{2Z_0} \quad \text{dB} \tag{11.84b}$$

Equation (11.84a) indicates complete transmission of the incident sound ($R_N^{\min} = 0$ dB) at compressional wave resonances if $\eta = 0$. However, Eq. (11.84a) indicates that even a small loss factor of $\eta = 0.001$ ensures that the normal-incidence transmission loss of a concrete partition does not dip below 24 dB at compressional wave resonances. Note, however, that no such dips are observed for random incidence because different incident angles correspond to different frequencies where such dips occur. As indicated in Eq. (11.84b), the maximum achievable normal-incidence sound transmission loss of homogeneous, isotropic single plates of any thickness is limited by the ratio of the characteristic impedances and is 80 dB for dense concrete, 94 dB for steel, and 68 dB for wood.

Transmission of Oblique-Incidence Plane Sound Waves through an Infinite Plate

The transmission of oblique-incidence sound through infinite plane plates can be formulated either in terms of shear and compressional waves, where the bending of the plate is considered as a superposition of these two wave types, or by utilizing the bending-wave equation of the plate. Both formulations are discussed below.

Combined Compressional and Shear Wave Formulation. The combined shear and compression wave formulation, which is valid also for thick plates, has been treated in Chapter 9 of the 1992 edition. It is too involved to present it here. The interested reader is referred to references 27 and 28 or to pages 248–288 of the 1992 edition of this book.

Separation Impedance Formulation. Sound transmission through plates can be conveniently characterized by the separation impedance Z_s defined as[1,26]

$$Z_s = \frac{p_s - p_{\text{rec}}}{v_n} \quad \text{N}\cdot\text{s/m}^3 \tag{11.85}$$

where p_s is the complex amplitude of the sound pressure at the source-side face of the plate representing the sum of the incident and reflected pressures ($p_s = p_{\text{inc}} + p_{\text{refl}}$), p_{rec} is the complex amplitude of sound pressure on the receiver-side face, and v_n is the complex amplitude of normal velocity of the receiver-side face of the plate. In the case of single panels, it is generally assumed that both faces of the panel vibrate in phase with the same velocity. The sound transmission coefficient of the plate τ and its sound transmission loss R are given by[1,26]

$$\tau(\phi) = \frac{I_{\text{trans}}}{I_{\text{inc}}} = \left|1 + \frac{Z_s \cos \phi}{2\rho_0 c_0}\right|^2 \tag{11.86}$$

$$R(\phi) = 10 \log \frac{1}{\tau(\phi)} = 10 \log \left(\left|1 + \frac{Z_s}{2\rho_0 c_0/\cos \phi}\right|^2\right) \quad \text{dB} \tag{11.87}$$

where ϕ is the angle of sound incidence ($\phi = 0$ for normal incidence).

The formulation of the sound transmission in terms of the separation impedance of Eq. (11.87) lends itself exceptionally well to predicting the sound transmission loss of multilayer partitions where the constituent layers may be thin plates separated by air spaces with and without porous sound-absorbing material (such as double and triple walls), windows, and so on. The reason for this is that the separation impedance of such multilayer partitions can be expeditiously obtained by appropriately combining the separation impedances of the constituent layers.

The separation impedance for a thick isotropic plate is obtained by solving the bending-wave equation of the plate as formulated by Mindlin[29]:

$$\left(\nabla^2 - \frac{\rho_M}{G}\frac{\partial^2}{\partial t^2}\right)\left(B\nabla^2 - \frac{\rho_M h}{12}\frac{\partial^2}{\partial t^2}\right)\xi(x,y) + \rho_M h \frac{\partial^2}{\partial t^2}\xi(x,y)$$
$$= \left(1 - \frac{B}{Gh}\nabla^2 + \frac{\rho_M h^2}{12G}\frac{\partial^2}{\partial t^2}\right)\Delta p(x,y,0) \tag{11.88}$$

where $\Delta p(x,y,0)$ is the pressure differential across the plate, $\zeta(x,y)$ is the plate displacement in the z direction normal to the plate surface, $\nabla^2 = \partial^2/\partial x^2 + \partial^2/\partial y^2$ is the Laplacian operator, ρ_M is the density, G is the shear modulus, B is the bending stiffness, and h is the thickness of the panel. The solution of Eq. (11.88) for a plane sound wave of incidence angle ϕ yields[30]

$$Z_s = \frac{j\{[\rho_M h + (\rho_M h^3/12 + \rho_M B/G)(\omega^2/c_0^2)\sin^2 \phi]\omega - [(B/c_0^4)\sin^4 \phi + \rho_M^2 h^3/(12G)]\omega^3\}}{1 + B\omega^2 \sin^2 \phi/(Gc_0^2 h) - \rho_M h^2 \omega^2/(12G)} \quad \text{N} \cdot \text{s/m}^3 \tag{11.89}$$

Equation (11.89) can be approximated by

$$Z_s \cong Z_m + \left(\frac{1}{Z_B} + \frac{1}{Z_{\text{sh}}}\right)^{-1} \quad \text{N} \cdot \text{s/m}^3 \tag{11.90}$$

where $Z_m = j\omega\rho_M h$ is the mass impedance of the plate per unit area and $Z_B = -jB\omega^3 \sin^4 \phi/c_0^4$ and $Z_{\text{sh}} = -jGh\omega \sin^2 \phi/c_0^2$ are the bending-wave and shear wave impedances of the plate per unit area.

If $|Z_{sh}| \ll |Z_B|$, the plate responds predominantly in shear and $Z_s \cong Z_m + Z_{sh}$. In this case, Eq. (11.89) yields

$$R_{sh}(\phi) = 20 \log \left| 1 + \frac{j\omega \rho_M h[1 - (c_s/c_0)^2 \sin^2 \phi]}{2\rho_0 c_0 / \cos \phi} \right| \quad \text{dB} \tag{11.91}$$

Equation (11.91) indicates that trace coincidence between the incident sound wave and the free shear waves in the plate (which would lead to complete transmission) can be avoided at all incident angles provided that the panel is specifically designed to yield a low shear wave speed such that $c_s^2 = G/\rho < c_0^2 = P_0\kappa/\rho_0$. Unfortunately, this desirable condition cannot be met with homogeneous plates made of any construction-grade material. As we will discuss later, however, it is possible to satisfy the $c_s < c_0$ criterion with specially designed[31] inhomogeneous sandwich panels without compromising the static strength of the panel and thereby preserving the mass law behavior described by

$$R_{sh}(\phi) \cong R_{mass}(\phi) = 20 \log \left(1 + \frac{\rho_2 \, \omega \cos \phi}{2\rho_0 c_0} \right) \quad \text{dB} \tag{11.92}$$

which is representative of a limp panel of the same mass per unit area $\rho_s = \rho_M h$ as the shear panel.

Thin homogeneous panels are easier to bend than to shear so that $Z_B \ll Z_{sh}$. It follows that $Z_s \cong Z_m + Z_B$ and Eqs. (11.87) and (11.89) yield

$$R(\phi) = 10 \log \frac{1}{\tau(\phi)} = 10 \log \left| 1 + \frac{j\rho_s \omega[1 - (f/f_c)^2 \sin^4 \phi]}{2\rho_0 c_0 / \cos \phi} \right|^2 \quad \text{dB} \tag{11.93}$$

where f_c is the critical frequency where the speed of the free bending waves in the plate, $c_B(f_c)$, equals the speed of sound in the media. It is given by

$$f_c = \left(\frac{\omega_c}{2\pi} \right) = \left(\frac{c_0^2}{2\pi} \right) \left(\frac{\rho_s}{B} \right)^{1/2} \quad \text{Hz(s)}^{-1} \tag{11.94}$$

where ρ_s = mass per unit are, $\rho_M h$ kg/m²
h = plate thickness, m
B = bending stiffness of plate, $Eh^3/[12(1 - v^2)]$ N · m

Consequently, the product $f_c \rho_s$ is a constant for a specific plate material and liquid (usually air). It is given by

$$f_c \rho_s = \left(\frac{c_0^2}{2\pi} \right) \sqrt{12(1 - v^2)} \sqrt{\frac{\rho_M}{E}} \quad \text{Hz(s)}^{-1} \tag{11.95}$$

This product is listed in Table 11.2 for air as a surrounding medium. It is used to determine f_c as $f_c = \rho_s f_c / f_c$.

The coincidence frequency in air of homogeneous isotropic plates can be easily determined by cutting a narrow strip of length l in meters, supporting the two ends of the strip on a knife edge (to simulate simple supported edge conditions), and measuring the sag d in millimeters at the midpoint of the strip. The coincidence frequency is computed as

$$f_c = 1.65(10^2)\sqrt{\frac{d/\text{mm}}{l/\text{m}}} \quad \text{Hz(s)}^{-1} \quad (11.96)$$

Equation (11.93) indicates that above the critical frequency f_c given in Eq. (11.94), trace coincidence between the incident sound wave and the free bending waves in the plate occurs when $f = f_c/\sin^2\phi$ (which is called the coincidence frequency) and would result in complete transmission if the plate had no internal damping.

It is customary to account for internal damping in Eq. (11.88) by introducing a complex Young's modulus $E' = E(1 + j\eta)$ that results in complex wavenumbers $k'_c = k_c(1 + \frac{1}{2}j\eta)$ and $k'_s = k_s(1 + \frac{1}{2}j\eta)$ and yields the following modified form of the sound transmission loss:

$$R(\phi) = 10\log\frac{1}{\tau(\phi)}$$
$$= 10\log\left\{\left|1 + \frac{\rho_s\omega}{2\rho_0 c_0/\cos\phi}\right.\right.$$
$$\left.\left.\cdot\left\{\eta\left(\frac{f}{f_c}\right)^2\sin^4\phi + j\left[1 - \left(\frac{f}{f_c}\right)^2\sin^4\phi\right]\right\}\right|^2\right\} \quad \text{dB} \quad (11.97)$$

Equation (11.97) indicates that in the vicinity of $f = f_c/\sin^2\phi$ the curves of sound transmission loss versus frequency exhibit a minimum that is controlled by the damping.

Figure 11.20 shows the curve of sound transmission loss versus frequency for a 4.7-mm-($\frac{3}{16}$-in.-) thick glass plate for normal ($\phi = 0°$), $\phi = 45°$, and near-grazing ($\phi = 85°$) angles of incidence computed according to Eq. (11.97). Figure 11.20 illustrates the decrease in sound transmission loss that occurs with increasing angle of incidence (owing to the $\cos\phi$ term) and the trace-matching dip that occurs at $f = f_c/\sin^2\phi$.

Transmission of Random-Incidence (Diffuse) Sound through an Infinite Plate

A plane wave impinging on the plate at one particular angle is not a typical problem. The sound field in a room is better modeled as a diffuse sound field, which is an ensemble of plane sound waves of the same average intensity traveling with equal probability in all directions. A region of unit area on the plate will be exposed, at any instant, to plane sound waves incident from all areas on

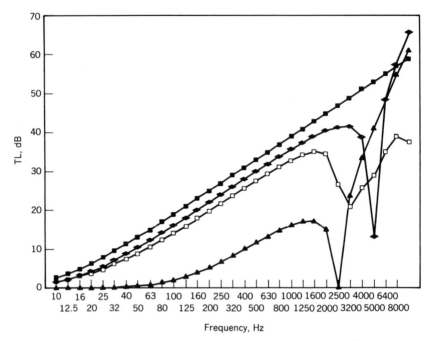

FIGURE 11.20 Computed TL-versus-frequency curve of a 3.7-mm-($\frac{3}{16}$-in.-) thick infinite glass plate for various angles of incidence: (■) normal ($\phi = 0°$); (♦) $\phi = 45°$; (▶) $\phi = 85°$; (□) random.

a hemisphere whose center is the area on the plate. These waves are uncorrelated and have equal intensity. The sound intensity incident on the unit area of the plate from any particular angle will be the intensity of the plane wave at angle I_{inc} multiplied by the cosine of the angle of incidence. The total transmitted intensity is then

$$I_{\text{trans}} = \int_\Omega \tau(\phi) I_{\text{inc}} \cos\theta \, d\Omega \quad \text{W/m}^2 \tag{11.98}$$

The integration is over a hemisphere of solid angle Ω, where $d\Omega = \sin\phi \, d\phi \, d\theta$. Because I_{inc} is the same for all plane waves and τ is independent of the polar angle θ, an average transmission coefficient may be defined by

$$\bar{\tau} = \frac{\int_0^{\phi\,\text{lim}} \tau(\phi) \cos\phi \sin\phi \, d\phi}{\int_0^{\phi\,\text{lim}} \cos\phi \sin\phi \, d\phi} \tag{11.99}$$

where ϕ_{lim} is the limiting angle of incidence of the sound field. For random incidence, ϕ_{lim} is taken as $\frac{1}{2}\pi$, or $90°$. The sound transmission coefficient $\tau(\phi)$ is that given in Eq. (11.97), and the random-incidence sound transmission loss is given by

$$R_{\text{random}} = 10 \log \frac{1}{\bar{\tau}} \quad \text{dB} \tag{11.100a}$$

At low frequencies ($f \ll f_c$) the random-incidence sound transmission loss (for $TL_N > 15$ dB) is found by averaging the argument of Eq. (11.97) over a range of ϕ from 0° to 90° to yield[32,33]

$$R_{\text{random}} \cong R_0 - 10\log(0.23 R_0) \quad \text{dB} \tag{11.100b}$$

which is commonly referred to as the random-incidence mass law.

It has become common practice to use the *field-incidence mass law*, which is defined (for $R_0 \geq 15$ dB) as[34]

$$R_{\text{field}} \cong R_0 - 5 \quad \text{dB} \tag{11.101}$$

This result, which yields better agreement with measured data than Eq. (11.100b), approximates a diffuse incident sound field with a limiting angle ϕ_{lim} of about 78° in Eq. (11.99).[34]

The mass law transmission losses R_0, R_{field}, and R_{random}, valid for frequencies well below coincidence, are plotted versus $f\rho_s$ in Fig. 11.21.

Field-Incidence Sound Transmission for Thin Isotropic Plates, R_{field}.
Equations (11.98) and (11.99) must be solved by numerical integration.

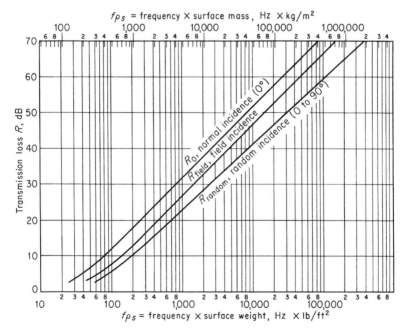

FIGURE 11.21 Theoretical sound transmission loss of large panels for frequencies well below coincidence ($f \leq 0.5 f_c$). Field incidence assumes a sound field that allows all angles of incidences up to 78° from normal.

The results of such an integration of the transmission coefficient between the angles 0° and 78° and application of Eq. (11.100a) to give the *field-incidence transmission loss* are presented in Fig. 11.22 for all values of f/f_c. The ordinate is the difference between the field-incidence transmission loss R_field and the normal-incidence mass law transmission loss at the critical frequency $R_0(f_c)$. The latter is easily determined from Eq. (11.83c) or from Fig. 11.21 when the mass per unit area and the critical frequency of the panel are known. Note that predicted transmission losses of less than about 15 dB for $f \ll f_c$ or less than 25 dB for $f \simeq f_c$ from Fig. 11.22 are not accurate.

Example 11.2. Calculate the normal-incidence mass law for an aluminum panel weighing 10 lb/ft² at a frequency of 500 Hz. Also determine the random-incidence and field-incidence mass laws. What is R_field at 2800 Hz when $\eta = 10^{-2}$?

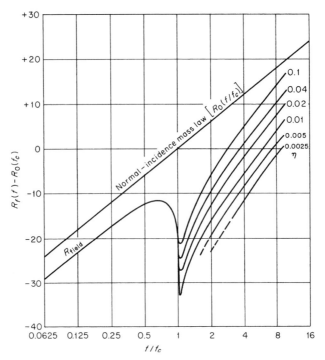

FIGURE 11.22 Field-incidence forced-wave transmission loss. The ordinate is the difference between the field-incidence transmission loss at the frequency f and the normal-incidence transmission loss at the *critical frequency* ($f/f_c = 1$). Note that for a predicted transmission loss of less than 15 dB or for the dashed areas on the figure, the transmission loss depends on both the surface weight and the loss factor, and the curves provide only a lower bound estimate to the actual transmission loss. Use of curve: (1) determine $\rho_s f_c$ from Table 11.2, (2) determined f_c, (3) determine $R_0(f_c)$ from Fig. 11.21 or Table 11.2, (4) read $R_f(f) - R_0(f_c)$ from Fig. 9.21 at the required η, and (5) $R_f(f) = [R_f(f) - R_0(f_c)] + R_0(f_c)$. Top curve is the normal-incidence mass law defined in Eq. (11.83c).

Solution The normal-incidence mass law is given by Eq. (11.83c) and the upper curve of Fig. 11.21. We have $f\rho_s = 500 \times 10 = 5000$ Hz · lb/ft². From Fig. 11.21, $R_0 = 45.5$ dB. The random-incidence mass law is given by Eq. (11.100b) and the lower curve of Fig. 11.21; that is, $R_{\text{random}} = 35$ dB. The field-incidence mass law is given by Eq. (11.101) and the middle curve of Fig. 11.21, that is, $R_{\text{field}} = 45.5 - 5 = 40.5$ dB.

From Table 11.2, $\rho_s f_c = 7000$ Hz · lb/ft²; $f_c = 7000/10 = 700$ Hz and $R_0(f_c) = 48$ dB from Fig. 11.21. Evaluating Fig. 11.22 at $f/f_c = 2800/700 = 4$ and $\eta = 0.01$, we get $R_f(f) - R_0(f_c) = -6$ dB, yielding $R_r(f) = [R_r(f) - R_0(f_c)] + R_0(f) = -6 + 48 = 42$ dB.

Sound Transmission for Orthotropic Plates. Sound transmission for orthotropic plates differs from that of isotropic plates because orthotropic plates have markedly different bending stiffnesses in the different principal directions. The difference in bending stiffness for plane plates may result from the anisotropy of the plate material (such as for wood caused by grain orientation) or from the construction of the plate such as corrugations, ribs, cuts, and so on. Consequently, the speed of free bending waves is different for these two directions and the orthotropic panel has two coincidence frequencies given by[35]

$$f_{c1} = \frac{c_0^2}{2\pi}\sqrt{\frac{\rho_s}{B_x}} \quad \text{Hz} \qquad (11.102a)$$

$$f_{c2} = \frac{c_0^2}{2\pi}\sqrt{\frac{\rho_s}{B_y}} \quad \text{Hz} \qquad (11.102b)$$

where B_x is the bending stiffness for the stiffest direction and B_y the direction perpendicular to this. The random-incidence sound transmission loss of an orthotropic plate is predicted by[35]

$$R_{\text{random}} \cong \begin{cases} 10\log\left[\left(\dfrac{\rho_s\omega}{2\rho_0 c_0}\right)^2\right] - 5 & \text{for } f \ll f_{c1} \\[2mm] 10\log\left[\left(\dfrac{\rho_s\omega}{2\rho_0 c_0}\right)^2\right] - 10\log\left[\dfrac{1}{2\pi^3\eta}\dfrac{f_{c1}}{f}\sqrt{\dfrac{f_{c1}}{f_{c2}}}\left(\ln\dfrac{4f}{f_{c1}}\right)^4\right] & \text{for } f_{c1} < f < f_{c2} \\[2mm] 10\log\left[\left(\dfrac{\rho_s\omega}{2\rho_0 c_0}\right)^2\right] - 10\log\dfrac{\pi f_{c2}}{2\eta f} & \text{for } f > f_{c2} \end{cases}$$

(11.103)

where η is the loss factor. For corrugated plates, as shown in Fig. 11.23, the bending stiffnesses can be approximated by

$$B_y = \frac{Eh^3}{12(1-v^2)} \quad \text{N} \cdot \text{m} \qquad (11.104a)$$

$$B_x = B_y \left(\frac{s}{s'}\right) \quad \text{N} \cdot \text{m} \tag{11.104b}$$

where s and s' are defined in Fig. 11.23. Note that the increase in bending stiffness caused by corrugations, ribs, and stiffeners always results in a reduction of the sound transmission loss, while measures such as partial-depth saw cuts, which decrease bending stiffness, result in an increase of the sound transmission loss of plates.

Sound Transmission Loss for Inhomogeneous Plates. Sound transmission loss for inhomogeneous plates, such as appropriately designed sandwich panels, can be substantially higher than for homogeneous panels of the same mass per unit area, provided that such plates favor the propagation of the free shear waves (with frequency-independent propagation speed) rather than free bending waves for which the propagation speed increases with increasing frequency. However, they must be designed such that the shear wave speed remains below the speed of sound in air so that no trace coincidence occurs. Consequently, the sound transmission loss of such so-called shear wall panels closely approximates the field-incidence mass law. Information for designing such panels is given in reference 31. However, ordinary sandwich panels are very poor sound barriers because

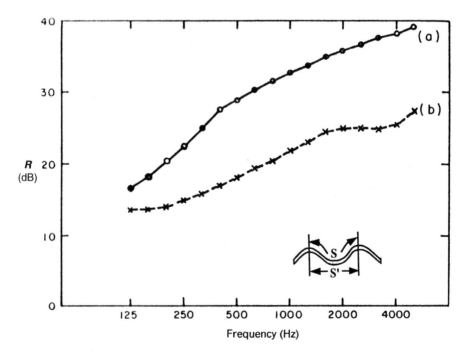

FIGURE 11.23 Measured sound transmission loss of steel plates (after Ref. 28): (*a*) plane plate, $\rho_s = 8$ kg/m^2; (*b*) corrugated plate, $\rho_s = 11$ kg/m^2; s, distance between corrugations along surface; s', distance along straight line.

of their low mass and high bending stiffness that result in a coincidence frequency that usually falls in the middle of the audio frequency range. Dilation resonance, which occurs at the frequency where the combined stiffness impedance of the face plate and that of the enclosed air equals the mass impedance of the plate, also leads to further deterioration of the sound transmission loss of sandwich panels.

Sound Transmission through a Finite-Size Panel

For most architectural applications, where the first resonance frequency of typical platelike partitions is well below the frequency range of interest and the plate size is much larger than the acoustical wavelength, Eq. (11.97) or Fig. 11.22 (which are strictly valid only for infinitely large panels) can be used to predict the sound transmission loss of finite panels. In many industrial applications, the finite size of the panel must be taken into account.[36]

In finite panels the sound-forced bending waves encounter the edges of the plate and generate free bending waves, such that the sum of the incident forced bending wave and the generated free bending wave satisfies the particular plate edge condition (e.g., zero displacement and angular displacement at a clamped edge). Consequently, the sound-forced bending waves continuously feed free bending wave energy into the finite panel and build up a reverberant, free bending wave field. The mean-square vibration velocity of this free bending wave field, $\langle v_{FR}^2 \rangle$ can be obtained using a power balance for the finite plate. The power introduced into the finite plate at the edges equals the power lost by the plate owing to viscous losses in the plate material, energy flow into connected structures, and sound radiation. The transmitted sound radiated by the finite panel is given by

$$W_{\text{rad}} = \langle v_{FO}^2 \rangle \, \rho_0 c_0 S \sigma_{FO} + \langle v_{FR}^2 \rangle \, \rho_0 c_0 S \sigma_{FR} \quad \text{W} \tag{11.105}$$

where $\langle v_{FO}^2 \rangle$ is the mean-square velocity of the sound-forced supersonic bending waves, $\sigma_{FO} \geq 1$ is the radiation efficiency of the forced waves, S is the surface area of the panel, and σ_{FR} is the radiation efficiency of the free bending waves. Since $\sigma_{FR} \ll 1$ below the critical frequency of the panel ($f \ll f_c$), it is frequently the case that the vibration response of the panel is controlled by the free bending waves (i.e., $\langle v_{FR}^2 \rangle \gg \langle v_{FO}^2 \rangle$) but the sound radiation is controlled by the less intense but more efficiently radiating forced waves.

The classical definition of sound transmission loss is $R = 10 \log(W_{\text{inc}}/W_{\text{trans}})$, where W_{inc} is the sound power incident at the source side and W_{trans} is that radiated from the receiver side of the panel. If the incident sound is a plane wave arriving at an incident angle ϕ ($\phi = 90°$ for grazing incidence), then it is assumed that

$$W_{\text{inc}} = \frac{0.5 |\hat{p}_{\text{in}}|^2 S \, \cos \phi}{\rho_0 c_0} \quad \text{W} \tag{11.106}$$

This assumption leads to the dilemma that at grazing incidence ($\phi = 90°$) no power is incident on the panel. It is common knowledge that grazing incidence sound excites the panel to forced vibrations, and the panel radiates sound into the receiver room when the forced bending waves in the panel and the sound wave at the receiver side, which run parallel to the panel, encounter the edges of the finite panel. This unresolved conceptual problem has been avoided[34] by limiting the incident angle range to 78°, in computing the field incidence sound transmission loss for the infinite panel according to Eq. (11.101), so as to yield reasonable agreement with laboratory measurements for panel sizes typically employed in such tests.

Obviously, it is not the incident sound power but the mean-square sound pressure on the source side that is forcing the panel. Since this quantity is proportional to the sound energy density in the source room $E_s = \langle p_s^2 \rangle / \rho_0 c_0^2$, it has been suggested[37,38] that the sound transmission loss of a finite partition be defined as

$$R_E \equiv 10 \log \left(\frac{E_s}{E_R} \frac{S}{A} \right) \quad \text{dB} \quad (11.107)$$

where $E_R = 4 W_{\text{trans}} / c_0 A$ is the energy density in the receiver room, S is the surface area of the panel (one side), and A is the total absorption in the receiver room. The transmitted sound power $W_{\text{trans}} = \frac{1}{4} c_0 A E_R$ is owing to the velocity of the plate. The forced response is dominated by the mass-controlled separation impedance $Z_s \simeq j\omega\rho_s$. The sound radiation of the sound-forced finite plate is controlled by its radiation impedance $Z_{\text{rad}} \cong \text{Re}\{Z_{\text{rad}}\} = \rho_0 c_0 \sigma_F$. Consequently, the low-frequency sound transmission loss of the finite partition is predicted is predicted as[38]

$$R_E \cong R_0 - 3 - 10 \log \sigma_F \quad \text{dB} \quad (11.108)$$

where R_0 is the normal-incidence mass law sound transmission loss given in Eq. (11.83c) and σ_F is the forced-wave radiation efficiency of the finite panel given in Table 11.8. Note that σ_F depends on panel size as well as on incidence angle and can be smaller or larger than unity. This implies that the sound transmission loss of finite panels can be larger than the normal-incidence mass law even for grazing incidence if the size of the panel is small compared with the wavelength. When the panel size is much larger than the acoustical wavelength, σ_F approaches $1/\cos\phi$. For predicting the sound transmission loss of finite partitions over the entire low-frequency range ($f \ll f_c$), Eq. (11.108) should be used.

According to reference 38, the classical sound transmission formulas for infinite panels can be used to approximate the sound transmission loss of finite partitions by substituting $1/\sigma_F$ instead of $\cos\phi$ and $\sqrt{1 - 1/\sigma_F^2}$ instead of $\sin\phi$ in Eq. (11.97) and carrying out the integration in Eq. (11.99) from $\phi = 0$ to $\phi = 90°$ to obtain an estimate of the random-incidence sound transmission loss that agrees well with laboratory measurements. There is no need to resort to limiting the incident angle range to 78°. The radiation efficiency for random-incidence (diffuse-field) sound-forced excitation is given in Table 11.8, and this expression

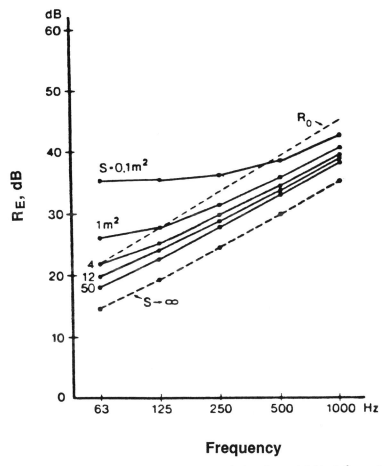

FIGURE 11.24 Random-incidence sound transmission loss of 25-kg/m² panels as a function of frequency with panel surface area S as a parameter, mass-controlled low-frequency region; computed according to Eq. (11.110). (After References 37 and 38.)

should be used in Eq. (11.108). As shown in Fig. 11.24, the random-incidence sound transmission loss of partitions of approximately 4 m² surface area, which are typically used in laboratory measurements, yield predicted values that are 5 dB below the normal-incidence mass law, giving theoretical justification for using the field-incidence mass law defined in Eq. (11.101) for partitions of this size. Note, however, that for small partitions Eq. (11.108) is more accurate. It should be pointed out that all of the sound transmission loss prediction formulas are valid only in the frequency region well above the first bending wave resonance of the panel where the forced response is mass controlled. For small, very stiff partitions the frequency range of interest may extend into the stiffness-controlled region below the frequency of the first bending wave resonance.* In this case,

*For prediction in this frequency region use Eqs. (11.70)–(11.73).

the volume compliance of the panel, as described in the beginning of this section in Chapter 12, should be used to predict sound transmission.

Empirical Method for Predicting Sound Transmission Loss of Single Partitions. An alternate technique useful in preliminary design is illustrated in Fig. 11.25. In essence, it considers the loss factor of the material to be determined completely by the material selection and substitutes a "plateau" or horizontal line for the peak and valley of the forced-wave analysis in the region of the critical frequency.[34,39] Its use will be demonstrated in Example 11.3.

Example 11.3. Calculate the transmission loss of a $\frac{1}{8}$-in.- (3-mm-) thick, 5×6.5-ft (1.52×2-m) aluminum panel by the alternate (plateau) method.

Solution From Table 11.2, the product of surface mass and critical frequency is $\rho_s f_c = 34,700$ Hz · kg/m^2. The $\frac{1}{8}$-in.-thick aluminum plate has a surface density

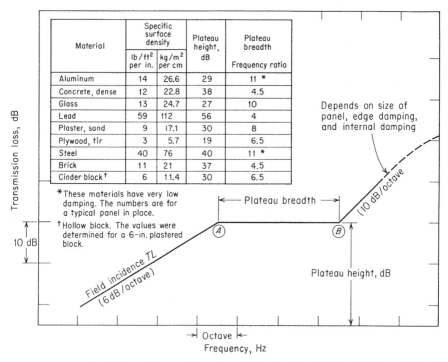

FIGURE 11.25 Approximate design chart for estimating the sound transmission loss of single panels. The chart assumes a reverberant sound field on the source side and approximates the behavior around the critical frequency with a horizontal line or plateau. The part of the curve to the left of A is determined from the field-incidence mass law curve (Fig. 11.21). The plateau height and length of the line from A to B are determined from the table. The part above B is an extrapolation. This chart is fairly accurate for large panels. Length and width of the panel should be at least 20 times the panel thickness.[34]

of 8.5 kg/m². From Fig. 11.21 we find that the normal-incidence transmission loss at the critical frequency is $R_0(f_c) = 48.5$ dB and that the field-incidence transmission loss at 1000 Hz is R_{field} (8500 Hz · kg/m²) = 31.5 dB.

The procedure by the plateau method is as follows[34,39]:

1. Using semilog paper (with coordinates decibels versus log frequency), plot the field-incidence mass law transmission loss as a line with a 6-dB/octave slope through the point 31.5 dB at 1000 Hz.
2. From Fig. 11.25, the plateau height for aluminum is 29 dB. Plotting the plateau gives the intercept of the plateau with the field-incidence mass law curve at approximately 750 Hz.
3. From Fig. 11.25, the plateau width is a frequency ratio of 11. The upper frequency limit for the plateau is therefore 11×750 Hz = 8250 Hz.
4. From the point 29 dB, 8250 Hz, draw a line sloping upward at 10 dB/octave. This completes the plateau method estimate (see curve b in Fig. 11.34).

Sound Transmission through Double- and Multilayer Partitions

The highest sound transmission loss obtainable by a single partition is limited by the mass law. The way to "break this mass law barrier" is to use multilayer partitions such as double walls, where two solid panels are separated by an air space that usually contains fibrous sound-absorbing material and double windows where light transparency requirements do not allow the use of sound-absorbing materials.

The transmission of sound through a multilayer partition can be computed in a similar manner as the sound absorption coefficient of multilayer sound absorbers treated in Chapter 8. Figure 11.26 shows the situation where a plane sound wave of frequency $f = \omega/2\pi$ is incident at an angle ϕ on a panel that has N layers and $N+1$ interfaces. The important boundary conditions are that the wavenumber component parallel to the panel surface $k_x = k \sin \phi$ must be the same in all of the layers and that the acoustical pressure and particle velocity at the interfaces of the layers must be continuous.[40–42]

The layers are characterized by their wave impedance formula, which relates the complex wave impedance at the input-side interface Z_I with that at the termination-side interface Z_T and by their pressure formula that relates the complex sound pressure at the input-side interface p_I to that at the terminal-side interface p_T.

The impedance and pressure formulas for an impervious orthotropic thin plate are given by[42]

$$Z_I = Z_T + Z_S$$
$$= Z_T + j \left[\omega \rho_s - \frac{1}{\omega}(B_x k_x^4 + 2B_{xy} k_x^2 k_y^2 + B_y k_y^4) \right] \quad \text{N·s/m}^3 \quad (11.109)$$

$$p_I = p_T \frac{Z_I}{Z_T} \quad \text{N/m}^2 \quad (11.110)$$

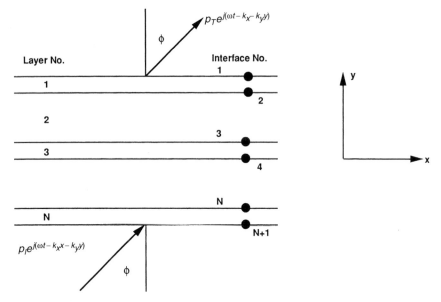

FIGURE 11.26 Transmission of a plane oblique incident sound wave through an infinite lateral dimension multilayer panel.

where ρ_s is the mass per unit area and B is the bending stiffness of the orthotropic plate.

For isotropic plates the second term in Eq. (11.109) reduces to $Z_s = Z_m + Z_B = j(\omega \rho_s - B k_x^4/\omega)$. If the impervious layer is a composite of an isotropic homogeneous plate with a bonded damping material of thickness h_D, Young's modulus E_D, Poisson ratio ν_D, and damping loss factor η_D and the plate characteristics are thickness h_p, Young's modulus E_p, Poisson ratio ν_p, and loss factor η_p, then the complex bending stiffness $B = B_{comp}(1 + j\eta_{comp})$ is obtained[1,42] from

$$B_{comp} = \frac{E_p h_p^3}{12(1 - \nu_p^2)} + \frac{E_D h_D (h_p + h_D)^2}{4(1 - \nu_D^2)} \quad \text{Nm} \quad (11.111)$$

$$\eta_{comp} = \tfrac{1}{4} B_{comp}(\eta_p E_p h_p + \eta_D E_D h_D)(h_p + h_D)^2 \quad (11.112)$$

For a porous sound-absorbing layer of thickness h the impedance and pressure formulas are given by[42]

$$Z_I = Z_a \frac{k_a}{k_{ay}} \frac{(1 + Z_a \Gamma_a / Z_T \Gamma_{ay}) e^{j\Gamma_{ay} h} + (1 - Z_a \Gamma_a / Z_T \Gamma_{ay}) e^{-j\Gamma_{ay} h}}{(1 + Z_a \Gamma_a / Z_T \Gamma_{ay}) e^{j\Gamma_{ay} h} - (1 - Z_a \Gamma_a / Z_T \Gamma_{ay}) e^{-j\Gamma_{ay} h}} \quad \text{N·s/m} \quad (11.113)$$

$$p_T = \frac{p_I}{2}\left[\left(1 + \frac{Z_a \Gamma_a}{Z_I \Gamma_{ay}}\right) e^{-j\Gamma_{ay} h} + \left(1 - \frac{Z_a \Gamma_a}{Z_I \Gamma_{ay}}\right) e^{j\Gamma_{ay} h}\right] \quad \text{N/m}^2 \quad (11.114)$$

where $Z_a = \rho_0 c_0 Z_{an}$ is the complex characteristic acoustical impedance of the porous bulk material for plane waves and Z_{an} is its normalized value while $\Gamma_a = \Gamma_{an} k_0$ is the complex wavenumber of plane sound waves in the bulk porous material; $\Gamma_{ay}^2 = \Gamma_a^2 - k_x^2$. Formulas for computing Γ_{an} and Z_{an} on the basis of the flow resistivity of the porous material R_1 are given in Chapter 8 [see Eqs. (8.22) and (8.19)]. The simpler approximate formulas for Γ_a and Z_a given in reference 42 are less accurate and their use is not recommended.

In the special case of an air layer without porous sound-absorbing material, substitute $Z_a = \rho_0 c_0$ and $k_a = k_0$ into Eqs. (11.113) and (11.114).

The computation of the sound transmission loss of a layered partition proceeds as follows:

1. Set the termination impedance at the receiver-side interface (interface 1 in Fig. 11.26) to $Z_T = \rho_0 c_0 / \cos\phi$.
2. Apply the appropriate impedance formula for layer 1 and compute the input impedance at interface 2.
3. Using the impedance computed in step 2 as the termination impedance for layer 2, compute the input impedance at interface 3 and proceed down the chain of impedance calculations until the input impedance at the source-side interface (interface $N+1$ in Fig. 11.26), Z_{N+1}, is obtained.
4. Compute the sound pressure at the source-side interface as the sum of the incident and reflected pressures $p_{N+1} = p_I[2\alpha/(\alpha+1)]e^{j(\omega t - k_x x - k_y y)}$, where $\alpha = Z_{N+1}/(\rho_0 c_0 / \cos\phi)$.
5. Apply the appropriate pressure formulas in succession until the sound pressure at the receiver-side interface, p, is obtained.
6. Determine the transmission coefficient $\tau(\omega, \phi) = p_1^2/p_I^2$ for all frequencies of interest.
7. Perform computational steps 1–6 for the incident angle range from $\phi = 0°$ to $\phi = 90°$ in one-third-degree increments.
8. Compute the random-incidence sound transmission coefficient for isotropic layers (see ref. 42 for orthotropic impervious layers) as

$$\tau_R(\omega) = \int_0^{\pi/2} \tau(\omega, \phi) \sin 2\phi \, d\phi$$

9. Compute the random-incidence sound transmission loss of the infinite layered partition as

$$R_{\text{random}}(\omega) = 10 \log \frac{1}{\tau_R(\omega)}$$

Figure 11.27 shows the random-incidence sound transmission loss of an infinite three-layer partition as well as that of its constituent layers, computed according to Eqs. (11.109)–(11.114). The partition consists of two 1-mm-thick steel plates and a 100-mm-thick air space that may or may not contain a fibrous sound-absorbing material. Figure 11.27 illustrates the benefit of using sound-absorbing

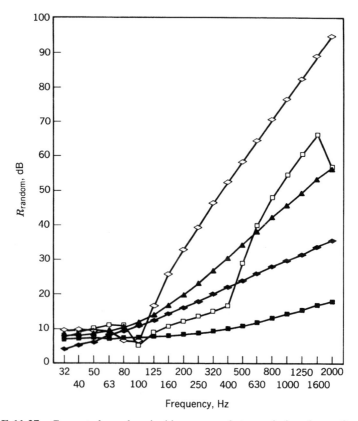

FIGURE 11.27 Computed random-incidence sound transmission loss of a double wall and its constituent layers: (■) 100-mm-thick fibrous absorber, flow resistivity $R_1 = 16,700$ N/s · m^4; (♦) 1-mm steel plate; (▲) 1-mm steel plate and 100-mm fibrous absorber. Double wall consisting of two 1-mm-thick steel plates: (□) 100-mm air space, no fibrous absorber; (◇) 100-mm air space with fibrous absorber.

layers. Note that below 500 Hz the double wall without sound-absorbing material in the air space provides substantially lower random-incidence sound transmission loss than a single 1-mm steel plate or the single plate combined with the sound-absorbing layer. The highest sound transmission loss is obtained when the air space is filled with the porous sound-absorbing material. In practical situations, leaks and structure-borne connections between the face plates at edges of the partition usually limit the maximally achievable sound transmission loss at high frequencies to a range of 40–70 dB. At low frequencies the finite size of the partition results in higher values than predicted for infinite partitions. The effect of the sound-absorbing material in the air space results in refraction of the oblique-incidence sound toward the normal, thereby reducing the dynamic stiffness of the air between the plates. The sound-absorbing material also prevents high sound energy buildup in the cavity. These result in a substantial increase in sound transmission loss. The flow resistivity of the sound-absorbing material should be about $R_1 = 5000$ N · s/m^4.[43] Higher values of R_1 yield only diminishing returns.

The filling of the air space with a gas of 50% lower speed of sound than air (such as SF$_6$ or CO$_2$) has the same effect as the sound-absorbing material,[44,45] as illustrated in Fig. 11.28. Using a light gas such as helium, which has three times higher speed of sound than air, also improves sound transmission loss to the same extent as a heavy-gas fill. In this case, the improvement is due to the higher speed of sound in the gas fill, which makes it easier to push the gas tangentially than to compress it. Double windows, which can be hermetically sealed and must be light transparent, are partitions where this beneficial effect can be exploited.

Empirical Method for Predicting Sound Transmission Loss of Double Partitions.

Goesele[46] has proposed a simplified method to predict the sound transmission loss R of a double partition when the measured sound transmission losses of the two constituent single partitions R_I and R_{II} are available, there are no structure-borne connections, and the gap is filled with porous sound-absorbing material. The prediction is given as

$$R \cong R_I + R_{II} + 20 \log \left(\frac{4\pi f \rho_0 c_0}{s'} \right) \quad \text{dB} \qquad (11.115)$$

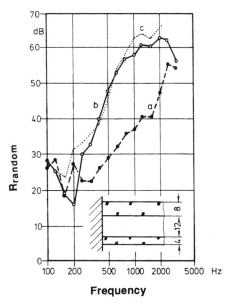

FIGURE 11.28 Improvement of the sound transmission loss of a double glass partition (no contact at the edges) owing to heavy gas (SF$_6$) fill of the gap: a, measured with air-filled gap; b, measured with SF$_6$-filled gap; c, computed for mineral wool fill. (After Ref. 44.)

where

$$s' = \begin{cases} \dfrac{\rho_0 c_0^2}{d} & \text{for } f < f_d = \dfrac{c_0}{2\pi d} \quad \text{N/m}^3 \quad (11.116a) \\ 2\pi f \rho_0 c_0 & \text{for } f > f_d \quad\quad\quad\quad \text{N/m}^3 \quad (11.116b) \end{cases}$$

is the dynamic stiffness per unit area of the gap and d is the gap thickness.

If no measured sound transmission loss data for the constituent single partitions are available, the sound transmission loss of the double wall made of two identical panels can be predicted on the basis of material properties as

$$R(f_R < f < f_c) \cong 20 \log \frac{\pi f \rho_{s1}}{\sqrt{2}\rho_0 c_0} + 40 \log \left(\frac{\sqrt{2}f}{f_R}\right) \quad \text{dB} \quad (11.117a)$$

$$R(f > f_R, f > f_c) \cong 40 \log \left[\frac{\pi f \rho_{s1}}{\rho_0 c_0}\sqrt{2\eta}\left(\frac{f}{f_c}\right)^{1/4}\right]$$

$$+ 20 \log \frac{4\pi f \rho_0 c_0}{s'} \quad \text{dB} \quad (11.117b)$$

where f_c is the critical frequency of the panels and f_R is the double-wall resonance frequency

$$f_R = \frac{1}{2\pi}\sqrt{\frac{2\sqrt{2}s'}{\rho_{s1}}} \quad \text{Hz} \quad (11.118)$$

Figure 11.29 shows that Eq. (11.115) yields good agreement with measured data in the entire frequency region, while Eqs. (11.117a) and (11.117b) give good agreement only well below and well above the critical frequency but fail in the frequency region near the critical frequency. Prediction methods for the sound transmission loss of double walls with point and line bridges are given by reference 30 and by Bies and Hansen (see Bibliography).

Sound Transmission Loss of Ducts and Pipes

Pipes and ducts that carry high-intensity internal sound are excited into vibration and radiate sound to the outside. This sound transmission in the breakout direction (i.e., from inside to outside) is characterized by breakout sound transmission loss R_{io}, which is a measure of the rate at which sound energy from the interior of the duct radiated to the outside. When pipes and ducts traverse areas of high-intensity sound such as found in mechanical equipment rooms, the exterior sound field excites ductwall vibrations and the vibrating walls generate an internal sound field that can travel to distant quiet areas. This sound transmission from the outside to the inside direction is characterized by the breakin sound transmission loss R_{oi}, which is a measure of the rate at which sound energy from the exterior sound field enters the duct.

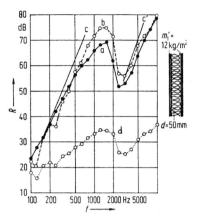

FIGURE 11.29 Sound transmission loss of a double partition consisting of two identical, 12.5-mm-thick gypsum boards separated by a 50-mm-thick gap filled with fibrous sound-absorbing material: a, measured values; b, predicted by Eq. (11.117); c, predicted by Eqs. (11.116a) and (11.116b); d, measured sound transmission loss of a single gypsum board wall. (After reference 46.)

The sound power level radiated by a duct or pipe of length l, L_w^{io}, is predicted as[47]

$$L_w^{io}(l) = L_w^i(0) - R_{io} + 10\log\left(\frac{Pl}{S}\right) + 10\log(C) \quad \text{dB re} 10^{-12} \text{ W} \quad (11.119a)$$

where

$$C = \frac{1 - e^{-(\tau+\beta)l}}{(\tau+\beta)l} \quad (11.119b)$$

$$\tau = \frac{P}{S} \times 10^{-R_{io}/10} \quad (11.119c)$$

$$\beta = \frac{\Delta L_1}{4.34} \quad (11.119d)$$

where $L_w^i(0)$ is the sound power level in the duct at the source side, S is the cross-sectional area in square meters, P is the perimeter of the duct cross section in meters, and ΔL_1 is the sound attenuation in decibels per unit length inside the duct due to porous lining. Equation (11.119a) contains only measurable quantities and is used as a basis for the experimental evaluation of the breakout sound transmission loss R_{io}, by measuring the sound power $W_{io}(l)$ radiated by a test duct of length l into a reverberation room, the sound power in the duct at the source side, $W_i(0)$, and the sound attenuation inside the duct, ΔL_1, and solving Eq. (11.119a) for R_{io} by iteration.

The sound power level of the sound propagating in one direction, $L_w^{io}(l)$ when a duct of length l traverses through a noisy area (and sound breaks into the duct is predicted by[47]

$$L_w^{io}(l) = L_w^{\text{inc}} - R_{io} - 3 + C \quad \text{dB re}10^{-12} \text{ W} \qquad (11.120)$$

where L_w^{inc} is the sound power level of the sound incident on the duct of length l, R_{io} is the breakin sound transmission loss, and C is as defined in Eq. (11.119b). On the basis of reciprocity,[48] the following relationship exists between breakout and breakin sound transmission loss:

$$R_{oi} \cong R_{io} - 10\log\left\{4\gamma\left[1 + 0.64\frac{a}{b}\left(\frac{f_{\text{cut}}}{f}\right)^2\right]\right\} \quad \text{dB} \qquad (11.121)$$

where a and b are the larger and smaller sides of a rectangular duct cross section, f is the frequency, f_{cut} is the cutoff frequency of the duct, and γ is 1 below cutoff and 0.5 above cutoff. Empirical methods for predicting the breakout sound transmission loss of unlined, unlagged rectangular sheet metal ducts are given in references 48–50. Chapter 17 contains predicted values of octave-band breakout sound transmission loss versus frequency for rectangular sheet metal ducts of sizes most frequently used in low-velocity HVAC systems. Sound transmission loss predictions for round and flat-oval ducts are given elsewhere.[49,50] Figure 11.30 shows the breakin sound transmission loss of an unlined, unlagged

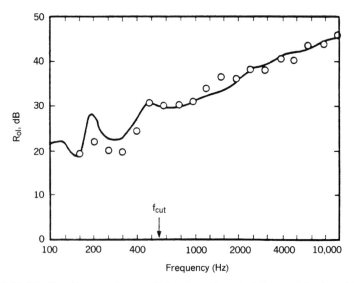

FIGURE 11.30 Breakin sound transmission loss R_{io} of a 0.3 m × 0.91 mm thick sheet metal duct (after Ref. 48): (———) measured: (O O O) predicted from measured R_{io} using Eq. (11.121).

sheet metal duct. The solid curve was obtained by directly measuring the breakin sound transmission loss while the open circles represent data points obtained by applying the reciprocity relationship embodied in Eq. (11.121) for the measured breakout sound transmission loss. The importance of Eq. (11.121) is that it makes it unnecessary to measure both R_{oi} and R_{io} separately because if one has been measured, the other can be predicted.

Sound Transmission through a Composite Partition

Partitions that separate adjacent rooms frequently consist of areas that have different sound transmission losses, such as a wall that contains a door with an uncovered key hole. If all parts of the composite partition are exposed to the same average sound intensity, I_{inc}, on the source side, then the sound power transmitted is

$$W_{\text{trans}} = I_{\text{inc}} \sum_{i=1}^{n} S_i \times 10^{-R_i/10} = I_{\text{inc}} \sum_{i=1}^{n} S_i \tau_i \quad \text{W} \tag{11.122a}$$

and the transmission loss of the composite partition is

$$R_{\text{comp}} = 10 \log \frac{W_{\text{inc}}}{W_{\text{trans}}} = -10 \log \sum_{i=1}^{n} \frac{S_i}{S_{\text{tot}}} \times 10^{-R_i/10} \quad \text{dB} \tag{11.122b}$$

where S_i is the surface area of each component, S_{tot} is the total area of all components, and R_i is the sound transmission loss of the ith component. The sound transmission loss of a small hole radius $a \ll \lambda_0$ in a thin plate of thickness h is well approximated by[26]

$$R_{\text{hole}} \cong 20 \log \frac{h + 1.6a}{\sqrt{2a}} \quad \text{dB} \tag{11.123}$$

indicating that small holes in thin partitions ($h < a$) yield a frequency-independent sound transmission loss of $R_{\text{hole}} \simeq 0$ dB. Note, however, that Eq. (11.123) is valid only for small round holes. Long narrow slits can have a "negative sound transmission loss."[51] The sound transmission loss of holes and slits can be increased substantially by sealing them with either porous sound-absorbing material or an elastomeric material or designing them as silencer joints. Prediction of the sound transmission loss of such acoustically sealed openings are given in references 52 and 53.

Flanking Sound Transmission

The sound transmission loss of partitions is measured in acoustical laboratories where sound transmission from the source room to the receiver room occurs only through the partition under test. However, if the same partition constitutes a part of a building, then sound can be transmitted through many paths, as shown schematically in Fig. 11.31. Path 1 represents the primary path, which is

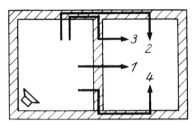

FIGURE 11.31 Sound transmission paths between two adjacent rooms in a typical building (after Reference 54): 1, primary path; 2–4, flanking paths.

characterized by sound transmission loss of the separating partition R_1, usually available from laboratory measurements. The sound transmission loss for each of the $n = 4$ paths (1 the direct and 2, 3, and 4 the flanking paths) is defined as[54]

$$R_n \equiv 10 \log \frac{W_1^{\text{inc}}}{W_{\text{trans}}^n} \quad \text{dB} \tag{11.124}$$

and the composite sound transmission loss, which combines sound transmission along each of the four paths, as

$$R_{\text{comp}} \equiv -10 \log \left(10^{-R_1/10} + \sum_{n=2}^{4} \sum_{m=1}^{4} 10^{-R_{mn}/10} \right) \quad \text{dB} \tag{11.125}$$

where $m = 4$ represents the four sound-excited flanking partitions (i.e., the two sidewalls, floor, and ceiling of the receiver room) each of which transmit sound along each of the three flanking paths $n = 2, 3, 4$. Usually flanking path $n = 2$ contributes as much to the receiver room sound power as do the two other flanking paths $n = 3$ and $n = 4$ together. The contribution of the back wall of the source and receiver room is usually negligible. The process of flanking transmission along flanking paths $n = 2, 3, 4$ is as follows: (1) the sound field in the source room excites the flanking walls to vibration, (2) the vibration is transmitted through the wall junctions to the receiver room walls, and (3) the receiver room walls radiate sound power into the receiver room that adds to that transmitted by the separating partition through the direct path. If the source and receiver rooms have no common wall, the entire sound transmission takes place through flanking paths. For adjacent rooms with a common wall, the composite sound transmission loss given in Eq. (11.125) should be used to predict the sound pressure level in the receiver room. The component flanking transmission losses R_{mn} for homogeneous isotropic single-wall construction can be approximated by[54]

$$R_{mn} = 10 \log \left[\frac{1}{4\pi \sqrt{12}} \left(\frac{\rho_M}{\rho_0} \right)^2 \frac{c_L h^3 \omega^2}{c_0^4} \eta_m \right] + \Delta L_{\text{junct}}$$

$$+ 10 \log \frac{S_1}{S_{\text{rad}}} - 10 \log(\sigma_m \sigma_{\text{rad}}) \quad \text{dB} \tag{11.126}$$

where ρ_M is density and c_L the longitudinal wave speed of the wall material, h is the thickness, η_m is the composite loss factor of the mth partition in the source room, and ΔL_{junct} is the attenuation of the structure-borne sound amplitude at the wall junction along the transmission path n. The symbol S_1 is the surface area of the separating partition and S_{rad} is the surface area of the partition in the receiver room involved in the transmission along path n. The symbols σ_m and σ_{rad} are the radiation efficiencies of the mth partition in the source room and the radiation efficiency of the partition in the receiver room that radiates sound owing to the sound-induced vibration of the mth partition in the source room transmitted through the nth path. A more accurate prediction of the effect of flanking paths on sound transmission loss can be made utilizing the statistical energy analysis method discussed in the next section.

11.8 STATISTICAL ENERGY ANALYSES

Statistical energy analysis (SEA) is a point of view in dealing with the vibration of complex resonant systems. It permits calculation of the energy flow between connected resonant systems, such as plates, beams, and so on, and between plates and the reverberant sound field in an enclosure.[55–59]

System of Modal Groups

In respect to the energy E stored in a structure or in an acoustical volume, may be thought of as a *system of* resonant modes or resonators. First, let us consider the power flow between two groups of resonant modes of two coupled structures having their modal resonance frequencies within the same narrow frequency band $\Delta\omega$ (see Fig. 11.32).

We assume that each resonant mode of the first system (box 1 in Fig. 11.32) has the same energy. Also, assume that the coupling of the individual resonant modes of the first system with each resonance mode of the second system is approximately the same.

If we further assume that the waves carrying the energy in one system are uncorrelated with the waves carrying the energy gained through coupling to the

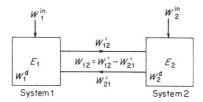

FIGURE 11.32 Block diagram illustrating power flow between two nondissipatively coupled systems.

other system, we can separate the power flow (each equation is written for a narrow frequency band $\Delta\omega$) as

$$W'_{12} = E_1 \omega \eta_{12} \quad W \tag{11.127}$$

$$W'_{21} = E_2 \omega \eta_{21} \quad W \tag{11.128}$$

where W'_{12} = power system 1 transmits to system 2, W
W'_{21} = power system 2 transmits to system 1, W
E_1 = total energy in system 1, m · kg/s
E_2 = total energy in system 2, m · kg/s
ω = center frequency of band, rad/s
η_{12} = coupling loss factor from system 1 to system 2, as defined in Eq. (11.127)
η_{21} = coupling loss factor from system 2 to system 1, as defined in Eq. (11.128)

The net power flow between the two systems is, accordingly,

$$W_{12} = W'_{12} - W'_{21} = E_1 \omega \eta_{12} - E_2 \omega \eta_{21} \quad W \tag{11.129}$$

Modal Energy E_m

Let us define modal energy as

$$E_m = \frac{E(\Delta\omega)}{n(\omega)\,\Delta\omega} \quad W \cdot s/Hz \tag{11.130}$$

where $E(\Delta\omega)$ = total energy in system in angular frequency band $\Delta\omega$
$n(\omega)$ = modal density, = number of modes in unit bandwidth ($\Delta\omega = 1$) centered on ω, the angular frequency
$\Delta\omega$ = bandwidth, rad/s

If the previously made assumptions about the equal distribution of energy in the modes and the same coupling loss factor are valid, then it may be shown that

$$\frac{\eta_{21}}{\eta_{12}} = \frac{n_1(\omega)}{n_2(\omega)} \tag{11.131}$$

where $n_1(\omega)$ = modal density of system 1 at frequency ω, s
$n_2(\omega)$ = modal density of system 2 at frequency ω, s

Equation (11.131) implies that for equal total energies in the two systems, $E_1 = E_2$, the system that has the *lower* modal density [lower $n(\omega)$] transfers more energy to the second system than is transferred from the second to the first system.

Combining Eqs. (11.129) and (11.131) yields*

$$W_{12} = \omega \eta_{12} n_1(\omega)[E_{m1} - E_{m2}] \, \Delta\omega \quad \text{W} \qquad (11.132)$$

where W_{12} = net power flow between systems 1 and 2 in band $\Delta\omega$, centered at ω, W
E_{m1}, E_{m2} = modal energies for systems 1 and 2, respectively [see Eq. (11.130)], W · s/Hz

This equation is positive if the first term in the brackets is greater than the second.

The principle of the SEA method is given by Eq. (11.132), which is a simple algebraic equation with energy as the independent dynamic variable. It states that *the net power flow between two coupled systems in a narrow frequency band, centered at frequency ω, is proportional to the difference in the modal energies of the two systems at the same frequency. The flow is from the system with the higher modal energy to that with the lower modal energy.*

It may help to understand Eq. (11.132) if we use the thermodynamical analogy of heat transfer between two connected bodies of different temperature, where the heat flow is from the body of higher temperature to that of lower temperature and the net heat flow is proportional to the difference in temperature of the two bodies. Consequently, the modal energy E_m is analogous to temperature, and the net power flow W_{12} is analogous to heat flow. The case of equal modal energies of the two systems where the net power flow is zero is analogous to the equal temperature of the two bodies.

Equal Energy of Modes of Vibration

Equal energy of the modes within a group will usually exist if the wave field of the structure is diffuse. Also, since the frequency-adjacent resonance modes of a structure are coupled to each other by scattering and damping, there is always a tendency for the modal energy of resonant modes to equalize within a narrow frequency band even if the wave field is not diffuse.

Noncorrelation between Waves in the Two Systems

In sound transmission problems usually only one system is excited. The power W'_{12} transmitted to the nonexcited system builds up a semidiffuse vibration field in that system. Accordingly, the waves that carry the transmitted power W'_{21} back to the excited system are almost always sufficiently delayed and randomized in phase with respect to the waves carrying the incident power W'_{12} that there is little correlation between the two wave fields.

---

*Note that Eq. (11.131) is a necessary requirement if Eq. (11.132) is to obey the consistency relationship $W_{12} = -W_{21}$.

Realization of Equal Coupling Loss Factor

Equality of the coupling loss factors between individual modes within a group is a matter of grouping modes of similar nature. If the coupled system is a plate in a reverberant sound field, the *acoustically slow edge and corner modes* and *acoustically fast surface modes* are grouped separately.

Composite Structures

Composite structures generally consist of a number of elements such as plates, beams, stiffeners, and so on. We may divide a complex structure into its simpler member provided the wavelength of the structure-borne vibration is small compared with the characteristic dimensions of the elements. Where this is true, the modal density of a complex structure is approximately that of the sum of the modal densities of its elements. If the power input, the various coupling loss factors, and the power dissipated in each element are known, the power balance equations will yield the vibrational energy in the respective elements of the structure.

The dissipative loss factor for an element of a structure is obtained by separating that element from the rest of the structure and measuring its decay rate, as discussed in Chapter 14. The coupling loss factor can be determined experimentally from Eq. (11.132); however, the procedure is difficult. Theoretical solutions are available for the coupling loss factors of a few simple structural connections.[59] When the coupling loss factor between a sound field and a simple structure is desired, such a loss factor can be calculated from Eq. (11.131) if the radiation ratio of the structure is known, as shown in the next section.

Power Balance in a Two-Structure System

The power balance of the simple two-element system of Fig. 11.32 is given by the following two algebraic equations:

$$W_1^{in} = W_1^d + W_{12} \quad W \tag{11.133}$$

$$W_2^{in} = W_{12} + W_2^d \quad W \tag{11.134}$$

where W_1^{in} = input power to system 1, W
W_1^d = power dissipated in system 1, W
W_{12} = net power lost by system 1 through coupling* to system 2, = $W'_{12} - W'_{21}$, W
W_2^{in} = input power to system 2, W
W_2^d = power dissipated in system 2, W

---

*As in our previous analysis, we assume that the coupling is nondissipative.

The power dissipated in a system is related to the energy stored by that system, E_i, through the dissipative loss factor η_i, namely,

$$W_i^d = E_i \omega \eta_i \quad \text{W} \tag{11.135}$$

where E_i is energy stored in system i in newton-meters.

Assuming that the second system does not have direct power input ($W_2^{in} = 0$), the combination of Eqs. (11.130)–(11.132), (11.134), and (11.135) (with $i = 2$) yields the ratio of the energies stored in the two respective systems:

$$\frac{E_2}{E_1} = \frac{n_2}{n_1} \frac{\eta_{21}}{\eta_{21} + \eta_2} \tag{11.136}$$

If the coupling loss factor is very large compared to the loss factor in system 2, that is, if $\eta_{21} \gg \eta_2$, Eq. (11.136) yields the equality of the modal energies ($E_1/n_1 \Delta\omega = E_2/n_2 \Delta\omega$).

Diffuse Sound Field Driving a Freely Hung Panel

Let us now examine the special case of the excitation of a homogeneous panel (system 2) that hangs freely, exposed to the diffuse sound field of a reverberant room (system 1).

For this case, the total energies for each system are given by

$$E_1 = DV = \frac{\langle p^2 \rangle}{\rho_0 c_0^2} V \quad \text{N} \cdot \text{m} \tag{11.137}$$

$$E_2 = \langle v^2 \rangle \rho_s S \quad \text{N} \cdot \text{m} \tag{11.138}$$

where D = average energy density in reverberant room, N/m^2
  $\langle p^2 \rangle$ = mean-square sound pressure (space–time average), N^2/m^4
  V = room volume, m^3
  $\langle v^2 \rangle$ = mean-square plate vibration velocity (space–time average), m^2/s^2
  S = plate surface area (one side), m^2
  ρ_s = mass per unit area of panel, kg/m^2

To find the coupling loss factor η_{21}, we must first recognize that W'_{21} equals the power that the plate, having been excited into vibration, radiates back into the room. Thus

$$W_{rad} = W'_{21} = 2\langle v^2 \rangle \rho c \sigma_{rad} S \equiv E_2 \omega \eta_{21} = \langle v^2 \rangle \rho_s S \omega \eta_{21} \quad \text{W} \tag{11.139}$$

where σ_{rad} = radiation ratio for plate, dimensionless
  W_{rad} = acoustical power radiated by both sides of plate, which accounts for factor 2, W

Solving for η_{21} yields

$$\eta_{21} = \frac{2\rho_0 c_0 \sigma_{rad}}{\rho_s \omega} \qquad (11.140)$$

The modal density of the reverberant sound field in the room $n_1(\omega)$ and that of the thin homogeneous plate $n_2(\omega)$ are given in Table 11.7, namely,

$$n_1(\omega) \simeq \frac{\omega^2 V}{2\pi^2 c_0^3} \quad s \qquad (11.141)$$

$$n_2(\omega) = \frac{\sqrt{12}\, S}{4\pi c_L h} \quad s \qquad (11.142)$$

where c_L = propagation speed of longitudinal waves in plate material, m/s
h = plate thickness, m

Inserting Eqs. (11.137)–(11.142) into Eq. (11.135) yields the desired relation between the sound pressure and plate velocity:

$$\langle v^2 \rangle = \langle p^2 \rangle \frac{\sqrt{12}\pi c_0^2}{2\rho_0 c_0 h c_L \rho_s \omega^2} \frac{1}{1 + \rho_s \omega \eta_2 / 2\rho c \sigma_{rad}} \quad m^2/s^2 \qquad (11.143)$$

The mean-square (space–time average) acceleration of the panel is simply

$$\langle a^2 \rangle = \omega^2 \langle v^2 \rangle = \langle p^2 \rangle \frac{\sqrt{12}\pi c_0^2}{2\rho_0 c_0 h c_L \rho_s} \frac{1}{1 + \rho_s \omega \eta_2 / 2\rho_0 c_0 \sigma_{rad}} \quad m^2/s^4 \qquad (11.144)$$

It can be shown that as long as the power dissipated in the plate is small compared with the sound power radiated by that plate ($\rho_s \omega \eta_2 \ll 2\rho_0 c_0 \sigma_{rad}$), the equality of the modal energies of the sound field and the plate yields the proper plate velocity and acceleration. Also, under this condition, the ratio of the mean-square plate acceleration to the mean-square sound pressure is independent of frequency.

In general, *the plate response is always smaller than that calculated by the equality of the modal energies by the last factor on the right of Eq. (11.143) or (11.144), which equals the ratio of the power loss by acoustical radiation to the total power loss.* In dealing with the excitation of structures by a sound field, the concept of equal modal energy often enables one to give a simple estimate for the upper bound of the structure's response.

Example 11.4. Calculate the rms velocity and acceleration of a 0.005-m- ($\frac{1}{10}$-in.-) thick homogeneous aluminum panel resiliently suspended in a reverberant room. The space-averaged sound pressure level $\overline{L}_p = 100$ dB($\sqrt{\langle p^2 \rangle} = 2$ N/m^2) as measured in a one-third-octave band centered at a frequency $f = \omega/2\pi = 1000$ Hz. The appropriate constants of the panel and surrounding media are $\rho_s = 13.5$ kg/m^2; $c_L = 5.2 \times 10^3$ m/s; $h = 5 \times 10^{-3}$ m; $\rho_0 = 1.2$ kg/m^3; $c_0 = 344$ m/s; and $\eta_2 = 10^{-4}$.

Solution First, calculate the factor

$$\frac{\rho_s \omega \eta_2}{2\rho_0 c_0 \sigma_{\text{rad}}} = \frac{13.5 \times 2\pi \times 10^3 \times 10^{-4}}{2 \times 1.2 \times 344 \times \sigma_{\text{rad}}} = \frac{8.5}{820\sigma_{\text{rad}}} \ll 1$$

According to the above inequality the mean-square acceleration of the panel given by Eq. (11.144) simplifies to

$$\langle a^2 \rangle \approx \langle p^2 \rangle \frac{\sqrt{12}\pi c_0^2}{2\rho_0 c_0 h c_L \rho_s} = 19 \ \text{m}^2/\text{s}^4$$

or $a_{\text{rms}} = \sqrt{\langle a^2 \rangle} = 4.34$ m/s^2, an acceleration level of 113 dB re 10^{-5} m/s^2. The mean-square velocity is

$$\langle v^2 \rangle = \frac{\langle a^2 \rangle}{\omega^2} = \frac{19}{4\pi^2 \times 10^6} = 4.76 \times 10^{-7} \ \text{m}^2/\text{s}^2$$

or $v_{\text{rms}} = \sqrt{\langle v^2 \rangle} = 6.9 \times 10^{-4}$ m/s, a velocity level of 97 dB re 10^{-8} m/s.

Sound Transmission Loss of a Simple Homogeneous Structure by the SEA Method

The SEA method may be used to analyze the transmission of sound between two rooms coupled to each other by a single common, thin homogeneous wall.[58] (i.e., there are no flanking paths). System 1 is the ensemble of modes of the diffuse, reverberant sound field in the source room, resonant within the frequency band $\Delta\omega$. System 2 is an appropriately chosen group of vibration modes of the wall. System 3 is the ensemble of modes of the diffuse reverberant sound field in the receiving room, resonant within the frequency band $\Delta\omega$. A loudspeaker in the source room is the only source of power, and the power dissipated in each system is assumed to be large compared with the power lost to the other two systems through the coupling (see Fig. 11.33).

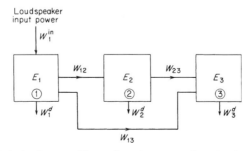

FIGURE 11.33 Block diagram illustrating the power flow in three-way coupled systems: W_{13}, transmission of sound by those modes whose resonance frequency lies outside the source band. The "nonresonant" modes are important below the critical frequency.

The procedure is as follows:

1. Relate W_1^{in} to the power lost by sound absorption in the room. This yields $\langle p_1^2 \rangle$, the space-averaged mean-square sound pressure in the source room.
2. Calculate $E_1 = \langle p_1^2 \rangle V_1 / \rho c^2$, where V_1 is the volume of the source room.
3. The reverberant sound power incident on the dividing wall of area S_2 is $W_{inc} = E_1 c S_2 / 4 V_1 = p_1^2 / 4 \rho c$.
4. From the power balance of the wall for resonant modes within the bandwidth, $\Delta \omega$ is next determined, that is, $W_{12} = W_2^d + W_{23}$, so that

$$W_{12} = \text{Eq.}(11.135) + 0.5\,[\text{Eq.}(11.139)]$$

This sum equals Eq. (11.132). Because η_{12} of Eq. (11.32) is not well known, it is replaced by using Eq. (11.131) and the definition of loss factor η_{21}, which yields $\eta_{21} = \rho_0 c_0 \sigma_{\text{rad}} / \rho_s \omega$.

5. The vibrational energy of the wall is $E_2 = \langle v^2 \rangle \rho_s S_2$.
6. Combining steps 3, 4, and 5 yields the mean-square wall velocity $\langle v^2 \rangle$ as a function of mean-square source room pressure $\langle p_1^2 \rangle$.
7. The power radiated into the receiving room is

$$W_{23} = \rho_0 c_0 S_2 \sigma_{\text{rad}} \langle v^2 \rangle$$

8. Finally, the resonance transmission coefficient τ_r is found by dividing step 7 by step 3.
9. The *resonance transmission loss*, defined as $R_r = 10 \log(1/\tau_r)$, is computed from step 8 using Eq. (11.97) and assuming that $\rho_s \omega \eta_2 \gg 2 \rho_0 c_0 \sigma_{\text{rad}}$ to yield

$$R_r = 20 \log \left(\frac{\rho_s \omega}{2 \rho_0 c_0} \right) + 10 \log \left(\frac{f}{f_c} \frac{2}{\pi} \frac{\eta_2}{\sigma_{\text{rad}}^2} \right) \quad \text{dB} \qquad (11.145)$$

The first term in Eq. (11.145) is approximately the normal-incidence mass law transmission loss R_0, so that Eq. (11.145) becomes

$$R_r = R_0 + 10 \log \left(\frac{f}{f_c} \frac{2}{\pi} \frac{\eta_2}{\sigma_{\text{rad}}^2} \right) \quad \text{dB} \qquad (11.146)$$

where f_c = critical frequency [see Eq. (11.94)], Hz
  η_2 = total loss factor of wall, dimensionless
  σ_{rad} = radiation efficiency for wall, dimensionless

Thus we have obtained the transmission loss between two rooms separated by a common wall using the SEA method.

Below the critical frequency and when the dimensions of the wall are large compared with the acoustical wavelength, the radiation factor σ_{rad} can be taken from Table 11.8.

It is important to note that if the sound transmission loss of an equivalent *infinite* wall is compared with the data measured and predicted by the SEA method, it is found that *above the critical frequency*, the transmission loss R for the infinite wall yields the same results as Eq. (11.146), which takes into account only the resonance transmission of a finite wall.

Below the critical frequency the sound transmission loss of a finite panel is more controlled by the contribution of those modes that have their resonance frequencies outside of the frequency band of the excitation signal than by those with resonance frequencies within that band. Since only the contributions of the latter are included in the previous SEA calculation, Eq. (11.146) usually overestimates the sound transmission loss of a finite panel below the critical frequency. Figure 11.34 shows that below the critical frequency for a $\frac{1}{8}$-in.-thick aluminum panel the sound transmission loss of the resonant modes alone (curve a) is approximately 10 dB higher than that measured on the actual panel (curve d).

A *composite transmission factor* that approximately takes into account both the forced and resonance waves is closely approximated by

$$\frac{1}{\tau} = \frac{W_{\text{inc}}}{W_{\text{forced}} + W_{\text{res}}}$$

$$= \frac{(\langle p^2 \rangle / 4\rho c) S_2}{\langle p^2 \rangle (\pi \rho c S_2 / \rho_s^2 \, \omega^2) + \langle p^2 \rangle (\sqrt{12} \pi c^3 \rho \sigma_{\text{rad}}^2 S_2 / 2 \, \omega^3 \rho_s^2 c_L h \eta_2)} \quad (11.147)$$

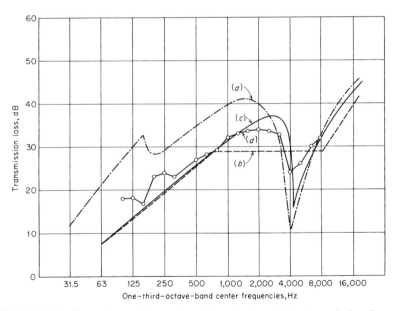

FIGURE 11.34 Comparison of experimental and theoretical transmission loss of a 5 ft × 6.5 ft × $\frac{1}{8}$ in. aluminum panel. The theoretical calculations are based on (*a*) resonance mode calculation, (*b*) plateau calculation, and (*c*) forced-wave calculation. Curve *d* shows the experimental results. (After Reference 58.)

At low frequencies where the first term in the denominator dominates, the transmission factor becomes

$$\frac{1}{\tau} \approx \frac{1}{\pi}\left(\frac{\rho_s \omega}{2\rho c}\right)^2 \qquad (11.148)$$

and the sound transmission loss is [see Eq. (11.101)]

$$R = 10 \log \frac{1}{\tau} \approx R_0 - 5 = R_{\text{field}} \quad \text{dB} \qquad (11.149)$$

At high frequencies where the second term in the denominator becomes dominant, the sound transmission loss is given by Eq. (11.146).

11.9 EQUIVALENCY BETWEEN THE EXCITATION OF A STRUCTURE SOUND FIELD AND POINT FORCE

In the mechanical equipment room of buildings, ships, and engine compartment of vehicles the boundaries (such as the walls, floor, and ceiling) are excited simultaneously by airborne noise and by dynamic forces. The airborne noise might emanate from the casing of machines and the dynamic forces might be those acting at the rigid or resilient attachment points of the machine to the floor or at the attachment points of pipes, conduits, or ducts (which are rigidly connected to the vibrating machine) to the wall or to the ceiling.

The noise control engineer is faced with the dilemma of predicting whether the airborne noise or the dynamic forces at the attachment points control the vibration response of the structure. The type of excitation that controls the vibration response of the structure will also control the airborne noise and vibration at distant noise- and vibration-sensitive receiver locations.

In his noise control engineering practice the author has been called upon to predict whether the ramble (low-frequency random noise) generated by the passage of subway trains in a nearby tunnel will be above or below the threshold of human hearing in a planned concert hall. In another project, he had to predict whether eye surgeons would be able to perform retina operations in a hospital located near another subway tunnel.

The dynamic forces acting in the tunnel floor owing to the wheel–rail interaction can be reduced substantially by mounting the track on a "floating slab" consisting of a thick concrete slab that is supported by resilient rubber mounts laid on the tunnel floor. However, the floating slab has no beneficial effect on reducing the airborne noise exposure of the walls and the roof of the tunnel. Actually, the floating slab results in an increase of the airborne noise, especially in the frequency range near its coincidence frequency where the propagation speed of bending waves in the floating slab coincides with the propagation speed of sound in air and the slab becomes a very efficient sound radiator. Consequently, the airborne sound excitation of the tunnel structure might be controlling the low-frequency vibration response of the tunnel walls and the noise and vibration at a distant observer location.

To enable him to make a quantitative judgment on the relative importance of the airborne noise versus the force excitation of a platelike structure, the author has derived the relationship[60]

$$F_{eq} = p\left(\frac{c_0}{f}\right)\sqrt{\frac{\sigma_{rad}S_f}{\pi}} = p\lambda\sqrt{\frac{\sigma_{rad}S_f}{\pi}}[2(10^{-5})]10^{L_p/20} \quad N \quad (11.150)$$

where F_{eq} is the point force in newtons that generates the same free bending wave response on a partition of surface area S_f as a random-incidence sound field with a space–time average sound pressure p. The symbol σ_{rad} is the radiation efficiency of the partition. It is unity at frequencies above the coincidence frequency. The principle of reciprocity requires that σ_{rad} is also a measure of the degree of coupling of the sound waves to the vibration response of the structure. The symbol $\lambda = c_0/f$ is the acoustical wavelength in meters, c_0 is the speed of sound in air in meters per second, and f is the frequency in hertz (reciprocal seconds). The symbol L_p is the sound pressure level in decibels re 2×10^{-5} N/m^2.

Equation (11.150) can be written in a form that is easy to remember:

$$F_{eq}^2 = \left(\frac{4}{\pi}\right)\left(\frac{(pS)^2}{S/(\lambda/2)^2}\right) \simeq \left(\frac{(pS)^2}{S/(\lambda/2)^2}\right) \quad N^2 \quad (11.151)$$

The numerator of Eq. (11.151) is a force squared (the product sound pressure and the area) and the denominator is the number of areas, each a half wavelength squared, that would fill the entire surface area S. The area $(\lambda/2)^2$ is where the sound pressure on the surface of the partition is in phase.

Equations (11.150) and (11.151) are extremely simple and universally useful because:

1. The force/sound pressure equivalency does not implicitly depend on the material properties of the partition such as the density, Young's modulus, loss factor, and geometry.
2. They do not depend on the density of the fluid:
3. They are also valid for partitions with fluid loading such as concrete slabs embedded in soil or steel plates with a liquid on the other side.

The reason for these unique properties of Eqs. (11.150) and (11.151) is that the vibration response to both sound and point force depends the same way on these properties. For example, a lightly damped structure will respond equally vigorously to sound and point force excitation, a plate made out of a material with high density will respond equally less vigorously to both sound and force excitation than one that is made of a less dense material, and so on. Equations (11.150) and (11.151) have been derived by assuming that the partition is large compared with the bending wavelength and the room where the sound field is generated is large compared with the acoustical wavelength. Both the structure and the sound field respond in a multimodal fashion and their response is dominated by the resonant modes.

In noise control design the airborne noise strength of machines and equipment is given in the form of their sound power $W(f)$ or sound power level spectrum $L_w(f)$. The diffuse-field sound pressure in a room, resulting from the injected sound power, is given by

$$\frac{p^2}{4\rho_0 c_0} S_{tot}\bar{\alpha} = W_f \quad \text{N/m}^2 \tag{11.152}$$

Combining Eqs. (11.150) and (11.152) yields

$$F_{eq}(f) = W(f) \frac{\rho_0 c_0^3 \sigma_{rad}(f)}{f^2} \left(\frac{1}{\alpha}\right)\left(\frac{S_i}{S_{tot}}\right)\left(\frac{4}{\pi}\right) \quad \text{N}^2 \tag{11.153}$$

where $W(f)$ = sound power of machine, W
  $F_{eq}(f)$ = equivalent point force acting on solid boundary that produces same resonant (i.e., free bending wave) vibration response of boundary as the sound field produced in room owing to sound power output of the machine, N/m²,
  S_i = surface area (one side) of boundary directly excited by force, m²,
  S_{tot} = total surface area of all boundaries of room, m²,
  α = sound absorption coefficient of internal boundary surfaces

The validity of Eqs. (11.150), (11.151), and (11.153) have been briefly checked[61] experimentally by exciting the boundaries of an underground fan room. First one wall was excited with a shaker and the point force $F(f)$ was measured with a force gauge built into the impedance head that connected the shaker to the fan room wall. Then, a sound field was generated in the fan room by a loudspeaker and the sound pressure level SPL(f) was measured with a calibrated microphone. For both types of excitation the force and the sound pressure level was generated at an identical series of pure-tone frequencies to maximize the signal–noise ratio. The response was measured by a geophone in the form of the ground vibration at a distant location.

Example 11.5. Predict whether a planned concert hall can be located on a site near an existing subway line without imposing speed limits on the subway train and without supporting the entire concert hall on resilient vibration isolation pads. The design goal is to keep the ramble noise in the concert hall below the threshold of human hearing, which is 35 dB in the 63-Hz center-frequency octave band.

The subway tunnel has a 10 ft × 10 ft cross section and the subway train is 100 ft long. The tunnel walls, roof, and floor are poured concrete of 0.4 m thickness. Analytical predictions carried out have indicated that mounting the rails on a floating slab would reduce the total force acting on the tunnel floor to 200 N in the 63-Hz center-frequency octave band and that this dynamic force was expected to produce a sound pressure level of 30 dB in the concert hall. This is 5 dB lower than the human threshold of hearing. Sound-level measurements carried out in the tunnel during the passage of a subway train yielded a sound pressure level of 110 dB in the 63-Hz center-frequency octave band. Predict whether this airborne excitation will produce noise levels in the concert hall above the 35-dB threshold of hearing.

Solution The critical frequency f_c is predicted from Table 11.2. The density of concrete $\rho = 2300$ kg/m³, the mass per unit area $\rho_s = \rho h = 2300 \times 0.4 = 920$ kg/m², $f_c \rho_s = 43,000$, and $f_c = 43,000/920 = 47$ Hz; consequently, $\sigma_{\text{rad}(f=63 \text{ Hz})} \simeq 1$. The surface area of the tunnel exposed to the high-level airborne noise excitation $S_{\text{tot}} = 10 \times 10 \times 100 = 10,000$ ft² = 929 m². With these values Eq. (11.150) yields

$$F_{eq} = \left(\frac{c_0}{f}\right)\sqrt{\frac{S_{\text{tot}}}{\pi}} 2(10^{-5})10^{\text{SPL}/20} = \left(\frac{340}{63}\right)\sqrt{\frac{929}{\pi}} 2(10^{-5})10^{110/20} = 586 \text{ N}$$

Consequently, the noise level in the concert hall that is attributable to the airborne sound excitation of the tunnel walls is predicted to be

$$\text{SPL}_{\text{Hall}}(f = 63 \text{ Hz}) = 30 + 20\log\left(\frac{586}{200}\right) \simeq 39 \text{ dB}$$

This is 4 dB above the design goal, indicating that the design goal cannot be achieved without restricting the train speed or putting the concert hall on vibration isolators that would need to provide a high degree of isolation at 63 Hz.

11.10 RECIPROCITY AND SUPERPOSITION

The principles of reciprocity and superposition apply to linear systems with time-invariant parameters. Not only solid structures but fluid volumes at rest fall into this category. Consequently, reciprocity and superposition apply to systems that consist of solid structures surrounded by acoustical spaces and can be used to great advantage not only in structure-borne noise and airborne noise but also in structural acoustics, which deals with the interaction of sound waves with solid structures.

The principle of superposition, illustrated in Fig. 11.35, allows the use of the simplest excitation sources such as a point force source or a point-monopole sound source to explore the response to more complex excitation sources such as a moment acting on a structural element or an acoustical dipole radiating into an acoustical volume. The principle of reciprocity, which can be traced back to Lord Rayleigh,[62] is illustrated in the upper three sketches in Fig. 11.36. It states that

$$F_2 v_2 = F_1 v_1 \quad \text{W} \quad (11.154a)$$

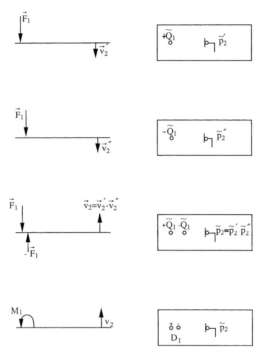

FIGURE 11.35 Use of superposition to predict the response of structures and acoustical spaces (v_2 and p_2) to complex excitation sources such as moments (M_1) and dipoles (D_1) on the basis of point force (F_1) and acoustical monopole (Q_1). Note that the structure and the acoustical space may be of arbitrary shape and the acoustical space may be unbound or bound by elastic or sound-absorbing boundaries and may contain an arbitrary number and size of rigid or elastic scatterers.

The symbol F_1 is (generalized) force when point 1 is source and point 2 is receiver and v_1 is (generalized) velocity when point 1 is receiver and point 2 is source. The vector product $F_1 v_1$ must yield the instantaneous power, or in complex notation $\text{Re}\{\frac{1}{2} F_1 v_1^*\} = \text{Re}\{\frac{1}{2} F_2 v_2^*\}$ must yield the time-averaged power. Note that F and v are vector quantities as signified in the figure by the arrow above the symbols. If v_1 and F_1 and v_2 and F_2 are measured or applied in the same direction, as illustrated by the sketch in the lower left side of Fig. 11.36, the vector notation can be exchanged for the less complicated scalar notation, where force and velocity are characterized by a magnitude and phase (i.e., $\tilde{F} = F e^{j\phi_f}$, $\tilde{v} = v e^{j\phi_r}$). In this case, the reciprocity takes the form of the equality of transfer functions

$$\frac{\tilde{v}_2}{\tilde{F}_1} = \frac{\tilde{v}_1}{\tilde{F}_2} \quad \text{m/N} \cdot \text{s} \qquad (11.154b)$$

This should be kept in mind in our latter deliberations, where the special vector notation is not carried through. Since monopole strength $\tilde{Q}$ and sound pressure

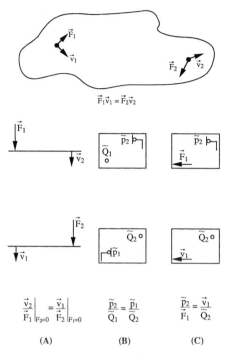

FIGURE 11.36 Principle of reciprocity as applied to a complex structure. Top sketch illustrates general principle. Lower sketches apply in special situations: (*a*) structure; (*b*) an acoustical space; (*c*) a structure coupled to an acoustical space. Symbols: F, point force vector; v, velocity response vector (measured by the same direction as F); $\tilde{Q}$, volume velocity of acoustical point source; $\tilde{p}$, sound pressure.

$\tilde{p}$ are scalar quantities (defined by their magnitude and phase), no directional restraints exist in the acoustical case illustrated in Fig. 11.36*b*.[63,64] However, dipole and quadrupole sound sources (constructed from adjacent out-of-phase monopoles) have highly directional radiation characteristics and must be connected to directional quantities of the sound field such as pressure gradients dp/dx and d^2p/d^2x measured in the same direction relative to the orientation of the dipole or quadrupole sound source for the reciprocity to apply.

Table 11.9 contains useful reciprocity relationships applicable to higher order excitation sources and responses. These relationships automatically follow from the joint application of superposition and reciprocity to the appropriate combination of simple sources such as point forces and acoustical monopoles.

The principle of reciprocity can be used to considerable advantage in both experimental and analytical work. In experimental work it is difficult and cumbersome to excite complex structures by point forces and moments and to measure the sound pressure such excitation causes in the interior of a vehicle. It is almost always easier to obtain the sought transfer function between the acoustical pressure and the exciting force or moment by placing a small acoustical source of known volume velocity at the microphone location and measuring the vibration response at the position and in the direction of the applied force and applying the reciprocity relationship illustrated in Fig. 11.36c.

The form of reciprocity illustrated in Fig. 11.36c is the most useful in structural acoustics. Its practical application is described in reference 64. In the direct experiment the force $\tilde{F}_1$ is applied to the structure by a shaker (and its magnitude and phase are measured by a force gauge inserted between the shaker and the structure), and the magnitude and phase of the sound pressure p_2 are measured by a microphone. The phase of F_1 and p_2 are referenced to the voltage U applied to the shaker. In the reciprocal experiment—which is much easier to perform than the direct experiment—a small omnidirectional sound source (an enclosed loudspeaker whose diameter is smaller than one-quarter acoustical wavelength) with calibrated volume velocity response Q is placed at the former microphone location and the velocity response of the structure at the former excitation point v_2 is measured (in the same direction as the force was applied) by a small accelerometer. The phases of Q and v_2 are referenced to the voltage U applied to the loudspeaker sound source. The volume velocity calibration of the sound source (Q/U) is obtained by placing it in an anechoically terminated rigid tube, baffling it so it radiates only toward the anechoic termination, sweeping the loudspeaker voltage through the frequency range of interest, and measuring the transfer function (p/U), where U is the voltage applied to the loudspeaker and p is the sound pressure measured by a microphone located two tube diameters or further away from the source. The sought volume velocity calibration of the source is then computed as[65] $|Q/U| = |P/U|(S/\rho_0 c_0)$, where S is the cross-sectional area of the tube, ρ_0 is the density of air, and c_0 is the speed of sound in air. When phase information is important, the sound source can be calibrated in an anechoic chamber by measuring the sound pressure $p(r)$ at a large distance $r \gg \lambda_0$ away from the source and computing the volume velocity calibration $Q/U = [p(r)/U](4\pi r^2/\rho_0 c_0)e^{-j2\pi f r/c_0}$.

Figure 11.37 illustrates the application of reciprocity on a complex structural acoustical problem, namely, the prediction of the interior noise of an automobile to point force excitation of the shock tower. First, the shock tower was excited by a point force, and sound pressure generated at the driver's head position was measured to obtain the direct transfer function $\tilde{p}/F$ identified by the solid line. Next, the reciprocal experiment was carried out by placing a point sound source of known volume velocity $\tilde{Q}$ at the former location of the microphone and measuring the vibration velocity response of the shock tower. The transfer function $v/\tilde{Q}$ obtained this way is shown as the dotted curve in Fig. 11.37a. Referencing the phases of $\tilde{p}$ and F to the voltage applied to the shaker and that of v and $\tilde{Q}$ to the voltage applied to the loudspeaker source, not only the magnitude but also the phase of the reciprocal transfer functions $\tilde{p}/F$ and $v/\tilde{Q}$ can be retained so that the interior noise caused by many simultaneously acting forces and moments can be predicted. Figure 11.37b shows the unrolled phase of the transfer function pair.

RECIPROCITY AND SUPERPOSITION **469**

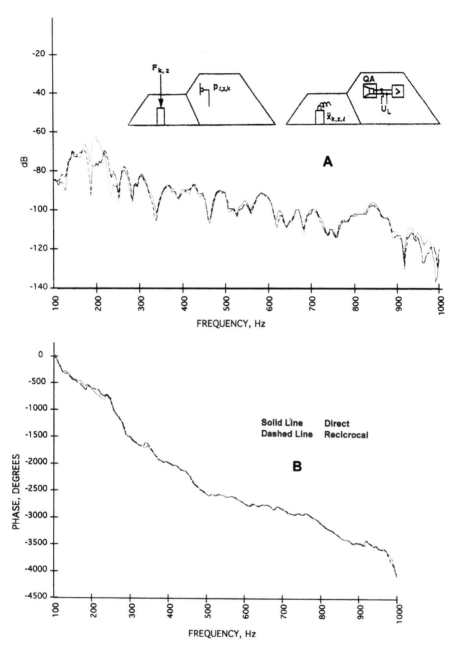

FIGURE 11.37 Pair of reciprocal transfer functions measured in a passenger vehicle: (a) direct, $\tilde{p}/\tilde{F}$; (b) reciprocal, $\tilde{v}/\tilde{Q}$.

Prediction of Noise Caused by Multiple Correlated Forces

The use of reciprocity and superposition in the case of multiple correlated force input is illustrated in Fig. 11.38. The problem is to predict the sound pressure p_R in the receiver room below after installation of a vibrating machine in the source room above. The building is constructed, and the machine manufacturer provides the magnitude, the direction, and the mutual phase ϕ_{12} of forces F_1 and F_2 the machine—when installed on soft springs—will impart to the floor. The prediction of p_R proceeds as illustrated in the lower part of Fig. 11.38 by measuring the reciprocal transfer functions $-v_1/\tilde{Q}_R = \tilde{p}_R/F_1$ and $-v_2/\tilde{Q}_R = \tilde{p}_{R2}/F_2$ and utilizing the principle of superposition. The forces and velocities must be measured in the same direction. Performing the reciprocity prediction for a number of different loudspeaker positions in the receiver room, the spatial

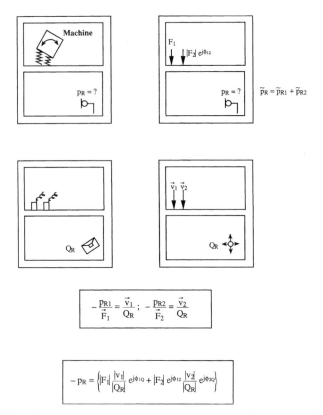

FIGURE 11.38 Use of reciprocity and superposition to predict the sound pressure in a room caused by two correlated forces $\tilde{F}_1$ and $\tilde{F}_2$ acting on the building structure. Top sketch represents actual situation and lower sketch the reciprocity prediction: ϕ_{1Q} and ϕ_{2Q} represent the phases of the transfer functions $\tilde{v}_1/\tilde{Q}_R$ and $\tilde{v}_2/\tilde{Q}_R$, respectively; each is conveniently referenced to the excitation voltage of the loudspeaker sound source.

variation of p_R can also be predicted. The methodology can be easily extended to more than two simultaneously acting, correlated forces.[66]

The reciprocity is most useful in the early stages of design of aircraft and ground vehicles, well before a flightworthy version of the aircraft or a roadworthy version of the ground vehicle is available. Reciprocity helps the noise control engineer to find answers to some difficult questions:

1. What will be the contribution of the structure-borne noise at the engine firing rate to the noise in the passenger compartment?
2. Which engine mount transmits most of the structure-borne noise?
3. Which direction of vibration force is most critical?
4. Which mutual phasing of forces acting on the individual engine-mounting points is most critical?
5. Most importantly, what is the effect of changing the design of engine-mounting brackets on cabin noise?

All these questions can be answered without applying known forces in three orthogonal directions to each of the engine-mounting brackets.

Source Strength Identification by Reciprocity

When it is not feasible to measure directly the strength of noise and vibration sources during the operation of vehicles, equipment, and machinery, reciprocity can be used to obtain them indirectly. This is accomplished by measuring the noise or vibration during the operation of the equipment at a distant, accessible receiver location and—when the equipment is not operating—exciting it at these distant receiver locations and measuring the acoustical or structural response at the source location, which is now accessible. The principle is illustrated in Fig. 11.39. Common in the three problems shown in Fig. 11.39 is the knowledge of the location and nature of the excitation sources. Unknown are their magnitude and mutual phase, which must be determined by observation of response to these sources at distant locations and by reciprocity experiments as described below.

The upper left-hand side of Fig. 11.39 represents the case where the sound field in an enclosure is excited by two monopole sound sources of unknown strength and mutual phase $\tilde{Q}_1(?)$ and $\tilde{Q}_2(?)$ (e.g., the openings of the inlet pipe leading to two cylinders of a reciprocating compressor). The first experiment, illustrated in the upper sketch, is the measurement of the magnitude and mutual phase of the sound pressure $\tilde{p}_3$ and $\tilde{p}_4$ at accessible distant locations 3 and 4 obtained when both sources were operating simultaneously. The reciprocal experiments, illustrated in the two lower sketches, are performed when the sources are not operational by placing a monopole sound source of known volume velocity Q at the former microphone locations 3 and 4 and measuring the magnitude and phase of the sound pressure produced at the two former source locations $\tilde{p}_{13}$, $\tilde{p}_{23}$,

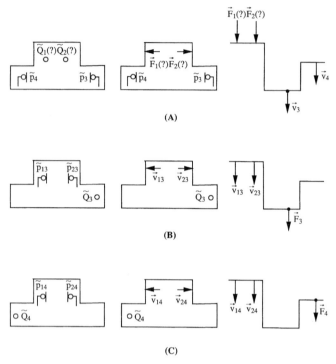

FIGURE 11.39 Source strength identification utilizing reciprocity and superposition: (a) response measurements at distant locations with sources active and (b, c) reciprocity measurements, sources inactive. (See text for explanation of symbols.)

$\tilde{p}_{14}$, and $\tilde{p}_{24}$. The phase of the sound pressure is referenced to the loudspeaker voltage.

The sketches in the middle column in Fig. 11.39 illustrate a case where the sound field in an enclosure is produced by two forces $\vec{F}_1$ and $\vec{F}_2$ of known location and direction but unknown magnitude and mutual phase, both acting simultaneously at the enclosure wall (e.g., forces caused by a vibration isolation-mounted reciprocating or rotating machine). In this case, the reciprocal experiment yields the vibration velocity responses at the former force application points $\tilde{v}_{13}$, $\tilde{v}_{23}$, $\tilde{v}_{14}$, and $\tilde{v}_{24}$, measured in the same direction as the force.

The situation shown in the sketch on the upper right-hand side in Fig. 11.39 illustrates the situation when the determination of the unknown magnitude and mutual phase of the two forces $\vec{F}_1(?)$ and $\vec{F}_2(?)$ must be diagnosed (e.g., forces transmitted to a structural floor by a vibration-isolated machine). In this case, the reciprocal experiment is carried out by exciting the building structure at the two distant observation points 3 and 4 by known forces $\vec{F}_3$ and $\vec{F}_4$ and measuring the velocity responses $\tilde{v}_{13}$, $\tilde{v}_{23}$, $\tilde{v}_{14}$, and $\tilde{v}_{24}$ at the former force excitation points 1 and 2. Utilizing the principle of reciprocity and superposition yields the following pairs of linear equations:

$$\tilde{p}_3 = \tilde{Q}_1(?) \left[\frac{\tilde{p}_{13}}{\tilde{Q}_3} \right] + \tilde{Q}_2(?) \left[\frac{\tilde{p}_{23}}{\tilde{Q}_3} \right] \quad \text{N/m}^2 \quad (11.155a)$$

$$\tilde{p}_4 = \tilde{Q}_1(?) \left[\frac{\tilde{p}_{14}}{\tilde{Q}_4} \right] + \tilde{Q}_2(?) \left[\frac{\tilde{p}_{24}}{\tilde{Q}_4} \right] \quad \text{N/m}^2 \quad (11.155b)$$

which can be solved for the two unknowns $\tilde{Q}_1(?)$ and $\tilde{Q}_2(?)$. Similarly, the set of equations for obtaining $\tilde{F}_1(?)$ and $\tilde{F}_2(?)$ in the situation illustrated in the center sketches in Fig. 11.39 are

$$\tilde{p}_3 = \tilde{F}_1(?) \left[\frac{\tilde{v}_{13}}{\tilde{Q}_3} \right] + \tilde{F}_2(?) \left[\frac{\tilde{v}_{23}}{\tilde{Q}_3} \right] \quad \text{N/m}^2 \quad (11.155c)$$

$$\tilde{p}_4 = \tilde{F}_1(?) \left[\frac{\tilde{v}_{14}}{\tilde{Q}_4} \right] + \tilde{F}_2(?) \left[\frac{\tilde{v}_{24}}{\tilde{Q}_4} \right] \quad \text{N/m}^2 \quad (11.155d)$$

and for that illustrated in the sketches on the right

$$\tilde{p}_3 = \tilde{F}_1(?) \left[\frac{\tilde{v}_{13}}{\tilde{F}_3} \right] + \tilde{F}_2(?) \left[\frac{\tilde{v}_{23}}{\tilde{F}_3} \right] \quad \text{N/m}^2 \quad (11.155e)$$

$$\tilde{p}_4 = \tilde{F}_1(?) \left[\frac{\tilde{v}_{14}}{\tilde{F}_4} \right] + \tilde{F}_2(?) \left[\frac{\tilde{v}_{24}}{\tilde{F}_4} \right] \quad \text{N/m}^2 \quad (11.155f)$$

In the case of n unknown excitation sources, the prediction equations represent an $n \times n$ matrix.

Extension of Reciprocity to Sound Excitation of Structures

As illustrated in Fig. 11.40, the reciprocity relationship can be extended for surface excitation of structures (e.g., by an incident sound wave). Consider first a small part of the surface of a cylindrical body (such as an aircraft fuselage) with surface area dA exposed to a local sound pressure of $\tilde{p}_1$ as illustrated in the upper left sketch in Fig. 11.40 resulting in a local force of $\tilde{F}_1 = \tilde{p}_1 \, dA$. For this force the reciprocity relationship shown in Fig. 11.36c yields $\Delta \tilde{p}_{R1}/\tilde{F}_1 = \Delta \tilde{v}_{1R}/Q_R$, which for $\tilde{F}_1 = \tilde{p}_1 \, dA$ and $\Delta \tilde{Q}_{1R} = \Delta \tilde{v}_{1R} \, dA$ becomes

$$\frac{\Delta \tilde{p}_R}{\tilde{p}_1} = \frac{\Delta \tilde{Q}_{1R}}{\tilde{Q}_R} \quad (11.156a)$$

When the structure is exposed to a complex sound field distribution $\tilde{p}_1, \tilde{p}_2, \ldots, \tilde{p}_n$, as illustrated in the lower sketch, the resulting interior sound pressure at the receiver location, $\tilde{p}_R$, is given by

$$\tilde{p}_R = \sum_{i=1}^{n} \tilde{p}_i \left(\frac{\Delta \tilde{Q}_{iR}}{\tilde{Q}_R} \right) \quad \text{N/m}^2 \quad (11.156b)$$

474 INTERACTION OF SOUND WAVES WITH SOLID STRUCTURES

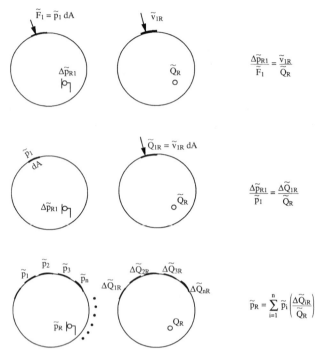

FIGURE 11.40 Extension of reciprocity to sound excitation of structures. Top: Reciprocity relationship for external pressure, $\tilde{p}_1$, acting on a small area (dA) of the structure and the resulting internal sound pressure, $\Delta \tilde{p}_{R1}$, and structural response v_{1R} produced by an internal point sound source of volume velocity Q_R located at the former receiver position. Middle: Substitution of volume velocity response $\Delta \tilde{Q}_{1R} = d\tilde{v}_{1R} dA$ into the reciprocity relationship. Bottom: Reciprocity relationship for external incident sound excitation p_i and internal sound pressure $\tilde{p}_R$ and volume velocity responses of the structure $\Delta \tilde{Q}_{1R}$ produced by an internal point sound source of volume velocity $\tilde{Q}_R$.

where the transfer functions $\Delta \tilde{Q}_{1R}/\tilde{Q}_R$ represent the reciprocity calibration of the structure as a transducer. This extension of reciprocity by Fahy[67] has the advantage that the reciprocity calibration of the structure (in the form of discretized transfer functions $\Delta \tilde{Q}_{iR}/\tilde{Q}_R$) can be carried out with a capacitive transducer that directly measures the structure's local volume displacement $\Delta \tilde{Q}_{iR}/j\omega$. To obtain sufficient resolution, the side length of the square-shaped capacitive transducer used in measuring the volume displacement must not exceed one-eighth of the acoustical wavelength. Note that the bending wavelength in thin, plate-like structures is usually much smaller than the transducer size so that the capacitive transducer acts as a wavenumber filter, accounting only for those components of the vibration field that results in a net volume displacement. The high-wavenumber components, which result only in local near fields, are "averaged out." The additional advantage of the capacitive transducer is that it does not influence the vibration response of the structure.

Reciprocity can also be used in predicting the sound pressure attributable to the complex vibration pattern of a vibrating body in case where the radiated sound cannot be measured directly (e.g., other correlated vibration sources dominate the sound field). The reciprocity prediction proceeds in two steps. First the vibration pattern of the body is mapped during the operation of the equipment by measuring the vibration velocity $\tilde{v}_i$ at a large number of locations. The phase of the velocity responses is referenced to the velocity measured at a designated reference location. Next the machine is shut off and a point sound source of known volume velocity Q is placed at the receiver location where the sound pressure should be predicted and the sound pressure $\tilde{p}_i$ produced by the point sound source at the various locations along the stationary surface of the body is measured. The phase of the pressure responses is conveniently referenced to the voltage applied to the loudspeaker sound source. The sound pressure at the receiver location, $\tilde{p}_R$, attributable to the periodic vibration of body is predicted as

$$\tilde{p}_R = \sum_{i=1}^{n} \tilde{v}_i \, dA_i \frac{\tilde{p}_i}{Q} \cong \sum_{i=1}^{n} \Delta \tilde{Q}_i \frac{\tilde{p}_i}{Q} \quad \text{N/m}^2 \tag{11.156c}$$

where dA_i is the area and $\Delta \tilde{Q}_i = \tilde{v}_i dA_i$ is the volume velocity of the ith sample of the vibrating surface.

The solid line in Fig. 11.41 represents the directly measured sound pressure $\tilde{p}_R$ at a specific location in a room when a thin plate was excited by a shaker to a

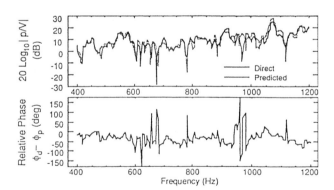

FIGURE 11.41 Sound pressure response at a specific location in a room produced by a thin plate excited by a shaker to complex vibration pattern. Solid line: Directly measured transfer function between sound pressure $\tilde{p}_R$ and excitation force F, phase is referenced to shaker voltage $\tilde{U}$. Dotted line: Transfer function predicted by measuring the magnitude and relative phase of the plate response, $\Delta \tilde{Q}_i = \tilde{v}_i \, dA$ at $i = 81$ position during shaker excitation and the magnitude and phase of the sound pressure $\tilde{p}_i$ at the surface of the stationary plate when a point sound source of volume velocity $\tilde{Q}_R$ is placed at the former receiver position and applying the reciprocity relationship $\tilde{p}_R = \Sigma_{i=1}^{81} \Delta \tilde{Q}_i (\tilde{p}_i / \tilde{Q}_R)$ (after Reference[67]). Top curve, magnitude of pressure; lower curve, phase spectrum of directly measured minus phase spectrum of predicted pressure.

complex vibration pattern. The dotted curve represents the reciprocity prediction according to Eq. (11.156a) utilizing $n = 81$ sampling points on the plate.[67] The experimental results indicate the feasibility of using reciprocity predictions in engineering applications.

Reciprocity in Moving Media

Reciprocity requires that exchanging the function of the source and receiver should not result in any change in the sound propagation path. This is true only if the acoustical medium is at rest. As illustrated in Fig. 11.42, the propagation path between source and receiver remains the same if the exchange of source and receiver positions is accompanied by a reversal of the direction of the uniform mean flow. In this case, or in the case of low-Mach-number potential flow where the shear layer is small compared with the acoustical wavelength, the reciprocity also applies in moving media.[68] Reversal of potential flow is usually easy to accomplish in analytical calculations. However, in many experimental situations where the shear layer is not small compared with the acoustical wavelength, reciprocity does not apply.

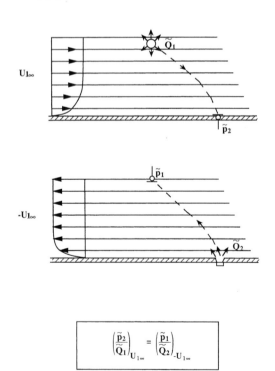

FIGURE 11.42 Reciprocity in moving media. Reversal of the function of source and receiver must be accompanied by reversing the flow direction. Streamlines must remain unchanged to assure that the propagation path between source and receiver remains the same.

Fundamental questions regarding reciprocity are dealt with in references 69–71. Analytical applications of reciprocity and superposition for predicting power input structures excited by complex airborne or structure-borne sources is presented in references 66, 72, and 73. Ship acoustics application of reciprocity are treated in references 74 and 75.

11.11 IMPACT NOISE

There are many practical cases where the excitation of a structure can be represented reasonably well by the periodic impact of a mass on its surface. Footfall in dwellings, punch presses, and forge hammers fall into this category. This section deals only with footfall noise in buildings. For the prediction and control of impact noise of machines and equipment, the reader is referred to a series of 10 papers[76–85] that covers all aspects of impact noise of machinery.

Standard Tapping Machine

A standard tapping machine[86] is used to rate the impact noise isolation of floors in dwellings. This machine consists of five hammers equally spaced along a line, the distance between the two end hammers being about 40 cm. The hammers successively impact on the surface of the floor to be tested at a rate of 10 times per second. Each hammer has a mass of 0.5 kg and falls with a velocity equivalent to a free-drop height of 4 cm. The area of the striking surface of the hammer is approximately 7 cm^2; the striking surface is rounded as though it were part of a spherical surface of 50 cm radius. The impact noise isolation capability of a floor is rated by placing the standard tapping machine on the floor to be tested and measuring the one-third-octave-band sound pressure level L'_p averaged in space in the room below.

$$L_n \equiv L_p - 10 \log \frac{A_0}{S\bar{\alpha}_{S,\text{ab}}} \quad \text{dB re} 2 \times 10^{-5} \text{ N/m}^2 \qquad (11.157a)$$

where L_p = one-third-octave-band sound pressure level as measured, dB
$S\bar{\alpha}_{S,\text{ab}}$ = total absorption in receiving room (see Chapter 7), m^2
A_0 = reference value of absorption, $=10$ m^2

The physical formulation of the problem of impact noise is that of the excitation of a plate by periodic force impulses. Such periodic forces can be presented by a Fourier series consisting of an infinite number of discrete-frequency components, each with amplitude F_n, given by

$$F_n = \frac{2}{T_r} \int_0^{T_r} F(t) \cos \frac{2\pi n}{T_r} t \, dt \quad \text{N} \qquad (11.157b)$$

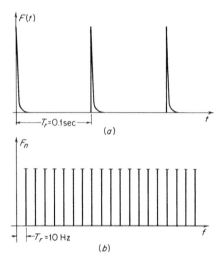

FIGURE 11.43 (a) Time function and (b) Fourier components of the force that a standard tapping machine exerts on a massive rigid floor.

where $T_r = 1/f_r = 0.1$ is the time interval between hammer strikes and $n = 1, 2, 3, \ldots$. The curve in Fig. 11.43a shows the time function of the force $F(t)$, and that in Fig. 11.43b shows the amplitude of its Fourier components.

It is an experimental fact that when the hammer strikes a hard concrete slab, the duration of the force impulse is small compared even with the period of the highest frequency of interest in impact testing. For less stiff structures, like wooden floors, this assumption is not valid and the exact shape of $F(t)$ has to be determined and used in Eq. (11.157b). For a thick concrete slab the effective length of the force impulse is short enough so that $\cos[(2\pi n/T_r)t] \approx 1$, and all components have the same amplitude. Because the integral in Eq. (11.157b) is the momentum of a single hammer blow (assuming no rebound) equal to mv_0 (in kg · m/s), the amplitudes of the Fourier components of the force for a repetition frequency f_r are

$$F_n = 2f_r m v_0 \quad \text{N} \tag{11.158}$$

The velocity of the hammer at the instant of impact is

$$v_0 = \sqrt{2gh} \quad \text{m/s} \tag{11.159}$$

where $h =$ falling height of hammer, m
$g =$ acceleration of gravity (9.8 m/s^2)

Let us define a mean-square-force spectrum density S_{f_0} that when multiplied by the bandwidth will yield the value of the mean-square force in the same bandwidth,

$$S_{f_0} = \tfrac{1}{2} T_r F_n^2 = 4 f_r m^2 g h \quad \text{N}^2/\text{Hz} \tag{11.160}$$

For the standard tapping machine the numerical value of S_{f_0} is 4 N²/Hz. Accordingly, the mean-square force in an octave band $\Delta f_{\text{oct}} = f/\sqrt{2}$ is

$$F^2_{\text{rms}}(\text{oct}) = \frac{4}{\sqrt{2}} f \quad \text{N}^2 \tag{11.161}$$

The octave-band sound power level radiated by the impacted slab (which is assumed to be isotropic and homogeneous) into the room below is calculated by inserting Eq. (11.161) into Eq. (11.67a), which yields

$$L_w(\text{oct}) \approx 10 \log_{10}\left(\frac{\rho c \sigma_{\text{rad}}}{5.1 \rho_p^2 c_L \eta_p t^3}\right) + 120 \quad \text{dB re} 10^{-12} \text{ W} \tag{11.162}$$

where ρ = density of air, kg/m³
c = speed of sound in air, m/s
σ_{rad} = radiation factor of slab
ρ_p = density of slab material, kg/m³
c_L = propagation speed of longitudinal waves in slab material, m/s
η_p = composite loss factor of slab
t = thickness of slab, m

Note that the *sound power level is independent of the center frequency of the octave*, that *doubling the slab thickness decreases the level of the noise radiated into the room below by 9 dB*, and that the *sound power level decreases with increasing loss factor*.

Improvement of Impact Noise Isolation by an Elastic Surface Layer

Experience has shown that the impact noise level of even an 8–10-in.-thick dense concrete slab is too high to be acceptable. A further increase of thickness to reduce impact noise is not economical.

Impact noise may be reduced effectively by an elastic surface layer, much softer than the surface of the slab, applied to the structural slab. The resilient layer changes the shape of the force pulse and the amount of mechanical power introduced into the slab by the impacting hammer, as shown in Fig. 11.44.

We would expect, if the elastic layer is linear and nondissipative, that the velocity will be at its maximum v_0 at the instant of impact $t = 0$. It will then decrease to zero and the mass will rebound to nearly the same velocity (it is assumed the hammer is not permitted to bounce a second time) according to the function shown by curve *a* of Fig. 11.44. The force function is shown by curve *b*.

The improvement in impact noise isolation achieved by the addition of the soft surface layer is defined in terms of the logarithmic ratio[87]

$$\Delta L_n = 20 \log \frac{F}{F'} = 20 \log \left(\left|\frac{1 - n f_r/f_0}{\cos[(\pi/2)n(f_r/f_0)]}\right|\right) \quad \text{dB} \tag{11.163}$$

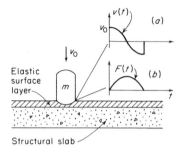

FIGURE 11.44 Velocity and force pulse of a single hammer blow on an elastic surface layer over a rigid slab: (a) velocity pulse; (b) force pulse.

where

$$n = 1, 2, 3, \ldots \quad (11.164)$$

$$f_0 = \frac{1}{2\pi}\sqrt{\frac{A_h}{m}}\sqrt{\frac{E}{h}} \quad \text{Hz} \quad (11.165)$$

F', F = forces acting on slab with and without resilient surface layer, respectively, N
A_h = striking area of hammer, m^2
m = mass of hammer, kg
E = dynamic Young's modulus of elastic material, N/m
h = thickness of layer, m

The characteristic frequency f_0 of an elastic surface layer for the standard tapping machine is plotted in Fig. 11.45 as a function of E/h.

Equation (11.163), which assumes no damping, is plotted in Fig. 11.46 as a function of the normalized frequency f/f_0. Below $f/f_0 = 1$, the improvement is zero. Above $f/f_0 = 1$ the improvement increases with an asymptotic slope of 40 dB/decade.

Figures 11.45 and 11.46 (use the 40-dB/decade asymptote) permit one to select an elastic surface layer to achieve a specified ΔL_n.

Example 11.6. The required improvement in impact noise isolation should be 20 dB at 300 Hz. Design a resilient covering for the concrete slab.

Solution From Fig. 11.46 we obtain $f/f_0 \approx 3$, which gives $f_0 = 100$ Hz. Entering Fig. 11.45 with this value of f_0 yields $E/h = 2.8 \times 10^8$ N/m^3 (or $E/h \approx$ 1000 psi/in.). Any material having this ratio of Young's modulus to thickness will provide the required improvement. If we wish to select a 0.31-cm- ($\frac{1}{8}$-in.-) thick layer, the dynamic modulus of the material should be 8.7×10^5 N/m^2 (8000 psi). Since the dynamic modulus of most elastic materials is about twice the statically

IMPACT NOISE **481**

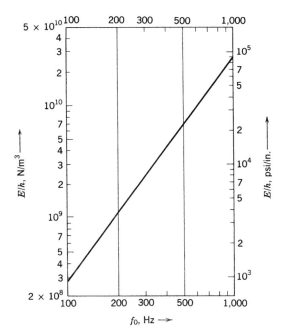

FIGURE 11.45 Chart for the selection of an elastic surface layer, where f_0 = characteristic frequency, E = Young's modulus, and h = thickness of layer.

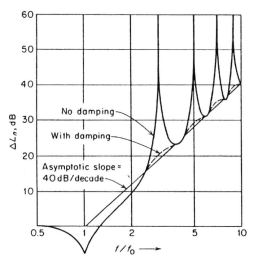

FIGURE 11.46 Improvement in impact noise isolation ΔL_n versus normalized frequency for a resilient surface layer (select f_0 to yield desired improvement).

measured Young's modulus,[88] a material with $E \leq 4.35 \times 10^5$ N/m² (4000 psi) should be selected.

Frequently used materials for the elastic surface layer are rubberlike materials, vinyl-cork tile, or carpet. The impact isolation improvement curve (ΔL_n vs. frequency) has been measured and reported[89] for a large variety of elastic surface layer configurations.

The expected normalized impact sound level in the room below (see Fig. 11.1) for a composite floor (with a heavy structural slab) is that of the bare concrete structural floor minus the improvement caused by the elastic surface layer, that is,

$$L_{n,\text{comp}} = L_{n,\text{bare}} - \Delta L_n \quad (11.166)$$

where

$$L_{n,\text{bare}}(\text{oct}) = 116 + 10 \log \left(\frac{\rho c \sigma_{\text{rad}}}{5.1 \rho_p^2 c_L \eta_p t^3} \right) \quad \text{dB re2} \times 10^{-5} \text{ N/m}^2 \quad (11.167)$$

for homogeneous isotropic slabs.

Measured values of impact noise isolation of a large number of floor constructions and improvement of impact noise isolation by various surface layers are presented in reference 89.

Improvement through Floating Floors. It is often more practical to use a floating floor above a structural slab than a soft resilient surface layer. The advantages are that (1) both the impact noise isolation and the airborne sound transmission loss of the composite floor are improved and (2) the walking surface is hard. For analysis, floating floors can be categorized as either (1) locally reacting or (2) resonantly reacting, as defined below.

Locally Reacting Floating Floors. A locally reacting floor is one where the impact force of the hammer on the upper slab (slab 1) is transmitted to the structural slab (slab 2), primarily in the immediate vicinity of the excitation point, and where there is no spatially homogeneous reverberant vibration field on slab 1. In this case the bending waves in the floating slab are highly damped. If the Fourier amplitude of the force acting on plate 1 is given by Eq. (11.158), the reduction in transmitted sound level is[87]

$$\Delta L_n = 20 \log \left[1 + \left(\frac{f}{f_0} \right)^2 \right] \approx 40 \log \frac{f}{f_0} \quad (11.168)$$

where

$$f_0 = \frac{1}{2\pi} \sqrt{\frac{s'}{\rho_{s_1}}} \quad (11.169)$$

and ρ_{s_1} = mass per unit area of floating slab, kg/m²
s' = dynamic stiffness per unit area of resilient layer between slab 1 and slab 2 including trapped air, N/m³

Resonantly Reacting Floating Floors. If the floating slab is thick, rigid, and lightly damped, the impact force of the hammers excites a more-or-less spatially homogeneous reverberant bending wave field.

The improvement in impact noise isolation at higher frequencies, where the power dissipated in slab 1 exceeds the power transmitted to slab 2, can be approximated by[87,90]

$$\Delta L_n \approx 10 \log \frac{2.3 \rho_{s_1}^2 \, \omega^3 \eta_1 c_{L_1} h_1}{n' s^2} \qquad (11.170)$$

where h_1 = thickness of floating slab, m
c_{L_1} = propagation speed of longitudinal waves in floating slab, m/s
ρ_{s_1} = mass per unit area of floating slab, kg/m²
η_1 = loss factor of floating slab
n' = number of resilient mounts per unit area of slab, m²
s = stiffness of mount, N/m

Equation (11.170) indicates that, in contrast to the locally reacting case where ΔL_n increases at a rate of 40 dB for each decade increase in frequency, the increase is only 30 dB/decade if the loss factor of the floating slab η_1 is frequency independent. Another difference is the marked dependence of ΔL_n on this loss factor. The loss factor is determined both by the energy dissipated in the slab material itself and by the energy dissipated in the resilient mounts.

Figure 11.47 shows the improvement of a floating-floor system under impacting by a standard tapping machine and high-heel shoes, respectively.[91] The negative improvement in the vicinity of the resonance frequency f_0 can be observed.

Impact Noise Isolation versus Sound Transmission Loss

Whether a floor is excited by the hammers of a tapping machine or by an airborne sound field in the source room, it will in both cases radiate sound into the receiving room. There is a close relation between the airborne sound transmission loss R and the normalized impact noise level L_n for a given floor.

In the case of acoustical excitation, the sound power transmitted to the receiver room comprises the contribution of forced waves and of resonance waves. The forced waves usually dominate below the critical frequency of the slab and the resonance waves above. The sound power transmitted by exciting the slab by a standard tapping machine is made up of the contributions of the near-field component and of the reverberant component.

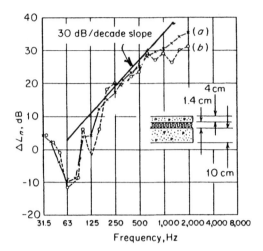

FIGURE 11.47 Improvement in impact noise isolation ΔL_n for a resonantly reacting floating floor for excitation by (a) a standard tapping machine and (b) high-heeled shoes. Note the negative ΔL_n in the vicinity of the resonance frequency. (After Ref. 91.)

The relation between sound transmission loss R and normalized impact noise level, assuming measurement in octave bands, is[92]

$$L_n + R = 84 + 10 \log \left[\frac{S_{f0} f}{\sqrt{2}} \left(\frac{\rho/(2\pi \rho_s^2 c) + \rho c \sigma_{\text{rad}}/(2.3 \rho_s^2 c_L \omega \eta_p h)}{\pi \rho c/(\omega^2 \rho_s^2) + \pi \sqrt{12} c^3 \rho \sigma_{\text{rad}}^2/(2\rho_s^2 c_L \omega^3 \eta_p h)} \right) \right] \text{dB} \quad (11.171)$$

where ρ_s = mass per unit area of slab, kg/m²
 c_L = propagation speed of longitudinal waves in slab, m/s
 σ_{rad} = radiation factor of slab
 h = thickness of slab, m
 η_p = composite loss factor of slab
 S_{f0} = mean-square force spectrum density as given in Eq. (11.160), N²/Hz

In the special case of a thick, lightly damped slab,

$$L_n + R = 43 + 30 \log f - 10 \log \sigma_{\text{rad}} - \Delta L_n \quad (11.172)$$

where ΔL_n represents the effect of the surface layer only. For a bare structural slab, by definition, $\Delta L_n = 0$.

Equation (11.172) states that the *sum of the airborne sound transmission loss and the normalized impact noise level is independent of the physical characteristics of the structural slab above the critical frequency of the slab where* $\sigma_{\text{rad}} \approx 1$.

Below the coincidence frequency where the forced waves control the airborne sound transmission loss but the impact noise isolation is still controlled by the resonant vibration of the impacted slab, Eq. (11.171) yields[92]

$$R + L_n = 39.5 + 20 \log f - \Delta L_n - 10 \log \frac{\eta_p}{f_c \sigma_{\text{rad}}} \quad \text{dB} \quad (11.173)$$

where ΔL_n = effect of surface layer only (zero for structural slab), dB
f_c = critical frequency of structural slab, Hz

In this case, the sum $R + L_n$ decibels depends on the physical characteristics of the slab and frequency. Figure 11.48 shows the measured sound transmission loss R and normalized impact sound level L_n as well as their sum for a typical floating floor. The measured and predicted values for the sum are in good agreement,

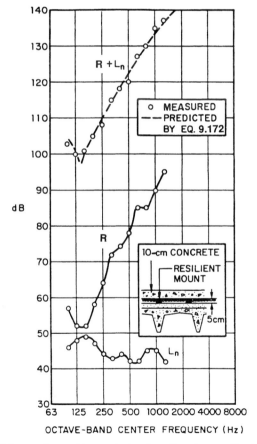

FIGURE 11.48 Measured sound transmission loss R and normalized impact sound level L_n and their sum ($R + L_n$) of a resonantly reacting floating floor assembly. Dotted curve: $R + L_n$ predicted by Eq. (11.172). (After reference 92.)

indicating that the precautionary measures taken to eliminate flanking have been successful.

In checking out the performance of floating floors in the field, it is advisable to measure both R and L_n. The discrepancy between the measured R and that calculated from Eq. (11.173) is a direct indication of flanking. By measuring the acceleration level on the wall surfaces in the source and receiving rooms during acoustical and impact excitation, the flanking paths can be immediately identified.

The measurement and rating of the impact noise isolation of floor assemblies is prescribed in ASTM E492–90 (1996), ASTM E989–89 (1999), and ASTM E1007–97.

REFERENCES

1. L. Cremer, M. Heckl, and E. E. Ungar, *Structureborne Sound*, 2nd ed., Springer Verlag, Berlin 1988.
2. M. Heckl, "Compendium of Impedance Formulas," Report No. 774, Bolt Beranek and Newman, Cambridge, MA, May 26, 1961.
3. I. Dyer, "Moment Impedance of Plate," *J. Acoust. Soc. Am.*, **32**, 247–248 (1960).
4. E. Eichler, "Plate Edge Admittances," *J. Acoust. Soc. Am.*, **36**(2), 344–348 (1964).
5. H. G. D. Goyder and R. G. White, "Vibrational Power Flow from Machines into Builtup Structures, Part I: Introduction and Approximate Analyses of Beam and Plate-Like Foundations," *J. Sound Vib.*, **68**(1), 59–75 (1980); "Part II: Wave Propagation and Power Flow in Beam-stiffened Plates," *J. Sound Vib.*, **68**(1), 77–96 (1980); "Part III: Power Flow through Isolation Systems," *J. Sound Vib.*, **68**(1), 97–117 (1980).
6. U. J. Kurze, *Laermarm Konstruieren XII; Mechanische Impedance* [*Low Noise Design XII; Mechanical Impedance*], Bundesanstalt fuer Arbeitsschutz, FB Nr. 398, Wirtschaftverlag NW, Bremerhaven, 1985.
7. K. -P. Schmidt *Laermarm Konstruieren (III); Aenderung der Eingangsimpedanz als Massnahme zur Laermminderung* [*Low Noise Design (III); Reducing Noise by Changing the Input Impedance*] Bundesanstalt fuer Arbeitsschutz, FB Nr. 169, Wirtschaftsverlag NW, Bremerhaven, 1979.
8. VDI 3720, Blatt 6 (Preliminary, July 1984), *Noise Abatement by Design; Mechanical Input Impedance of Structural Elements, Especially Standard-Section Steel*, VDI-Verlag, Dusseldorf, 1984 (in German).
9. M. Heckl, "Bending Wave Input Impedance of Beams and Plates," in *Proceedings 5th ICA*, Liége, Paper L67, 1965 (in German).
10. M. Heckl, "A Simple Method for Estimating the Mechanical Impedance," in *Proceedings DAGA*, VDI Verlag, 1980, pp. 828–830 (in German).
11. M. Heckl, "Excitation of Sound in Structures," *Proc. INTER-NOISE*, 1988, pp. 497–502.
12. R. J. Pinnington, "Approximate Mobilities of Builtup Structures," Report No. 162, Institute of Sound and Vibration Research, Southampton, England, 1988.
13. H. Ertel and I. L. Vér, "On the Effect of the Vicinity of Junctions on the Power Input into Beam and Plate Structures," *Proc. INTER-NOISE*, 1985, pp. 684–692.

14. P. W. Smith Jr., "The Imaginary Part of Input Admittance: A Physical Explanation of Fluid-Loading Effects on Plates," *J. Sound Vib.*, **60**(2), 213–216 (1978).
15. M. Heckl, "Input Impedance of Plates with Radiation Loading," *Acustica*, **19**, 214–221 (1967/68) (in German).
16. D. G. Crighton, "Force and Moment Admittance of Plates under Arbitrary Fluid Loading," *J. Sound Vib.*, **20**(2), 209–218 (1972).
17. D. G. Crighton, "Point Admittance of an Infinite Thin Elastic Plate under Fluid Loading," *J. Sound Vib.*, **54**(3), 389–391 (1977).
18. R. H. Lyon, "Statistical Analysis of Power Injection and Response in Structures and Rooms," *J. Acoust. Soc. Am.*, **45**(3), 545–565, (1969).
19. R. D. Blevins, *Formulas for Natural Frequency and Mode Shape*, Van Nostrand Reinhold, New York, 1979.
20. T. Kihlman, "Transmission of Structureborne Sound in Buildings," Report No. 9. National Swedish Institute of Building Research, Stockholm, 1967.
21. M. Paul, "The Measurement of Sound Transmission from Bars to Plates," *Acustica*, **20**, 36–40 (1968) (in German).
22. E. J. Richards et al. "On the Prediction of Impact Noise, II: Ringing Noise," *J. Sound Vib.*, **65**(3), 419–451 (1979).
23. R. Timmel, "Investigation on the Effect of Edge Conditions on Flexurally Vibrating Rectangular Panels on Radiation Efficiency as Exemplified by Clamped and Simply Supported Panels," *Acustica*, **73**, 12–20 (1991) (in German).
24. ASTM E90-02 1990, "Laboratory Measurement of Airborne-Sound Transmission Loss of Building Partitions," American Society for Testing and Materials, Philadelphia, PA.
25. ISO 140-1 : 1997 "Laboratory Measurement of Airborne Sound Insulation of Building Elements," American National Standards Institute, New York.
26. L. Cremer, *Vorlesungen ueber techische Akustik*, [*Lectures on Technical Acoustics*], Springer-Verlag, Berlin, 1971.
27. H. Reissner, "Der senkrechte und schräge Durchtritt einer in einem fluessigen Medium erzeugten ebenen Dilations(Longitudinal)–Welle durch eine in diesem Medium benfindliche planparallele feste Platte" ["Transmission of a Normal and Oblique-Incidence Plane Compressional Wave Incident from a Fluid on a Solid, Plane, Parallel Plate"] *Helv. Phys. Acta*, **11**, 140–145 (1938).
28. M. Heckl, "The Tenth Sir Richard Fairey Memorial Lecture: Sound Transmission in Buildings," *J. Sound Vib.*, **77**(2), 165–189 (1981).
29. R. D. Mindlin, "Influence of Rotary Inertia and Shear on Flexural Motion of Elastic Plates," *J. Appl. Mech.*, **18**, 31–38 (1951).
30. B. H. Sharp, "A Study of Techniques to Increase the Sound Insulation of Building Elements," Report WR-73-5, HUD Contract No. H-1095, Wyle Laboratories, Arlington, VA June 1973.
31. G. Kurtze and B. G. Watters, "New Wall Design for High Transmission Loss or High Damping," *J. Acoust. Soc. Am.*, **31**(6), 739–748 (1959).
32. H. Feshbach, "Transmission Loss of Infinite Single Plates for Random Incidence," Report No. TIR 1, Bolt Beranek and Newman, Cambridge Mass. October 1954.

33. L. L. Beranek, "The Transmission and Radiation of Acoustic Waves by Structures" (the 45th Thomas Hacksley Lecture of the British Institution of Mechanical Engineers), *J. Inst. Mech. Eng.*, **6**, 162–169 (1959).
34. L. L. Beranek, *Noise Reduction*, McGraw-Hill, New York, 1960.
35. M. Heckl, "Untersuchungen an Orthotropen Platten" ["Investigations of Orthotropic Plates"] *Acustica*, **10**, 109–115 (1960).
36. M. Heckl, "Die Schalldaemmung von homogene Einfachwaenden endlicher Flaeche" ["Sound Transmission Loss of the Homogeneous Finite Size Single Wall"], *Acustica*, **10**, 98–108 (1960).
37. H. Sato, "On the Mechanism of Outdoor Noise Transmission through Walls and Windows; A Modification of Infinite Wall Theory with Respect to Radiation of Transmitted Wave," *J. Acoust. Soc. Jpn.*, **29**, 509–516 (1973) (in Japanese).
38. J. H. Rindel, "Transmission of Traffic Noise through Windows; Influence of Incident Angle on Sound Insulation in Theory and Experiment," Ph.D. Thesis, Technical University of Denmark, Report No. 9, 1975 (also see *Proc. DAGA*, pp. 509–512), 1975; also private communications.
39. B. G. Watters, "Transmission Loss of Some Masonry Walls," *J. Acoust. Soc. Am.*, **31**(7), 898–911 (1959).
40. Y. Hamada and H. Tachibana, "Analysis of Sound Transmission Loss of Multiple Structures by Four-Terminal Network Theory," *Proc. INTER-NOISE*, 1985, pp. 693–696.
41. A. C. K. Au and K. P. Byrne "On the Insertion Loss Produced by Plane Acoustic Lagging Structures," *J. Acoust. Soc. Am.*, **82**(4), 1325–1333 (1987).
42. A. C. K. Au and K. P. Byrne, "On the Insertion Loss Produced by Acoustic Lagging Structures which Incorporate Flexurally Orthotropic Impervious Barriers," *Acustica*, **70**, 284–291 (1990).
43. K. Goesele and U. Goesele, "Influence of Cavity-Volume Damping on the Stiffness of Air Layer in Double Walls," *Acustica*, **38**(3), 159–166 (1977) (in German).
44. K. W. Goesele, W. Schuele, and B. Lakatos. *(Glasfuellung bei Isolierglasscheiben) Forschunggemeinschaft Bauen und Wohen* [*Gas Filling of Isolated Glass Windows*], FBW Blaetter, Vol. 4, 1982 Stuttgart.
45. H. Ertel and M. Moeser, *Effects of Gas Filling on the Sound Transmission Loss of Isolated Glass Windows in the Double Wall Resonance Frequency Range*, FBW Blaetter, Stuttgart, Vol. 5, 1984 (in German).
46. K. Goesele, "Prediction of the Sound Transmission Loss of Double Partitions (without Structureborne Connections)," *Acustica*, **45**, 218–227 (1980) (in German).
47. I. L. Vér, "Definition of and Relationship between Breakin and Breakin Sound Transmission Loss of Pipes and Ducts," *Proc. INTER-NOISE*, 1983, pp. 583–586.
48. L. L. Vér, "Prediction of Sound Transmission through Duct Walls: Breakout and Pickup," ASHRAE Paper 2851 (RP-319), *ASHRAE Trans.*, **90**(Pt. 2), 391–413 (1984).
49. Anonymous, "Sound and Vibration Control," in *ASHRAE Handbook, Heating, Ventilating, and Air-Conditioning Systems and Applications*, American Society of Heating, Refrigerating and Air Conditioning Engineers, Atlanta, GA, 1987, Chapter 52.
50. A. Cummings, "Acoustic Noise Transmission through Duct Walls," *ASHRAE Trans.*, **91**(Pt. 2A), 48–61 (1985).

51. M. C. Gomperts and T. Kihlman, "Transmission Loss of Apertures in Walls," *Acustica*, **18**, 140 (1967).
52. F. P. Mechel, "The Acoustic Sealing of Holes and Slits in Walls," *J. Sound Vib.*, **111**(2), 297–336 (1986).
53. F. P. Mechel, "Die Schalldaemmung von Schalldaempfer-Fugen" ["Acoustical Insulation of Silencer Joints"], *Acustica*, **62**, 177–193 (1987).
54. W. Fasold, W. Kraak, and W. Schirmer, *Pocketbook Acoustic*, Part 2, Section 6.2, VEB Verlag Technik, Berlin, 1984 (in German).
55. R. H. Lyon and G. Maidanik, "Power Flow between Linearly Coupled Oscillators," *J. Acoust. Soc. Am.*, **34**(5), 623–639 (1962).
56. E. Skudrzyk, "Vibrations of a System with a Finite or Infinite Number of Resonances," *J. Acoust. Soc. Am.*, **30**(12), 1114–1152 (1958).
57. E. E. Ungar, "Statistical Energy Analysis of Vibrating Systems," *J. Eng. Ind. Trans. ASME Ser. B*, **87**, 629–632 (1967).
58. M. J. Crocker and A. J. Price, "Sound Transmission Using Statistical Energy Analysis," *J. Sound Vib.*, **9**(3), 469–486 (1969).
59. R. H. Lyon, *Statistical Energy Analysis of Dynamical Systems; Theory and Application*, MIT Press, Cambridge, MA, 1975.
60. I. L. Ver, Equivalent Dynamic Force that Generates the same Vibration Response as a Sound Field, Proceedings of Symposium on International Automotive Technology, SIAT 2005-SAE Conference, Pune, India, 19–22 January 2005, pp 81–86.
61. Private communications. J. Barger of BBN LLC.
62. Lord Rayleigh, *Theory of Sound*, Vol. I, Dover, New York, 1945, pp. 104–110.
63. M. Heckl, "Some Applications of Reciprocity Principle in Acoustics," *Frequenz*, **18**, 299–304 (1964) (in German).
64. I. L. Vér, "Uses of Reciprocity in Acoustic Measurements and Diagnosis," *Proc. INTER-NOISE*, 1985, pp. 1311–1314.
65. I. L. Vér, *Reciprocity as a Prediction and Diagnostic Tool in Reducing Transmission of Structureborne and Airborne Noise into an Aircraft Fuselage*, Vol. 1: *Proof of Feasibility*, BBN Report No. 4985 (April 1982); Vol. 2: *Feasibility of Predicting Interior Noise Due to Multiple, Correlated Force Input*, BBN Report No. 6259 (May 1986), NASA Contract No. NAS1-16521, Bolt Beranek and Newman.
66. I. L. Vér, "Use of Reciprocity and Superposition in Predicting Power Input to Structures Excited by Complex Sources," *Proc. INTER-NOISE*, 1989, pp. 543–546.
67. F. J. Fahy, "The Reciprocity Principle and Its Applications in Vibro-Acoustics," *Proc. Inst. Acoust. (UK)*, **12**(1), 1–20 (1990).
68. L. M. Lyamshev, "On Certain Integral Relations in Acoustics of a Moving Medium," *Dokl. Akad. Nauk SSSR*, **138**, 575–578 (1961); *Sov. Phys. Dokl.*, **6**, 410 (1961).
69. Yu I. Belousov and A. V. Rimskii-Korsakov, "The Reciprocity Principle in Acoustics and Its Application to the Calculation of Sound Fields of Bodies," *Sov. Phys. Acoust.*, **21**(2), 103–109 (1975).
70. L. Cremer, "The Law of Mutual Energies and Its Application to Wave-Theory of Room Acoustics," *Acustica*, **52**(2), 51–67 (1982/83) (in German).
71. M. Heckl, "Application of the Theory of Mutual Energies," *Acustica*, **58**, 111–117 (1985) (in German).

72. J. M. Mason and F. J. Fahy, "Development of a Reciprocity Technique for the Prediction of Propeller Noise Transmission through Aircraft Fuselages," *Noise Control Eng. J.*, **34**, 43–52 (1990).
73. P. W. Smith, "Response and Radiation of Structures Excited by Sound," *J. Acoust. Soc. Am.*, **34**, 640–647 (1962).
74. H. F. Steenhock and T. TenWolde, "The Reciprocal Measurement of Mechanical-Acoustical Transfer Functions," *Acustica*, **23**, 301–305 (1970).
75. T. TenWolde, "On the Validity and Application of Reciprocity in Acoustical, Mechano-Acoustical and other Dynamical Systems," *Acustica*, **28**, 23–32 (1973).
76. E. J. Richards, M. E. Westcott, and R. K. Jeyapalan, "On the Prediction of Impact Noise, I: Acceleration Noise," *J. Sound Vib.*, **62**, 547–575 (1979).
77. E. J. Richards, M. E. Westcott, and R. K. Jeyapalan, "On the Prediction of Impact Noise, II: Ringing Noise," *J. Sound Vib.*, **65**, 419–451 (1979).
78. E. J. Richards, "On the Prediction of Impact Noise, III: Energy Accountancy in Industrial Machines," *J. Sound Vib.*, **76**, 187–232 (1981).
79. J. Cuschieri and E. J. Richards, "On the Prediction of Impact Noise, IV: Estimation of Noise Energy Radiated by Impact Excitation of a Structure," *J. Sound Vib.*, **86**, 319–342 (1982).
80. E. J. Richards, I. Carr, and M. E. Westcott, "On the Prediction of Impact Noise, Part V: The Noise from Drop Hammers," *J. Sound Vib.*, **88**, 333–367 (1983).
81. E. J. Richards, A. Lenzi, and J. Cuschieri, "On the Prediction of Impact Noise, VI: Distribution of Acceleration Noise with Frequency with Applications to Bottle Impacts," *J. Sound Vib.*, **90**, 59–80 (1983).
82. E. J. Richards and A. Lenzi, "On the Prediction of Impact Noise, VII: Structural Damping of Machinery," *J. Sound Vib.*, **97**, 549–586 (1984).
83. J. M. Cuschieri and E. J. Richards, "On the Prediction of Impact Noise, VIII: Diesel Engine Noise," *J. Sound Vib.*, **102**, 21–56 (1985).
84. E. J. Richards and G. J. Stimpson, "On the Prediction of Impact Noise, IX: The Noise from Punch Presses," *J. Sound Vib.*, **103**, 43–81 (1985).
85. E. J. Richards and I. Carr, "On the Prediction of Impact Noise, X: The Design and Testing of a Quietened Drop Hammer," *J. Sound Vib.*, **104**(1), 137–164 (1986).
86. ASTM E492-90, "Standard Test Method for Laboratory Measurement of Impact Sound Transmission through Floor-Ceiling Assemblies Using the Tapping Machine," American Society for Testing and Materials, Philadelphia, PA, 1996.
87. I. L. Vér, "Impact Noise Isolation of Composite Floors," *J. Acoust. Soc. Am.*, **50**(4, Pt. 1), 1043–1050 (1971).
88. K. Gösele, "Die Bestimmung der Dynamischen Steifigkeit von Trittschall-Dämmstoffen," Boden, Wand und Decke, Heft 4 and 5, Willy Schleunung, Markt Heidenfeld, 1960.
89. I. L. Vér and D. H. Sturz "Structureborne Sound Isolation," in C. M. Harris (ed.), *Handbook of Acoustical Measurements and Noise Control*, 3rd ed., McGraw-Hill, New York, 1991, Chapter 32.
90. I. L. Vér, "Acoustical and Vibrational Performance of Floating Floors," Report No. 72, Bolt Beranek and Newman, Cambridge, MA, October 1969.

91. R. Josse and C. Drouin, "Étude des impacts lourds a l'interieur des bâtiments d'habitation," Rapport de fin d'étude, Centre Scientifique et Technique du Bâtiment, Paris, February 1, 1969, DS No. 1, 1.24.69.
92. I. L. Vér, "Relationship between Normalized Impact Sound Level and Sound Transmission Loss," *J. Acoust. Soc. Am.*, **50**(6, Pt. 1), 1414–1417 (1971).
93. L. Cremer, "Theorie des Kolpfschalles bei Decken mit Schwimmenden Estrich," *Acustica*, **2**(4), 167–178 (1952).

BIBLIOGRAPHY

Beranek, L. L., *Acoustical Measurements*, rev. ed., Acoustical Society of America, Woodbury, NY, 1988.

Bies, D. A., and C. H. Hansen, *Engineering Noise Control*, Unwin Hyman, London/Boston/Sydney, 1988.

Cremer, L., M. Heckl, and E. E. Ungar, *Structureborne Sound*, 2nd ed., Springer-Verlag, Berlin, Heidelberg, New York, London, Paris, Tokyo, 1988.

Fahy, F., *Sound and Structural Vibration*, Academic, London, 1989, revised, corrected paperback edition.

Fasold, W., W. Kraak, and W. Shirmer, *Taschenbuch Akustik* [*Pocket Book Acoustics*] Vols. 1 and 2, VEB Verlag Technik, Berlin, 1984.

Harris, C. M. (Ed.), *Handbook of Acoustical Measurements Noise and Control*, 3rd ed., McGraw-Hill, New York, 1991.

Junger, M. C., and D. Feit, *Sound Structures and Their Interaction*, 2nd ed., MIT Press, Cambridge, MA, 1986.

Reichardt, W., *Technik-Woerterbuch, Technische Akustik* [*Dictionary of Technical Acoustics*; English, German, French, Russian, Spanish, Polish, Hungarian, and Slowakian], VEB Verlag, Technik, Berlin, 1978.

Mechel, F. P. (Ed.), *Formulas of Acoustics*, Springer, Berlin, 2002.

Heckl, M., and H. A. Mueller, *Taschenbuch der Technischen Akustik* [*Pocketbook of Technical Acoustics*], 2nd ed., Springer, Berlin, 1997.

TABLE 11.1 Speeds of Sound in Solids, Deformation of Solids for Different Wave Types, and Formulas for Propagation Speed

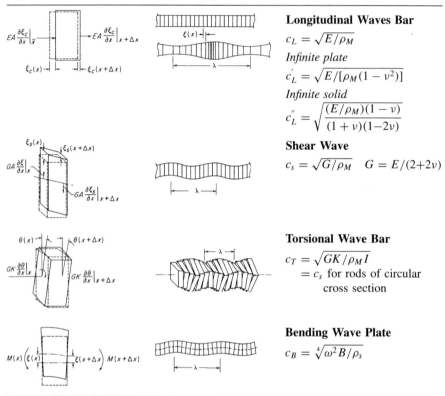

Longitudinal Waves Bar

$$c_L = \sqrt{E/\rho_M}$$

Infinite plate

$$c_L' = \sqrt{E/[\rho_M(1-\nu^2)]}$$

Infinite solid

$$c_L'' = \sqrt{\frac{(E/\rho_M)(1-\nu)}{(1+\nu)(1-2\nu)}}$$

Shear Wave

$$c_S = \sqrt{G/\rho_M} \quad G = E/(2+2\nu)$$

Torsional Wave Bar

$$c_T = \sqrt{GK/\rho_M I}$$
$$= c_S \text{ for rods of circular cross section}$$

Bending Wave Plate

$$c_B = \sqrt[4]{\omega^2 B/\rho_s}$$

Young's modulus E, N/m², relates the stress S (force per unit area) to the strain (change in length per unit length). *Poisson's ratio* ν is the ratio of the transverse expansion per unit length of a circular bar to its shortening per unit length, under a compressive stress, dimensionless. It equals about 0.3 for structural materials and nearly 0.5 for rubberlike materials. The *density* of the material is ρ_M, kg/m³. ρ_s is mass per unit area (kg/m²) for plates and mass per unit length (kg/m) for bars, rods, or beams. The *shear modulus* G is the ratio of shearing stress to shearing strain, N/m². I is the *polar moment* of inertia, m⁴. The *torsional stiffness factor* K relates a twist to the shearing strain produced, m⁴. The *bending stiffness* per unit width B equals $Eh^3/[12(1-\nu^2)]$ for a homogeneous plate, N·m, where h is the thickness of the bar (or plate) in the direction of bending, m. For rectangular rods $B = Eh^3w/12$, where h is the cross-sectional dimension in the plane of bending and w that perpendicular to it (and width), m.

TABLE 11.2 Key Acoustical Parameters of Solid Materials

Material	Density ρ_M, kg/m³	Young's Modulus E, N/m²	Poisson Ratio v	Speed of Sound c_L, m/s	Product of Surface Density and Critical Frequency $\rho_s f_c$		TL at Critical Frequency $R(f_c)$, dB	Internal Damping Factor for Bending at 1000 Hz, η^a	Acoustical–Mechanical Conversion Efficiency η_{am}, Eq. (11.66)
					Hz·kg/m²	Hz·lb/ft²			
Aluminum	2,700	7.16×10^{10}	0.34	5,150	34,700	7,000	48.5	10^{-4}–10^{-2b}	2.5×10^{-3}
Copper	8,900	1.3×10^{11}	0.35	3,800					3.3×10^{-4}
Glass	2,500	6.76×10^{10}		5,200	38,000	7,800	49.5	0.001–0.01^b	2.7×10^{-3}
Lead (chemical or tellurium)	11,000	1.58×10^{10}	0.43	1,200	605,000	124,000	73.5	0.015	1.4×10^{-4}
Plexiglas or Lucite	1,150	3.73×10^9	—^{c}	1,800	35,400	7,250	49.0	0.002^b	2×10^{-3}
Steel	7,700	1.96×10^{11}	0.31	5,050	97,500	20,000	57.5	10^{-4}–10^{-2b}	8.5×10^{-4}
Brick	1,900–2,300	—^{c}	—^{c}	—^{c}	34,700–58,600	7,000–12,000	48.5 to 53	0.01	—^{c}
Concrete, dense poured	2,300	2.61×10^{10}	—^{c}	3,400	43,000	9,000	50.5	0.005–0.02	1.9×10^{-3}
Concrete (Clinker) slab, plastered on both sides 5 cm thick	1,500	—		—	48,800	10,000	51.5	0.005–0.02	—^{c}
Masonry block	750	—		—	23,200	4,750	45.0	0.005–0.02	

(continued overleaf)

TABLE 11.2 (*continued*)

Material	Density ρ_M, kg/m³	Young's Modulus E, N/m²	Poisson Ratio ν	Speed of Sound c_L, m/s	Product of Surface Density and Critical Frequency $\rho_s f_c$		TL at Critical Frequency $R(f_c)$, dB	Internal Damping Factor for Bending at 1000 Hz, η^a	Acoustical–Mechanical Conversion Efficiency η_{am}, Eq. (11.66)
					Hz·kg/m²	Hz·lb/ft²			
Hollow cinder with 1.6 cm sand plaster, nominal thickness 15 cm (6 in.)	900	—	—	—	25,500	5,220	46.0	0.005–0.02	—[c]
Hollow dense concrete, nominal 15 cm (6 in.) thick	1100	—	—	—	23,000	4,720	45.0	0.007–0.02	—[c]
Hollow dense concrete, sand-filled voids, nominal 15 cm (6 in.) thick	1,700	—[c]	—[c]	—[c]	42,200	8,650	50.0	Varies with frequency	—[c]
Solid dense concrete, nominal 10 cm (4 in.) thick	1,700	—[c]	—[c]	—[c]	54,100	11,100	52.5	0.012	—[c]

Material								
Gypsum board 1.25–5 cm ($\frac{1}{2}$–2 in.) thick	650	—[c]	6,800	20,000	4,500	45.0	0.01–0.03	—[c]
Plaster, solid, on metal or gypsum lathe	1,700	—[c]	—[c]	24,500	5,000	45.5	0.005–0.01	—[c]
Fir timber	550	—[c]	3,800	4,880	1,000	31.5	0.04	9×10^{-3}
Plywood 0.6–3.12 cm ($\frac{1}{2}$–2 in.) thick	600	—[c]	—[c]	12,700	2,600	40	0.01–0.04	—[c]
Wood waste material bonded with plastic 23 kg/m² (5 lb/ft²)	750	—[c]	—[c]	73,200	15,000	55.0	0.005–0.01	—[c]

[a] The range in values of η are based on limited data. The lower values are typical for material alone while the higher values are the maximum observed on panels in place.

[b] The loss factors for structures of these materials are sensitive to construction techniques and edge conditions.

[c] The parameter either is not meaningful or is not available.

TABLE 11.3 Driving Force Point Impedance of Infinite Structures

Element	Picture	Equivalent Mass [see Eq. (11.19)]	Driving Point Force Impedance	Range of Validity	Auxiliary Expressions and Notes
Beam, in compression					
Semi-infinite		$\lambda_L/2\pi$	$Z_F = \rho_M c_L S$	$S < (\lambda_s/4)^2$	$c_L = \sqrt{\dfrac{E}{\rho_M}}$
Infinite		λ_L/π	$Z_F = 2\rho_M c_L S$		S = cross-sectional area $\lambda_s = \dfrac{1}{f}\sqrt{\dfrac{E}{2(1+\nu)\rho_M}}$
Beam in bending					r = radius of contact area
Semi-infinite		$\dfrac{\sqrt{2}}{2\pi}\lambda_B$	$Z_F = \tfrac{1}{2}(1+j)\rho_M S c_B$	$S < (\lambda_B/6)^2$ $2r > 9S/\lambda_s$	$c_B = \left(\dfrac{EI\omega^2}{\rho_M S}\right)^{1/4}$
Infinite		$\dfrac{\sqrt{2}}{\pi}\lambda_B$	$Z_F = 2(1+j)\rho_M S c_B$		ν = Poisson Ratio $\lambda_B = c_B/f$ I = polar moment of inertia

Thin infinite plate Vertical force Horizontal force (in plane)		$Z_F = 8\sqrt{B'\rho_M c_L h}$ $= 2.3 \rho_M c_L h^2$ $\dfrac{1}{Z_F} = \dfrac{\pi f}{4Gh}$ $\times \left(\dfrac{3-\mu}{2} + jH \right)$ $\operatorname{Re}\{1/Z_F\} = [n_L(f) + n_s(f)]/8M$	$h < \lambda_B/6$ $2r > 3h$ $\lambda_s \gg 2\pi r > 10h$
Thin semi-infinite plate	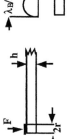 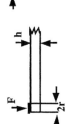	$Z_F = 3.5\sqrt{B'\rho_M h}$ $\approx \rho_M c_L h^2$	$h < \lambda_B/6$ $2r > 3h$

$B' = \left(\dfrac{Eh^3}{12(1-\nu^2)} \right)$

$\lambda_s = \dfrac{1}{f}\sqrt{\dfrac{E}{2(1+\nu)\rho_M}}$

$H = K + L$

$K = (1-\mu)\ln\left(\dfrac{\lambda_l}{\pi r}\right)$

$L = 2\ln\left(\dfrac{\lambda_s}{\pi r}\right)$

$n_L(f)$ = longitudinal modal density
$n_L(f)$ = shear modal density
M = total mass
$\lambda_l = c_L/f$
$\lambda_s = c_s/f$
$\lambda_B = c_B/f$

(continued overleaf)

TABLE 11.3 (*continued*)

Element	Picture	Equivalent Mass [see Eq. (11.19)]	Driving Point Force Impedance	Range of Validity	Auxiliary Expressions and Notes
Beam-stiffened infinite thin plate		**High Frequency Approximation**	$Z_F = \dfrac{(1-j)k'}{4\rho'_s \omega}$ $Z_F \cong Z_{FB}$ $= 2(1+j)\rho_M S_b c_B$	$S_b < (\lambda_B/6)^2$ $2r > 9S_b/\lambda_B$	$\rho'_s = \rho_M S_B + 2\rho_M h/k_p$ $k' = \left(\dfrac{\rho'_s}{B}\right)^{1/4} \omega^{1/2} A$ $A = 1 - j\dfrac{\rho_M h}{2\rho'_s k_p}$ k_p = plate-bending wavenumber $B = Eh^3b/3$ $s = k_b/k_p$
Infinite string			$Z_F = 2\sqrt{\rho_l F_T}$		$c_s = \sqrt{F_T/\rho_l}$ ρ_l = mass per unit length F_T = tension force

Semi-infinite plate, in-plane edge force

$$\frac{1}{Z_F} = -\frac{j\omega^2}{\pi E h} \log a + \frac{\omega}{4}\left(\frac{1}{D} + \frac{1}{S}\right)$$

$D = \dfrac{Eh(1-v)}{(1-v)(1-2v)}$

$S = Gh$

G = shear modulus

Infinite corrugated plate

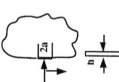

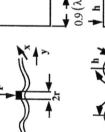

$Z_F = 8[(\rho_M h)^2 \times B_x B_y]^{1/4}$

$S \ll \lambda_B$

$B_y = \dfrac{Eh^3}{12(1-v^2)}$

$B_x \cong \left(\dfrac{1-v}{1+v}\right)^2 \dfrac{S}{S'} B_y$

$f_{c1} = \dfrac{c_0^2}{2\pi}\sqrt{\dfrac{\rho_M h}{B_x}}$

$f_{c2} = f_{c1}\sqrt{\dfrac{B_x}{B_y}}$

(*continued overleaf*)

TABLE 11.3 (*continued*)

Element	Picture	Equivalent Mass [see Eq. (11.19)]	Driving Point Force Impedance	Range of Validity	Auxiliary Expressions and Notes
Elastic half space			$Z = \dfrac{-j0.64Gr}{f(1-\nu/2)}$ $+ 1.79r^2\sqrt{\dfrac{4G\rho_M}{1-\nu}}$	$2r < \lambda_s/6$	ρ_M = density G = shear modulus ν = Poisson's ratio
Infinite cylindrical shell		—	$Z_\infty \cong 2.3\rho_M c_L h$ $\operatorname{Re}\left(\dfrac{1}{Z_\infty}\right) = \dfrac{\sqrt{12}}{8\rho_M c_L h^2}$ $\times \left(\dfrac{f}{f_R}\right)^{2/3}$	$f > 1.5 f_R$ $f < 0.7 f_R$	$f_R = c_L/\pi D$ ring frequency

TABLE 11.4 Moment Impedance Z_M of Infinite and Semi-Infinite Structures

Element	Picture	Driving Point Moment Impedance
Semi-infinite beam Free end		$\dfrac{(1-j)\rho_l c_B^3(f)}{8\pi^2 f^2}$
Pinned end		$\dfrac{(1-j)\rho_l c_B^3(f)}{4\pi^2 f^2}$
Infinite beam		$\dfrac{(1-j)\rho_l c_B^3(f)}{2\pi^2 f^2}$
Infinite homogeneous isotropic plate		$\dfrac{16\rho_M h c_L^2 k_B^{-2}(f)}{2\pi f[(1-j)1.27\,\ln(k_B a/2.2)]}$
Semi-infinite homogeneous isotropic plate		$\dfrac{12\rho_M h c_L^2 k_B^{-2}(f)}{2\pi f[(1-j)3.35\ln(kr/3.5)]}$
At joint of homogeneous, isotropic plates		$\dfrac{\rho_M c_L^2 h^3}{75.4 f}\left\{16\left[\dfrac{(1+j)1.27\ln(kb/2.2)}{1+(1.27\ln kb/2.2)^2}\right]\right.$ $\left.+12\left[\dfrac{(1+j)3.35\ln(kr/3.5)}{1+(3.35\ln kr/3.5)^2}\right]\right\}$

Auxiliary expressions and notes:

ρ_l = mass per unit length (kg/m)

$c_B(f)$ = bending wave speed (m/s) in bending, $=\sqrt{2\pi f}(EI/\rho_l)^{1/4}$

E = Young's modulus (N/m^2)

I = area moment of inertia in bending (m^4)

f = frequency

ρ_M = material density (kg/m^3)

$k_B(f) = 2\pi f/c_B(f)$ bending wavenumber

h = plate thickness

TABLE 11.5 Effective Length l_M Connecting Force and Moment Impedance, $Z_M = l_M^2 Z_f$ for Infinite Beams and Plates

Element	Picture	Equivalent Force Pair	Effective Length l_M	Auxiliary Expressions
Beam in bending			$\dfrac{\lambda_B}{2\pi}/\sqrt{j}$	$j = \sqrt{-1}$
Beam in torsion			$0.79i$	For hollow beams: $i = \sqrt{\dfrac{d_o^2 - d_i^2}{2}}$ d_o = outside diameter d_i = inside diameter
Plate, vertical moment			$\dfrac{\lambda_B/\sqrt{8\pi j}}{\sqrt{\ln(\lambda_B/\pi r)}}$	Bending waves: $3h < 2r \ll \lambda_B/\pi$
Plate, in-plane torsional moment			$\dfrac{r}{\sqrt{\ln(\lambda_s/\pi r)}}$	Shear waves: $3h < 2r \ll \lambda_s/\pi$ λ_s = shear wavelength

Source: After reference 6.

TABLE 11.6 Power Input to Structures

Element	Picture	Power Input to Infinite Element		Finite Elements		Auxiliary Expressions
		Force or Moment Excitation	Velocity or Angular Velocity Excitation	Onset of Infinite Behavior	$W_{\text{fin}}/W_{\text{inf}}$	
Beam in longitudinal wave motion; force or velocity excitation	$\hat{F},\hat{v}$ S	$\dfrac{|\hat{F}|^2}{4\rho_M S c_L}$	$4|\hat{v}|^2 S \rho_M c_L$	$\omega > \dfrac{\pi c_L}{\eta l}$	$\dfrac{4}{\pi \eta}$	$c_L = \sqrt{E/\rho_M}$ l = length η = loss factor Q = torsion constant G = shear modulus J = mass moment of inertia per unit length
Beam in torsion moment or angular velocity excitation	$\hat{M},\hat{\theta}$ S	$\dfrac{|\hat{M}|^2}{4GQJ}$	$4|\hat{\theta}|^2 \sqrt{GQJ}$	$\omega > \dfrac{\pi c_T}{\eta l}$	$\dfrac{4}{\pi \eta}$	$c_T = \sqrt{\dfrac{E/\rho_M}{GQ/J}}$ = torsional wave speed ρ_M = density E = Young's modulus I = second moment of inertia
Beam in bending; force or velocity excitation	$\hat{F},\hat{v}$ S	$\dfrac{|\hat{F}|^2}{8 \rho_M S c_B(f)}$	$|\hat{v}|^2 S \rho_M c_B(f)$	$\omega > \dfrac{4\pi c_B(f)}{\eta l}$	$\dfrac{4\sqrt{2}}{\pi \eta}$	$\dot\theta$ = angular velocity $c_B = \sqrt{\omega}\sqrt{\dfrac{E/I}{\rho_M S}}$
Beam in bending; moment or angular velocity excitation	$\hat{M},\hat{\theta}$ S	$\dfrac{|\hat{M}|^2 c_B(f)}{8EI}$	$\dfrac{|\hat{\theta}|^2 EI}{c_B(f)}$	$\omega > \dfrac{4\pi c_B(f)}{\eta l}$	$\dfrac{2\sqrt{2}}{\pi \eta}$	= bending wave speed $B_p = \dfrac{h^3 E}{12(1-\nu^2)}$ S = area
Plate in bending; force or velocity excitation	$\hat{F},\hat{v}$ h	$\dfrac{|\hat{F}|^2}{16\sqrt{B_p \rho_M h}}$ $=\dfrac{|\hat{F}|^2}{4.6 \rho h^2 c_L}$	$4\hat{v}^2 \sqrt{B_p \rho_M h}$ $=1.15\hat{v}^2 \rho_M h^2 c_L$	$\omega > \dfrac{8}{\eta l_1 l_2}\sqrt{\dfrac{B_p}{\rho_M h}}$	$\dfrac{32 l_1 l_2}{\pi^2 \eta (l_1^2 + l_2^2)}\dfrac{\omega}{c_B}$	

(continued overleaf)

TABLE 11.6 (continued)

Element	Picture	Power Input to Infinite Element — Force or Moment Excitation	Power Input to Infinite Element — Velocity or Angular Velocity Excitation	Finite Elements — Onset of Infinite Behavior	Finite Elements — $W_{\text{fin}}/W_{\text{inf}}$	Auxiliary Expressions
Plate in bending; moment or angular velocity excitation	$\widehat{M}, \hat{\theta}$; h; $2r$	$\sim \dfrac{\omega \lvert \hat{M} \rvert^2}{16 B_p}$ for $r > h$	$\dfrac{4\lvert\hat{\theta}\rvert^2 B_p}{\omega\left\{1+\left[\dfrac{4}{\pi}\ln\left(\dfrac{\omega r}{c_B}\right) - \dfrac{8}{\pi(1-\nu)}\left(\dfrac{h}{\pi r}\right)^2\right]^2\right\}}$	$\omega > \dfrac{8}{\eta l_1 l_2}\sqrt{\dfrac{B_p}{\rho_M h}}$		
Thin-walled pipe in bending; force excitation	$\hat{F}$; h; $2r$	$\hat{F}^2/(16\pi\rho_M r h\sqrt{c_L r \omega})$ for $f < 0.123 c_L h/r$ $\hat{F}^2\sqrt{V/(2+V)}/(\omega\rho_s 2\lambda^2/\pi^2)$ for $f > 0.123 c_L h/r$				$V = \omega r / c_L$ λ_p = bending wavelength in equivalent thickness plate $\rho_s = \rho_M h$
Plate in bending; multiple force excitation, equally spaced	$\hat{F}_1 \ldots \hat{F}_n$; h; $2a$; $\hat{F}=\sum_{i=1}^{n}\hat{F}_i$	$\hat{F}^2[2J_1(z)/z]^2/(16\sqrt{B_p \rho_M h})$				$Z = 2\pi a/\lambda_B$ J_1 = Bessel function of order 1

TABLE 11.6 (continued)

Element	Picture	Power Input to Infinite Element		Finite Elements		Auxiliary Expressions
		Force or Moment Excitation	Velocity or Angular Velocity Excitation	Onset of Infinite Behavior	$W_\text{finite}/W_\text{inf}$	
Plate in bending; large area velocity excitation		$\hat{v}^2 \dfrac{\pi}{2} \rho_M c_B (r + 0.8\lambda_B) h$				
Elastic half space; single force		$\dfrac{48\hat{F}^2}{\omega \rho_M \pi \lambda_s^3}$				$\lambda_s = \sqrt{G/\rho_M}/f$ shear wavelength
Elastic half space equal multiple forces along a line, equally spaced		$\dfrac{16\hat{F}^2}{\omega \rho_M l \lambda_s^2}, l > \lambda_s/2$				

(continued overleaf)

TABLE 11.6 (continued)

Element	Picture	Power Input to Infinite Element Nearby Monopole and Dipole Excitation	Auxiliary Expressions
Infinite plate excited by nearby:			
Monopole		$W_{\text{mon}} \cong \dfrac{Q^2(\rho_o c_0)^2 \, \text{Re}\{Y_\infty\}}{(f_c/f) - 1}$ $\times \exp\left(\dfrac{4\pi f}{c_0} y \sqrt{f_c/f - 1}\right), f < f_c$ $y \ll \lambda_0/4$ light fluid loading	Q = rms volume velocity Y_∞ = point force admittance k_B = free plate bending wavenumber f_c = critical coincidence frequency
Lateral dipole		$W_{\text{LD}} \cong 0.5(k_B d)^2 \, W_{\text{mon}}, \, f < f_c$	
Perpendicular dipole		$W_{\text{PD}} \cong W_{\text{mon}}(k_B d)^2 [1 - (f/f_c)]^2$ (for $f < f_c$ and light fluid loading)	

TABLE 11.7 First Resonance Frequency, Mode Shape and Modal Density of Finite Structures

Element	Picture	Boundary Conditions[a]	First Resonance Frequency	Mode Shape $\phi(x, y, z)$	Modal Density[b] $n(\omega)$	Auxiliary Formulas
Beam in compression		f–f c–c	$c_L/2l$	$\cos(n\pi x/l)$ $\sin(n\pi x/l)$	$l/\pi c_L$	$\kappa = \sqrt{I/S}$ radius of gyration
Beam in bending		p–p f–f c–c c–f	$(\pi/2)(\kappa c_L/l^2)$ $(1/2\pi)(4.73/l)^2 \kappa c_L$ $(1/2\pi)(4.73/l)^2 \kappa c_L$ $(1/2\pi)(1.875/l)^2 \kappa c_L$	$\sqrt{2}\sin(k_n X)$ See ref. 19	$\dfrac{l}{2\pi}\dfrac{1}{\sqrt{\omega\kappa c_L}}$	$k_n = \sqrt{2\pi f_n/c_L\kappa}$ a/b = aspect ratio
Rectangular plate in bending		ffff ssss cccc	$\begin{array}{c|cccc} a/b & 1 & 1.5 & 2.5 \\ \hline & 3.33 & 3.31 & 2.13 \\ C_1 & 4.88 & 5.28 & 7.1 \\ & 8.89 & 10.0 & 14.6 \end{array}$ $f_1 = 10^3 C_1 (h/S)(c_L/c_{L\text{st}})$	See ref. 19	$\dfrac{\sqrt{12}S}{4\pi c_L h}$	h = plate thickness, m S = area, m² $c_{L\text{st}}$ = 5050 m/s c_L = longitudinal wave speed

(continued overleaf)

TABLE 11.7 (continued)

Element	Picture	Boundary Conditions[a]	First Resonance Frequency	Mode Shape $\phi(x, y, z)$	Modal Density[b] $n(\omega)$	Auxiliary Formulas
Membrane			See ref. 18	See ref. 19	$\dfrac{S}{2\pi c_m}$	F' = tension per unit length; ρ_s = mass per unit area; $c_m = \sqrt{F'/\rho_s}$; $c_s = \sqrt{F'/\rho_l}$
String		c–c	$\pi c_s / l$	$\sin(n\pi x / l)$	$l/\pi c_s$	F = tension force; ρ_l = mass per unit length
Rectangular air volume		Hard walls	$c_0/2l_{\max}$	$\cos\left(\dfrac{n_x \pi x}{l_x}\right) \cos\left(\dfrac{n_y \pi y}{l_y}\right) \cos\left(\dfrac{n_z \pi z}{l_z}\right)$ $n(\omega) = \dfrac{\omega^2 V}{2\pi^2 c_0^3} + \dfrac{S\omega}{4\pi c_0} + \dfrac{L}{16\pi c_0}$		$V = l_x l_y l_z$; c_0 = speed of sound; $n = 1, 2, 3, \ldots$

[a] f = free; s = simply supported; c = clamped; p = pinned.
[b] $n(\omega) = n(f)/2\pi$, $S = 2(l_x l_y + l_x l_z + l_y l_z)$, $L = 4(l_x + l_y + l_z)$.

Source: After reference 19.

TABLE 11.8 Radiation Efficiency of Vibrating Bodies

Body	Picture	σ_{rad}	Auxiliary Expressions
Small pulsating body		$\dfrac{(ka)^2}{1+(ka)^2}$	c_0 = speed of sound $k_0 = 2\pi f/c_0$ a = source radius
Small oscillating rigid body		$\dfrac{(ka)^4}{4+(ka)^4}$, see Fig. 11.14	
Pulsating pipe		$2/\pi k_0 a |H_1(k_0 a)|^2$ for $(\pi/2)k_0 a \leq 2/\pi$	H_1 = Hankel function, second kind, order 1 $k_0 = 2\pi f/c_0$
Oscillating pipe or rod		$2/\pi k_0 a |H_1'(k_0 a)|^2$, see Fig. 11.15	H_1' = first derivative of H_1 in respect of its argument

(*continued overleaf*)

TABLE 11.8 (continued)

Body	Picture	σ_{rad}	Auxiliary Expressions
Circular pipes in bending		zero; $f < f_c$ $(k_0 a)^3 [1 - (f_c/f)]$; $f > f_c$; $k_d a \ll 1$ 1; $f > f_c$; $k_0 a \gg 1.5$	$k_d^2 = k_0^2 - k_B^2$ $k_B = 2\pi/\lambda_B$ f_c = critical frequency where $k_B = k_0$
Rectangular and elliptic beams in bending		See ref. 22	
Infinite thin plate supporting free bending waves		0 for $f < f_c$ $1/[1 - (f_c/f)]^{1/2}$ for $f > f_c$	
Finite thin plate supporting free bending waves; plate surrounded by rigid baffle		$\dfrac{P c_0}{\pi S f_c} \sqrt{f/f_c}$; $f < f_c$ $0.45 (P/\lambda_c)^{1/2} (L_{\min}/L_{\max})^{1/4}$; $f = f_c$ $(1 - f_c/f)^{-1/2}$; $f > 1.3 f_c$ 1; $f \geq 1.3 f_c$	f_c = critical frequency. See Eq. 11.94 $\lambda_c = c_0/f_c$ $S = L_{\max} L_{\min}$ = area (one side) $P = 2(L_{\max} + L_{\min})$ = perimeter $\beta = (f/f_c)^{1/2}$

TABLE 11.8 (continued)

Body	Picture	σ_{rad}	Auxiliary Expressions
Thick finite plate supporting free bending waves		$g_1(\beta) = \begin{cases} (4/\pi^4)[(1-2\beta^2)/\beta(1-\beta^2)^{1/2}]) & ; f < 0.5 f_c \\ 0 & ; f > 0.5 f_c \end{cases}$ $g_2(\beta) = \left(\dfrac{1}{4\pi^2}\right) \dfrac{(1-\beta^2)\ln[(1+\beta)/(1-\beta)] + 2\beta}{(1-\beta^2)^{3/2}}$; $C_1 = \begin{cases} 1 \text{ for simple supported edges} \\ \beta^2 \exp(10\lambda_c/P) \text{ for clamped edges} \end{cases}$ $\sigma_{\text{rad}} = \dfrac{P}{S}\dfrac{c_0}{\pi^2}\sqrt{\dfrac{f}{f_c^3}}$ $\begin{array}{l} 0.45\sqrt{P/\lambda_0} \quad \text{for } f \leq f_b \\ 1 \qquad\qquad \text{for } f \gg f_b \end{array}$	$f_b = f_c + \dfrac{5c_0}{P}$ $P = \text{perimeter}$
Infinite plate sound-forced waves		$\sigma_F = 1/\cos\phi$	$\phi = \text{incidence angle, degrees}$
Finite square-plate oblique-incidence plane sound wave excitation		$\sigma_F = \min\begin{cases} A[(k_o/2)\sqrt{S}] \\ 1/\cos\phi \end{cases}$ for $0.1\lambda_0^2 < S < 0.4\lambda_0^2$ $\sigma_F = \min\begin{cases} [(0.5)^{(\phi/90)}\sqrt{k_0/2\sqrt{S}}] \\ 1/\cos\phi \end{cases}$ for $S > 0.4\lambda_0^2$	$A = (0.5)(0.8)^{(\phi/90)}$ $\alpha = 1 - 0.34\phi/90$ $k_o = 2\pi/\lambda_0 = 2\pi f/c_0$
Finite square-plate diffuse-sound-field excitation		$\sigma_F = 0.5[0.2 + \ln(k_0\sqrt{S})]$ for $k_0\sqrt{S} > 1$	

Source: After references 1, 23, and 28.

TABLE 11.9 Reciprocity Relationships for Higher Order Excitation Sources and Responses

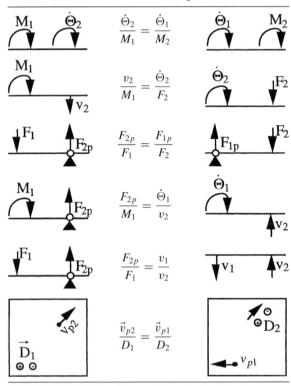

Note: M = moment, v = velocity, v_p = particle velocity, $\dot{\Theta}$ = angular velocity, F = force, F_p = force acting at pinned restraints, D = dipole sound source.
Source: After reference 63.

TABLE 11.10 USS Gauges and Weights of Steel Plates

Gauge	Steel USS Gauge Rev. Thickness in.	mm	Surface Weight lb/ft²	kg/m²	Galvanized Steel USS Gauge Thickness in.	mm	Surface Weight lb/ft²	kg/m²	Stainless Chrome Alloy USS Gauge Thickness in.	mm	Surface Weight lb/ft²	kg/m²	Stainless Chrome Nickel USS Gauge Surface Weight lb/ft²	kg/m²	Monel USS Gauge Thickness in.	mm	Surface Weight lb/ft²	kg/m²
32	0.01	0.254			0.013	0.330	0.563	2.75	0.01	0.254	0.418	2.04	0.427	2.08				
31	0.011	0.279			0.014	0.356	0.594	2.90	0.0109	0.277	0.45	2.20	0.459	2.24				
30	0.012	0.305	0.5	2.44	0.0157	0.399	0.656	3.20	0.0125	0.318	0.515	2.51	0.525	2.56				
29	0.0135	0.343	0.563	2.75	0.0172	0.437	0.719	3.51	0.014	0.356	0.579	2.83	0.591	2.89				
28	0.0149	0.378	0.625	3.05	0.187	4.750	0.781	3.81	0.0156	0.396	0.643	3.14	0.656	3.20				
27	0.0164	0.417	0.688	3.36	0.0202	0.513	0.844	4.12	0.0171	0.434	0.708	3.46	0.721	3.52				
26	0.0179	0.455	0.75	3.66	0.0217	0.551	0.906	4.42	0.0187	0.475	0.772	3.77	0.787	3.84	0.0187	0.475	0.827	4.04
25	0.0209	0.531	0.875	4.27	0.0247	0.627	1.031	5.03	0.0218	0.554	0.901	4.40	0.918	4.48	0.0218	0.554	0.965	4.71
24	0.0239	0.607	1	4.88	0.0276	0.701	1.156	5.64	0.025	0.635	1.03	5.03	1.05	5.13	0.025	0.635	1.148	5.60
23	0.0269	0.683	1.125	5.49	0.0306	0.777	1.281	6.25	0.0281	0.714	1.158	5.65	1.181	5.77	0.0281	0.714	1.1286	5.51
22	0.0299	0.759	1.25	6.10	0.0336	0.853	1.406	6.86	0.0312	0.792	1.287	6.28	1.312	6.41	0.0312	0.792	1.424	6.95
21	0.0329	0.836	1.375	6.71	0.0366	0.930	1.531	7.47	0.0343	0.871	1.416	6.91	1.443	7.04	0.0343	0.871	1.562	7.63
20	0.0359	0.912	1.5	7.32	0.0396	1.006	1.656	8.08	0.0375	0.953	1.545	7.54	1.575	7.69	0.0375	0.953	1.7	8.30
19	0.0418	1.062	1.75	8.54	0.0456	1.158	1.906	9.31	0.0437	1.110	1.802	8.80	1.837	8.97	0.0437	1.110	1.975	9.64
18	0.0478	1.214	2	9.76	0.0516	1.311	2.156	10.53	0.05	1.270	2.06	10.06	2.1	10.25	0.05	1.270	2.297	11.21
17	0.0538	1.367	2.25	10.98	0.0575	1.461	2.406	11.75	0.0562	1.427	2.317	11.31	2.362	11.53	0.0562	1.427	2.572	12.56

(*continued overleaf*)

TABLE 11.10 (*continued*)

Gauge	Steel USS Gauge Rev.				Galvanized Steel USS Gauge						Stainless Chrome Alloy USS Gauge						Stainless Chrome Nickel USS Gauge				Monel USS Gauge			
	Thickness		Surface Weight		Thickness		Surface Weight				Thickness		Surface Weight				Surface Weight				Thickness		Surface Weight	
	in.	mm	lb/ft²	kg/m²	in.	mm	lb/ft²	kg/m²			in.	mm	lb/ft²	kg/m²			lb/ft²	kg/m²			in.	mm	lb/ft²	kg/m²
16	0.0598	1.519	2.5	12.21	0.0635	1.613	2.656	12.97			0.0625	1.588	2.575	12.57			2.625	12.82			0.0625	1.588	2.848	13.90
15	0.0673	1.709	2.812	13.73	0.071	1.803	2.969	14.49			0.0703	1.786	2.896	14.14			2.953	14.42			0.0703	1.786	3.216	15.70
14	0.0747	1.897	3.125	15.26	0.0785	1.994	3.281	16.02			0.0781	1.984	3.218	15.71			3.281	16.02			0.0781	1.984	3.583	17.49
13	0.0897	2.278	3.75	18.31	0.0934	2.372	3.906	19.07			0.0937	2.380	3.862	18.85			3.937	19.22			0.0937	2.380	4.272	20.86
12	0.1046	2.657	4.375	21.36	0.1084	2.753	4.531	22.12			0.1093	2.776	4.506	22.00			4.593	22.42			0.1093	2.776	5.007	24.44
11	0.1196	3.038	5	24.41	0.1233	3.132	5.156	25.17			0.125	3.175	5.15	25.14			5.25	25.63			0.125	3.175	5.742	28.03
10	0.1345	3.416	5.625	27.46	0.1382	3.510	5.781	28.22			0.1406	3.571	5.793	28.28			5.906	28.83			0.1406	3.571	6.431	31.40
9	0.1497	3.802	6.25	30.51	0.1532	3.891	6.406	31.27			0.1562	3.967	6.437	31.43			6.562	32.04			0.1562	3.967	7.166	34.98
8	0.1644	4.176	6.875	33.56	0.1681	4.270	7.031	34.33			0.1718	4.364	7.081	34.57			7.218	35.24			0.1718	4.364	7.855	38.35
7	0.1793	4.554	7.5	36.62							0.1875	4.763	7.59	37.05			7.752	37.85			0.1875	4.763	8.59	41.94

TABLE 11.11 Percentage Open Area of Perforated Plates

Diameter Divided by Centers or side of square divided by centers D/C or S/C	Round Holes, Standard Staggered	Round Holes, Standard Staggered	Square Holes, Straight Or
.200	3.6	3.1	4.0
.225	4.6	4.0	5.1
.250	5.7	4.9	6.3
.275	6.9	5.9	7.6
.300	8.1	7.1	9.0
.325	9.6	8.3	10.6
.350	11.1	9.6	12.3
.375	12.8	11.0	14.1
.400	14.5	12.6	16.0
.425	16.4	14.2	18.1
.450	18.4	15.9	20.3
.475	20.5	17.7	22.6
.500	22.7	19.6	25.0
.525	25.0	21.6	27.6
.550	27.4	23.8	30.3
.575	30.0	26.0	33.1
.600	32.7	28.3	36.0
.625	35.4	30.7	39.1
.650	38.3	33.2	42.3
.675	41.3	35.8	45.6
.700	44.4	38.5	49.0
.725	47.7	41.3	52.6
.750	51.0	44.2	56.3
.775	54.4	47.2	60.0
.800	58.0	50.3	64.0
.825	61.7	53.5	68.0
.850	65.5	56.7	72.3
.875	69.5	60.1	76.6
.900	73.5	63.6	81.0
.925	77.6	67.2	85.6
.950	81.9	70.9	90.3

CHAPTER 12

Enclosures, Cabins, and Wrappings

ISTVÁN L. VÉR

Consultant in Acoustics, Noise, and Vibration Control
Stow, Massachusetts

This chapter deals with the acoustical design of enclosures, cabins, and wrappings. Acoustical enclosures are structures that house a noise source (usually a machine) for protection of the environment from the noise emitted by the source. Cabins are fully enveloping structures specifically designed for protecting a human being from environmental noise. The human being protected by the cabin may be the operator of a noisy machine, supervisor of a manufacturing operation, or attendant of a road toll booth. Wrappings are acoustical structures which closely envelop the casing of machines, valves, and connected piping to provide a high degree of noise reduction at high frequencies, practically no reduction at low frequencies, and modest reduction in the frequency range in-between.

The recently issued international standard ISO 15667:2000E,[1] entitled "Guidelines for Noise Control by Enclosures and Cabins," defines a large number of acoustical performance ratings and specifies the measurement procedure of how to obtain them in the laboratory or in situ. These performance measures, which include single-number and spectral and directivity information, are listed in Table 12.1 for enclosures and in Table 12.2 for cabins. For purposes of easy comparison, these tables retain the somewhat cumbersome nomenclature used in the standard.

The technical content of ISO 15667:2000E is based, to a large extent, on the content of Chapter 13 of the 1992 edition of this book[2] and, consequently, the reader will find much of the information in this chapter. The standard contains excellent suggestions about the collection of input information, planning, and performance verification of acoustical enclosures and cabins.

Annex A provides a large number of sketches depicting (1) details for joining wall panels, (2) mounting the enclosure airtight to the floor, (3) seals around doors and observation windows, (4) vibration isolation of the enclosed machines, (5) pipe and shaft penetrations, (6) ventilation possibilities, and so on. Though the information contained in these sketches is useful, the author cautions the purchaser of acoustical enclosures to contract with an experienced provider of noise control hardware who has similar proven details and the necessary experience to

TABLE 12.1 Acoustical Performance Measures of Enclosures

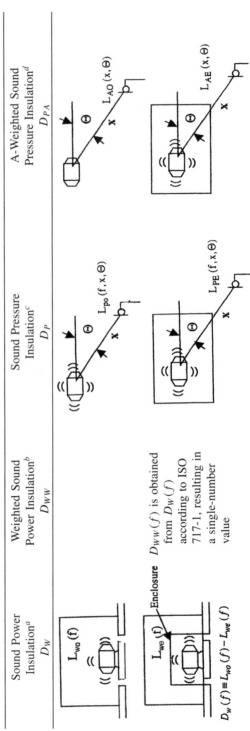

Sound Power Insulation[a] D_W	Weighted Sound Power Insulation[b] D_{WW}	Sound Pressure Insulation[c] D_P	A-Weighted Sound Pressure Insulation[d] D_{PA}

$D_W(f) \equiv L_{wo}(f) - L_{we}(f)$

$D_{WW}(f)$ is obtained from $D_W(f)$ according to ISO 717-1, resulting in a single-number value

L_{w0} = Sound power level of unenclosed source

L_{we} = Sound power level of enclosed source according to ISO 11546-1 or ISO 110546-2

$D_w(f) = R(f) + 10\log[\alpha(f)]$

$R(f)$ = sound transmission loss of enclosure wall according to ISO 140-3

$\alpha(f)$ = average absorption coefficient of internal side of enclosure panels

$D_p(f, x, \theta) = L_{po}(f, x, \theta) - L_{pe}(f, x, \theta)$

$L_{po}(f, x, \theta)$ = sound-pressure-level spectrum of unenclosed source at distance x, direction θ

$L_{pe}(f, x, \theta)$ = sound-pressure-level spectrum of enclosed source at distance x, direction θ

$D_{PA}(X, \theta) = L_{AO}(X, \theta) - L_{AE}(X, \theta)$

L_{AO} = dBA for unenclosed source

L_{AE} = dBA for enclosed source

[a] Contains no information about directivity. Limited usefulness if (1) source is highly directional and (2) source is much nearer to one wall than to others.
[b] Used for rough comparison of different enclosures in cases where source spectrum is not known.
[c] Best suited for detailed analyses of enclosure performance for different directions.
[d] Single-number rating for particular machine of known sound power spectrum.

Note: According ISO 15667:2000.

TABLE 12.2 Acoustical Performance Measures of Cabins

	Sound Pressure Insulation[a] D_P	A-Weighted Sound Pressure Insulation[b] D_{PA}	Apparent Sound Pressure Insulation[c] D'_P
Room	$\langle L_{P_o}(f) \rangle$	$L_{PA}(x)$	$L_{P_o}(f,x)$
Cabin	$\langle L_{P_c}(f) \rangle$	$L_{PAC}(x)$	$L_{pc}(f,x)$

$D_P \equiv \langle L_{P_o}(f) \rangle - \langle L_{P_c}(f) \rangle$

$\langle L_{P_o}(f) \rangle$ = space-average sound pressure level in a room with semidiffuse sound field

$\langle L_{P_c}(f) \rangle$ = space-average sound pressure level inside the cabin

$D_{PA} \equiv L_{PA}(x) - L_{PAC}(x)$

$L_{PA}(x)$ = A-weighted sound pressure level in room at position x, without cabin

$L_{PAC}(x)$ = A-weighted sound pressure level in cabin at the same position x

$D'_P \equiv L_{P_o}(f,x) - L_{pc}(f,x)$

$L_{P_o}(f,x)$ = sound pressure level at frequency f in position x, arbitrary sound field distribution

$L_{pc}(f,x)$ = Sound pressure level at frequency f in position x inside cabin

[a] $\langle L_{P_o}(f) \rangle$ assumed to be semidiffuse field.
[b] $L_{PA}(x)$ is the actual A-weighted level in the Room.
[c] L_{po} is produced by an arbitrary sound pressure distribution.
Note: According ISO 15667:2000.

integrate them into the total design and successfully install them onsite. Giving the construction details included in Annex A to a local sheet metal shop without prior experience in producing noise control hardware may result in a low bid but it is very likely that the end result will be disappointing.

In Annex B the standard provides informative case studies where the non-expert reader can find typical ranges of acoustical performance of functional enclosures and cabins which have operable doors, inlet and discharge openings for ventilation, and so on.

The key difference between enclosures and wrapping is that in the case of enclosures, the sound-absorbing layer is not in contact with the surface of the vibrating equipment, while in the case of wrappings, the porous sound-absorbing layer is in full surface contact with the vibrating body it surrounds. Because the porous sound-absorbing material of wrappings provides a full-surface, structure-borne connection between the vibrating equipment and the exterior layer, it must not only be a good sound absorber but also be highly resilient so as to prevent the transmission of vibration to the outer impervious layer where it can be radiated as sound. Sound-absorbing materials used in enclosures, where there is no contact between the porous material and the vibrating equipment, can have a fairly rigid skeleton. Wrappings are most frequently used to decrease the sound radiation of vibrating surfaces such as ducts and pipes and sometimes also to gain extra sound attenuation of acoustical enclosures. Since fibrous sound-absorbing materials, such as glass fiber and mineral wool, are good heat insulators, properly designed wrappings can provide both substantial acoustical and heat insulation. On the other hand, the heat-insulating properties of the enclosure may be detrimental and require that provision be made for auxiliary cooling of the interior of the enclosure to prevent the buildup of excessively high temperatures. Only the acoustical design aspects of enclosures and wrappings are treated in this chapter.

12.1 ACOUSTICAL ENCLOSURES

Depending on their size (compared with the acoustical and bending wavelength) acoustical enclosures can be termed either small or large. The enclosure is considered small* if both the bending wavelength is large compared with the largest wall panel dimensions and the acoustical wavelength is large compared with the largest interior dimension of the enclosure volume. In small acoustical enclosures, the interior volume has no acoustical resonances. If the largest dimension of the acoustical volume is $L_{\max} \leq \frac{1}{10}\lambda$, the sound pressure is evenly distributed within the volume. The enclosure is considered large if all of its interior dimensions are large compared with the acoustical wavelength and there are a large number of acoustical resonances in the interior volume in the frequency range of interest. Accordingly, even enclosures with large physical dimensions are acoustically small at very low frequencies while enclosures of small physical size are acoustically large at very high frequencies. In almost all acoustical enclosures the

*Small enclosures are also treated in Chapter 6.

enclosure walls already exhibit numerous structural resonances in the frequency range where the first acoustical resonance occurs.

If the enclosure has no mechanical connections to the enclosed equipment, it is termed *free standing*. If there are mechanical connections, then the enclosure is *equipment mounted*. Enclosures that very closely surround the enclosed equipment and the volume of the machine is comparable to the volume of the enclosure are called *close fitting*. Enclosures without acoustically significant openings are referred to as *sealed enclosures*, and those with significant acoustical leaks (intentional or unintentional) as *leaky* acoustical enclosures. Figure 12.1 shows various configurations of sealed acoustical enclosures. This chapter deals with the acoustical design of enclosures. Nonacoustical aspects—such as ventilation, safety, and economy—are treated in a handbook by Miller and Montone (see Bibliography). Construction details and advice for writing purchase specifications are given in a VDI guideline (also see Bibliography).

Insertion Loss as Acoustical Performance Measure

The insertion loss is the most appropriate descriptor for the acoustical performance of enclosures of all types. The operational definition of the insertion loss (IL) of an acoustical enclosure is illustrated in Fig. 12.2. For noise sources that will be positioned indoors, such as machinery in factory spaces, the sound-power-based insertion loss of the enclosure as indicated in Fig. 12.2a is the most meaningful. It is defined as

$$\text{IL}_w = 10 \log \left(\frac{W_0}{W_E} \right) = L_{w0} - L_{WE} \quad \text{dB} \tag{12.1}$$

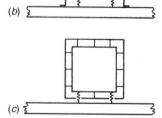

FIGURE 12.1 Enclosure types: (*a*) free standing, large; (*b*) free standing, close fitting; (*c*) equipment mounted, close fitting.

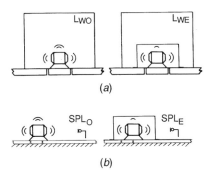

FIGURE 12.2 Operational definition of enclosure insertion loss IL: (a) power based; $IL_w \equiv L_{w0} - L_{WE}$, (b) sound pressure based; $IL_P \equiv SPL_O - SPL_E$.

where W_0 is the sound power radiated by the unenclosed source and L_{W0} the corresponding sound power level and W_E is the sound power radiated by the enclosed source and L_{WE} the corresponding sound power level. Both W_0 and W_E are measured in a reverberation room (see Chapter 4) or with the aid of a sound intensity meter.

For enclosures used with equipment deployed outdoors, a less precise but more easily implemented definition of insertion loss, the so-called pressure-based insertion loss, or IL_p, illustrated in Fig. 12.2b is most appropriate. It is defined as

$$IL_p = SPL_O - SPL_E \quad dB \qquad (12.2)$$

where SPL_O is the average sound pressure level measured at a number of locations around the source without the enclosure and SPL_E is that measured with the source surrounded by the enclosure. The measurement positions may be chosen on a circle that is centered at the source location. The measurement distance should be at least three times the longest dimension of the enclosure. Equations (12.1) and (12.2) represent definitions readily implemented in the field or laboratory. Using these definitions one can readily determine whether a specific enclosure will or will not meet specific performance requirements stated in the form of a sound-power-level reduction, a sound-pressure-level reduction for a specified distance, or a sound-pressure-level reduction at a specified distance and direction. If the laboratory facilities are available, the sound-pressure-based insertion loss may also be measured in a large hemianechoic chamber.

Note that if the radiation patterns of both the unenclosed and enclosed source is omnidirectional, the two measures yield the same result ($IL_w = IL_p$).

Qualitative Description of Acoustical Performance

Figure 12.3 shows the typical shape of a curve of the acoustical insertion loss versus frequency of a free-standing, sealed acoustical enclosure. Region I is the small-enclosure region where neither the interior air volume nor the enclosure

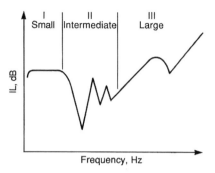

FIGURE 12.3 Typical curve of insertion loss versus frequency of a sealed, free-standing acoustical enclosure. Region I: panel stiffness controlled; damping and interior absorption are ineffective. Region II: resonance controlled; damping and interior absorption are effective. Region III: controlled by sound transmission loss; sound transmission loss and interior absorption are effective; usually limited by leaks.

wall panels exhibit any resonances. In this frequency region the insertion loss is frequency independent and is controlled by the ratio of the volume compliance (inverse of volume stiffness) of the enclosure walls and that of the enclosed air volume.

Region II in Fig. 12.3 is the intermediate region where the insertion loss is controlled by the resonant interaction of the enclosure structure with the enclosed acoustical volume. The region is characterized by a number of alternating maxima and minima in the insertion loss. Typically, the first and most important minima in the insertion loss occurs when the combined volume compliance of the interior air volume and the volume compliance of the wall panels matches the mass compliance of the wall panels. The insertion loss at this resonance frequency usually controls the low-frequency insertion loss of the enclosure and in some instances can become negative, signifying that the equipment with the enclosure may radiate more noise than without the enclosure. Additional minima of the insertion loss occur at acoustical resonances of the enclosure volume. Further minima in insertion loss occur when the frequencies of structural resonances of a wall panel and the frequencies of acoustical resonances of the enclosure volume coincide. It is imperative that enclosures designed for sound sources that radiate noise of predominantly tonal character (such as transformers, gears, reciprocating compressors and engines, etc.) should have no structural or acoustical resonances that correspond to the frequencies of the predominant components of the source noise.

Region III in Fig. 12.3 is the large-enclosure region where both the enclosure wall panels and the interior air volume exhibit a very large number of acoustical resonances. Here, statistical methods of room acoustics can be used to predict the sound field inside the enclosure (see Chapter 7) and the sound transmission through the enclosure walls (see Chapter 11). In this frequency region, the insertion loss is controlled by the interior sound absorption and by the sound transmission loss (R) of the enclosure wall panels. The dip in the curve of insertion loss versus frequency in region III corresponds to the coincidence frequency of the wall panel (see Chapter 11), which for the most frequently used wall panels (1–2 mm steel or aluminum) falls above the frequency region of interest. The coincidence frequency may fall into the region of interest if the ratio of stiffness of the panel to its mass per unit area is high (e.g., a honeycomb panel).

The prediction of the insertion loss of small and large acoustical enclosures is treated in the following sections. Performance prediction in the intermediate frequency region requires a detailed finite-element analysis of the coupled mechanical acoustical system such as outlined in Chapter 6.

Small Sealed Acoustical Enclosures

For a small, sealed acoustical enclosure, where the sound pressure inside the cavity is evenly distributed, the insertion loss is given by[4–5]

$$\text{IL}_{SM} = 20 \log \left(1 + \frac{C_v}{\sum_{i=1}^{n} C_{wi}} \right) \quad \text{dB} \tag{12.3}$$

where

$$C_v = \frac{V_0}{\rho c^2} \quad \text{m}^5/\text{N} \tag{12.4}$$

is the compliance of the gas volume inside the enclosure, V_0 is the volume of the gas in the enclosure volume, ρ is the density of the gas, c is the speed of sound of the gas, and C_{wi} is the *volume compliance* of the ith enclosure wall plate defined as

$$C_{wi} = \frac{\Delta V_{pi}}{p} \quad \text{m}^5/\text{N} \tag{12.5}$$

where ΔV_{pi} is the volume displacement of the ith enclosure wall plate in response to the uniform pressure p. It is assumed here that the enclosure is rectangular and is made of n separate, homogeneous, isotropic plates, each with its own volume compliance.

At frequencies below the first mechanical resonance of the isotropic enclosure wall panel, the volume compliance of a homogeneous, isotropic panel C_{pi} is given by[3]

$$C_{wi} = \frac{10^{-3} A_{wi}^3 F(\alpha)}{B_i} \quad \text{m}^5/\text{N} \tag{12.6}$$

where A_{wi} is the surface area of the ith wall panel and $F(\alpha)$ is given in Fig. 12.4 as a function of the aspect ratio $\alpha = a/b$ of the panel, where a is the longest and b is the smallest edge dimension of the wall panel. For homogeneous, isotropic wall panels, the bending stiffness per unit length of the panel is

$$B = \frac{Eh^3}{12(1-\nu^2)} \quad \text{N} \cdot \text{m} \tag{12.7}$$

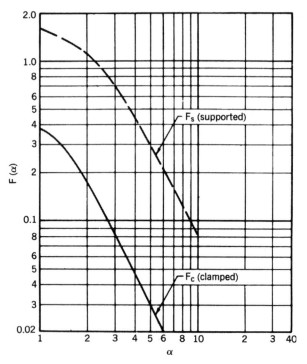

FIGURE 12.4 Plate volume compliance function $F(\alpha)$ plotted versus the aspect ratio $\alpha = a/b$ for homogeneous isotropic panels either with clamped or simply supported edges. (After Ref. 3).

where E is Young's modulus and h is the thickness and ν the Poisson ratio of the wall panel. For a rectangular enclosure combining Eqs. (12.3)–(12.7) yields

$$\mathrm{IL_{SM}} = 20 \log \left[1 + \frac{V_0 E h^3}{12 \times 10^{-3}(1 - \nu^2)\rho c^2} \sum_{i=1}^{6} \frac{1}{A_{wi}^3 F(\alpha_i)} \right] \quad \mathrm{dB} \quad (12.8)$$

For the special case of a cubical enclosure with clamped walls of edge length a Eq. (12.8) yields

$$\mathrm{IL}_s = 20 \log \left[1 + 41 \left(\frac{h}{a} \right)^3 \frac{E}{\rho c^2} \right] \quad \mathrm{dB} \quad (12.9)$$

Equation (12.8) indicates that high insertion loss is achieved if the enclosure has small edge length, high aspect ratio, and large wall thickness, edge conditions are clamped, and the panels are made of a material of high Young's modulus. In short, for an enclosure with high insertion loss at low frequencies, the wall must be made as stiff as possible. According to Eq. (12.9), the low-frequency insertion loss of a cubical steel enclosure of edge length $a = 300$ mm, $E = 2 \times 10^{11}$ N/m^2, $\rho = 1.2$ kg/m^3, and $c = 340$ m/s is for two different wall thicknesses h:

Wall Thickness h	Clamped		Simply Supported	
	IL$_s$	f_0	IL$_s$	f_0
3 mm	35.5 dB	296 Hz	24 dB	162 Hz
1.5 mm	18.5 dB	148 Hz	9 dB	81 Hz

Here f_0 is the resonance frequency of each of the identical homogeneous isotropic clamped panels. The insertion loss values listed in the table above are valid only at frequencies well below f_0. Note in the table that much higher insertion loss is achieved with a clamped-edge condition than with simply supported edge conditions. In practice, clamped edges are almost impossible to achieve. Consequently, one should use the simply supported edge condition in the initial design to ensure that the designed performance will be achieved.

It is of considerable practical interest to know which material should be used for a small sealed cubical enclosure to yield the highest insertion loss at low frequencies for the same enclosure volume and same total weight. Considering that the total mass M of a cubical enclosure of edge length a is $M = 6\rho_M a^2 h$, Eq. (12.9) can be expressed as

$$\text{IL}_s = 20 \log \left[1 + 0.19 \frac{M^3}{a^9 \rho c^2} \left(\frac{c_L}{\rho_M} \right)^2 \right] \text{ dB} \quad (12.10)$$

where $c_L = \sqrt{(E/\rho_M)}$ is the speed of longitudinal waves in the bulk enclosure material and ρ_M is the density of the enclosure material. Equation (12.10) indicates that for all materials giving the same enclosure mass M, the material with the highest c_L/ρ_M ratio yields the highest insertion loss. Table 12.3 lists the c_L/ρ_M ratio and the normalized low-frequency insertion loss ΔIL for frequently used materials, where ΔIL is defined as the low-frequency insertion loss of an enclosure made out of a specific material minus the insertion loss of an enclosure of the same volume and weight built of steel. Table 12.3 indicates that, in regard to low-frequency insertion loss, aluminum and glass are superior to steel and lead is the worst possible choice! Note, however, that this conclusion can be exactly opposite for a large enclosure if the coincidence frequency falls within the frequency range of interest.

Formstiff Small Enclosures. Because the insertion loss of small, sealed enclosures at very low frequencies ($L_{\max} < \frac{1}{10}\lambda$) is controlled by the volume compliance of the enclosure walls, it is desirable to select constructions that provide the highest wall stiffness for the allowable enclosure weight. An enclosure consisting of a round, cylindrical body and two half-spherical end caps yields a very

TABLE 12.3 Difference in Low-Frequency Insertion Loss ΔIL and c_L/ρ_M Ratio for Different Construction Materials[a]

	Lead	Steel	Concrete	Plexiglass	Aluminum	Glass
c_L/ρ_M, m⁴/kg·s	0.11	0.65	1.5	1.6	1.9	2.1
ΔIL, dB	−31	0	+14	+15	+19	+20

[a] ΔIL is for a cubical enclosure with identical sides.

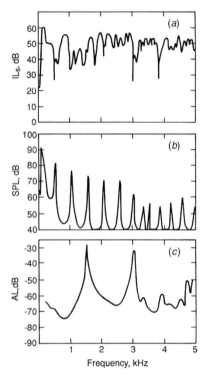

FIGURE 12.5 Acoustical characteristics of a small, sealed, round cylindrical enclosure without interior absorption: (a) insertion loss IL; (b) resonant acoustical response of interior volume, SPL; (c) sound-induced vibration acceleration response of enclosure wall, AL.

stiff construction and much higher insertion loss than a rectangular enclosure of the same volume and weight. Figure 12.5 shows the acoustical characteristics of such a particular small, cylindrical enclosure with no internal sound-absorbing treatment. Curve A represents insertion loss versus frequency measured with an external source, indicating that insertion loss at most frequencies exceeds 55 dB. Curve B represents the sound pressure levels in the enclosed air volume when excited with an external source, indicating the presence of strong acoustical resonances at 550 Hz and above. Curve C represents the sound-induced vibration acceleration of the cylindrical shell caused by an external sound source. Strong structural resonances are seen at 1.5, 3, and 3.75 kHz. Observing the three curves in Fig. 12.5, it is noted that the frequencies where the insertion loss is minimum coincide either with acoustical resonances of the interior volume or with structural resonances of the enclosure shell. The strong minima of the curve of insertion loss versus frequency in the vicinity of 1.5, 3, and 3.75 kHz are caused by coincidence of the structural resonance of the shell with acoustical resonances of the interior air volume.

Another way to achieve a low wall compliance for a given total weight is to make the enclosure walls out of a composite material consisting of a honeycomb core sandwiched between two lightweight plates. A special form of such high-stiffness, low-weight panel, described by Fuchs, Ackermann, and Frommhold,[6] also provides high sound absorption at low frequencies in addition to the low volume compliance. In this case, the panel on its interior side has two double membranes. The first, thicker membrane, which is rigidly bonded to the honeycomb core, has round openings to make an inward-facing Helmholtz resonator out of each honeycomb cavity. A second, thin, membrane that is covering the first one is free to move over the resonator openings. The presence of this second, thin membrane lowers the resonance frequency of the Helmholtz resonators owing to the mass of the membrane covering the opening and increases the dissipation owing to air pumping. Such Helmholtz plates can be constructed from stainless steel, aluminum, or light-transparent plastic. They are fully sealed and can be hosed down for cleaning. Figure 12.6 shows the curve of insertion loss versus frequency of a cubical enclosure of 1 m edge length. The solid curve was obtained with enclosure walls made of the above-described Helmholtz plates of

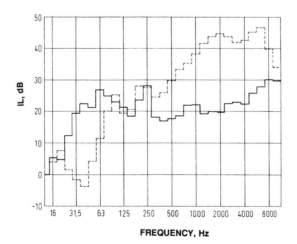

FIGURE 12.6 Measured insertion loss of two $1 \times 1 \times 1$-m enclosures. (After Ref. 6.) Solid curve: Helmholtz plate walls, 10 cm thick, 8.5 kg/m^2, no sound-absorbing treatment. Dashed curve: Woodchip board walls, 1.3 cm thick, 5-cm-thick sounding-absorbing layer, 10 kg/m^2.

10 cm thickness with a mass per unit area of 8.5 kg/m² without any additional sound-absorbing lining. The dotted curve was obtained with an enclosure of the same size, made of 1.3-cm-thick wood chip board with 5-cm sound-absorbing lining, yielding a mass per unit area of 10 kg/m². The Helmholtz plate enclosure yields superior low-frequency performance and does not require extra sound-absorbing lining. However, the insertion loss of this enclosure at midfrequencies (above 250 Hz) is low because of the low coincidence frequency of the light and stiff panels.

Small Leaky Enclosures. Except for special constructions, such as hermetically sealed compressors, all enclosures are likely to be leaky and provide zero insertion loss as the frequency approaches zero. If an enclosure has a leak in the form of a round opening of radius a_L in a wall of thickness h, the leak represents a compliance $C_L = 1/j\omega Z_L$, where Z_L is the acoustical impedance of the leak given as

$$C_L = \frac{(\pi a_L^2)^2}{-\omega^2 \rho (h + \Delta h) \pi a_L^2 + j\omega R} \quad \text{m}^5/\text{N} \quad (12.11)$$

where h is the plate thickness, $\Delta h \simeq 1.2 a_L$ represents the end correction, and R is the real part of the impedance of the leak. The sound pressure inside the small, leaky enclosure owing to the operation of a source of volume velocity q_0 is

$$p_{\text{ins}}^{\text{leaky}} = \frac{q_0}{j\omega} \frac{1}{C_L + C_v + \sum_i C_{wi}} \quad \text{N/m}^2 \quad (12.12)$$

and the insertion loss of the small leaky enclosure is

$$\text{IL}_{\text{leaky}} = \text{IL}_L = 20 \log \frac{C_L + C_v + \sum_i C_{wi}}{C_L + \sum_i C_{wi}} \quad \text{dB} \quad (12.13)$$

where C_v and C_{wi} are given in Eqs. (12.4) and (12.5), respectively. Since C_L approaches infinity as the frequency approaches zero, the insertion loss of leaky enclosures approaches zero for any form of leak. The insertion loss becomes negative in the vicinity of the Helmholtz resonance frequency of the compliant leaky enclosure given by

$$f_{0L} = \frac{a_L c}{2} \sqrt{\frac{1}{\pi (h + \Delta h)(V_0 + \rho c \sum_i C_{wi})}} \quad \text{Hz} \quad (12.14)$$

As $C_L / \Sigma C_{wi}$ becomes small with increasing frequency, the insertion loss of the small, leaky enclosure, IL_L, approaches that of the sealed enclosure, IL_s. This behavior is illustrated in Fig. 12.7.

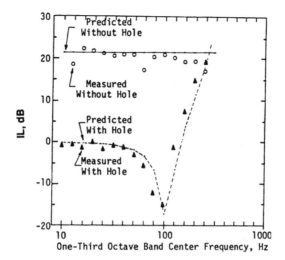

FIGURE 12.7 Measured and predicted sound-pressure-based insertion loss (outside–inside direction) of a 50 × 150 × 300-mm aluminum enclosure of 1.5-mm-thick wall with a single 9.4-mm-diameter hole (after Ref. 5): (———) predicted, sealed; (- - - -) predicted, leaky; (○) measured, sealed; (▲) measured, leaky.

Close-Fitting, Sealed Acoustical Enclosures

A close-fitting acoustical enclosure is one in which a considerable portion of the enclosed volume is occupied by the equipment being quieted. Such enclosures are used in cases where the radiated noise must be reduced using a minimum of added volume. Vehicle engine enclosures, portable compressors, and transportable transformers fall into this category.

Free-standing, close-fitting enclosures, such as shown in Fig. 12.1b, have no mechanical connections to the vibrating source and the enclosure walls are excited to vibration only by the airborne path. Conversely, the walls of machine-mounted, close-fitting enclosures, such as depicted in Fig. 12.1c, are excited by both airborne and structure-borne paths.

The characteristic property of a close-fitting enclosure is that the air gap (the perpendicular distance between the vibrating surface of the machine and the enclosure wall) is small compared with the acoustical wavelength in most of the frequency range. The walls of the close-fitting enclosures are usually made of thin, flat sheet metal with a layer of sound-absorbing material such as glass fiber or acoustical foam on the interior face. The purpose of the sound-absorbing lining is to damp out the acoustical resonances in the air space.

Free-Standing, Close-Fitting Acoustical Enclosures. Free-standing, close-fitting enclosures, which have no structure-borne connection to the vibrating equipment, always achieve higher insertion loss than a similar machine-mounted enclosure can provide. If the enclosure is perfectly sealed, then its insertion loss depends on (1) thickness, size, and material of the enclosure wall; (2) edge-joining conditions of the wall plates and their loss factors; (3) vibration pattern of the machine surface; (4) average thickness of the air gap between the wall and the machine; and (5) type of sound-absorbing material in the air gap.

Figure 12.8 shows the measured dependence of the power-based insertion loss of a model close-fitting enclosure consisting of 0.6 × 0.4-m enclosure walls of thickness h and a variable spacing gap of thickness T. To obtain these data, an array of loudspeakers, which represented the vibrating surface of the enclosed machine, was phased to simulate various vibration patterns.[8] Figure 12.8a shows

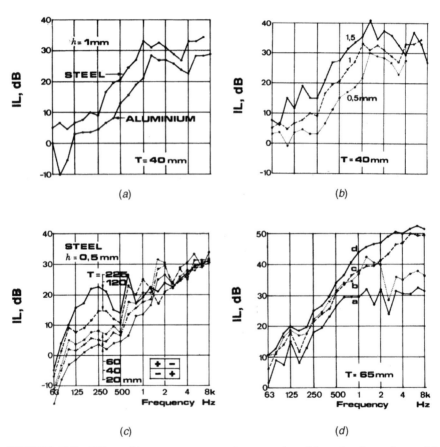

FIGURE 12.8 Effect of key parameters on the power-based insertion loss of a model close-fitting enclosure. Wall panel 0.6 × 0.4. (After Ref. 8.) (a) Effect of the wall material; 4-cm air gap, in-phase excitation. (b) Effect of wall panel thickness, 4-cm air gap, in-phase excitation. (c) Effect of average air gap thickness; 0.5-mm steel plate, quadrupole excitation. (d) Effect of sound-absorbing material and structural damping; 1-mm steel plate, 6.5-cm air gap, in-phase excitation; a, no damping, no absorption in air space; b, with damping treatment; c, no damping, but absorption in air space; d, with damping and absorption.

the dependence of the measured insertion loss on frequency obtained for 1-mm-thick aluminum and steel wall plates, respectively. Because the steel and aluminum plates have practically the same dynamic behavior, the higher insertion loss of the steel wall panel is attributable to its higher mass. In the frequency range between 300 Hz and 1 kHz, the 9-dB average difference corresponds to the difference in density of the two materials. Figure 12.8b shows the effect of wall thickness. Here again the insertion loss increases with increasing wall thickness. Below 250 Hz, where the insertion loss is controlled by the stiffness of the air and the stiffness and damping of the wall panels, the IL_w changes little with increasing frequency. In the frequency range between 200 Hz and 1 kHz, where the IL_w is controlled by the volume stiffness of the air and by the mass per unit area of the wall, the IL_w increases with a slope of 40 dB/decade. Above 1 kHz, the insertion loss is limited by the acoustical resonances in the air space. Figure 12.8c shows that the insertion loss increases with increasing air gap thickness. Figure 12.8d shows the effect of sound-absorbing material in the air space and damping treatment of the enclosure wall on the achieved insertion loss. The sound-absorbing treatment in the air space prevents the acoustical resonances in the air space and results in a substantial increase of the insertion loss above 1 kHz. Structural damping helps to reduce the deleterious effect of the efficiently radiating plate resonance at 160 Hz. A combination of sound-absorbing treatment and structural damping results in a smooth, steeply increasing insertion loss with increasing frequency and provides a very high degree of acoustical performance.

One-Dimensional Model. According to Bryne, Fischer, and Fuchs,[9] the insertion loss of free-standing, sealed, close-fitting acoustical enclosures can be predicted with reasonable accuracy with the simple one-dimensional model shown in Fig. 12.9. It is assumed that the machine wall (A) vibrates in phase (like a rigid piston) with velocity v_M and that the motion of the machine is not affected by the presence of the enclosure. The vibrating machine is surrounded by a

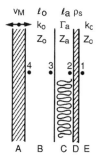

FIGURE 12.9 One-dimensional model for predicting the power-based insertion loss IL of free-standing, sealed close-fitting acoustical enclosures; A machine wall vibrating with velocity v_M; B, air gap thickness l_0; C, sound-absorbing layer thickness l_a; D, impervious enclosure wall of mass per unit area ρ_s; E, free space into which sound is radiated. (After Ref. 9).

flat, limp, impervious enclosure wall (D) which has an internal sound-absorbing lining (C) of thickness l_a and there is an air gap (B) of thickness l_0 between the machine and the sound-absorbing layer. The sound pressure on the outside surface of the enclosure (point 1 in Fig. 12.9) is given by

$$p_1 = p_2 \left(1 - \frac{j\omega\rho_s}{Z_0}\right) \quad \text{N/m}^2 \tag{12.15}$$

where $Z_0 = \rho c$ is the characteristic impedance of air. The following equations are used to obtain p_1 as a function of the machine wall vibration velocity v_M:

$$p_3 = 0.5 p_2 \left[\left(1 + \frac{Z_a}{Z_3}\right) e^{-j\Gamma_a l_a} + \left(1 - \frac{Z_a}{Z_3}\right) e^{j\Gamma_a l_a}\right] \quad \text{N/m}^2 \tag{12.16}$$

$$p_4 = v_M Z_4 = 0.5 p_3 \left[\left(1 + \frac{Z_0}{Z_4}\right) e^{-jk_0 l_0} + \left(1 - \frac{Z_0}{Z_4}\right) e^{jk_0 l_0}\right] \quad \text{N/m}^2 \tag{12.17}$$

$$Z_2 = Z_0 + j\omega\rho_s \quad \text{N} \cdot \text{s/m}^3 \tag{12.18}$$

$$Z_3 = Z_a \frac{(1 + Z_a/Z_2) e^{j\Gamma_a l_a} + (1 - Z_a/Z_2) e^{-j\Gamma_a l_a}}{(1 + Z_a/Z_2) e^{j\Gamma_a l_a} - (1 - Z_a/Z_2) e^{-j\Gamma_a l_a}} \quad \text{N} \cdot \text{s/m}^3 \tag{12.19}$$

$$Z_4 = Z_0 \frac{(1 + Z_0/Z_3) e^{jk_0 l_0} + (1 - Z_0/Z_3) e^{-jk_0 l_0}}{(1 + Z_0/Z_3) e^{jk_0 l_0} - (1 - Z_0/Z_3) e^{-jk_0 l_0}} \quad \text{N} \cdot \text{s/m}^3 \tag{12.20}$$

where $k_0 = 2\pi f/c$ is the wavenumber, ρ the density, and c the speed of sound in air; Z_a is the complex characteristic impedance and Γ_a the complex propagation constant of the porous sound-absorbing material as given in Chapter 8.

The sound power radiated by the unenclosed machine wall is given by

$$W_0 = v_M^2 \rho c A \quad \text{W} \tag{12.21a}$$

and when it is enclosed with a free-standing, close-fitting enclosure

$$W_A = \frac{p_1^2}{\rho c} A \quad \text{W} \tag{12.21b}$$

yielding for the insertion loss

$$\text{IL}_F = 10 \log \frac{W_0}{W_A} = 20 \log \frac{v_M \rho c}{p_1^2} \quad \text{dB} \tag{12.22}$$

In deriving this equation, no account is taken of possible change in surface area or radiation efficiency between the enclosure surface and machine surface.

If the angle of sound incidence on the enclosure wall is known (i.e., a specific oblique angle or random), the power-based insertion loss of free-standing, close-fitting enclosures can be predicted on the basis of the two-dimensional model of sound transmission through layered media presented in Chapter 11.

Machine-Mounted, Close-Fitting Acoustical Enclosure. If the close-fitting enclosure is machine mounted, the rigid or resilient connections between the machine and the enclosure wall give rise to additional sound radiation of the enclosure wall. Assuming that the vibration velocity of the machine v_M is not affected by the connected enclosure, the additional sound power W_M radiated owing to the structural connections is

$$W_M = \sum_{i=1}^{n} F_i^2 \left(\frac{\rho}{\rho_s^2 c} + \frac{\rho c \sigma}{2.3 \rho_s^2 c_L h \omega \eta} \right) \quad \text{W} \tag{12.23}$$

where

$$F_i^2 = \frac{v_{mi}^2}{|1/2.3 \rho_s^2 c_L h + j\omega/s|^2} \quad \text{N}^2 \tag{12.24}$$

In Eq. (12.24), F_i is the force transmitted by the ith attachment point, v_{Mi} is the vibration velocity of the machine at the ith attachment point, n is the number of point attachments between the machine and the homogeneous, isotropic enclosure wall, ρ_s is the mass per unit area of the enclosure wall, h is the wall thickness, c_L is the speed of longitudinal waves in the wall material, η is the loss factor, σ is the radiation efficiency, and s is the dynamic stiffness of the resilient mount connecting the enclosure wall to the machine ($s = \infty$ for rigid point connections). To minimize structure-borne transmission, it is advantageous to select attachment points at those locations on the machine that exhibit the lowest vibration. The insertion loss of the machine-mounted, close-fitting enclosure is

$$\text{IL}_{\text{MM}} = 10 \log_{10} \frac{W_0}{W_A + W_M} \quad \text{dB} \tag{12.25}$$

where W_0 and W_A are given in Eqs. (12.21a) and (12.21b).

Figure 12.10 shows a close-fitting enclosure investigated experimentally and analytically by Byrne, Fischer, and Fuchs[9] in three configurations: (1) free standing, (2) rigidly machine mounted, and (3) resiliently machine mounted. Figure 12.11 shows the measured insertion loss obtained for each of the three configurations. The machine-mounted enclosure, which was supported from the machine at four points at each side, yield substantially lower insertion loss at high frequencies than the free-standing enclosure. Elastic mounting yields higher insertion loss than rigid mounting.

As shown by reference 9, the simple one-dimensional analytical model based on Eqs. (12.15)–(12.25) yields predictions that are in good agreement with measured data. The reason for this surprisingly good agreement is that the sound-absorbing treatment effectively prevents sound propagation (and the occurrence of acoustical resonances) in the plane parallel to the enclosure wall and also provides structural damping to the thin enclosure wall. Replacing the effective glass fiber sound-absorbing treatment with a thinner, less effective acoustical foam has resulted in substantial decrease in the insertion loss, indicating the crucial importance of an effective sound-absorbing treatment. Figure 12.12 shows

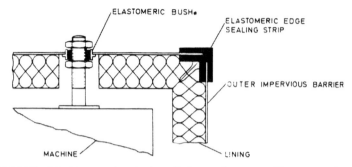

FIGURE 12.10 Machine-mounted, close-fitting enclosure; resilient mounting, resilient panel edge connections. (After Ref. 9).

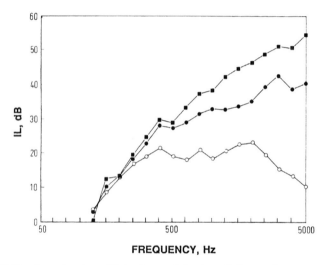

FIGURE 12.11 Insertion loss (IL_s) of a sealed, close-fitting enclosure made of 1-mm steel plates. (After Refs. 7 and 10.) Dimensions: $1 \times 0.6 \times 0.8$ m; machine, enclosure wall distance 50 mm; interior lining 50 mm thick; flow resistivity $R_1 = 2 \times 10^4 \text{N} \cdot \text{s/m}^4$. (■) Free standing; (●) machine mounted, resilient mounting (see Fig. 12.10); (○) machine mounted, rigid mounting.

how the specific choice of edge connections of the enclosure plates affected the insertion loss of the machine-mounted, close-fitting enclosure. Elastically sealed edges, such as shown in Fig. 12.10, yield higher insertion loss than rigidly connected (welded) edges. The better performance obtained with elastically sealed edges is due to the reduced coupling between the in-plane motion of one plate (which radiates no sound) with the normal motion of the connected plate (that radiates sound efficiently). The elastic edge seal also increases the loss factor of the enclosure plates, thereby reducing their resonant vibration response.

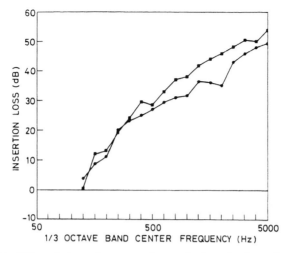

FIGURE 12.12 Effect of plate edge connection on the insertion loss of sealed, resiliently machine-mounted, close-fitting enclosure: (■) elastically sealed edge; (●) welded edge.

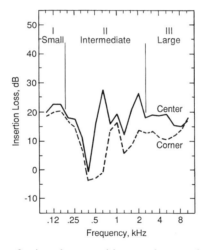

FIGURE 12.13 Effect of microphone position on the sound-pressure-based insertion loss measured in the outside–inside direction of a 50 × 150 × 300-mm (1.6-mm-thick) empty, plane, sealed, unlined aluminum acoustical enclosure. (After Ref. 11.) Solid line: center microphone location. Dashedline: corner microphone location.

Intermediate-Size Enclosures

The intermediate frequency region, designated as region II in Figs. 12.3 and 12.13, is defined as the frequency region where the enclosure walls, the enclosure air volume, or both exhibit resonances; but the resonances do not overlap so that statistical methods are not yet applicable.

Comparison of the solid and dashed curves in Fig. 12.13 show how the insertion loss, measured in the outside–inside direction (such as in a quiet control room in a noisy test facility), depends on the location of the microphone. Below 400 Hz, the sound pressure is uniformly distributed in the enclosure and the microphone in the corner measures the same sound pressure as that placed at the center. The first acoustical resonance occurs at 565 Hz, which results in a very low insertion loss at the enclosure corner in the 500- and 800-Hz center-frequency one-third-octave bands, as illustrated by the dashed curve, but in a high insertion loss in the same bands for the enclosure center location, as depicted by the solid curve. Fluctuations owing to structural resonances can be reduced by application of damping treatment to the panels (see Chapter 14) and those owing to acoustical resonances by internal sound-absorbing lining (see Chapter 8).

Because the insertion loss in this intermediate frequency range fluctuates widely with frequency and position, it is very difficult to make accurate analytical predictions of the insertion loss. Frequently, involved finite-element analysis (see Chapter 6), model-scale or full-scale experiments, or crude approximations must be employed. For crude approximations one connects the low- and high-frequency predictions to cover the intermediate range.

Enclosures without Internal Sound-Absorbing Treatment

There are cases where the danger of bacterial growth or the accumulation of combustible particles precludes the use of porous or fibrous sound-absorbing materials inside of acoustical enclosures. In such cases bare enclosures must be used. Simple analytical models yield zero insertion loss for such enclosures. Consequently, it is also of theoretical interest to treat this extreme case.

Acoustical Performance Prediction of Large, Bare Enclosures. Even if the enclosure is built from homogeneous, isotropic panels without any interior sound-absorbing treatment, the interior sound pressure will not be infinitely high and the insertion loss of the enclosure will not be zero.

The interior sound pressure does not go to infinity because the interior sound field loses power through (1) sound radiation of the walls by the forced bending waves, (2) sound radiation of the walls by the free bending waves, (3) energy dissipation in the panel by the free bending waves, and (4) the inevitable acoustical energy dissipation at the interior surfaces of the impervious enclosure panels owing to the acoustical shear layer and heat conduction losses.

The insertion loss of the enclosure will be finite because of the energy dissipation by processes 3 and 4 above.

The power balance of the interior sound field is given as

$$W_{\text{source}} = W_{\text{rad}}^{\text{forced}} + W_{\text{rad}}^{\text{free}} + W_{d}^{\text{free}} + W_{d}^{\text{shear}} \quad \text{W} \qquad (12.26)$$

where W_{source} = sound power of enclosed source
$W_{\text{rad}}^{\text{force}}$ = sound power radiated by forced bending waves

$W_{\text{rad}}^{\text{free}}$ = sound power radiated by free bending waves
W_d^{free} = sound power dissipated in plate by free bending waves
W_d^{shear} = sound power lost through shear and heat conduction at interior surface of impervious, solid walls

Unless the panel is very heavily damped, the power dissipated by the forced bending waves is small compared with that dissipated by the free bending waves and can neglected.

Inserting into Eq. (12.26) the appropriate relationships from Section 11.8 and for $\alpha_{\min}$ from reference 12 yields for the random-incidence sound absorption coefficient of the bare enclosure walls,

$$\bar{\alpha}_{\text{rand}} = \frac{W_{\text{loss}}^{\text{tot}}}{W_{\text{inc}}} = A + B + C + D \tag{12.27}$$

for the space-average mean-square sound pressure in the enclosure,

$$\overline{p^2} = \frac{4\rho_0 c_0 \, W_{\text{source}}}{S_{\text{tot}}(A + B + C + D)} \quad \text{N}^2/\text{m}^4 \tag{12.28}$$

and for the insertion loss of the bare enclosure,

$$\text{IL}_{\text{bare}} = 10 \log \left(\frac{A + B + D}{A + BC} \right) \quad \text{dB} \tag{12.29}$$

where

$$A = \frac{1}{1 + (1/52)[\omega \rho_s/(\rho_0 c_0)]^2} \qquad B = \frac{2\pi \sqrt{12} c_0^2 \rho_0 c_0 \sigma_{\text{rad}}}{c_L \rho_M h^2 \omega^2}$$

$$C = \frac{\rho_0 c_0 \sigma_{\text{rad}}}{\rho_0 c_0 \sigma_{\text{rad}} + \rho_s \omega \eta} \qquad D = \alpha_{\min} = 0.72(10^{-4})\sqrt{\omega}$$

where S_{tot} = total interior surface area of enclosure, m² (assuming that all six partitions are made from the same panels)
ρ_s = mass per unit area of plate, $h\rho_M$, kg/m²
ρ_M = density of plate material, kg/m³
h = plate thickness, m
ω = radian frequency, $2\pi f$, Hz
ρ_0 = density of air, kg/m³
c_0 = speed of sound, m/s
σ_{rad} = radiation efficiency of free bending waves on plate (see Chapter 11)
η_c = loss factor of plate (see Chapter 14)

Inspecting Eq. (12.29), note that for $\eta = 0$, $C = 1$, and $D \approx 0$, Eq. (12.29) yields IL = 0, as it should be. In the vicinity of the coincidence frequency (i.e., in the plateau region of the transmission loss), where $B \gg A$, the insertion loss of the bare enclosure approaches

$$\text{IL} \approx 10 \log \left(\frac{1}{C}\right) = 10 \log \left(\frac{\rho_0 c_0 \sigma_{\text{rad}} + \rho_s \omega \eta}{\rho_0 c_0 \sigma_{\text{rad}}}\right) \simeq 10 \log \left(\frac{\rho_s \omega \eta}{\rho_0 c_0 \sigma_{\text{rad}}}\right) \quad (12.30)$$

Figure 12.14 compares the insertion loss predicted for an unlined, sealed enclosure [using Eq. (12.8) at low frequencies and Eq. (12.27) at high frequencies] with experimental data obtained by reference 11. There is a reasonably good agreement between the predicted and measured values at both low and high frequencies not only for this specific enclosure but also for a large variety of unlined sealed enclosures of different size and wall panel thickness, indicating that the prediction formulas embodied in Eqs. (12.8) and (12.27) yield reasonable estimates for engineering design. Figure 12.14 also indicates that without internal sound-absorbing treatment the insertion loss remains very modest even at high frequencies, highlighting the importance of sound absorption in enclosure design.

Large Acoustical Enclosures with Interior Sound-Absorbing Treatment

Acoustical enclosures are termed large at frequencies where both the enclosure wall panels and the enclosure volume exhibit a large number of resonant modes in a given frequency band and statistical methods for predicting the level of the interior sound field and the vibration response and sound radiation of the enclosure wall panels can be applied. Large acoustical enclosures used in industrial noise control have many paths for transmitting acoustical energy, as illustrated

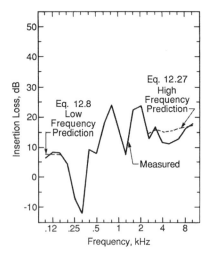

FIGURE 12.14 Comparison of measured and predicted pressure-based IL of a 50 × 150 × 300-mm (0.8-mm-thick) sealed, unlined aluminum enclosure, with source outside. (After Ref. 11).

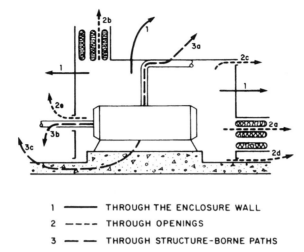

1 ——— THROUGH THE ENCLOSURE WALL
2 ----- THROUGH OPENINGS
3 — — THROUGH STRUCTURE-BORNE PATHS

FIGURE 12.15 Paths of noise transmission from a typical acoustical enclosure.

schematically in Fig. 12.15. These paths can be grouped into three basic categories: (1) through the enclosure walls, (2) through openings, and (3) through structure-borne paths. The first group is characterized by the sound excitation of the enclosure wall by the interior sound field, resulting in sound radiation from the exterior wall surfaces. Sound transmission by this path is relatively well understood, and in most cases the sound power transmitted can be predicted with good engineering accuracy (see Chapter 11). The second group is characterized by sound energy escaping through openings in the enclosure wall such as air intake and exhaust ducts, gaps between panels, and gaps around the gasketing at the floor and doors. These are also fairly well understood[13–15] but not so easily controlled. The third group is characterized by radiation of solid surfaces excited to vibration by dynamic forces, such as the enclosure walls when rigidly connected to the enclosed vibrating equipment, shafts and pipes that penetrate the enclosure wall, and the vibration of the uncovered portion of the floor. This path is difficult to predict without detailed information about the motion of the enclosed machine and its dynamic characteristics. Consequently, all possible efforts should be made to prevent any solid connections to the source. To obtain a balanced design, one must control each of these transmission paths and avoid overdesigning any one of them.

Analytical Model for Predicting Insertion Loss at High Frequencies. A

sound source enclosed in a large, acoustically lined enclosure will radiate approximately the same sound power as it would in the absence of the enclosure. To achieve a high power-based insertion loss, a high percentage of this radiated sound power must be dissipated (i.e., converted into heat) within the enclosure proper. This is accomplished by providing walls of high sound transmission loss to contain the sound waves and by covering the interior with sound-absorbing treatment to convert the trapped acoustical energy into heat.

In analyzing the acoustical behavior of an enclosure at high frequencies, the first step is to predict the space–time average mean-square pressure of the diffuse sound field, $\langle p^2 \rangle$, within the enclosure proper. Once the interior sound field is predicted, one can determine the sound power escaping from the enclosure through the various paths. The mean-square value of the diffuse sound pressure in the interior of the enclosure is obtained by balancing (a) the sound power injected into the enclosure from all the noise sources with (b) the power loss through dissipation (in the sound-absorbing lining, in the air, and in the wall structure) and the sound transmission through the walls of the enclosure and through diverse openings. According to reference 4, a sound source of power output W_0 generates in the enclosure a reverberant field with a space–time average mean-square sound pressure $\langle p^2 \rangle$ given by

$$\langle p^2 \rangle = W_0 \frac{4\rho c}{\left\{ \begin{array}{l} S_w \left[\alpha_w + \sum_i (S_{wi}/S_w) 10^{-R_{wi}/10} + D \right] \\ + S_i \alpha_i + \sum_k S_{sk} + mV + \sum_j S_{Gj} 10^{-R_{Gj}/10} \end{array} \right\}} \quad \text{N}^2/\text{m}^4 \qquad (12.31)$$

where

$$D = \left(\frac{4\pi \sqrt{12} \rho c^3 \sigma}{c_L h \rho_s \omega^2} \right) \left(\frac{\rho_s \omega \eta}{\rho_s \omega \eta + 2\rho c \sigma} \right) \qquad (12.32)$$

and S_w is the total interior wall surface, S_{wi} is the surface area of the ith wall, α_w is the average energy absorption coefficient of the walls, R_{wi} is the sound transmission loss of the ith wall, ρ_s is the mass per unit area of a typical wall panel, η is the loss factor of a typical wall panel in place, σ is the radiation efficiency of a typical panel, c_L is the propagation speed of longitudinal waves in the plate material, h is the wall panel thickness, $S_i \alpha_i$ is the total absorption in the interior in excess of the wall absorption (i.e., the machine body itself), S_{Gj} is the area of the jth leak or opening, R_{Gj} is the sound transmission loss of the jth leak or opening, S_{sk} is the face area of the kth silencer opening (assumed completely absorbing), m is the attenuation constant for air absorption, and V is the volume of the free interior space. The sound power incident on the unit surface area is given by

$$W_{\text{inc}} = \frac{\langle p^2 \rangle}{4\rho c} = \frac{W_0}{\{\cdots\}} \quad \text{W} \qquad (12.33)$$

where $\{\cdots\}$ represents the expression in the denominator in Eq. (12.31) and W_0 is the sound power output of the enclosed machine.

The terms in the denominator of Eq. (12.31) from left to right stand for (1) power dissipation by the wall absorption ($S_w \alpha_w$), (2) power loss through sound radiation of the enclosure walls ($\sum_i S_{wi} \times 10^{-R_{wi}/10}$), (3) power dissipation in the walls through viscous damping effects ($S_w D$), (4) power dissipation by sound-absorbing surfaces in addition to the walls ($S_i \alpha_i$), (5) sound power loss to silencer terminals ($\sum_k S_{sk}$), (6) sound absorption in air (mV), and (7) sound transmission to the exterior through openings and gaps ($\sum_j S_{Gj} \times 10^{-R_{Gj}/10}$).

The relative importance of the various terms may differ widely for different enclosures and even for the same enclosure in the different frequency ranges. For example, for a small airtight enclosure without any sound-absorbing treatment, the power dissipation in the wall panels may be of primary importance, while it is usually negligible for enclosures with proper interior sound-absorbing treatment. Air absorption is important in large enclosures at frequencies above 1000 Hz. The sound power transmitted through the enclosure walls, W_{TW}, that through gaps and openings, W_{TG}, and that through silencers, W_{TS}, are given in Eqs. (12.34)–(12.36), respectively:

$$W_{TW} = \frac{W_0 (\sum_i S_{wi} \times 10^{R_{wi}/10})}{\{\cdots\}} \quad W \qquad (12.34)$$

$$W_{TG} = \frac{W_0 (\sum_j S_{Gj} \times 10^{-R_{Gj}/10})}{\{\cdots\}} \quad W \qquad (12.35)$$

$$W_{TS} = \frac{W_0 (\sum_k S_{sk} \times 10^{-\Delta L_k/10})}{\{\cdots\}} \quad W \qquad (12.36)$$

where ΔL_k is the sound attenuation through the silencer over opening k. The power-based insertion loss of the enclosure in the inside–outside direction is defined as

$$\text{IL} \equiv 10 \log_{10} \frac{W_0}{W_{TW} + W_{TG} + W_{TS} + W_{SB}} \quad \text{dB} \qquad (12.37)$$

which takes the form

$$\text{IL} = 10 \log_{10} \frac{\{\cdots\}}{\sum_i S_{wi} \times 10^{-R_{wi}/10} + \sum_k S_{sk} \times 10^{-\Delta L_k/10}} \qquad (12.38)$$

$$\times \frac{1}{\sum_j S_{Gj} \times 10^{-R_{Gj}/10} + W_{SB}(\{\cdots\}/W_0)} \quad \text{dB}$$

where W_{SB} is the sound power transmitted through structure-borne paths considered separately according to Eqs. (12.23) and (12.24).

The denominator of Eq. (12.38) reveals that if full advantage is to be taken of the high transmission loss of the enclosure walls, the air paths through gaps, openings, and air intake and exhaust silencers and structure-borne paths must be controlled to a degree that their contribution to the sound radiation is small compared with the sound radiation of the walls. If these paths are well controlled and the dissipation is achieved by the sound-absorbing treatment of the interior wall surfaces and by the absorption in the interior, then for $R_{wi} = R_w$ the insertion loss of the enclosure is well approximated by

$$\text{IL} \cong 10 \log \left(1 + \frac{S_w \alpha_w + S_i \alpha_i}{S_w} \times 10^{R_w/10} \right) \quad \text{dB} \qquad (12.39)$$

If we further assume that the second term in Eq. (12.39) is much larger than unity and call the total absorption in the interior of the enclosure $S_w\alpha_w + S_i\alpha_i = A$, Eq. (12.39) simplifies to

$$\text{IL} \cong R_w + 10 \log \frac{A}{S_w} \quad \text{dB} \tag{12.40}$$

This approximate formula would yield negative insertion loss for very small values of interior absorption. It has been widely promulgated in the acoustical literature without mention of its limited validity. Note that the insertion loss of a large enclosure can approach the sound transmission loss of the enclosure wall panels only if there are no leaks and α_w approaches unity.

Key Parameters Influencing the Insertion Loss of Acoustical Enclosures.
The primary paths for the transmission and dissipation of sound in large acoustical enclosures are shown in the block diagram in Fig. 12.16. The key parameter affecting enclosure insertion loss are summarized in Table 12.4. Fisher and Veres[10] and Kurtze and Mueller[16] have carried out systematic experimental investigations on how (1) choice of wall panel parameters, (2) internal sound-absorbing lining, (3) leaks, and (4) vicinity of the enclosed machine to the enclosure walls and its vibration pattern influence the acoustical performance of enclosures. The model–machine sound source employed in reference 10 consisted of $1 \times 1.5 \times 2$-m steel boxes of 1 mm wall thickness and were combined to yield a small ($3 \times 2 \times 1.5$-m) and a large ($4 \times 2 \times 1.5$-m) model machine. The

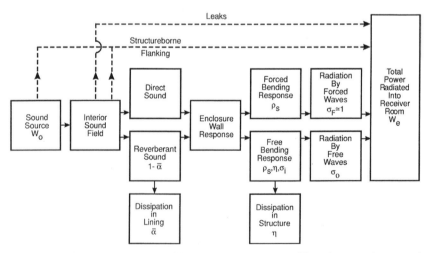

FIGURE 12.16 Block diagram of key components controlling the sound attenuation process in acoustical enclosures: W_0, source sound power output; $\bar{\alpha}$, average sound absorption coefficient; ρ_s, mass per unit area; η, loss factor; σ_i, radiation efficiency of wall panels for radiation for inside direction; σ_0, radiation efficiency of wall panels for radiation for outside direction; σ_F, 1 radiation efficiency of forced waves; W_e, total sound power radiated by the enclosure.

TABLE 12.4 Effect of Key Parameters on Sound Insertion Loss of Large Enclosures

Parameter	Symbol	Effect on Insertion Loss
Absorption coefficient of lining	$\bar{\alpha}$	Increases IL by reducing reverberant buildup
Distance between machine and enclosure wall	d	If d decreases beyond a certain limit, IL decreases at low frequencies (close-fitting enclosure behavior)
Thickness of wall panel	h	Increases IL
Density of wall panel material	ρ_M	Increases IL
Speed of longitudinal waves in panel material	$c_L = \sqrt{E/\rho_M}$	Decreases IL below critical frequency
Loss factor of wall panel	η	Increases IL, especially near and above critical frequency and at first panel resonance
Critical frequency of wall panel	$f_c = c^2/1.8 c_L h$	The higher is f_c for a given mass per unit area, the higher is IL
Radiation efficiencies of wall panels	σ_i, σ_0	The higher is the radiation efficiency for inside direction, σ_i, and for outside direction, σ_0, the smaller is IL
Stiffeners	—	Decrease IL by increasing σ
Leaks	—	Limit achievable IL; watch out for door gaskets and penetrations
Structure-borne flanking	—	Limit achievable IL; watch out for solid connections between vibrating machine and enclosure and for floor vibration

walls of the model machines could be excited by either an internal loudspeaker or an internal tapping machine to simulate both sound and structure-borne excitation. The investigated rectangular walk-in enclosure had 4.5 × 2.5 × 2 m inside dimensions and was equipped with an operational personal-access door. The thickness and material of the wall panels as well as the thickness of the interior lining could be varied.

Wall Panel Parameters. Figure 12.17 shows the measured insertion loss versus frequency for three different values of the wall thickness h, indicating that above 1.5 mm, a further increase of the wall thickness brings only slight improvement at low frequencies and a slight deterioration at high frequencies near coincidence. Figure 12.17 also includes, as the dashed line, the field-incidence

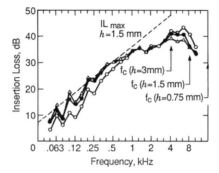

FIGURE 12.17 Curves of power-based insertion loss versus frequency measured (inside–outside direction) for a large walk-in enclosure for various thicknesses of steel plate, h. (After Ref. 10.) Sound-absorbing treatment, 70 mm thick, door sealed; f_c, coincidence frequency; $IL_{max} = R_f$ represents the maximum achievable insertion loss. (△) $h = 3$ mm; (●) $h = 1.5$ mm; (○) $h = 0.75$ mm.

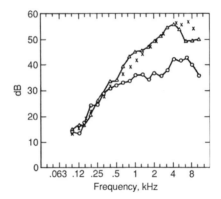

FIGURE 12.18 Comparison of measured enclosure insertion loss IL and panel sound transmission loss R (after Ref. 10): 1.5-mm-thick steel plate, 70-mm-thick sound-absorbing treatment; (△) R, measured; (×) R, predicted according to Chapter 11; (○) IL measured.

sound transmission loss for a 1.5-mm-thick steel plate for comparison, indicating that up to 500 Hz, the insertion loss matches well the field-incidence sound transmission loss of the panel.

Figure 12.18 compares the measured sound transmission loss of a typical 2.1 × 1.2-m wall panel and predicted sound transmission loss of an identical infinite-enclosure panel with the measured insertion loss obtained with an enclosure built with the same panels. The data indicate that the prediction formulas presented in Chapter 11 for infinite panels can be used to predict accurately the sound transmission loss of the finite panels of this specific size and that above 500 Hz the insertion loss is dominated by unintentional small leaks. According to reference 10, the sound-absorbing interior liner provides sufficient structural damping of the wall panels so that additional damping treatment did not result in improved performance.

Stiffening the wall panels with exterior L-channels, which increase the radiation efficiency, resulted in slight deterioration of the acoustical performance. Using steel and wood chipboard of equal mass per unit area yielded different transmission loss but resulted in practically the same insertion loss because in the frequency range where the chipboard had its coincidence frequency the insertion loss was already controlled by leaks.

As a general rule, wall panels should be selected to have *high enough coincidence frequency* to be above the frequency region of interest and be *large enough that their first structural resonance occurs below the frequency region of interest* but heavy enough to yield a field-incidence mass law sound transmission loss that matches the insertion loss requirements. Steel plates 1.5 mm thick usually fulfill most of these requirements.

Stiff, lightweight panels (such as honeycombs) usually yield a critical frequency as low as 500 Hz and provide a very low sound transmission loss in this frequency region. Consequently, such panels should not be used in enclosure design unless the enclosure is required to provide high insertion loss at only very low frequencies and mid- and high-frequency performance is not required.

Effect of Sound-Absorbing Treatment. The proper choice of sound-absorbing treatment plays a more crucial role than the specific choice of wall material or wall thickness, as shown in Fig. 12.19. The sound-absorbing treatment helps to increase insertion loss by (1) reducing reverberant buildup in the enclosure at mid- and high frequencies, (2) increasing the transmission loss of the enclosure walls at high frequencies, and (3) covering up some of the unintentional leaks between adjacent panels and between panels and frames. Whenever feasible, the thickness of the interior sound-absorbing treatment should be chosen to yield a normal-incidence absorption coefficient $\alpha \geq 0.8$ in the frequency region of interest. As shown in Chapter 8, this can be achieved by a layer thickness $d \geq \frac{1}{10}\lambda$, where λ_L is the acoustical wavelength at the lower end of the frequency region, and by choosing a porous material of normalized flow resistance of $1.5 \leq R_1 d/\rho c < 3$.

Leaks. Leaks reduce the insertion loss of large enclosures just as they reduce it for small enclosures [see Eq. (12.13)]. The reduction in insertion loss due to leaks, ΔIL_L, is defined as

$$\Delta \text{IL}_L \equiv \text{IL}_s - \text{IL}_L \cong 10 \log(1 + \beta \times 10^{R_w/10}) \quad \text{dB} \qquad (12.41)$$

where IL_s and IL_L are the insertion losses of the sealed and leaky enclosures, respectively, R_w is the sound transmission loss of the enclosure wall, and $\beta = (1/S_w)\Sigma_j S_{Gj} \times 10^{-R_{Gj}/10}$ is the leak ratio factor, S_{Gj} is the face area, and R_{Gj} is the sound transmission loss of the jth leak. The sound transmission loss of leaks can be positive or, in the case of longitudinal resonances in wide, rigid-walled gaps, also negative. For preliminary calculations it is customary to assume

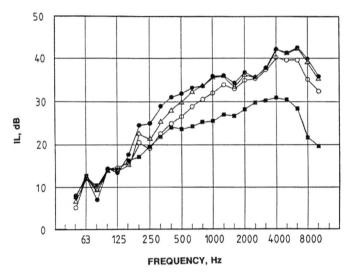

FIGURE 12.19 Measured insertion loss of the enclosure with 1.5-mm-thick steel plates for various thicknesses of the sound-absorbing material (after Ref. 10): (■) 0 mm (no absorption); (○) 20 mm; (△) 40 mm; (•) 70 mm.

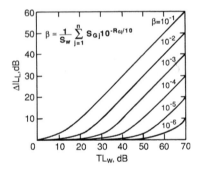

FIGURE 12.20 Decrease of enclosure insertion loss, ΔIL, as a function of the wall sound transmission loss R_w with the leak ratio factor β as parameter; S_w is the wall surface area and R_{Gj} and S_{Gj} are the sound transmission loss and face area of the jth leak, respectively.

$R_{Gj} = 0$. In this case, $\beta = \sum_j S_{Gj}/S_w$ is the ratio of the total face area of the leaks and gaps and the surface area (one side) of the enclosure walls.

Figure 12.20 is a graphical representation of Eq. (12.41) indicating that the higher is the sound transmission loss of the enclosure walls, the higher is the reduction in insertion loss caused by leaks. For example, for an enclosure assembled from wall panels that provide a R_w of 50 dB, the total area of all leaks must be less than 1/1000 percent ($\beta = 10^{-5}$) of the total enclosure surface area if the insertion loss is not to be decreased by more than 3 dB!

Machine Position. Experimental investigations by Fischer and Veres[10] have shown that the insertion loss of an enclosure at low frequencies also depends on the specific position of the enclosed machine within the enclosure. At low frequencies, where the distance between the flat machine wall and the flat enclosure wall panel, d, becomes smaller than one-eighth of the acoustical wavelength ($d < \frac{1}{8}\lambda$), the observed decrease of the insertion loss (from its value obtained when the machine was positioned centrally so that all wall distances were large compared with $\frac{1}{8}\lambda$) was as high as 5 dB. Note that the machine–wall distance d includes the thickness of the internal lining. As a general rule, machines with omnidirectional sound radiation should be positioned centrally. It is especially important to avoid positioning the noisy side of machines very close to enclosure walls, doors, windows, and ventilation openings.

Machine Vibration Pattern. The insertion loss of an enclosure also depends on the specific vibration pattern of the machine, although that dependence appears to not be very strong. Experimental investigations[10] have shown that variation of the insertion loss was about ± 2.5 dB when measured with a loudspeaker inside the model machine and with excitation of the model machine by an ISO tapping machine.

Flanking Transmission through the Floor. The potential insertion loss of an enclosure can be limited severely by flanking if the floor is directly exposed to vibration forces, the internal sound field of the enclosure, or both. Figure 12.21 shows typical enclosure installations ranging from the best to the worst with regard to flanking transmission via the floor. For most equipment it is imperative to provide vibration isolation, a structural break in the floor, or both to reduce the transmission of structure-borne excitation of the floor. Applicable prediction tools and isolation methods are described in Chapters 11 and 13. For the cases depicted in Figs. 12.21c,d, where the interior sound field directly impinges on the floor, the sound-induced vibration of the floor sets a limit to the achievable insertion loss of the enclosure. This limit can be estimated roughly from

$$\text{IL}_L \cong R_F + 10 \log \sigma_F \quad \text{dB} \tag{12.42}$$

where R_F is the sound transmission loss and σ_F is the radiation efficiency of the floor slab.

Relationship between Inside–Outside and Outside–Inside Transmission. Acoustical enclosures are most frequently used to surround a noisy equipment for reducing the noise exposure of a receiver located outside of the enclosure. Less frequently, the enclosure surrounds the receiver (i.e. it is a cabin) to reduce its noise exposure to a sound source located outside of the enclosure. The acoustical performance of the enclosure in the former case is given by its insertion loss in the inside–outside direction, IL_{io}, while the latter is given by its insertion loss in the outside–inside direction, IL_{oi}. All of

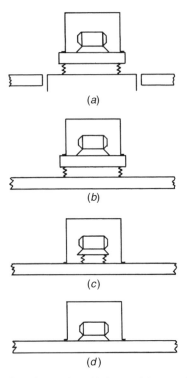

FIGURE 12.21 Rating of enclosure installations with regard to flanking transmission of the floor: (a) best; (b) good; (c) bad; (d) worst.

our considerations in this chapter have dealt with sound transmission from the inside to the outside. According to the principle of reciprocity (see Chapter 11), exchanging the position of the source and receiver does not affect the transfer function. Though it is strictly true only for point sound sources and receivers, it also can be applied to extended sound sources and averaged observer positions in an enclosure, yielding $IL_{io} = IL_{oi}$.

Example. To demonstrate the use of the previously provided design information, predict the insertion loss of the large, acoustically lined, walk-in enclosure investigated experimentally in reference 10. The enclosure is $4.5 \times 2.5 \times 2.0$ m high and is constructed of 1.5-mm-thick steel plates and has a 70-mm-thick interior sound-absorbing lining. The sound transmission loss of the wall panels, R_w, given in Fig. 12.18 by the triangular data points, and the sound absorption coefficient of the interior panel surfaces, $\bar{\alpha}$, taken from reference 10, as a curve that monotonically increases from 0.16 at 100 Hz and reaches a plateau of 0.9 at 600 Hz. The insertion loss is predicted by Eq. (12.38) using the following values: $S_{wi} = S_w = 39$ m^2, $\sum_j S_{Gj} \times 10^{-R_{Gj}/10} = 10^4$, $S_{Gj} = 3.9 \times 10^{-3}$, assuming that leaks account for 1/100 percent of the wall surface area, that there are no silencer openings ($S_s = 0$) and no structure-borne connections ($W_{SB} = 0$), and

that air absorption (mV), the sound absorbed by the machine ($S_i\alpha_i$), and power lost due to dissipation in the panels ($S_w D$) and due to sound radiation ($10^{-R_w/10}$) are small compared with the power dissipated in the sound-absorbing treatment on the interior wall surfaces.
Then Eq. (12.38) simplifies to

$$\text{IL}_L \cong 10 \log \frac{S_w \overline{\alpha}}{\sum_i S_{wi} \times 10^{-R_{wi}/10} + \sum_j S_{Gj} \times 10^{-R_{Gj}/10}} \quad \text{dB} \qquad (12.43)$$

The curve of insertion loss versus frequency predicted by Eq. (12.43) is represented by the open circles in Fig. 12.22, while the solid curve represents the measured data. The measured sound transmission loss of the wall panels, R_w, is the dashed line and represents the limiting insertion loss that could be achieved only if there would be no leaks and the random-incidence sound absorption coefficient $\overline{\alpha}_w$ would approach unity. Observing Fig. 12.22, note that the assumption of 1/100 percent leaks ($\beta = 10^{-4}$) yielded a prediction that matches reasonably well the measured data, which indicates that even such a very small percentage of leaks, which can be realized only if extreme care is exercised during erecting the enclosure and careful caulking of all leaks detected during the initial performance checkout and careful adjustment of door gaskets, can substantially decrease the potential insertion loss of an enclosure.

Partial Enclosures

When the work process of the machine or safety and maintenance requirements do not allow a full enclosure, a partial enclosure (defined as those with more than 10% open area) is used to reduce radiated noise. When a partial enclosure is far enough from the sound source not to cause an increase in source sound

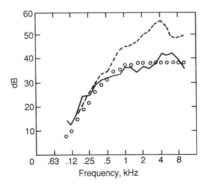

FIGURE 12.22 Typical acoustical performance of a leaky enclosure: (———) IL, measured by Ref. 10; (- - - -) R_w, measured by Ref. 10; (O) IL, predicted by Eq. (12.43) for

$$\beta = \left(\sum_j \left(\frac{S_{Gj}}{S_w} \right) \times 10^{-R_{Gj}/10} \right) = 10^{-4}$$

power by hydrodynamic interactions of the near field of the source with edges of the opening and the walls of the partial enclosure are not rigidly connected to the vibrating machine to serve as a sounding board, a partial enclosure can provide a modest insertion loss of 5 dB or less. According to reference 17, the power-based insertion loss of partial enclosure can be estimated by

$$\text{IL} = 10 \log \left[1 + \alpha \left(\frac{\Omega_{\text{tot}}}{\Omega_{\text{open}}} - 1 \right) \right] \quad \text{dB} \quad (12.44)$$

where Ω_{tot} is the solid angle of sound radiation of the unenclosed source and $\Omega_{\text{open}} = S_{\text{open}}/r$ is the solid angle at which the enclosed source (located at a distance r) "sees" the opening of area S_{open}. Reference 17 contains construction details, curves of measured insertion loss versus frequency, and cost information of 54 different partial enclosures.

12.2 WRAPPINGS

Many machines and pipes need thermal insulation to provide protection of operating personnel or to prevent excessive heat loss. Typically, they need a minimum of 25-mm- (1-in.-) thick glass or ceramic fiber blanket to achieve proper protection. Since many hot equipment, such as turbines, boiler feed pumps, compressor valves, and pipelines, are also sources of intense noise, it is frequently feasible to achieve both heat and acoustical insulation by providing a (preferably) limp, impervious surface layer on top of the porous blanket. The impervious surface layer also provides protection for the porous blanket. Acoustically, the combination of the resilient blanket and the heavy, limp, impervious surface layer provide a spring–mass isolation system. At frequencies where the mass impedance of the surface layer exceeds the combined stiffness impedance of the skeleton of these flexible blankets and the entrapped air, the vibration amplitude of the surface layer becomes smaller than that of the machine surface and wrapping results in an insertion loss that monotonically increases with increasing frequency. The insertion loss of the wrapping is defined the same way as the insertion loss of a close-fitting enclosure and for flat wrappings can be predicted in a manner similar to that of the close-fitting enclosure.

The major difference between wrappings and close-fitting enclosures is that for wrappings the skeleton of the porous layer is in full surface contact with the vibrating surface of the machine, but for the close-fitting enclosures there is no such contact. Consequently, for wrappings the vibrating machine surface transmits a pressure to the impervious surface layer not only through sound propagation in the voids between the fibers but also through the skeleton of the porous material. At low frequencies where the thickness of the porous layer, L, is much smaller than the acoustical wavelength, the porous layer can be represented by a stiffness per unit area, S_{tot}, that according to Mechel[18] can be estimated as

$$S_{\text{tot}} = S_M + S_L \left(1 - \frac{P}{A} \sqrt{\frac{\rho c^2}{L \pi f \gamma R_1 h'}} \right) \quad \text{N/m}^3 \quad (12.45)$$

where S_L is the stiffness of the trapped air,

$$S_L \cong \frac{\rho c^2}{\gamma L h'} \quad \text{N/m}^3 \qquad (12.46)$$

γ the adiabatic exponent, h' the porosity, f the frequency, and R_1 the flow resistivity of the material, and S_M is given by

$$S_M = (2\pi f_0)^2 \rho_s \quad \text{N/m}^3 \qquad (12.47)$$

In Eq. (12.47) the symbol f_0 represents the measured resonance frequency of a mass–spring system consisting of a small rectangular or square-shaped sample of the porous material covered with a metal plate of mass per unit area ρ_s. The resonance frequency is measured by putting the mass–spring system on top of a shaker table and performing a frequency sweep. The second term in Eq. (12.45) accounts for the air that escapes along the perimeter P of a test sample of surface area A, used in the experiment to determine the resonance frequency f_0 according to Eq. (12.47). Young's modulus of the porous material is then determined as $E = S_{\text{tot}}/L$, the speed of longitudinal waves in the porous layer as $c_L = \sqrt{E/\rho_M}$, where ρ_M is the density of the porous material, and the wavenumber as $k = 2\pi f/c_L$. Characteristics of some frequently used thermal insulation materials reported by Wood and Ungar[19] are given below:

Material	Dynamic Stiffness per Unit Area, S_M (N/m^3)	Density, ρ_M (kg/m^3)
Erco-Mat[a]	1.3×10^7	138
Erco-Mat F[a]	2×10^6	104
Glass fiber[b]	4×10^4	12

[a] Nedled glass fiber insulation.
[b] Low-density Owens Corning Fiberglas blanket.

Considering the porous heat insulating layer as a wave-bearing medium of density ρ and a complex Young's modulus $E' = E(1 + j\eta)$ and characterizing the impervious, limp surface layer by its mass per unit area ρ_s, Wood and Ungar[19] derived the analytical formula

$$\text{IL} = 20 \log \left| \cos(kL) - \frac{\rho_s}{\rho_M L}(kL)\sin(kL) \right| \quad \text{dB} \qquad (12.48)$$

where $k = \omega/\sqrt{E(1+j\eta)/\rho_M}$ is the complex propagation constant of the porous layer, $\eta = 1/Q$ the loss factor, and $\rho_s \omega$ the mass impedance per unit area of the covering layer.

At very low frequencies, where $kL \ll 1$, Eq. (12.48) can be approximated by

$$\text{IL}_L = 20 \log \left[1 - \left(\frac{\omega}{\omega_n}\right)^2 \right] \quad \text{dB} \qquad (12.49)$$

where $\omega_n = E/\rho_s L$ is the resonance frequency of a system consisting of a massless spring of stiffness E/L and a mass ρ_s (per unit area). According to Eq. (12.49), the cladding provides no insertion loss for $\omega \ll \omega_n$ and yields a negative insertion loss (amplifies the radiated sound) if $\omega \cong \omega_n$.

For frequencies above the resonance frequency ($\omega \gg \omega_n$) the blanket provides an attenuation that increases monotonically with increasing frequency. Noting that in Eq. (12.48) neither the cosine nor the sine term can exceed unity, an upper bound for the curve of insertion loss versus frequency at high frequencies is obtained by[19]

$$\text{IL}_L \leq 20 \log \left[1 + \frac{\rho_s}{\rho_L} (kL)^2 \right] \quad \text{dB} \tag{12.50}$$

Pipe Wrappings

Pipe wrappings consist of a resilient porous layer and an impervious jacket. The impervious jacket is usually sheet metal or loaded plastic. They achieve their acoustical performance in a similar manner as flat wrappings. However, there is an important difference. While flat wrappings do not increase the sound-radiating surface, when a wrapping is applied to a small-diameter pipe, the diameter of the impervious jacket can be substantially larger than that of the bare pipe. This increases both the radiating surface and the radiation efficiency. Accordingly, at low frequencies, below or slightly above the resonance frequency of the pipe wrapping, the insertion loss is negative. Positive insertion loss is usually achieved only above 200 Hz.

According to Michelsen, Fritz, and Sazenhofen,[20] the maximal achievable insertion loss of wrappings can be estimated by the empirical formula

$$\text{IL}_{\max} = \frac{40}{1 + 0.12/D} \log \frac{f}{2.2 f_0} \quad \text{dB} \tag{12.51}$$

where

$$f_0 = 60/\sqrt{\rho_s L} \quad \text{Hz} \tag{12.52}$$

where L is the thickness of the porous resilient layer and D is the pipe diameter, both in meters, and ρ_s is the mass per unit area of the impervious jacket in kilograms per square meter.

Equation (12.51) is valid only if there are no structure-borne connections between the pipe and jacket and for frequencies $f \geq 2 f_0$. Figure 12.23 shows curves of measured insertion loss versus frequency obtained in reference 20 for typical pipe wrapping consisting of a galvanized steel jacket, thickness 0.75–1 mm, and a porous resilient layer with the following parameters:

Thickness L	30, 60, 80, and 100 mm
Density	85–120 kg/m^3
Flow resistivity	3×10^4 N · s/m^4
Dynamic Young's modulus	2×10^5 N/m^2

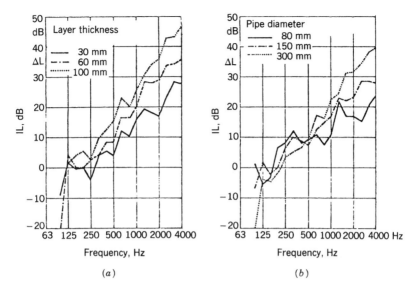

FIGURE 12.23 Curves of measured insertion loss versus frequency of pipe wrappings. (After Ref. 20.) (*a*) Effect of layer thickness. (*b*) Effect of pipe diameter.

Figure 12.23*a* shows the effect of layer thickness L on the insertion loss for a pipe of diameter $D = 300$ mm while Fig. 12.23*b* illustrates the effect of pipe diameter D for a constant layer thickness $L = 60$ mm. The insertion loss above 250 Hz increases with increasing thickness of the porous layer and with increasing diameter of the bare pipe. The standard deviation of the measured insertion loss around that predicted by Eq. (12.52) was 4 dB.

Note that spacers between the pipe and jacket result in insertion loss values that may be substantially lower than those predicted by Eq. (12.50) unless the spacers are less stiff dynamically than the porous layer.

REFERENCES

1. ISO 15667: 2000E, "Guidelines for Noise Control by Enclosures and Cabins," International Organization for Standardization, Geneva, Switzerland, 2000.
2. I. L. Ver, "Enclosures and Wrappings," in L. L. Beranek and I. L. Ver (Eds.), *Noise and Vibration Control Engineering*, J. Wiley, New York, 1992.
3. R. H. Lyon, "Noise Reduction of Rectangular Enclosures with One Flexible Wall," *J. Acoust. Soc. Am.*, **35**(11), 1791–1797 (1963).
4. I. L. Vér, "Reduction of Noise by Acoustic Enclosures," in *Isolation of Mechanical Vibration, Impact and Noise*, Vol. 1, American Society of Mechanical Engineers, New York, 1973, pp. 192–220.
5. J. B. Moore, "Low Frequency Noise Reduction of Acoustic Enclosures," *Proc. NOISE-CON*, (1981) pp. 59–64.

6. H. M. Fuches, U. Ackermann, and W. Frommhold, "Development of Membrane Absorbers for Industrial Noise Abatement," *Bauphysik*, **11**(H.1), 28–36 (1989) (in German).
7. M. H. Fischer, H. V. Fuchs, and U. Ackermann, "Light Enclosures for Low Frequencies," *Bauphysik*, **11**(H.1), 50–60 (1989) (in German).
8. I. L. Vér and E. Veres, "An Apparatus to Study the Sound Excitation and Sound Radiation of Platelike Structures," *Proc. INTER-NOISE*, (1980) pp. 535–540.
9. K. P. Byrne, H. M. Fischer, and H. V. Fuchs "Sealed, Close-Fitting, Machine-Mounted Acoustic Enclosures with Predictable Performance," *Noise Control Eng. J.*, **31**, 7–15 (1988).
10. M. H. Fischer and E. Veres, "Noise Control by Enclosures" Research Report 508, Bundesanstalt fuer Arbeitschutz, Dortmund, 1987 (in German); also Fraunhofer Institute of Building Physics Report, IBP B5 141/86, 1986.
11. J. B. Moreland, Westinghouse Electric Corporation, private communication.
12. L. Cremer, M. Heckl, and E. E. Ungar, *Structureborne Sound*, Berlin, Springer, 1988.
13. K. Goesele, *Berichte aus der Bauforschung [Sound Transmission of Doors]*, H. 63 Wilhelm Ernst & Sohn Publishers, Berlin, 1969, pp. 3–21.
14. F. P. Mechel, "Transmission of Sound through Holes and Slits Filled with Absorber and Sealed," *Acustica*, **61**(2), 87–103 (1986).
15. F. P. Mechel, "The Acoustic Sealing of Holes and Slits in Walls," *J. Sound Vib.*, **III**(2), 297–336 (1986).
16. G. Kurtze and K. Mueller "Noise Reducing Enclosures," Research Report BMFT-FB-HA 85-005, Bundesministerium fuer Forschung und Technologie, Berlin, 1985 (in German).
17. U. J. Kurze et al., "Noise Control by Partial Enclosures; Shielding in the Nearfield," Research Report No. 212, Bundesanstalt fuer Arbeitschutz und Unfallforshung, 1979 (in German).
18. F. P. Mechel, "Sound Absorbers and Absorber Functions" in *Reduction of Noise in Buildings and Inhabited Regions*, Strojnizdat, Moscow, 1987 (in Russian).
19. E. W. Wood and E. E. Ungar, BBN Systems and Technologies Corporation, private communications.
20. R. Michelsen, K. R. Fritz, and C. V. Sazenhofen, "Effectiveness of Acoustic Pipe Wrappings" in *Proc. DAGA '80*, VDE-Verlag, Berlin, 1980, pp. 301–304 (in German).

BIBLIOGRAPHY

Miller, R. K., and W. V. Montone, *Handbook of Acoustical Enclosures and Barriers*, Fairmont, Atlanta, GA, 1978.

"Noise Reduction by Enclosures," VDI Guideline 2711, VDI-Verlag GmbH, Dusseldorf, Germany (in German).

CHAPTER 13

Vibration Isolation

ERIC E. UNGAR AND JEFFREY A. ZAPFE

Acentech, Inc.
Cambridge, Massachusetts

13.1 USES OF ISOLATION

Vibration isolation refers to the use of comparatively resilient elements for the purpose of reducing the vibratory forces or motions that are transmitted from one structure or mechanical component to another. The resilient elements, which may be visualized as springs, are called vibration isolators. Vibration isolation is generally employed (1) to protect a sensitive item of equipment from vibrations of the structure on which it is supported or (2) to reduce the vibrations that are induced in a structure by a machine it supports. Vibration isolation may also be used to reduce the transmission of vibrations to structural components whose attendant sound radiation one wishes to control.

13.2 CLASSICAL MODEL

Mass–Spring–Dashpot System

Many aspects of vibration isolation can be understood from analysis of an ideal, linear, one-dimensional, purely translational mass–spring–dashpot system like that sketched in Fig. 13.1. The isolator is represented by the parallel combination of a massless spring of stiffness k (which produces a restoring force proportional to the displacement) and a massless damper with viscous damping coefficient c (which produces a force proportional to the velocity and opposing it). The rigid mass m, which here is taken to move only vertically and without rotation, corresponds either to an item to be protected (Fig. 13.1a) or to a machine frame on which a vibratory force acts (Fig. 13.1b).

Transmissibility

Although the mass–spring–dashpot diagrams of Figs. 13.1a and b are similar, they describe physically different situations. In Fig. 13.1a, the support S is

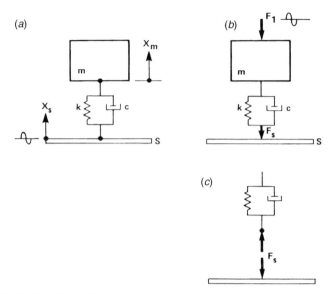

FIGURE 13.1 One-dimensional translational mass–spring–dashpot system: (*a*) excited by support motion; (*b*) excited by force acting on mass; (*c*) force F_s between isolator and support.

assumed to vibrate vertically with a prescribed amplitude X_s at a given frequency; the purpose of the isolator is to keep the displacement amplitude X_m of mass m acceptably small. In Fig. 13.1b, a force of a prescribed amplitude F_1 at a given frequency is assumed to act on mass m; the purpose of the isolator is to keep the amplitude F_s of the force that acts on the support within acceptable limits, thereby also keeping the support's resulting motion adequately small.

For the situation represented by Fig. 13.1a, the ratio $T = X_m/X_s$ of the amplitude of the mass' displacement to that of the disturbing displacement of the support S is called the (motion) transmissibility. For the situation represented by Fig. 13.1b, the ratio $T_F = F_s/F_1$ of the amplitude of the force transmitted to the support S to that of the disturbing force is called the force transmissibility. In many practical instances corresponding to force excitation as represented by Fig. 13.1b, the support is so stiff or massive that its displacement may be taken to be zero. It turns out that the force transmissibility T_{F0} obtained with an immobile support (for the case of Fig. 13.1b) is given by the same expression as the motion transmissibility (for the case of Fig. 13.1a),* namely,[1,2]

$$T = T_{F0} = \sqrt{\frac{1 + (2\zeta r)^2}{(1 - r^2)^2 + (2\zeta r)^2}} \tag{13.1}$$

*This equality holds not only for the simple system of Fig. 13.1, but also for any mathematically linear system.[3] Because of this equality, the literature generally does not differentiate between the two types of transmissibility.

where $r = f/f_n$ is the ratio of the excitation frequency to the natural frequency of the mass–spring system and $\zeta = c/c_c$, which is called the *damping ratio*, is the ratio of the system's viscous damping coefficient c to its critical damping coefficient c_c.

The critical damping coefficient is given by $c_c = 2\sqrt{km} = 4\pi f_n m$. The natural frequency f_n of a spring–mass system obeys[1,2]

$$2\pi f_n = \sqrt{\frac{k}{m}} = \sqrt{\frac{kg}{W}} = \sqrt{\frac{g}{X_{st}}} \qquad (13.2)$$

where g denotes the acceleration of gravity, W the weight associated with the mass m, and X_{st} the static deflection of the spring due to this weight. In customary units,

$$f_n(\text{Hz}) \approx \frac{15.76}{\sqrt{X_{st}(\text{mm})}} \approx \frac{3.13}{\sqrt{X_{st}(\text{in.})}} \qquad (13.3)$$

The relation $X_{st} = k/W$ and Eq. (13.3) hold only if the spring is mathematically linear—that is, if the slope of the spring's force–deflection curve (which slope corresponds to the stiffness k) is constant. For small-amplitude vibrations about an equilibrium deflection one may take k in the first form of Eq. (13.2) as the slope dF/dx of the spring's force–deflection curve* at the spring's static deflection under an applied load W. However, for some practical isolators, notably those incorporating elastomeric materials, the effective dynamic stiffness may be considerably greater than that obtained from a quasi-static force–deflection curve.[4]

Figure 13.2 shows curves based on Eq. (13.1) for several values of the damping ratio. For small frequency ratios, $r \ll 1$, the transmissibility T is approximately equal to unity; the motion or force is transmitted essentially without attenuation or amplification. For values of r near 1.0, T becomes large (at $r = 1$ or $f = f_n$, $T = 1/2\zeta$); the system responds at resonance, resulting in amplification of the motion or force. All curves pass through unity at $r = \sqrt{2}$.

For $r > \sqrt{2}$, T is less than unity and decreases continually with increasing r. In this high-frequency range, which may be called the *isolation range*, the inertia of the mass plays the dominant role in limiting the mass excursion and thus in limiting the mass' response to support displacement or to forces acting on the mass. Thus, if one desires to achieve good isolation, which corresponds to small transmissibility, one needs to choose an isolator with the smallest possible k (i.e., with the largest practical static deflection X_{st}) in order to obtain the smallest f_n and the greatest value of $r = f/f_n$ for a given excitation frequency.

---

*Nonlinear spring elements for which $k = dF/dx$ is proportional to the applied load W in a given range are the basis for isolators which have the practical advantage of providing the same natural frequency for all loads in the given range.[4]

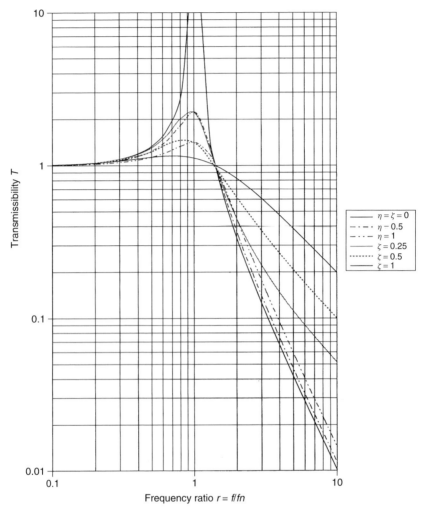

FIGURE 13.2 Transmissibility of mass–spring–damper system of Fig. 13.1 from Eqs. (13.1) and (13.4).

Isolation Efficiency

The performance of an isolation system is sometimes characterized by the isolation efficiency I, which is given by $I = 1 - T$. Whereas the transmissibility indicates the fraction of the disturbing motion or force that is transmitted, the isolation efficiency indicates the fraction by which the transmitted disturbance is less than the excitation. Isolation efficiency is often expressed in percent. For example, if the transmissibility is 0.0085, the isolation efficiency is 0.9915, or 99.15%, indicating that 99.15% of the disturbance does not "get through" the isolator.

Effect of Damping

In the isolation range, that is, for frequency ratios $r > \sqrt{2}$, increased viscous damping results in increased transmissibility, as evident from Fig. 13.2 or Eq. (13.1). Although this fact is mentioned in many texts, it is of little practical consequence for two reasons. First, in practice one rarely encounters systems with damping ratios that are greater than 0.1 unless high damping is designed into a system on purpose—and small values of the damping ratio, and certainly small changes in that damping ratio, have little effect on transmissibility. Second, Eq. (13.1) and Fig. 13.2 pertain only to viscously damped systems, in which a damper produces a retarding force that is proportional to the velocity. Although such systems have been studied most extensively (largely because they are relatively easy to analyze mathematically), the retarding forces in practical systems generally have other parameter dependences.

The effect of damping in practical isolators is generally better represented by *structural damping* than by viscous damping. In structural damping, the retarding force acts to oppose the motion, as in viscous damping, but is proportional to the displacement. For structurally damped systems,[2]

$$T = F_{F0} = \sqrt{\frac{1+\eta^2}{(1-r^2)^2 + \eta^2}} \qquad (13.4)$$

where η denotes the *loss factor* of the system.*

If a structurally damped system has the same amplification at the natural frequency as a similar viscously damped system, then $\eta = 2\zeta$. The transmissibility of a structurally damped system in the isolation range increases much less rapidly with increasing damping than does that of a system with viscous damping, as evident from Fig. 13.2.

In practical isolation arrangements the damping typically is very small, that is, $\zeta < 0.1$. For such small amounts of damping, the transmissibility in the isolation range differs little from that for zero damping, so that one may approximate the transmissibility in that range by

$$T = T_{F0} = \frac{1}{|r^2 - 1|} \approx \left(\frac{f_n}{f}\right)^2 \qquad (13.5)$$

where the rightmost expression applies for $r^2 = (f/f_n)^2 \gg 1$.

Effects of Inertia Bases

An isolated machine is often mounted on a massive support, generally called an *inertia base*, to increase the isolated mass. If the isolators are not changed as

---

*The loss factor η in Eq. (13.4) may be taken to vary with frequency as determined from experimental data. See Chapter 14.

the mass is increased, then the natural frequency of the system is reduced, the frequency ratio r corresponding to a given excitation frequency is increased, and the transmissibility is reduced; that is, isolation is improved.

However, practical considerations, such as the load-carrying capacities of isolators, usually dictate that the total isolator stiffness be increased* as the mass is increased, with the static deflection changing little. (In fact, conventional commercial isolators typically are specified in terms of the loads they can carry and the corresponding static deflection.) As evident from Eq. (13.3), the natural frequency remains unchanged if the static deflection remains unchanged, and addition of an inertia base then does not change the transmissibility.

Thus, inertia bases in practice typically do not improve isolation significantly. They are of some benefit, however: With a greater mass supported on stiffer springs, forces that act directly on the isolated mass produce smaller static and vibratory displacements of that mass.

Effect of Machine Speed

In a rotating or reciprocating machine, the dominant excitation forces generally are due to dynamic unbalance and occur at frequencies that correspond to the machine's rotational speed and/or multiples of that speed. Greater speed thus corresponds to increased excitation frequencies and larger values of r—and therefore to reduced transmissibility (or improved isolation). As evident from Eq. (13.1) or (13.5), the transmissibility—the ratio of the transmitted to the excitation force—becomes smaller as the speed (and the excitation frequency) is increased; in the isolation range, the transmissibility is very nearly inversely proportional to the square of the excitation frequency. However, the excitation forces associated with unbalance vary as the square of speed (and frequency) and thus increase with speed about as much as the transmissibility decreases. The net result is that the magnitudes of the forces that are transmitted to the supporting structure are virtually unaffected by speed changes, although speed changes do affect the frequencies at which these forces occur.

Limitations of Classical Model

Although the linear single-degree-of-freedom model provides some useful insights into the behavior of isolation systems, it obviously does not account for many aspects of realistic installations. Clearly, real springs are not massless and may be nonlinear, and real machine frames and supporting structures are not rigid. Resiliently supported masses generally move not only vertically but also horizontally—and they also tend to rock.

In simple classical analyses, furthermore, the magnitude of the exciting force or motion is taken as constant and independent of the resulting response, whereas

---

*Note that the single spring and dashpot in the schematic diagrams of Fig. 13.1 represent the entire isolation system, which in reality may consist of many isolators. Addition of isolators amounts to an increase in the stiffness of the isolation system.

in actual situations the excitation often depends significantly on the response, as discussed in Section 13.4 under Loading of Sources.

13.3 ISOLATION OF THREE-DIMENSIONAL MASSES

General

Unlike the simple model shown in Fig. 13.1, where the mass can only move vertically without rotation and the system has only one natural frequency, an actual rigid three-dimensional mass has six degrees of freedom; it can translate in three coordinate directions and rotate about three axes. An elastically supported rigid mass thus has six natural frequencies. A nonrigid mass has many additional ones associated with its deformations. Obtaining effective isolation here requires that all of the natural frequencies fall considerably below the excitation frequencies of concern. Descriptions of the natural frequencies and responses of general isolated rigid masses are available[3,5] but are so complex that they provide little practical insight and tend to be used only rarely for design purposes.

Coupling of Vertical Motion and Rocking

Figure 13.3 is a schematic diagram of a mass m supported on two isolators in a plane through its center of gravity, parallel to the plane of the paper (or on two rows of isolators extending in the direction perpendicular to the plane of the paper) with stiffnesses k_1 and k_2 located at distances a_1 and a_2 from the mass' center of gravity. One may visualize easily that a downward force applied at the center of gravity of the mass in general would produce not only a downward displacement of the center of gravity but also a rotation of the mass, the latter due to the moment resulting from the isolator forces. Similarly, purely vertical up-and-down motion of the support S would in general result in rocking of the mass, in addition to its vertical translation. Vertical and rocking motions here are said to be "coupled."

The natural frequency f_v at which pure vertical vibration would occur if the mass would not rock (i.e., the "uncoupled" vertical natural frequency) is given by

$$2\pi f_v = \sqrt{\frac{k_1 + k_2}{m}} \tag{13.6}$$

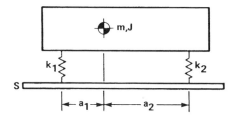

FIGURE 13.3 Mass m with moment of inertia J supported on two isolators.

The natural frequency f_r at which the mass would rock if its center of gravity would not move vertically (i.e., the uncoupled rocking frequency) is given by

$$2\pi f_r = \sqrt{\frac{k_1 a_1^2 + K_2 a_2^2}{J}} \tag{13.7}$$

where J denotes the mass polar moment of inertia about an axis through the mass' center of the gravity and perpendicular to the plane of the paper.

The rocking frequency f_r may be either smaller or larger than f_v. In the general case, the two natural frequencies of the system differ from both of these frequencies and correspond to motions that combine rotation and vertical translation (but without rotation out of the plane of the paper). These "coupled" motion frequencies always lie above and below the foregoing uncoupled motion frequencies; that is, one of the coupled motion frequencies lies below both f_v and f_r and the other lies above both f_v and f_r; coupling in effect increases the spread between the natural frequencies.[6,7]

The presence of coupled motions complicates the isolation problem because one needs to ensure that both of the coupled natural frequencies fall considerably below the excitation frequencies of concern. To eliminate this complication, one may select the locations and stiffnesses of the isolators so that the forces produced by them when the mass moves downward without rotation result in zero net moment about the center of gravity. This situation occurs if the isolators are linear and designed so that they have the same static deflection under the static loads to which they are subject. (Selection of isolators that have the same unloaded height as well as the same static deflection results also in keeping the isolated equipment level.) Then f_v and f_r of Eqs. (13.6) and (13.7) are the system's actual natural frequencies, and vertical forces at the center of gravity or vertical support motions produce no rocking.

Effect of Horizontal Stiffness of Isolators

In the foregoing discussion, horizontal translational motions and the effects of horizontal stiffnesses of the isolators were neglected. However, any real isolator that supports a vertical load also has a finite horizontal stiffness, and in some systems separate horizontally acting isolators may be present. Figure 13.4 is a schematic diagram of a mass on two vertical isolators (or on two rows of such isolators extending perpendicular to the plane of the paper) with stiffnesses k_1 and k_2. The effects of the horizontal stiffnesses of these isolators or of separate horizontally acting isolators are represented by the horizontally acting spring elements with stiffnesses h_1 and h_2, here assumed to act in the same plane at a distance b below the center of gravity of the mass.

The system of Fig. 13.4 has three degrees of freedom and therefore three natural frequencies. If the vertically acting isolators are selected and positioned so that they have the same static deflection, then (as discussed in the foregoing section) there is very little coupling between the vertical translational and the

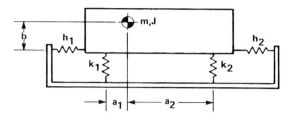

FIGURE 13.4 Mass m with moment of inertia J supported by two vertically and two collinear horizontally acting isolators.

rocking motions, and the natural frequency corresponding to vertical translation is given by Eq. (13.6). The other two natural frequencies, which are associated with combined rotation and horizontal translation, are given by the two values of f_H one may obtain from[6,7]

$$\frac{f_H}{f_v} = (N \pm \sqrt{N^2 - SB})^{1/2} \qquad (13.8)$$

where

$$N = \frac{1}{2}\left[S\left(1 + \frac{b^2}{r^2}\right) + B\right] \qquad B = \frac{a_1^2 k_1 + a_2^2 k_2}{r^2(k_1 + k_2)} \qquad S = \frac{h_1 + h_2}{k_1 + k_2} \qquad (13.8a)$$

Here, $r^2 = J/m$ represents the square of the radius of gyration of the mass about an axis through its center of gravity.

For a rectangular mass of uniform density, the center of gravity is in the geometric center and the radius of gyration r obeys $r^2 = \frac{1}{12}(H^2 + L^2)$, where H and L represent the lengths of the vertical and horizontal edges of the mass (in the plane of Fig. 13.4).

Nonrigidity of Mass or Support

If the mass of Fig. 13.1a is flexible instead of rigid and has a resonance at a certain frequency, then it will deflect (by deforming) considerably in response to excitation at that frequency, resulting in large transmissibility and thus in poor isolation. An analogous situation occurs if the system of Fig. 13.1b is excited at a resonance frequency of the support.

Figure 13.5 is a schematic diagram representing the vertical translational motion of a machine isolated from a nonrigid support, where the support is represented by a spring–mass system, corresponding, for example, to the static stiffness of a building's floor and to the effective mass that participates in the floor's vibration at its fundamental resonance. Isolator and floor damping have little effect on the off-resonance vibrations and may be neglected for the sake of simplicity. If f_M denotes the natural frequency of the isolated machine on a

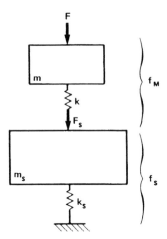

FIGURE 13.5 Schematic representation of isolated machine on nonrigid support structure. Machine is represented by mass m subject to oscillatory force F; isolator is represented by spring k. Support structure is represented by effective mass m_s on spring k_s.

rigid support and f_S represents the natural frequency of the support without the machine in place, so that

$$2\pi f_M = \sqrt{\frac{k}{m}} \qquad 2\pi f_S = \sqrt{\frac{k_s}{m_s}} \qquad (13.9)$$

then the force transmissibility, that is, the ratio of the magnitude of the force F_S transmitted to the support to that of the excitation force F, may be written as

$$T_F = \frac{F_S}{F} = \left| \frac{1 - R^2}{(1 - R^2)(1 - R^2 G^2) - R^2/M} \right| \qquad (13.10)$$

where $R = f/f_S$, $G = f_S/f_M$, and $M = m/m_s$ with f representing the excitation frequency.

Figure 13.6 shows a plot of the transmissibility calculated from Eq. (13.10) for the illustrative case where $M = \frac{1}{2}$, $G = 2$, together with a corresponding plot of the transmissibility that would be obtained with an immobile (infinitely rigid) support. For excitation frequencies that fall between the two resonance frequencies of the system with the nonrigid support, the transmissibility obtained with the nonrigid support is less than that with the rigid support; the reverse is true for excitation frequencies above the upper of the two resonance frequencies. For sufficiently high excitation frequencies, the difference between the two transmissibilities becomes negligible.

The two resonance frequencies f_c of the system with the nonrigid support may be found from

$$\left(\frac{f_c}{f_M}\right)^2 = P \pm \sqrt{P^2 - G^2} \qquad P = \tfrac{1}{2}(1 + G^2 + M) \qquad (13.11)$$

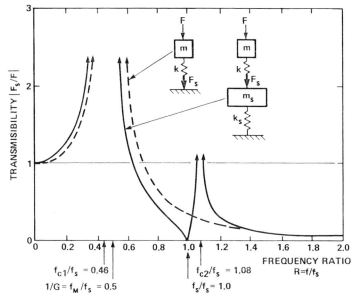

FIGURE 13.6 Comparison of transmissibilities obtained with rigid and nonrigid supports. Curves were obtained from Eqs. (13.1) and (13.10) for zero damping and for mass ratio $M = m/m_s = \frac{1}{2}$ and system frequency ratio $G = f_S/f_M = 2$. Resonance frequencies f_{c1} and f_{c2} of coupled system and f_M and f_S of uncoupled machine and support are indicated on the frequency scale.

with G and M defined as indicated after Eq. (13.10). The higher resonance frequency, obtained with the plus sign in Eq. (13.11), always lies above both f_M and f_S; the lower resonance frequency, obtained with the minus sign, always lies below both f_M and f_S.

13.4 HIGH-FREQUENCY CONSIDERATIONS

Although the relations discussed in the foregoing sections usually suffice for analysis and design in cases where only relatively low frequencies are of concern (i.e., for frequencies that are generally below the resonance frequencies of the machine and support structures themselves), at higher frequencies there occur complications that may need to be taken into account. At high frequencies, the isolated item and the support structure cease to behave as rigid masses, isolators may exhibit internal resonances, and vibration sources tend to be affected by "loading."

Loading of Sources

Loading of a source refers to the reduction of the source's vibratory motion that results from a force that opposes this motion. For example, the vibrations of a sheet metal enclosure tend to be reduced by the opposing forces produced when a hand is placed against it. Similarly, a factory floor that vibrates with a certain amplitude tends to vibrate less when a (nonvibrating) machine is bolted to it as the result of reaction forces produced by the machine.

In the latter example, if the machine is isolated from the floor, it is likely to produce smaller reaction forces on the floor than it would produce if it is bolted rigidly to the floor, and the floor may be expected to vibrate more if the machine is isolated than if it is rigidly fastened to the floor. Since greater floor vibrations then are transmitted to the machine, use of isolation will protect the machine less than anticipated from transmissibility considerations.

To be able to evaluate how much protection of the machine the isolator provides in this case, one needs a quantitative description of the source's response to loading. One may obtain such a description by measuring the motion produced by the source as it acts on several different masses or structures with different dynamic characteristics (impedances), which produce different (known or measurable) reaction forces. For prediction purposes, it often suffices to assume that the source's response to loading at any given frequency is "linear"; that is, the motion amplitude produced by the source decreases in proportion to the amplitude F_0 of the reaction force (which is equal to the amplitude of the force that the source generates). One may describe source motions at a given frequency equally well in terms of acceleration, displacement, or velocity, but use of velocity has become customary. For a linear source one may express the dependence of the source's velocity amplitude V_0 on the force as[8]

$$V_0 = V_{\text{free}} - M_s F_0 \qquad (13.12)$$

Here V_{free} represents the velocity amplitude that the source generates if it is free of reaction forces, that is, if it produces zero force. (In general, the parameters V_{free} and M_s for a given source may be different for different frequencies.) The quantity M_s, which indicates how rapidly V_0 decreases with increasing F_0, is known as the *source mobility* and may be found from $M_s = V_{\text{free}}/F_{\text{blocked}}$, where F_{blocked} denotes the force amplitude obtained if the source is "blocked" so that it has zero velocity V_0.

One may readily verify that $M_s = 0$ corresponds to a *velocity source*, that is, to a source whose output velocity amplitude is constant, regardless of the magnitude of its output force F_0. Similarly, infinite M_s corresponds to a *force source*, whose output force F_0 is constant and independent of its output velocity.[8] (For example, a rotating unbalanced mass generates forces that are virtually independent of its support motions and thus acts essentially like a force source. On the other hand, a piston driven by a shaft with a large flywheel moves at essentially the same amplitude regardless of the force acting on it and thus behaves like a velocity source.)

Isolation Effectiveness[3,8]

In the presence of significant source loading, one cannot evaluate the performance of an isolation system on the basis of transmissibility because in the definition of transmissibility the magnitudes of the disturbances are prescribed and thus, in effect, are taken as constant. A measure of isolation performance that is useful in the presence of loading is the so-called *isolation effectiveness E*. Isolation effectiveness is defined as the ratio of the magnitude of the vibrational velocity of the item to be protected (called the "receiver") that results if the item is rigidly connected to the source to the magnitude of the receiver's velocity that is obtained if the isolator is inserted between the source and the receiver in place of the rigid connection. The definition of isolation effectiveness is analogous to that of insertion loss in airborne acoustics.

If the receiver velocity V_R is proportional to the force F_R that acts on the receiver, so that $V_R = M_R F_R$, where M_R is called receiver mobility,* then the isolation effectiveness may be expressed in terms of a ratio of forces acting on the receiver as well as of a ratio of receiver velocities, namely,

$$E = \frac{V_{Rr}}{V_{Ri}} = \frac{F_{Rr}}{F_{Ri}} \qquad (13.13)$$

where the added subscript r refers to the case in which a rigid connection replaces the isolator and the subscript i refers to the situation where the isolator is present.

Whereas small transmissibility T corresponds to good isolation, it is large values of E that imply effective isolation. For this reason, the reciprocal of the effectiveness is sometimes used to characterize the performance of an isolation system. Although this reciprocal differs from the transmissibility T in the general situation where the source is affected by loading, in the special case of sources that are unaffected by loading (i.e., for sources that generate load-independent velocity or force amplitudes), $E = 1/T$.

Effectiveness of Massless Linear Isolator

An isolator may be considered as "massless" if it transmits whatever force is applied to it. (Equal and opposite forces must act on the two sides of a massless isolator if it is not to accelerate infinitely.) A "linear" isolator is one whose deflection is proportional to the applied force. The velocity difference across such an isolator at any frequency then is also proportional to the applied force, and the ratio of the magnitude of this velocity difference to that of the applied force is called the isolator's mobility M_I. Like the reciprocal of isolator stiffness, M_I is large for a soft isolator and zero for a rigid isolator.

For a massless linear isolator, the effectiveness obeys[8,11]

$$E = \left| 1 + \frac{M_I}{M_S + M_R} \right| \qquad (13.14)$$

---

*Mobilities of receivers at their attachment points may be estimated analytically for simple configurations (e.g., refs. 9 and 10) or may be measured (as functions of frequency). The velocities and forces usually are expressed in terms of complex numbers, or *phasors*, which indicate both the magnitudes and the relative phases of sinusoidally varying quantities. Mobilities then also are complex quantities in general.

570 VIBRATION ISOLATION

At a resonance of the receiver, the receiver's vibratory velocity V_R resulting from a given force F_R is large, so that the receiver's mobility is large. In view of Eq. (13.14), the effectiveness of an isolator is small in the presence of such a resonance.

Equation (13.14) also indicates that the effectiveness is small if the source mobility M_S is large. For a force source, for which M_S is infinite, the effectiveness is equal to unity, implying that the receiver vibrates just as much with the isolator in place as it does if the isolator is replaced by a rigid connection. This initially somewhat surprising result is correct: After all, the force source generates the same force, regardless of the velocity or displacement it produces, and the isolator transmits all of this force, with a softer isolator merely leading to greater displacement of the source at its output point.

Consequences of Isolator Mass Effects

Isolator mass effects may be neglected—that is, an isolator may be considered as massless—as long as the frequencies under consideration are appreciably lower than the first internal or *standing-wave* resonance frequency of the isolator.* Such standing-wave resonances tend to reduce the isolator's effectiveness severely, as illustrated by Fig. 13.7. This figure shows the calculated transmissibility of a leaf spring modeled as a uniform cantilever beam. The upper left-hand corner of the plot may be recognized as the usual transmissibility curve (similar to Fig. 13.2) in the vicinity of the resonance frequency $f_n = (1/2\pi)\sqrt{k/m}$ obtained for a massless isolator. With increasing excitation frequency the transmissibility does not decrease monotonically, as it would for a massless spring (see curve for $m_{\rm sp} = 0$); instead, there occur secondary peaks associated with standing-wave resonances of the beam. The frequency at which these peaks begin to occur increases as the ratio of the isolated mass m to the mass $m_{\rm sp}$ of the spring increases. Although in the figure only two peaks are shown for each mass ratio, there actually occurs a succession of peaks that become more closely spaced with increasing frequency. The magnitude of these peaks decreases with increasing damping.

To reduce the effects of standing-wave resonances, one thus needs to select an isolator with relatively high damping and a configuration for which the onset of standing-wave resonances occurs at comparatively high frequencies. This implies use of a material with high stiffness-to-weight ratio or, equivalently, with a high longitudinal wave velocity $\sqrt{E/\rho}$ (where E denotes the material's modulus of elasticity and ρ its density), and also use of a configuration with small overall dimensions.

---

*At lower frequencies, the only effect of the mass of the isolator is to reduce slightly the fundamental resonance frequency of the system. The modified resonance frequency may be calculated from the isolator stiffness and a mass consisting of the isolated mass plus a fraction of the mass of the isolator. If the isolator consists of a uniform spring or pad in compression or shear, the fraction is $\frac{1}{3}$; if the isolator consists of a uniform cantilever beam, the fraction is approximately 0.24.

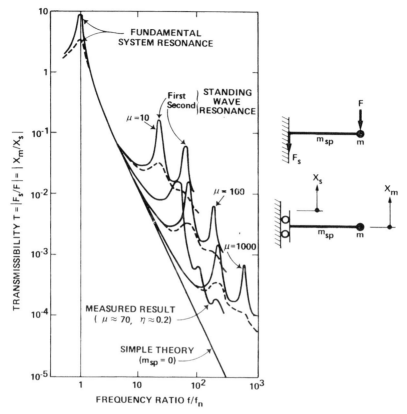

FIGURE 13.7 Effect of isolator mass and damping on transmissibility of uniform cantilever for three values of ratio $\mu = m/m_{sp}$ of isolated mass to mass of cantilever spring. (After Refs. 11 and 12.) Solid calculated lines are for loss factor $\eta = 0.1$; dashed lines for $\eta = 0.6$. (Measured result shown is from Ref. 13.) Frequency is normalized to f_n, the fundamental resonance frequency obtained with a massless spring.

The isolation effectiveness at any specified frequency of a system in which isolator mass effects are not negligible is given by[8]

$$E = \left| \frac{\alpha}{M_S + M_R} \right| \left| 1 + \frac{M_S}{M_{lsb}} + \frac{M_R}{M_{lrb}} \left(1 + \frac{M_S}{M_{lsf}} \right) \right|$$

$$\frac{1}{\alpha^2} = \frac{1}{M_{lrb}} \left(\frac{1}{M_{lsb}} - \frac{1}{M_{lsf}} \right)$$

(13.15)

where M_{lsb} denotes the isolator mobility (i.e., the velocity-to-force ratio) measured on the source side of the isolator if the receiver side of the isolator is "blocked" (i.e., prevented from moving), M_{lrb} denotes the mobility measured on the receiver side of the isolator if the source side is blocked, and M_{lsf} denotes the mobility measured on the source side if the receiver side is "free" or unconstrained. One may readily verify that Eq. (13.15) reduces to Eq. (13.14) for massless isolators, for which $M_{lsb} = M_{lrb} = M_I$ and M_{lsf} is infinite.

13.5 TWO-STAGE ISOLATION

A force applied to one side of a massless isolator, as has been mentioned, must be balanced by an equal and opposite force at the other side of the isolator. This is not the case for an isolator that incorporates some mass, because the force applied to one side then is balanced by the sum of the inertia force and the force acting on the other side of the isolator. Thus, unlike for a massless isolator, the force transmitted by an isolator with mass can be less than the applied force.

One may realize such a force reduction benefit, even at low frequencies at which mass effects in isolators themselves are negligible, by adding a lumped mass "inside" an isolator. One may visualize this concept by considering a spring that is cut into two lengths with a rigid block of mass welded in place between the two parts, resulting in an isolator consisting of two lengths of spring with a mass between them. If a mass m is mounted atop this isolator, one obtains a system that may be represented by a diagram like that of Fig. 13.8a. Because this system consists of a cascade of two spring–mass systems, it is said to have two stages of isolation.

Transmissibility

The system of Fig. 13.8a has two natural frequencies f_b that may be found from

$$\left(\frac{f_b}{f_0}\right)^2 = Q \pm \sqrt{Q^2 - B^2} \qquad Q = \frac{1}{2}\left(B^2 + 1 + \frac{k_2}{k_1}\right) \qquad (13.16)$$

where

$$B = \frac{f_I}{f_0} \qquad 2\pi f_I = \sqrt{\frac{k_1 + k_2}{m_I}} \qquad 2\pi f_0 = \frac{1}{\sqrt{m(1/k_1 + 1/k_2)}} \qquad (13.16a)$$

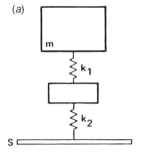
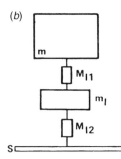

FIGURE 13.8 Two-stage isolation with intermediate mass m_I: (a) springs; (b) general isolation elements.

The frequency f_0 is the natural frequency of the system in the absence of any included mass m_I; that is, it is the natural frequency of a conventional simple single-stage system. The frequency f_I is the natural frequency of mass m_I moving between the two springs, with mass m held completely immobile. The upper frequency f_b, which one obtains if one uses the plus sign before the square root, always is greater than both f_0 and f_I; the lower frequency f_b, corresponding to the minus sign, always falls below both f_0 and f_I.

The transmissibility of a two-stage system like that of Fig. 13.8a obeys

$$\frac{1}{T} = \frac{1}{T_{F0}} = \frac{1}{B^2}\left(\frac{f}{f_0}\right)^4 - \left[1 + \frac{1 + k_2/k_1}{B^2}\right]\left(\frac{f}{f_0}\right)^2 + 1 \approx \left(\frac{f^2}{f_0 f_I}\right)^2 \quad (13.17)$$

where the last approximate expression applies for high frequencies, namely, for excitation frequencies that are much greater than both f_0 and f_I.

Figure 13.9 shows an illustrative plot of the transmissibility of an undamped two-stage system with $B = f_I/f_0 = 5$ and $k_2/k_1 = 1$, together with a plot of the transmissibility of an (undamped) single-stage system. The second natural frequency of the two-stage system (at $f/f_0 \approx 5.1$) is clearly evident, as is the subsequent rapid decrease in that system's transmissibility with increasing frequency. At high frequencies, that is, a little above the aforementioned second natural frequency, the transmissibility of the two-stage system may be seen to be smaller than that of a single-stage system with the same fundamental natural frequency. As evident from Eq. (13.17), the high-frequency transmissibility of a two-stage system varies inversely as the fourth power of the excitation frequency, whereas Eq. (13.5) indicates that the transmissibility of a single-stage system varies inversely as only the second power of the excitation frequency.

The advantage of a two-stage system over a single-stage system is that it results in greatly reduced transmissibility at high frequencies (i.e., above the higher of its two natural frequencies). It has the disadvantage that it introduces an additional transmissibility peak at a low frequency (i.e., at its second natural frequency). Thus, two-stage isolation is beneficial in general only if the aforementioned second natural frequency is somewhat lower than the lowest excitation frequency of concern. Thus, to reap the benefit of a two-stage system, one typically needs a relatively large intermediate mass m_I. Where several items need to be isolated, it often is advantageous to support these on a common massive platform (often called a "subbase" or "raft"), to isolate each item from the platform, and to isolate the platform from the structure that supports it. In this arrangement the platform serves as a relatively large intermediate mass for each of the isolated items, resulting in efficient two-stage isolation performance with comparatively small weight penalty.

Isolation Effectiveness

Although the foregoing results apply strictly only to isolators without damping, they also provide a reasonable approximation to the behavior of lightly damped systems, except near the natural frequencies. To account for high damping or more

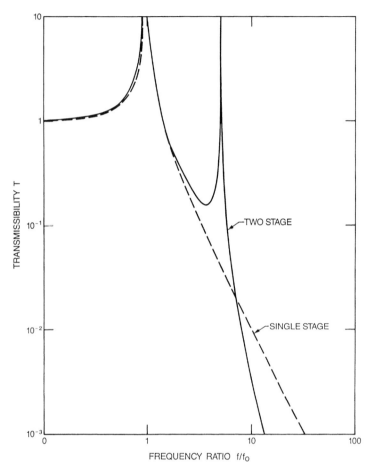

FIGURE 13.9 Transmissibility of two-stage system. Calculated from Eq. (13.17) for $k_2/k_1 = 1$ and $B = f_1/f_0 = 5$.

complicated linear isolator configurations (e.g., where each isolator is modeled by various series and parallel combinations of springs and dampers), it is convenient to represent each isolator by its mobility. A corresponding diagram appears in Fig. 13.8b. As has been discussed in Section 13.4, the source mobility M_S is a measure of a vibration source's susceptibility to loading effects, and the isolation effectiveness E is a measure of the isolation performance, which, unlike transmissibility, takes loading effects into account. If one takes S in Fig. 13.8b to represent a general linear source and replaces the mass m by a general linear receiver with mobility M_R, one obtains a general linear two-stage system whose effectiveness one may write as[8]

$$E = |E_1 + \Delta E| \qquad E_1 = 1 + \frac{M_I}{M_S + M_R} \qquad (13.18)$$

where

$$M_I = M_{I1} + M_{I2} \qquad \Delta E = \frac{(M_{I1} + M_S)(M_{I2} + M_R)}{M_m(M_S + M_R)} \qquad (13.18a)$$

One may recognize E_1 as corresponding to the effectiveness of a single-stage system [see Eq. (13.14)], that is, to a two-stage system with zero included mass, with M_I denoting the mobility of the two partial isolators in series. Thus, ΔE represents the effectiveness increase obtained by addition of the included mass m_I, whose mobility is M_m. Note that ΔE is inversely proportional to M_m, indicating that greater included masses generally result in greater effectiveness increases.

Optimization of Isolator Stiffness Distribution

Once one has selected the mobility M_I of the total isolator (or, equivalently, its compliance or stiffness), one needs to consider how to allocate this mobility among the components M_{I1} and M_{I2}. If one lets r_1 denote the fraction of the total mobility on the source side of the included mass, so that $M_{I1} = r_1 M_I$ and $M_{I2} = (1 - r_1) M_I$, it turns out that one may obtain the largest value of ΔE, namely,

$$\Delta E_{\max} = \frac{(M_I + M_s + M_R)^2}{4 M_m (M_s + M_R)} \qquad (13.19)$$

by making r_1 equal to its optimum value,*

$$r_{\text{opt}} = \frac{1}{2}\left(1 + \frac{M_R - M_S}{M_I}\right) \qquad (13.20)$$

In view of Eq. (13.18a), in an efficient isolation system M_{I1} must be considerably greater than M_S, and also M_{I2} must be considerably greater than M_R. For such a system, one finds that $r_{\text{opt}} \approx \frac{1}{2}$ and that $\Delta E_{\max}$ may be approximated by replacing the expression in the parentheses of the numerator of Eq. (13.19) by M_I. Thus, if the total mobility of the isolator is sufficiently great, that is, if the total stiffness of the isolator is sufficiently small, one may generally obtain the greatest improvement $\Delta E_{\max}$ by allocating the same mobility or stiffness to the two isolator components.

It may be shown[8] that placing a given mass "within" the isolator as described above, so as to obtain a two-stage system with two like isolator mobility components, results in greater effectiveness than placement of the mass directly at the receiver as long as $M_I \gg M_R$. This inequality usually is likely to be satisfied in practice, except at resonances of the receiver at which M_R is very large. Similarly, as long as $M_I \gg M_S$, placing the mass within the isolator results in greater

---

*This result applies strictly only if the various mobilities or mobility ratios are real quantities. It suffices for development of an intuitive understanding, although more complicated expressions apply in the general case where the mobilities are represented by complex quantities.

effectiveness than placement of the mass directly at the source. The foregoing inequality generally is likely to be satisfied in practice, except for sources that behave essentially like force sources and thus have very high mobility.

13.6 PRACTICAL ISOLATORS

A great many different isolators are available commercially in numerous sizes and load capacities, with various attachment means, and with many types of specialized features.[4] Details concerning such isolators typically may be found in suppliers' catalogs.

Most commercial isolators incorporate metallic or elastomeric resilient elements. Metallic elements most often are in the form of coil springs but also occur in the form of flexural configurations such as leaf springs or conical Belleville washers. Coil springs are predominantly used in compression, usually because such tensile springs tend to involve configurations that give rise to stress concentrations and thus have lesser fatigue life. Coil spring isolator assemblies often involve parallel and/or series arrangements of springs in suitable housings, are designed to have the same stiffness in the lateral directions as in the axial direction, and may also incorporate friction devices (such as wire mesh inserts), snubbers to limit excursions due to large disturbances (such as earthquakes), and in-series elastomeric pads for enhanced damping and isolation at high frequencies. Spring systems in housings need to be installed with care to avoid binding between the housing elements and between these elements and the springs.

Some commercial metallic isolators employ pads or woven assemblies of wire mesh to provide both resilience and damping. Others use arrangements of coils or loops of wire rope not only to provide damping but also to serve as springs.

A great many commercially available isolators that employ elastomeric elements have these elements bonded or otherwise attached to support plates or sleeves that incorporate convenient means for fastening to other components. The isolators may be designed so that the elastomeric element is used in shear, torsion, compression, or a combination of these modes. There are also available a variety of elastomeric gaskets, grommets, sleeves, and washers, intended to be used with bolts or similar fasteners to provide both connection and isolation.

Elastomeric pads often are used as isolators by themselves, as are pads of other resilient materials, such as cork, felt, fiberglass, and metal mesh. Such pads often are convenient and relatively inexpensive; their areas can be selected to support the required loads, and their thicknesses can be chosen to provide the desired stiffness.

In the design and selection of pads of solid (in contrast to foamed) elastomeric materials, one needs to take into account that a pad's stiffness depends not only on its thickness and area but also on its shape and constraints. This behavior is due to the incompressibility of elastomeric materials, which essentially prevents a pad from changing its volume as it is compressed and thus in essence does not permit a pad to be compressed if it is confined so that its edges cannot bulge outward. A pad's freedom to compress may be characterized by its shape factor, defined as the ratio of its loaded area to the total area of the edges that are free to bulge; the greater the shape factor, the greater is the pad's effective stiffness. However, a pad's freedom to deform while maintaining constant volume also is affected by how easily the loaded surfaces can slip relative to the adjacent surfaces; the more restricted this slippage, the greater the pad's effective stiffness. Some commercial isolation pads are furnished with top and bottom load-carrying surfaces bonded to metal or other stiff plates to eliminate the stiffness uncertainties due to unpredictable slippage.

To avoid the need for considering the shape factor in pad selection, many commercial isolation pad configurations have a multitude of cutouts (e.g., closely spaced arrays of holes) or ribs, which provide roughly constant amounts of bulging area per unit surface area. If ribbed or corrugated pads or pads with cutouts are used in stacks, plates of a stiff material (e.g., metal sheets) generally are used between pads to distribute the load on the load-bearing surfaces and to avoid having protrusions on one pad extending into openings on the adjacent pad.

So-called pneumatic, or air spring, isolators, which have found considerable use, obtain their resilience primarily from the compressibility of confined volumes of air. They may take the form of air-filled pillows of rubber or plastic, often with cylindrical or annular shapes, or they may consist essentially of piston-in-cylinder arrangements. Air springs can be designed to have small effective stiffnesses while supporting large loads and to have smaller heights than metal springs of equal stiffness. Practical air springs typically can provide fundamental resonance frequencies that may be as low as about 1 Hz. Some air spring configurations are laterally unstable under some load conditions and require the use of lateral restraints; some are available with considerable lateral stability.

Air springs of the piston-and-cylinder type can be provided with leveling controls, which automatically keep the isolated item's static position at a predetermined distance from a reference surface and (by use of several air springs and a suitable control system) at a predetermined inclination. The stiffness of a piston-type air spring is proportional to PA^2/V, where P denotes the air pressure, A the piston face area, and V the cylinder volume. The product $(P - P_0)A$, where P_0 represents the ambient atmospheric pressure, is equal to the static load carried by the spring. Lower stiffnesses may be obtained with a given area at a given air pressure by use of larger effective volumes; for this reason, some commercial air spring isolators are available with auxiliary tanks that communicate with the cylinder volume via piping. In some instances, a flow constriction in this piping is used to provide low-frequency damping. If the pressure in a piston-and-cylinder-type air spring is considerably greater than the atmospheric pressure, the pressure in the spring is nearly proportional to the load it supports. Because the spring's stiffness is proportional to this pressure, the natural frequency one obtains with such an air spring isolation system is essentially independent of the load, making air springs (like other constant-natural-frequency systems[4]) particularly useful for applications in which the loads are variable or uncertain.

Pendulum arrangements often are convenient means for obtaining horizontally acting isolation systems with low natural frequencies. One may calculate the horizontal natural frequency of a pendulum system from Eq. (13.3) if one replaces X_{st} in that equation by the pendulum length. Some commercial isolation systems combine pendulum action for horizontal isolation with spring action for vertical isolation.

Various exotic isolation systems have also been investigated or employed for special applications. These include systems in which the spring action is provided by magnetic or electrostatic levitation or by streams or thin films of gases or liquids.

Active isolation systems (see Chapter 18) have recently received increased attention. Such systems essentially are dynamic control systems in which the vibration of the item to be protected is sensed by an appropriate transducer whose suitably processed output is used to drive an actuator that acts on the item so as to reduce its vibration. Active systems are relatively complex, but they can provide better isolation than passive systems under some conditions, notably, in the presence of low-frequency disturbances, attenuation of which by passive means generally tends to be most difficult.

REFERENCES

1. W. T. Thomson, *Theory of Vibration with Applications*, 2nd ed., Prentice-Hall, Englewood Cliffs, NJ, 1981.
2. J. C. Snowdon, *Vibration and Shock in Damped Mechanical Systems*, Wiley, New York, 1968.
3. D. J. Mead, *Passive Vibration Control*, Wiley, New York, 1998.
4. E. I. Rivin, *Passive Vibration Isolation*, American Society of Mechanical Engineers, New York, 2003.
5. H. Himelblau, Jr. and S. Rubin, "Vibration of a Resiliently Supported Rigid Body," in C. M. Harris and C. E. Crede (Eds.), *Shock and Vibration Handbook*, 2nd ed. McGraw-Hill, New York, 1976.
6. J. N. Macduff and J. R. Curreri, *Vibration Control*, McGraw-Hill, New York, 1958.
7. C. E. Crede and J. E. Ruzicka, "Theory of Vibration Isolation," in C. M. Harris and C. E. Crede (Eds.), *Shock and Vibration Handbook*, 2nd ed., McGraw-Hill, New York, 1976.
8. E. E. Ungar and C. W. Dietrich, "High-Frequency Vibration Isolation," *J. Sound Vib.*, **4**, 224–241 (1966).
9. L. Cremer, M. A. Heckl, and E. E. Ungar, *Structure-Borne Sound*, 2nd ed., Springer-Verlag, Berlin, 1988.
10. E. E. Ungar, "Mechanical Vibrations," in H. A. Rothbart (Ed.), *Mechanical Design and Systems Handbook*, 2nd ed., McGraw-Hill, New York, 1985, Chapter 5.
11. D. Muster and R. Plunkett, "Isolation of Vibrations," in L. L. Beranek (Ed.), *Noise and Vibration Control*, McGraw-Hill, New York, 1971, Chapter 13.
12. E. E. Ungar, "Wave Effects in Viscoelastic Leaf and Compression Spring Mounts," *Trans. ASME Ser. B*, **85**(3), 243–246 (1963).
13. D. Muster, "Resilient Mountings for Reciprocating and Rotating Machinery," Eng. Rept. No. 3, ONR Contract N70NR-32904, July 1951.

CHAPTER 14

Structural Damping

ERIC E. UNGAR AND JEFFREY A. ZAPFE

Acentech, Inc.
Cambridge, Massachusetts

14.1 THE EFFECTS OF DAMPING

The dynamic responses and sound transmission characteristics of structures are determined by essentially three parameters: mass, stiffness, and damping. Mass and stiffness are associated with storage of kinetic and strain energy, respectively, whereas damping relates to the dissipation of energy, or, more precisely, to the conversion of the mechanical energy associated with a vibration to a form (usually heat) that is unavailable to the mechanical vibration.

Damping in essence affects only those vibrational motions that are controlled by a balance of energy in a vibrating structure; vibrational motions that depend on a balance of forces are virtually unaffected by damping. For example, consider the response of a classical mass–spring–dashpot system to a steady sinusoidal force. If this force acts at a frequency that is considerably lower than the system's natural frequency, the response is controlled by a quasi-static balance between the applied force and the spring force. If the applied force acts at a frequency that is considerably above the system's natural frequency, the response is controlled by a balance between the applied force and the mass's inertia. In both of these cases, damping has practically no effect on the responses. However, at resonance, where the excitation frequency matches the natural frequency, the spring and inertia effects cancel each other and the applied force supplies some energy to the system during each cycle; as a result, the system's energy (and amplitude) increases until steady state is reached, at which time the energy input per cycle is equal to the energy lost per cycle *due to damping*.

In light of energy considerations like the foregoing, one finds that increased damping results in (1) more rapid decay of unforced vibrations, (2) faster decay of freely propagating structure-borne waves, (3) reduced amplitudes at resonances of structures subject to steady periodic or random excitation with attendant reductions in stresses and increases in fatigue life, (4) reduced response to sound and increased sound transmission loss (reduced sound transmission) above the coincidence frequency (at which the spatial distribution of the disturbing pressure matches that of the structural displacement), (5) reduced rate of buildup of vibrations at resonances, and (6) reduced amplitudes of "self-excited" vibrations, in which the vibrating structure accepts energy from an external source (e.g., wind) as the result of its vibratory motion.

14.2 MEASURES AND MEASUREMENT OF DAMPING

Most measures of damping are based on the dynamic responses of simple systems with idealized damping behaviors. Damping measurements typically involve observation of some characteristics of these responses.

Decay of Unforced Vibrations with Viscous Damping

Many aspects of the behaviors of vibrating systems can be understood in terms of the simple ideal linear mass–spring–dashpot system shown in Fig. 14.1. If this system is displaced by an amount x from its equilibrium position, the massless spring produces a force of magnitude kx tending to restore the mass m toward its equilibrium position, and the massless dashpot produces a retarding force of magnitude $c\dot{x}$. Here k and c are constants of proportionality; k is known as the spring constant and c as the *viscous damping coefficient*.

If this system is displaced from its equilibrium position by an amount X_0 and then released, the resulting displacement varies with time t as[1]

$$x = X_0 e^{-\zeta \omega_n t} \cos(\omega_d t + \phi) \qquad (14.1)$$

provided that $\zeta < 1$. Here ϕ represents a phase angle, which depends on the velocity with which the mass is released, and ω_n and ω_d represent the undamped and damped radian natural frequencies of the system. These obey

$$\omega_n = \sqrt{\frac{k}{m}} = 2\pi f_n \qquad \omega_d = \omega_n \sqrt{1 - \zeta^2} \qquad (14.2)$$

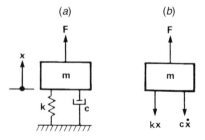

FIGURE 14.1 System with single degree of freedom: (*a*) schematic representation of mass–spring–dashpot system or of a vibrational mode of a structure; (*b*) free-body diagram of mass. Spring produces restoring force kx; dashpot produces retarding force $c\dot{x}$.

with f_n representing the cyclic (undamped) natural frequency. The constant ζ is called the *damping ratio* or *fraction of critical damping*; it is defined as

$$\zeta = \frac{c}{c_c} \qquad c_c = 2\sqrt{km} = 2m\omega_n \qquad (14.3)$$

where c_c is known as the *critical damping coefficient*. For the small values of ζ one usually encounters in practice, ω_d is sufficiently close to ω_n so that one rarely needs to distinguish between the damped and the undamped natural frequencies. Furthermore, the foregoing expression for ω_d applies only for viscous damping; other relations hold for other damping models.

The right-hand side of Eq. (14.1) represents a cosine function with an amplitude $X_0 e^{-\zeta \omega_n t}$ that decreases as time t increases (see Fig. 14.2); its rate of decrease is $\zeta \omega_n$ and thus is proportional to ζ. However, Eq. (14.1) does not apply for values of ζ that equal or exceed unity (or for values of c that equal or exceed c_c). For such large values of ζ or c one obtains a nonoscillatory decay represented by pure exponential expressions instead of the decaying oscillation represented by Eq. (14.1). The critical damping coefficient c_c constitutes the boundary between oscillatory and nonoscillatory decays.

The *logarithmic decrement* δ is a convenient, time-honored representation of how rapidly a free oscillation decays. It is defined by[1]

$$\delta = \frac{1}{N} \ln \frac{X_i}{X_{i+N}} \qquad (14.4)$$

where X_i represents the value of x at any selected peak and X_{i+N} represents the value at the peak at N cycles from the aforementioned one. It follows from Eq. (14.1) that $\delta = 2\pi \zeta$.

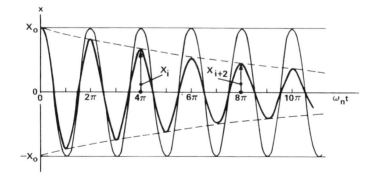

FIGURE 14.2 Time variation of displacement of mass–spring–dashpot system released with zero velocity from initial displacement X_0. *Light curve*: Undamped system ($c = \zeta = 0$); amplitude remains constant at X_0. *Heavy curve*: Damped system ($0 < c < c_c$; $0 < \zeta < 1$); amplitude decreases according to $x = X_0 e^{-\zeta \omega_n t}$, which is represented by upper dashed curve. Lower dashed curve corresponds to $x = -X_0 e^{-\zeta \omega_n t}$. Amplitudes X_i and X_{i+2} illustrate values that may be used to calculate logarithmic decrement from Eq. (14.4) for $N = 2$.

The utility of logarithmic measures of oscillatory quantities has long been recognized in acoustics, and definitions analogous to acoustical levels have come into use in the field of vibrations, particularly in regard to measurement. For example, one may define the displacement level L_x, in decibels, corresponding to an oscillatory displacement $x(t)$ in analogy to sound pressure level, as

$$L_x = 10 \log_{10} \frac{x^2(t)}{x_{\text{ref}}^2} \tag{14.5}$$

where x_{ref} denotes a (constant) reference value of displacement. One may then define a decay rate Δ, in decibels per second, and find that for a viscously damped system[2]

$$\Delta = -\frac{dL_x}{dt} = 8.69 \zeta \omega_n = 54.6 \zeta f_n \tag{14.6}$$

Also in analogy to acoustics, one may define the reverberation time T_{60} as the time it takes for the displacement level to decrease by 60 dB; thus,

$$T_{60} = \frac{60}{\Delta} = \frac{1.10}{\zeta f_n} \tag{14.7}$$

Because velocity and acceleration levels may be defined in full analogy to the definition of displacement level in Eq. (14.5), the decay rate and reverberation time expressions of Eqs. (14.6) and (14.7) also apply to these other vibration levels.

If any extended structure that is not too highly damped vibrates in the absence of external forces at one of its natural frequencies, all points on that structure move either in phase or in opposite phase with each other, and the structure is said to vibrate in one of its modes. In addition to the modal natural frequency, there corresponds to each mode a modal mass, a modal stiffness, and a modal damping value. With the aid of these parameters the behavior of a modal vibration may be described in terms of that of an equivalent simple mass–spring–dashpot system.[1,3,4] Thus, all of the foregoing discussion concerning this simple system also applies to structural modes.

Of course, extended structures also can exhibit wave motions at a given frequency in which all points are not in or out of phase with each other. Such motions, which can be described in terms of freely propagating waves, also decrease due to damping. For flexural waves on a beam or for nonspreading (straight-crested) flexural waves on a plate, the *spatial decay rate* Δ_λ, defined as the reduction in vibration level per wavelength, obeys[2] $\Delta_\lambda = 27.2 \zeta$ in decibels per wavelength.

Steady Forced Vibrations

If the system of Fig. 14.1 is subject to a sinusoidal force $F(t) = F \cos \omega t$, then its equation of motion may be written as

$$m\ddot{x} + c\dot{x} + kx = F(t) = F \cos \omega t \tag{14.8}$$

One may find its steady-state solution by substituting $x(t) = X\cos(\omega t - \phi)$ and solving for X and ϕ. Alternatively, one may obtain this solution by taking $x(t) = \text{Re}\lfloor \overline{X} e^{j\omega t} \rfloor$, where $j = \sqrt{-1}$ and the complex amplitude or *phasor* $\overline{X} = X e^{-j\phi}$ indicates both the amplitude and the phase of the vibration.* In terms of phasors (and omitting the writing of "Re"), the equation of motion may be written as

$$(-m\omega^2 + j\omega c + k)\overline{X} = (\overline{k} - m\omega^2)\overline{X} = F \tag{14.9}$$

In the foregoing there has been introduced the complex stiffness

$$\overline{k} = k + j\omega c = k + k_i = k\left(1 + \frac{jk_i}{k}\right) \tag{14.10}$$

which includes information about the system's damping as well as about it stiffness. By means of either solution approach one may determine that

$$\frac{X}{F/k} = \frac{X}{X_{st}} = \frac{1}{\sqrt{(1-r^2)^2 + (2r\zeta)^2}} \qquad \tan\phi = \frac{2r\zeta}{1-r^2} \tag{14.11}$$

where $r \equiv \omega/\omega_n$.

The deflection $X_{st} = F/k$, which was introduced in order to obtain a nondimensional expression, is the quasi-static or zero-frequency deflection that the system would experience due to a statically applied force of magnitude F. The ratio X/X_{st}, which indicates by what factor the amplitude under dynamic excitation exceeds the quasi-static deflection, is called the *amplification*.

The foregoing response expressions (and related ones that involve velocity V or acceleration A instead of displacement) depend on damping; thus, damping data may be extracted from corresponding measurements. Two widely used measures of damping may be derived readily from the amplification expressions of Eq. (14.11), a plot of which is shown in Fig. 14.3. One is the *amplification at resonance*, conventionally represented by the letter Q and often simply called "the Q" of the system.[2] This corresponds to the value of X/X_{st} that results if the excitation frequency ω is equal to the natural frequency ω_n and is related to the viscous damping ratio by $Q = 1/2\zeta$. The second commonly used measure is the *relative bandwidth* $b = \Delta\omega/\omega_n \approx 1/Q$, where $\Delta\omega$ represents the difference between the two frequencies† (one below and one above ω_n; see Fig. 14.3) at which the amplification is equal to $Q/\sqrt{2}$.

In view of Eq. (14.11), the phase lag ϕ also provides a measure of damping. It is particularly convenient to use phase information in "Nyquist plots," that is, in plots of the real and imaginary parts of responses at a number of frequencies, as illustrated in Fig. 14.4. These plots are circles or nearly circles, with a diameter that is equal to Q if the plots are appropriately nondimensionalized.[5,6]

---

*Note that $e^{jz} = \cos z + j\sin z$ for any real number z. The amplitude X is equal to the absolute value of the phasor; that is, $X = |\overline{X}|$.

†These frequencies often are called the *half-power points* because at these the energy stored in the system (and that dissipated by it), which is proportional to the square of the amplitude [see Eq. (14.6)], is half of the maximum value.

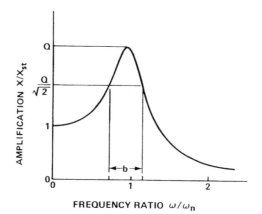

FIGURE 14.3 Steady-state response of mass–spring–dashpot system to sinusoidal force. See Eq. (14.11): Q = value of amplification X/X_{st} at resonance. Relative bandwidth $b = \Delta\omega/\omega_n$ is determined from "half-power points," that is from frequencies at which response amplitude is $1/\sqrt{2}$ times the maximum.

If the system of Fig. 14.1 (or a structural mode modeled by it) is subject to a broadband force (or broadband modal force) rather than a single-frequency sinusoidal force, then the mean-square displacement $\overline{x^2}$ of the mass is given by[7]

$$\frac{\overline{x^2}}{\pi S_F(\omega)\omega_n/k^2} = \frac{1}{2\zeta} \tag{14.12}$$

where $S_F(\omega)$ denotes the spectral density of the force in terms of radian frequency (i.e., the value of excitation force squared per unit radian frequency interval). Note that the spectral density in cyclic frequency obeys $S_F(f) = 2\pi S_F(\omega)$. The foregoing equation is exact for excitations with spectral densities that are constant for all frequencies. It is a good approximation for excitations with spectral densities that vary only slowly in the vicinity of the system's natural frequency ω_n; the spectral density value to be used in the equation then is that corresponding to ω_n.

Energy; Complex Stiffness

All of the measures of damping discussed so far are based on the motions of simple systems. However, since damping pertains to the dissipation of energy, damping measures that relate to energy are more basic and more general.

The *damping capacity* ψ is defined as the ratio of the energy that is dissipated per cycle to the total energy present in the vibrating system. The *loss factor* η is defined similarly as the ratio of the energy that is dissipated per radian to the total energy. If D denotes the energy dissipated per cycle and W the total energy in the system, then

$$\eta = \frac{\psi}{2\pi} = \frac{D}{2\pi W} \tag{14.13}$$

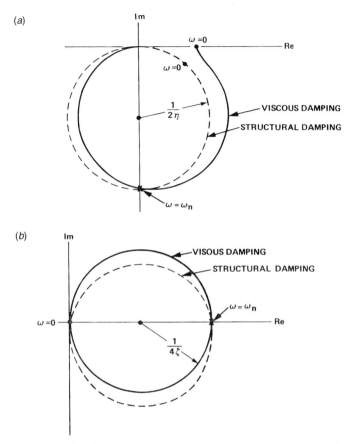

FIGURE 14.4 Nyquist plots of nondimensional responses of viscously and structurally damped mass–spring–damper systems. (*a*) Real and imaginary parts of amplification X/X_{st}. (*b*) Real and imaginary parts of mobility $Vk/F\omega_n = Vc_c/2F$. Amplification plot for structural damping and mobility plot for viscous damping are exact circles; others are approximate circles that become more nearly circular with decreased damping. Diameter is exactly or approximately equal to Q. Figure plots correspond to $\zeta = 0.2$, $\eta = 0.4$.

These expressions apply for any damping mechanism. However, it is interesting to relate them to the special case of a viscously damped mass–spring–dashpot system like that of Fig. 14.1. In this system the energy dissipated corresponds to the work that is done on the dashpot, and one may readily determine that the energy dissipated per cycle in a steady vibration at radian frequency ω and displacement amplitude X obeys $D = \pi \omega c X^2$. The total energy W stored by the system consists of the kinetic energy W_{kin} of the mass and the potential (or strain) energy W_{pot} in the spring. If the energy dissipated is small compared to the total energy stored, then W is approximately equal to the energy stored in the spring when the kinetic energy is zero—that is, when the spring is displaced to the full extent of its amplitude—and $W = W_{kin} + W_{pot} \approx kX^2/2$.

Thus, one finds by use of Eqs. (14.3) and (14.13) that for a viscously damped system

$$\eta \approx \frac{\omega c}{k} = \frac{2\zeta\omega}{\omega_n} = 2r\zeta \tag{14.14}$$

From Eq. (14.10) one finds that

$$\overline{k} = k(1 + j\eta) \qquad k_i/k = \eta \tag{14.15}$$

Equation (14.14) indicates that for the particular case of a viscously damped system the loss factor is proportional to the frequency. In cases where the loss factor exhibits some other frequency dependence, Eqs. (14.13) and (14.15) still apply—and Eq. (14.11) holds if $2r\zeta$ is replaced by η. These equations then permit one not only to consider the case of *structural damping*, which is characterized by a constant loss factor, but also to take into account experimentally determined frequency variations of the loss factor.

Interrelation of Measures of Damping

The following relations apply at all frequencies:

$$\eta = \frac{\psi}{2\pi} = \frac{k}{k_i} = |\tan \phi|_{r=0} \tag{14.16a}$$

but the relations of Eq. (14.16b) are exact only at resonance:

$$\eta = \frac{1}{Q} = 2\zeta \quad \text{and} \quad \eta \approx b \text{ for small damping} \tag{14.16b}$$

For systems with viscous damping,

$$2\zeta = \frac{\delta}{\pi} = \frac{2.20}{f_n T_{60}} = \frac{\Delta}{27.3 f_n} = \frac{\Delta_\lambda}{13.6} \tag{14.16c}$$

For systems with small damping, one may take $\eta \approx 2\zeta$, consider that the system behaves approximately like a viscously damped one, and use the relations of Eq. (14.16c).

Measurement of Damping

Most approaches to measurement of the damping of structures are based on the previously discussed responses of simple systems, which, as has been mentioned, also correspond to those of structural modes. However, unlike mass–spring–dashpot systems, structures have a multiplicity of modes and corresponding natural frequencies. Therefore, many of the approaches applicable to simple systems can be applied only to structural modes whose responses can be separated adequately from those of all others because of differences in their natural frequencies or mode shapes.

Measurement of logarithmic decrement δ typically is applicable only to the fundamental modes of structures for which a clear record of the amplitude-versus-time trace can be obtained. If more than one mode is present, their decaying responses are superposed and the record becomes difficult to interpret.

The counting of peaks that is required for determination of the logarithmic decrement from Eq. (14.4) is not needed if one focuses on the decaying signal's envelope. For purposes of evaluating this envelope it is particularly convenient to use a display of the logarithm of the rectified amplitude versus time. Rectification is needed because the logarithm of negative numbers is undefined. In such a logarithmic display the envelope becomes a straight line whose slope is proportional to $\zeta\omega_n$ and thus to the decay rate. Not only does measurement of the slope enable one to evaluate the damping, but also observation of deviations of the envelope from a straight line permit one to judge whether the structure's damping is indeed viscous and amplitude-independent and whether a superposition of responses with different decay rates is present.

Determination of decay rates is useful also in frequency bands in which a multitude of modes are excited. A typical measurement here involves excitation of a structure by a broadband force in a given frequency band, cutting off the excitation, then observing the envelope of the logarithm of the rectified signal obtained by passing the output of a transducer (usually an accelerometer) through a bandpass filter* tuned to the excitation band. The center frequency of this passband may be taken to represent ω_n for all modes in the band. Some judgment in interpreting the resulting envelopes and averaging of results from repeated measurements is generally required because different modes in the band may exhibit somewhat different decays.

A conceptually straightforward approach to measuring the damping of a structure involves the application of Eq. (14.13) to observed values of the energy dissipation and total vibrational energy W present in a structure in the steady state.[8] The structure is excited via an impedance head or a similar transducer arrangement that measures the force and motion at the excitation point. The instantaneous force and velocity values are multiplied and the product is time averaged to yield the average energy input per unit time, which is equal to the energy dissipated per unit time under steady-state conditions. For a given excitation frequency f, the energy D dissipated per cycle is equal to $1/f$ times the energy dissipated per unit time.

The energy W stored in the structure may be determined from its kinetic energy, which may be calculated from information on its mass distribution and from velocity values measured by a suitable array of accelerometers or other motion transducers. This measurement approach requires particular care in instrumentation selection and calibration, but it has a significant advantage: Because it involves direct measurement of the dissipated energy, it does not rely on any

---

*The filter's response must be fast enough so that it can follow the decaying signal; otherwise one observes the decay of the filter response instead of that of the structural vibration. Filters with wider passbands need to be used to observe the more rapid decays associated with greater damping.

particular model of dissipation. It also permits one, for example, to investigate how the loss factor varies with amplitude.

Force-and-motion transducer combinations may also be employed for the direct measurement of complex impedance or mobility (or of other force–motion ratios), from which damping information may be extracted, typically on the basis of Nyquist plots. Much corresponding specialized *modal testing* or *modal parameter extraction* instrumentation and software has recently become available.

Simple steady sinusoidal response measurements may also be made to measure Q and the half-power point bandwidth b directly on the basis of their definitions. These measurements require particular care to ensure that the near-resonance response of the mode of interest is not affected significantly by the responses of other modes with resonance frequencies near that of the mode of interest.

14.3 DAMPING MODELS

Analytical Models

Most of the foregoing discussion has dealt with "viscous" damping, where energy dissipation results from a force that is proportional to the velocity of a vibrating system and that acts opposite to the velocity. This viscous model of damping action has been used most widely because it results in relatively simple linear differential equations of system motion and because it yields a reasonable approximation to the action of some real systems, particularly at small amplitudes.

Among the many other models that have received considerable attention, most also involve a motion-opposing force that is a function of velocity. In *dry friction* or *Coulomb* damping, the force is constant in magnitude (but changes its algebraic sign when the velocity does). In *square law* and *power law* damping the force magnitude is proportional to the square or to some other power of velocity. Of course, modern numerical methods also permit one readily to analyze models involving other velocity dependences, such as one might obtain from corresponding experiments.

Whereas in viscous damping the retarding force is proportional to the velocity, in *structural damping* the retarding force is proportional to the displacement. As has been mentioned, structural damping is characterized by a constant loss factor η and the dimensionless response relation for a system with structural damping is given by Eq. (14.11) with $2r\zeta$ replaced by η. Because of the difference in the retarding forces, a structurally damped system behaves differently from a viscously damped one, except near the natural frequency. The structural damping model is often extended to loss factors that vary with frequency as determined from experimental data. With the appropriate frequency dependence of the loss factors, the structural damping model can be made to represent the sinusoidal response of viscously damped systems—and, indeed, of any system whose loss factor is independent of amplitude.*

---

*Caution is required if one wants to undertake transformations from the frequency to the time domain. Not all conceivable frequency variations of the loss factor lead to physically realizable results; e.g., some imply system motions that begin before application of a force.[9]

Applicability of Models

If one desires to determine the precise response of a system to a prescribed excitation, one generally needs to have a complete description of all forces, including the damping forces; that is, one needs a damping model that corresponds to the actual system. This is true, for example, if one wants to study the "wave shape" of the motion of a screeching brake or a chattering tool or if one needs to determine the response of a system to a transient, such as a shock.

In many practical instances, however, the details of the system's motion are of no interest and only the amplitudes are of concern. As has been mentioned, under steady resonant or free decay conditions (and in a few other situations), the amplitudes are established essentially by the energy in the system. For such conditions, the details of the damping model are unimportant as long as the model gives the correct energy dissipation per cycle. It is for this reason that measures of damping that involve only energy considerations have found wide acceptance.*

14.4 DAMPING MECHANISMS AND MAGNITUDES

Since damping involves the conversion of energy associated with a vibration to other forms that are unavailable to the vibration, there are as many damping mechanisms as there are ways to remove energy from a vibrating system. These include mechanisms that convert mechanical energy into heat, as well as others that transport energy away from the vibrating system of concern.

Energy Dissipation and Conversion

Material damping, mechanical hysteresis, and internal friction refer to the conversion of mechanical energy into heat that occurs within materials due to deformations that are imposed on them. This conversion may result from a variety of effects on the molecular, crystal lattice, or metal grain level, including magnetic, thermal, metallurgical, and atomic phenomena.[10] Figure 14.5 indicates the ranges of the loss factors reported for some common materials.

Damping of a vibrating structure may also result from friction associated with relative motion between the structure and solids or fluids that are in contact with it. Also, an electrically conductive structure moving in a magnetic field is subject to damping due to eddy currents that result from the motion and that are converted into heat.

A granular material, such as sand, placed in contact with a vibrating structure tends to produce damping by two different mechanisms. At small amplitudes, damping results predominantly from interaction of asperities on adjacent grains and the attendant energy loss due to mechanical hysteresis. At large amplitudes,

---

*Equivalent viscous damping, defined as viscous damping that results in the same energy dissipation as the damping actually present in the system, is often used in analyses. This damping model obviously should not be used where details of the system motion are of concern.

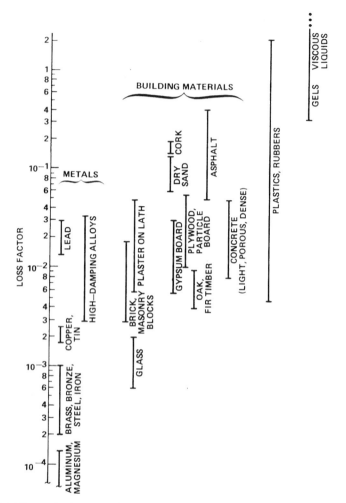

FIGURE 14.5 Typical ranges of material loss factors at small strains, near room temperature, at audio frequencies. The loss factors of metals tend to increase with strain amplitude, particularly near the yield point, but the loss factors of plastics and rubbers tend to be relatively independent of strain amplitude up to strains of the order of unity. The loss factors of some materials, particularly those that can flow or creep, tend to vary markedly with temperature and frequency.

damping results predominantly from impacts between the structure and the grains or between grains; these impacts produce high-frequency vibrations of the structure and of the granular material, and the energy that goes into these vibrations (which eventually is converted to heat) is no longer available to the structural vibrations of concern.[11] Impact dampers, in which a small element is made to rattle against a vibrating structure, similarly rely on conversion of the energy of the vibrating structure to higher frequency vibrations.

Damping due to Boundaries and Reinforcements

For panels or other structural components that may be considered as uniform plates, one may estimate the loss factor η_b associated with energy loss at the panel boundaries, due to both energy transport to adjacent panels and dissipation at its boundaries, from information on the boundary absorption coefficients.[12] For a panel of area A vibrating at a frequency at which the flexural wavelength λ on the panel is considerably shorter than a panel edge, this loss factor is given by

$$\eta_b = \frac{\lambda}{\pi^2 A} \sum \gamma_i L_i \qquad (14.17)$$

where γ_i denotes the absorption coefficient of the ith boundary increment whose length is L_i and where the summation extends over all boundary increments.

At frequency f the flexural wavelength on a homogeneous plate of thickness h of a material with longitudinal wave speed c_L and Poisson's ratio ν is given by $\lambda = \sqrt{(\pi/\sqrt{3})hc_L/f(1-\nu^2)}$.

The absorption coefficient γ of a boundary element is defined as the fraction of the panel bending-wave energy impinging on the boundary element that is not returned to the panel. Although the absorption coefficient values associated with a given boundary element rarely can be predicted well analytically, they can often be determined experimentally. For example, one might add a boundary element of length L_0 to a panel, measure the resulting loss factor increase $\Delta\eta$ at various frequencies, and calculate the absorption coefficient values γ_0 of this boundary element from $\gamma_0 = (\Delta\eta)\pi^2 A/\lambda L_0$, which follows from Eq. (14.17). Both Eq. (14.17) and the expression for γ_0 are based on the assumption that the absorption coefficient of a boundary element is independent of its length, an assumption that generally holds true if the wavelength λ is considerably smaller than the element length.

Equation (14.17) also permits one to account for the damping effects of linear discontinuities, such as seams or attached reinforcing beams, on panels, provided one knows the corresponding absorption coefficients. Since plate waves can impinge on both sides of a discontinuity located within the panel area, for such a discontinuity location one needs to use in Eq. (14.17) twice the actual discontinuity length.

The energy that beams or reinforcements attached to a panel can dissipate, and thus the damping they can produce, depends markedly on the fastening method used. Metal beams attached to metal panels or seams in such panels generally produce little damping if they are continuously welded or joined by means of a rigid adhesive. However, they can contribute significant damping if they are fastened by a flexible, dissipative adhesive or if they are fastened at only a number of points, for example, by rivets, bolts, or spot welds. At high frequencies, at which the flexural wavelength on the panel is smaller than the distance between fastening points, damping results predominantly not from interface friction but from an "air-pumping" effect produced as adjacent surfaces (at locations between the connection points) move away from and toward each other. Energy loss here is due to the viscosity of the air or other fluid present between the surfaces.[13–15]

The high-frequency absorption coefficient corresponding to a beam that is fastened to a panel at a multitude of points may be estimated from the experimental results summarized in Fig. 14.6, which shows how the beam's reduced absorption coefficient γ_r varies with reduced frequency f_r or with the ratio d/λ of fastener spacing to plate flexural wavelength. These reduced quantities, definitions of which are given in the figure, account for the absorption coefficients' dependence on beam width w, plate thickness h, fastener spacing d, and longitudinal wave speed c_L in the plate material by relating these to reference values of these parameters (indicated by the subscript 0). It should be noted that Fig. 14.6 pertains to panels immersed in air at atmospheric pressure; lesser absorption coefficient values apply for panels at reduced pressure and higher values for panels at greater pressures. A theory is available[13] to account for the effects of atmospheric pressure changes and for other gases or liquids present between the contacting surfaces.

Damping due to Energy Transport

Structural Transmission. Energy that is transported away from a vibrating structure constitutes energy lost from that structure and thus contributes damping. Energy transport may occur to neighboring structural elements or to fluids in contact with the vibrating structure.

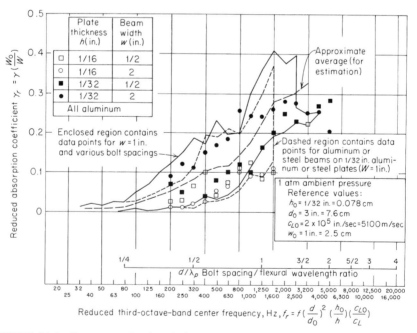

FIGURE 14.6 Summary of reduced absorption coefficient data for beams fastened to plates by rows of rivets, bolts, or spot welds.[15]

For example, a panel that is part of a multipanel array (such as an aircraft fuselage) is damped not only due to energy dissipation within the panel but also due to energy transport to adjacent panels. Energy transport makes it very difficult in practice to measure the dissipative damping of a structural component that is connected to others. Energy transported via the supports of test samples also tends to contaminate laboratory measurements of dissipative damping, potentially introducing large errors in measurements on samples with small inherent damping.

If a structural element is attached to a vibrating structure at a given point, then the energy D that is transported to the attached structure per cycle at frequency f is given by

$$D = \frac{V_S^2 \operatorname{Re}[Z_A]}{2f} \left| 1 + \frac{Z_A}{Z_S} \right|^{-2} \tag{14.18}$$

Here V_S denotes the amplitude of the velocity of the vibrating structure at the attachment point before the added structure is attached; Z_A denotes the driving-point impedance of the attached structure, and Z_S denotes the impedance of the vibrating structure at the attachment point (with both impedances measured in the direction of V_S). The loss factor contribution due to an attached structure then may be found by use of Eq. (14.13).

A *waveguide absorber* is a damping device that is intended to conduct energy away from its attachment point and to dissipate that energy. Such an absorber consists essentially of a structural element along which waves can travel and which includes means for dissipating the energy transported by these waves. For example, a long slender beam (which may be straight or coiled in some fashion) that is made of a highly damped plastic or coated with a high-damping material may serve as a waveguide absorber at frequencies considerably above its fundamental resonance. To be effective, a waveguide absorber must support waves in the frequency range of interest, it must be attached at a point where the vibrating structure moves with considerable amplitude, and its impedance must be such that it does not excessively reduce the vibrating structure's motion at the attachment point.[16]

A *tuned damper*, often also called a *dynamic absorber* or *neutralizer*, may be visualized as a mass–spring–damper system whose spring base is attached to a point on a vibrating structure. In the region near its natural frequency a tuned damper tends to impede the motion of the attachment point and to dissipate considerable energy; outside this frequency region, the damping effect of such a damper typically is small. Any system, such as a beam or plate, which exhibits a resonance at the frequency of concern, can act as a tuned damper at that frequency. Considerable damping of a plate over a relatively wide frequency range can be obtained by distributing a number of small tuned dampers with slightly different natural frequencies over the plate surface.[17]

Sound Radiation. Sound radiated by a vibrating structure transports energy from the structure and thus contributes damping. For a homogeneous panel of thickness h and material density ρ_p, one may calculate the panel's loss factor η_R at frequency f due to sound radiation from one side of the panel from its radiation efficiency σ by use of the relation[14]

$$\eta_R = \frac{\rho}{\rho_p} \frac{c}{2\pi f h} \sigma \qquad (14.19)$$

where ρ and c denote the density of the ambient medium and the speed of sound in it, respectively. If the panel can radiate from both of its sides, η_R is twice as great as indicated by Eq. (14.19). The magnitude of the radiation efficiency σ depends on the vibratory velocity distribution on the panel, as well as on frequency, and thus generally is different for different excitation distributions.

For a plate that has little inherent damping and that is excited at a single point, the radiation efficiency σ obeys[18]

$$\sigma = \begin{cases} (Uc/\pi^2 A f_c)\sqrt{f/f_c} & \text{for } f \ll f_c \\ 0.45\sqrt{U f_c/c} & \text{for } f = f_c \\ 1.0 & \text{for } f \gg f_c \end{cases} \qquad (14.20)$$

where A denotes the panel's surface area (one side), U its circumference, and f_c the coincidence frequency. This frequency, which is defined as that at which the plate flexural wavelength is equal to the acoustical wavelength in the ambient medium, is given by $f_c \approx c^2/1.8 h c_L$, where c_L represents the longitudinal wave speed in the plate material. More detailed information on radiation efficiency is provided in Chapter 11. Equation (14.20) may be used for the general estimation of radiation efficiency values for plates that are not too highly damped. This equation also provides a reasonable estimate of the radiation efficiency of rib-stiffened plates if twice the total rib length is included in the circumference U.

14.5 VISCOELASTIC DAMPING TREATMENTS

Viscoelastic Materials and Material Combinations

Materials that have both damping (energy dissipation) and structural (strain energy storage) capability are called "viscoelastic." Although virtually all materials fall into this category, the term is generally applied only to materials, such as plastics and elastomers, that have relatively high ratios of energy dissipation to energy storage capability.

Structural materials with high strength-to-weight ratios typically have little inherent damping, as is evident from Fig. 14.5, whereas plastics and rubbers that are highly damped tend to have relatively low strength. This circumstance has led to the consideration of combinations of high-strength materials and high-damping viscoelastic materials for applications where both strength and damping are required. Additions of viscoelastic materials to structural elements have come to be known as viscoelastic damping treatments.

If a composite structure is deflected, it stores energy via a variety of deformations (such as shear, tension and compression, flexure) in each structural element. If η_i denotes the loss factor corresponding to the ith element deformation and W_i represents the energy stored in that deformation, then the loss factor η of the entire structure obeys[19]

$$\eta = \sum \eta_i \frac{W_i}{W_T} \tag{14.21}$$

where $W_T = \sum W_i$ denotes the total energy stored in the structure. The foregoing expression indicates that the loss factor η of the composite structure is equal to a weighted average of the loss factors corresponding to all of the element deformations, with the energy storages serving as the weighting factors. This expression also leads to an important conclusion: An element deformation can make a significant contribution to the total loss factor only if (1) the loss factor associated with it is significant *and* (2) the energy storage associated with it is a significant fraction of the total energy storage.*

Mechanical Properties of Viscoelastic Materials

Because viscoelastic materials have both energy storage and energy dissipation capability, it is convenient to describe their behavior in terms of elastic and shear moduli that are complex quantities, in analogy to the definition of the complex stiffness introduced in Eq. (14.10). The complex Young's modulus $\overline{E}$ of a material, defined as the ratio of the stress phasor to the strain phasor, may be written as[10] $\overline{E} = E_R + jE_I = E_R(1 + j\eta_E)$, where the real part E_R is called the storage modulus, the imaginary part E_I is called the loss modulus, and the loss factor η_E associated with Young's modulus is equal to E_I/E_R. A completely analogous definition applies to the complex shear modulus.†

For plastics and elastomers, the viscoelastic materials of greatest practical interest, the real and imaginary moduli as well as the loss factors vary considerably with frequency and temperature. However, these parameters usually vary relatively little with strain amplitude, preload, and aging.[10] The loss factor associated with the shear modulus typically is equal to that associated with the Young's modulus for all practical purposes, so that one generally need not distinguish between the two. Also, since most of the viscoelastic materials of practical interest are virtually incompressible, the shear modulus value is nearly equal to one-third of the corresponding Young's modulus value. One may also note that often $\eta_E^2 \ll 1$, so that $|\overline{E}| \approx E_R$.

---

*Equation (14.21) applies precisely only for cases where all energy storage elements are deflected in phase, so that they reach their maximum energy storages at the same instant.

†An advantage of the complex modulus representation is the ease with which it enables one to incorporate damping in an analysis. One merely needs to replace the real moduli in the undamped formulation of a problem by the corresponding complex moduli—or, equivalently, to replace the real stiffnesses by the corresponding complex stiffnesses—to obtain a formulation of the problem that includes damping. This approach applies for lumped-parameter dynamic systems as well as for continuous systems and can take account of different values of damping in different elements and materials.

Figure 14.7 shows how the (real) shear modulus and loss factor of a typical viscoelastic material vary with frequency and temperature. At low frequencies and/or high temperatures, the material is soft and mobile enough for the strain to follow an applied stress without appreciable phase shift so that the damping is small; the material is said to be in its "rubbery" state. At high frequencies and/or low temperatures, the material is stiff and immobile, may tend to be brittle, is relatively undamped, and behaves somewhat like glass; it is said to be in its "glassy" state. At intermediate frequencies and temperatures, the modulus takes on intermediate values and the loss factor is highest; the material is said to be in its "transition" state.

This material behavior may be explained on the basis of the interactions of the long-chain molecules that constitute polymeric materials. At low temperatures, the molecules are relatively inactive; they remain "locked together," resulting in high stiffness, and because they move little relative to each other, there is little intermolecular "friction" to produce damping. At high temperatures, the molecules become active; they move easily relative to each other, resulting in low stiffness, and because they interact little, there is again little energy dissipation due to intermolecular friction. At intermediate temperatures, where the molecules have intermediate relative motion and interaction, the stiffness also takes on an intermediate value and the loss factor is greatest. A similar discussion applies to the effect of frequency on the material properties, with the inertia of the molecules leading to their decreasing mobility and interaction with increasing frequency.

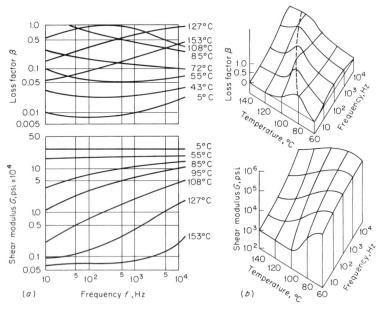

FIGURE 14.7 Dependence of shear modulus and loss factor of a polyester plastic on frequency and temperature[20]: (a) functions of frequency at constant temperature; (b) isometric plots on temperature–frequency plane. Note logarithmic frequency axes.

The observation that there exists a *temperature–frequency equivalence*, namely that an appropriate temperature decrease produces the same effect as a given frequency increase, has led to the development of convenient plots in which data for the frequency and temperature variations of each material modulus collapse onto single curves.[10,21] This collapse is achieved by plotting the data against a reduced frequency $f_R = f\alpha(T)$, where $\alpha(T)$ is an appropriately selected function of temperature T. In presentations of data in this form the function $\alpha(T)$ may be given analytically, in a separate plot or, as has recently been standardized[22,23], in the form of a nomogram that is superposed on the reduced data plot. Figure 14.8 is an illustration of such a plot and nomogram; its use is explained in the figure's legend.

Data on the properties of damping materials are available from knowledgeable suppliers of these materials. Compilations of data appear in references 10, 21, and 24. Key information on some of these materials appears in Table 14.1 in a form that is useful for preliminary material comparison and selection for specific applications in keeping with the concepts discussed in the later portions of this chapter. For each listed material the table shows the greatest loss factor value η_{max} exhibited by the material and the temperatures at which this value is obtained at three frequencies. The table also lists three values of the modulus of elasticity: E_{max}, the greatest value of Young's modulus, applies at low temperatures (i.e.,

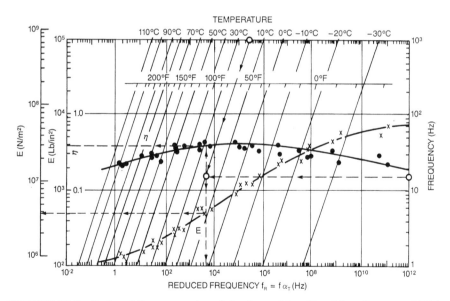

FIGURE 14.8 Reduced frequency plot of elastic modulus E and loss factor η of "Sylgard 188" silicone potting compound. (After Ref. 25) Points indicate measured data to which curves were fitted. Nomograph superposed on data plot facilitates determination of reduced frequency f_R corresponding to frequency f and temperature T. Use of nomogram is illustrated by dashed lines: For $f = 15$ Hz and $T = 20°C$, one finds $f_R = 5 \times 10^3$ Hz and $E = 3.8 \times 10^6$ N/m², $\eta = 0.36$.

TABLE 14.1 Properties of Some Commercial Damping Materials[a]

Material	Maximum Loss Factor η_{max}	Temperature (°F)[b] for η_{max} at			Elastic Moduli (psi)[c]			
		10 Hz	100 Hz	1000 Hz	E_{max}	E_{min}	E_{trans}	$E_{l,max}$
Antiphon-13	1.8	25	75	120	3e5	1.2e3	1.9e4	3.e3e
blachford Aquaplas	0.5	50	80	125	1.6e6	3e4	2.2e5	1.1e5
Barry Controls H-326	0.8	−40	−25	−10	6e5	3e3	4.2e4	3.4e4
Dow Corning Sylgard 188	0.6	60	80	110	2.2e4	3e2	2.6e3	1.5e3
EAR C-1002	1.9	23	55	90	3e5	2e2	7.7e3	1.5e4
EAR C-2003	1.0	45	70	100	8e5	6e2	2.2e4	2.2e4
lord LD-400	0.7	50	80	125	3e6	3.3e3	1e5	7e4
Soundcoat DYAD 601	1.0	15	50	75	3e5	1.5e2	6.7e3	6.7e3
Soundcoat DYAD 606	1.0	70	100	130	3G5	1.2c2	6e3	6e3
Soundcoat DYAD 609	1.0	125	150	185	2e5	6e2	1.1e4	1.1e4
Soundcoat N	1.5	15	30	70	3e5	7e1	4.6e3	6.9e3
3M ISD-110	1.7	80	115	150	3e4	3e1	1e3	1.7e3
3M ISD-112	1.2	10	40	80	1.3e5	8e1	3.2e3	3.9e3
3M ISD-113	1.1	−45	−20	15	1.5e5	3e2	2.1e2	2.3e2
3m 468	0.8	15	50	85	1.4e5	3e1	2e3	1.6e3
3M ISD-830	1.0	−75	−50	−20	2e5	1.5e2	5.5e3	5.5e3
GE SMRD	0.9	50	80	125	e35	5e3	3.9e4	3.5e4

[a] Approximate values taken from curves in Ref. 11.
[b] To convert to °C, use the formula °C = ($\frac{5}{9}$)(°F − 32) or the approximate table below:

°F	−80	−60	−40	−20	0	20	40	60	80	100	120	140	160	180	200
°C	−62	−51	−40	−29	−18	−7	4	16	27	38	49	60	71	82	93

[c] Numbers shown correspond to storage (real) values of Young's modulus, except that $E_{l,max}$ represents the maximum values of the loss (imaginary) modulus. E_{max} applies for low temperatures and/or high frequencies. E_{min} applies for high temperatures and/or low frequencies. E_{trans} and $E_{l,max}$ applies in the range of η_{max}. Divide by 3 to obtain the corresponding shear modulus values. To convert to N/m², multiply tabulated values by 7×10^3. The number following e represents the power of 10 by which the number preceding e is to be multiplied; e.g., 1.2e3 represents 1.2×10^3.

temperatures considerably below those corresponding to η_{max}); E_{min}, the smallest value of E, applies at high temperatures; the transition value E_{trans} applies in the η_{max} range; and $E_{l,max} \approx \eta_{max} E_{trans}$, the maximum value of the loss modulus, applies in the transition range.

It is important to keep in mind that the mechanical properties of polymeric materials, including plastics and elastomers, tend to be more variable than those of metals and other classical structural materials. Some of this variability results from a polymer's molecular structure and molecular weight distribution, which depend not only on the material's chemical composition but also on its processing. Additional variability results from the various types and amounts of plasticizers and fillers that are added to most commercial materials for a number of practical purposes. Thus, it is quite common for nominally identical polymeric materials to exhibit considerably different mechanical behaviors. It may also occur that even material samples from the same production run have considerably different loss factors and moduli at frequencies and temperatures at which they are intended to be used, pointing toward the need for careful quality control and performance verification for critical applications.

Structures with Viscoelastic Layers

One may calculate the loss factor of a structure vibrating in a given mode by use of Eq. (14.21) if one knows the energies W_i stored in the various deformations of all of the component elements. Indeed, modern finite-element analysis methods[26–28] proceed by calculating the modal deflections, applying these to evaluate all the energy storage components, and then using Eq. (14.21) to find the loss factor.

Analytical results have been developed for flexure of uniform beam and plate structures under conditions that are often approximated in practice. These results, which are extremely useful for design guidance and for development of an understanding of the important parameters, apply to structures whose deflection distributions are sinusoidal.*

Two-Component Beams

In flexure of a uniform beam with an insert or added layer of viscoelastic material, as illustrated by Fig. 14.9, the energy storage (and dissipation) associated with shear and torsional deformations may generally be neglected. If contact between the components is maintained without slippage at all surfaces and if the loss factor of the basic structural (nonviscoelastic) component is negligible, then the

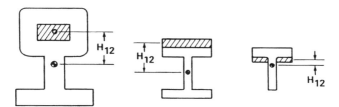

FIGURE 14.9 End views of beams with viscoelastic inserts or added layers. Structural material is unshaded, viscoelastic material is shown shaded; H_{12} represents distance between neutral axis of structural component and that of viscoelastic component. Beam deflection is vertical, with wave propagation along the beam length, perpendicular to the plane of the paper.

---

*The deflection distribution of any beam (or plate) vibrating in one of its natural modes is at least approximately sinusoidal (in one dimension for beams and in two dimensions for plates) at locations that are one wavelength or more from the boundaries, regardless of the boundary conditions. Therefore, the assumption of a sinusoidal deflection distribution is valid for a larger fraction of the structure as the frequency increases.

loss factor η of the composite beam is related to the loss factor β of the material of the viscoelastic component by[29,30]

$$\frac{\eta}{\beta} = \left[1 + \frac{k^2(1+\beta^2) + (r_1/H_{12})^2\alpha}{k[1 + (r_2/H_{12})^2\alpha]}\right]^{-1} \quad (14.22)$$

where $\alpha = (1+k)^2 + (\beta k)^2$ and H_{12} denotes the distance between the neutral axes of the two components. With subscript 1 referring to the structural (undamped) component and subscript 2 referring to the viscoelastic component, $k = K_2/K_1$, where $K_i = E_i A_i$ denotes the extensional stiffness of component i, expressed in terms of its Young's modulus (real part) E_i and cross-sectional area A_i. Furthermore, $r_i = \sqrt{I_i/A_i}$ represents the radius of gyration of A_i, where I_i denotes the centroidal moment of inertia of A_i.

For the often-encountered case where the structural component's extensional stiffness is much greater than that of the viscoelastic component, $k \ll 1$ and Eq. (14.22) reduces to

$$\eta \approx \frac{\beta E_2 I_T}{E_1 I_1 + E_2 I_T} \approx \frac{\beta E_2}{E_1} \frac{I_T}{I_1} \quad (14.23a)$$

where $I_T = I_2 + H_{12}^2 A_2 = A_2(r_2^2 + H_{12}^2)$ denotes the moment of inertia of A_2 about the neutral axis of A_1.

The last expression in Eq. (14.23a) applies for $E_2 I_T \ll E_1 I_1$, which is generally true in practical structures where the area and elastic modulus of the viscoelastic component are small compared to those of the structural component. In this case the composite structure's neutral axis coincides very nearly with that of the structural component and the dominant energy storage is associated with flexure of the structural component (whose flexural stiffness is $E_1 I_1$). The dominant energy dissipation is associated with extension and compression of the viscoelastic component, with the average extension (equal to the extension at the viscoelastic component's neutral axis) resulting from the flexural curvature and the distance H_{12} between the neutral axes of the viscoelastic and the structural components.* The flexural curvature is greatest at the antinodes of the vibrating structure; most of the damping action thus occurs at these locations, with little damping resulting from the material near the nodes.

The second form of Eq. (14.23a) contains two ratios; the first involves only material properties and the second only geometric parameters. It indicates that the most important dynamic mechanical property of the viscoelastic material is its extensional loss modulus $E_l = \beta E_2$. In keeping with the conclusions based on the general energy expression [Eq. (14.21)], good damping of the composite structure can be obtained only from a viscoelastic material that has not only a high loss factor but also a considerable energy storage capability.

---

*A "spacer," a layer that is stiff in shear and soft in extension (e.g., like honeycomb), inserted between the structural and the viscoelastic component can increase H_{12} and thus the damping obtained with a given amount of viscoelastic material.[30]

Plates with Viscoelastic Coatings

A strip of a plate (see insert of Fig. 14.10) may be considered as a special case of a two-component beam, where the two components have rectangular cross sections. Thus Eqs. (14.22) and (14.23a) apply, with $r_i = H_i/\sqrt{12}$ and $H_{12} = \frac{1}{2}(H_1 + H_2)$, where H_i denotes the component thickness. The energy storage and dissipation considerations that were discussed in the foregoing section, as well as the foregoing remarks concerning the dominant damping material properties, apply here also.

Figure 14.10 is a plot based on Eq. (14.22) for $\beta^2 \ll 1$. It shows that for small relative thicknesses $h_2 = H_2/H_1$, the loss factor ratio η/β is proportional to the viscoelastic layer thickness, whereas for very large relative thicknesses the loss factor ratio approaches unity; that is, the loss factor of the coated plate approaches that of the viscoelastic coating, as one would expect.* As also is evident from

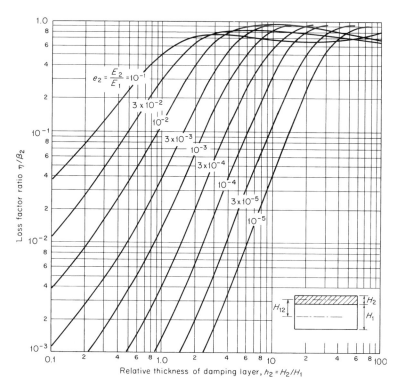

FIGURE 14.10 Dependence of loss factor η of plate strip with added viscoelastic layer on relative thickness and relative modulus of layer.[30] Curves apply for loss factors β of viscoelastic material that are small compared to unity.

---

*For very thick viscoelastic coatings, deformations in the thickness direction (which are not considered in this simplified analysis) also may play a significant role, particularly at frequencies at which standing-wave resonances may occur in the viscoelastic material.[31]

the figure, at small relative thicknesses the loss factor ratio is proportional to the modulus ratio $e_2 = E_2/E_1$. For small relative thicknesses, that is, in the regions where the curves of Fig. 14.10 are nearly straight, the loss factor of a coated plate may be estimated from[30]

$$\eta \approx \frac{\beta E_2}{E_1} h_2 \, (3 + 6h_2 + 4h_2^2) \tag{14.23b}$$

where $h_2 = H_2/H_1$.

If two viscoelastic layers are applied to a plate, one layer on each side, then the loss factor of the coated plate may be taken as the sum of the loss factors contributed by the individual layers, with each contribution calculated as if the other layer were absent, provided that each viscoelastic layer has low relative extensional rigidity, that is, that $E_2 H_2 \ll E_1 H_1$. If this inequality is not satisfied, a more complex analysis is required.

Three-Component Beams with Viscoelastic Interlayers

Figure 14.11 illustrates uniform beams consisting of two structural (nonviscoelastic) components interconnected via a relatively thin viscoelastic component. Such three-component beams may be preferable to two-component beams for practical reasons because the viscoelastic material is exposed only at its edges; however, such beams can also be designed to have higher damping than two-component beams of similar weight.*

In flexure of a three-component beam with a viscoelastic layer whose extensional and flexural stiffnesses are small compared to those of the structural components, the dominant energy dissipation is associated with shear in the viscoelastic component and the most significant energy storage occurs in connection with extension/compression and flexure of the two structural components. The shear in the viscoelastic component is greatest at the vibrating structure's nodes. Thus, most of the energy dissipation occurs in the viscoelastic material near

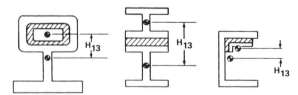

FIGURE 14.11 End views of composite beams made up of two structural components (unshaded) joined via a viscoelastic component (shaded); H_{13} is distance between neutral axes of structural components. Beam deflection is vertical, with wave propagation along the beam length, perpendicular to the plane of the paper.

---

*It should be noted that design changes to obtain increased damping generally also result in mass and stiffness changes, which tend to affect a structure's vibratory response and should be considered in the design process.[17,32,33]

the nodes, with relatively little resulting in that near the antinodes. For efficient damping, it is important that the shearing action in the viscoelastic material not be restrained (particularly at and near nodes) by structural interconnections, such as bolts.

The loss factor η corresponding to a spatially sinusoidal deflection shape of such a three-component beam is related to the loss factor β of the viscoelastic material by[29,30]

$$\eta = \frac{\beta Y X}{1 + (2 + Y)X + (1 + Y)(1 + \beta^2)X^2} \qquad (14.24)$$

where

$$X = \frac{G_2 b}{p^2 H_2} S \qquad \frac{1}{Y} = \frac{E_1 I_1 + E_3 I_3}{H_{13}^2} S \qquad S = \frac{1}{E_1 A_1} + \frac{1}{E_3 A_3} \qquad (14.25)$$

Here subscripts 1 and 3 refer to the structural components and 2 to the viscoelastic component; E_i, A_i, and I_i represent, respectively, the Young's modulus, cross-sectional area, and moment of inertia of component i; H_{13} denotes the distance between the neutral axes of the two structural components; and G_2 represents the shear modulus (real part) of the viscoelastic material, H_2 the average thickness of the viscoelastic layer, and b its length as measured on a cross section through the beam. The wavenumber p of the spatially sinusoidal beam deflection obeys

$$\frac{1}{p^2} = \left(\frac{\lambda}{2\pi}\right)^2 = \frac{1}{\omega}\sqrt{\frac{B}{\mu}} \qquad (14.26)$$

where λ represents the bending wavelength and B denotes the flexural rigidity and μ the mass per unit length of the composite beam.

The *structural parameter* Y of three-component structures depends only on the geometry and Young's moduli of the two structural components, whereas the *shear parameter* X depends also on the properties of the viscoelastic layer and on the wavelength of the beam deflection. The shear parameter X is proportional to the square of the ratio of the beam flexural wavelength to the *decay distance*,[34] that is, the distance within which a local shear disturbance decays by a factor of e, where $e \approx 2.72$ denotes the base of natural logarithms; thus, X also is a measure of how well the viscoelastic layer couples the flexural motions of the two structural components.

The (complex) flexural rigidity of a three-component beam is given by

$$\overline{B} = (E_1 I_1 + E_3 I_3)\left(1 + \frac{X^* Y}{1 + X^*}\right) \qquad X^* = X(1 - j\beta) \qquad (14.27)$$

Its magnitude is $B = |\overline{B}|$. Thus, for small X, the flexural rigidity B of the composite beam is equal to the sum of the flexural rigidities of the structural components, that is, to the total flexural rigidity that the two components exhibit if they are not interconnected. For $X \gg 1$, however B approaches $1+Y$ times the foregoing value, which is equal to the flexural rigidity of a beam with rigidly interconnected structural components 1 and 3.

For a given value of β and Y, the loss factor η of the composite beam takes on its greatest value,

$$\eta_{\max} = \frac{\beta Y}{2+Y+2/X_{\text{opt}}} \tag{14.28}$$

at the optimum value of X, which is given by

$$X_{\text{opt}} = [(1+Y)(1+\beta^2)]^{-1/2} \tag{14.29}$$

With the aid of these definitions, one may rewrite Eq. (14.24) in terms of the ratio $R = X/X_{\text{opt}}$ in the following form, which provides a convenient view of the damping behavior of three-component beams:

$$\frac{\eta}{\eta_{\max}} = \frac{2(1+N)R}{1+2NR+R^2} \qquad N = (1+\tfrac{1}{2}Y)X_{\text{opt}} \tag{14.30}$$

Figure 14.12, which is based on Eqs. (14.28) and (14.29), shows how $\eta_{\max}/\beta$ increases monotonically with Y, indicating the importance of selecting a configuration with a large value of Y in the design of highly damped composite structures.* Figure 14.13 gives approximate values of Y for some often encountered

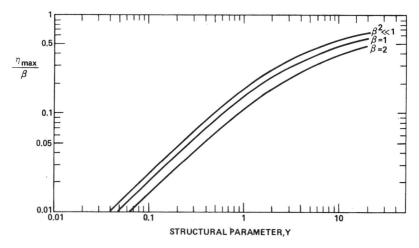

FIGURE 14.12 Dependence of maximum loss factor $\eta_{\max}$ of three-component beams or plates on structural parameter Y and loss factor β of viscoelastic material.

*A shear-stiff, extensionally soft "spacer" (e.g., honeycomb) inserted between the viscoelastic and one or both structural components can serve to increase H_{13} and, therefore, the value of Y. See Eq. (14.25). For a given deflection of the composite structure, a spacer increases the damping by increasing the shear strain—and thus the energy storage and dissipation—in the viscoelastic component.[30,34]

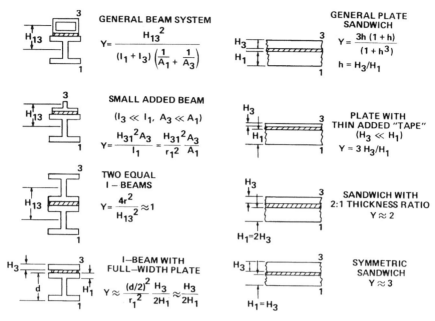

FIGURE 14.13 Values of structural parameter Y for three-component beam and plate configurations with thin viscoelastic components and with structural components of the same material; viscoelastic component is shown cross-hatched. I, moment of inertia; A, cross-sectional area; r, radius of gyration; H_{13}, distance between neutral axes.

configurations. Figure 14.14 shows how $\eta/\eta_{\max}$ varies with X/X_{opt}, indicating the importance of making the operating value of X of a given design match X_{opt} as closely as possible in order to obtain a loss factor that approaches $\eta_{\max}$.

If one knows the wavenumber $p = 2\pi/\lambda$ and the frequency associated with a given beam vibration, one may calculate the loss factor η of a composite beam simply by substituting the beam parameters and the material properties at any frequency (and temperature) of interest into Eqs. (14.24) and (14.25) or (14.28)–(14.30). The latter set of equations is particularly useful for judging how far from the optimum a given configuration may be operating.

If one knows only the frequency and not the wavenumber p corresponding to a beam vibration, one needs to use Eq. (14.26) to determine p. Substitution of B as calculated from Eq. (14.27) into Eq. (14.26), followed by substitution of the result into the first of Eqs. (14.25), leads to a cubic equation in X. Although one may solve this numerically, it is often more convenient to determine X by use of an iteration procedure like that indicated in Fig. 14.15.

In contrast to the previously discussed two-component beam, the loss factor η of a three-component beam does not depend primarily on the loss modulus (i.e., on the product of the loss factor β and storage modulus E_2 or G_2) of the viscoelastic material. The loss factor of a three-component beam depends on β and G_2 separately, and the separate dependences must be taken into account in

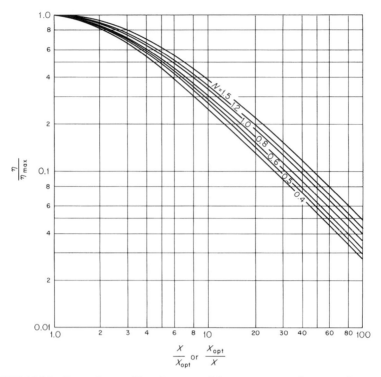

FIGURE 14.14 Dependence of loss factor η of three-component beam or plate on shear parameter X [from Eq. (14.30)].

the design of such a beam. Design of a highly damped three-component beam structure requires (1) choice of a configuration with a large structural parameter Y, (2) selection of a damping material with a large loss factor β in the frequency and temperature range of interest, and (3) adjustment of the damping material thickness H_2 and length b so as to make X [as calculated from Eq. (14.25) for the value of G_2 applicable to the frequency and temperature of concern] approximately equal to X_{opt} [given by Eq. (14.29)]. The resulting design then will have a loss factor approximately equal to η_{max} as given by Eq. (14.28).

The expected performance of any design should be checked for the frequency and temperature ranges of interest by means of the procedure described in the previous paragraph. Note that a given design may be expected to perform optimally—that is, to have X under operating conditions approximately equal to X_{opt}—only in a limited range of frequencies and temperatures, with reduced performance outside this range.

Plates with Viscoelastic Interlayers

A strip of a plate consisting of a viscoelastic layer between two structural layers may be considered as a special case of a three-component beam. In a plate strip,

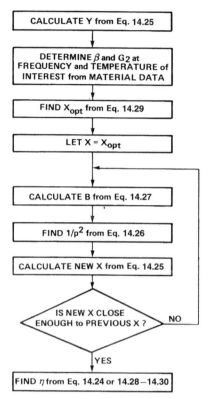

FIGURE 14.15 Iteration procedure for determination of loss factor of three-component beams or plates for which wavelength is not known initially.

all components have rectangular cross sections of the same width; Eqs. (14.25) accordingly become

$$X = \frac{G_2}{p^2 H_2} S \qquad \frac{1}{Y} = \frac{E_1 H_1^3 + E_3 H_3^3}{12 H_{31}^2} S \qquad S = \frac{1}{E_1 H_1} + \frac{1}{E_3 H_3} \qquad (14.31)$$

If, furthermore, $E_1 I_1 + E_3 I_3$ in Eq. (14.27) is replaced by $\frac{1}{12}(E_1 H_1^3 + E_3 H_3^3)$ and if μ in Eq. (14.26) is interpreted as the mass per unit surface area of the plate, then all of the foregoing discussion pertaining to beams also applies to plates.

REFERENCES

1. W. T. Thomson, *Theory of Vibration with Applications*, 2nd ed., Prentice-Hall, Englewood Cliffs, NJ, 1981.
2. R. Plunkett, "Measurement of Damping," in J. F. Ruzicka (Ed.), *Structural Damping*, American Society of Mechanical Engineers, New York, 1959.

3. K. N. Tong, *Theory of Mechanical Vibration*, Wiley, New York, 1960.
4. E. E. Ungar, "Mechanical Vibrations," in H. A. Rothbart (Ed.), *Mechanical Design and Systems Handbook*, 2nd ed., McGraw-Hill, New York, 1985, Chapter 5.
5. D. J. Ewins, *Modal Testing: Theory and Practice*, Research Studies Press, Letchworth, Hertfordshire, England, 1986.
6. V. H. Neubert, *Mechanical Impedance: Modelling/Analysis of Structures*, Josens Printing and Publishing, State College, PA, distributed by Naval Sea Systems Command, Code NSEA-SSN, 1987.
7. B. L. Clarkson and J. K. Hammond, "Random Vibration," in R. G. White and J. G. Walker (Eds.), *Noise and Vibration*, Wiley, New York, 1982, Chapter 5.
8. B. L. Clarkson and R. J. Pope, "Experimental Determination of Modal Densities and Loss Factors of Flat Plates and Cylinders," *J. Sound Vib.*, **77**(4), 535–549 (1981).
9. S. H. Crandall, "The Role of Damping in Vibration Theory," *J. Sound Vib.*, **11**(1), 3–18 (1970).
10. A. D. Nashif, D. I. G. Jones, and J. P. Henderson, *Vibration Damping*, Wiley, New York, 1985.
11. G. Kurtze, "Körperschalldämpfung durch Körnige Medien" [Damping of Structure-Borne Sound by Granular Media], *Acustica*, **6**(Beiheft 1), 154–159 (1956).
12. M. A. Heckl, "Measurements of Absorption Coefficients on Plates," *J. Acoust. Soc. Am.*, **34**, 308–808 (1962).
13. G. Maidanik, "Energy Dissipation Associated with Gas-Pumping at Structural Joints," *J. Acoust. Soc. Am.*, **34**, 1064–1072 (1966).
14. E. E. Ungar, "Damping of Panels Due to Ambient Air," in P. J. Torvik (Ed.), *Damping Applications in Vibration Control*, AMD-Vol 38, American Society of Mechanical Engineers, New York, 1980, pp. 73–81.
15. E. E. Ungar and J. Carbonell, "On Panel Vibration Damping Due to Structural Joints," *AIAA J.*, **4**, 1385–1390 (1966).
16. E. E. Ungar and L. G. Kurzweil, "Structural Damping Potential of Waveguide Absorbers," *Trans. Internoise 84*, December 1985, pp. 571–574.
17. D. J. Mead, *Passive Vibration Control*, J Wiley, New York, 2002.
18. L. Cremer, M. Heckl, and E. E. Ungar, *Structureborne Sound*, 2nd ed., Springer-Verlag, New York, 1988.
19. E. E. Ungar and E. M. Kerwin, Jr., "Loss Factors of Viscoelastic Systems in Terms of Energy Concepts." *J. Acoust. Soc. Am.*, **34**(7), 954–957 (1962).
20. E. E. Ungar, "Damping of Panels," in L. L. Beranek (Ed.), *Noise and Vibration Control*, McGraw-Hill, New York, 1971, Chapter 14.
21. D. I. G. Jones, *Handbook of Viscoelastic Vibration Damping*, Wiley, New York, 2002.
22. ISO 10112, "Damping Materials—Graphic Presentation of the Complex Modulus," International Organization for Standardization, Geneva, Switzerland, 1991.
23. ANSI S2.24-2001, "Graphical Presentation of the Complex Modulus of Viscoelastic Materials," American National Standards Institute, Washington, DC, 2001.
24. J. Soovere and M. L. Drake, "Aerospace Structures Technology Damping Design Guide," AFWAL-TR-84-3089, Flight Dynamics Laboratory, Wright-Patterson Air Force Base, OH, December 1985.

25. D. I. G. Jones and J. P. Henderson, "Fundamentals of Damping Materials," Section 2 of Vibration Damping Short Course Notes, M. L. Drake (Ed.), University of Dayton, 1988.
26. M. L. Soni and F. K. Bogner, "Finite Element Vibration Analysis of Damped Structures." *AIAA J.*, **20**(5), 700–707 (1982).
27. C. D. Johnson, D. A. Kienholz, and L. C. Rogers, "Finite Element Prediction of Damping in Beams with Constrained Viscoelastic Layers," *Shock Vib. Bull.*, **51**(Pt. 1), 71–82 (1981).
28. M. F. Kluesner and M. L. Drake, "Damped Structure Design Using Finite Element Analysis," *Shock Vib. Bull.*, **52**(Pt. 5), 1–12 (1982).
29. E. E. Ungar, "Loss Factors of Viscoelastically Damped Beam Structures," *J. Acoust. Soc. Am.*, **34**(8), 1082–1089 (1962).
30. D. Ross, E. E. Ungar, and E. M. Kerwin, Jr., "Damping of Plate Flexural Vibrations by Means of Viscoelastic Laminae," in J. Ruzicka (Ed.), *Structural Damping*, American Society of Mechanical Engineers, New York, 1959, Section 3.
31. E. E. Ungar and E. M. Kerwin, Jr., "Plate Damping Due to Thickness Deformations in Attached Viscoelastic Layers," *J. Acoust. Soc. Am.*, **36**, 386–392 (1964).
32. D. J. Mead, "Criteria for Comparing the Effectiveness of Damping Materials," *Noise Control*, **1**, 27–38 (1961).
33. D. J. Mead, "Vibration Control (I)," in R. G. White and J. G. Walker (Eds.), *Noise and Vibration*, Ellis Horwood, Chichester, West Sussex, England, 1982.
34. E. M. Kerwin, Jr., "Damping of Flexural Waves in Plates by Spaced Damping Treatments Having Spacers of Finite Stiffness," in L. Cremer (Ed.), *Proc. 3rd Int. Congr. Acoust., 1959*, Elsevier, Amsterdam, Netherlands, 1961, pp. 412–415.

CHAPTER 15

Noise of Gas Flows

H. D. BAUMANN

Consultant
Rye, New Hampshire

W. B. CONEY

BBN Systems and Technologies Corporation
Cambridge, Massachusetts

15.1 INTRODUCTION

The sound produced by unsteady gas flows and by interactions of gas flows with solid objects is called *aerodynamic sound*. The same flows often excite structural modes of vibration in surfaces bounding the flow and are then said to generate *structure-borne sound*. Unwanted flow-generated sound, noise, is a common by-product of most industrial processes. It also accompanies the operation of ships, automobiles, aircraft, rockets, and so on, and can adversely affect structural stability and be an important source of fatigue.

A practical understanding of the sources of aerodynamic sound is necessary over the whole range of mean-flow Mach numbers, from the very lowest (0.01 or less) associated with flows in air conditioning systems and underwater applications to the high supersonic range occurring in jet engines and high-pressure valves. In subsonic flows, the sound may be attributed to three basic aerodynamic source types: monopole, dipole, and quadrupole.[1]

These fundamental source types are discussed in this chapter, together with a survey of noise mechanisms associated with turbulent jets, spoilers and airfoils, boundary layers and separated flow over wall cavities, combustion, and valves.

15.2 AEROACOUSTICAL SOURCE TYPES

Aerodynamic Monopole

Monopole radiation is produced by the unsteady introduction of mass or heat into a fluid. Typical examples are pulse jets (where high-speed air is periodically ejected through a nozzle), turbulent flow over a small aperture in a large wall (where the flow induces pulsating motion in the aperture), unsteady combustion processes, and heat release from boundaries or from a pulsed laser beam.

The radiation from a monopole source in an otherwise stationary fluid is equivalent to that produced by a pulsating sphere (Fig. 15.1a). Both the amplitude and phase of the acoustical pressure are spherically symmetric. When the monopole sound is generated by unsteady flow velocities, the dimensional relation between the radiated sound power and the flow parameters is

$$W_{\text{monopole}} \propto \frac{\rho L^2 U^4}{c} = \rho L^2 U^3 M \qquad (15.1)$$

where W_{monopole} = radiated sound power, W
ρ = mean speed of gas, kg/m^3
c = speed of sound in gas, m/s
U = flow velocity in source region, m/s
L = length scale of flow in source region, m
M = Mach number, $= U/c$, dimensionless

Aerodynamic Dipole

Dipole sources arise when unsteady flow interacts with surfaces or bodies, when the dipole strength is equal to the force on the body, or when there are significant variations of mean fluid density in the flow. This source type is found in compressors where turbulence impinges on stators, rotor blades, and other control surfaces. Similarly, the unsteady shedding of vorticity from solid objects,

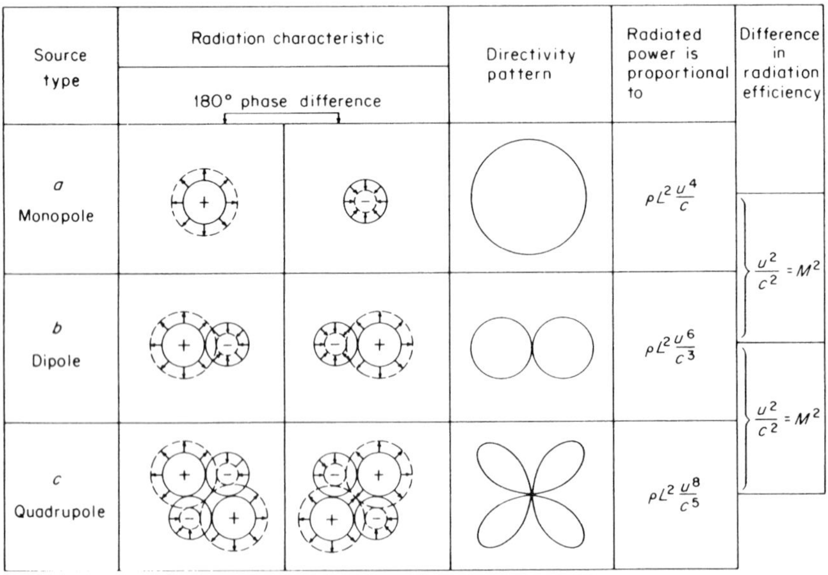

Source type	Radiation characteristic (180° phase difference)		Directivity pattern	Radiated power is proportional to	Difference in radiation efficiency
a Monopole	(+)	(−)	○	$\rho L^2 \frac{U^4}{c}$	
b Dipole	(+)(−)	(−)(+)	∞	$\rho L^2 \frac{U^6}{c^3}$	$\frac{U^2}{c^2} = M^2$
c Quadrupole	(+)(−)(−)(+)	(−)(+)(+)(−)	✿	$\rho L^2 \frac{U^8}{c^5}$	$\frac{U^2}{c^2} = M^2$

FIGURE 15.1 Aeroacoustical source types and their dimensional properties in fluid of uniform mean density. See also Fig. 1.2.

such as telegraph wires, struts, and airfoils, generates "singing" tones that are also attributable to dipole sources. Other examples include the noise generated by hot jets exhausting into a cooler ambient medium and by the acceleration of temperature (or "entropy") inhomogeneities in a mean pressure gradient, as in a duct contraction.

The dipole is equivalent to a pair of equal monopole sources of opposite phase separated by a distance that is much smaller than the wavelength of the sound. Destructive interference between the radiations from the monopoles reduces the efficiency with which sound is generated by the dipole relative to a monopole and produces a double-lobed, figure-eight radiation field shape proportional to the cosine of the angle measured from the dipole axis (Fig. 15.1b). In fluid of uniform mean density, the dimensional dependence of the aerodynamic dipole sound power is

$$W_{\text{dipole}} \propto \frac{\rho L^2 U^6}{c^3} = \rho L^2 U^3 M^3 \qquad (15.2)$$

This differs by a factor M^2 from the power output of the monopole. In subsonic flow ($M < 1$) the dipole is a less efficient source of sound.

If the specific entropy or temperature of the flow in the source region is not uniform (e.g., in the shear layer of a hot jet exhausting into a cooler ambient atmosphere), the density must also be variable. The relatively strong pressure fluctuations in the turbulent flow are then scattered by the density variations and produce sound of the dipole type. The dipole strength is proportional to the difference between the actual acceleration of the density inhomogeneity in the turbulent pressure field and that which it would have experienced had the density been uniform.[2] The dimensional dependence of the corresponding sound power is

$$W_{\text{entropy}} \propto \frac{\rho L^2 (\delta T/T)^2 U^6}{c^3} = \rho L^2 \left(\frac{\delta T}{T}\right)^2 U^3 M^3 \qquad (15.3)$$

where $(\delta T/T)^2$ is the mean-square fractional temperature fluctuation.

Aerodynamic Quadrupole

Quadrupole radiation is produced by the Reynolds stresses in a turbulent gas in the absence of obstacles. These arise from the convection of fluid momentum by the unsteady flow. The Reynolds stress forces must occur in opposing pairs, since the net momentum of the fluid is constant. Force pairs of this type are called quadrupoles and are equivalent to equal and opposite dipole sources (see Fig. 15.1c).

Aerodynamic quadrupoles and entropy dipoles are the dominant source types in high-speed, subsonic, turbulent air jets. The quadrupole strength is larger where both the turbulence and mean velocity gradients are high, for example, in the turbulent mixing layer of a jet.

The dimensional dependence of the radiated quadrupole sound power is

$$W_{\text{quadrupole}} \propto \frac{\rho L^2 U^8}{c^5} = \rho L^2 U^3 M^5 \tag{15.4}$$

This differs from the dipole power by a factor M^2. At subsonic speeds ($M < 1$), the quadrupole radiation efficiency is lower than that of the dipole because of the double cancellation illustrated in Fig. 15.1c.

The monopole, dipole, and quadrupole sources decrease in their respective radiation efficiencies for subsonic flows, but the dependencies of their radiated sound powers on flow velocity show the opposite trend, that is, the total radiated sound power varies as the fourth, sixth, and eighth powers of the flow speed U for monopole, dipole, and quadrupole sources, respectively. Thus, if the flow velocity is high enough, the radiation from the quadrupole sources may be the dominant source of sound even though the efficiency with which the sound is produced is small. This is usually the case for a jet engine at high subsonic exhaust velocities, although other internal sources caused by unsteady combustion or rough burning (predominantly monopole) or compressor noise (predominantly dipole) can also make a significant contribution to the total noise.

The value of the constant of proportionality for each type of source depends on both the sound-generating mechanism and the flow configuration. Thus, the constant for a singing wire differs from that of an edge tone, although both arise from unsteady surface forces (dipole sources). However, the proportionality relations (15.1)–(15.4) can be used to estimate the influence on radiated sound power of changes in one or more of the source parameters. A twofold increase in the exhaust velocity U of a jet (quadrupole-type source) causes the sound power level to increase by 24 dB (eighth power of flow velocity), whereas a doubling of the exhaust nozzle area A (proportional to L^2) increases the sound power level by only 3 dB. Because the thrust of a jet engine is proportional to AU^2, the increase in radiated sound will be smaller if a doubling of the thrust is achieved by increasing the nozzle area by a factor of 2 rather than the exhaust velocity by a factor of 1.4.

Aerodynamic Sources of Fractional Order

The efficiency with which sound is produced by aerodynamic dipole sources on surfaces whose dimensions greatly exceed the acoustical wavelength frequently differs from those implied by Eqs. (15.2) and (15.3). For example, the net dipole strength associated with turbulent flow over a smooth, plane wall is zero: The radiation is the same as that generated by the quadrupole sources in the flow when the wall is regarded as a plane reflector of sound. Similarly, the net strength of the dipoles induced by turbulence near the edge of a large wedge-shaped body varies with angle ν—being equivalent to that of a quadrupole when $\nu = 180°$ (i.e., for a plane wall) and to a *fractional* multipole order $\frac{3}{2}$ when $\nu = 0$ (knife edge). The latter case is important in estimating the leading and trailing edge noise of an airfoil, where at high enough frequency the sound power is proportional to $\rho L U^3 M^2$.

Influence of Source Motion

The directional characteristics of the radiation from acoustical sources are changed if the source moves relative to the fluid. Both the frequency and intensity of the sound are increased ahead of the source and decreased to its rear. It is usual to refer the observer to a coordinate system based on the source location at the time of emission of the received sound.

For a monopole source of strength $q(t)$ kilograms per second (equal to the product of the volume velocity and the fluid density) moving at constant speed U in the direction illustrated in Fig. 15.2, the acoustical pressure p in the far field is

$$p_{\text{monopole}} = \pm \left[\frac{dq/dt}{4\pi r (1 - M_{fr})^2} \right] \tag{15.5}$$

where the square brackets denote evaluation at the time of emission of the sound, (r, Φ) are coordinates defining the observer position relative to the source at the time of emission of the sound, and the plus and minus sign is taken according as $M_{fr} < 1$ or $M_{fr} > 1$, where

$$M_{fr} = \left(\frac{U}{c} \right) \cos \Phi \quad \text{(dimensionless)}$$

The frequency of the received sound is $f/(1 - M_{fr})$, where f, per second, is the frequency in a frame fixed relative to the source. The term $1/(1 - M_{fr})$ is called the *Doppler factor*. Its effect is to modify both the frequency and the amplitude of the acoustic field.

For a uniformly translating dipole of strength $f_i(t)$ (newtons), which is equivalent to an applied force in the i direction, and for a quadrupole $T_{ij}(t)$ (newtons-meters), specified by directions i, j (and equivalent to a force pair applied to the fluid), the respective acoustical pressure fields become

$$p_{\text{dipole}} = \pm \left[\frac{df_r/dt}{4\pi c r (1 - M_{fr})^2} \right] \qquad p_{\text{quadrupole}} = \left[\frac{d^2 T_{rr}/dt^2}{4\pi c^2 r |1 - M_{fr}|^3} \right] \tag{15.6}$$

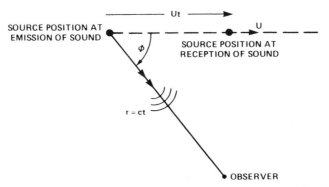

FIGURE 15.2 Coordinates defining the observer position at the time of emission of the received sound from a moving source.

where f_r, T_{rr} denote the components of the dipole and quadrupole source strengths in the observer direction at the time of emission of the sound and the plus and minus sign is taken according as $M_{fr} < 1$ or $M_{fr} > 1$.

These expressions are valid for ideal, point sources moving at constant velocity. The influence of source motion on real, aerodynamic sources is often much more complicated. For example, a pulsating sphere is a monopole source when at rest. In uniform translational motion, however, the radiation is amplified by a Doppler exponent of $\frac{7}{2}$ instead of the 2 of Eq. (15.5). This is because the monopole is augmented due to the motion by a dipole whose strength is proportional to the convection Mach number M_f. The interaction of the volume pulsations with the mean flow over the sphere produces a net fluctuating force on the fluid.

15.3 NOISE OF GAS JETS

The sound generated by high-speed jets is usually associated with several different sources acting simultaneously. *Jet mixing noise*, caused by the turbulent mixing of the jet with the ambient medium, and for imperfectly expanded supersonic jets *shock-associated noise*, produced by the convection of turbulence through shock cells in the jet, are the principal components of the radiation. The properties of sound sources in real jets differ considerably from those of the idealized models described in Section 15.2. Sources located within the "jet pipe" can also make a significant contribution to the noise. In the case of a gas turbine engine, the additional radiation includes combustion noise as well as tonal and broadband sound produced by interactions involving fans, compressors, and turbine systems. The following discussion is based on experimental data obtained and validated by several independent investigators and collated for prediction purposes by the Society of Automotive Engineers.[3] Formulas are given for predicting the *free-field* radiation from a jet in an ideal acoustical medium. In many applications it will be necessary to modify these predictions to account for atmospheric attenuation and interference caused by reflections from surfaces.

Jet Mixing Noise

Mixing noise is the most fundamental source of sound produced by a jet. The simplest free jet is an air stream issuing from a large reservoir through a circular convergent nozzle (Fig. 15.3). The gas accelerates from near-zero velocity in the reservoir to a peak velocity in the narrowest cross section of the nozzle. Sonic flow occurs at the nozzle exit when the pressure ratio p_0/p_s exceeds 1.89, where p_0 is the steady reservoir pressure and p_s is the ambient pressure downstream of the nozzle. An increase above this critical pressure ratio leads to the appearance of a shock cell structure downstream of the nozzle and "choking" of the flow unless the convergent part of the nozzle is followed by a divergent section in which the pressure decreases smoothly to p_s.

For the idealized, shock-free jet, no interaction is assumed between the gas flow and the solid boundaries. The noise is produced entirely by turbulent mixing

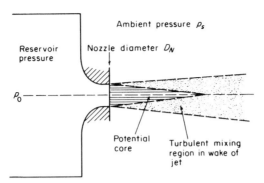

FIGURE 15.3 Subsonic turbulence-free jet.

in the shear layer. The sources extend over a considerable distance downstream of the nozzle. The high-frequency components of noise are generated predominantly close to the nozzle, where eddy sizes are small. Lower frequencies are radiated from sources further downstream where eddy sizes are much larger. The sources may be regarded as quadrupoles, whose strength and directivity are modified by the influences of nonuniform fluid density (temperature) and convection by the flow.

The total radiated sound power W in watts of the mixing noise may be expressed in terms of the mechanical stream power of the jet, equal to

$$W_{\text{mech stream}} = \tfrac{1}{2}mU^2 \tag{15.7}$$

where m = mass flow of gas, kg/s
U = fully expanded mean jet velocity, m/s

The dependence of $W/\tfrac{1}{2}mU^2$ on jet Mach number $M = U/c$ and density ratio ρ_j/ρ_s, where c is the ambient sound speed and ρ_j, ρ_s are respectively the densities (in kilograms per cubic meter) of the fully expanded jet and the ambient atmosphere, is illustrated in Fig. 15.4. For $M < 1.05$ the sound power increases as ρ_j/ρ_s decreases (i.e., as the jet temperature increases) and decreases at higher Mach numbers. When $M < 1$, W may be estimated from

$$\frac{W}{W_{\text{mech stream}}} = \frac{4 \times 10^{-5}(\rho_j/\rho_s)^{(w-1)}M^{4.5}}{(1 - M_c^2)^2} \qquad M < 1 \tag{15.8}$$

where $M_c = 0.62U/c$ and w is the jet density exponent given by[4]

$$w = \frac{3M^{3.5}}{0.6 + M^{3.5}} - 1 \tag{15.9}$$

Equation (15.9) is applicable for M greater than about 0.35, including the supersonic region and is also used in the formulas given below.

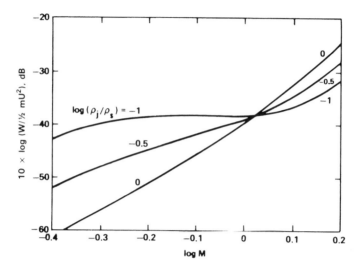

FIGURE 15.4 Ten times the logarithm of the ratio of acoustical power of jet mixing noise to mechanical stream power as a function of the fully expanded jet Mach number $M = U/c$ for different values of ρ_j/ρ_s.

The overall sound pressure level (OASPL) at any point in space is defined as

$$\text{OASPL} = 20 \times \log_{10} \frac{p_{\text{rms}}}{20 \ \mu\text{Pa}} \quad \text{dB} \quad (15.10)$$

where p_{rms} is the rms sound pressure in pascals (newtons per square meter) (see Chapter 2). Source convection and refraction in the shear layers causes the sound field to be directive, the maximum noise being radiated in directions inclined at angles θ to the jet axis between about 30° and 45°.

Typical field directivities at a fixed observer distance for an air jet at three different jet Mach numbers $M = U/c$ are shown in Fig. 15.5. The curves give the variation of $\text{OASPL}(\theta) - \text{OASPL}(90°)$, the sound pressure level relative to its value at $\theta = 90°$, and are approximated by the formula

$$\text{OASPL}(\theta) = \text{OASPL}(90°) - 30 \times \log\left[1 - \frac{M_c \cos\theta}{(1 + M_c^5)^{1/5}}\right]$$

$$- 1.67 \times \log\left[1 + \frac{1}{10^{(40.56-\theta')} + 4 \times 10^{-6}}\right] \quad (15.11)$$

where $M_c = 0.62 U/c$
$\theta' = 0.26(180 - \theta) M^{0.1}$, degrees

The final term on the right of (15.11) is negligible unless $\theta < 180° - 150°/M^{0.1}$, that is, except in directions close to the jet axis.

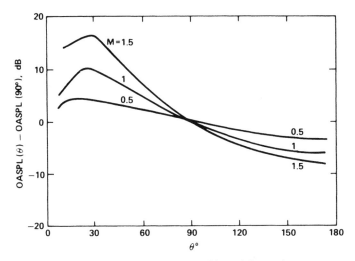

FIGURE 15.5 Directivity of jet mixing noise.

At $\theta = 90°$ the overall sound pressure level of jet mixing noise can be calculated from the formula

$$\text{OASPL} = 139.5 + 10 \times \log \frac{A}{R^2} + 10 \times \log\left[\left(\frac{p_s}{p_{\text{ISA}}}\right)^2 \left(\frac{\rho_j}{\rho_s}\right)^w\right]$$

$$+ 10 \times \log\left(\frac{M^{7.5}}{1 - 0.1M^{2.5} + 0.015M^{4.5}}\right) \tag{15.12}$$

where A = fully expanded jet area, $= \frac{1}{4}\pi D_N^2$ for a subsonic jet, m²
D_N = nozzle exit diameter, m
R = distance from center of nozzle exit, m
p_{ISA} = international standard atmospheric pressure at sea level, = 10.13 μPa

The dependence of $\text{OASPL}(90°) - 10 \times \log(A/R^2)$ on jet Mach number M and density ratio ρ_j/ρ_s is illustrated in Fig. 15.6.

The sound-pressure-level (SPL) frequency spectrum of jet mixing noise is broadband and peaks at a frequency $f = f_p$ that is a function of the radiation direction θ, jet Mach number M, and temperature ratio T_j/T_s, where T_j (in kelvin) is the fully expanded jet temperature and T_s (kelvin) is the ambient gas temperature. The spectrum has a broad peak that occurs in the range $0.3 < S < 1$, where

$$S = \frac{f_p D_N}{U} \tag{15.13}$$

is the *Strouhal number* (dimensionless), tabulated in Table 15.1 for $\theta \geq 50°$ and for different values of T_j/T_s. For practical purposes the difference $\Delta =$

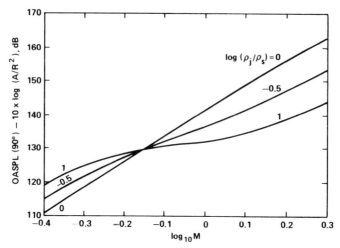

FIGURE 15.6 Overall sound pressure level at $\theta = 90°$.

TABLE 15.1 Values of Strouhal Number and $\Delta(\cdot)$ for Different Values of T_j/T_s[a]

T_j/T_s	$\theta = 50°$	$\theta = 60°$	$\theta = 70°$	$\theta = 80°$	$\theta \geq 90°$
1	0.7 (11 dB)	0.8 (11 dB)	0.8 (11 dB)	1.0	0.9
2	0.5 (10 dB)	0.4 (10 dB)	0.6 (11 dB)	0.5	0.6
3	0.3 (9 dB)	0.4 (10 dB)	0.4 (10 dB)	0.4	0.5

[a] Strouhal number $S = fD_N/U$ at the peak of the one-third-octave-band spectrum of jet mixing noise for $U/c \leq 2.5$ (left columns); $\Delta = \text{OASPL} - \text{SPL}_{\text{peak}} \cong 11$ dB for all $\theta \geq 80°$ except as indicated in parentheses (right columns).

OASPL $-$ SPL$_{\text{peak}}$ between the peak of the one-third-octave-band spectrum and the overall SPL at the same value of θ may be taken to be 11 dB except as indicated in the table.

Figure 15.7 shows a typical relative one-third-octave-band SPL spectrum SPL(f) $-$ OASPL for $\rho_j = \rho_s$ and angles $\theta \geq 90°$. The characteristic shape is the same for all temperatures and angles, although there is significant dependence on temperature and Mach number when θ is smaller than about 50° and $f > f_p$ when refraction of sound by the jet shear layer becomes important. For subsonic jets, the spectrum varies roughly as f^3 at low frequencies and decays like $1/f$ for $f > f_p$.

Figures 15.4–15.7 and Table 15.1, supplemented when necessary by Eqs. (15.8) and (15.9), comprise a general procedure for estimating the SPL, directivity, and spectrum of jet mixing noise.

Example 15.1. Find the total jet mixing noise sound-power- and sound-pressure-level spectrum at 60° from the jet axis and 3 m from the nozzle of a 0.03-m-diameter air jet. The jet exhausts into ambient air (15°C, $p_s = 10.13$ μPa, $\rho_s = 1.225$ kg/m^3) at sonic ($U/c = 1$) exit velocity, that is, 340 m/s.

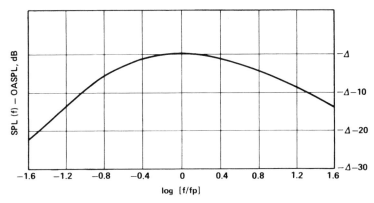

FIGURE 15.7 One-third-octave-band SPL − OASPL (dB) for radiation directions $\theta \geq 90°$ and $\rho_j = \rho_s$.

Solution The mechanical stream power $\frac{1}{2}mU^2$ is 1.70×10^4 W. The ratios ρ_j/ρ_s and U/c are unity. Therefore, from Fig. 15.3 [or Eqs. (15.8) and (15.9)] the ratio of sound power to mechanical stream power is 1.0×10^{-4}. The resulting overall sound power is 1.7 W. From Fig. 15.6 [or Eq. (15.12)] the OASPL at 90° to the jet axis and $R = 3$ m is 98.8 dB re 20 µPa. From Eq. (15.11) the OASPL at 60° to the jet axis and $R = 3$ m is 103.5 dB, and the peak Strouhal number is 0.8. The one-third-octave-band SPL spectrum is found from Fig. 15.6 by adding 103.5 dB to the ordinate with $\Delta = 11$ dB (from Table 15.1).

Realistic jets exhausting from pipes and engine nozzles do not provide smooth, low-turbulence entrainment of flow but rather flow that has been disturbed or spoiled before leaving the nozzle. In this case, the above procedure is not valid unless the jet velocity U exceeds about 100 m/s. Below this speed major portions of the aerodynamically generated noise emanate from internal sources.

Noise of Imperfectly Expanded Jets

Supersonic, underexpanded, or "choked" jets contain shock cells through which the flow repeatedly expands and contracts (see inset to Fig. 15.8). Seven or more distinct cells are often visible extending up to 10 jet diameters downstream of the nozzle. They are responsible for two additional components of jet noise: *screech tones* and broadband *shock-associated noise*. Screech is produced by a feedback mechanism in which a disturbance convected in the shear layer generates sound as it traverses the standing system of shock waves. The sound propagates upstream through the ambient atmosphere and causes the release of a new flow disturbance at the nozzle exit. This is amplified as it convects downstream, and the feedback loop is completed when it encounters the shocks. A notable feature of the screech tones is that the frequencies are independent of the radiation direction, the fundamental occurring at approximately $f = U_c/L(1 + M_c)$, where $M_c = U_c/c$, U_c is the convection velocity of the disturbance in the shear layer, L is the axial length

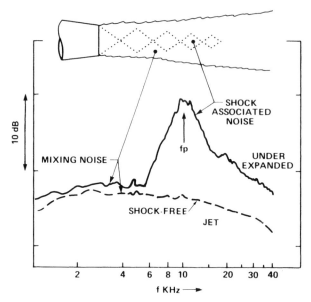

FIGURE 15.8 Jet noise spectra for shock-free and underexpanded jets at $\theta = 90°$, $\beta = 1$.

of the first shock cell, and c is the ambient speed of sound. Although screech is often present in model-scale experiments and can be important for choked jet engines, it is usually easy to eliminate by minor modifications to nozzle design, for example, by notching.

In practice, it is the broadband, shock-associated noise that is important for supersonic jet engines. It can be suppressed by proper nozzle design, but it is always present for a convergent nozzle. The dominant frequencies are usually higher than the screech tones and can range over several octave bands. For predictive purposes it is permissible to estimate separately the respective contributions of mixing noise and shock-associated noise to the overall sound power generated by the jet. The mixing noise predictions are the same as for a shock-free jet.

The overall SPL of shock-associated noise is approximately independent of the radiation direction and may be estimated from

$$\text{OASPL} = C_0 + 10 \times \log \frac{\beta^n A}{R^2}$$

$$C_0 = 156.5 \quad (\text{dB}) \quad n = \begin{cases} 4 & (\beta < 1) \\ 1 & (\beta > 1) \end{cases} \quad \frac{T_j}{T_s} < 1.1 \quad (15.14)$$

$$C_0 = 158.5 \quad (\text{dB}) \quad n = \begin{cases} 4 & (\beta < 1) \\ 2 & (\beta > 1) \end{cases} \quad \frac{T_j}{T_s} > 1.1$$

where T_j = total (reservoir) jet temperature, K
T_s = total ambient temperature, K
$\beta = \sqrt{M_j^2 - 1}$
M_j = fully (ideally) expanded jet Mach number, based on sound speed in jet

The shock-associated noise spectrum exhibits a well-defined peak in the vicinity of

$$f_p = \frac{0.9 U_c}{D_N \beta (1 - M_c \cos \theta)} \quad (15.15)$$

where the convection velocity U_c meters per second is equal to $0.7 U$ and θ is measured from the jet axis. When $T_j / T_s > 1.1$, the one-third-octave-band SPL (re 20 µPa) is given by

$$\text{SPL} = 143.5 + 10 \times \log \frac{\beta^n A}{R^2} - 16.13 \times \log \left(\frac{5.163}{\sigma^{2.55}} + 0.096 \sigma^{0.74} \right)$$

$$+ 10 \times \log \left\{ 1 + \frac{17.27}{N_s} \cdot \sum_{i=0}^{N_s - 1} \left[C(\sigma)^{(i^2)} \right. \right.$$

$$\left. \left. \times \sum_{j=1}^{N_s - i - 1} \frac{\cos(\sigma q_{ij}) \sin(0.1158 \sigma q_{ij})}{\sigma q_{ij}} \right] \right\} \quad (15.16)$$

where $C(\sigma) = 0.8 - 0.2 \times \log_{10}(2.239/\sigma^{0.2146} + 0.0987\sigma^{2.75})$
$\sigma = 6.91 \beta D_N f / c$, dimensionless
c = ambient speed of sound, m/s
$N_s = 8$ (number of shocks)
$q_{ij} = (1.7ic/U)\{1 + 0.06[j + \frac{1}{2}(i+1)]\}[1 - 0.7(U/c) \cos \theta]$

and n is defined as in (15.14). In the case of a "cold jet," for which $T_j / T_s < 1.1$, the prediction (15.16) should be reduced by 2 dB.

The prediction procedure for shock-associated noise is applicable over all angles θ where shock cell noise is important (say, $\theta > 50°$). When used in conjunction with the prediction procedure for jet mixing noise, the one-third-octave-band sound pressure spectrum for the overall jet noise may be estimated by adding the respective contributions to each frequency band of the mean-square acoustical pressures of the mixing noise and the shock-associated noise. The influence of shock-associated noise is illustrated in Fig. 15.8, which compares the spectrum of the jet mixing noise of a fully expanded, shock-free supersonic jet from a convergent–divergent nozzle at the same pressure ratio. In the latter case the spectral peak occurs at the frequency f_p given by Eq. (15.15).

Flight Effects

The noise generated by a jet engine in flight is modified because of the joint effects of Doppler amplifications (discussed in Section 15.2) and the reduction in mean shear between the jet flow and its environment. From an extensive series of experimental studies[3] empirical formulas have been developed to predict the SPL in flight in terms of the corresponding static levels.

In the Mach number range of $1.1 < M < 1.95$ jet mixing noise $\text{OASPL}(\theta)_{\text{flight}}$ is related to $\text{OASPL}(\theta)_{\text{static}}$ by

$$\text{OASPL}(\theta)_{\text{flight}} = \text{OASPL}(\theta)_{\text{static}} - 10 \times \log\left[\left(\frac{U}{U - V_f}\right)^{m(\theta)} (1 - M_f \cos \Phi)\right]$$

$$m(\theta) = \left[\left(\frac{6959}{|\theta - 125|^{2.5}}\right)^7 + \frac{1}{[31 + 18.5M - (0.41 - 0.37M)\theta]^7}\right]^{-1/7}$$

(15.17)

where θ = angle between jet axis and observer at retarded time of emission of sound, degrees
Φ = angle between flight direction and observer direction at retarded time of emission of sound, defined in Fig. 15.2
U = fully expanded jet velocity (relative to nozzle), m/s
V_f = aircraft flight speed, m/s
M_f = flight Mach number relative to sound speed in air, $= V_f/c$

The formula is applicable for $20° < \theta < 160°$. For $M < 1.1$ and $M > 1.95$, the relative velocity exponent $m(\theta)$ should be taken to be given by the above formula at $M = 1.1, 1.95$, respectively.

Estimates of the one-third-octave-band SPL spectrum in flight can be made by using Fig. 15.6 with the OASPL determined by (15.17) and the Strouhal number fD_N/U replaced by that based on the jet velocity: $fD_N/(U - V_f)$.

For shock-associated noise the influence of flight on both the spectrum and the OASPL is approximately given by

$$\text{SPL}_{\text{flight}} = \text{SPL}_{\text{static}} - 40 \log \times (1 - M_f \cos \Phi) \qquad (15.18)$$

15.4 COMBUSTION NOISE OF GAS TURBINE ENGINES

Jet mixing noise and shock-associated noise of high-speed jet engines are the result of sources in the flow downstream of the nozzle. Their importance has progressively diminished in recent years with the introduction of large-diameter, high-bypass-ratio turbofan engines with much reduced mass efflux velocities. In consequence, greater attention has been given to noise generated within the engine (termed *core noise*), which tends to be predominant at frequencies less than 1 kHz. Combustion processes are a significant component of core noise, both directly in the form of thermal monopole sources and indirectly through the creation of temperature and density inhomogeneities ("entropy spots"), which behave as dipole sources when accelerated in nonuniform flow.

The noise prediction scheme outlined below is based on an analysis of combustion noise of turbojet, turboshaft, and turbofan engines as well as model-scale data[3]. Annular, can-type, and "hybrid" combusters were all included in the valuation studies.

The overall sound power level (OAPWL, in decibels) is a function of the operating conditions of the combuster and the turbine temperature extraction and may be estimated from the equation

$$\text{OAPWL} = -60.5 + 10 \times \log \frac{mc^2}{\Pi_{\text{ref}}} + 20 \times \log \frac{(\Delta T/T_I)(P_I/p_{\text{ISA}})}{[(\Delta T)_{\text{ref}}/T_s]^2} \quad (15.19)$$

where
m = combuster mass flow rate, kg/s
P_I = combuster inlet total pressure, Pa
ΔT = combuster total temperature rise, K
$(\Delta T)_{\text{ref}}$ = reference total temperature extraction by engine turbines at maximum takeoff conditions, K
T_I = inlet total temperature, K
T_s = sea-level atmospheric temperature, 288.15 K
Π_{ref} = reference power, 10^{-12} W
p_{ISA} = sea-level atmospheric pressure, 10.13 kPa = 1.013×10^4 N/m²
c = sea-level speed of sound, 340.3 m/s

This formula is expected to yield predictions that are accurate to within ±5 dB.

The one-third-octave-band power level spectrum PWL(f) is given in terms of the OAPWL by

$$\text{PWL}(f) = \text{OAPWL} - 16 \times \log[(0.003037 f)^{1.8509} + (0.002051 f)^{-1.8168}] \, \text{dB} \quad (15.20)$$

where f is frequency (in hertz) and the formula is applicable for 100 Hz $\leq f \leq$ 2000 Hz. The spectrum is essentially symmetric about a peak at $f = 400$ Hz (Fig. 15.9). In applications where the observed peak frequency f_p differs slightly from this value, the spectrum in the figure should be shifted, retaining its shape, so that the peak coincides with observation. In (15.20), f would be replaced by $400 f/f_p$.

The one-third-octave-band sound power pressure spectrum may now be determined from the formula

$$\text{SPL} = -10.8 + \text{PWL}(f) - 20 \times \log R$$
$$- 2.5 \times \log \left[\frac{1}{(10^{(1.633-0.0567\theta)} + 10^{(19.43-0.233\theta)})^{0.4}} + 10^{(4.333-0.115\theta)} \right]$$
$$(15.21)$$

626 NOISE OF GAS FLOWS

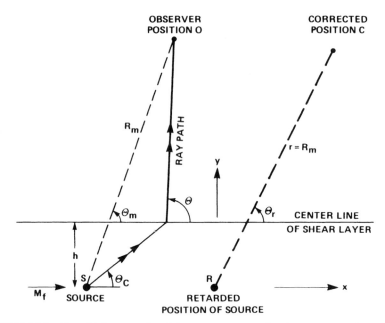

FIGURE 15.9 One-third-octave-band frequency power level spectrum of combuster noise.

where θ = angle from jet exhaust axis, degrees
R = observer distance, m

The peak radiation is predicted by (15.21) to occur at $\theta \approx 60°$, although at very low frequencies (≤ 200 Hz) the peak may be shifted slightly toward the jet axis.

For an aircraft in flight at speed V_f, the modification of the predicted SPL may be estimated from Eq. (15.18).

15.5 TURBULENT BOUNDARY LAYER NOISE

The intense pressure fluctuations that can occur beneath a turbulent boundary layer are a source of sound and structural vibrations. The sound is produced directly by aerodynamic sources within the flow and indirectly by the *diffraction* at discontinuities in the wall (e.g., corners, ribs, support struts) of structural modes excited by the boundary layer pressures.

The pressure developed beneath a boundary layer on a *hard* wall is called the *blocked pressure* and is twice the pressure that a nominally identical flow would produce if the wall were absent. The rms wall pressure p_{rms} (in pascals) of the turbulent boundary layer can be estimated by[5]

$$\frac{p_{rms}}{q} \approx \sigma \varepsilon^* = \frac{\sigma}{\frac{1}{2}(1 + T_w/T) + 0.1(\gamma - 1)M^2} \quad (15.22)$$

where $\sigma = 0.006$, dimensionless
q = local dynamic pressure, $= \frac{1}{2}\rho U^2$, Pa
ρ = fluid density at outer edge of boundary layer, kg/m^3
U = free-stream velocity at outer edge of boundary layer, m/s
M = free-stream Mach number, $= U/c$, dimensionless
c = speed of sound at outer edge of boundary layer, m/s
T = temperature at outer edge of boundary layer, K
T_w = temperature of the wall, K
γ = ratio of specific heats of gas, dimensionless

The quantity $\sigma \equiv (p_{\text{rms}}/q)_{\text{incompressible}}$, and Eq. (15.22) represents a mapping from an incompressible flow to the compressible state by means of the compressibility factor, ε^*. Recent measurements[6] suggest that the currently accepted value $\sigma = 0.006$ may be too low and that a better approximation is $\sigma = 0.01$.

In many applications wall pressure fluctuations are substantially higher if the flow *separates*. For example, when separation occurs at the compression corner at a ramp or at an expansion corner, the rms wall pressure can typically exceed 2% of the local dynamic pressure q.

The structural response of a flexible wall to forcing by the boundary layer depends on both the temporal and spatial characteristics of the pressure fluctuations. When the wall is locally plane, these can be expressed in terms of the *wall pressure wavenumber–frequency spectrum* $P(\mathbf{k}, \omega)$. This is the two-sided Fourier transform $(1/2\pi)^3 \int_{-\infty}^{\infty} R_{pp}(x_1, x_3, t) \exp[-i(\mathbf{k} \cdot \mathbf{x} - \omega t)] \, dx_1 \, dx_3 \, dt$ of the space–time correlation function of the wall pressure R_{pp}. By convention, coordinate axes (x_1, x_2, x_3) are taken with x_1, x_3 parallel and transverse to the mean flow, respectively, x_2 measured outward of the wall, and the wavenumber $\mathbf{k} = (k_1, k_3)$ (in reciprocal meters) has components parallel to the x_1, x_3 axes only. The principal properties of the blocked-pressure spectrum $P_0(\mathbf{k}, \omega)$, say, are understood only for low-Mach-number flows ($M \ll 1$).

Wall Pressure Spectrum at Low Mach Number

Here, $P_0(\mathbf{k}, \omega)$ is an even function of ω, and its general features are shown in Fig. 15.10[7] for $\omega > 0$ The strongest pressure fluctuations are produced by eddies in the *convective ridge* of the wavenumber plane that convect along the wall at about 70% of the free-stream velocity. The region $k \equiv |\mathbf{k}| < \omega/c$ (c = speed of sound, in meters per second) is called the acoustical domain, and $k_0 = \omega/c$ is the acoustical wavenumber. The adjacent *subconvective* domain ($k_0 < k \ll \omega/U_c$) is important to determining the structural response of the wall to forcing by the turbulence pressures. In these regions $10 \times \log[P_0(\mathbf{k}, \omega)]$ is typically 30–60 dB below the levels in the convective domain.

Pressure fluctuations in the acoustical domain correspond to sound waves in the fluid. When the wall is smooth and flat, the generation of sound is dominated by quadrupole sources in the flow,[7] and the acoustical pressure frequency spectrum

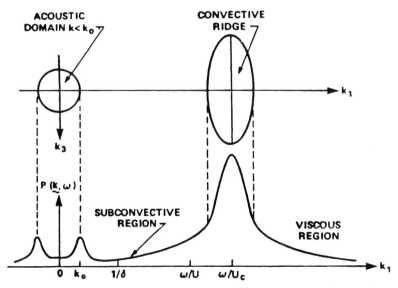

FIGURE 15.10 Turbulent boundary layer wall pressure spectrum at low Mach number for $\omega\delta/U \gg 1$, where δ is the boundary layer thickness (m).

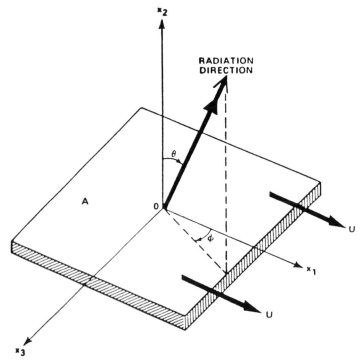

FIGURE 15.11 Coordinates defining the radiation of sound from a region of the wall of area A.

of sound produced by a fixed area of the wall (Fig. 15.11) is

$$\Phi(\omega) = 2\frac{A}{R^2}k_0^2 \cos^2\theta \, P_0(k_0 \sin\theta \cos\phi, k_0 \sin\theta \sin\phi, \omega) \tag{15.23}$$

where OASPL $= 10 \times \log\left[\int_0^\infty \Phi(\omega)\,d\omega/(20\ \mu\text{Pa})^2\right]$, dB
A = area of wall region, m^2
R = observer distance from center of A, m
(θ, ϕ) = polar angles of observer defined in Fig. 15.11

Chase[8] has proposed the following representation of the blocked-pressure spectrum that is applicable over the whole range of wavenumbers and is based on an empirical fit to experimental data in the convective and subconvective domains:

$$\frac{P_0(\mathbf{k}, \omega)}{\rho^2 v_*^2 \delta^3} = \frac{1}{[(k_+\delta)^2 + 1.78]^{5/2}} \left\{ \frac{0.1553(k_1\delta)^2 k^2}{|k_0^2 - k^2| + \beta^2 k_0^2} + 0.00078 \right.$$

$$\left. \times \frac{(k\delta)^2[(k_+\delta)^2 + 1.78]}{(k\delta)^2 + 1.78} \left(4 + \frac{|k_0^2 - k^2|}{k^2} + \frac{k^2}{|k_0^2 - k^2| + \beta^2 k_0^2}\right) \right\}$$

$$\frac{\omega\delta}{U} > 1 \tag{15.24}$$

where δ = boundary layer thickness (distance from wall at which mean-flow velocity is equal to $0.99U$), m
ρ = mean fluid density, kg/m^3
v_* = friction velocity, $\approx 0.035U$, m/s
$U_c \approx 0.7U$, m/s
U = mainstream velocity at outer edge of boundary layer, m/s
ω = radian frequency, $= 2\pi f$, rad/s
$k_+ = \sqrt{(\omega - U_c k_1)^2/9v_*^2 + k^2}$, m^{-1}
$\beta \approx 0.1$, dimensionless

The first term in the curly brackets determines the behavior near the convective ridge and the second the low-wavenumber and acoustical domains. The numerical coefficient β controls the height of the spectral peak (Fig. 15.10) at the boundary $k = k_0$ of the acoustical domain and determines the intensity of the sound waves propagating parallel to the plane of the wall.

Example 15.2. Estimate the frequency spectrum $\Psi(\omega)$ of the power dissipated by flexural motion of a plane wall when the wall is excited by a low-Mach-number turbulent flow over a region of area A whose normal impedance $\mathbf{Z}(\mathbf{k}, \omega)$ (kg/m$^2 \cdot$ s) is independent of the orientation of $\mathbf{k}$.

Solution The impedance satisfies $Z(\mathbf{k}, \omega) = -p(\mathbf{k}, \omega)/v_2(\mathbf{k}, \omega)$, where p and v_2, respectively, denote the Fourier components of the pressure and normal velocity on the wall. Let $Z = R - iX$, where R and X are the resistive and reactive components of the impedance, respectively. The total power delivered to the flexural motions is equal to $\int_0^\infty \Psi(\omega)\,d\omega$, where

$$\Psi(\omega) = 2A \int_{k>|k_0|}^\infty \frac{R(\mathbf{k},\omega) P_0(\mathbf{k},\omega)\, d^2\mathbf{k}}{R(\mathbf{k},\omega)^2 + [X(\mathbf{k},\omega) + \rho\omega/\sqrt{k^2-k_0^2}]^2}$$

The flexural-mode wavenumbers are the roots $\mathbf{k} = \mathbf{k}_n, n = 1, 2, 3, \ldots$, of $X(\mathbf{k},\omega) + \rho\omega/\sqrt{k^2-k_0^2} = 0$. In practice, these usually lie in the low-wavenumber region where $P_0(\mathbf{k},\omega) \equiv P_0(k,\omega)$, so that, when $R \ll X$ and Z is a function of $k \equiv \mathbf{k}$ and ω only,

$$\Psi(\omega) \approx \sum_n \frac{4\pi^2 A k_n P_0(k_n,\omega)}{\left|(\partial/\partial k)\left[X(k,\omega) + \rho\omega/\sqrt{k^2-k_0^2}\right]\right|_{k_n}}$$

When the wall consists of a vacuum-backed, thin elastic plate of bending stiffness B (kg·m^2/s^2), mass density m (kg/m^2) per unit area, and negligible damping $R = 0$, we have $X = -(Bk^4 - m\omega^2)/\omega$, and there is only one flexural mode that occurs at $k = k_* > |k_0|$ such that

$$\Psi(\omega) = \frac{4\pi^2 A\omega(k_*^2 - k_0^2) P_0(k_*,\omega)}{5Bk_*^4 - 4B(k_0 k_*)^2 - m\omega^2}$$

where k_* is the positive root of $Bk^4 - m\omega^2 - \rho\omega^2/\sqrt{k^2-k_0^2} = 0$. Numerical estimates may be made by substituting for $P_0(k_*,\omega)$ from (15.24).

15.6 NOISE FROM FLUID FLOW IN PIPES

Piping system noise typically comes mainly from control valves (see Section 15.11) and high-speeds machines (see Chapter 18). However, under some circumstances the noise induced by flow through milder discontinuities (bends, tees, swags, and other components) may be significant. Here we describe an approximate approach for estimating this type of flow-induced noise.

Pressure fluctuations in the turbulent boundary layer drive the wall of the pipe, causing it to radiate sound. As described in Section 15.5, the structural response of a flexible wall to forcing by the boundary layer depends in a complex manner on both the temporal and spatial characteristics of the pressure fluctuations. Seebold[9] suggested the following approximate method for estimating the noise produced by fluid flow in piping systems. The method estimates the noise that results from boundary layer pressure fluctuations in fully developed turbulent flow in uninterrupted straight pipes[10] and then applies a loss factor correction for local discontinuities. This loss factor correction is based on the notion that the sound power radiated at a discontinuity is approximately proportional to the square of the pressure drop Δp there and can therefore be related to the pressure head loss factor K, where $\Delta p = K(\tfrac{1}{2}\rho U^2)$.

The sound pressure level at a distance of 1 m from the pipe in the octave band centered on frequency f_c is approximated by

$$\text{SPL} = -3.5 + 40 \log U + 20 \log \rho + 20 \log K$$
$$- 10 \log \left[\frac{t}{D} \left(1 + \frac{1.83}{D} \right) \right] - 5 \log \left[\frac{f_c}{f_r} \left(1 - \frac{f_c}{f_r} \right) \right] + \Delta L(f_c) \quad (15.25)$$

where U = gas flow velocity, m/s
ρ = fluid density, kg/m^3
t = pipe wall thickness, m
D = inside diameter of pipe, m
f_c = octave-band center frequency, Hz
f_r = ring frequency of pipe, Hz; = 1608/D for steel
K = total loss factor per 10-diameter pipe section, dimensionless

The spectral correction $\Delta L(f_c)$ depends on the ratio of the octave-band center frequency f_c to the Strouhal frequency of the pipe flow, $f_p = 0.2U/D$, as[11]

$$\Delta L = \begin{cases} 10.4 + 11.4 \log \dfrac{f_c}{f_p} & \text{for } \dfrac{f_c}{f_p} < 0.5 \\ 7 & \text{for } 0.5 \leq \dfrac{f_c}{f_p} < 5 \\ 14 - 10 \log \dfrac{f_c}{f_p} & \text{for } 5 \leq \dfrac{f_c}{f_p} < 12 \\ 41.9 - 36.1 \log \dfrac{f_c}{f_p} & \text{for } \dfrac{f_c}{f_p} \geq 0.5 \end{cases} \quad (15.26)$$

When employing (15.25), the piping system should be thought of as being composed of segments 10 diameters in length. The total loss factor K is determined by adding the individual loss factors K_i for the flow fittings and elements present within each 10-diameter segment of pipe. This total loss factor is then entered in Eq. (15.25) to estimate the flow noise emanating from that segment (at a distance of 1 m from the pipe wall). Loss factors for various piping components are provided in Table 15.2. For example, the loss factor for a straight pipe is taken to be 0.12 velocity heads per 10-diameter segment. On the other hand, a 10-diameter-long segment of piping containing an expander (1.5 : 1), an elbow (90°, $R/D = 1.5$), and a tee (flow-through run) would have a total loss factor of approximately 1, $K \approx 0.1 + 0.33 + 0.5 \approx 1$.

15.7 SPOILER NOISE

The term *spoiler noise* is used to characterize the noise produced by an obstacle or other obstruction that spans a duct carrying a mean flow. We shall discuss the case of air flowing in a duct or pipe that may be regarded as "semi-infinite" when

TABLE 15.2 Pressure Head Loss Factors[a] K

Straight pipes (10-diameter-long segment)		0.12					
45° Elbow	Screwed	0.42	Welded, $R/D = 1$	0.29	Welded, $R/D = 1.5$	0.21	
90° Elbow	Screwed	0.92	Welded, $R/D = 1$	0.45	Welded, $R/D = 1.5$	0.33	
180° Elbow	Screwed	2.00	Welded, $R/D = 1$	0.60	Welded, $R/D = 1.5$	0.43	
Tees (screwed)	Thru branch	1.80		Thru run	0.50		
Tees (welded)	Thru branch	1.40		Thru run	0.40		
Reducer	$D_2/D_1 = 0.3$	0.25	$D_2/D_1 = 0.5$	0.17	$D_2/D_1 = 0.7$	0.07	
Expander	$D_2/D_1 = 3$	0.80	$D_2/D_1 = 2$	0.56	$D_2/D_1 = 1.25$	0.10	
Sudden Contraction	$D_2/D_1 = 0.1$	0.48	$D_2/D_1 = 0.33$	0.41	$D_2/D_1 = 0.8$	0.12	
Sudden Expansion	$D_2/D_1 = 10$	0.98	$D_2/D_1 = 3$	0.79	$D_2/D_1 = 1.25$	0.12	

[a] After Ref. 9.

it is desired to estimate the acoustical radiation from the open end. A spoiler-noise-generating system is shown schematically in Fig. 15.12. In practice, the spoiler may be a strut, stringer, guide vane, or other flow control device. The nominally steady mean flow exerts unsteady lift and drag forces on the spoiler, which accordingly behaves as an acoustical source of *dipole* type.

The following idealizations will be made:

1. The spoiler can be of arbitrary shape, but its cross-flow dimensions are small relative to the pipe cross section. This ensures that the flow speeds near the spoiler are not significantly greater than the pipe mean flow speed and therefore that turbulence mixing noise (quadrupole) can be neglected. Thus, the case of a valve, which produces a severe throttling of the flow, is excluded (see Section 15.11).

2. The peak frequency of the noise spectrum is below the "cutoff" frequency f_{co} of the pipe, which is given by

$$f_{co} = \begin{cases} 0.293c/r & \text{Hz (circular pipe)} \\ 0.5c/w & \text{Hz (rectangular pipe)} \end{cases} \quad (15.27)$$

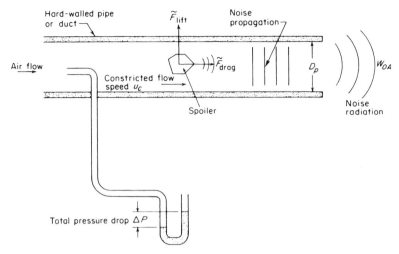

FIGURE 15.12 Flow spoiler system. The sound power is generated by the unsteady drag force on the spoiler and radiated from the open end of the pipe.

where c = sound speed, m/s
r = radius of circular pipe, m
w = largest transverse dimension of rectangular pipe, m

If this condition is not fulfilled, the propagated sound power near and above the peak frequency (which determines most of the overall sound power) will be influenced by transverse modes of the pipe.

3. The pipe wall is acoustically rigid. The fluctuating lift forces on the spoiler are then canceled by images in the wall, and sound is produced solely by the unsteady drag, which corresponds to an acoustical dipole whose axis is parallel to the mean flow. A low-noise spoiler system should therefore use bodies of low drag, such as airfoils, rather than grids or other oddly shaped bodies.

For several elementary flow spoiler configurations, such as the one shown in Fig. 15.12, the *fluctuating drag is directly proportional to the steady-state drag experienced by the body*. This depends on the pressure drop ΔP across the spoiler (as measured by an upstream and a downstream total-pressure probe), and the broadband noise radiated from the open end of the pipe is given by[12]

$$W_{\text{OA}} = \frac{k(\Delta P)^3 D_p^3}{\rho^2 c^3} \tag{15.28}$$

where W_{OA} = overall radiated sound power, W
k = constant of proportionality, dimensionless

ΔP = total pressure drop across spoiler, Pa
D_p = pipe diameter, m
ρ = atmospheric density, kg/m³
c = atmospheric sound speed, m/s

Specific information about spoiler geometry is absent from Eq. (15.28) but is implicit in the pressure drop ΔP. From a variety of experimental spoiler configurations the constant k is found to be about 2.5×10^{-4} for air.

Equation (15.28) determines the broadband sound power. It should strictly be regarded as a lower bound that is applicable only if there are no *discrete-frequency* components of the noise. These are present in certain conditions (e.g., excitation of edge tones) and may well stand out against the broadband levels.[13] When such a mechanism has been identified, the discrete tones can usually be eliminated or controlled by detuning or rounding off sharp corners or edges or cutting feedback paths by treating reflecting surfaces with sound-absorbent materials.

The frequency spectrum measured outside the pipe for the noise generated by the spoiler exhibits a haystack structure (Fig. 15.13) with a peak frequency given by

$$f_p \approx \frac{u_c}{db} \quad \text{Hz} \tag{15.29}$$

where u_c = constricted flow speed, m/s
d = projected width of spoiler, m
b = constant that equals 0.2 for pressure differences ΔP of the order 4000 Pa and 0.5 for ΔP of the order of 40,000 Pa

The constricted flow speed u_c for cold air is given in Table 15.3 as a function of ΔP. Values in between those in the table can be interpolated.

Example 15.3. Find the sound-power-level spectrum of noise produced by a spoiler consisting of a flat plate of width 2 cm stretching across a circular pipe of internal diameter $D_p = 5$ cm. The pressure drop across the spoiler is $\Delta P = 10,000$ Pa.

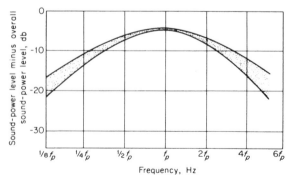

FIGURE 15.13 Generalized octave-band spectrum for in-pipe-generated spoiler noise referenced to the total power radiated from the open end and the peak frequency.

TABLE 15.3 Constricted Flow Speed u_c for Cold Air as Function of Pressure Drop

ΔP, Pa	2500	5000	10,000	20,000	30,000	40,000
u_c, m/s	63	90	124	173	209	238

Solution From Eq. (15.28) the total sound power $L_w \approx 101$ dB re 10^{-12} W. When $\Delta P = 10{,}000$ Pa, the peak frequency of the spectrum can be expected to be at $f_p \approx 0.35 u_c/d$.

From Table 15.3, we find $u_c = 124$ m/s; hence $f_p \approx 2170$ Hz. The sound power spectrum in octave bands is obtained form Fig. 15.13 with 101 dB added to the ordinate.

15.8 GRID OR GRILLE NOISE

The characteristics of noise produced by flow through grids, grilles, diffusers, guide vanes, or porous plates, which often terminate air conditioning ducts, are similar to those of spoiler-generated sound. The principal differences are that (1) the grid is located at the open end of the duct, (2) the duct cross-sectional area is typically quite large (say, 0.04–1 m²), and (3) the velocity of the air in the duct is usually small, rarely exceeding 30 m/s. The speed of the "air jets" exhausting from the individual air passages (or orifices) in a diffuser is generally low enough (<100 m/s) that jet mixing noise can be neglected. The dominant sound source is of dipole type and is associated with the interaction of the flow with the diffuser elements (e.g., guide vanes).

When a duct is terminated by a grid of circular rods (Fig. 15.14a), periodic shedding of vortices can occur, producing a fluctuating lift force on each rod and an associated *tonal* component of the radiated sound. This source is a dipole (whose axis is parallel to the lift direction) and is in addition to the drag dipole. The frequency of the tone is given approximately by

$$f = \frac{0.2u}{D_R} \tag{15.30}$$

where u = mean flow speed, m/s
D_R = diameter of rods, m

Tonal oscillations produced by a feedback mechanism can also arise when the duct termination consists of a plate perforated with sharp-edged circular cylindrical apertures (Fig. 15.14b).

In general, however, the noise produced by typical air conditioner grids is almost always broadband and of conventional dipole character, with the sound power varying as the sixth power of the velocity. As in the case of spoiler noise, the overall sound power can be related to the pressure drop ΔP across the grid,

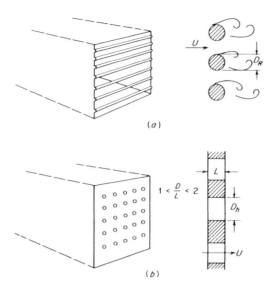

FIGURE 15.14 Special cases of duct terminations: (*a*) circular rods; (*b*) sharp-edge circular cylindrical holes.

independently of the specific grid geometry. To do this, we introduce the pressure drop coefficient

$$\xi = \frac{\Delta P}{\frac{1}{2}\rho u^2} \qquad (15.31)$$

where ρ = density of air, kg/m³
u = mean flow speed in duct prior to grid, m/s

Three typical diffuser configurations together with their pressure drop coefficients ξ are illustrated in Fig. 15.15.[14] The values of ξ for similar diffuser configurations are usually available from the manufacturer. If not, they may be estimated from this figure.

If not given by the manufacturer, the overall sound power level L_w from air conditioning diffusers can be estimated from the empirical formula[14]

$$L_w = 10 + 10 \times \log(S\xi^3 u^6) \quad \text{dB re } 10^{-12} \text{ W} \qquad (15.32)$$

where S = area of duct cross section prior to diffuser, m²
ΔP = pressure drop through diffuser, Pa

The noise spectra of different diffusers do not exhibit identical shapes even when normalized to similar flow speeds and exhaust areas. Construction differences tend to emphasize different frequency regimes, and poorly designed diffusers will radiate discrete-frequency sound. In practical noise control problems, however, a general spectrum shape $L_w - 10 \times \log_{10}(S\xi^3)$ (dB re 10^{-12} W) can be used for each duct velocity u (in meters per second) that fits most diffuser noise spectra to within about ±5 dB, as shown in Fig 15.16.

GRID OR GRILLE NOISE

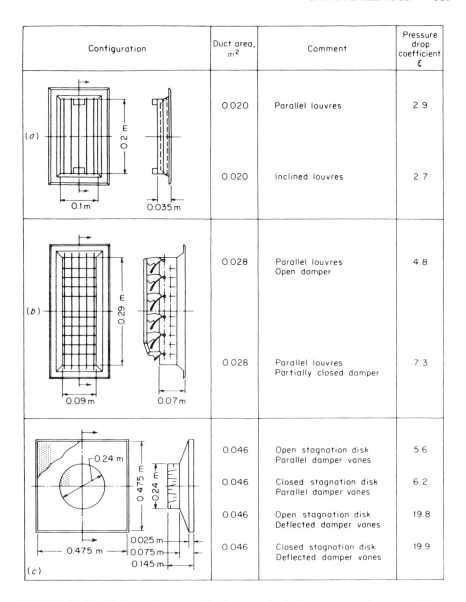

FIGURE 15.15 Various duct terminations and their pressure drop coefficients $\xi = \Delta P / \frac{1}{2}\rho u^2$. The duct area S is in square meters. (After Ref. 14.)

To estimate the sound power spectrum of the radiation from a given diffuser, first determine the relevant curve from Fig. 15.16 for the particular flow speed. The ordinate in the figure is then increased by $10 \times \log_{10}(S\xi^3)$ decibels to yield the desired one-third-octave-band spectrum with a margin of error of about ±5 dB.

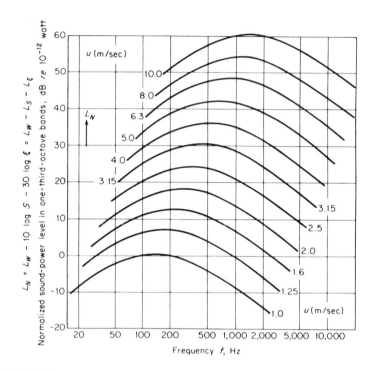

FIGURE 15.16 Normalized one-third-octave-band sound power level spectra for noise radiated from diffusers of various flow velocities. (After Ref. 14.)

15.9 SOUND GENERATION BY AIRFOILS AND STRUTS

Noise of a Strut in a Turbulent Stream

Long struts, airfoils, guide vanes, and so on, which offer negligible drag to a mean flow, frequently experience significant fluctuations in lift when exposed to a turbulent stream and behave as *dipole-type* sources of broadband sound. The strength of the radiation depends on the dimensions of the airfoil and on the turbulence intensity and correlation length of the velocity fluctuations.

A simple description of the radiation can be given when the mean flow is of low Mach number (say, <100 m/s in air) and when the distance to the nearest boundary (e.g., duct wall) is at least of the order of the acoustical wavelength. In these circumstances the *frequency spectral density* of the total radiated sound power $\Pi(\omega)$ is given approximately by

$$\Pi(\omega) = \frac{\pi l a^2 \rho u^2 M^3 (\omega \Lambda / U)^4}{4(1 + \pi \omega a/U)[1 + (\omega \Lambda/U)^2]^{5/2}[1 + (\omega a/3c)^3]^{1/3}} \quad (15.33)$$

where ω = radian frequency, $= 2\pi f$, f in Hz
 a = chord of airfoil, m

l = span of airfoil, m
ρ = density of mean stream, kg/m³
u = rms turbulence velocity in lift direction, m/s
U = velocity of mean stream, m/s
M = mean flow Mach number, $= U/c$, dimensionless
c = sound speed, m/s
Λ = integral scale of turbulence velocity, m

The total radiated sound power $W = \int_0^\infty \Pi(\omega)\,d\omega$, in watts, can be estimated from

$$W = \frac{1.78(la^2/\Lambda)\rho u^2 U M^3}{(1 + 10.71 a/\Lambda)(1 + 1.79\, Ma/\Lambda)} \qquad 0 \leq M \leq 0.3 \qquad (15.34)$$

At low frequencies the acoustical intensity exhibits the characteristic dipole *field shape*, proportional to $\cos^2 \theta$, where θ is measured from the direction of the mean lift. At higher frequencies (when the acoustical wavelength is much smaller than the chord of the airfoil but still exceeds the airfoil thickness), the field shape assumes a cardioid form, with a null in the forward direction and peaks close to the direction of the mean flow.

Airfoil Self-Noise

In addition to gust/inflow turbulence-induced noise, an airfoil can also generate *self-noise* when turbulence that arises from the natural instability of the boundary layers on the surface of the airfoil is swept past the trailing edge. This is usually important when the hydrodynamic wavelength of the turbulence eddies is larger than the airfoil thickness at the trailing edge. The mechanism of noise production tends to be weaker than at the leading edge, because the violence of the unsteady motion at the trailing edge is alleviated by vortex shedding into the wake, and to be more prominent at higher frequencies.[15]

The noise may be ascribed to a distribution of lift dipoles near the trailing edge. At low frequencies the acoustical intensity exhibits the characteristic dipole field shape, proportional to $\cos^2 \theta$, with a null in the plane of mean motion of the airfoil. At higher frequencies (when the acoustical wavelength is smaller than the chord of the airfoil) the field shape assumes a cardioid form, with a null in the downstream direction and the peak in the forward direction. In either case, when the trailing edge is at right angles to the mean flow, the frequency spectral density of the overall sound power $\Pi(\omega)$ (in watt-seconds) is given approximately by

$$\Pi(\omega) = \frac{\pi l a \omega^2}{24\rho c^3 [1 + (\omega a/3c)^3]^{1/3}} \int_{-\infty}^{\infty} \frac{P_0(k_1, 0, \omega)}{|k_1|}\, dk_1 \qquad (15.35)$$

where $P_0(\mathbf{k}, \omega)$ = blocked-pressure wavenumber–frequency spectrum on airfoil just upstream of trailing edge (see Section 15.5)
k_1 = wavenumber component parallel to mean flow, m⁻¹

If the flow approaching the trailing edge is turbulent on both sides of the airfoil, $P_0(k_1, 0, \omega)$ in Eq. (15.35) should be replaced by the sum of the wall pressure spectra on the two sides.

When the Chase model [Eq. (15.24)] is used to estimate the integral in Eq. (15.35), the frequency spectral density of the sound power radiated by a section of the airfoil of spanwise length l (in meters) is

$$\Pi(\omega) = \frac{0.08 a l \delta \rho v_*^4 U_c (\omega \delta / U_c)^3}{c^3 [1 + (\omega a / 3c)^3]^{1/3} [(\omega \delta / U_c)^2 + 1.78]^2} \quad (15.36)$$

where δ = boundary layer thickness, m
 v_* = friction velocity, m/s
 $U_c \approx 0.7 \times$ velocity of mean stream, m/s

and other quantities are defined after Eq. (15.35).[16] However, measured levels of $\Pi(\omega)$ frequently exceed predictions of this formula. More detailed empirical formulas are given in the references.[16,17]

When the frequency is high enough [beyond the range of applicability of Eq. (15.36)] that the *Strouhal number* $\omega h / U \approx 1$, where h is the thickness of the trailing edge of the airfoil, the shedding of discrete vortices from the trailing edge can occur, provided the Reynolds number Ua/ν based on the chord a of the airfoil (dimensionless; ν is kinematic viscosity in square meters per second) does not exceed $10^6 – 10^7$. This produces a distinct contribution to trailing-edge noise, often called "airfoil singing," whose amplitude is not easily predicted.[16]

15.10 FLOWS PAST CAVITIES

Sheared flow passing over a hole or opening in a wall can be the source of intense acoustical tones. Instability of the mean flow over such a wall cavity not only can excite hydrodynamic self-sustained oscillations but also can couple with resonant acoustical modes of the cavity, as in the Helmholtz resonator. Cavity tones are a frequent source of unwanted noise and vibration in branched duct and piping systems, turbomachinery, and exposed openings on automobiles, aircraft, and other high-speed vehicles. In this section we provide relationships for determining when flow over an opening that can be approximated as either a rectangular cavity (Fig. 15.17a) or a Helmholtz resonator (Fig. 15.17b) might be expected to exhibit strongly resonant behavior.

Flow over Cavities of Uniform Cross Section

For cavities of the type shown in Fig. 15.17a, resonant cavity oscillation occurs when the frequency associated with the shedding of vorticity from the upstream edge of the cavity is sufficiently high and the acoustical wavelengths within the cavity are sufficiently short as to allow standing waves inside the cavity. Longitudinal wave resonance is possible when the acoustical wavelength is less

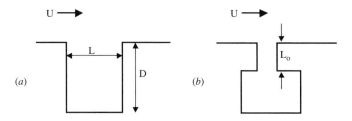

A_o = Orifice Cross-Sectional Area
V_c = Cavity Volume
A_c = Cavity Cross-Sectional Area

FIGURE 15.17 Examples of flows that sustain cavity resonances: (a) rectangular cavity of uniform cross section; (b) non-axisymmetric Helmholtz resonator.

than twice the cavity length, $c/f \le 2L$, and depth resonance is observed when the acoustical wavelength is less than four times the cavity depth, $c/f \le 4D$. For shallow cavities, where cavity length-to-depth ratio is large ($L/D > 1$), the longitudinal standing waves are dominant, while for deep cavities ($L/D < 1$) the depth resonance dominates.

Rossiter[18] derived the following equation for the Strouhal number associated with longitudinal cavity resonance modes:

$$\frac{fL}{U_\infty} = \frac{m - \xi}{M + 1/k_v} \tag{15.37}$$

where k_v = ratio of shear layer velocity to free-stream velocity, = $U_c/U_\infty \cong 0.57$
m = mode number, = 1, 2, 3, ..., dimensionless
ξ = empirical constant, = 0.25, dimensionless
L = cavity length, m
M = free-stream Mach number, = U_∞/c, dimensionless

Block[19] developed the following relation for the longitudinal cavity modes that accounts for the dependence of the Strouhal number on the cavity L/D ratio as observed in experimental data for shallow ($L/D > 1$) cavities:

$$\frac{fL}{U_\infty} = \frac{m}{1.75 + M(1 + 0.514/(L/D))} \tag{15.38}$$

The following relation for depthwise modes of cavity oscillation was developed by East[20]:

$$\frac{fL}{U_\infty} = \left(\frac{1}{M}\right)\left(\frac{L}{D}\right)\left(\frac{0.25}{1 + 0.65(L/D)^{0.75}}\right) \tag{15.39}$$

Lucas,[21] after comparing the ranges of validity of various cavity oscillation models, recommends, for low to moderate Mach number, the use of Block's formula (15.38) for shallow ($L/D > 1$) cavities. For deep cavities he suggests use of East's formula (15.39) to determine the resonance of the depth mode and Rossiter's formula (15.37) for the longitudinal resonance mode. For such deep cavities the depth resonance will be the dominant mode.

Block[19] combined Eqs. (15.38) and (15.39) so that the frequency for the shallow acoustical waves coincides with the depthwise acoustical modes, providing the following formula for estimating the Mach number at which a cavity begins to oscillate in a given mode m:

$$M = \frac{1.75 L/D}{4m[1 + 0.65(L/D)^{0.75}] - [(L/D) + 0.514]} \qquad (15.40)$$

Equation (15.40) was shown by Block[19] to give adequate agreement with experimental data for Mach numbers in the range of 0.1–0.5 and for $L/D < 2$.

Flow over a Helmholtz Resonator

The natural frequency f_n of the Helmholtz resonator shown in Fig. 15.17b is given by

$$f_n = \frac{c}{2\pi} \sqrt{\frac{A_o}{V_c L_{\text{eff}}}} \qquad (15.41)$$

where c = sound speed, m/s
A_o = orifice cross-sectional area, m^2
V_c = cavity volume, m^3
L_{eff} = effective neck length of orifice, m

The following approximate expression[22] for the effective neck length L_{eff} in the presence of a mean flow over the Helmholtz resonator accounts for an end correction due to the entrained mass of the virtual piston forming the interface between the neck and the exterior space (length ΔL_o):

$$L_{\text{eff}} = L_o + \Delta L_o$$
$$= L_o + 0.48\sqrt{A_o} \qquad (15.42)$$

where L_o is the length of the neck length of the orifice in meters.

An experimental study by Panton and Miller[23] indicates a strong response of the Helmholtz resonator excited by a turbulent boundary layer when

$$\frac{2 d_o f_n}{U_c} \cong 1 \qquad (15.43)$$

where d_o = Helmholtz resonator neck diameter, m
U_c = boundary layer convection velocity, $\approx 0.7U$, m/s
U = mainstream velocity at outer edge of boundary layer, m/s

In Eq. (15.43) f_n is either the Helmholtz resonance frequency in Eq. (15.41) or the fundamental standing-wave (organ pipe) resonance. For the latter, the natural frequency is approximately given by

$$f_n \cong \frac{c}{2L_{\text{eff}}} \qquad (15.44)$$

Example 15.4. Many automobiles have a sunroof opening in their roof. Under some conditions the interior volume of the automobile cabin can act in combination with such an opening to create a Helmholtz resonator. If no measures are taken to prevent such an occurrence, at what speed might we expect the automobile's passengers to experience intense pressure fluctuations given a cabin interior volume of approximately 3 m², a square opening in the roof 0.4 m on a side, and a combined thickness of the roof and edging surrounding the opening of approximately 10 cm.

Solution Assuming an air temperature of 20°C, the speed of sound $c = 331.5 + 0.58T$ (in °C) $= 343.1$ m/s. The orifice cross sectional area $A_o = 0.4 \times 0.4 = 0.16$ m². The effective length of the opening is given by Eq. (15.42) as $L_{\text{eff}} = 0.1 + 0.48\sqrt{0.16} = 0.29$ m. Applying the above values and the cavity volume of $V_c = 3$ m², from Eq. (15.41) we expect the Helmholtz resonance to occur at $f_n = 23.4$ Hz. We take the neck diameter to be given by the length of the opening, $d_o = 0.4$ m. Substituting this value and that of the resonance frequency into Eq. (15.43), we can solve for the convection velocity, $U_c \approx 2 \times 0.4 \times 23.4 = 18.7$ m/s. The corresponding velocity of the mainstream, and thus our estimate of the vehicle speed at which we expect to strongly excite the Helmholtz resonance, is $U = U_c/0.7 = 26.7$ m/s, or 96 kph.

15.11 AERODYNAMIC NOISE OF THROTTLING VALVES

The noise of gas control or throttling valves may generally be associated with two sources: (1) mechanical vibration of the trim and (2) aerodynamic throttling. Noise generation by these mechanisms rarely occurs simultaneously, but when it does, the cure of one is usually the cure of the other.

Aerodynamic Noise

The prediction of the noise radiated from the body of a valve and from the downstream piping connected to it involves the following conceptual steps: (1) prediction of the magnitude and the spatial and spectral distribution of the noise inside the pipe based on the aerodynamic characteristics of the valve and the cross sectional area of the pipe, (2) prediction of the vibration response of the pipe to the internal sound field, (3) prediction of the sound radiation of the vibrating pipe.

In Step 1, the sound power generated by the valve has two components. The first is owing to the dynamic drag and lift forces (dipole) generated by the interaction of the turbulent flow with the valve, which are roughly proportional to the sixth power of the flow velocity. The second is sound power generated by the jet noise component (quadrupole), which is roughly proportional to the eighth power of the jet exit velocity. At very high pressure ratios, when the flow in the valve passage is supersonic, a shock pattern emerges downstream of the valve and generates intense screech noise. The dipole mechanism dominates the internal sound field at low Mach numbers and the quadrupole at high Mach numbers.

Based on the scaling of experimental data,[24] the lowest pipe wall attenuation occurs at $f_0 = (f_R/4)(c_{gas}(T)/c_0)$, where f_R is the ring frequency (Eq.15.58) and $c_{gas}(T)$ and $c_0 = 340$ m/s are the speed of sound of the gas in the pipe at local temperature and the that of air at room temperature respectively.

A fully analytical prediction procedure is too cumbersome. Consequently, the valve industry has adopted a prediction procedure based on both analytical and empirical methods that predicts the sound pressure level of the valve noise at 1 meter from the pipe wall directly from the aerodynamic properties of the valve and the geometric and material properties of the pipe. The rest of this chapter describes this semi-empirical prediction procedure.

Investigations of noise-induced pipe failures[25] have enabled maximum safe sound power levels to be established for given pipe sizes and wall thicknesses, as indicated in Fig. 15.18. The power levels shown in the figure correspond approximately to a sound level of 130 dBA at 1 m from the pipe wall downstream of the valve. Exceeding this level will most probably lead to piping failure, and a limit of 110 dBA at 1 m is recommended for safety and to maintain the structural integrity of valve-mounted accessories.

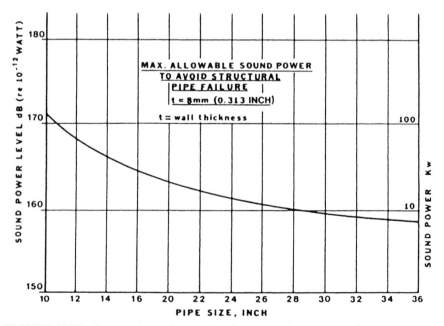

FIGURE 15.18 Suggested sound power level and sound power (Kw) limits inside pipe to avoid structural pipe failure based on actual occurrence.[24]

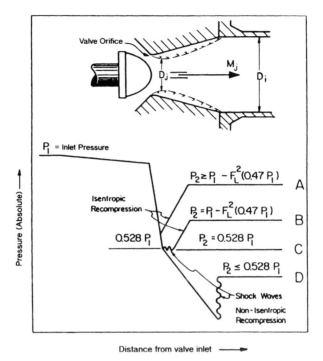

FIGURE 15.19 Schematic flow profile and static pressure diagram of throttling valve for various downstream pressures (P_2).

The aerodynamic noise is determined by the mechanical stream power $W_{\text{mech strm}} = \frac{1}{2}mU^2$ (m being the mass flow and U the velocity; see Section 15.3) that is converted from potential energy (inlet pressure) to kinetic energy (velocity head) within the valve and subsequently to thermal energy (corresponding to reduced downstream pressure and an increase in entropy).[26] The pressure reduction (i.e., the conversion of kinetic into thermal energy) occurs via the generation of turbulence or, when the flow is supersonic, through shock waves. The sound power is equal to $\eta W_{\text{mech strm}}$, where η is an acoustical efficiency factor. Unlike the case of a jet discharging into the atmosphere, the jet from a valve cannot expand freely, and only a fraction of the kinetic energy is converted by turbulence into thermal energy. This is illustrated in Fig. 15.19, which is a schematic view of a throttling valve and the associated pressure profile for various downstream pressures P_2.* On curve A the flow is subsonic, and turbulence-generated sound is predominantly of dipole type. At the critical pressure ratio ($P_1/P_2 = 1.89$ for air), curve B shows the presence of incipient shock waves followed by subsonic recompression, with both turbulence and shocks as noise sources. At higher pressure ratios recompression tends to be nonisentropic, and sound is produced predominantly by shock waves (curve D).

*In the rest of this chapter all pressures are absolute static pressures (in pascals).

Downstream of the vena contracta (D_j in Fig. 15.19) a portion of the velocity head is always recovered. The fraction lost is equal to F_L^2, where F_L is an experimentally determined pressure recovery coefficient that depends on valve type and size. The pressure drop $P_1 - P_2$ across the valve required to reach sonic velocity in the vena contracta is $F_L^2(P_1 - P_0)$, where P_0 is the static pressure in the vena contracta.[26] In subsonic flow the remaining portion of the jet power is recovered through isentropic recompression and is not converted to sound. If there were no pressure recovery at the valve orifice and assuming sonic flow at speed in c_0 meters per second (curve D in Fig. 15.19),

$$W_{\text{mech strm}} = \tfrac{1}{2} m c_0^2 \tag{15.45}$$

where m is the mass flow in kilograms per second, which also determines the valve flow coefficient C_v [see Eq. (15.50) or Table 15.4] together with the specific gravity G_f of the vapor or gas (relative to air $= 1$) and the valve inlet pressure P_1 (in pascals) such that

$$W_{\text{mech strm}} = 7.7 \times 10^{-11} C_v F_L c_0^3 P_1 G_f \quad \text{W} \tag{15.46}$$

The total sound power is then

$$W = \eta W_{\text{mech strm}} \quad \text{W} \tag{15.47}$$

$$L_W = 10 \times \log \frac{\eta W}{10^{-12}} \quad \text{dB} \tag{15.48}$$

When the pressure ratio is not sonic, η should be replaced by the acoustical power coefficient η_m determined below and (to simplify the calculations) should be used in conjunction with the sonic mechanical stream power given by Eq. (15.45). The acoustical power coefficient is further used in Eq. (15.53).

It may be remarked that the pressure recovery coefficient F_L can be used to predict the pressure P_0 in the vena contracta and also, to a satisfactory approximation, the area A_v of the vena contracta by

$$P_0 = P_1 - \frac{P_1 - P_2}{F_L^2} \tag{15.49a}$$

$$A_v = \frac{C_v F_L}{5.91 \times 10^4} \quad \text{m}^2 \tag{15.49b}$$

The valve's flow coefficient C_v can be calculated as follows:

$$C_v = \begin{cases} 2.14 \times 10^7 \dfrac{m}{\sqrt{\Delta P(P_1 + P_2) G_f}} & \Delta P \leq \tfrac{1}{2} F_L^2 P_1 \\ 1.95 \times 10^7 \dfrac{m}{F_L P_1 \sqrt{G_f}} & \Delta P \geq \tfrac{1}{2} F_L^2 P_1 \end{cases} \tag{15.50}$$

TABLE 15.4 Factors for Valve Noise Prediction (Typical Values)

Valve Type	Flow To	Percentage of Capacity or Angle of Travel	$C_v/D^{2,a}$	F_L	F_d
Globe, single-port parabolic plug	Open	100%	0.020	0.90	0.46
Globe, single-port parabolic plug	Open	75%	0.015	0.90	0.36
Globe, single-port parabolic plug	Open	50%	0.010	0.90	0.28
Globe, single-port parabolic plug	Open	25%	0.005	0.90	0.16
Globe, single-port parabolic plug	Open	10%	0.002	0.90	0.10
Globe, single-port parabolic plug	Close	100%	0.025	0.80	1.00
Globe, V-port plug	Open	100%	0.016	0.92	0.50
Globe, V-port plug	Open	50%	0.008	0.95	0.42
Globe, V-port plug	Open	30%	0.005	0.95	0.41
Globe, four-port cage	Open	100%	0.025	0.90	0.43
Globe, four-port cage	Open	50%	0.013	0.90	0.36
Globe, six-port cage	Open	100%	0.025	0.90	0.32
Globe, six-port cage	Open	50%	0.013	0.90	0.25
Butterfly valve, swing-through vane	—	75° open	0.050	0.56	0.57
Butterfly valve, swing-through vane	—	60° open	0.030	0.67	0.50
Butterfly valve, swing-through vane	—	50° open	0.016	0.74	0.42
Butterfly valve, swing-through vane	—	40° open	0.010	0.78	0.34
Butterfly valve, swing-through vane	—	30° open	0.005	0.80	0.26
Butterfly valve, fluted vane	—	75° open	0.040	0.70	0.30
Butterfly valve, fluted vane	—	50° open	0.013	0.76	0.19
Butterfly valve, fluted vane	—	30° open	0.007	0.82	0.08
Eccentric rotary plug value	Open	50° open	0.020	0.85	0.42
Eccentric rotary plug value	Open	30° open	0.013	0.91	0.30
Eccentric rotary plug value	Close	50° open	0.021	0.68	0.45
Eccentric rotary plug value	Close	30° open	0.013	0.88	0.30
Ball valve, segmented	Open	60° open	0.018	0.66	0.75
Ball valve, segmented	Open	30° open	0.005	0.82	0.63

[a] D is the internal pipe diameter in mm.

Acoustical Efficiency

At Mach 1 a free jet has an acoustical efficiency $\eta \approx 10^{-4}$. If it is assumed that the sources of valve noise are dipoles and quadrupoles of equal magnitude (5×10^{-5}) at the Mach 1 reference point, then their respective efficiencies are represented by curves A and B in Fig. 15.20, and both curves can be combined to yield a subsonic slope C that originates at 1×10^{-4} and varies as $U^{3.6}$. This curve can be modified further to account for the decrease in $W_{\text{mech strm}}$ ($\propto U^3$) with subsonic velocities, leading to the final curve D for the effective efficiency (i.e., acoustical power coefficient) η_m that varies as $U^{6.6}$ or $(P_1/P_0 - 1)^{2.57}$ when $U \propto (P_1/P_0 - 1)^{0.39}$, P_0 being the static pressure in the vena contracta. This approximation gives satisfactory results for $0.38 \leq M \leq 1$ (i.e., down to $P_1/P_2 = 1.1$).

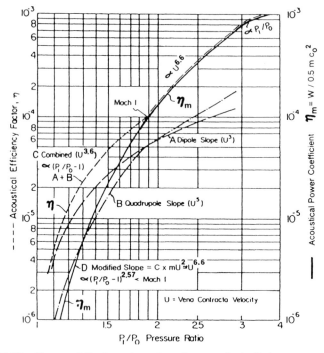

FIGURE 15.20 Slopes of dipole and quadrupole acoustical efficiency curves, assuming each has equal strength at Mach 1; their combined slope and the η_m curve after subtracting the effects of change in mass flow m and orificial velocity U for regions below Mach 1 from the combined curve.

For practical use it is convenient to express η_m in terms of $P_1/P_2 =$ inlet pressure/outlet pressure. The following formulas[28,29] may be used for this purpose:

regime I: $P_1/P_2 < P_1/P_{2\text{critical}}$ (subsonic):

$$\eta_{m\text{I}} = 10^{-4} F_L^2 \left(\frac{P_1 - P_2}{P_1 F_L^2 - P_1 + P_2} \right)^{2.6} \quad (15.51a)$$

regimes II and III: $P_1/P_{2\text{critical}} < P_1/P_2 < 3.2\alpha$:

$$\eta_{m\text{II}} = 10^{-4} F_L^2 \left(\frac{P_1/P_2}{P_1/P_{2\text{critical}}} \right)^{3.7} \quad (15.51b)$$

regime IV: $3.2\alpha < P_1/P_2 < 22\alpha$ ($M_j > 1.4$):

$$\eta_{m\text{IV}} = 1.32 \times 10^{-3} F_L^2 \left(\frac{P_1/P_2}{P_1/P_{2\text{break}}} \right) \quad (15.51c)$$

regime V: $P_1/P_2 > 22\alpha$ (constant efficiency):

$$\eta_{mV} = \text{maximum value from regime IV} \qquad (15.51d)$$

where
$$\alpha = (P_1/P_{2\text{critical}})/1.89$$
$$P_1/P_{2\text{break}} = \alpha\gamma^{\gamma/(\gamma-1)} \text{ at } M = \sqrt{2}$$
$$\gamma = \text{ratio of specific heat of gas}$$
$$P_1/P_{2\text{critical}} = P_1/(P_1 - 0.5 P_1 F_L^2)$$

The values of these parameters may be taken from Table 15.5.

When the downstream piping is straight and there are no sudden changes in cross-sectional area, the highest internal $\frac{1}{3}$-octave band sound pressure level L_{pi} (re 2×10^{-5} Pa) downstream of the valve is given by

$$L_{pi} = -61 + 10 \times \log \frac{\eta_m C_v F_L P_1 P_2 c_0^4 G_f^2}{D_i^2} \quad \text{dB} \qquad (15.52)$$

where D_i = internal diameter of downstream pipe, m
G_f = specific gravity of gas relative to air at 20°C
$c_0 = \sqrt{\gamma P_0/\rho_0}$, m/s (speed of sound)
γ = ratio of specific heats
ρ_0 = gas density at vena contracta, kg/m³

Pipe Transmission Loss Coefficient

Knowledge of the peak internal sound frequency f_p is crucial for a proper prediction of the pipe transmission loss coefficient T_L. The coefficient T_L does not vary significantly between the first cutoff frequency f_0 [see Eq. (15.58)] and the ring frequency f_R of the pipe (see lower curve in Fig. 15.21), but variations can be large at other frequencies. The slope of T_L is about -6 dB per octave below f_0 and $+6$ dB per octave above f_r.[28]

For Mach numbers in the pipe less than about 0.3 and for relatively heavy pipes (as found in typical process plants) it may be assumed that the *minimum*

TABLE 15.5 Important Pressure Ratios for Air[a]

F_L	0.5	0.6	0.7	0.8	0.9	1.0
P_1/P_2 critical	1.13	1.20	1.30	1.43	1.61	1.89
P_1/P_2 break	1.95	2.07	2.24	2.40	2.76	3.25
α Ratio	0.60	0.64	0.68	0.76	0.85	1.0
22α	13	14	15	16	19	22

Note: P_1/P_0 critical $= 1.89$.
[a] May be used for other gases with reasonable accuracy.

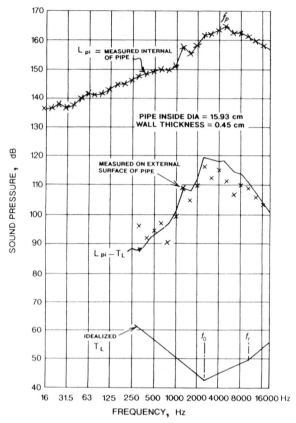

FIGURE 15.21 Test data taken on a 150-mm globe valve showing good agreement using the stated transmission loss equation. Notice the resultant shift in the peak sound frequency (f_p) from ~4.8 kHz inside the pipe to the first cutoff frequency of the pipe (f_0) at 2.5 kHz outside the pipe.

transmission loss occurs at f_0 and is given by[28]

$$T_{Lf_0} = 10 \times \log \left[9 \times 10^6 \frac{rt_p^2}{D_i^3} \left(\frac{P_2}{P_a} + 1 \right) \right] \quad \text{dB} \qquad (15.53)$$

where r = distance from pipe wall to observer, m
t_p = thickness of pipe wall, m
D_i = internal pipe diameter, m
P_2 = internal static pressure downstream of valve, Pa
P_a = external static pressure at same downstream distance, Pa

Typical values are given in Table 15.6. For fluids with higher sonic velocities and for thinner pipe walls, the minimum value of T_{Lf_0} shifts toward the ring frequency f_R [see Eq. (15.58)].

TABLE 15.6 Minimum Pipe Transmission Loss Coefficient $T_{Lf_0}{}^a$

Nominal Pipe Size, m	Pipe Schedule	
	40	80
0.025	72	76
0.050	65	69
0.080	64	68
0.100	60	64
0.150	57	61
0.200	54	58
0.250	52	57
0.300	51	56
0.400	50	55
0.500	48	53

[a] At 1 m distance from steel pipe (dB) for 200 kPa internal and 100 kPa external air pressure. Pipe schedules per ANSI B36.10.

The total transmission loss coefficient for the valve is now expressed in the form

$$T_L = T_{Lf_0} + \Delta T_{Lf_p} \tag{15.54}$$

where the correction ΔT_{Lf_p} (in decibels) is determined by the peak noise frequency according to

$$\Delta T_{Lf_p} = \begin{cases} 20 \times \log \dfrac{f_0}{f_p} & f_p \leq f_0 \\ 13 \times \log \dfrac{f_p}{f_0} & f_p \leq 4f_0 \\ 20 \times \log \dfrac{f_p}{4f_0} + 7.8 & f_p > 4f_0 \end{cases} \tag{15.55}$$

The peak frequency is estimated from[29]

$$f_p = \begin{cases} \dfrac{0.2 M_j c_0}{D_j} & M_j < 4 \\ \dfrac{0.28 c_0}{D_j \sqrt{M_j^2 - 1}} & M_j > 4 \end{cases} \tag{15.56}$$

where c_0 = speed of sound at vena contracta, m/s
 D_j = jet diameter at valve orifice

and

$$M_j = \left\{ \dfrac{2}{\gamma - 1} \left[\left(\dfrac{P_1}{\alpha P_2} \right)^{(\gamma - 1)/\gamma} - 1 \right] \right\}^{1/2} \tag{15.57}$$

$$f_R = \frac{c_L}{\pi D_i} = 4f_0; \quad \text{for steel pipes and air} \quad f_R \approx \frac{5000}{\pi D_i} \quad (15.58)$$

For a gas other than air f_0 must be multiplied by $c_{0,\text{gas}}/c_{0,\text{air}}$.

The diameter D_j is difficult to determine because of the complex flow geometry in many valves. A reasonable approximation is to use the hydraulic diameter D_H as the jet diameter and make use of a valve-style modifier F_d (see Table 15.4) to obtain

$$D_j \approx 4.6 \times 10^{-3} F_d \sqrt{C_v F_L} \quad \text{m} \quad (15.59)$$

This formula is *not* applicable if jet portions combine downstream (usually at higher pressure ratios). This is common with multiple-hole valve cages,[30] short-stroke parabolic valve plugs in the "flow to close" direction, and butterfly valves at larger openings.

The external sound level (measured at 1 m from the wall) is now given by

$$L_a = 5 + L_{pi} - T_L - L_g \quad \text{dBA} \quad (15.60)$$

where L_g is a correction for fluid velocity in the pipe:

$$L_g = -16 \times \log\left(1 - \frac{1.3 \times 10^{-5} P_1 C_v F_L}{D_i^2 P_2}\right) \quad (15.61)$$

The laboratory test data shown in Fig. 15.21, taken with air on a DN150 globe valve, exhibits good agreement between predicted and measured external valve sound levels using T_L calculated from Eq. (15.54). In the absence of unusual phenomena such as jet "screech," Eq. (15.60) is found to be accurate to within ± 3 dB assuming that all the valve coefficients are well established.

Accounting for Additional Noise due to High Velocities in Valve Outlet

Gas expanding in volume due to pressure reduction in a valve can cause high valve outlet velocities. These, in turn, can create secondary noise sources in the downstream pipe or pipe expander.[31,32]

The first step is to estimate the valve's outlet velocity (limited to sonic c_2):

$$U_R = 4\frac{m}{\pi \rho_2 (d_i^2)} \quad (15.62)$$

where d_i is the valve diameter in meters.

The internal sound pressure level produced by this velocity is

$$L_{piR} = 65 + 10 \log\left(M_R^3 \frac{W_{mR}}{D_i^2}\right) \quad (15.63)$$

Here

$$W_{mR} = \frac{m}{2} U_R^2 \left[\left(1 - \frac{d_i^2}{D_i^2}\right) + 0.2 \right] \qquad (15.64)$$

and

$$M_R = \frac{U_R}{c_2} \qquad (15.65)$$

Use Eq. (15.60) to calculate the external sound level L_R, substituting L_{piR} for L_{pi}, and using Eq. (15.54) to calculate TL, here substituting f_{pR} for f_p, where

$$f_{pR} = 0.2 \frac{U_R}{d_i} \qquad (15.66)$$

The combined estimated external sound level at 1 m from the pipe from both the valve and the downstream piping now is

$$L_{ps} = 10 \log(10^{L_a/10} + 10^{L_R/10}) \qquad (15.67)$$

The reader in encouraged to review reference 24, which illustrates the use of the prediction procedure in numerous examples.

Methods of Valve Noise Reduction

The two basic variables controlling the valve noise are the jet Mach number M_j and the internal peak frequency f_p. In subsonic flows lower jet velocities can be achieved by use of a valve trim that is less streamlined ($F_L \geq 0.9$, say).

At higher pressure ratios the only remedy is to use orifices or resistance paths in series, so that each operates subsonically. Such a "zig-zag" path valve is shown in Fig. 15.22. For example, for natural gas with an inlet pressure of 10,000 kPa that is to be throttled down to 5000 kPa and assuming a valve pressure recovery factor $F_L = 0.9$, the internal pressure ratio P_1/P_0 for a single-stage reduction would be $P_1/P_2 \alpha = 2.35 : 1$. However, using the trim from Fig. 15.22 with 10 reduction steps, the orificial pressure ratio for each step is now only $1.603 : 1$ with an acoustical power coefficient of only 1.1×10^{-7}. Ignoring the partial addition of 10 separate power sources, this represents a noise reduction of about 30 dB.

A less costly way to reduce "audible" valve noise is to increase the peak frequency by using multiorifice trim, as shown in Fig. 15.23. The throttling is still single stage, but the jets are divided into a number of parallel ports. This reduces the value of the valve style modifier $F_d \approx 1/\sqrt{n_o}$, where n_o is the number of equally sized parallel ports. Since F_d is proportional to jet diameter D_j, f_p can be shifted to higher levels. Thus if $n_o = 16$, as in Fig. 15.23, $F_d = 0.25$ and f_p is four times higher than for a single orifice ($F_d = 1$) with the same total flow area. If the peak frequency occurs in the mass-controlled region, an overall reduction in the external (audible) sound level of $\Delta T_{Lf_p} = 20 \times \log(f_{p_1}/f_{p_0}) = 20 \times \log(1/0.25) \approx 12$ dB is achieved.

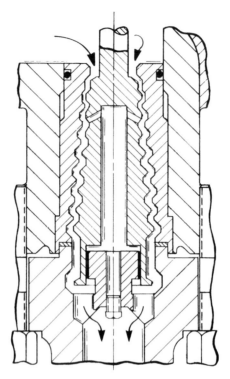

Single flow area, multi-step valve plug

FIGURE 15.22 High-pressure reducing valve using multiple-step trim with a single flow path. Ideally each step will have subsonic orifice velocity.

The beneficial effects of a reduction in velocity combined with higher transmission loss caused by increased frequency are obtained with the elaborate arrangement shown in Fig. 15.24. This combines the multistep and multipath approach by use of a layer of disks having individually cast or etched channels.

A more economical solution is to couple a static downstream pressure-reducing device, such as a multihole restrictor, with a throttling valve with a fluted disk as indicated in Fig. 15.25. The fluted disk generates multiple jets with increased noise frequencies, although most of the energy is converted by the static pressure plate. At maximum design flow the valve is sized to have a low pressure ratio of 1.1 or 1.15 and the remainder of the pressure reduction occurs across the static plate having inherent low-noise throttling due to multistage, multihole design.

The most advanced and most compact form of a low noise valve trim is shown in Fig. 15.26. It consists of identical stamped or cut plates and constitutes a two-stage device, hence the narrow envelope. The device combines a supersonic first stage with a subsonic second stage for lowest overall acoustic efficiency, while retaining the final high frequency peak outlet sound level for high transmission losses.

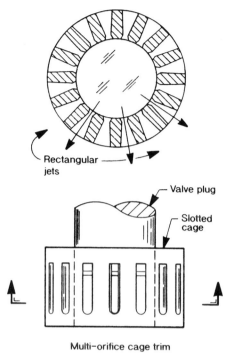

FIGURE 15.23 Slotted valve cage subdivides single-step flow path in 16 separate openings ($F_d = 0.25$) to increase peak frequency and thereby transmission loss.

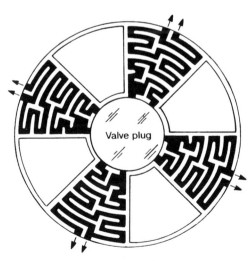

FIGURE 15.24 Labyrinth-type flow path cast or machined into each of a stack of metal plates surrounding a valve plug to provide a combination of multistep and multichannel flow paths in order to reduce throttling velocity and increase peak frequency.

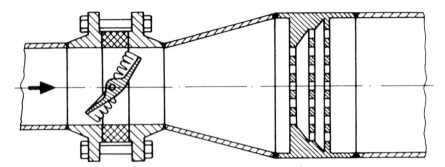

FIGURE 15.25 Modulating fluted butterfly valve cooperates with triple-stage static resistance plate for purposes of pressure reduction. Static plates see 90% of pressure drop at maximum design flow. Balance is handled by the valve.

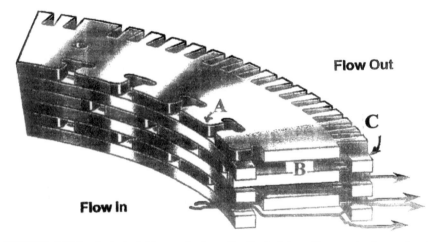

FIGURE 15.26 A stacked trim made from identical plates offers most advanced noise reduction. Here gas enters inlet orifices (A) configured as supersonic diffusers. These orifices convert about 95% of the stream power of the gas through a mechanism of shock waves; the gas then expands into sound-absorbing settling chambers (B) and finally exits the trim through a series of small subsonic passages (C).

Placing a silencer downstream of a valve is not cost effective because a good portion of the sound power travels upstream or radiates through the valve body and actuator, yielding noise reductions that typically do not exceed about 10 dB. Similar remarks apply to acoustical insulation of the pipe wall, where attenuations are limited to about 15 dB. The effectiveness of various abatement procedures is summarized below:

1. One-inch-thick pipe insulation: 5–10-dB reduction
2. Doubling thickness of pipe wall: 6-dB reduction

3. Silencer downstream: 10-dB reduction
4. Silencers upstream and downstream: 20-dB reduction
5. Multiport resistance plate downstream of valve: 15–20-dB reduction (use only 5–10% of valve inlet pressure as valve ΔP at maximum flow)
6. Special low-noise valve: 15–30-dB reduction

ACKNOWLEDGMENT

The authors wish to acknowledge the contribution of Dr. M. H. Howe for the portions of this chapter that have been retained from previous editions of this book.

REFERENCES

1. M. J. Lighthill, "On Sound Generated Aerodynamically. Part I: General Theory," *Proc. Roy. Soc. Lond.*, **A211**, 564–587 (1952).
2. M. S. Howe, "Contributions to the Theory of Aerodynamic Sound with Applications to Excess Jet Noise and the Theory of the Flute," *J. Fluid Mech.*, **71**, 625–673 (1975).
3. Society of Automotive Engineers, *"Gas Turbine Jet Exhaust Noise Prediction,"* Report No. SAE ARP 876C, Society of Automotive Engineers, Warrendale, PA, 1985.
4. J. R. Stone, "Prediction of In-Flight Exhaust Noise for Turbojet and Turbofan Engines," *Noise Control Eng.*, **10**, 40–46 (1977).
5. A. L. Laganelli and H. Wolfe, *"Prediction of Fluctuating Pressure in Attached and Separated Turbulent Boundary Layer Flow,"* Paper 89, American Institute of Aeronautics and Astronautics, Washington, DC, 1989, p. 1064.
6. M. S. Howe, "The Role of Surface Shear Stress Fluctuations in the Generation of Boundary Layer Noise," *J. Sound Vib.*, **65**, 159–164 (1979).
7. M. S. Howe, "Surface Pressures and Sound Produced by Low Mach Number Turbulent Flow over Smooth and Rough Walls," *J. Acoust. Soc. Am.*, **90**, 1041–1047 (1991).
8. D. M. Chase, "The Character of Turbulent Wall Pressure Spectrum at Subconvective Wavenumbers and a Suggested Comprehensive Model," *J. Sound Vib.*, **112**, 125–147 (1987).
9. J. G. Seebold, "Smooth Piping Reduces noise—Fact or Fiction?" *Hydrocarbon Process.*, September 1973, pp. 189–191.
10. D. A. Bies, *"A Review of Flight and Wind Tunnel Measurements of Boundary Layer Pressure Fluctuations and Induced Structural Response,"* NASA CR-626, NASA Langley Research Center, Hampton, VA, May 1966.
11. D. A. Nelson, *Reduced-Noise Gas Flow Design Guide*, NASA Glenn Research Center, Cleveland, OH, April 1999.
12. H. H. Heller and S. E. Widnall, "Sound Radiation from Rigid Flow Spoilers Correlated with Fluctuating Forces," *J. Acoust. Soc. Am.*, **47**, 924–936 (1970).
13. P. A. Nelson and C. L. Morfey, "Aerodynamic Sound Production in Low Speed Flow Ducts," *J. Sound Vib.*, **79**, 263–289 (1981).

14. M. Hubert, "Untersuchungen ueber Geraeusche durchstroemter Gitter," Ph.D. Thesis, Technical University of Berlin, 1970.
15. M. S. Howe, "The Influence of Vortex Shedding on the Generation of Sound by Convected Turbulence," *J. Fluid Mech.*, **76**, 711–740 (1976).
16. W. K. Blake, *Mechanics of Flow-Induced Sound and Vibration*, Vol. 2: *Complex Flow-Structure Interactions*, Academic, New York, 1986.
17. T. F. Brooks, D. S. Pope, and M. A. Marcolini, *"Airfoil Self-Noise and Prediction,"* NASA Reference Publication No. 1218, NASA Langley Research Center, Hampton, VA, 1989.
18. J. E. Rossiter, *"Wind Tunnel Experiments on the Flow over Rectangular Cavities at Subsonic and Transonic Speeds,"* RAE TR 64037, Royal Aircraft Establishment, Farnsborough, U.K., 1964.
19. P. J. W. Block, *"Noise Response of Cavities of Varying Dimensions at Subsonic Speeds,"* Technical Paper D-8351, NASA Langley Research Center, Hampton, VA, 1976.
20. L. F. East, "Aerodynamically Induced Resonance in Rectangular Cavities," *J. Sound Vib.*, **3**(3), 277–287 (1966).
21. M. J. Lucas, "Impinging Shear Layers," in M. J. Lucas et al., *Handbook of the Acoustic Characteristics of Turbomachinery Cavities*, American Society of Mechanical Engineers, New York, 1997, pp. 205–241.
22. M. C. Junger, "Cavity Resonators," in M. J. Lucas et al., *Handbook of the Acoustic Characteristics of Turbomachinery Cavities*, American Society of Mechanical Engineers, New York, 1997, pp. 13–38.
23. R. L. Panton and J. M. Miller, "Excitation of a Helmholtz Resonator by a Turbulent Boundary Layer," *J. Acoust. Soc. Am.*, **58**, 533–544 (1975).
24. IEC 60534-8-3, *"Control Valve Aerodynamic Noise Prediction Method,"* International Electrical Commission, Geneva, Switzerland, 2000.
25. V. A. Carucci and R. T. Mueller, *"Acoustically Induced Piping Vibration in High Capacity Pressure Reducing Systems,"* Paper No. 82-WA/PVP-8, American Society of Mechanical Engineers, New York, 1982.
26. H. D. Baumann, *"On the Prediction of Aerodynamically Created Sound Pressure Level of Control Valves,"* Paper No. WA/FE-28, American Society of Mechanical Engineers, New York, 1970.
27. ANSI/ISA S75.01-1985, *"Flow Equations for Sizing Control Valves,"* Instrument Society of America, Research Triangle Park, NC, 1985.
28. ANSI/ISA S75.17, *"Control Valve Aerodynamic Valve Noise Prediction,"* Instrument Society of America, Research Triangle Park, NC, 1989.
29. H. D. Baumann, *"A Method for Predicting Aerodynamic Valve Noise Based on Modified Free Jet Noise Theories,"* Paper No. 87-WA/NCA-7, American Society of Mechanical Engineers, New York, 1987.
30. C. Reed, "Optimizing Valve Jet Size and Spacing Reduces Valve Noise," *Control Eng.*, **9**, 63–64 (1976).
31. H. D. Baumann, "Predicting Control Valve Noise at High Exit Velocities," *INTECH*, February 1997, pp. 56–59.

CHAPTER 16

Prediction of Machinery Noise

ERIC W. WOOD AND JAMES D. BARNES
Acentech, Inc.
Cambridge, Massachusetts

Engineering information and prediction procedures are presented describing the sound power emission characteristics of various industrial machinery. The types of equipment addressed in this chapter include air compressors, boilers, coal-handling equipment, cooling towers, air-cooled condensers, diesel-engine-powered equipment, fans, feed pumps, gas turbines, steam turbines, steam vents, transformers, and wind turbines. The information has been extracted from consulting project files and the results of field measurement programs. The machinery noise prediction procedures are based on studies of empirical field data sponsored by numerous clients, including the Edison Electric Institute and the Empire State Electric Energy Research Corporation, for whom references 1–3 were prepared with the authors' colleagues: Robert M. Hoover, Laymon N. Miller, Susan L. Patterson, Anthony R. Thompson, and István L. Vér.

Characteristics identified and described for the noise produced by the machines discussed in this chapter include the estimated overall, A-weighted, and octave-band sound power levels (L_w) in dB re 1 pW, general directivity and tonal characteristics observed in the far field of the source, and temporal characteristics associated with the source operation. In addition, noise abatement concepts that have proven to be useful to the authors during previous consulting projects are described briefly. Successful noise control treatments often require a detailed understanding of site-specific operating and maintenance requirements as well as pertinent acoustical conditions at the site. We suggest that acoustical design handbooks or experienced acoustical engineering professionals be consulted when guidance is needed for specific design applications.

The information and procedures presented here can be used for numerous engineering applications requiring predictions of the approximate acoustical characteristics of new machinery installations and estimates of the noise attenuation that may be needed to meet local requirements. The prediction procedures are expected to provide A-weighted sound-power-level estimates that are generally accurate to within about ±3 dB. Individual octave-band sound-power-level estimates will necessarily be somewhat less accurate than A-weighted sound-power-level estimates. The prediction procedures estimate the equivalent (energy average) L_{eq} sound power level during operation of one machine. For equipment with intermittent operating cycles, the estimated sound power levels may be reduced by 10 log (operating cycle) if long-term equivalent values are required. When several identical machines are operating simultaneously, the sound power levels can be increased by 10 log (number of machines). When equipment is located within a large plant, other nearby large equipment may provide some shielding that will reduce the noise radiated to distant neighbors in shielded directions (see Chapter 7). When equipment is located within an enclosed plant, a portion of the acoustical energy will be dissipated inside the plant, leading to further reductions in the far-field noise. One should note that doors, windows, and ventilation openings can substantially diminish the acoustical performance (composite sound transmission loss) of building exterior walls (see Chapter 12).

The reader should try to obtain actual field data and/or manufacturers' information whenever possible for machines being studied. Information about the noise produced by gas flows and fans in buildings is presented in Chapters 15 and 17. Noise abatement information is provided in several other chapters. Indoor sound field characteristics are described in Chapters 6 and 7, and outdoor sound propagation is discussed in Chapter 5. The noise emission information provided in this chapter is representative of equipment presently available in the United States. Equipment in use in other countries may produce noise levels that are somewhat different than presented here. Prediction formulas for the overall sound power level are provided in the text of this chapter. Adjustments to predict the A-weighted and octave-band sound power levels are provided in Table 16.1.

16.1 AIR COMPRESSORS

The noise produced by air compressors is radiated from the filtered air inlet, the compressor casing, the interconnected piping and interstage cooler, as well as the motor or engine used to drive the compressor. The noise produced by air compressors can exhibit a distinctive duty cycle based on the demand for compressed air and is generally considered omnidirectional and broadband and absent prominent discrete tones. The air inlet noise of certain vacuum pumps sometimes includes high levels of low-frequency pulsating noise.

Noise-estimating procedures presented in this section are applicable to conventional industrial-, utility-, and construction-type air compressors without special noise abatement packages. Air compressors are often available from their manufacturer with special noise abatement packages that include various types of inlet mufflers, laggings, insulations, and enclosures that provide varying degrees of noise reduction. Acoustical data for these low-noise compressors should be obtained directly from the manufacturer's literature.

TABLE 16.1 Adjustments Used to Estimate A-Weighted and Octave-Band L_w [a]

Line	Equipment[b]	A-Weighted	31.5	63	125	250	500	1000	2000	4000	8000
1	RR compressor	2	11	15	10	11	13	10	5	8	15
2	C Compressor case	2	10	10	11	13	13	11	7	8	12
3	C Compressor inlet	0	18	16	14	10	8	6	5	10	16
4	S boilers	9	6	6	7	9	12	15	18	21	24
5	L boilers	12	4	5	10	16	17	19	21	21	21
6	CC shakers	9	5	6	7	9	12	15	18	21	24
7	R car open	12	4	5	10	13	17	19	21	21	21
8	R car enclosed	10	3	5	11	14	14	14	18	19	19
9	BL unloaders	9	8	5	8	10	12	13	18	23	27
10	CS unloaders	11	3	8	12	14	14	16	19	21	25
11	Coal crushers	9	6	6	6	10	12	15	17	21	30
12	T towers	7	7	7	7	9	11	12	14	20	27
13	C mill (13–36)	5			4	6	8	11	13	15	
14	C mill (37–55)	5			6	7	4	11	14	17	
15	ND C towers	0			12	13	11	9	7	5	7
16	MD C towers	10	9	6	6	9	12	16	19	22	30
17	MD C towers 1/2S	5	9	6	6	10	10	11	11	14	20
18	A-C condenser	12	5	6	6	10	14	17	24	29	34
19	D engine	5		11	6	3	8	10	13	19	25
20	C fan I + O	5	11	9	7	8	9	9	13	17	24
21	C fan case	13	3	6	7	11	16	18	22	26	33
22	R fan case	12	10	7	4	7	18	20	25	27	31
23	Axial fan	3	11	10	9	8	8	8	10	14	15
24	S Fd pumps	4	11	5	7	8	9	10	11	12	16
25	L Fd pumps	1	19	13	15	11	5	5	7	19	23
26	C-C plant	11	7	3	7	13	15	17	19	23	21
27	S St turbines	5	11	7	6	9	10	10	12	13	17
28	L St turbines	12	9	3	5	10	14	18	21	29	35
29	SL blow-out	11		2	6	14	17	17	21	18	18
30	Transformers	0		−3	−5	0	0	6	11	16	23

[a] Subtract values shown from estimated overall L_w; except for transformers, subtract values from A-weighted sound level.
[b] RR: rotary and reciprocating; C: casing; S: small, L: large, CC: coal car; R: rotary car dumpers; BL: bucket ladder; CS: clamshell; T: transfer; C: coal; ND C: natural-draft cooling; MD C: mechanical-draft cooling; A-C: air cooled; D: diesel; C-C plant: inside combined-cycle power plant main building; SL: steam line.

The overall sound power level radiated by the casing and air inlet of rotary and reciprocating air compressors can be estimated using Eq. (16.1) (see also line 1 of Table 16.1). These relations for rotary and reciprocating compressors assume that the air inlet is equipped with a filter and small muffler as normally provided by the manufacturer:

Rotary and reciprocating compressor, overall $L_w = 90 + 10 \log(\text{kW})$ (dB) (16.1)

where kW is the shaft power in kilowatts. The casing and unmuffled air inlet overall sound power levels for centrifugal air compressors in the power range of 1100–3700 kW can be estimated using Eqs. (16.2) and (16.3) (see also lines 2

662 PREDICTION OF MACHINERY NOISE

and 3 of Table 16.1). The insertion loss of any muffler installed at the air inlet should be deducted from the calculated inlet power levels:

$$\text{Centrifugal compressor casing, overall } L_w = 79 + 10 \log(\text{kW}) \quad (\text{dB}) \quad (16.2)$$
$$\text{Centrifugal compressor inlet, overall } L_w = 80 + 10 \log(\text{kW}) \quad (\text{dB}) \quad (16.3)$$

The term $\log(x)$ represents the common logarithm to base 10 of the value of x.

Equations (16.2) and (16.3) do not include the noise radiated by the motor or engine used to drive the compressor. Sound power levels calculated for the driving equipment should be added to the calculated sound power levels for the compressor.

Mufflers, laggings, barriers, and enclosures have all been used with varying degrees of success to control the noise radiated from compressors to nearby workspaces or neighborhoods. However, it is suggested that it is often most practical to specify and purchase a reduced-noise compressor from the manufacturer.

16.2 BOILERS

Two procedures are provided to predict the noise from boilers; one is for relatively small boilers in the range of about 50–2000 boiler horsepower (one boiler horsepower equals 15 kg of steam per hour). Another procedure is provided for large boilers of the type that serve electric-power-generating stations in the range of 100–1000 megawatts (MWe).

Small Boilers

The sound power output of small boilers is only weakly related to the thermal rating of the boiler. The combustion air fans and the burners probably radiate more noise than do the insulated sidewalls of small boilers. The overall sound power level of small boilers can be estimated using Eq. (16.4) (also see line 4 of Table 16.1):

$$\text{Small boiler, overall } L_w = 95 + 4 \log(\text{bhp}) \quad (\text{dB}) \quad (16.4)$$

Large Boilers

Noise levels measured in work spaces adjacent to large central-station boilers that are not enclosed in a building (open boilers) are often in the range of 80–85 dBA at the lower half and 70–80 dBA at the upper half of the boiler. Noise levels measured in the vicinity of enclosed boilers (generally found in areas with cold climates) are often about 5 dBA higher. The overall sound power output of large central-station boilers can be estimated by the relation given in Eq. (16.5) (also see line 5 of Table 16.1):

$$\text{Large boiler, overall } L_w = 84 + 15 \log(\text{MWe}) \quad (\text{dB}) \quad (16.5)$$

The noise from boilers is essentially omnidirectional and broadband in character. However, the combustion air or forced-draft fans, induced-draft fans, gas recirculation fans, overfire air fans, and drive motors sometimes produce tonal noise and may increase the noise levels on their side of the boiler. A small number of boilers have also been found to exhibit a strong discrete tone, usually at a frequency between about 20 and 100 Hz during operation at particular boiler loads. This can be caused by vortex shedding from heat exchanger tubes that excite an acoustical resonance within the boiler that in turn causes coherent (rather than random) vortex shedding to occur at a boiler resonance frequency. The low-frequency tonal noise that radiates from the boiler sidewalls has caused strong adverse reactions from residential neighbors. Boiler operators have also expressed concern about possible fatigue failure of vibrating tube sheets. Installation of one or more large metal plates to subdivide the interior of the boiler volume and thereby change its resonance frequency has successfully corrected this problem. Additional information about this resonant condition and its control is provided in references 4 and 5.

The typical broadband noise radiated by boiler sidewalls has been successfully reduced with the use of a well-insulated exterior enclosure, which also serves as weather protection for equipment and workers. The noise produced by fans serving the boiler can also be controlled as described in Section 16.6.

16.3 COAL-HANDLING EQUIPMENT

A wide variety of noise-producing equipment, such as railcar, ship and barge unloaders, transfer towers, conveyors, crushers, and mills, are used to unload, transport, and condition coal for use at large industrial and utility boilers. The results of numerous noise-level surveys have been studied and condensed to prepare the following general noise prediction procedures for this equipment.

Coal Car Shakers

Coal car shakers vibrate coal cars during bottom unloading. The unloading operation of each car often occurs for about 2–5 min when the coal is not frozen. When frozen coal is located in the car, the shake-out operation can last for up to 10 min or longer. Car shakers are often located inside a sheet metal or masonry building with little or no interior insulation for sound absorption. Shaker buildings include large openings at both ends and sometimes smaller openings and windows along the sidewalls. They also include dust collection and ventilation systems with fans that produce noise. The overall sound power produced during a coal car shake-out can be estimated using the following relation (also see line 6 of Table 16.1):

$$\text{Coal car shake-out, overall } L_w = 141 \text{ dB} \qquad (16.6)$$

Noise produced by the dust collection and ventilation fans can be estimated using the relations provided in Section 16.6 and in Chapter 17.

The far-field noise produced by car shake-outs can be reduced by enclosing the operation inside a metal or masonry building or by reducing the openings in any existing building. In addition, modern car-shaking equipment is sometimes available that produces measured A-weighted sound levels about 10 dB lower than are estimated with Eq. (16.6). Careful train movements can reduce the impact noise associated with indexing the train through the shaker operation. Limiting the shake-out operation to daytime hours also reduces the community noise impacts.

Rotary Car Dumpers

Rotary car dumpers are often used to unload long unit trains equipped with special rotary couplings. It typically takes 2–3 min to unload unfrozen coal from each car. A 100-car unit train can be unloaded in about 3–5 h unless delays are encountered. The overall sound power produced during rotary car unloading cycles can be estimated using Eqs. (16.7) and (16.8) for open and enclosed facilities (also see lines 7 and 8 of Table 16.1). However, short-duration impact sounds can be as much as 10–25 dB greater than the equivalent values given by these relations:

$$\text{Rotary car open unloading, overall } L_w = 131 \text{ dB} \quad (16.7)$$

$$\text{Rotary car enclosed unloading, overall } L_w = 121 \text{ dB} \quad (16.8)$$

Noise abatement methods for rotary car dumpers are similar to those discussed previously for car shakers.

Bucket Ladder and Clamshell Bucket Unloaders

Bucket ladder and clamshell bucket unloaders are often used to unload coal delivered by ship or barge. This equipment is located outdoors along the dock line with the electric motors and speed reduction gears often inside metal enclosures for weather protection. Acoustical data studied to prepare the prediction relations provided here were obtained at bucket ladder unloaders with free-digging rates in the range of 1800–4500 metric tons per hour and clamshell unloaders with free-digging rates in the range of 1600–1800 metric tons per hour. The energy-averaged overall sound power level produced during operation of bucket ladder and clamshell bucket unloaders can be estimated using Eqs. (16.9) and (16.10) (see also lines 9 and 10 of Table 16.1):

$$\text{Bucket ladder unloader, overall } L_w = 123 \text{ dB} \quad (16.9)$$

$$\text{Clamshell bucket unloader, overall } L_w = 131 \text{ dB} \quad (16.10)$$

Noise control treatments for bucket ladder and clamshell bucket unloaders generally employ insulated enclosures for the drive equipment and limiting unloading operations to daytime hours.

Coal Crushers

Coal is sometimes conveyed to a crusher building where the chunks are reduced in size in preparation for firing in a power plant boiler. Noise produced during crushing operations is a composite of the crusher, metal chutes, conveyors, drive motors, and speed reducers. Coal crushing is usually an intermittent operation and the noise is often omnidirectional without major tonal components. The overall sound power level produced during operation of the crusher can be estimated in accordance with the following relation (also see line 11 of Table 16.1):

$$\text{Coal crusher, overall } L_w = 127 \text{ dB} \quad (16.11)$$

The far-field noise associated with coal-crushing operations can be reduced by using a well-insulated and ventilated building to enclose the operation.

Coal Transfer Towers

Transfer towers reload coal from one conveyor to another as the coal is relocated within the coal yard on its way to the plant. Transfer tower noise is comprised of coal impacts, local conveyors, drive motors, and dust collection and ventilation system fans. Coal transfer is usually an intermittent operation, and the noise is often omnidirectional without major tonal components. The overall sound power level produced during coal transfer in open buildings can be estimated in accordance with the following relation (also see line 12 of Table 16.1):

$$\text{Transfer tower, overall } L_w = 123 \text{ dB} \quad (16.12)$$

Coal transfer operations can be enclosed in a well-ventilated building to reduce the far-field community noise. In this case, the estimated far-field noise can be mitigated by the composite sound transmission loss of the exterior building walls. Fan noise associated with the ventilation system should be estimated with the relations given in Section 16.6 or Chapter 17.

Coal Mills and Pulverizers

Coal mills and pulverizers crush and size coal in preparation for burning in a boiler. The resulting noise is associated with internal impacts, drive motors, speed reducers, and fans. The noise is essentially broadband, omnidirectional, reasonably steady, and continuous while the plant is operating. The overall sound power level produced by a coal mill in a building with large openings can be estimated in accordance with the following relations for mills in the size ranges of 13–36 and 37–55 metric tons per hour (also see lines 13 and 14 of Table 16.1). When coal mills are located in a closed building, the estimated far-field noise can be reduced by the composite sound transmission loss of the exterior building walls:

$$\text{Coal mill, overall } L_w = \begin{cases} 110 \text{ dB} & 13\text{–}36 \text{ metric tons per hour} \quad (16.13) \\ 112 \text{ dB} & 37\text{–}55 \text{ metric tons per hour} \quad (16.14) \end{cases}$$

16.4 COOLING TOWERS

The sound power output of wet cooling towers is caused primarily by the water splash in the fill and in the basin as well as by the fans, motors, and gears used to provide draft in mechanical towers. The noise is usually continuous and somewhat directive for rectangular towers. The noise produced by a specific pair of large natural-draft hyperbolic cooling towers has been observed to include a modest low-frequency discrete tone associated with aerodynamic vortex shedding at the base of each tower.

Methods that can be used to estimate the sound power output of natural-draft and mechanical induced-draft cooling towers are provided below. Major tower manufacturers can provide additional useful information regarding noise estimates and noise abatement.

Natural-Draft Cooling Towers

The overall sound power radiated from the rim of large hyperbolic natural-draft wet cooling towers can be estimated with Eq. (16.15) (see line 15 in Table 16.1). Noise radiated by the top of the tower is usually not significant at ground elevations compared to the rim-radiated noise:

Natural-draft cooling tower rim, overall $L_w = 86 + 10 \log Q$ (dB)

(16.15)

where Q is the water flow rate in gallons per minute (one cubic meter per minute is equal to 264 U.S. gallons per minute).

Inlet mufflers and fan assistance have been installed to control rim-radiated noise for hyperbolic cooling towers in Europe when located near residential neighbors. However, noise abatement is expensive and not common for large hyperbolic towers.

Mechanical Induced-Draft Cooling Towers

The sound power produced by mechanical induced-draft wet cooling towers operating at full fan speed and at half fan speed can be estimated using the following relations (also see lines 16 and 17 of Table 16.1):

Mechanical-draft tower full speed, overall

$$L_w = 96 + 10 \log(\text{fan kW}) \quad \text{(dB)} \quad (16.16)$$

Mechanical-draft tower half speed, overall

$$L_w = 88 + 10 \log(\text{fan kW}) \quad \text{(dB)} \quad (16.17)$$

where kW is the full-speed fan power rating in kilowatts for both relations.

The above relations apply in all horizontal directions from round towers and most directions away from the inlet face of rectangular towers. For directions away from the enclosed ends of rectangular towers, the far-field noise is several decibels less than estimated above due to the effects of shielding by the solid closed ends. For a line of as many as 6–10 rectangular towers, the far-field noise from the enclosed ends could be as much as 5–6 dB less than estimated, based on omnidirectional radiation using Eqs. (16.16) and (16.17).

Reduced fan speed is a common method used to mitigate the noise from mechanical-draft cooling towers during evening and nighttime hours when excess cooling capacity may be available due to reduced ambient air temperatures. In multiple-cell tower installations, it is preferable to reduce the fan speed for all cells rather than to shut down unneeded cells; the fans' speeds should be properly selected to avoid introducing a strong acoustical beat. Another common noise control method is to install wide-chord high-efficiency fan blades that can provide the necessary fan performance at reduced speeds. In addition, mechanical-draft towers with centrifugal fans may be selected because they can sometimes be designed to produce less noise than towers with propeller fans, although at some energy cost. Mufflers have also been installed at the air inlet and discharge to reduce both the fan and water noise. However, the mufflers must be protected from the wet environment, can become coated with ice during freezing weather, and introduce an additional aerodynamic restriction that must be overcome by the fan. Barrier walls and partial enclosures have successfully shielded neighbors from cooling-tower noise; however, barrier walls that avoid excessive restrictions to the airflow often limit their acoustical benefit for protecting far-field locations.

16.5 AIR-COOLED CONDENSERS

The sound power output of air-cooled condensers (dry cooling systems) is caused primarily by the fans used to move air across the condenser with some additional noise contribution from the motors and gears. The noise is usually continuous and somewhat directive for rectangular installations. The sound power output of dry air-cooled condensers does not, of course, include the water splash noise associated with wet cooling towers.

The need to reduce water consumption is causing the increasing trend to install air-cooled condensers even at large industrial facilities. When installed at a moderate to large new combined-cycle power plant, the fans for the forced-draft air-cooled condenser can contribute approximately the same sound power output as the balance of equipment at the plant. Fortunately, low-noise fans are now readily available for air-cooled condensers that produce comparable airflow rates and 3–6 dBA less noise than do conventional fans. These low-noise fans include high-efficiency wide-chord blades operating at reduced tip speeds. For some installations, noise reductions greater than 6 dBA can be available.

To avoid water consumption, power consumption, and fan noise generation, air-cooled systems are installed in hyperbolic natural-draft towers.

The sound power produced by air-cooled condensers with low-noise fans operating at full speed can be estimated using the following relation (also see line 18 of Table 16.1):

$$\text{Air-cooled condenser full speed, overall } L_w = 84 + 10 \log(\text{fan kW}) \quad \text{(dB)} \quad (16.18)$$

where kW is the full-speed fan power rating in kilowatts.

The above relation applies in all horizontal directions from a low-noise unit; a standard unit will produce greater noise. It is suggested that fan- and air-cooled condenser manufacturers be contacted to obtain useful information regarding noise estimates and noise abatement for specific applications. See also references 6 and 7 for information about air-cooled condenser noise generation and propagation. Summary comparisons of the operating performance of wet, dry, and parallel condensing systems are provided in reference 8.

16.6 DIESEL-ENGINE-POWERED EQUIPMENT

When machinery such as compressors, generators, pumps, and construction equipment is powered by a diesel engine, it is usually the diesel engine that is the dominant noise source.

Mobile construction and coal yard equipment powered by diesel engines, such as dozers, loaders, and scrapers, often produces in-cab A-weighted noise levels as high as 95–105 dB. Methods to retrofit and reduce in-cab noise levels for several loaders and dozers have been developed, field tested, and documented in references 9 and 10. Well-designed cabs and noise control features are now often available from major manufacturers when purchasing new diesel-engine-powered mobile equipment.

A simple relation for estimating the maximum exterior overall sound power level for naturally aspirated and turbocharged diesel engines used to power equipment is provided in Eq. (16.19) (also see line 19 in Table 16.1):

$$\text{Diesel engine equipment, overall } L_w = 99 + 10 \log(\text{kW}) \quad \text{(dB)} \quad (16.19)$$

The above relation assumes that the engine has a conventional exhaust muffler in good working condition as typically provided by the engine manufacturer and further assumes that the engine is operating at rated speed and power. Noise associated with material impacts during equipment operation is not included.

Equipment used on construction sites often operates at part power. Measurements obtained at the operating equipment indicate that work-shift-long equivalent L_{eq} sound levels are therefore typically about 2–15 dB less than the maximum values provided by Eq. (16.19). It is assumed that the following values could be subtracted from the Eq. (16.19) maximum levels to obtain work-shift-long equivalent L_{eq} levels. When project-long equivalent L_{eq} levels are required, the estimated values can be further reduced to account for the percentage of time that the equipment will actually operate on the construction site [10 log(operating time/project time)]:

3–4 dB	Backhoes, rollers
5–6 dB	Dozers, graders, haulers, loaders, scrapers
7–8 dB	Air compressors, concrete batch plants, mobile cranes, trucks
12–13 dB	Derrick cranes, paging systems

When stationary diesel-engine-driven equipment is located inside masonry or metal buildings, it is necessary to consider the sound power radiated by the engine inlet, the engine casing, the engine exhaust, and the engine cooling fan as well as the driven equipment. It is also necessary to account for the attenuation expected from the building walls and openings, the inlet filter/muffler, the exhaust muffler, and mufflers to be installed at building openings that serve ventilation and cooling systems. Additional information about the propagation of noise from equipment located inside buildings and about muffler systems is provided in Chapters 9 and 17.

16.7 INDUSTRIAL FANS

The noise produced by industrial fans is caused by the dynamic interaction of the gas flow with rotating and stationary surfaces of the fan. Noise produced by shear flow is usually not considered important. Broadband fan noise results from the random aerodynamic interactions between the fan and the gas flow. The prominent discrete tones produced by centrifugal fans result from the periodic interaction of the outlet flow and the cutoff located directly downstream of the blade trailing edges. Tonal noise produced by axial-flow fans results from periodic interactions between distorted inflow and the rotor blades as well as the rotor wakes and nearby downstream surfaces, including struts and guidevanes. This tonal noise is usually most prominent at the harmonics of the passing frequency of the fan blade (number of blades times the rotation rate in revolutions per second). Detailed information about the noise associated with gas flows and fans is provided in Chapters 15 and 17.

The noise produced by new forced-draft fans can be controlled most easily with the use of ducted and muffled inlets. Sometimes the ducted inlet is extended to a location in the plant where warm air is available and additional ventilation is needed. To provide further noise reduction, acoustical insulation lagging should be considered for the inlet and outlet ducts and the fan housing. Open-inlet forced-draft fans can be located within an acoustically treated fan room that contains the fan and its noise. However, workers inside the fan room during inspections and maintenance will be exposed to high levels of fan noise and should wear ear protection.

Discharge noise of large induced-draft fans radiated from the top of the stack to residential neighbors has caused serious community noise problems. The fan and duct system design should include provisions to control this noise if residential areas are located within $\frac{1}{2}-1$ mile of the plant. Mufflers used for induced-draft fan service at boilers and furnaces should be of the dissipative, parallel-baffle, open-cavity type tuned to the harmonics of fan blade passing frequency. These mufflers are built to avoid fly-ash clogging and include erosion protection to ensure adequate long-term performance. They can usually be designed to introduce a pressure loss in the range of about 10–40 mm water gauge. Consideration should also be given to the noise attenuation that will be provided by any flue gas scrubbers, filters, or precipitators that are to be located between the induced-draft fan and its stack. Alternatively, variable-speed fan drives can be installed to reduce nighttime noise at cycling plants that shed load during nighttime hours. This has the added benefit of also reducing the power consumed by the fans and reducing the erosion rates and stresses of rotating components. When multiple fans are installed, they should be operated at the same speed to avoid acoustical beats between the fan tones. (see references 3 and 11).

When an induced-draft centrifugal fan discharges directly to a stack, it is not uncommon for the tonal fan noise to radiate relatively long distances from the top of the stack. This tonal noise is generated by the fluid dynamic interaction of the gas flow leaving rotating fan blades and the stationary cutoff of the fan. This noise propagates with little attenuation up the stack and radiates from the stack top. Some success in reducing the magnitude of this tonal noise at large industrial centrifugal fan installations has been achieved by modifying the geometry of the cutoff. The fabrication and installation of "slanted" and rounded cutoffs, as described in reference 12, have been shown to reduce the tonal noise level by about 3–10 dB with no observed loss in fan performance.

Well-designed and sized inlet and outlet ducts that properly manage the flow are essential to avoid excessive fan noise. Inlet swirl, distorted inflow, and excessive turbulence can result in high noise levels and reduced system efficiency. Useful design guidelines have been established by the Air Movement and Control Association (AMCA) for large industrial fan installations.

Procedures are provided in this section to estimate the overall sound power output of large industrial fans operating at peak efficiency conditions with undistorted inflow. Experience indicates that fans operating with highly distorted inflow or at off-peak efficiency conditions often produce sound power levels that may be 5–10 dB higher than indicated below. Centrifugal- and axial-flow fans produce less sound power when operating at low speeds and reduced working points than during full-load, high-speed operation. For part-speed operation, the sound power levels estimated below can be reduced by about 55 log(speed ratio). Additional information about fan noise generation and attenuation is provided in reference 13.

Centrifugal-Type Forced-Draft and Induced-Draft Fans with Single-Thickness, Backward-Curved or Backward-Inclined Blades or Airfoil Blades

The overall sound power level radiated from the inlet of forced-draft centrifugal fans and the outlet of induced-draft centrifugal fans (with single-thickness, backward-curved or backward-inclined blades or airfoil blades) can be estimated with the relationship provided in Eq. (16.20) (also see line 20 of Table 16.1):

$$\text{Centrifugal fan, overall } L_w = 10 + 10 \log Q + 20 \log(\text{TP}) \quad \text{(dB)} \quad (16.20)$$

where Q is fan flow rate in cubic meters of gas per minute and TP is fan total pressure rise in newtons per square meter at rated conditions.

To account for the tonal noise, 10 dB should be added to the octave bands containing the fan blade passing frequency and its second harmonic. For multiple-fan installations, the estimated sound power levels should be increased by $10 \log N$, where N is the number of identical fans.

The overall sound power level radiated from the uninsulated casing of centrifugal fans can be estimated with the relationship provided in Eq. (16.21) (also see line 21 of Table 16.1). To account for the tonal noise, 5 dB should be added to the octave band containing the fan blade passing frequency:

$$\text{Centrifugal fan casing, overall } L_w = 1 + 10 \log Q + 20 \log(\text{TP}) \quad \text{(dB)} \quad (16.21)$$

where Q and TP are as defined for Eq. (16.20).

The sound power radiated by the uninsulated discharge breaching is about 5 dB less than the sound power estimated using Eq. (16.21) for the fan casing.

The overall sound power level radiated from the casings of radial-blade centrifugal fans with ducted inlets and outlets, such as are sometimes used for gas recirculation and dust collection service, can be estimated with the relationship provided in Eq. (16.22) (also see line 22 of Table 16.1):

$$\text{Radial fan casing, overall } L_w = 13 + 10 \log Q + 20 \log(\text{TP}) \quad \text{(dB)} \quad (16.22)$$

where again Q and TP are as defined for Eq. (16.20).

To account for the tonal noise, 10 dB should be added to the octave bands containing the fan blade passing frequency and its second harmonic.

Axial-Flow Forced-Draft and Induced-Draft Fans

The overall sound power level radiated from the inlet of forced-draft axial-flow fans and the outlet of induced-draft axial-flow fans can be estimated with the following relation:

$$\text{Axial-flow fan, overall } L_w = 24 + 10 \log Q + 20 \log(\text{TP}) \quad \text{(dB)} \quad (16.23)$$

where Q and TP have been defined in Eq. (16.20).

To account for the tonal noise, 6 dB should be added to the octave band containing the fan blade passing frequency and 3 dB should be added to the octave band containing its second harmonic. For multiple-fan installations, the above estimated values should be increased by $10 \log N$, where N is the number of identical fans.

Ventilation Fans

The sound power output of industrial ventilation fans can be estimated using Eq. (16.20) for centrifugal ventilation fans or Eq. (16.23) for axial-flow ventilation fans; also see Chapter 17.

16.8 FEED PUMPS

It is common for the noise radiated by a pump motor set to be dominated by the motor noise. One exception is the relatively high-flow, high-head pumps used for boiler and reactor feedwater service in large modern power-generating stations that operate in the United States. These feed pumps are usually driven by an electric motor, an auxiliary steam turbine, or the main turbine-generator shaft. Midsize pumps generally produce broadband noise without strong tonal components. The larger pumps, however, commonly produce both broadband noise and strong midfrequency tonal noise. Feed pump noise is omnidirectional and continuous during plant operation.

A high-performance thermal-acoustical blanket insulation, described in reference 14, has been developed and evaluated at field installations specifically for use with equipment such as noisy feedwater pumps, valves, and turbines. This thermal-acoustical insulation provides far better noise attenuation than conventional blanket insulations. It is also easier to remove and reinstall during maintenance and inspections than most rigid insulations. It has been successfully used as a retrofit insulation at installed pumps, turbines, valves, and lines and has been used by equipment manufacturers at new installations requiring reduced noise levels. The use of flexible thermal-acoustical insulation to control noise avoids the mechanical, structural, and safety problems associated with the large rigid enclosures that have sometimes been used at boiler feedwater pump installations. Note that experience in Germany indicates that it is possible to design and operate boiler feedwater pumps that produce relatively low noise levels without external means for controlling the noise.

The overall sound power level radiated by boiler and reactor feedwater pumps can be estimated using the information provided below for pumps in the power range of 1–18 MWe:

Rated Power (MWe)	Overall Sound Power (dB)
1	108
2	110
4	112
6	113
9	115
12	115
15	119
18	123

Octave-band and A-weighted sound power levels for boiler and reactor feedwater pumps can be estimated by subtracting the values provided in line numbers 24 and 25 of Table 16.1 from the overall power level estimated above.

16.9 INDUSTRIAL GAS TURBINES

Industrial gas turbines are often used to provide reliable, economic power to drive large electric generators, gas compressors, pumps, or ships. A prediction scheme for the combustion noise of flightworthy gas turbine engines used in aircraft operations (turbojet, turboshaft, and turbofan engines) is provided in Chapter 15. The noise produced by industrial gas turbine installations is radiated primarily from four general source areas: the inlet of the compressor; the outlet of the power turbine; the casing and/or enclosure of the rotating components; and various auxiliary equipment, including cooling and ventilating fans, the exhaust-heat recovery steam generator, poorly gasketed or worn access openings, generators, and electric transformers. Each of the above areas should receive considerable attention if an industrial gas turbine is to be located successfully near a quiet residential area.

Operation of unmuffled industrial gas turbines of moderate size would produce sound power levels of 150–160 dB or greater. Because the resulting noise levels would be unacceptable, essentially all industrial gas turbine installations include at least some noise abatement provided by the manufacturer. Noise abatement performance requirements may call for reducing the sound power output to levels as low as 100 dB for installations located near sensitive residential areas.

Noise radiated from the compressor inlet includes high levels of both broadband and tonal noise, primarily at frequencies above 250–500 Hz. This mid- and high-frequency noise is controlled with the use of (a) a conventional parallel-baffle or tubular muffler comprised of closely spaced thin baffles and (b) inlet ducting, plenum walls, expansion joints, and access hatches that are designed to contain the noise and avoid flanking radiation.

Noise radiated by the casings of rotating components is typically contained with the use of thermal insulation laggings and close-fitting steel enclosures that include interior sound absorption and well-gasketed access doors. It is not uncommon for the enclosure to be mounted directly on the structural steel base that supports the rotating components. This encourages the transmission of structure-borne noise to the enclosure and can limit its effective acoustical performance at installations requiring high degrees of noise reduction. Off-base mounting of the enclosure can be considered for installations where noise abatement performance must be improved. In addition, many large industrial gas turbines are being installed within a turbine-generator building that provides weather protection and additional reduction of noise radiated through the enclosure from the rotating components.

Noise radiated from the power turbine outlet includes high levels of both broadband and tonal noise at low, mid, and high frequencies, with the low-frequency noise being the most difficult and expensive to control at simple-cycle installations (without heat recovery boilers) located near residential areas. The wavelength of sound at about 30 Hz is on the order of 15–20 m in the gas stream at the back end of the power turbine, thereby increasing the required size of the discharge mufflers. Furthermore, low-frequency exhaust noise has caused significant community noise problems when the sound level in the 31-Hz octave band exceeded about 70 dB at sensitive residential areas. This low-frequency exhaust noise might be best controlled by the manufacturer with improved gas management designs behind the power turbine. The outlet noise is typically controlled with the use of (a) conventional parallel-baffle or tubular mufflers that include thick acoustical treatments; (b) additional dissipative or reactive muffler elements tuned to attenuate low-frequency noise; and (c) outlet ducts, plenum chamber, expansion joints, and access hatches that include adequate acoustical treatments to avoid flanking along these paths and their radiation of excessive noise.

At combined-cycle and cogeneration plants, heat recovery steam generators (HRSGs) are installed downstream of the power turbine to absorb waste heat from the exhaust stream. The HRSG also serves to attenuate the stack-radiated turbine exhaust noise, often by as much as 15–30 dBA, depending on the configurations and sizes of the gas turbine and HRSG. The HRSG can eliminate, or at least reduce, the performance requirements of the exhaust muffler. Also, the HRSG reduces the exhaust gas stream temperature, the wavelengths of the exhaust sounds, and the required size of an exhaust muffler. It is necessary to consider and account for the noise radiated from the surfaces and various steam vents of the HRSG. It is suggested that most HRSG manufacturers can provide technical information describing the noise radiated from their equipment. See also references 15–19 for additional information.

Noise produced by auxiliary cooling and ventilation fans can be controlled effectively with the use of high-performance low-speed fan blades, mufflers, or partial enclosures, depending on specific site and installation requirements. The control of transformer noise, if required for a specific installation, is discussed in Section 16.11. The control of noise radiated from the sidewalls of heat recovery steam generators can be accomplished with large barrier walls or an enclosing building.

Combined-cycle gas turbine power plants produce relatively steady noise as well as relatively brief intermittent loud noises often associated with steam venting and steam bypass operations. Noise levels of 75–80 dBA or greater at 300 m are not uncommon; however, these intermittent sources can usually be reduced 5–10 dBA or more. Further information about such intermittent loud noise from power plants is provided in reference 20.

Sound-power-level prediction procedures are not provided here specifically for industrial gas turbines because of the rather wide range of noise control treatments available within the industry. Many manufacturers offer their machines with a variety of optional noise abatement treatments, ranging from modest to highly effective. For example, 50–100-MWe installations, with or without heat recovery boilers, are readily available with noise abatement treatments resulting in A-weighted noise levels in the range of 50–60 dB at 120 m. Installations with lower noise levels are available from certain manufacturers and can be designed by several independent professionals who have specialized in the control of noise from industrial gas turbines. Combined-cycle plants in the range of 500–600 MWe have been designed and constructed to meet 43 dBA at 300 m.

The overall sound power level of the combined-cycle equipment located within the main power building, including the combustion turbine, HRSG, steam turbine, generators, and support equipment can be estimated with the following relation (also see line 26 of Table 16.1):

Inside main power building, overall $L_w = 96 + 10 \log(A)$ (dB)   (16.24)

where A is the wall and roof area of the building in square meters.

This relationship is based on field measurements obtained near the outer wall and roof areas at numerous combined-cycle and cogeneration power plants. Equation (16.24) assumes that the interior surfaces of the building have sound absorption, with a resulting average sound level of 85 dBA at the outer wall and roof areas. For buildings without significant sound absorption treatment, the estimates should be increased by 5 dB. The interior sound level and the building shell transmission loss will determine the amount of noise that radiates to the outdoors from the main plant equipment. For additional discussion on these plants, see reference 21.

One early but important step in the design of a new gas turbine installation is the preparation of a reasonable and well-founded technical specification by the buyer or buyer's acoustical consultant that fully describes the site-specific noise requirements. Methods and procedures for preparing gas turbine procurement specifications that describe expected noise-level limits are included in ANSI/ASME B133.8-l977 (R1996). If residential neighbors are near the site, the reader is cautioned to consider fully the site-specific needs when planning control of the low-frequency exhaust noise. Several independent consultants with many years of gas turbine experience have suggested that the noise-level threshold of complaints resulting from low-frequency noise in the 31-Hz octave band is about 65–70 dB, measured at residential wood-frame homes. Higher levels of low-frequency noise can sometimes vibrate the walls and rattle the windows and doors of wood-frame homes and result in varying degrees of annoyance. Gas turbine manufacturers, however, correctly indicate that low-frequency noise control treatments are expensive and that many industrial gas turbines are operating that produce higher levels of low-frequency noise in residential areas without causing complaints.

16.10 STEAM TURBINES

Two procedures are provided to predict the sound power radiation from steam turbines; one is for the relatively small turbines in the power range of 400–8000 kW that operate at about 3600–6000 rpm and are often used to drive auxiliary equipment at a plant where steam is readily available. Another procedure is provided for large steam turbine generators in the range of about 200–1100 MWe used at central electricity-generating stations. Thermal-acoustical blanket insulation that has been developed to control noise radiated from equipment including steam turbines is discussed in Section 16.8 and reference 14.

Auxiliary Steam Turbines

The overall sound power output of auxiliary steam turbines, with common thermal insulation installed, can be estimated using Eq. (16.25) (see also line 27 of Table 16.1). This noise is considered to be omnidirectional, generally nontonal, and continuous when the driven equipment is operating:

$$\text{Auxiliary steam turbine, overall } L_w = 93 + 4 \log(\text{kW}) \quad \text{(dB)} \quad (16.25)$$

Large Steam Turbine-Generators

The overall sound power output of large steam turbine-generators can be estimated using Eq. (16.26). This includes the noise radiated by the low-pressure, intermediate-pressure, and high-pressure turbines as well as the generator and shaft-driven exciter. The turbine-generator produces both tonal and broadband noise. The tonal components produced by the generator are typically most noticeable at 60 and 120 Hz for 3600-rpm machines and 30, 60, 90, and 120 Hz for 1800-rpm machines:

$$\text{Large steam turbine-generator, overall } L_w = 113 + 4 \log(\text{MW}) \quad \text{(dB)} \quad (16.26)$$

Octave-band and A-weighted sound power levels for large steam turbine-generators can be calculated by subtracting the values provided in line 28 of Table 16.1 from the overall sound power level estimated using Eq. (16.26).

Most major manufacturers of large steam turbine-generators will provide their equipment with additional noise abatement features that reduce the noise by about 5–10 dBA, when required by a well-written purchase specification. The reverberant noise inside a turbine building can also be reduced through the proper selection of the building siding. The use of building siding with a fibrous insulation sandwiched between a perforated metal inner surface and a solid exterior surface will provide improved midfrequency sound absorption and thereby reduce the reverberant buildup of noise within the turbine building. It can also reduce the noise radiated to the outdoors.

16.11 STEAM VENTS

Atmospheric venting of large volumes of high-pressure steam is probably one of the loudest noise sources found at industrial sites. The overall sound power produced during steam line blow-outs at large central stations prior to starting a new boiler can be estimated using Eq. (16.27) (see also line 29 of Table 16.1).

This noise is broadband and only occurs for a few minutes during each blow-out for the first few weeks of boiler operation:

$$\text{Steam line blow-out, overall } L_w = 177 \text{ dB} \qquad (16.27)$$

The actual sound power level produced during the venting of high-pressure gas is, of course, related to various factors, including the conditions of the flowing gas and the geometry of the valve and pipe exit. However, the above relationship will provide a reasonable estimate of the noise associated with the blow-out of steam lines at large boilers used for utility service.

Large heavy-duty mufflers are sometimes purchased or rented that reduce the noise by about 15–30 dB during steam line blow-outs. The noise produced by the more common atmospheric vents and commonly encountered valves can be estimated using prediction procedures available from many valve manufacturers and their representatives. Many manufacturers offer valves with special low-noise trims, orifice plates, and inline mufflers that effectively reduce noise generation and radiation. The low-noise trim can also reduce the vibration and maintenance that are sometimes associated with valves used in high-pressure-drop service.

16.12 TRANSFORMERS

The noise radiated by electrical transformers is composed primarily of discrete tones at even harmonics of line frequency, that is, 120, 240, 360, ... Hz when the line frequency is 60 Hz and 100, 200, 300, ... Hz when the line frequency is 50 Hz. This tonal noise is produced by magnetostrictive forces that cause the core to vibrate at twice the electrical line frequency. The cooling fans and oil pumps at large transformers produce broadband noise when they operate; however, this noise is usually less noticeable and therefore less annoying to nearby neighbors. The tonal core noise should be considered omnidirectional and continuous while the transformer is operating. The broadband fan and pump noise occurs only during times when additional cooling is required.

The technical literature includes numerous relations and guidelines for the prediction of noise produced by transformers. Reference 22 reports the results of measurements obtained at 60 transformer banks and indicates that the space-averaged A-weighted sound level produced by the core of the average transformer (without built-in noise abatement) at an unobstructed distance of about 150 m is well represented by the relationship given in Eq. (16.28). Ninety-five percent of the A-weighted noise data reported in reference 22 lies within about ±7 dB of this relation for transformers with maximum ratings in the range of 6–1100 MVA:

Average A-weighted core sound level at 150 m,

$$L_p = 26 + 8.5 \log(\text{MVA}) \qquad (\text{dBA}) \quad (16.28)$$

where MVA is the maximum rating of the transformer in million volt-amperes.

The space-averaged sound pressure levels of the transformer core tones at 120, 240, 360, and 480 Hz at 150 m can be estimated by adding the following values to the A-weighted sound level of Eq. (16.28):

120	240	360	480
17	5	−4	−8

Another relation for transformer A-weighted sound levels versus distance, extracted from reference 23, is

Space-averaged far-field sound level $L_p = L_n - 20(\log d/S^{1/2}) - 8$ (dBA)   (16.29)

where L_n = circumferential average sound pressure levels measured at National Electrical Manufacturers Association (NEMA) close-in measurement positions (A-weighted or tonal)
S = total surface area of four sidewalls of transformer tank
d = distance from transformer tank (in units that are compatible with tank sidewall area), must be greater than $S^{1/2}$

Equation (16.29) can be used if L_p and L_n represent the A-weighted sound levels or the sound pressure levels of the discrete tones produced by the transformer tank. The octave-band sound pressure levels of the transformer core noise can be obtained directly from the sound pressure levels of the discrete tones estimated above. The octave-band sound pressure levels of the total transformer noise with the cooling fans operating can be estimated by subtracting the values provided in line 30 of Table 16.1 from the A-weighted sound level estimated with Eq. (16.29). This applies to conventional cooling-fan systems with motors in the power range of about 0.15–0.75 kW operating at about 1000–1700 rpm with two- or four-bladed propeller fans. Further information about special low-noise cooling systems should be obtained from the manufacturer. Excess attenuation should be considered (see Chapter 5) when estimating sound levels at distances beyond about 150 m.

NEMA has published standard tables of close-in noise levels for transformers; see reference 24. Experience indicates that the noise near most operating transformers is often equal to or somewhat less than the NEMA standard values, and the noise near new high-efficiency transformers can be significantly less than the NEMA values. However, the noise produced by converter transformers operating at AC-to-DC converter terminals can include discrete tones at frequencies up to about 2000 Hz and can be about 5–10 dB greater than the NEMA standard values. This additional high-frequency noise has been found to be unusually noticeable and disturbing to residential neighbors when the converter transformer is located in quiet rural-suburban areas.

Two basic methods are available for reducing the far-field noise produced by transformers. First, manufacturers are able to respond to custom noise requirements and produce transformers with reduced flux density that generate noise levels as much as 10–20 dB lower than the NEMA standard values. New high-efficiency transformers with low electrical losses usually produce core noise levels that are less than the NEMA standard values. For the quieted-design transformers, reductions of the higher frequency tones (e.g., 480 and 600 Hz) are typically 1–3 dB more than the reductions of the lower frequency tones (e.g., 120, 240, and 360 Hz). High-efficiency low-speed cooling fans and cooling-fan mufflers are also available from some manufacturers when needed for special siting applications. In some cases, the cooling fans are eliminated from the transformer tank and replaced with oil-to-water heat exchangers.

Second, barrier walls, partial enclosures, and full enclosures can be provided to shield or contain the transformer noise. They are usually fabricated with masonry blocks or metal panels. The interior surfaces of the barrier or enclosure walls should usually include sound absorption that is effective at the prominent transformer tones. Care must be used to ensure adequate strike distances and space for cooling-air flows. If the fin-fan oil coolers are to be located outside of the enclosure, some attention should be given to the structure-borne and oil-borne noise from the core that can be radiated by the coolers. Space and provisions must be provided for inspections, maintenance, and transformer removal. It is also important to ensure that the enclosure walls are structurally isolated from the transformer foundation to avoid radiation of structure-borne noise resulting transformer-induced vibration. Further information about the control of transformer noise is available from most major transformer manufacturers as well as from reference 25. And a promising passive means of canceling tonal noise radiated from transformer sidewalls by means of an attached mechanical oscillator is discussed in reference 26.

As energy efficiency becomes increasingly important, new transformers are being purchased with low electrical losses reducing operating costs. Experience at numerous large electrical facilities indicates that new high-efficiency low-loss transformers also often produce core noise levels 5–10 dB less than the NEMA standard values.

Instead of, or in addition to, reducing the noise radiated from a transformer substation, it is sometimes possible to site a transformer far from residential neighbors or within an existing noisy area, such as close to a well-traveled highway, where the ambient sounds partially mask the transformer noise.

16.13 WIND TURBINES

The installation of modern wind energy systems is growing rapidly throughout many areas of the world. Installed capacity by 2001 exceeded 30,000 MWe as the reliability and efficiency of modern wind turbines increased and the costs declined. It has been estimated that wind turbines will generate more than 10% of the world's energy, more than 1200 GW, by year 2020. Wind turbine capacity is generally rated based on rotor diameter, rotor swept area, or generator capacity in watts. Small-capacity wind turbines are generally rated at less than about 5 kW, medium turbines at about 5–300 kW, and large wind turbines at about 500 kW and above. Large turbines are now available rated at up to 1–5 MW, and with multiple turbines on a single tower, ratings can continue to increase. Wind turbine parks or farms are being developed onshore and offshore with rated capacities of 50–150 MWe and greater. The development of wind turbine rotors to operate with low wind speeds will increase the available resource area and will bring development closer to more people.

The main types of noise from wind turbines are aerodynamic, mechanical, and electrical. The principal sources of noise include the rotor blades, the gearbox, and the generator. Additional sources include the brakes, the electronics, and the tower. Earlier wind turbine installations produced both prominent tonal noise from the gear box and a strongly modulated "thumping" noise most often associated with the rotor blades passing through and interacting with the turbulent tower wake flow at downwind machines. Fortunately, the manufacturers of modern wind turbines have learned to avoid what were these most noticeable and annoying aspects of wind turbine noise.

Considerable progress in reducing noise generation and radiation continues to be made by the equipment manufacturers with ongoing technical support from research organizations (reference 27). Noise abatement for modern wind turbines includes increasingly efficient rotor blade profiles, variable-speed and variable-pitch rotors, advanced electronics, and low-noise gearboxes, as well as vibration isolation mounts and sound absorption within turbine nacelles. Noise from modern wind turbines is generally dominated by broadband rotor noise that is directly related to tip speed. Information about aeroacoustical noise generation is provided in Chapter 15.

Modern wind turbines are generally available that produce broadband noise without strong tonal components. The midfrequency aerodynamic rotor noise generally includes some noticeable time-varying amplitude modulation at the passing frequency of the rotor blade (number of blades times the rotation rate in revolutions per second). For a three-blade constant-speed turbine operating at 25 rpm, this is 1.25 Hz. For variable-rotor-speed turbines, the modulation frequency will often be in the range of 0.5–1 Hz.

The approximate A-weighted sound power level of modern medium to large wind turbines can be estimated with the relationships provided in Eq. (16.30) or (16.31), although the sound power level of some turbines will be as much as 10 dBA greater:

$$\text{Wind turbine, A-weighted sound power level} = \begin{cases} 86 + 10 \log D & \text{(dBA)} \quad (16.30) \\ 73 + 10 \log \text{kW} & \text{(dBA)} \quad (16.31) \end{cases}$$

where D is the rotor diameter in meters and kW is the rated turbine power.

Alternatively, the approximate footprint area surrounding a modern wind turbine within which the A-weighted sound level equals or exceeds 40 dBA can be estimated with the relationship provided in Eq. (16.32):

$$\text{Wind turbine 40 dBA footprint area in square meters} = 800 \times \text{kW} \quad (\text{m}^2) \quad (16.32)$$

The operating characteristics, including noise generation under specific conditions, of most modern wind turbines have been certified in accordance with national and international standards. The special requirements associated with reliable measurements and specification of wind turbine noise are defined by International Standard IEC 61400-11, 1998, "Wind Turbine Generator Systems—Part 11: Acoustic Noise Measurement Techniques."

16.14 SUMMARY

The machinery noise prediction procedures and relations presented in this chapter are based primarily on extensive field measurement data collected by the authors and their colleagues during many years of consulting projects. The results obtained when using these relations should be useful for many engineering applications. The reader is cautioned, however, that site-specific installation conditions and individual equipment characteristics can cause noise levels to be somewhat higher or lower than predicted, and detailed knowledge of these exceptions can be important for critical applications. Also, many items of equipment can be purchased with reduced noise, can be installed so as to reduce noise, or can be fitted with effective noise control treatments.

The authors continue to add to and update their library of equipment noise emission data used in the preparation of this chapter. Readers with access to new or useful data or information on equipment noise characteristics or control and wanting to share their information are encouraged to send copies to the authors.

REFERENCES

1. L. N. Miller, E. W. Wood, R. M. Hoover, A. R. Thompson, and S. L. Patterson, *Electric Power Plant Environmental Noise Guide,* rev. ed., Edison Electric Institute, Washington, DC, 1984.
2. J. D. Barnes, L. N. Miller, and E. W. Wood, *Power Plant Construction Noise Guide*, BBN Report No. 3321, Empire State Electric Energy Research Corporation, New York, 1977.
3. I. L. Vér and E. W. Wood, *Induced Draft Fan Noise Control: Technical Report and Design Guide*, BBN Report Nos. 5291 and 5367, Empire State Electric Energy Research Corporation, New York, 1984.
4. I. L. Vér, "Perforated Baffles Prevent Flow-Induced Resonances in Heat Exchangers," in *Proceedings of DAGA*, Deutschen Arbeitsgemeinschaft fur Akustik, Göttingen, Germany, 1982, pp. 531–534.
5. R. D. Blevins, *Flow-Induced Vibrations*, Krieger Publishing Company, Melbourne, FL, 2001.
6. W. E. Bradley, "Sound Radiation from Large Air-Cooled Condensers," in *Proceedings of NoiseCon 2003*, June 2003, The Institute of Noise Control Engineering of the USA, Inc., Washington, DC, 2003.
7. J. Mann, A. Fagerlund, C. DePenning, F. Catron, R. Eberhart, and D. Karczub, "Predicting the External Noise from Multiple Spargers in an Air-Cooled Condenser Power

Plant," in *Proceedings of NoiseCon 2003*, June 2003, The Institute of Noise Control Engineering of the USA, Inc., Washington, DC, 2003.
8. L. De Backer and W. Wurtz, "Why Every Air-Cooled Steam Condenser Needs a Cooling Tower," *CTI J.*, **25** (1), 52–61 (Winter 2004).
9. *Bulldozer Noise Control,* manual prepared by Bolt Beranek and Newman for the U.S. Bureau of Mines Pittsburgh Research Center, Pittsburgh, PA, May 1980.
10. *Front-End Loader Noise Control,* manual prepared by Bolt Beranek and Newman for the U.S. Bureau of Mines Pittsburgh Research Center, Pittsburgh, PA, May 1980.
11. R. M. Hoover and E. W. Wood, "The Prediction, Measurement, and Control of Power Plant Draft Fan Noise," Electric Power Research Institute Symposium on Power Plant Fans—The State of the Art, Indianapolis, IN, October 1981.
12. J. D. Barnes, E. J. Brailey, Jr., M. Reddy, and E. W. Wood, "Induced Draft Fan Noise Evaluation and Control at the New England Power Company Salem Harbor Generating Station Unit No. 3," in *Proceedings of ASCE Environmental Engineering*, Orlando, FL, July 1987.
13. *Proceedings of Fan Noise Symposiums*, Senlis, France, 1992 and 2003.
14. C. C. Thornton, C. B. Lehman, and E. W. Wood, "Flexible-Blanket Noise Control Insulation-Field Test Results and Evaluation," in *Proc. INTER-NOISE* 84, June 1984, Noise Control Foundation, Poughkeepsie, NY, 1984, pp. 405–408.
15. G. F. Hessler, "Certifying Noise Emissions from Heat Recovery Steam Generators (HRSG) in Complex Power Plant Environments," in *Proceedings of NoiseCon 2000*, December 2000, The Institute of Noise Control Engineering of the USA, Inc., Washington, DC, 2000.
16. W. E. Bradley, "Integrating Noise Controls into the Power Plant Design," in *Proceedings of NoiseCon 2000*, December 2000, The Institute of Noise Control Engineering of the USA, Inc., Washington, DC, 2000.
17. J. R. Cummins and J. B. Causey, "Issues in Predicting Performance of Ducts and Silencers Used in Power Plants," in *Proceedings of NoiseCon 2003*, June 2003, The Institute of Noise Control Engineering of the USA, Inc., Washington, DC, 2003.
18. *Noise Prediction—Guidelines for Industrial Gas Turbines*, Solar Turbines, San Diego, CA, 2004.
19. R. S. Johnson, "Recommended Octave Band Insertion Losses of Heat Recovery Steam Generators Used on Gas Turbine Exhaust Systems," in *Proceedings of NoiseCon 1994*, May 1994, Noise Control Foundation, Poughkeepsie, NY, 1984, pp. 181–184.
20. D. Mahoney and E. W. Wood, "Intermittent Loud Sounds from Power Plants," in *Proceedings of NoiseCon 2003*, June 2003, The Institute of Noise Control Engineering of the USA, Inc., Washington, DC, 2003.
21. G. F. Hessler, "Issues In HRSG System Noise," in *Proceedings of NoiseCon 1997*, June 1997, The Institute of Noise Control Engineering of the USA, Inc., Washington, DC, 1997, pp. 297–304.
22. R. L. Sawley, C. G. Gordon, and M. A. Porter, "Bonneville Power Administration Substation Noise Study," BBN Report No. 3296, Bolt Beranek and Newman, Cambridge, MA, September 1976.
23. L. Vér, D. W. Anderson, and M. M. Miles, "Characterization of Transformer Noise Emissions," Vols. 1 and 2, BBN Report No. 3305, Empire State Electric Energy Research Corporation, New York, July 1977.

24. NEMA Standards Publication No. TR 1–1993 (R2000), *"Transformers, Regulators and Reactors,"* National Electrical Manufacturers Association, Rosslyn, VA, 2000.
25. C. L. Moore, A. E. Hribar, T. R. Specht, D. W. Allen, I. L. Vér, and C. G. Gordon, "Power Transformer Noise Abatement," Report No. EP 9–14, Empire State Electric Energy Research Corporation, New York, October 1981.
26. J. A. Zapfe and E. E. Ungar, "A Passive Means for Cancellation of Structurally Radiated Tones," *J. Acoust. Soc. Am.*, **113** (1), 320–326 (January 2003).
27. P. Migliore, J. van Dam, and A. Huskey, "Acoustic Tests of Small Wind Turbines," AIAA-2004-1185, 23rd ASME Wind Energy Symposium, U.S. DOE National Renewable Energy Laboratory, Reno, NV, January 2004.

CHAPTER 17

Noise Control in Heating, Ventilating, and Air Conditioning Systems

ALAN T. FRY

Consultant
Colchester, Essex, United Kingdom

DOUGLAS H. STURZ

Acentech, Inc.
Cambridge, Massachusetts

This chapter provides guidelines, recommendations, and design tools useful in assessing and controlling noise and vibration stemming from mechanical systems serving buildings. Discussed are the transmission of noise along duct systems, airflow velocities at various points in duct systems that are consistent with particular noise goals, fan noise, terminal box noise, and special design features for especially quiet heating, ventilating, and air conditioning (HVAC) systems. Sound isolation for mechanical rooms and special consideration for the application of vibration isolation for mechanical equipment in buildings are also addressed. Further discussion and additional detailed design information can be found in references 1–5, and in Chapters 9–15 of this book.

17.1 DUCT-BORNE NOISE TRANSMISSION

It is of significant practical interest to predict the noise that transmits via a ducted system from a source to a receiver space to achieve a desired noise goal. This is typically done by starting with the noise of a known source (typically and primarily a fan) and subtracting from this the attenuation provided by each of the various duct elements that the noise encounters as it propagates along the duct path. Thus, from a known noise source level, one can predict the noise level that reaches the occupant of a space and can engineer any special attenuation treatments that may need to be incorporated in the duct system to help achieve a desired result. Alternatively, one can work backward from a desired noise goal in the receiving space to determine the permissible sound power level of a fan serving the system (or other source in the system). There are many acoustical complexities that are difficult to model precisely without exhaustive knowledge of the duct construction and the nature of the sound field entering a duct element, and such details are generally not well known. The prediction procedure is really a guide to make decisions in order to achieve designs that are in the correct neighborhood of the goal, and generally the procedure is a good guide in the moderately lower frequency range that is of greatest interest for ventilation systems. There are many aspects of ducted systems which are strongly frequency dependent in the higher frequency range, and the precision of the calculations may be questionable for extensive systems in this frequency range. Fortunately, in designing noise control for ducted systems, when the lower frequencies are suitably controlled, the predicted noise levels at higher frequencies are dramatically lower than the goal and there is no need to make precise predictions in this higher frequency range. The following presents some of the data necessary to assess attenuation in the path of sound propagation from source to receiver. Considerations for some relevant noise sources are presented later.

Sound Attenuation in Straight Ducts

Even the attenuation of sound propagating in simple, straight ducts is quite complex. The sound attenuation in straight ducts of uniform cross section and wall construction is usually given in attenuation per unit length, decibels per meter or decibels per feet, assuming that every unit of duct length provides the same amount of attenuation. This is a considerable simplification because the attenuation is a function of the character of the sound field entering the duct section and the character of the sound field is constantly changing along the length of even a straight-duct run. At frequencies high enough for the duct to support higher order modes, these higher order modes are much more rapidly attenuated by duct attenuation treatments than the plane-wave fundamental mode. Attenuation of sound along a duct is a function of the sound dissipation when the sound wave interacts with the walls of the duct and this is determined by the impedance of the duct wall. Sound attenuation in lined ducts and silencers is treated in detail in Chapter 9. At high frequencies, energy loss due to sound transmission through the bare sheet metal duct walls yields only very little sound attenuation for the duct path. However, with sound-absorptive lining of the walls, high-frequency sound attenuation along the ducts can be quite high. At low frequencies, a substantial portion of the sound attenuation along the duct path is provided by energy transmission through the sheet metal walls of the duct—breakout. Form, stiffness, and surface weight control the low-frequency sound attenuation. Consequently, unlined form-stiff round ducts yield much less sound attenuation (and also much higher breakout sound transmission loss) than do ducts with rectangular or oval cross section and with large aspect ratio. The sound attenuation values represented herein are for typical duct construction within the range of constructions typically allowed by the Sheet Metal and Air Conditioning National Association (SMACNA) for low-pressure ductwork.

Estimation of the attenuation provided by unlined and lined ducts of round and rectangular duct constructions typically found in the field is presented in Tables 17.1–17.6.

TABLE 17.1 Sound Attenuation in Unlined Rectangular Sheet Metal Ducts

Duct Size, in. × in.	P/A, ft^{-1}	Attenuation by Octave-Band Center Frequency, dB/ft			
		63 Hz	125 Hz	250 Hz	>250 Hz
6 × 6	8.0	0.30	0.20	0.10	0.10
12 × 12	4.0	0.35	0.20	0.10	0.06
12 × 24	3.0	0.40	0.20	0.10	0.05
24 × 24	2.0	0.25	0.20	0.10	0.03
48 × 48	1.0	0.15	0.10	0.07	0.02
72 × 72	0.7	0.10	0.10	0.05	0.02

Source: 1999 ASHRAE *Applications Handbook*, © American Society of Heating, Refrigerating and Air Conditioning Engineers, Inc., www.ashrae.org.

Sound Attenuation by Duct Divisions

When sound propagating along a duct encounters a duct branch, the acoustical energy divides into the two branches according to the fraction of the area that each represents to the total area leaving the division. This is probably a reasonable approximation of what happens at lower frequencies, but at higher frequencies there are likely to be directional effects for which this simple model does not account. Fortunately, for most practical HVAC noise control concerns, the low-frequency range is of the greatest interest. The attenuation in decibels that occurs at a duct division is given by

$$\text{Attenuation} = 10 \log \left(\frac{A_1}{A_T} \right) \quad \text{dB} \tag{17.1}$$

where A_1 is the area of the duct leaving the division on the path being studied and A_T is the total area of all branches leaving the division.

Sound Attenuation by Duct Cross-Sectional Area Changes

Sound propagating along a duct is reflected back if it encounters a sudden cross-sectional area change. The strength of reflection at low frequencies, where only the fundamental plane wave can propagate in the duct, depends only on the ratio of the cross-sectional areas. Whether the sound is coming from the duct of the larger or smaller cross-sectional area, the resulting sound attenuation is given by

$$\text{Attenuation} = 10 \log \left\{ 0.25 \left[\left(\frac{A_1}{A_2} \right)^{0.5} + \left(\frac{A_2}{A_1} \right)^{0.5} \right]^2 \right\} \quad \text{dB} \tag{17.2}$$

TABLE 17.2 Insertion loss for Rectangular Sheet Metal Ducts with 1-in. Fiberglass Lining[a]

Dimensions, in. × in.	Insertion Loss by Octave Band Center Frequency, dB/ft					
	125 Hz	250 Hz	500 Hz	1000 Hz	2000 Hz	4000 Hz
6 × 6	0.6	1.5	2.7	5.8	7.4	4.3
6 × 10	0.5	1.2	2.4	5.1	6.1	3.7
6 × 12	0.5	1.2	2.3	5.0	5.8	3.6
6 × 18	0.5	1.0	2.2	4.7	5.2	3.3
8 × 8	0.5	1.2	2.3	5.0	5.8	3.6
8 × 12	0.4	1.0	2.1	4.5	4.9	3.2
8 × 16	0.4	0.9	2.0	4.3	4.5	3.0
8 × 24	0.4	0.8	1.9	4.0	4.1	2.8
10 × 10	0.4	1.0	2.1	4.4	4.7	3.1
10 × 16	0.4	0.8	1.9	4.0	4.0	2.7
10 × 20	0.3	0.8	1.8	3.8	3.7	2.6
10 × 30	0.3	0.7	1.7	3.6	3.3	2.4
12 × 12	0.4	0.8	1.9	4.0	4.1	2.8
12 × 18	0.3	0.7	1.7	3.7	3.5	2.5
12 × 24	0.3	0.6	1.7	3.5	3.2	2.3
12 × 36	0.3	0.6	1.6	3.3	2.9	2.2
15 × 15	0.3	0.7	1.7	3.6	3.3	2.4
15 × 22	0.3	0.6	1.6	3.3	2.9	2.2
15 × 30	0.3	0.5	1.5	3.1	2.6	2.0
15 × 45	0.2	0.5	1.4	2.9	2.4	1.9
18 × 18	0.3	0.6	1.6	3.3	2.9	2.2
18 × 28	0.2	0.5	1.4	3.0	2.4	1.9
18 × 36	0.2	0.5	1.4	2.8	2.2	1.8
18 × 54	0.2	0.4	1.3	2.7	2.0	1.7
24 × 24	0.2	0.5	1.4	2.8	2.2	1.8
24 × 36	0.2	0.4	1.2	2.6	1.9	1.6
24 × 48	0.2	0.4	1.2	2.4	1.7	1.5
24 × 72	0.2	0.3	1.1	2.3	1.6	1.4
30 × 30	0.2	0.4	1.2	2.5	1.8	1.6
30 × 45	0.2	0.3	1.1	2.3	1.6	1.4
30 × 60	0.2	0.3	1.1	2.2	1.4	1.3
30 × 90	0.1	0.3	1.0	2.1	1.3	1.2
36 × 36	0.2	0.3	1.1	2.3	1.6	1.4
36 × 54	0.1	0.3	1.0	2.1	1.3	1.2
36 × 72	0.1	0.3	1.0	2.0	1.2	1.2
36 × 108	0.1	0.2	0.9	1.9	1.1	1.1
42 × 42	0.2	0.3	1.0	2.1	1.4	1.3
42 × 64	0.1	0.3	0.9	1.9	1.2	1.1
42 × 84	0.1	0.2	0.9	1.8	1.1	1.1
42 × 126	0.1	0.2	0.9	1.7	1.0	1.0
48 × 48	0.1	0.3	1.0	2.0	1.2	1.2
48 × 72	0.1	0.2	0.9	1.8	1.0	1.0
48 × 96	0.1	0.2	0.8	1.7	1.0	1.0
48 × 144	0.1	0.2	0.8	1.6	0.9	0.9

[a] Add to attenuation of bare sheet metal ducts.
Source: 1999 ASHRAE *Applications Handbook*, © American Society of Heating, Refrigerating and Air Conditioning Engineers, Inc., www.ashrae.org.

TABLE 17.3 Insertion loss for Rectangular Sheet Metal Ducts with 2-in. Fiberglass Lining[a]

Dimensions, in. × in.	Insertion Loss by Octave-Band Center Frequency, dB/ft					
	125 Hz	250 Hz	500 Hz	1000 Hz	2000 Hz	4000 Hz
6 × 6	0.8	2.9	4.9	7.2	7.4	4.3
6 × 10	0.7	2.4	4.4	6.4	6.1	3.7
6 × 12	0.6	2.3	4.2	6.2	5.8	3.6
6 × 18	0.6	2.1	4.0	5.8	5.2	3.3
8 × 8	0.6	2.3	4.2	6.2	5.8	3.6
8 × 12	0.6	1.9	3.9	5.6	4.9	3.2
8 × 16	0.5	1.8	3.7	5.4	4.5	3.0
8 × 24	0.5	1.6	3.5	5.0	4.1	2.8
10 × 10	0.6	1.9	3.8	5.5	4.7	3.1
10 × 16	0.5	1.6	3.4	5.0	4.0	2.7
10 × 20	0.4	1.5	3.3	4.8	3.7	2.6
10 × 30	0.4	1.3	3.1	4.5	3.3	2.4
12 × 12	0.5	1.6	3.5	5.0	4.1	2.8
12 × 18	0.4	1.4	3.2	4.6	3.5	2.5
12 × 24	0.4	1.3	3.0	4.3	3.2	2.3
12 × 36	0.4	1.2	2.9	4.1	2.9	2.2
15 × 15	0.4	1.3	3.1	4.5	3.3	2.4
15 × 22	0.4	1.2	2.9	4.1	2.9	2.2
15 × 30	0.3	1.1	2.7	3.9	2.6	2.0
15 × 45	0.3	1.0	2.6	3.6	2.4	1.9
18 × 18	0.4	1.2	2.9	4.1	2.9	2.2
18 × 28	0.3	1.0	2.6	3.7	2.4	1.9
18 × 36	0.3	0.9	2.5	3.5	2.2	1.8
18 × 54	0.3	0.8	2.3	3.3	2.0	1.7
24 × 24	0.3	0.9	2.5	3.5	2.2	1.8
24 × 36	0.3	0.8	2.3	3.2	1.9	1.6
24 × 48	0.2	0.7	2.2	3.0	1.7	1.5
24 × 72	0.2	0.7	2.0	2.9	1.6	1.4
30 × 30	0.2	0.8	2.2	3.1	1.8	1.6
30 × 45	0.2	0.7	2.0	2.9	1.6	1.4
30 × 60	0.2	0.6	1.9	2.7	1.4	1.3
30 × 90	0.2	0.5	1.8	2.6	1.3	1.2
36 × 36	0.2	0.7	2.0	2.9	1.6	1.4
36 × 54	0.2	0.6	1.9	2.6	1.3	1.2
36 × 72	0.2	0.5	1.8	2.5	1.2	1.2
36 × 108	0.2	0.5	1.7	2.3	1.1	1.1
42 × 42	0.2	0.6	1.9	2.6	1.4	1.3
42 × 64	0.2	0.5	1.7	2.4	1.2	1.1
42 × 84	0.2	0.5	1.6	2.3	1.1	1.1
42 × 126	0.1	0.4	1.6	2.2	1.0	1.0
48 × 48	0.2	0.5	1.8	2.5	1.2	1.2
48 × 72	0.2	0.4	1.6	2.3	1.0	1.0
48 × 96	0.1	0.4	1.5	2.1	1.0	1.0
48 × 144	0.1	0.4	1.5	2.0	0.9	0.9

[a] Add to attenuation of bare sheet metal ducts.
Source: 1999 ASHRAE *Applications Handbook*, © American Society of Heating, Refrigerating and Air Conditioning Engineers, Inc., www.ashrae.org.

TABLE 17.4 Sound Attenuation in Straight Unlined Round Ducts

Diameter, in.	Attenuation by Octave-Band Center Frequency, dB/ft						
	63 Hz	125 Hz	250 Hz	500 Hz	1000 Hz	2000 Hz	4000 Hz
$D \leq 7$	0.03	0.03	0.05	0.05	0.10	0.10	0.10
$7 < D \leq 15$	0.03	0.03	0.03	0.05	0.07	0.07	0.07
$15 < D \leq 30$	0.02	0.02	0.02	0.03	0.05	0.05	0.05
$30 < D \leq 60$	0.01	0.01	0.01	0.02	0.02	0.02	0.02

Source: 1999 ASHRAE *Applications Handbook*, © American Society of Heating, Refrigerating and Air Conditioning Engineers, Inc., www.ashrae.org.

TABLE 17.5 Insertion Loss for Acoustically Lined Round Ducts with 1-in. Lining[a]

Diameter, in.	Insertion Loss by Octave-Band Center Frequency, dB/ft							
	63 Hz	125 Hz	250 Hz	500 Hz	1000 Hz	2000 Hz	4000 Hz	8000 Hz
6	0.38	0.59	0.93	1.53	2.17	2.31	2.04	1.26
8	0.32	0.54	0.89	1.50	2.19	2.17	1.83	1.18
10	0.27	0.50	0.85	1.48	2.20	2.04	1.64	1.12
12	0.23	0.46	0.81	1.45	2.18	1.91	1.48	1.05
14	0.19	0.42	0.77	1.43	2.14	1.79	1.34	1.00
16	0.16	0.38	0.73	1.40	2.08	1.67	1.21	0.95
18	0.13	0.35	0.69	1.37	2.01	1.56	1.10	0.90
20	0.11	0.31	0.65	1.34	1.92	1.45	1.00	0.87
22	0.08	0.28	0.61	1.31	1.82	1.34	0.92	0.83
24	0.07	0.25	0.57	1.28	1.71	1.24	0.85	0.80
26	0.05	0.22	0.53	1.24	1.59	1.14	0.79	0.77
28	0.03	0.19	0.49	1.20	1.46	1.04	0.74	0.74
30	0.02	0.16	0.45	1.16	1.33	0.95	0.69	0.71
32	0.01	0.14	0.42	1.12	1.20	0.87	0.66	0.69
34	0	0.11	0.38	1.07	1.07	0.79	0.63	0.66
36	0	0.08	0.35	1.02	0.93	0.71	0.60	0.64
38	0	0.06	0.31	0.96	0.80	0.64	0.58	0.61
40	0	0.03	0.28	0.91	0.68	0.57	0.55	0.58
42	0	0.01	0.25	0.84	0.56	0.50	0.53	0.55
44	0	0	0.23	0.78	0.45	0.44	0.51	0.52
46	0	0	0.20	0.71	0.35	0.39	0.48	0.48
48	0	0	0.18	0.63	0.26	0.34	0.45	0.44
50	0	0	0.15	0.55	0.19	0.29	0.41	0.40
52	0	0	0.14	0.46	0.13	0.25	0.37	0.34
54	0	0	0.12	0.37	0.09	0.22	0.31	0.29
56	0	0	0.10	0.28	0.08	0.18	0.25	0.22
58	0	0	0.09	0.17	0.08	0.16	0.18	0.15
60	0	0	0.08	0.06	0.10	0.14	0.09	0.07

[a] Add to attenuation of bare sheet metal ducts.
Source: 1999 ASHRAE *Applications Handbook*, © American Society of Heating, Refrigerating and Air Conditioning Engineers, Inc., www.ashrae.org.

TABLE 17.6 Insertion Loss for Acoustically Lined Round Ducts with 2-in. Lining[a]

	Insertion Loss by Octave-Band Center Frequency, dB/ft							
Diameter, in.	63 Hz	125 Hz	250 Hz	500 Hz	1000 Hz	2000 Hz	4000 Hz	8000 Hz
6	0.56	0.80	1.37	2.25	2.17	2.31	2.04	1.26
8	0.51	0.75	1.33	2.23	2.19	2.17	1.83	1.18
10	0.46	0.71	1.29	2.20	2.20	2.04	1.64	1.12
12	0.42	0.67	1.25	2.18	2.18	1.91	1.48	1.05
14	0.38	0.63	1.21	2.15	2.14	1.79	1.34	1.00
16	0.35	0.59	1.17	2.12	2.08	1.67	1.21	0.95
18	0.32	0.56	1.13	2.10	2.01	1.56	1.10	0.90
20	0.29	0.52	1.09	2.07	1.92	1.45	1.00	0.87
22	0.27	0.49	1.05	2.03	1.82	1.34	0.92	0.83
24	0.25	0.46	1.01	2.00	1.71	1.24	0.85	0.80
26	0.24	0.43	0.97	1.96	1.59	1.14	0.79	0.77
28	0.22	0.40	0.93	1.93	1.46	1.04	0.74	0.74
30	0.21	0.37	0.90	1.88	1.33	0.95	0.69	0.71
32	0.20	0.34	0.86	1.84	1.20	0.87	0.66	0.69
34	0.19	0.32	0.82	1.79	1.07	0.79	0.63	0.66
36	0.18	0.29	0.79	1.74	0.93	0.71	0.60	0.64
38	0.17	0.27	0.76	1.69	0.80	0.64	0.58	0.61
40	0.16	0.24	0.73	1.63	0.68	0.57	0.55	0.58
42	0.15	0.22	0.70	1.57	0.56	0.50	0.53	0.55
44	0.13	0.20	0.67	1.50	0.45	0.44	0.51	0.52
46	0.12	0.17	0.64	1.43	0.35	0.39	0.48	0.48
48	0.11	0.15	0.62	1.36	0.26	0.34	0.45	0.44
50	0.09	0.12	0.60	1.28	0.19	0.29	0.41	0.40
52	0.07	0.10	0.58	1.19	0.13	0.25	0.37	0.34
54	0.05	0.08	0.56	1.10	0.09	0.22	0.31	0.29
56	0.02	0.05	0.55	1.00	0.08	0.18	0.25	0.22
58	0	0.03	0.53	0.90	0.08	0.16	0.18	0.15
60	0	0	0.53	0.79	0.10	0.14	0.09	0.07

[a] Add to attenuation of bare sheet metal ducts.
Source: 1999 ASHRAE *Applications Handbook*, © American Society of Heating, Refrigerating and Air Conditioning Engineers, Inc., www.ashrae.org.

Equation (17.2) yields an attenuation of only 0.5 dB for a cross-sectional area change of $A_1/A_2 = 2$ or $A_2/A_1 = 0.5$ and that of 2 dB for a 1:4 or 4:1 change. Since large abrupt changes in duct cross-sectional area do not occur very often for aerodynamic reasons, the attenuation attributable to duct cross-sectional area changes usually encountered in HVAC duct design is very small.

Sound Attenuation by Elbows

When sound propagating along a duct encounters an elbow, some of the sound is reflected back in the direction from which it was coming, some is dissipated, and some continues to propagate along the duct path. The fraction of sound energy that is reflected or dissipated determines the attenuation. The attenuation is a function of frequency and size of the elbow. The attenuation is also impacted by the type of elbow (mitered, mitered with vanes, and radiused) and the presence (or absence) of sound-absorbing lining in the duct (or on the turning vanes). Mitered elbows with vanes and radiused elbows tend to transmit more higher frequency energy around the bend than mitered elbows without vanes and so they provide less attenuation for the duct path. However, they yield lower pressure drop and therefore are commonly used. Tables 17.7–17.9 present the estimated sound attenuation for various elbow types.

TABLE 17.7 Insertion Loss of Unlined and Lined Rectangular Elbows without Turning Vanes

	Insertion Loss, dB	
	Unlined Elbows	Lined Elbows
$fw < 1.9$	0	0
$1.9 \leq fw < 3.8$	1	1
$3.8 \leq fw < 7.5$	5	6
$7.5 \leq fw < 15$	8	11
$15 \leq fw < 30$	4	10
$fw > 30$	3	10

Note: $fw = f \times w$, where f = center frequency, kHz, and w = width, in.
Source: 1999 ASHRAE *Applications Handbook*, © American Society of Heating, Refrigerating and Air Conditioning Engineers, Inc., www.ashrae.org.

TABLE 17.8 Insertion Loss of Unlined and Lined Rectangular Elbows with Turning Vanes

	Insertion Loss, dB	
	Unlined Elbows	Lined Elbows
$fw < 1.9$	0	0
$1.9 \leq fw < 3.8$	1	1
$3.8 \leq fw < 7.5$	4	4
$7.5 \leq fw < 15$	6	7
$fw > 15$	4	7

Note: $fw = f \times w$, where f = center frequency, kHz, and w = width, in.
Source: 1999 ASHRAE *Applications Handbook*, © American Society of Heating, Refrigerating and Air Conditioning Engineers, Inc., www.ashrae.org.

TABLE 17.9 Insertion Loss of Unlined Round Elbows

	Insertion Loss, dB
$fw < 1.9$	0
$1.9 \leq fw < 3.8$	1
$3.8 \leq fw < 7.5$	2
$fw > 7.5$	3

Note: $fw = f \times w$, where $f =$ center frequency, kHz, and $w =$ width, in.
Source: 1999 ASHRAE *Applications Handbook*, © American Society of Heating, Refrigerating and Air Conditioning Engineers, Inc., www.ashrae.org.

Prefabricated Sound Attenuators

Prefabricated sound-attenuating devices (duct silencers) are available from a variety of manufacturers in a variety of styles, configurations, and constructions to fit virtually every HVAC system application and to meet every sound attenuation requirement. The sound-absorptive media may be conventional dissipative materials such as glass fiber, mineral wool, or nylon wool covered by a perforated metal facing sheet to provide physical protection for the fibrous sound-absorbing material from erosion by the turbulent flow. For low-velocity flows the sound-dissipating material may be protected only by a thin surface layer of flow-resistive facing such as applied on the surface of duct liners. For special applications, to prevent any fibrous material from getting into the air stream, the fill material can be sealed in thin plastic bagging. This typically will slightly increase the acoustical attenuation at specific low and midfrequencies and diminish it substantially at high frequencies compared to unfaced fill. Special construction details are typically required to prevent the film from sealing the holes in the perforated metal protective facing and to avoid chafing of the film on sharp edges formed on the back of the perforated metal in the punching process. Detailed information on the design and prediction of dissipative silencers is given in Chapter 9. There are also silencers without traditional dissipative fill but with special acoustically reactive surfaces lining the air channels which remove acoustical energy from the air stream.

Basic silencers have a straight-through air path with aerodynamic inlet and discharge geometry to help minimize pressure losses which are greatest at the flow transition points. Silencers are also available with special flow configurations for special applications. For HVAC system applications, elbow configuration silencers are particularly notable since these can help to avoid difficult flow conditions that could otherwise exist or even be caused by aerodynamically poor application of a straight silencer.

Manufacturers' information should be used for the prediction of the sound attenuation, pressure drop, and flow-generated noise. For a given application there are typically several silencers that might be able to provide the desired acoustical performance. The noise control and HVAC design engineers must select one that also yields a suitably low pressure drop and trade-offs will need to be considered between the length and the resistance class of the silencer. For most typical HVAC system applications the pressure drop across the silencer should be limited to about 0.30 in. water gauge (wg). The pressure loss can be higher, but the designer needs to be sure the extra pressure loss will not have undesirable impacts on the system. The energy consumed to overcome the pressure drop of the silencer should be considered in deciding what pressure loss is acceptable. Over the lifetime of the system, significant energy cost savings will result from having a silencer that has a particularly low pressure drop. The savings may be much more than the extra initial cost of the lower pressure drop model. Silencers should not be placed in the duct just upstream of a fan to avoid having the turbulent flow from the tail end of the baffles interact with the fan blade, which will increase the fan's broadband and blade passage tone noise.

Ideally, the airflow entering and leaving the silencer should be straight and smooth to approach the conditions under which the sound attenuation, pressure drop, and flow noise of the silencer are tested in accordance with ASTM E-477 or ISO 7235. In the field, such ideal conditions seldom occur, and there is some degree of excess pressure drop beyond the catalogued value for the application due to system effects. The system designer needs to consider these when making pressure drop predictions. To the extent that the silencers are placed in positions where the inflow is distorted and turbulent, there will be extra pressure drop. Figure 17.1 presents some common flow configurations and the associated extra pressure drop multiplier to be applied to the catalogued silencer pressure loss.

Because of the high velocity flows through the narrow silencer channels and the imperfect conversion of velocity pressure to static pressure at the discharge, turbulence is generated in the flow through silencers and this creates flow noise. The manufacturers typically tabulate the flow-generated noise of their silencers at various face velocities. The total flow-generated sound power is a function of the face velocity and the area of the silencers, and the noise generated by the silencer will typically be published for various face velocities and for a specific cross-sectional area. This data must be adjusted for the cross-sectional size of the silencer actually used when one predicts the flow-generated noise by adding $10 \log A_1/A_0$ to the published data (where A_1 is the actual silencer area and A_0 is the area that is the basis for the presented data, in consistent units). The noise generated by the silencer becomes another source of noise that propagates along the duct and needs to be addressed. The closer the silencer is placed to the receiving space and the quieter the receiving space, the more care needs to be taken in controlling flow-generated noise because there is less opportunity to attenuate it before it reaches the receiving space. For duct systems serving spaces where the noise goal is moderate (perhaps NC-35 to NC-40; NC = noise criterion) silencers located near the occupied space (such as on the room-side of a terminal box) typically need to be sized for less about 0.10 in. wg pressure drop. Where the noise goal is more stringent, the pressure loss through the silencer

DUCT-BORNE NOISE TRANSMISSION **695**

DUCT ELEMENT	SILENCER SYSTEM EFFECT FACTOR DUCT ELEMENT ON...	
	SILENCER INLET	SILENCER DISCHARGE
TRANSITIONS		
$7\tfrac{1}{2}$ degrees per side		
Distance of transition from silencer		
$D = 1$	1.0	1.0
$D = 2$	1.1	1.1
$D = 3$	1.2	1.1
25 degrees per side		
Distance of transition from silencer		
$D = 1$	1.3	1.1
$D = 2$	1.6	1.1
$D = 3$	1.8	1.1
45 degrees per side		
Distance of transition from silencer		
$D = 1$	1.7	1.1
$D = 2$	1.9	1.1
$D = 3$	2.0	1.1
ELBOW-RADIUS TYPE		
Distance of radius elbow from silencer		
$D = 0$	1.2	1.4
$D = 1$	1.1	1.2
ELBOW-MITERED TYPE WITH SHORT TURNING VANES		
Distance of mitered elbow from silencer		
$D = 0$	1.2	1.1
$D = 1$	1.2	1.1
$D = 2$	1.2	1.2
ELBOW-MITERED TYPE WITH NO TURNING VANES		
Distance of mitered elbow from silencer		
$D = 0$	1.2	2.9
$D = 1$	1.0	1.8
$D = 2$	1.1	1.4
ABRUPT ENTRY OR EXIT		
Smooth inlet or discharge		
Distance of entry or exit from silencer		
$D = 0$	1.1	1.8
$D = 1$	1.0	1.4
$D = 2$	1.0	1.1
$D = 3$	1.0	1.0
ABRUPT ENTRY OR EXIT		
Sharp inlet or discharge		
Distance of entry or exit from silencer		
$D = 0$	1.2	2.0
$D = 1$	1.1	1.5
$D = 2$	1.0	1.2
$D = 3$	1.0	1.0
CENTRIFUGAL FAN		
Distance of centrifugal fan from silencer		
$D = 0$	1.5	2.0
$D = 1$	1.2	1.7
$D = 2$	1.1	1.5
$D = 3$	1.0	1.2
AXIAL FAN		
Distance of axial fan from silencer		
$D = 0$	1.5	2.0
$D = 1$	1.2	1.7
$D = 2$	1.1	1.5
$D = 3$	1.0	1.2

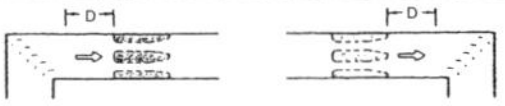

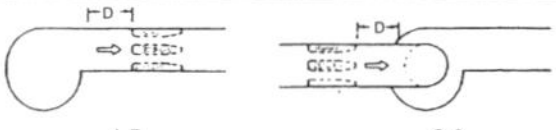

D is the diameter of round duct or equivalent diameter of rectangular duct.

FIGURE 17.1 Factor by which catalogue silencer pressure loss is multiplied for various applications. (Reproduced with permission from Vibro-Acoustics.)

needs to be commensurately lower. When silencers are located near the fans, at the beginning of systems, the noise of the fan on the building side of the silencer will often still be substantially higher than the flow-generated noise of the silencer (if selected for reasonable pressure loss), so that flow noise is not of concern. The lower the noise goal, the more carefully this needs to be checked.

Poorly selected and positioned silencers, which produce excessive pressure drop, can cause more noise problems than they solve.

Sound Attenuation by Plenums

Plenums in duct systems can often provide significant sound attenuation for the ducted path. The extent of the attenuation depends upon the size of the plenum compared to the connecting ducts, the orientation of the inlets and outlets, and the absorptivity of the inner walls. Equations (17.3)–(17.5) as applied to Fig. 17.2 provide a method for estimating the attenuation provided by plenums:

$$\text{TL} = -10 \log_{10} \left[S_{\text{out}} \left(\frac{Q \cos \theta}{4\pi r^2} + \frac{1 - \alpha_A}{S \alpha_A} \right) \right] \quad (17.3)$$

where (refer to Fig. 17.2)
where TL = transmission loss, dB
S_{out} = area of outlet section of plenum, ft^2
S = total inside surface area of plenum minus inlet and outlet area, ft^2
r = distance between centers of inlet and outlet sections of plenum, ft
Q = directivity factor, which may be taken as 4
α_A = average absorption coefficient of the plenum lining
θ = angle of vector representing r to long axis l of plenum

The average absorption coefficient α_A of the plenum lining is

$$\alpha_A = \frac{S_1 \alpha_1 + S_2 \alpha_2}{S} \quad (17.4)$$

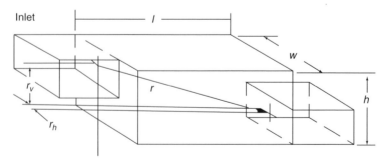

FIGURE 17.2 Schematic of plenum chamber. *Source*: 1999 ASHRAE *Applications Handbook*, © American Society of Heating, Refrigerating and Air Conditioning Engineers, Inc., www.ashrae.org.

where α_1 = sound absorption coefficient of any bare or unlined inside surfaces of plenum
S_1 = surface area of bare or unlined inside surface of plenum, ft^2
α_2 = sound absorption coefficient of any acoustically lined inside surfaces of plenum
S_2 = surface area of acoustically lined inside surface of plenum, ft^2

The value of cos θ is obtained from

$$\cos\theta = \frac{l}{r} = \frac{l}{(l^2 + r_v^2 + r_h^2)^{0.5}} \tag{17.5}$$

where l = length of plenum, ft
r_v = vertical offset between axes of plenum inlet and outlet, ft
r_h = horizontal offset between axes of plenum inlet and outlet, ft

The precision of plenum insertion loss predictions is generally not considered to be very good, and research is being conducted currently to improve the accuracy of the predicted attenuation of these duct elements. The results of the above equations are only valid where the wavelength of sound is small compared to the characteristic dimensions of the plenum. Equation (17.3) is not valid below the cutoff frequency of the entering and exiting ducts, given as

$$f_{co} = \frac{c_0}{2a} \quad \text{for rectangular ducts}$$

and

$$f_{co} = \frac{0.586 c_0}{d} \quad \text{for circular ducts}$$

where f_{co} = cutoff frequency, Hz
c_0 = speed of sound in air, m/s
a = larger cross-sectional dimension of duct, m
d = diameter of duct, m

End-Reflection Loss

When sound propagating along ducts reaches an abrupt change in cross-sectional area (such as the termination of a duct system at a diffuser or grille) and there is no disturbance of the sound field as it makes the transition, some low-frequency sound reflects back from the point of the abrupt change and does not propagate on toward the receivers of concern. The end-reflection loss is a function of the size of the termination compared with the acoustical wavelength and the specific location of the duct termination within the room. Typical values of end reflections are given in Table 17.10. Note that these are for idealized conditions. If there is a grille or diffuser at the termination of the duct system or if the grille or diffuser is not in the middle of a large plane (like the ceiling) but is in a two- or three-dimensional corner of the room, the end reflection will be lower. For each factor that deviates

TABLE 17.10 Duct End-Reflection Loss

Duct Diameter, in.	End Reflection Loss by Octave-Band Center Frequency, dB					
	63 Hz	125 Hz	250 Hz	500 Hz	1000 Hz	2000 Hz
Duct Terminated in Free Space						
6	20	14	9	5	2	1
8	18	12	7	3	1	0
10	16	11	6	2	1	0
12	14	9	5	2	1	0
16	12	7	3	1	0	0
20	10	6	2	1	0	0
24	9	5	2	1	0	0
28	8	4	1	0	0	0
32	7	3	1	0	0	0
36	6	3	1	0	0	0
48	5	2	1	0	0	0
72	3	1	0	0	0	0
Duct Terminated Flush with Surface						
6	18	13	8	4	1	
8	16	11	6	2	1	
10	14	9	5	2	1	
12	13	8	4	1	0	
16	10	6	2	1	0	
20	9	5	2	1	0	
24	8	4	1	0	0	
28	7	3	1	0	0	
32	6	2	1	0	0	
36	5	2	1	0	0	
48	4	1	0	0	0	
72	2	1	0	0	0	

Source: 1999 ASHRAE *Applications Handbook*, © American Society of Heating, Refrigerating and Air Conditioning Engineers, Inc., www.ashrae.org.

from the ideal case, one might consider the effective duct emitting area to be 4 times the actual duct size when entering the chart.

Note that end-reflection losses are not limited to the duct terminations but can be a useful technique within a ducted system to get low-frequency attenuation, for instance where a duct abruptly enters a large highly sound absorptive plenum volume.

Room Effect

Room effect is the translation of the sound power emitted into the room from the duct system to the sound pressure level that results in the space at a given location. The process of predicting the sound pressure level from the sound power level is treated in detail in Chapter 7.

The classical method to relate the sound pressure that results in a room to the sound power emitted from a source is modeled by a point sound source for small-size sources, such as round or rectangular diffusers, and by a line sound source in the case of strip diffusers or duct breakout. Note that the amplitude of the reverberant sound field, which dominates at large distances from the source, is independent of the nature of the source. However, the amplitude of the direct sound, which dominates in the vicinity of the sound source, strongly depends on how the sound energy spreads from the source.

For a point sound source radiating into the room the sound pressure level at a distance r (resulting from the power-based addition of the direct and reverberant fields) is given in the equation

$$L_p(r) = L_w + 10 \log_{10}\left(\frac{Q}{4\pi r^2} + \frac{4}{R}\right) \tag{17.6}$$

where L_w = source sound power level, dB re 10^{-12} W
L_p = sound pressure level, dB re 2×10^{-5} N/m^2
R = room constant, = $S\alpha_A/(1 - \alpha_A)$, m^2
S = total room absorption, m^2
α_A = average absorption coefficient of room surfaces
r = source–receiver distance, m
Q = inverse of fraction of sphere to which sound energy radiates

For a case where there is only one source of concern for design, using this equation is simple and sufficient. However, where there are multiple noise sources such as various diffusers or grills with close to the same noise emission level and producing close to the same noise level at the receiver position, some way to adjust the room effect for the multiple diffusers is needed. This might be done by adjusting the room effect by 10 log N decibels, where N is the effective number of outlets contributing to the sound at the critical receiver location. Other, more sophisticated, methods based of sound propagation modeling in rooms presented in Chapter 7 may be used when dealing with special situations.

17.2 FLOW NOISE IN DUCTED SYSTEMS

The noise generated by flow in duct systems is a concern for noise transmission inside the duct along the duct path and that radiated from the outside surface of the duct. Noise generated at particular fittings or duct elements can be estimated in accordance with the methods presented in Chapter 15. Flow-generated noise propagates down the ducts just like noise from any other sources in the duct system. It is typically in the mid- and higher frequency range and thus is relatively easily attenuated by dissipative treatments such as silencers or lining. When attenuating elements do not exist in the ductwork between the source and the occupied space, this noise can become a concern.

TABLE 17.11 Recommended Airflow Velocities in Lined Duct Systems[a,b]

	Airflow Velocities (fpm) Consistent with Indicated Noise Criterion (NC) through Net Free Area of Duct Section or Device							
	NC 15		NC 20		NC 25		NC 30	
Duct Element or Device	Supply	Return	Supply	Return	Supply	Return	Supply	Return
Terminal device[c] ($\frac{1}{2}$ in. minimum slot width)	250	300	300	360	350	420	425	510
First 8–10 ft of duct	300	350	360	420	420	490	510	600
Next 15–20 ft of duct	400	450	480	540	560	630	680	765
Next 15–20 ft of duct	500	570	600	685	700	800	850	970
Next 15–20 ft of duct	640	700	765	840	900	980	1080	1180
Next 15–20 ft of duct	800	900	960	1080	1120	1260	1360	1540
Maximum within space[b]	1000	1100	1200	1320	1400	1540	1700	1870

Note: Fan noise must be considered separately. Reduce duct velocities (not diffuser/grille velocities) by 20% if ductwork is unlined.
[a] All ducts with 1-in.-thick internal sound absorptive lining.
[b] Above mineral fiber panel ceiling. Lower velocities 20% if open or acoustically transparent ceiling.
[c] No dampers, straighteners, deflectors, equalizing grids, etc. behind terminal devices.

The following guidelines are offered for duct sizing based on the location, type, and class of ductwork:

<3000 feet per minute (fpm) velocity in round ducts in mechanical rooms and shafts,

<2500 fpm velocity in rectangular ducts in mechanical rooms and shafts,

<2000–2500 fpm velocity in the ceiling of occupied spaces with mineral fiber ceiling for NC-35 to NC-40 goal,

<1500–2000 fpm velocity in the ceiling of occupied spaces with open or acoustically transparent ceiling for NC-35 to NC-40 goal,

<1500 fpm velocity in larger final distribution ducts serving NC-35 to NC-40 spaces,

<a friction rate (pressure loss rate) of 0.10 in. wg/100 ft of duct run in smaller final duct distribution for NC-40, and

<a friction rate (pressure loss rate) of 0.08 in. wg/100 ft of duct run in smaller final duct distribution for NC-35.

Airflow velocity guidelines for the design of special low noise spaces (typically NC-30 and lower) are presented in Table 17.11 for lined duct systems.

17.3 SYSTEM DESIGN FOR ESPECIALLY QUIET SPACES

The lower the noise goal for a particular space, the more carefully the duct arrangement needs to be considered from the air balance and aerodynamic standpoints. In particular, the duct system needs to be designed to be as naturally balanced as possible. That is, if the system is turned on, the airflow naturally delivered from each outlet would match the desired airflow with little or no damper throttling to control the flow. Creating systematically branching duct arrangements to an array of diffusers or grilles is typically a good approach. There should be suitably long straight ducts between the branches for the flow to straighten out and divide, as desired at the next branch. However, not all buildings and conditions can accommodate this design concept. Often there is a need to have a series of diffusers or branches to diffusers off the side of a large main duct. Having at least a modest length (a duct diameter) of straight duct run out to the diffuser is always preferred to improve the airflow presentation to the back of the diffuser and to create a more desirable position for a damper that can be used to control the flow at a point that is away from the diffuser. For these cases, it is often best to have the main duct that feeds the diffusers remain a constant size so that the pressure drop along the main "header" duct becomes small and the flow distribution is controlled by the pressure drop of the runouts. To achieve a naturally balanced design, it is generally best to have the diffusers and grilles be in a reasonably cohesive grouping rather than be widely spaced out.

It is recommended to avoid having a tight cluster of diffusers serving a critical space and then a long "tail" duct extending downstream, because to balance the system, it will be necessary to throttle dampers near the diffusers serving the critical space, and this will cause noise to be generated.

Dampers should be provided in duct systems as needed to control the flow, but they must be located sufficiently far from the terminal devices so that the turbulence and noise they generate can be sufficiently attenuated before it reaches the occupied space. For especially quiet systems, dampers should not be placed at or near the face of the diffusers and grilles. It is essential to use dampers well back in the duct system to adjust for major deviations from a natural air balance design.

The airflow velocity near diffusers and grilles needs to be appropriately slow to avoid excessive noise generation. As one moves back in the duct system away from the critical space, flow velocities can progressively increase. Ideally, the velocities would step down in moving toward the critical space in a manner that is consistent with the desired systematic branching pattern. The best correlation between airflow velocities and the noise resulting in the space of interest is for noise from grilles and diffusers in the room and for the ducts closest to the room. The correlation between noise reaching the space of interest and the flow velocity becomes less direct as one moves further into the duct system, away from the receiving space. Thus, there is not much room for compromise in velocities close to the occupied space, but there may be room for compromise from an ideal design at greater distances from the occupied space.

Table 17.11 presents suggested velocities in duct systems serving quiet spaces to help ensure achieving particular noise goals with acoustically lined ductwork. These suggested flow velocities generally assume that the duct configuration will be only average in terms of its aerodynamics. If very desirable airflow geometry is achieved, somewhat higher velocities can be allowed. The suggested velocities through the net free area of the diffusers and grilles, given in Table 17.11, assumes a nominal slot width of about 1/2 in. To the extent that wider slot widths are used, higher velocities can be allowed for a given noise goal. Typically, there might be an allowable increase in slot velocity of 15–20% for each doubling of the nominal slot width up to about 2-in.-wide slots. Special nozzle-type diffusers are available to operate more quietly so that they can throw the air further for a given noise level. The manufacturer's noise data should be reviewed for special diffuser applications, being sure that the noise generation is adjusted from the conditions under which the devices are tested to the actual conditions of use. Because, at the indicated velocities, it is difficult to blow the air very far, for quiet systems, supply air is often delivered using a distributed array of diffusers at the ceiling rather than from the sidewalls of a space. This is not to say that sidewall blow applications need to be avoided; they just need to be limited to covering a small zone at the sides of the space rather than trying to condition the entire space from this position. Using a combination of ceiling diffusers and sidewall blow diffusers can be a useful technique to divide the main ducts into smaller pieces that will fit through the building more easily than handling all the airflow in fewer large ducts. Note that to help fit ducts into buildings it is often better to avoid having diffusers or grilles directly off the sides of a main duct because the entire airflow in the main duct will have to run at a very low velocity and this duct will be very large. However, in applications where there is ample space for large ducts or where ducts are exposed, this may be a simple, neat, and clean approach to the design. Developing the duct distribution in a systematic branching pattern naturally allows the ducts that have the large airflow quantity to be sized at a higher velocity, which helps them fit in the building more easily. For quiet systems, the space required to handle the airflow at the desired velocities can be considerable and this can translate into requiring more building volume in which to maneuver the ducts. Considering the system design concept and velocities early in the design can allow the building to be more economical from the standpoint of the sheet metal that is used and the building volume that is constructed. Not considering the duct system requirements early in the design may result in an undesirable compromise in the design and/or the noise levels that can be achieved.

Note that about a 20% change in velocity is generally associated with a 5-NC point change in noise level in the occupied spaces. Thus, a compromise of increasing the airflow velocities in a project design by 20% will generally increase the noise by about 5 NC points. Also, lowering the noise level by about 5 NC points requires about 20% lower velocities; this translates into only about 10–15% increase in the dimensions and surface area of round and (reasonably square) rectangular ducts. Increasing the noise goal by 5 NC points will only save about 10–15% of the sheet metal cost.

The airflow velocity guidelines in Table 17.11 assume that there will be lining in the duct system, and this is generally recommended to control the noise generated by flow in the ducts at fittings and unexpected flow obstacles in systems that are to be especially quiet. For lower noise goals, more things need to be done properly in the implementation of the design and there is less margin for error. If the ductwork does not have lining, it is suggested that the velocities in the ductwork (not at the diffusers and grilles) be about 20% lower than indicated to allow about a 5-dB margin for the lack of attenuation in the duct path.

17.4 DIFFUSER AND GRILLE SELECTION

The airflow velocities for low-noise spaces presented in Table 17.11 are only guidelines which include a number of assumptions, including that the nominal slot width is about 1/2 in. and the ductwork leading to the diffuser is only nominally configured. This also includes assumptions regarding the room effect and the number of diffusers that are contributing to the sound field at the receiving position. Note that the larger the room, the more likely it is that diffusers will be more numerous (and hence there will be more sources of noise), but in this case, it is also more likely that the diffusers will be further from the receiver and hence the room effect will be larger. These two factors tend to counter each other so that a single velocity guideline chart can cover a wide range of applications.

Manufacturers often provide data for the noise generated by their diffusers and typically these are boiled down to a single NC rating for the diffuser at a particular flow. Be careful in applying manufacturers' data because they are generated under ideal flow conditions and often include very favorable room effects so that the reported noise levels are as favorable as possible. Also, they typically are reported for one diffuser or for a short section of a linear diffuser where the application may have many more such elements contributing to the noise level at a receiver location. It is necessary to account for the sum of the noise that is generated by multiple diffusers/grilles which reach the receiver as well as to make other adjustments to correct the assumptions in the presented data to match the actual conditions of a project. It is also essential to make adjustments to the ideal flow condition used as the basis for the data to get to the actual applied flow condition. For especially quiet spaces, use of the velocity guidelines may be a preferable approach for establishing the selection parameters for the diffusers and grilles.

The noise produced by dampers needs to be assessed separately from diffuser and grille noise, but having two flow turbulence-producing devices closely spaced in series may add to the noise each device produces by itself.

Displacement ventilation systems which deliver air to the occupied zone through diffusers at the floor can often be very quiet. This is largely because air is introduced to the space in small quantities at any one location and the flow velocity at the diffuser needs to be particularly low for comfort reasons. The result can be a system with little diffuser-generated noise and this can be a favorable design approach for spaces that need to be very quiet.

17.5 NOISE BREAKOUT/BREAK-IN

When ducts containing high levels of noise pass though spaces with low or moderate noise goals, it is necessary to consider whether the noise that transmits out through the surface of the sheet metal duct (called the breakout noise) is suitably controlled. Similarly, when a duct in which the sound field inside has been quieted passes near a noisy piece of mechanical equipment prior to penetrating the mechanical room wall, noise may break into the duct and travel down the duct to a receiving space of concern. To minimize concern for noise breakout/break-in, it is usually best to position the high-insertion-loss elements in the duct system (the silencer) at or very near the penetration of the mechanical room boundary. However, it is not always possible to achieve this goal along with the goal of achieving good aerodynamics for the application of the silencer. Special elbow (or other configuration) silencers can help achieve the goal of the desired position of the silencer for breakout/break-in considerations while avoiding poor flow. This is a very good use of special silencers.

The noise that breaks out of and into ducts can be predicted using Eqs. (17.7) and (17.8), respectively:

$$L_{w(\text{out})} = L_{w(\text{in})} - \text{TL}_{\text{out}} + 10 \log\left(\frac{S}{A}\right) - C \qquad (17.7)$$

$$L_{w(\text{in})} = L_{w(\text{out})} - \text{TL}_{\text{in}} - 3 \qquad (17.8)$$

where $L_{w(\text{out})}$ = sound power level of sound radiated from outside surface of duct walls, dB
$L_{w(\text{in})}$ = sound power level of sound inside duct, dB
S = surface area of outside sound-radiating surface of duct, in.^{2}
A = cross-sectional area of inside of duct, in.^{2}
TL_{out} = duct breakout transmission loss, dB
C = correction factor as follows for values of
$\Delta = \text{TL}_{\text{out}} - 10 \log(S/A)$

Δ	C
> +10	0
+8 to +5	1
+4 to +2	2
+1 to −1	3
−2 to −3	4
−4	5
−5	6
−6	7
−7	8
−8	9
< −9	−Δ

Note: These breakout and break-in equations should not be used where $\Delta < -10$. Also, where Δ is a negative number, the precision of the result is questionable. The value of Δ tends to go negative with relatively long duct sections. This approach does not take into account any dissipative attenuation along the duct path, but at the low frequencies principally of concern for breakout and break-in issues, dissipative attenuation is usually small along the duct path in the length of duct sections for which the approach should be used.

The noise breakout formula can further lead to prediction of the sound pressure level in a receiving space by applying an appropriate room effect and including the sound isolation of any intervening construction. The noise break-in equation will determine the sound power that can then be used as a source level and the propagation of this noise can be traced along the duct system. Treatments for noise breakout/break-in typically include increasing the surface weight of the duct construction, stiffening the duct, applying a flexible mass barrier material on resilient backing (such as a glass fiber mat or foam), or enclosing the duct in gypsum board using a variety of possible details. Prediction of the insertion loss achieved by various types of wrappings is given in Chapter 13. The further the enclosing mass layer is from the duct surface and the heavier the enclosure material, the better this will control low-frequency sound, which is most typically the concern. Circular ducts are inherently stiffer than rectangular ducts made of flat metal sheets, and for this reason, distribution systems using circular ducts are often less prone to breakout/break-in noise problems. Note that there is a particular concern for breakout/break-in noise when glass fiber panel or other acoustically transparent ceilings are used in occupied spaces in buildings. Where there is a solid gypsum board ceiling, breakout noise is naturally less of a concern than where there is a conventional mineral fiber acoustical ceiling.

Ducts that drop out of the bottom of rooftop air-handling units into the ceiling plenums above occupied spaces are a classic case where duct breakout noise is a problem, and this typically is worst when the supply duct discharges directly down into the ducts rather than into a plenum within the unit. Elbow silencers with high-transmission-loss casings can be a convenient method of addressing this problem. If a straight silencer is used and is located well away from the elbow that turns in the ceiling space, there can be a significant amount of duct wall which can radiate noise to the occupied space and may need to be treated. This is discussed further under the rooftop unit section (Section 17.10). Most of the duct enclosure treatments to address this concern are very difficult to install, especially when there is limited space and the building is finished. Whenever possible, designs should be developed which avoid the need for this sort of treatment.

17.6 FANS

Fans are typically the major source of noise emission into duct systems. The various fan types have differing noise spectrum and magnitude characteristics and this is also a function of the duty. Certain fans are aerodynamically better suited for particular flow and pressure applications. This needs to be considered for the full operating range of the application. Many older references, such as older editions of the ASHRAE handbooks, include generic methods to predict the noise generated by fans, and such sources can provide useful guidance in the absence of more specific data from the manufacturer for the fans that are planned. If such data are used, remember that, due to changes in manufacturing methods, modern fans may not be made the same way as the fans used as the basis for the generic prediction methodology. Often new fans are made more economically and produce more noise that older fans. It is best to obtain the noise data for the fan from the prospective manufacturer, tested in accordance with the most current applicable standards. These data should be used as a guide in the selection of the fan and as a point of comparison between candidate fans for an application. In addition, the quietest of the reasonable fans should be selected for a given application. Such a process should help avoid selecting a particularly noisy fan, which could lead to unexpected noise problems or noise control treatment which could unduly burden the system.

The manufacturer's noise emission data for fans are based on a standard test which may or may not closely match the conditions of a particular field

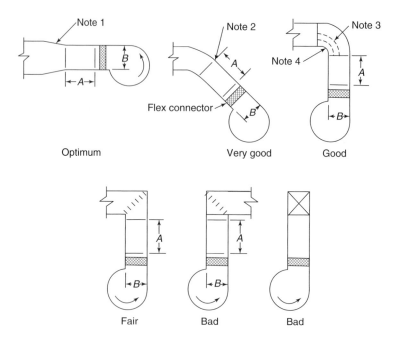

Notes:
1. Slopes of 1 in 7 preferred. Slopes of 1 in 4 permitted below 200 fpm.
2. Dimension A should be at least 1.5 times B, where B is the largest discharge duct dimension.
3. Rugged turning vanes should extend the full radius of the elbow.
4. Minimum 6-in, radius required.

FIGURE 17.3 Various outlet configurations for centrifugal fans and their possible rumble conditions. *Source*: 1999 ASHRAE *Applications Handbook*, © American Society of Heating, Refrigerating and Air Conditioning Engineers, Inc., www.ashrae.org.

application. The noise radiated by the fan into the duct will vary with the particular acoustical impedance of the connected ductwork. Consequently, the actual sound power of the fan that is emitted into a duct system may differ from the manufacturer's data.

Particular care should be given to the inlet and outlet flow conditions for fans. Poor inlet flow can unbalance the flow over a fan wheel, causing performance problems and excess noise generation at the anticipated speed for the application. The fan speed may need to be increased to compensate for unexpected pressure losses and thus generate more noise than expected. Inlet vanes or substantial flow distortion due to flexible connections protruding into the flow can easily cause an increase of noise on the order of 6 dB across the full frequency spectrum. The local airflow velocities on the discharge of fans can be quite high in some regions and probably will not be uniform over the outlet area. This needs to be taken into account in the system design and arrangement. It is often not possible to create ideal flow conditions, but it is important to create at least reasonable flow conditions and to avoid poor flow conditions which can create severe noise problems. Figure 17.3 illustrates some favorable fan discharge conditions as well as conditions that need to be avoided. It is always necessary to consider the flow arrangement on the discharge of a fan or air-handling unit to be sure that the duct and fan geometry work together as well as possible for the condition and, most essentially, to avoid poor conditions. Poor flow conditions at the fans can lead to low-frequency noise generation, either by the fan or due to large-scale turbulence in the ductwork which will drum the duct walls, causing a low-frequency noise problem that can be very difficult to solve.

17.7 TERMINAL BOXES/VALVES

Many HVAC systems serving buildings include a variety of terminal boxes or valves to control the amount of air delivered to the occupied zone. Some boxes are used to vary the airflow to the room in accordance with thermal load. Others provide a constant amount of flow to assure ventilation rates and control zone pressurization. Some boxes incorporate fans to mix plenum air with a variable amount of main system air to maintain a minimum amount of air movement while delivering variable heating/cooling. All of these boxes and valves have in common that they produce a pressure drop from the higher pressure in the main duct system to a low pressure in the final distribution ductwork on the room side of the box. In throttling the pressure from the main to the final distribution ductwork, noise is generated and this noise needs to be considered in the design to meet desirable noise levels in the occupied spaces. The greater the airflow quantity through the box, the greater the noise that is generated. However, the most significant factor is that the noise boxes generate is a strong function of the pressure drop across the valve. Boxes on a system that are near the fans naturally see a higher pressure in the main duct than boxes near the ends of the system, and boxes close to the beginning of the system will generate more noise than comparable boxes located toward the end of the system. Excess static pressure in the main ducts can be a significant factor in noisy systems as the terminal boxes in the system are throttling hard (have large pressure differential across them) to control the excess pressure.

The manufacturers of terminal boxes generally provide noise data for their boxes rated at various flow rates and pressure losses, and these data can be used for noise analysis of the ducted system just as fan noise propagation down the ducts is done. In such an analysis, the pressure loss that is of interest in determining the noise of the box is the actual pressure drop that the box is expected to produce in the field, not the minimum pressure at which the box will satisfactorily control flow or some arbitrary pressure loss that is assigned. Because of airflow control considerations, the boxes in many applications have inlet flow conditions that are reasonably close to the ideal flow condition under which the terminal box is tested for flow and noise. However, if there are adverse flow conditions, the noise generated by the boxes may be higher than the manufacturer's rating.

Noise radiation from the casings of the boxes can also impact the occupied spaces near them, and to enable assessment of this, manufacturers also publish casing radiated noise data. For fan-power boxes, these data include the noise that radiates from the fan out of the plenum air inlet and this can be a significant issue. In some cases it is necessary to provide attenuation for the plenum air inlet to the box. Most moderate-size terminal boxes operating at nominal pressures can be located in the ceilings of occupied spaces where the noise goal is NC-35 or higher when there is a conventional mineral fiber ceiling. Where boxes are in the ceilings of spaces with noise goals more stringent than NC-35, where system pressures are anticipated to be high, and/or where ceilings with lower transmission loss are used, special consideration should be given to the issue of casing radiated noise. In particular, be cautious of exposed box applications or where glass fiber, perforated, or open-slat ceilings are used.

The manufacturer's noise data should be reviewed for verification, but typically the frequency band of concern for flat-plate damper boxes is the 250-Hz octave band; for pneumatic-style boxes, the 500-Hz octave band; and for plunger-style valves, the 1000-Hz octave band. The type of noise control treatments that are applied in the duct systems should ideally be selected to treat the particular noise spectrum that is expected based on the box/valve type that is used.

For systems that utilize terminal boxes/valves it is usually a good idea to try to conceive the main duct system to be able to deliver the design airflow with the minimum overall pressure loss and the minimum differential in pressure arriving at the various boxes in the system. Consider gradually reducing the flow resistance in the main duct as the duct proceeds toward the end of the run and the magnitude of air being handled is reduced. The overall velocity in the duct system is often a significant factor determining the pressure drop from beginning to end, and the design velocities should be as low as reasonably possible. Use aerodynamically favorable fittings. Try to avoid particularly long duct runs by using more vertical shafts in the building to minimize the length of duct runs on the floors served. Where it is possible and acceptable to create ring ducts fed from both ends to serve the floor plate of a building, this can favorably reduce the pressure that is needed in the system. Consideration of these design features can help avoid noise problems and/or minimize the magnitude of attenuation treatments that are needed.

Always be sure that the final setup and balancing have the system operating at the lowest pressure that is consistent with delivering design airflow. This includes making sure that the system pressure cannot be further lowered without jeopardizing the required flow at the extremes of the duct system and making sure that the fan is operating at the lowest speed that will deliver design flow. If there are inlet vanes on the fan to control capacity, they should be nearly fully open when the system is demanding full design flow. The inlet vanes should not be constricting the airflow when the demand is full; if they are, the fan can probably be slowed, which should reduce noise emission.

17.8 MECHANICAL PLANT ROOM SOUND ISOLATION AND NOISE CONTROL

Good space planning is the best way to avoid sound isolation and noise problems at mechanical rooms. Avoid a common wall with a noise-sensitive space that would require a substantial and expensive wall construction. Very high sound isolation walls typically require discontinues in or special protection of the flanking structural paths; such treatments can be cumbersome and expensive. Having a common wall also invites penetration of that wall with ducts and pipes. Since creating a sufficiently tight seal around a single wall that is penetrated can often be difficult and risky, these should be avoided. Creating a buffer zone of less critical space between a mechanical room and noise-sensitive space allows simpler and less costly constructions to provide the desired isolation, and this minimizes the risk of noise leaks at penetrations. It is good practice to position particularly noisy equipment in mechanical rooms away from critical sound-isolating walls to avoid exposure of critical constructions to unnecessarily high levels of noise. Especially avoid locating equipment extremely close to walls where the sound has no opportunity to lose intensity owing to spreading losses.

Building constructions that provide a good deal of sound isolation can be constructed well, but there is always a door into the space, which is a weak sound isolation condition, and the location of doors needs special planning. Certainly it is necessary to avoid doors to mechanical rooms directly from noise-sensitive rooms. To the extent that it is possible to naturally create a double-door entrance vestibule ("noise-lock") to the mechanical room, such as by accessing the room off of a back-of-the-house corridor that is separated with another door from more public space, this is a good idea since good isolation can be reliably achieved with ordinary door construction.

Mechanical plant rooms often contain equipment which radiate relatively high levels of low-frequency noise, and it is necessary to have heavy constructions, such as masonry walls, surrounding the room. Lighter constructions such as gypsum board on stud can be used, but double-wall constructions with a substantial air space between the two sides of the wall will likely be required. Hence, light weight wall construction alternatives may take up more floor space. To the extent that there is auxiliary equipment associated with the main plant equipment that is wall mounted, it is often wise to have a double-wall construction so that minor vibration and structure-borne noise that might get coupled to the inner wall of the mechanical room will not get strongly transmitted to the side of the wall construction in the occupied space that is to be protected.

Creating an air-tight seal, as is always necessary for sound isolation, at the heads of walls can be particularly problematic. Necessary firestopping can be achieved for walls with fibrous packing and coating with intumescent material, but this is not a suitable seal for sound control purposes where the basic construction needs to close within a typical caulk joint of the abutting construction. Closing concrete block construction to the underside of fluted metal deck and beams can be very difficult. It is also difficult to create a seal when concrete block comes up under roof constructions that are sloped for drainage or need to accommodate structural deflection. The seals that are needed under these conditions need to be considered early in the design, and practical methods to accomplish the necessary seal in the field need to be detailed.

Floating concrete floors are sometimes required to separate mechanical spaces from noise-sensitive spaces that are above or below the mechanical room. Prediction of the insertion loss of floating floors is provided in Chapter 11. These constructions consist of a concrete slab (typically 4 in. thick) which is supported above the structural floor on resilient mounts. Systems to accomplish this are available from the various isolation product vendors. These constructions can and do provide effective control of sound transmission through floors, but they take up space that needs to be allocated from the start. They also require careful coordination and implementation in the field in order to function as planned. Floating floors are not generally designed as part of a vibration isolation scheme for plant equipment; their primary use is for airborne sound isolation. Concentrated loads such as from equipment can be accommodated on floated floors with appropriate planning and design of the floated floor, but it can also be acceptable to have such loads supported on small structural piers (housekeeping pedestals) that penetrate through the floated floor. To avoid a sound isolation concern at support points that penetrate the floors, these need to be kept as small as practical and typically complete housekeeping pads are not used. The sizes and positions of support piers need to be carefully coordinated in the field. Rather than use floating floors to provide high-level sound isolation to spaces above or below mechanical rooms, it is sometimes necessary to have special sound-isolating ceilings above or below the mechanical spaces. These can be complex to implement and require special coordination of the building services to avoid or minimize penetrations for supports through the ceiling. The timing of construction of the ceiling relative to installation of the building services needs forethought since it typically cannot follow the usual sequence.

17.9 VIBRATION ISOLATION CONSIDERATIONS FOR BUILDING MECHANICAL SYSTEMS

Technical aspects of isolation of vibration sources are treated in detail in Chapter 13. This section will provide some practical guidance regarding the application of vibration isolation systems and devices to building mechanical systems.

Numerous references, such as the sound and vibration control chapters of the ASHRAE *Applications Handbook*,[1] and vibration isolation equipment vendors provide guidance regarding the application of isolation devices for building equipment, and the reader is referred to such sources for specific isolation selections based on the type of equipment, the speed of the equipment, the location in the building, and the sensitivity of the application. For application to building equipment, the selection of vibration isolators is reduced to identifying the deflection of the isolation element under the static load. The deflection of the isolator under load is related to the natural frequency (the frequency at which resonance occurs) of the isolation system by the equation

$$f_n = 3.13 \left(\frac{1}{d}\right)^{0.5} \qquad (17.9)$$

where f_n = isolator natural (resonance) frequency, Hz
d = isolator deflection, in.

Generally, to provide effective isolation, the natural frequency of the isolation system needs to be no more than 0.30–0.10 times the drive frequency (RPM/60) of the equipment. At ratios greater than 0.30 isolation efficiency will be minimal, and at a ratio of 0.70 there is theoretically no isolation at all. Clearly, it is necessary to avoid having the equipment operate at or near the natural frequency of the equipment. There is generally little point in an isolation system that has a ratio of isolation system natural frequency to equipment drive speed smaller than 0.10 since little additional effectiveness is gained for practical installations and the stability of the equipment on the isolation system can be marginal.

Many pieces or equipment are equipped with variable-speed drives and, in this case, it is generally appropriate to select the isolation system for the normal full speed of the equipment. As the equipment slows down and operates closer to the natural frequency of the isolation system, the isolation system will provide less isolation efficiency, but fortunately, the magnitude of the machine vibration force will be reduced at approximately the same rate. So, there are generally no problems with this. The natural frequency of the isolator should typically be no more than about half of the lowest operating frequency of the equipment to prevent operation too close to resonance, but for more difficult applications, this might be as high as 0.7. If it is not feasible to select large deflection springs, sometimes it is necessary to limit the operating range of the equipment to avoid operating too close to the isolation system natural frequency. This possibility should be considered early in the design. In some cases it might be worthwhile to judiciously increase the deflection of the springs to lower the natural frequency to avoid impacting equipment operation and efficiency. However, an isolation system should not be so flexible that it creates instability or operational problems such as with alignment or stress transfer.

Inertia bases are used with many pieces of equipment to lower the center of gravity and to reduce the vibratory movement of the equipment when it is resiliently supported. Movement due to starting torque, resistance to forces created by the static pressure a fan develops, and turbulent flow in large fan systems are also reasons such bases are recommended for equipment. The inertia bases only reduce the motion of the equipment on the isolation system. They do not change the inherent unbalanced forces of the equipment, even though the magnitude of vibratory movement of the equipment on the base will be lower. The vibratory displacement of the base and that of the equipment rigidly mounted on it will be lower because the mass that the forces are now acting on is the combined mass of the equipment and the inertia base compared with the mass of the equipment alone. Note that the inertia base does not change the vibration forces that are transmitted to the building assuming that the deflection of the isolation system remains the same. This is because, for the same static deflection, the springs need to be stiffer to support the additional weight of the equipment plus its inertia base.

In theory, the weight of the inertia base would be determined by the desired reduction of the magnitude of the vibratory movement, but the magnitude of the motion of the equipment is not well known, so it is difficult to tell what mass ratio to use. In practice, the weight of the inertia base for building mechanical equipment is often determined by the weight of concrete fill in a steel frame that approximates the rectangular outline of the equipment footprint (with allowance for fastening), and the depth is determined by the need to make the base suitably stiff. The depth of the base is often related to the length of the base and is typically about $\frac{1}{10}$ to $\frac{1}{12}$ the longest dimension. Experience indicates that for many pieces of mechanical equipment, the base weight that results from this methodology is sufficient to control excessive movement and prevent undue forces from being transmitted to connected pipes and appurtenances.

Spring isolators most typically are manufactured to achieve rated deflections of approximately 1, 2, 3, and 4 in. This means that when the full design load for the spring is applied, the spring will approximately deflect this rated distance, and at this condition, the spring will also achieve other desirable design features such as for lateral stability, horizontal stiffness, and deflection reserve to bottoming out. The rated deflection has nothing to do with the deflection actually achieved in the field under the actual equipment load applied. For example, a spring that has a rated deflection of 1 in. for a 1000-lb load will only deflect 0.10 in. if it is supporting 100 lb. The only thing that is important in terms of the isolation efficiency of the mounting is the actual static deflection that results under the applied load and, in the case of this example, the natural frequency would be determined by 0.1 in. deflection. Isolation devices need to be properly selected to achieve the desired minimum static deflection and hence the desired isolation efficiency. Underloading an isolator can result in achieving less than the desired isolation efficiency and overloading can cause the spring to collapse (bottom out) so that it provides no isolation at all. If a spring is specified to achieve 1 in. minimum deflection under load, there is only one load that will exactly satisfy the requirement if a spring is selected from a 1-in. rated series of springs, and that is the maximum load that the mount can support; lower loads will achieve less than 1 in. deflection and will not meet the required minimum deflection, and greater loads will overload the spring. To achieve the required minimum deflection in this case, a 2-in. rated deflection spring would need to be used so that as long as it is loaded between half and full load, it will achieve the required minimum deflection. This can cause unnecessary expense for isolators that have substantially more capacity than needed. To clarify the performance desired and to avoid unnecessarily costly springs, it is often better to specify spring deflections of 0.75, 1.5, 2.5, and 3.5 in., as expected from 1, 2, 3, and 4-in. rated static deflection springs, respectively, so that there is a reasonable selection range available.

For applications where the weight of the isolated equipment may vary substantially from time to time, such as for a cooling tower or chiller if the water is drained for maintenance, the isolation devices that are used need to incorporate some method to restrain the upward movement of the equipment as the weight is removed because this can put excessive stress on connecting pipes. Equipment that is located outdoors may be exposed to wind and exhibit lateral movement which needs to be restrained. Special travel-limited isolators have been developed for this application which will snub the movement of the equipment under such conditions, typically at about 1/4 in.

Many pieces of equipment need to be specially restrained to avoid damage to the equipment or the building if a seismic event occurs. This applies to many pieces of equipment, whether resiliently mounted or not, but clearly, equipment that is resiliently supported could easily fall off its mounts if there is a seismic event. Seismic restraint of equipment is an entirely separate issue from vibration isolation; it is essentially a structural issue. Isolation and seismic restraint become related because so many pieces of equipment that are resiliently supported need restraint. There are seismic restraints available from vendors which are completely separate devices from vibration isolators. In some cases it is convenient to incorporate seismic restraint into the isolators, and products that do this are available from vibration isolation vendors, but it may not be necessary to use combination devices. Where the equipment has no other point of attachment but the mounting points at which suitable support and restraint can be achieved, a combination device is necessary. It is essential that the seismic restraint not compromise the performance of the vibration isolation system, and this is not an easy task considering the relatively tight installation tolerances that are required for both seismic restraints and vibration isolators. Often isolators with combined seismic restraint can help achieve the desired level of installation more reliably, and this is another benefit that such devices provide.

Electrical connections to vibration isolated equipment need to be made flexible to avoid creating a problematic vibration transmission path to the building structure. These are typically made with either a section of flexible conduit or a special flexible electrical coupling. For small-size equipment and conduits, practical lengths of flexible conduit are effective and sufficient. As the size of the equipment and the connecting conduit increase, the flexibility of these systems may diminish and it may be necessary to use special flexible electrical couplings. The flexible connections should be installed in whatever manner creates a slack condition.

Piping that connects to isolated equipment also needs to avoid creating a vibration transmission path to the building. There are flexible pipe connections available for many applications, but because vibration can travel through the fluid in the pipes (which is often highly incompressible), vibratory energy is often not stopped sufficiently by them. Such connections are best used (or not) based on considerations other than vibration control (e.g., thermal expansion/contraction or alignment). To control vibration transmission to the building via pipes, it is usually necessary to resiliently support the pipes for some prescribed extent, but that extent is not easy to describe and could vary substantially from project to project. Many schemes have been devised to describe the scope of pipe isolation in buildings, such as a prescribed length of pipe or a certain multiple of the pipe diameter, and this is somewhat dependent on the designer's philosophical approach. The fixed length of pipe isolation from a vibration source approach does not account for variations of pipe stiffness based on size. Clearly, a 1-in. pipe for a particular duty is much more flexible over a given distance than a 12-in. pipe for a similar duty. Prescribing the length of isolation in pipe diameters partly addresses this issue, but this approach can be difficult to work with in the field. In practice, there needs to be some compromise between a practical scheme that can be implemented in the field, a system that makes sense economically, and a scheme which will provide the isolation that is needed for the size of pipes and the vibration sources involved. There are many possible approaches to this specification depending upon what is workable and cost effective. There may be a generic scheme that is appropriate for many applications, but this should be carefully considered for each project before finalizing the scheme. The designer should keep in mind that a resilient connection between a pipe and the building structure provides a reduction of the vibratory force only if the local impedance of the supporting structure is high compared to the impedance of the isolation device. Little isolation will be gained if a moderately stiff isolator like a neoprene pad is used to resiliently support a pipe from a lightweight and flexible gypsum board on stud wall.

Classic vibration isolation theory assumes that isolators sit on supports that are infinitely stiff and massive. This is reasonably the case for support conditions at slab-on-grade floors, but this may not be the case for support conditions at all structures above grade. Most isolation application guides take this into account and provide recommendations for a variety of less than perfect support conditions such as on above-grade structures. The implications of structural flexibility on vibration isolation for most above-grade support conditions are rather modest where the structure is reasonably massive (such as a concrete floor) and is suitably strong to support mechanical service equipment. This accommodation involves only slightly softer isolators than would be used on grade, but the isolators are still within the range of conventional products that are economically available. When the supporting structure is particularly lightweight, such as wood frame construction or a simple metal deck roof, special attention needs to be given to the support condition, and it may be necessary to create special load transfer schemes to transfer the load to more substantial points of structure such as columns.

17.10 ROOFTOP AIR CONDITIONING UNITS

Many HVAC systems employ a variety of types of packaged rooftop air conditioning units with circulating fan systems and sometimes with integral compressor/condenser sections. Because such equipment is often noisy and typically supported on a lightweight structure directly above occupied space, this equipment is one of the most common sources of noise problems in buildings. Because of space limitations and the magnitude and frequency of excess noise that results, problems with this equipment are often very difficult and expensive to address, especially once the equipment is installed. Sometimes it is not practical and economically feasible to fully resolve the problems presented by this equipment at the initially chosen location and major system changes are needed. The best way to address noise from this equipment is to have it installed at the outset according to a well-conceived noise and vibration control scheme. The necessary installation features to control noise and vibration may be new and unusual to some installers, and it is wise to take the extra steps to clearly document all the necessary details so they can be properly implemented.

The roof construction needs to extend fully under the equipment and needs to close tightly around any penetrating elements such as pipes and ducts. The roof construction needs to be appropriate to control airborne sound transmission to the occupied space, including sound that might transmit both through the roof deck directly under the unit and off to the side of the unit (for some distance from the unit). If concrete is used in the roof deck, this is very good and can help to address, in a simple manner, both airborne noise and vibration isolation issues. Concrete is not always necessary and lighter weight sound isolation alternatives can be conceived, but they may be unfamiliar to some installers.

Having the supply fan discharging directly down into the supply duct entering the ceiling of an occupied space is often a problem and does not allow easy opportunities to incorporate silencing devices. If possible, it is usually much better to have the fan discharge horizontally within the unit, into a discharge plenum (also in the unit), before the air is delivered to the duct system. The discharge plenum presents an opportunity to incorporate sound attenuation treatment before the air stream enters the duct in the ceiling of the occupied space. In some cases it is acceptable to run the ductwork across the roof before penetrating the roof, and this is always a favorable noise control approach since it allows space for turbulent airflow to smooth out, allows room for substantial silencing treatments before the ductwork gets into the ceiling of the occupied space, avoids what are often poor duct flow geometries at the fan discharge which occur when the duct drops directly out of the bottom of the unit, and allows low-frequency fan and duct rumble noise in the initial duct sections to dissipate outdoors rather than radiate in the ceiling of occupied space. However, such ducts may present practical and aesthetic design issues. The return/relief air path associated with rooftop units can present similar issues as discussed for the supply air path, but typically conditions at the return are not as severe because the flow conditions are not as bad and there is less fan noise. Similar noise control treatments might be considered for the return as for the supply.

Be sure that the duct geometry on the discharge of the supply fan is as reasonably advantageous as possible and avoid poor geometries as discussed above.

Silencers are often needed in the supply and return ducts and this attenuation should ideally happen before the ducts enter the occupied space, but if the ductwork from the unit drops directly into the ceiling and then has to turn into the horizontal ceiling plane, there is no aerodynamically appropriate location to put the silencer close to the unit discharge. Often, an elbow silencer can work well for these applications. When the silencer is moved away from the roof penetration, it is often necessary to encase the ductwork between the roof penetration and the silencer in a special noise barrier construction. The silencers for these applications also often need to have special high-transmission-loss casings or be encased along with the adjacent ductwork. All of these details are awkward, expensive, and difficult to construct within the limited ceiling space that typically exists. Sometimes, special curbs are created under the units to allow the return air to flow back to the units within the curb space, such as shown in Figure 17.4. With appropriate design and detailing, this can facilitate very good and practical ways to incorporate noise control treatments for both the supply and return paths.

Vibration isolation of packaged rooftop HVAC equipment can present problems, but this most often results when the units have integral compressor/condenser sections. When the unit just has basic supply and return/relief fans in the cabinet, vibration isolation can usually be satisfactorily accomplished with conventional, suitably soft, spring isolation for the fans within the unit. In this case, such units can externally be rigidly supported to the building in whatever manner is desired based on other considerations. When a unit has a compressor/condenser section, there is a concern for vibration isolation even though the compressor may be "internally" isolated. This is because the refrigerant piping within the unit is typically not adequately internally isolated, allowing vibration transmission to the equipment casing and hence to the building. In this case it is necessary to externally vibration isolate the entire unit with a spring isolation system. This is done in either of two ways, depending upon how the unit is intended to be supported and how the building designer plans to transfer the load to the building structure. There are special curb isolation devices available for equipment that requires curb support and closure of the space below the unit,

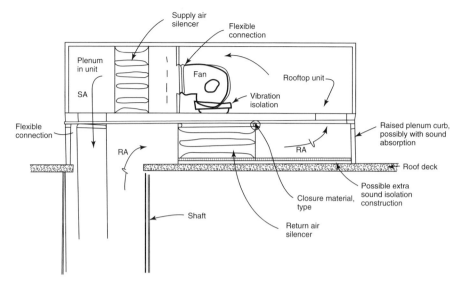

FIGURE 17.4 Rooftop unit with curb plenum return air. RA = return air, SA = supply air.

and some of these products can accommodate the aforementioned in-curb return airflow path scheme. If the load is to be transferred to the roof via a structural system raised above the roof, then point isolators are used. If the unit is intended for curb support but is raised above the roof in this manner, it may be necessary to mockup a structural curb for the unit to transfer its load to the point isolators.

17.11 OUTDOOR NOISE EMISSIONS

Many local authorities have noise regulations to control noise impact on neighboring properties and generally throughout the community. These need to be uncovered as part of the code searches at the outset of a building project so that they can be addressed in the mechanical system concept for the building. Typically, either fixed noise limits for specific receivers are established or the allowable noise level is based on the existing ambient noise level in the impacted area. If the noise requirement is based on the existing ambient noise condition, the existing noise has to be measured. Where neighbors are close by, requirements are stringent, and/or the receivers look down on the noise sources, meeting the noise requirements can be quite challenging.

Many pieces of equipment that are particularly economical (such as many air-cooled devices) are quite noisy and may not easily conform to noise regulations. Fortunately, equipment and mechanical schemes that produce less noise are often available, and, although they may be more expensive at the outset, they often provide some energy payback in the long run. Quieting the noise at the source is almost always the best approach to controlling outdoor noise emissions. Evaporative cooling equipment is usually quieter than air-cooled equipment. Equipment incorporating centrifugal fans is usually much quieter for a given duty than comparable equipment with propeller fans. Propeller fans can be quieted by using larger and/or slower moving fan blades, even if more fans are required for the duty. There are several degrees of propeller blade technology available (particularly for moderate to larger blade size applications) which can provide worthwhile noise reduction at an affordable cost. Typically, propeller blade noise reduction is achieved by having more aerodynamically efficient blades with wider chords so that they move more air with each rotation and can move a given quantity of air with lower fan speeds. Some equipment can be equipped with variable-speed drives so that when the demand on the equipment is low, the equipment (typically fans) can operate more slowly and therefore with lower noise emission levels. This is of course statistical noise reduction and will not produce noise reduction when full speed is required. However, this is particularly useful for equipment such as cooling towers where the heat rejection requirements at night are lower than the design condition and it is easier to reject heat to the cooler nighttime environment. The noise reduction that results with variable-speed drives often works naturally with the typical diurnal noise cycle which has lower noise conditions at night. It is best to avoid equipment which has a noise emission spectrum that is particularly tonal since many regulations prohibit or more strictly limit tonal sounds. Tonal sounds are typically judged to be more annoying than an equivalent level of broadband sound, so the likelihood of avoiding a noise complaint is better with broadband sources than with tonal sources. For a given level of satisfaction with the noise level achieved, a greater degree of noise control will need to be applied to tonal sources.

Sometimes equipment cannot be purchased suitably quiet within practical or economical limits, and it is necessary to apply noise control treatments. For many pieces of equipment and systems this can be reasonably and economically accomplished, such as with silencers or by building a noise barrier to block sound emissions in a particular direction. However, some pieces of equipment are not well suited to having noise control applied and treatments (such as barriers) can inhibit airflow. For instance, propeller fans produce dramatically lower airflow when working against even small degrees of extra pressure loss, and it is difficult to apply silencing devices to such equipment without degrading mechanical performance. Heat rejection equipment such as cooling towers and condensers need to avoid having the hot discharge air be recirculated back into the inlet because this can reduce efficiency and limit the mechanical capacity of the unit. Providing high levels of noise control to such equipment is often not possible and alternate equipment selection may become essential. The ability to adapt noise control treatments to equipment, if required, is a factor that should be considered in selecting the basic equipment and mechanical scheme. For small amounts of excess noise (typically less than about 10 dBA) it may be possible to create a noise barrier with sound-attenuating louvers which can allow some airflow to help mitigate the adverse airflow due to the barrier.

REFERENCES

1. *Applications Handbook*, American Society of Heating, Refrigerating and Air Conditioning Engineers, Inc., Atlanta, GA, 1999, Chapter 46.
2. C. Harris, *Handbook of Acoustical Measurements and Noise control*, 3rd ed., McGraw-Hill, New York, 1991.
3. A. T. Fry (Ed.), *Noise Control in Building Services*, Pergamon, Oxford, 1988.
4. M. E. Schaffer, *A Practical Guide to Noise and Vibration Control for HVAC Systems*, American Society of Heating, Refrigerating and Air Conditioning Engineers, Inc., Atlanta, GA, 1991.
5. R. S. Johnson, *Noise and Vibration Control in Buildings*, McGraw-Hill, New York, 1984.

CHAPTER 18
Active Control of Noise and Vibration

RONALD COLEMAN AND PAUL J. REMINGTON

BBN Technologies
Cambridge, Massachusetts

18.1 INTRODUCTION

The term active noise and vibration control (ANVC) refers to systems that use externally powered actuators to generate sound and/or vibration to reduce responses caused by unwanted disturbances. These systems consist of a set of reference sensors that monitor the source and a set of control, or residual, sensors that monitor system response. In some cases (e.g., feedback control) the reference and control sensors are the same. The signals from these sensors are inputs to an electronic controller, which filters the sensor signals to generate drive signals for the actuators. In a well-designed system the actuators excite a structure and/or sound field in such a way as to interfere with the disturbance and reduce the unwanted noise and/or vibration. The advent of modern digital signal processing hardware and software has allowed the controller in the system to be extremely flexible and even to adapt to changing conditions.

While ANVC has become a powerful tool for the noise control engineer, it is not a "silver bullet," and careful consideration must be given to passive mitigation approaches in any noise and vibration control problem. At low frequencies where the offending source consists of a small number of discrete frequencies and where the system to be controlled has only a small number of degrees of freedom, ANVC can be very cost effective. Passive treatments can also be utilized successfully on tonal sources at low frequency, but because the acoustic and structural wavelengths can be very long, the weight and size of the treatments must often be quite large to be effective.

Broadband sources of noise and vibration at high frequency, the other extreme, are often best controlled using passive techniques because many passive approaches are broadband in nature. In addition, because the acoustic and structural wavelengths in the system to be controlled are short, the treatments need not be large and heavy to be effective. On the other hand, active systems are usually less effective against broadband sources, and higher frequencies can place considerably higher demands on the controller hardware.

As the above discussion indicates, ANVC can be effectively employed against narrow-band disturbances at low frequencies where the system to be controlled has a small number of degrees of freedom. The term *degrees of freedom* refers to the number of ways that the system to be controlled can respond. The number of modes contributing significantly to the response of a system is an example of the number of degrees of freedom. It is important in ANVC because, as will be discussed later, the number of degrees of freedom corresponds to the number of actuators needed to provide effective control. As the number of input and output channels to the controller, the frequency range for control, and the performance goals increase, so do the risks associated with the stability, robustness, and costs of the resulting control system. In general, one wants to employ the system of minimum complexity to achieve the design goals. In many cases the use of a hybrid system employing both passive and active components is the best approach and the most likely to succeed.

This chapter is designed to provide the practicing noise control engineer with an overview of the technology currently employed in the application of ANVC. Enough detail is included in the hope of allowing readers to assess the applicability of the technology to their problem and to allow them to lay out an approach to the solution of that problem appropriate for ANVC. We begin with a brief discussion of some basic issues underlying ANVC technology, including conceptual overviews of several types of ANVC systems. Modern-day ANVC systems rely on digital signal processing technology, and the heart of digital signal processing technology is, of course, the digital filter. These are discussed in Section 18.3. The design of digital filters is a rich topic that has seen many years of development. Here we can only scratch the surface and hope to give the reader some appreciation of the power of the technology. Sections 18.4 and 18.5 deal with feedforward and feedback control architectures, respectively. The chapters compare the two architectures, provide guidelines for design of optimal controllers, including adaptation, and discuss system identification issues. For feedback a suboptimal heuristic design approach is also presented. The next section deals with practical design considerations. The topics include the determination of the number and location of actuators and sensors, selection of control architecture and hardware, performance simulations, and controller implementation and testing. The final section provides a detailed discussion of three ANVC systems that have been successfully implemented.

18.2 BASIC PRINCIPLES

Placement and Selection of Control Sources/Actuators

The basic concept behind active noise control is illustrated in Fig. 18.1. There a loudspeaker has been placed a distance L from an unwanted source of sound. The intent is to create sound from the *control* loudspeaker that will be out of phase with and will cancel the sound from the *disturbance* source. Ideally it would be desirable to control the sound radiation in all directions (global cancellation),

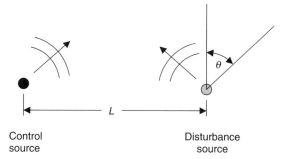

FIGURE 18.1 Disturbance source and control source.

and if the distance between the control source and disturbance source is small compared to a wavelength of sound, excellent global cancellation is possible. This comes about because the phases of the sound fields surrounding the two sources are very nearly the same at every location. Consequently, if one source is out of phase with the other two, the sound fields will cancel everywhere. On the other hand, if the separation distance is large compared to a wavelength, the two sound fields will have very different phases with the differences depending on location. As a result, the two sound fields will in some locations cancel and in others reinforce one another. This is illustrated in Fig. 18.2, where the change in the sound radiation from the disturbance source is shown as a function of angle θ. For $k_0 L = 0.1$ in the figure, where k_0 is the acoustical wavenumber and L the distance between the disturbance source and the control source, the change is a reduction of 20 dB or more for all angles. However, as the frequency increases and $k_0 L$ increases, the global reductions decrease. For example, in the figure, for $k_0 L = 1$ the reductions are limited to a small range of angles around $\pm 90°$ and for $k_0 L = 10$ there is a 6-dB increase for most angles. The figure clearly shows that it is desirable to locate the control source as close as possible to the disturbance source. In some cases, however, when the physical dimensions of the source are comparable to or larger than a wavelength at the frequencies of interest, the control source cannot be located close enough to the disturbing source. Figure 18.3 shows the effect of source separation on the overall radiated acoustic power (the integral over all angles in Fig. 18.2). The figure shows that for separation distances greater than one-quarter of an acoustic wavelength the total radiated power will actually increase. This is a fundamental limitation on ANVC that can only be overcome through the use of additional control sources. The use of multiple sources in a prototype system for controlling locomotive exhaust noise is discussed in Section 18.7 under MIMO Feedforward Active Locomotive Exhaust Noise Control System with Passive Component.

One of the earliest demonstrations of active noise control was carried out by Conover[1] and Conover and Ringlee[2] of General Electric in the early 1950s to control the noise from a high-voltage transmission line transformer. Using a single loudspeaker, he utilized the electrical line signal to drive the speaker and

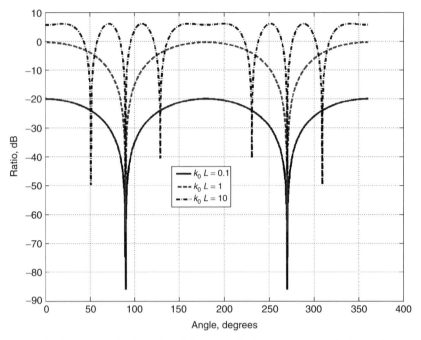

FIGURE 18.2 Change in the ratio of the controlled to the uncontrolled sound pressure level as a function of frequency and angle.

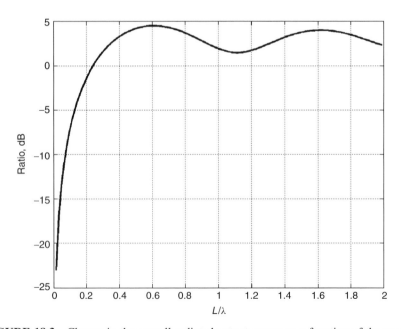

FIGURE 18.3 Change in the overall radiated output power as a function of the separation between sources.

appropriately amplified and phase shifted the harmonics of the line frequency to achieve the desired control. Because the system was not adaptive, occasional manual adjustment was required to correct for changes in sound propagation at the test site. In addition, because of the large size of the transformer, even at the low frequencies he was concerned with (120 Hz), he could not achieve global cancellation but had to be satisfied with ~10 dB of cancellation over a beam width of about 23°, corresponding to a point somewhere between the curves for $k_0 L = 1$ and $k_0 L = 10$ in Fig. 18.2. Remarkably, Conover was able to achieve this level of performance using purely analog hardware. Digital hardware did not come into use until the 1970s. Among the first to employ digital electronics were Kido[3] and Chaplin and Smith.[4] Kido used digital hardware to implement an active system to control transformer noise similar to Conover's and Chaplin and Smith developed a digital system for controlling tonal exhaust noise.

While the above discussion has focused on acoustical sources, similar principles apply to structural systems. Consider, for example, a plate with a point force disturbance applied as illustrated in Fig. 18.4 with a control actuator located as shown. In this example the plate is 10×5 ft and $\frac{1}{2}$-in-thick steel with a loss factor of 10% and the disturbance force and control actuator are about 1.4 ft apart. Because the plate in this example is finite in size, there will be resonant modes, which will add further complications to the selection of the number and location of the actuators. In this example, the control actuator is applying a force to the plate, the amplitude and phase of which is designed to control the first mode. Figure 18.5 shows the contribution of each mode to the plate response with and without activating the control actuator. Comparing the two plots in the figure, one can see that activating the control actuator has suppressed the first mode, and, in addition, the second and third modes have also been reduced. Higher

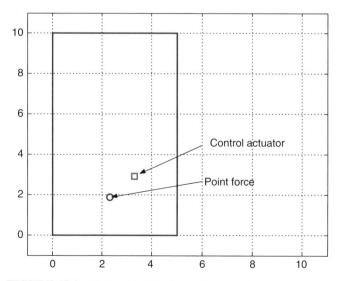

FIGURE 18.4 Plate with disturbance force and control actuator.

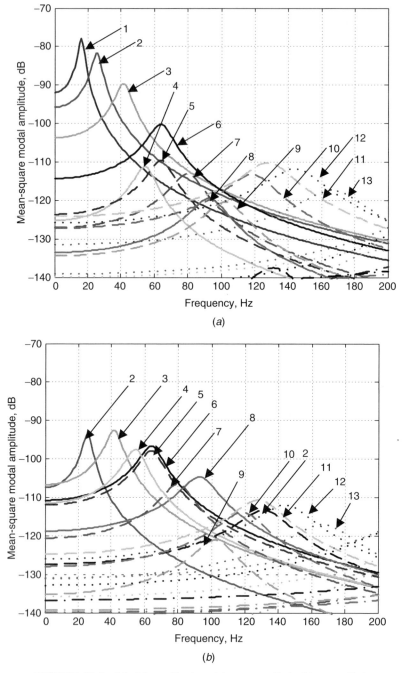

FIGURE 18.5 Modal amplitudes: (*a*) uncontrolled; (*b*) controlled.

order modes, however, have generally increased in amplitude. This increase in the amplitude of uncontrolled modes is referred to as *modal spillover*. Modes that are not targeted to be controlled can increase in amplitude. We can overcome this problem by moving the control actuator closer to the point of application of the disturbance force or we can add additional actuators, one for each mode that we wish to control. In general, placing the actuator as close as possible to the disturbance source will result in the best reduction in vibration. For example, in Fig. 18.6 the mean-square plate displacement averaged over the area of the plate is shown uncontrolled and controlled with a single actuator 1.4 ft away from the disturbance force and 0.28 ft away. When the actuator and disturbance are 1.4 ft apart, the vibration below 40 Hz is well controlled, but above about 50 Hz the control actuator increases the plate vibration due to modal spillover. When the control actuator is moved closer, spillover is virtually eliminated in the frequency range shown.

We can also increase the number of actuators, which will enable us to control additional modes and increase the frequency range over which the control will be effective. As an example consider the four control actuators located as shown in Fig. 18.7. These will be used to control the first four modes.* The result is shown in Fig. 18.8. While spillover has not been eliminated at all frequencies,

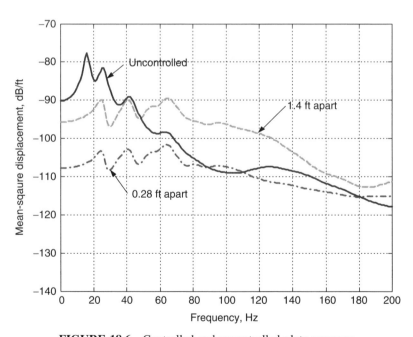

FIGURE 18.6 Controlled and uncontrolled plate response.

*Note that the actuator positions have been selected to ensure that at least one actuator is capable of exciting each of the four modes. If the actuators are placed near the node of a mode to be controlled, the actuator will be ineffective in controlling that mode.

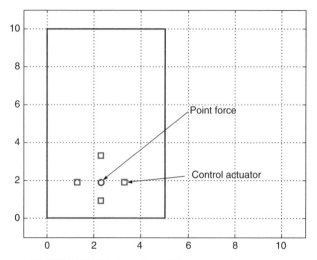

FIGURE 18.7 Location of four control actuators.

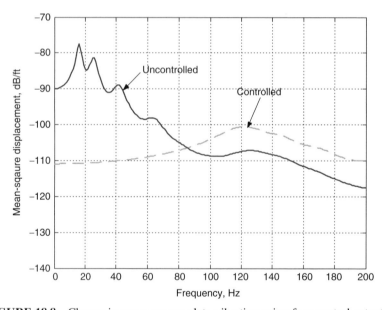

FIGURE 18.8 Change in mean-square plate vibration using four control actuators.

below 80 Hz the reduction in vibration is comparable to that achieved by bringing the control actuator very close to the disturbance point force. Good performance above 80 Hz can be achieved by continuing to increase the number of actuators and targeting more and more modes for control. Note that similar modal issues come into play in active noise control when the noise control is to be carried out in a confined space rather than in the free field in the first example.

BASIC PRINCIPLES   **729**

The lesson to be learned from all of this is that, when possible, placing the control sources as close as feasible to the offending sources will provide the best noise and vibration reduction performance. When this is not possible due to the size of the source, lack of access, and so on, the loss in performance can be made up, in principle, by employing a larger number of control sources. However, increasing the number of control sources will increase the complexity of the controller needed to drive them and will certainly increase the cost of the entire system.

Control sources come in a wide variety of forms. For acoustic applications, loudspeakers are typically employed. The active elements in the speakers may be electrodynamic (voice coil), piezoelectric, or electromagnetic, to mention a few. The active element might be used to move a speaker cone, modulate airflow, or deform a structure. The most common configuration is a speaker cone driven by a voice coil. Whatever the configuration, the speaker usually needs to be enclosed for protection against physical damage, weather protection, and performance enhancement.[5] Figure 18.9 shows an enclosure arrangement developed for the active control system discussed in Section 18.7 under MIMO Feedforward Active Locomotive Exhaust Noise Control System with Passive Component. These enclosures are bandpass enclosures designed to enhance the speaker output between 40 and 250 Hz. Each contains 2–12-in. loudspeakers and 10 were employed to control the locomotive exhaust noise. Despite these efforts the

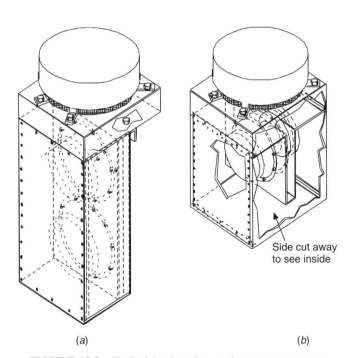

FIGURE 18.9 Typical loudspeaker enclosure arrangement.

performance of the active noise control system was limited by the output of these control sources. This is typically described as having insufficient *control authority*. Insufficient control authority is a problem often encountered in active noise control systems, and an assessment of the control source requirements is one of the first things that should be done before proceeding with any active noise and vibration control system design. If there is not sufficient space and power available to ensure adequate control authority, the active system is doomed to have inferior performance.

A wide variety of devices are available to be employed as structural control actuators. Typically these are devices that generate a force that is reacted against an inertial mass (inertial shaker) or against another part of the structure (interstructural shaker). The active element may be pneumatic,[6] hydraulic,[7] electrodynamic,[8] electromagnetic,[9] piezoelectric,[10] electrostatic,[11] or magnetostrictive,[12] to mention a few. A large-capacity electromagnetic inertially referenced shaker designed for use in an active control system is shown in Fig. 18.10.

In place of shakers piezoelectric[13] patches are sometimes applied directly to panels to control vibration or to radiate sound for noise control. Any of these devices can be very effective,* but the critical issue still remains. The device must provide sufficient control authority.

Control Sensors and Control Architectures

Sensors provide the input signals that the controller filters to drive the control sources and actuators. Sensors provide two functions, and to discuss those

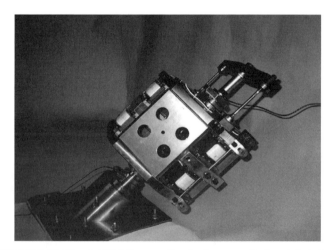

FIGURE 18.10 A 2000-lb-capacity electromagnetic inertial shaker.

*The inertial shaker usually provides better performance than the interstructural shaker, especially if the interstructural shaker creates a flanking path outside of the frequency range where the control system is functioning, but the interstructural device may be preferred when high control authority is needed with minimum weight.

functions, we need to describe the two commonly used control architectures: feedforward and feedback. The two architectures are illustrated in Fig. 18.11. In the feedforward system (Fig. 18.11a) one or more reference sensors measure a signal that is correlated with the disturbance to be controlled. The reference sensor output is sent to a control filter, which acts on the signal to generate an output signal that is used to drive the control actuators. The control actuators drive the system to be controlled (the plant function P in the figure) and the response of the plant is monitored by one or more residual sensors. The residual sensors are sometimes called the control sensors or error sensors. The purpose of the residual sensors is to monitor the performance of the control system. In later sections we will see how the reference and residual sensors are use to define and/or adapt the control filter.

In a feedback system (Fig. 18.11b) there are only residual sensors, or, to put it in other terms, the sensors in a feedback system provide both reference and residual functions. As in a feedforward system, the residual sensors monitor the performance of the control system. In addition, their signals are fed to the control filter, which acts on the signals to generate an output signal that is used to

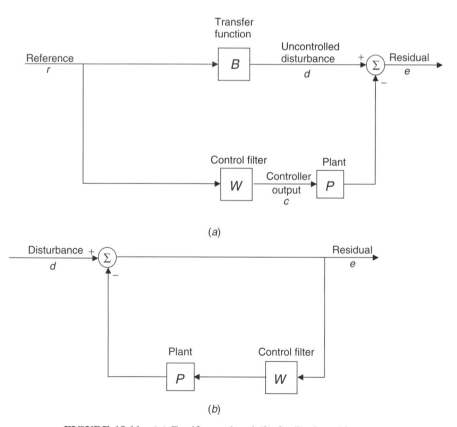

FIGURE 18.11 (a) Feedforward and (b) feedback architectures.

drive the control actuators. The control actuators drive the system and generate responses in the residual sensors through the plant transfer function, P.

The choice of whether to use feedback or feedforward architectures depends on a number of factors. In general, if reference sensors can be placed to measure a system response that is well correlated with the disturbance and residual sensors can be placed on the system such that there is sufficient delay between the reference and residual sensors (i.e., the reference sensors receive their signals before the residual sensors), then feedforward is the architecture of choice. This is especially true if the reference sensors are unaffected by the operation of the control actuators, such as would occur with a tachometer reference, for example. Feedforward is preferred (as we shall see in later sections of this chapter) because the algorithms are simple and easy to implement and the bandwidth requirements on a feedforward system are much less stringent than for feedback.

If a suitable set of reference and residual sensors satisfying the above requirements does not exist, then feedback must be used. As will be shown in later sections of this chapter, feedback control systems can present special problems. In particular, stability and out-of-band amplification requirements* can force the bandwidth of the system to be up to 100 times the control bandwidth.

As with actuators there is a wide variety of sensors used in active noise and vibration control systems. For systems designed to control only tones, a tachometer is often the reference sensor of choice, because in most cases the operation of the control system does not affect the tachometer, making the implementation of a feedforward system much easier. Other sensors that are commonly used include, accelerometers, force gauges, strain gauges and piezoelectric patches, proximity sensors, linear variable differential transformers (LVDTs), microphones, hydrophones, and pressure sensors. The requirements on the sensors are the same as for any high-quality instrumentation system: sufficient sensitivity to measure the phenomena of interest, low noise, stable operation, durability, and so on.

Performance Expectations

Active noise and vibration control systems can be powerful tools in the hands of a knowledgeable noise control engineer. In many cases, significant reductions in noise and vibration can be achieved with active systems that cannot be achieved using passive treatments with the same space and weight constraints. This is especially true at low frequencies. Despite the power of the technology, there are realistic limits on what reductions can be achieved. Figure 18.12 shows the performance achieved by the prototype active system described in Section 18.7 under Active Machinery Isolation. We present it here because it reflects the performance that is normally achievable in a properly designed active system. The figure shows the reduction in force transmitted by an active vibration isolation

---

*Outside of the frequency band where control is desired, the active system can amplify the disturbance unless great care is exercised in implementing the system filters to frequencies well beyond the frequency band for control.

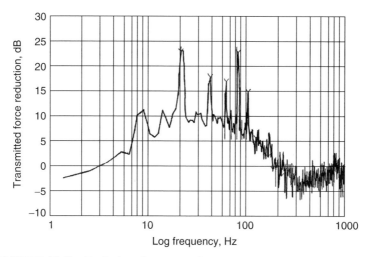

FIGURE 18.12 Typical performance of an active vibration control system.

mount. The system was a single-input, single-output (SISO) feedback control system designed for suppression of both broadband and narrow-band vibration. The narrow peaks in the figure with amplitudes of 15 to nearly 25 dB reflect the reduction in tonal components of the force transmitted by the mount. These values are what one would typically expect in tonal suppression from an active system. In addition to the peaks there is broad hill beneath them that in the frequency range of 10–100 Hz has an amplitude of from 6 to 10 dB. This too is typical of what one might expect from a system designed to provide control of broadband noise and vibration. Of course, there are many factors that go into determining the performance of an ANVC system, and there is no substitute for careful analytical/empirical modeling helping to direct the design process, but these values represent reasonable expectations for system performance when the technology is properly implemented.

Examples of Prototype ANVC Systems

There have been a vast number of ANVC prototype systems implemented over the past 50 years or so. Here we present a few examples of those systems to give the reader a sense of how the technology has been applied, where it has been most successful, and where additional technology and research is required. For a more detailed look at the early development of the technology the reader is referred to a number of review articles.[14–17]

The first patents proposing the use of a sound to cancel unwanted sound in the early 1930s proposed the type of system[18,19] illustrated in Fig. 18.13. These early proposals focused on the control of sound in a duct and involved the use of a reference microphone to sense the unwanted sound. In Fig. 18.13 the signal from that microphone was then acted on by appropriate electronics to synthesize

734 ACTIVE CONTROL OF NOISE AND VIBRATION

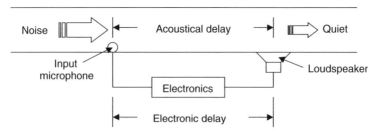

FIGURE 18.13 Schematic diagram of the active noise control concept for controlling sound in ducts.

a signal to drive a loudspeaker downstream from the microphone. The sound wave from the loudspeaker then cancels the unwanted sound. The use of active systems to control sound in ducts has received a great deal of attention over the years. Swinbanks[20] provided an extensive theoretical foundation in the early 1970s focusing on various arrangements of control sources at two different axial positions in a duct to minimize propagation of canceling sound back to the reference microphone. Other researchers[21,22] examined more complex control source configurations to deal with the control-source-to-reference coupling. Eventually laboratory experiments were carried out using analog electronics to confirm the predictions of the analysts[23] and to demonstrate both narrow-band and broadband control.[24] Eriksson's work in this area is particularly noteworthy.[25-28] His theoretical and experimental work eventually resulted in a commercially available active duct silencer utilizing digital electronics that was sold through Digisonix,* a subsidiary of Nelson Industries. The system configuration was very similar to that shown in Fig. 18.13.

Another early concept (1950s) for active noise control is illustrated in Fig. 18.14. This idea was developed by Olsen and May[29] while at the Radio

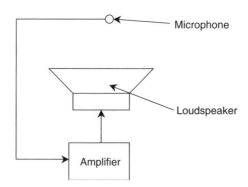

FIGURE 18.14 Olsen and May's "electronic sound absorber".

---

*Nelson Industries is currently owned by the Cummins Engine Company and Digisonix is no longer in business.

Corporation of America. The concept utilizes a microphone placed close to a loudspeaker in which the microphone detects the unwanted sound and the speaker generates a canceling signal that creates a zone of silence around the microphone. They assembled and successfully demonstrated their concept using only analog components. This basic concept has been extended and utilized in a number of applications, including control of automobile[30–34] and aircraft[35–38] interior noise and noise-reducing headsets.[39–41] The example in Section 18.7 under Active Control of Airborne Noise in a High-Speed Patrol Craft describes a system that uses this concept to reduce the noise in the berthing spaces of a high-speed patrol craft.

Figure 18.15 shows a schematic diagram of a notional active automotive interior noise control system. As shown, the system is designed to control noise generated by the propagation of vibration from the vehicle's suspension to the passenger compartment where it is radiated as sound to the vehicle interior (road noise) and engine noise. Feedforward active control architectures have usually been used to attack this problem. A number of reference accelerometers are placed on vehicle suspension components and in the engine compartment. Reference microphones might also be used. The reference signals are fed to a controller, which generates signals to drive the control loudspeakers in the passenger compartment. Residual microphones are located in the passenger compartment near the ears of the occupants so as to generate a zone of silence around the head of each occupant. Because of the nature of the interior noise, the control system must reduce both narrow-band and broadband noise. While the system concept is fairly simple, its successful implementation has proven difficult. First the number of residual sensors needed to sample all of the suspension sources can be quite large[42] (10–20). This can lead to significant costs due to the high channel count. In addition, the reductions achievable seem to be limited to about 6 dBA and the zone of silence around the control microphones is usefully large only at low

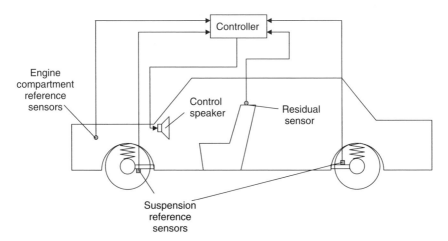

FIGURE 18.15 Schematic of an active automobile interior noise suppression system.

frequency (up to a few hundred hertz), limiting the useful frequency range of the system. Consequently, active systems providing broadband automobile interior noise control are not presently commercially available.

Systems for the control of aircraft interior noise have generally focused on tonal noise, that is, the propeller blade passage frequency and its harmonics. A schematic of a typical system is shown in Fig. 18.16. Experimental systems like that shown have generally provided about 10 dB of tonal interior noise control at the blade passage frequency and somewhat less at the higher harmonics. The system in the figure is similar to the automotive interior noise control system except that the reference is now a tachometer from the aircraft engine. The schematic shows control speakers as the actuators, but piezoelectric patches on the fuselage panels and inertial shakers on the ribs and stringers have also been tried, effectively turning the fuselage into a loudspeaker. Systems suitable for small propeller-driven business aircraft are presently commercially available from Ultra Electronics.

A number of manufacturers now provide reasonably priced active noise cancellation headsets. Figure 18.17 shows one arrangement. The figure shows the shell of one side of the headset in which there is a small speaker and colocated microphone. The signal that the wearer wishes to hear drives the speaker, the output of which is picked up by the microphone. The microphone also picks up any noise that passes through the shell. The microphone signal is then fed back, as shown in the block diagram in the figure. By increasing the gain of the filter,

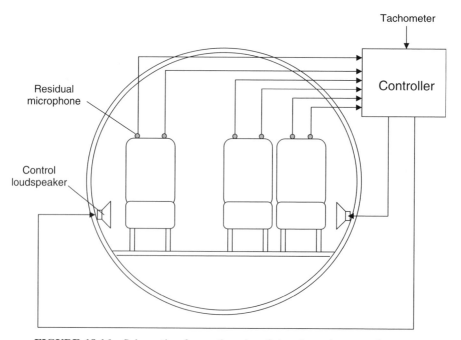

FIGURE 18.16 Schematic of an active aircraft interior noise control system.

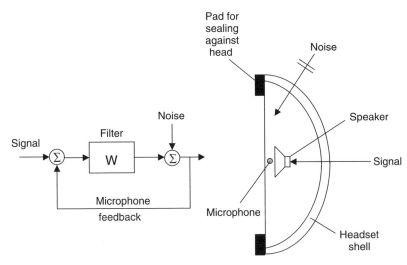

FIGURE 18.17 Typical noise cancellation headset.

W, in the block diagram* (typically an analog filter in commercial products rather than a digital one), the signal will pass through unchanged and the noise will be attenuated. Of course, this gain cannot be increased without limit, since too high a gain can result in instability and ringing. These devices produce noticeable improvement (10–15 dB) in the passive performance of hearing protection headsets above 100 Hz. The improvement gradually decreases with increasing frequency until at ~1 kHz most of the improvement has gone away. In very loud environments headsets can be limited in performance by the sound level the small speaker can generate within the shell.

Another area that has received considerable attention in recent years is the active control of tonal noise from turbofan engines.[43–50] Tonal noise from these engines is caused primarily by the interaction between the fan blade wakes and the stator vanes. The control systems for the most part have been feedforward systems utilizing a tachometer for a reference sensor. Since the fan and stator vanes are located in the bypass duct, the systems need to take into account the complex physics of the propagation of sound in ducts. The physics dictates that the fan–stator interaction noise will propagate down the duct as, so-called, spinning modes. These modes have a sinusoidal pressure variation in the circumferential directions that rotates as the mode propagates, hence the term *spinning mode*. Only a selected number of these modes are excited by the fan–stator interaction and a still smaller number propagate. Consequently, the control actuators must also generate spinning modes in order to cancel the fan–stator interaction noise without generating additional propagating modes in the duct. Figure 18.18 shows a schematic drawing of one such active noise control system. This particular

---

*In the block diagram W also contains the frequency response between the speaker input and the microphone output.

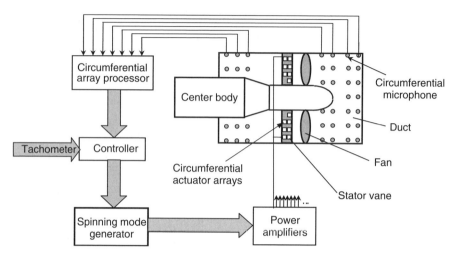

FIGURE 18.18 Schematic of the active noise control system for controlling fan-stator interaction noise.

system utilizes actuators in the stator vanes. Actuators have also been installed in the walls of the duct. The system must be designed with a sufficient number of actuators to ensure that the propagating modes to be controlled can be generated by the array without aliasing into other modes that will propagate.

Each array is driven by a single signal passed through analog or digital electronics (spinning-mode generator) that properly phase shifts the actuators in each array to produce a spinning mode of the desired circumferential order. Each actuator in the array must be the same. If there are large variations in amplitude and phase between actuators when driven by the same signal, then modes other than the desired mode will be generated, resulting in noise generation rather than suppression if the spurious modes propagate. The control sensors are typically wall-mounted microphones arranged in circumferential arrays and each array is steered (circumferential array processor) to accept a spinning mode of the circumferential mode order that is to be controlled. In a laboratory setting, systems of this type are capable of reducing the tonal noise from fan–stator interaction by 10–20 dB. In real high bypass turbofan engines, however, especially in modern-day large engines, none of the actuators tested to date have sufficient control authority to be able to overcome the disturbance. In addition, at the present time, while the controller channel count can be fairly low, typically equal to the number of modes to be controlled, the number of actuators and sensors needed to field a system can be quite large. It is very likely, however, that if active noise control is included early in the design process of an engine along with passive noise control treatments, the number of actuators and sensors needed could be substantially reduced. If more powerful actuation systems become available, commercially available active systems for the control of fan–stator interaction noise in turbofan engines may become a reality.

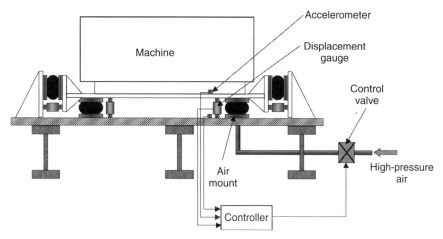

FIGURE 18.19 Notional activated vibration isolation mount.

In the 1950s Olsen[51] was the first to suggest extending active control to the control of machinery vibration. Since that time numerous studies have been carried out into the activation of vibration isolation mounts.[52–55] Figure 18.19 shows a schematic drawing of a notional active vibration isolation system. In this case, we envision a set of mounts supporting a vibrating machine on the deck of a ship. In the figure only one mount is shown activated. In actuality all (possibly 10 or 12 mounts) would be. It is desired to reduce the vibration excitation of the deck while maintaining the position of the machine, which is in a confined space, so that it does not strike any of the ship structure surrounding it during ship maneuvers. These two requirements are conflicting. Good vibration isolation performance requires a mount that is as compliant as possible while good position keeping requires a stiff mount. Two approaches can be taken to the activation of the mount. We can employ a mount that is stiff enough to ensure good position keeping and use an active system to increase the compliance of the mount at higher frequency where good vibration isolation performance is desired. We will refer to this type of system as the active isolation system. Another approach is to use a very compliant mount that provides the desired vibration isolation performance at high frequency and use the active system to provide position keeping at low frequency. We will refer to this system as the position-keeping system. For either concept to work the frequencies associated with ship maneuvers should be much lower than the frequencies where vibration isolation performance is desired. The notional system that we have shown in Fig. 18.19 incorporates air mounts. Since air mounts tend to be very compliant, they would be most appropriate for the position-keeping approach. The system would require 10–12 mounts to conveniently control all six rigid-body degrees of freedom. Each mount incorporates sensors for measuring relative displacement between the machine and the deck and for measuring the acceleration on the deck and the machine. These sensors would be used in the controller to generate a frequency-dependent linear combination of relative displacement, velocity, and acceleration at each mount. Those inputs would then be used by the controller in a feedback architecture to drive the control valves, which can inject high-pressure air into or vent it from the air mounts. Since the frequencies associated with ship maneuvers tend to be lie below 1 Hz, the machine would most likely act as a rigid body, significantly simplifying the system transfer functions and making the controller simpler as well. This approach does have some disadvantages. For example, it would require about 10–12 air mounts to conveniently implement it. In addition, all mounts must be coupled in the controller, making it a multiple-input, multiple-output (MIMO) system. On the other hand, there would only need to be six independent channels in the controller to control the six rigid-body degrees of freedom of the machine. Multi-degree-of-freedom systems of this type and of even greater complexity have been implemented in research programs, but none are commercially available. However, there remains considerable interest in active isolation systems and position-keeping systems for both military and civilian applications.

The above examples of types of ANVC systems represent only a taste of the many noise cancellation problems that have been addressed using the technology. In Section 18.7 we present three detailed examples of ANVC systems to provide the reader with a more in-depth look at the development, implementation, and performance of real prototype systems.

18.3 DIGITAL FILTERS

Advantages of Digital Filters

Active noise and vibration controllers are basically filters. The inputs are sensor signals and the outputs are the drive signals to the actuator electronics, which impart desired control forces to the physical system being controlled. For some ANVC applications, implementing these filters using analog electronics is an efficient and cost-effective approach. In particular, analog controllers should be considered for applications where the required magnitude and phase responses of the filters are a relatively simple function of frequency and when these filters are not required to change over time. An example of an application where analog controllers have been used effectively is active headsets. However, when the required magnitude and phase responses vary significantly as a function of frequency and the characteristics of these filters are required to change over time to maintain a desired level of performance, digital controllers are often the best choice. The filters implemented within these digital controllers are referred to as digital filters.

Digital controllers offer many advantages over analog controllers. As discussed by Nelson and Elliott,[56] these advantages include the following:

> *Flexibility.* Digital controllers implement digital filters by running programs written to operate on digital signal processor (DSP) chips. The codes to implement these filters can be written to allow for these filters to change

their characteristics over time (i.e., adapt). Further, these codes can be written to accommodate increases in the number of input and output channels. Additionally, the complexity of the digital filters can be increased or decreased by modifying the codes running on the DSPs.

Accuracy. Many DSPs support floating-point computations running in either single- or double-precision arithmetic. Although greater accuracy in the calculations can be achieved using double precision, this increased precision is seldom required and comes at the cost of increased memory requirements. The accuracy of single-precision controllers has been demonstrated for a wide variety of ANVC applications, including those with large numbers of input and output signals and digital filters requiring very complex magnitude and phase responses as a function of frequency.

Adaptability. Applications of ANVC often require the controller to alter filtering characteristics over time to maintain a desired level of performance. Controllers that satisfy this requirement are said to be "adaptive." An example of an adaptive controller is ANVC applied to systems whose dynamics (as a function of frequency) change over time. Digital controllers implemented using DSPs can be designed to adapt to these changes.

Cost. The cost of DSPs has continued to drop over the past 15 years. At the same time, the computational and memory capabilities of these chips have continued to increase. Currently available DSPs deliver up to 1×10^9 floating-point operations per second (1 gigaflop, abbreviated Gflop) and are equipped with up to 256 Mbytes of local, fast-access memory. In addition, the architecture of these chips has expanded to allow communications between DSPs to support efficient implementation of digital controllers employing multiple DSPs for large-scale applications. Both fixed- and floating-point DSPs are available. Fixed-point processors are cheaper than floating-point processors but are generally more difficult to program, resulting in increased software development costs.

In the following section the concepts of digital sampling and filtering are discussed. Subsequent sections use these concepts to summarize optimal digital and adaptive digital filter designs.

Description of Digital Filters

There are many good texts covering the concepts of digital sequences, sampling, and filtering.[56-59] For purposes here, we summarize these concepts following the developments of Nelson and Elliott[56] and Oppenheim and Schafer.[57] The interested reader should look to these references for more detailed discussions.

The input to a digital controller is typically an analog signal from a sensor that passes through an analog-to-digital (A/D) converter. The purpose of the A/D converter is to sample the input signal at regular time intervals (Δt) to produce a sequence of numbers corresponding to the amplitude of the input signal at each sample time. A schematic of the operation of an A/D converter is shown in

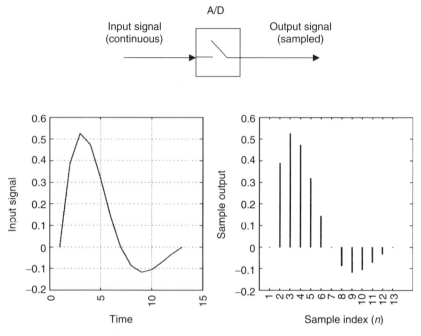

FIGURE 18.20 Digital sampling of an analog signal.

Fig. 18.20. The output of the A/D converter can be related to the continuous-time input signal through the following relationship:

$$f(n) = \sum_{i=-\infty}^{\infty} f(i)\delta(n-i) \qquad (18.1)$$

where $\delta(n-i)$ is called the Kronecker delta function, or unit-sample sequence, and is defined as

$$\delta(n-i) = \begin{cases} 1 & n=i \\ 0 & n \neq i \end{cases} \qquad (18.2)$$

The sampled signal in Eq. (18.1) is only defined for integer values of n. That is, $f(n)$ is only defined for $n = 0, \pm 1, \pm 2, \ldots$. An example of a continuous-time signal and the corresponding sampled signal $f(n)$ is shown in the lower portion of Fig. 18.20. The sample index n can be related to a discrete time by $t_n = n \, \Delta t$, where Δt is the fixed time interval between samples.

A digital system (e.g., digital filter) operates on a sequence of numbers to produce an output sequence. This is shown schematically in Fig. 18.21. The *impulse response* $w(n)$ of the digital system W is defined as the output corresponding to a unit-sample input [i.e., $\delta(n-i)$, $i = 0$]:

$$w(n) = W\{\delta(n)\} \qquad (18.3)$$

FIGURE 18.21 Digital filter operating on input $f(n)$ to produce an output $g(n)$.

where $w(i) = 0$ for $i < 0$ ensures that the output of the filter does not occur before the input sequence is applied. Filters that obey this constraint are referred to as *causal* filters.

The filter operation implied by the block diagram of Fig. 18.21 can be expressed mathematically as

$$g(n) = \sum_{i=0}^{I-1} w(i) f(n-i) \tag{18.4}$$

where the impulse response of the digital filter $w(i)$ in Eq. (18.4) is assumed to be a sequence of numbers of length I. It is common to refer to these numbers as the coefficients, taps, or weights of the digital filter. For example, $w(i)$ is called the ith coefficient of the digital filter w. Using this nomenclature, Eq. (18.4) states that the output of the digital filter w at time step n is equal to the summation over i of the ith coefficient of w times the input at time step $n - i$. In vector notation, this operation is equivalent to the inner product of a row vector and a column vector such that

$$g(n) = \mathbf{f}^T(n)\mathbf{w} \tag{18.5}$$

where $(\cdot)^T$ represents the vector transpose operator and $\mathbf{w}$ and $\mathbf{f}(n)$ are defined as

$$\begin{aligned}\mathbf{w}^T &= [w(0), w(1), w(2), \ldots, w(I-1)] \\ \mathbf{f}^T(n) &= [f(n), f(n-1), \ldots, f(n-I+1)]\end{aligned} \tag{18.6}$$

An example of this filtering process is shown schematically in Figs. 18.22 and 18.23. Figure 18.22 shows the input time sequence $f(n)$ and the impulse response of the digital filter $w(i)$. Figure 18.23 depicts the filtering of Eq. (18.4) at three increments in time during the filtering operation. The upper three charts plot the input sequence $f(n-i)$ for $n = 0, 4, 9$. The middle three charts plot the impulse response of the digital filter w. The lower three charts plot the output $g(n)$ at time step $n = 0, 4, 9$. As shown in these figures, the output of the filter is obtained by computing the summation of the time-reversed input signal as it slides past the impulse response of the filter.

The digital filter of Eq. (18.4) depends only on current and past values of the input sequence and is limited in length to I coefficients. These types of digital filters are referred to as finite impulse response (FIR) filters because their impulse response is zero after a finite number of taps or coefficients. These filters are also known as "tapped delay line," "moving-average" (MA), "nonrecursive," "all zero," or "transversal" filters.

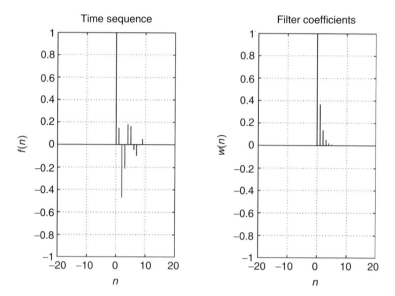

FIGURE 18.22 Example input sequence and impulse response.

A more general class of filters is defined as

$$g(n) = \sum_{i=0}^{I-1} a_i f(n-i) + \sum_{k=1}^{K} b_k g(n-k) \qquad (18.7)$$

In addition to depending on current and past values of the inputs, the current output of the digital filter represented by Eq. (18.7) depends on past values of the output. Digital filters of this type are referred to as infinite impulse response (IIR) filters because, in general, the impulse response of these filters never decays to zero. These filters are also known as "recursive," "pole-zero," or "autoregressive moving average" (ARMA) filters. The general form of Eq. (18.7) reduces to a FIR filter when $b_k = 0$ for all k.

Implementations of FIR and IIR filters can be represented schematically if we define a delay operator z^{-k} such that

$$f(n-k) = z^{-k}\{f(n)\} \qquad (18.8)$$

Using the property of the delay operator in Eq. (18.8) and the general filter representation in Eq. (18.7), a FIR filter can be implemented as shown in Fig. 18.24. A corresponding implementation for an IIR filter is shown in Fig. 18.25. A desired filter response can be approximated digitally using either the FIR or IIR filter structures discussed above. The choice between these two involves certain trade-offs, including the following:

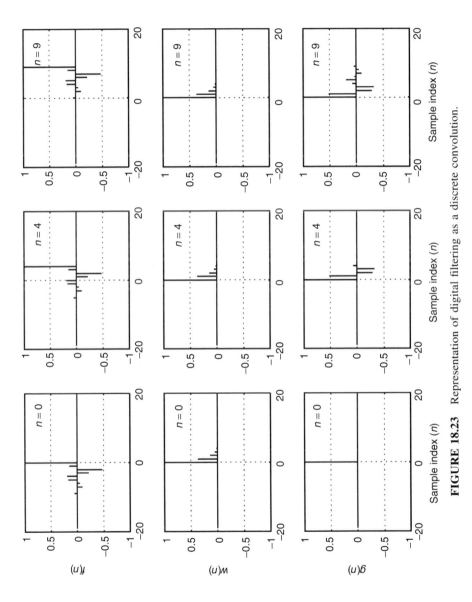

FIGURE 18.23 Representation of digital filtering as a discrete convolution.

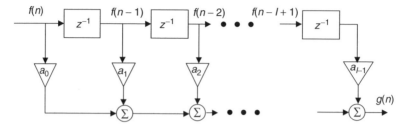

FIGURE 18.24 Implementation of a FIR filter using current and delayed values of the input.

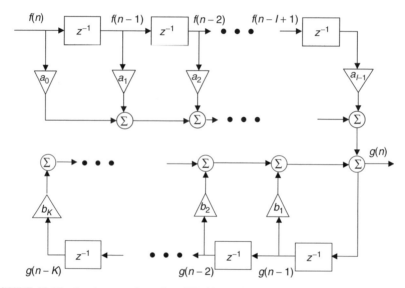

FIGURE 18.25 Implementation of an IIR filter using current input and delayed values of the inputs and outputs.

- FIR filters are guaranteed to be stable provided the coefficients are finite: Bounded inputs produce bounded outputs.
- IIR filters can be unstable even when all coefficients are finite: Bounded inputs can produce unbounded outputs.
- IIR filters may require fewer coefficients than FIR filters to approximate the frequency response necessary to provide active noise and vibration control: Typical transfer functions associated with structural acoustics include both poles and zeros, as do IIR filters, whereas FIR filters have only zeros.
- Design procedures to solve for FIR filter coefficients are generally numerically better behaved than those to solve for IIR filter coefficients.

Although there are advantages and disadvantages to both approaches, the vast majority of ANVC applications rely on the numerically stable design procedures and guaranteed stability associated with FIR filters and consequently implement the digital filters within ANVC controllers as FIR filters or composites of FIR filters.

Optimal Digital Filter Design

Having described the concept of digital filtering in the previous section, we focus in this section on design procedures to solve for the optimal coefficients of a digital filter for a particular class of problems. The class of problems considered is illustrated in Fig. 18.26. This problem is referred to as the electrical noise cancellation problem and is discussed in detail in references[56,58, and 59]. This is an important class of problems because it introduces the concept of minimizing a mean-square error metric and because it has strong parallels to problems encountered in ANVC applications; the main difference is that ANVC applications will have a nonunity transfer function relating the output of the control filter to the error signal response.

As shown in Fig. 18.26, the input signal $f(n)$, which is also referred to as the reference signal, is filtered by the FIR filter **w** to produce an output sequence $y(n)$. The desired signal, $\hat{d}(n)$, is contaminated by noise correlated with the input signal but presumably uncorrelated with the desired signal. The resulting signal plus noise is shown in the figure as $d(n)$. The correlation of the noise and the input signal is indicated by the unknown physical impulse response function h_i. The objective for this problem is to solve for the coefficients of the FIR filter that will take the input sequence $f(n)$ and generate the output sequence $y(n)$ that will remove the unwanted noise from the signal $d(n)$. In other words, we seek to design the optimal FIR filter to remove the part of $d(n)$ that is correlated with the input signal $f(n)$. If successful, the error signal $e(n)$ will be uncorrelated with the input signal when the filter is in place.

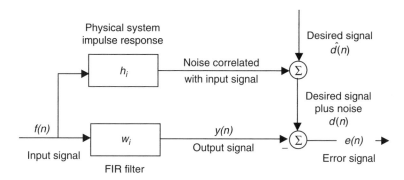

FIGURE 18.26 Electrical noise cancellation problem.

Following the developments in references 56 and 58 the error signal can be expressed in terms of the desired signal plus noise, the input signal, and the FIR filter coefficients:

$$e(n) = d(n) - \sum_{i=0}^{I} w(i) f(n-i)$$
$$= d(n) - \mathbf{w}^T \mathbf{f}(n) = d(n) - \mathbf{f}^T(n) \mathbf{w} \qquad (18.9)$$

where the lower equation has been used to express this relationship using vector inner products and $f(n)$ is defined in Eq. (18.6).

An expression for the mean-square error can be obtained by expanding terms in Eq. (18.9) and taking the expected value:

$$E\{e^2(n)\} = E\{d^2(n)\} - 2\mathbf{w}^T E\{\mathbf{f}(n)d(n)\} + \mathbf{w}^T E\{\mathbf{f}(n)\mathbf{f}^T(n)\} \mathbf{w} \qquad (18.10)$$

Equation (18.10) is a key result because it reveals that the mean-square error is a quadratic function of the weights of the FIR filter. The quadratic function will have a *global minimum* provided that the matrix $E\{\mathbf{f}(n)\mathbf{f}^T(n)\}$ in the third term is *positive definite*,[56,58–60] which means that all of its eigenvalues are greater than zero. Assuming this requirement is met, if one were to plot the mean-square error as a function of any two of the coefficients of the FIR filters, the "error surface" would be a bowl, as shown in Fig. 18.27. As a consequence, the FIR filter coefficients that correspond to the minimum mean-square error can be determined by differentiating Eq. (18.10) with respect to each coefficient of the FIR filter and setting the resulting set of equations equal to zero.

This process of differentiating and equating the results to zero will produce a set of linear equations in terms of the unknown coefficients of the FIR filter. This matrix equation can be expressed as

$$[A]\mathbf{w} = \mathbf{b} \qquad (18.11)$$

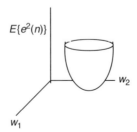

FIGURE 18.27 Error surface for a quadratic function in terms of coefficients w_1 and w_2.

where **[A]** is the *input correlation matrix* defined as

$$[\mathbf{A}] = \begin{bmatrix} R_{ff}(0) & R_{ff}(1) & \cdots & R_{ff}(I-1) \\ R_{ff}(1) & R_{ff}(0) & & \\ \vdots & & \ddots & \vdots \\ R_{ff}(I-1) & & & R_{ff}(0) \end{bmatrix} \quad (18.12)$$

where

$$R_{ff}(m) = E\{f(n)f(n-m)\} = E\{f(n)f(n+m)\} \quad (18.13)$$

The vector **b** in Eq. (18.11) corresponds to the lag values of the cross correlation between the reference sequence and the desired sequence given by

$$\mathbf{b} = [\, R_{fd}(0) \quad R_{fd}(1) \quad \cdots \quad R_{fd}(I-1) \,]^{\mathrm{T}} \quad (18.14)$$

where

$$R_{fd}(m) = E\{f(n)d(n+m)\} = E\{f(n-m)d(n)\} \quad (18.15)$$

With the above definitions for **[A]** and **b**, the coefficients of the optimal FIR filter can be expressed in terms of auto- and cross-correlation functions of the input and the desired signals. Provided that the input correlation matrix **[A]** is invertible, the optimal coefficients are given by

$$\mathbf{w}_{\text{opt}} = [\mathbf{A}]^{-1}\mathbf{b} \quad (18.16)$$

where $[\mathbf{A}]^{-1}$ is the inverse of the matrix **[A]**. As discussed by Widrow and Stearns[59] and Elliott,[58] the solution of Eq. (18.16) is called the *Wiener filter*. It can be shown that using this filter will cause the cross correlation between the error signal and input signal to be zero over the length of the filter $(I-1)$. For the problem posed in Fig. 18.26, this implies that the Wiener filter will remove any portion of the signal, $d(n)$, that is correlated with the input signal $f(n)$.

The residual mean-square error can be determined by substituting Eq. (18.16) into Eq. (18.10):

$$E\{e^2(n)\}_{\min} = E\{d^2(n)\} - \mathbf{b}^{\mathrm{T}}[\mathbf{A}]^{-1}\mathbf{b} \quad (18.17)$$

Solving for $\mathbf{w}_{\text{opt}}$ in Eq. (18.16), however, requires computing the inverse of the input correlation matrix **[A]**, which can be computationally expensive and may be numerically ill-conditioned (i.e., singular). As a consequence, procedures have been developed to search for the minimum of Eq. (18.10) using iterative techniques. One of the most computationally efficient iterative techniques, the least mean-square (LMS) algorithm, is introduced in the next section.

Adaptive Digital Filter Design

Gradient Descent Algorithms. As discussed in the previous section, the mean-square error is a quadratic function in terms of the optimal weights of the FIR filter. As such, gradient descent algorithms are guaranteed to converge to the global minimum, provided [**A**] is positive definite and the algorithm is itself stable.

The basic concept of a gradient descent algorithm is to estimate the gradient (i.e., derivative) of the mean-square error with respect to the FIR filter coefficients and update the filter coefficients by moving in the direction of the negative of the local gradient. There are many different algorithms that can be used to adaptively determine the filter coefficients corresponding to the global minimum. These algorithms include Newton-based algorithms, steepest-descent algorithms, and the LMS algorithm, which was introduced by Widrow and Stearns.[59] These algorithms typically trade off convergence speed versus numerical stability (for IIR filters) and computational complexity. Details of these algorithms can be found in texts written by Widrow and Stearns[59] and Elliott.[58] While sometimes having performance advantages over FIR filters, IIR filters can encounter numerical stability problems during design and can converge to nonglobal minima, resulting in less than optimal performance. On the other hand, the LMS algorithm discussed in the next section (for FIR filter design) approximates the gradient term, thus significantly reducing the required number of computations while ensuring convergence to the optimal coefficient vector.

LMS Algorithm. Widrow and Stearns[59] developed an elegant algorithm that greatly simplifies the computational complexity yet still converges to the optimal coefficient vector (i.e., Wiener filter). This algorithm estimates the gradient term using the instantaneous value of the error, as opposed to performing a matrix inversion or estimating expected values as required by Newton's algorithm and steepest-descent algorithm, respectively. As a consequence, the gradient term is approximated as [see Eq. (18.10)]

$$\frac{\partial E\{e^2(n)\}}{\partial \mathbf{w}} \approx 2e(n)\frac{\partial e(n)}{\partial \mathbf{w}} \approx -2\mathbf{f}(n)e(n) \tag{18.18}$$

Using this expression for the gradient, the update equation of the filter coefficients using the LMS algorithm is given by

$$\mathbf{w}(n+1) = \mathbf{w}(n) + 2\mu \mathbf{f}(n)e(n) \tag{18.19}$$

where μ is a constant known as the adaptation coefficient, which characterizes the speed of convergence of the algorithm. This equation states that the vector of filter coefficients at time step $n+1$ can be determined from the vector of coefficients at time step n plus a correction term in the direction of the negative of the gradient of the error surface as estimated by Eq. (18.18).

Widrow and others have shown that this algorithm will converge to the optimal solution and is guaranteed to be stable provided that the value of the convergence parameter satisfies the relationship

$$0 < \mu < \frac{2}{m\overline{f^2}} \tag{18.20}$$

The denominator of the right-hand term is an approximation to the trace of the input correlation matrix, where m is the number of FIR filter coefficients and $\overline{f^2}$ is equal to the mean-square value of $f(n)$.

Another important parameter associated with the LMS algorithm is termed the *misadjustment*, which is the ratio of the excess mean-square error to the minimum mean-square error. Essentially, this is a measure of how close the algorithm will converge toward the bottom of the quadratic performance surface (see, e.g., Fig. 18.27) for a particular value of μ. It can be shown that the *misadjustment* is equal to the convergence parameter times the trace of the input correlation matrix for the LMS algorithm.[59] If we invoke the same approximation to this trace as in Eq. (18.20), we obtain an expression for the convergence parameter in terms of the misadjustment and mean-square value of the input sequence:

$$\mu_{\max} = \frac{\rho}{m\overline{f^2}} \quad (18.21)$$

where ρ is the misadjustment factor. This final expression provides a useful procedure for selecting the convergence parameter. In fact, it is often useful to specify the desired misadjustment (e.g., $\rho = 0.01$) and let the adaptation coefficient be calculated from Eq. (18.21). In this way, the resulting adaptation coefficient will be normalized by the power of the reference signal. When the adaptation coefficient is adjusted to be inversely proportional to the power of the input signal, the algorithm is referred to as the *normalized* LMS algorithm. In practice, the value of the convergence coefficient obtained in this way serves as a starting point and manual adjustment is often necessary to obtain the desired performance.

To preserve clarity, we have presented the LMS algorithm assuming a SISO error signal. However, the algorithm is easily extendable to the MIMO case. Details for the MIMO implementation can be found in Widrow and Stearns,[59] Nelson and Elliott,[16] and Elliott.[58]

Because of its computational simplicity and good convergence properties, the LMS algorithm has been used extensively in a variety of applications (e.g., in-wire cancellation, system identification) and is the basis for the filtered-x LMS algorithm that is most often used in ANVC applications. In fact, the LMS algorithm is typically used to estimate the impulse response between the filter output and the error signal (i.e., plant), which is needed for the implementation of the filtered-x LMS algorithm. Details and examples of the filtered-x LMS algorithm are covered in the next section.

18.4 FEEDFORWARD CONTROL SYSTEMS

Basic Architecture

The basic architecture of a feedforward active control system is shown in Fig. 18.28. The figure shows a disturbance d exciting a system which is to be controlled. A transfer function T relates the disturbance to the system output o.

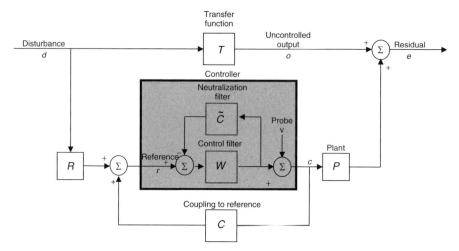

FIGURE 18.28 Basic feedforward architecture without adaptation.

A reference signal r correlated with the disturbance is used by the controller to generate a signal that drives a control actuator to excite the system and reduce the system response to the disturbance. The transfer function R is shown simply to indicate that the disturbance and the reference are correlated. The transfer function P, referred to as the plant, relates the controller output to the system output. The block diagram shows that the actuator, when driven by the controller output, is coupled not only to the system output but also to the reference through the transfer function C. The feedback to the reference signal is not desirable and the controller contains a filter $\tilde{C}$ the transfer function of which is a copy of the coupling transfer function between the control actuator and the reference sensor. The purpose of this neutralization filter is to cancel or neutralize the coupling between actuator and reference signal. In addition to $\tilde{C}$, the controller contains a second filter, W, the purpose of which is to take the reference signal and generate a signal to drive the actuator to reduce the system output o. If the neutralization filter performs properly, the coupling to the reference is removed and the block diagram simplifies to that shown in Fig. 18.29. Note that an alternate approach has been pioneered by Eriksson and his colleagues in which the designs of W and $\tilde{C}$ are combined in a slightly different architecture from that shown here. The interested reader is referred to references 25 and 61–63.

To see what the block diagram in Fig. 18.28 means physically, consider the duct problem described in Section 18.2. Figure 18.30 show a schematic of the duct with a reference microphone, controller, and control speaker. The disturbance, a sound wave of amplitude d, can be seen to the left propagating down the duct toward the control speaker. The transfer function T relates the disturbance pressure to the uncontrolled pressure, o, downstream of the speaker location in the duct. The reference microphone senses the disturbance pressure and sends a reference signal r to the controller. The control filter W operates

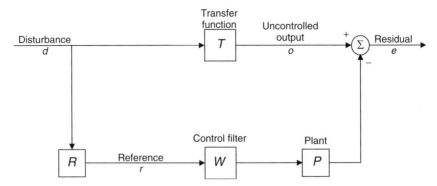

FIGURE 18.29 Simplified feedforward architecture with coupling to reference removed.

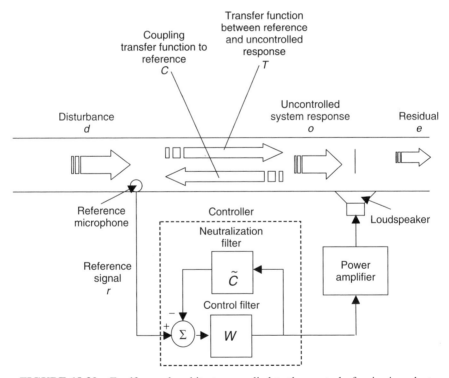

FIGURE 18.30 Feedforward architecture applied to the control of noise in a duct.

on the reference and sends the resulting signal to the power amplifier, which in turn drives the speaker. The transfer function relating the controller output signal to the pressure in the duct at the speaker is the plant transfer function P. The pressure generated by the speaker interacts with the disturbance propagating down the duct, reducing the amplitude and producing the residual sound wave of amplitude e. In an adaptive system there would be a microphone in the duct downstream of the control speaker to sense this residual. As we shall see later, this residual, or error, signal and the reference signal can then be used in the controller electronics to update, or adapt, the coefficients of the control filter. In addition to generating a sound wave that interferes with the sound propagating down the duct, the speaker also generates a sound wave which travels upstream to the reference microphone, contaminating the reference signal. The filter $\tilde{C}$ is a copy of the transfer function relating the controller output to the pressure at the reference microphone. It electronically removes the contamination from the control speaker in the reference microphone before it reaches the control filter W.

The architecture in Figs. 18.28 and 18.29 has been discussed as if there were only one reference signal and one residual. In fact, the controller need not be restricted to the SISO case. There commonly are cases with multiple references and multiple residuals. In fact, for feedforward systems the procedures for estimating the optimum control filter W for the MIMO case as well as the SISO case are well in hand. In addition the block diagrams in Figs. 18.28 and 18.29 are completely general and apply equally to SISO or MIMO systems. The only difference is that the blocks, which for SISO systems are transfer functions, for MIMO systems become transfer function matrices.

Optimal Control Filter Estimation

Tonal Disturbances. We will first consider the case of narrow-band, or tonal, disturbances. To do so, we will first modify the block diagram of Fig. 18.29 by showing a functional connection between the reference and the disturbance indicated by the transfer function B, as shown in Fig. 18.31. Mathematically this is equivalent to setting $T = 1$ and $B = R^{-1}$. A simple equation can be written for the residual e in terms of the controller transfer function in Fig. 18.31:

$$e(\omega) = d(\omega) - P(\omega)W(\omega)r(\omega) \qquad (18.22)$$

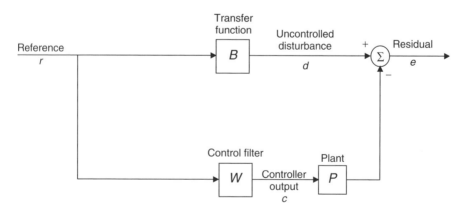

FIGURE 18.31 Simplified feedforward block diagram.

where ω is the frequency of the disturbance, the other variables are defined in Fig. 18.31, and all of the variables are scalars. We can use this equation to express the autospectrum of the residual as $S_{ee}(\omega) = E\{e(\omega)e^*(\omega)\}$, where $E\{\cdot\}$ means the expected value. Substituting Eq. (18.22) into this equation, we obtain

$$S_{ee}(\omega) = S_{dd}(\omega) - P(\omega)W(\omega)S_{rd}(\omega)$$
$$- P^*(\omega)W(\omega)^*S_{dr}(\omega) + |P^2(\omega)|S_{rr}(\omega)|W(\omega)^2| \quad (18.23)$$

where $(\cdot)^*$ means complex conjugate, S_{dd} is the autospectrum of the disturbance, S_{rr} is the autospectrum of the reference, $S_{dr} = E\{dr^*\}$ is the cross-spectra of the disturbance and the reference, and $S_{rd} = E\{rd^*\}$ is the complex conjugate of that cross-spectrum. If we take the derivative of Eq. (18.23) with respect to W and equate it to zero and solve for $W(\omega)$, we obtain

$$W(\omega) = \{|P^2(\omega)|S_{rr}(\omega)\}^{-1}P^*(\omega)S_{rd}(\omega) = P^{-1}(\omega)\left\{\frac{S_{rd}(\omega)}{S_{rr}(\omega)}\right\} = P^{-1}(\omega)B(\omega)$$
$$(18.24)$$

where we have taken advantage of the fact that $\{S_{rd}(\omega)/S_{rr}(\omega)\}$ is an estimate of the transfer function between the disturbance and the reference. In Eq. (18.24) W is the optimum control filter. It consists of the inverse of the plant times an estimate of the transfer function between the reference and the disturbance. Such a form for the control filter makes a lot of sense, because W is attempting to take the reference signal r and turn it into the disturbance d. To do so, the filter must first remove the influence of the plant transfer function, which is the purpose of the plant inverse. It then must include an estimate of the transfer function between the reference and the disturbance, which is the purpose of the term in brackets.

If we substitute Eq. (18.24) into Eq. (18.23), we can obtain an estimate of the spectrum of the residual when this control filter is used:

$$S_{ee}(\omega) = S_{dd}(\omega) - \frac{|S_{rd}(\omega)|^2}{S_{rr}(\omega)} \quad (18.25)$$

This equation shows that for perfect correlation between the reference and disturbance ($d = \gamma r$), where γ is constant, the spectrum of the residual goes to zero.

Note that this approach does not guarantee that the control filter will be causal. As discussed in Section 18.3, a causal filter is one whose impulse response is zero for $t < 0$. Causal filters can be formulated as FIR or stable IIR digital filters. Noncausal filters cannot. Consequently, a digital representation of a noncausal filter cannot be found for use in a control system. The filter is simply not realizable.

In some cases using the above control filter can result in excessive demands on the actuators, in which case it is useful to be able to limit the output of the control filter. Control effort weighting is one means of reducing the actuator effort. It can be included in the design equations for the filter by adding a term to Eq. (18.23), the equation for the residual. The term that is added is the weighted output of the controller given by

$$\alpha^2(\omega)|W(\omega)^2|S_{rr}(\omega) \tag{18.26}$$

where α^2 is a real positive scalar that might be a function of frequency. By selecting increasing values of α^2, the minimization of the residual can be adjusted. If α^2 is very small, the control filter will be designed as if there were no control effort weighting. If α^2 is large, the control effort will be reduced, and the performance of the control system will also be reduced. Including Eq. (18.26) in Eq. (18.23) results in a control filter given by

$$W(\omega) = \{[|P^2(\omega)| + \alpha^2(\omega)]S_{rr}(\omega)\}^{-1} P^*(\omega) S_{rd}(\omega) \tag{18.27}$$

and the residual is given by

$$S_{ee}(\omega) = S_{dd}(\omega) - \left\{\frac{|P^2(\omega)|}{|P^2(\omega)| + \alpha^2(\omega)}\right\} \left\{2 - \frac{|P^2(\omega)|}{|P^2(\omega)| + \alpha^2(\omega)}\right\} \frac{|S_{rd}(\omega)|^2}{S_{rr}(\omega)} \tag{18.28}$$

It is clear from this equation that as α^2 approaches zero the residual becomes the same as Eq. (18.25) with no control effort weighting. However, as α^2 becomes large, $S_{ee} \to S_{dd}$ and the control performance is reduced. Similarly the spectrum of the controller output c (see Fig. 18.31) is given by

$$S_{cc}(\omega) = |W^2(\omega)|S_{rr}(\omega) = \frac{|P^2(\omega)|}{\{|P^2(\omega)| + \alpha^2(\omega)\}^2} \frac{|S_{rd}(\omega)|^2}{S_{rr}(\omega)}$$

This equation shows that as α^2 increases the spectrum of the controller output will decrease, reducing the drive to the actuators. As a result increasing the control effort weighting will decrease the actuator drive at the cost of increasing the residual.

Broadband Disturbances. For a pure-tone disturbance, Eqs. (18.24) and (18.27) provide the proper control filter representation (amplitude and phase) at the frequency of the disturbance. For a broadband disturbance, if the broadband filter defined by these equations is noncausal, as it often is, then, a different approach that includes a constraint on causality needs to be employed to determine the optimum causal filter. We can ensure that the control filter is causal by inserting the form of a causal filter into Eq. (18.23) and carrying out the same optimization procedure. Because it results in a set of linear equations, we will use the form of an FIR digital filter given by

$$W(\omega) = \sum_{n=0}^{N-1} W_n e^{-j\omega n \Delta t} \tag{18.29}$$

where N is the number of taps in the filter, W_n is the real amplitude associated with each tap, and Δt is the sampling period (1/sampling rate). Before substituting Eq. (18.29) into (18.23) we need to modify Eq. (18.23) slightly. Since we are now concerned with a broadband disturbance, we need to integrate the residual spectrum over the frequency range where control is desired. Following the same procedure as for Eq. (18.23), we obtain for the optimum control filter the matrix equation

$$\begin{Bmatrix} W_0 \\ W_1 \\ \vdots \\ W_{N-1} \end{Bmatrix} = \begin{bmatrix} C_0 & C_{-1} & \cdots & C_{1-N} \\ C_1 & C_0 & \cdots & C_{2-N} \\ \vdots & & & \vdots \\ C_{N-1} & C_{N-2} & \cdots & C_0 \end{bmatrix}^{-1} \begin{Bmatrix} A_0 \\ A_1 \\ \vdots \\ A_{N-1} \end{Bmatrix} \quad (18.30)$$

where

$$A_p = \mathrm{Re}\left\{ \sum_{k=0}^{K-1} P(k) S_{rd}(k) e^{-jpk(2\pi/K)} \right\}$$

$$C_{(p-m)} = \mathrm{Re}\left\{ \sum_{k=0}^{K-1} \{|P^2(k)| + \alpha^2(k)\} S_{rr}(k) e^{-j(p-m)k(2\pi/K)} \right\} \quad (18.31)$$

and $p = 0, 1, \ldots, N-1$ and $m = 0, 1, \ldots, N-1$. In writing this equation, we have substituted a discrete sum for the integrals. The knowledgeable reader will recognize that Eq. (18.31) is in the form of the discrete Fourier transform (DFT), allowing for the efficient use of the fast Fourier transform algorithm in the evaluation of the vector and matrix elements in Eq. (18.30). In the equations K is the number of elements in the DFT.

Example. To illustrate these ideas, we will compute the noncausal and causal control filters for a simple example. We will use the following parameters:

$$S_{dd} = 1 \quad S_{rr} = 1$$

$$S_{rd} = e^{-j\omega T} \left\{ \frac{\omega_{\mathrm{BW}}^4}{\omega^4 + \omega_{\mathrm{BW}}^4} \right\} \quad (18.32)$$

$$P = \frac{\omega_0^2}{-\omega^2 + j\eta\omega_0\omega + \omega_0^2} \quad (18.33)$$

$$T = 0.02 \text{ s} \quad f_0 = \frac{\omega_0}{2\pi} = 25 \text{ Hz}$$

$$\eta = 0.1 \quad f_{\mathrm{BW}} = \frac{\omega_{\mathrm{BW}}}{2\pi} = 40 \text{ Hz}$$

$$\text{Sampling rate} = 200 \text{ Hz} \quad \text{Filter length} = 100 \text{ taps}$$

In this case the autocorrelations of the disturbance and reference are taken as unity. The cross-spectral density S_{rd} shows a delay of $T = 0.02$ s and includes a transfer function (in brackets) that decreases with increasing frequency. The plant P is a simple second-order system with a natural frequency of 25 Hz and a loss factor of 0.1. For the causal control filter estimation procedure we will use the indicated sampling rate and filter length.

Figure 18.32 compares the frequency responses of the control filters estimated using the noncausal and causal estimation procedures. The two estimation procedures give different filters but not markedly so. The difference is slightly more evident in Figs. 18.33a and 18.33b, where the impulse response of the control filter estimated using the noncausal approach and the filter tap coefficients of the filter from the causal procedure are compared. Bear in mind that if we plot the filter tap coefficients separated by the sampling period (0.005 s at 200 Hz sampling rate), we will obtain a graph corresponding to the impulse response of the filter. The two figures are similar except that in the noncausal impulse response there is a small increase near the end of the time interval. Because of the periodicity of the impulse response of the filter obtained using the DFT, the increase at the end of the time interval actually indicates that the filter is responding before $t = 0$ and consequently that the filter is noncausal. The effect in this

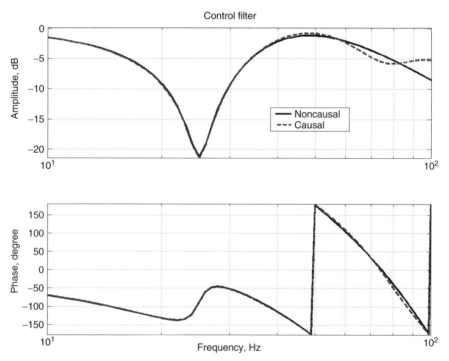

FIGURE 18.32 Control filter frequency response using the noncausal and causal estimation procedures.

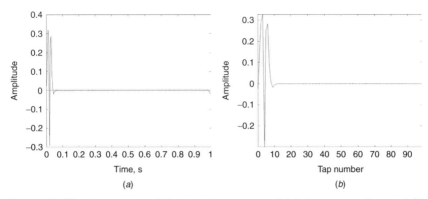

FIGURE 18.33 Comparison of the impulse response of (*a*) the noncausal control filter and (*b*) the tap coefficients of the causal filter.

instance is small; however, in later examples where we bring more realism into the calculation by including anti-aliasing filters and other practical considerations we will see much larger noncausal effects. Note that the tap coefficients of the filter estimated using the causal procedure show no such increase and the higher tap coefficients decrease to zero.

The performance of this system using the causal control filter is shown in Fig. 18.34. The figure shows the ratio of the spectrum of the controlled residual to the spectrum of the uncontrolled residual $\{S_{ee}(\omega)/S_{dd}(\omega)\}$ as a function of

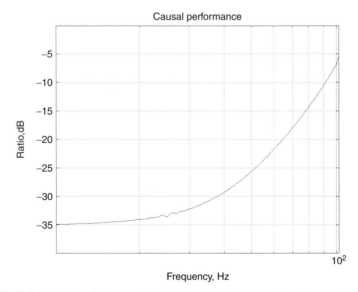

FIGURE 18.34 Ratio of the residual to the disturbance using the control filter in Fig. 18.32.

frequency out to the Nyquist frequency. Note that the performance of the non-causal filter (not shown in the graph) would be predicted to be perfect, that is, the residual would be zero for all frequencies.

The performance shown in the Fig. 18.34 is unrealistically large for a broadband active control system. This is a consequence of the fact that, at this stage, we have neglected to include other elements in the analysis that would add more realism to the estimate, such as system delays, anti-aliasing filters, sensor noise, and so on. In addition the figure only shows system performance up to the Nyquist frequency (one-half the sampling frequency). As we shall see in a later section, the digital control filter is periodic in frequency with a period equal to the sampling frequency. Consequently, the performance above the Nyquist frequency may also need to be controlled, especially if there is significant energy in the disturbance at those frequencies. We will see how this is done in a later section.

Adaptive Control

In the previous section we discussed the design of a control filter to reduce optimally a chosen residual in a least-squares sense. In that derivation the system to be controlled was assumed to be fixed. In reality the transfer functions defining dynamic systems often change with time. This means that those transfer functions must be measured periodically and the new results used to update the control filter coefficients. We will discuss this process of system identification in the next section. Here we examine an adaptive control algorithm based on the LMS algorithm that continuously updates the control filter coefficients in order to minimize the chosen residual sensor signals. The filtered-x algorithm, as it has come to be called, updates the control filter coefficients at each time step.

The filtered-x LMS algorithm, a modification of the LMS algorithm[59] commonly used for cancellation of electronic noise, was first proposed by Morgan.[64] If used unmodified in an active cancellation system, the LMS algorithm can lead to instabilities because the signal from the controller must pass through the plant dynamics where it will experience amplitude changes and phase shifts. The solution to this problem proposed by Morgan was also proposed independently by Widrow et al.[65] and Burgess.[66] Morgan's solution, which has come to be called the filtered-x LMS algorithm, is used extensively today and has been generalized to MIMO systems by Elliott and Nelson.[67]

Figure 18.35a shows a block diagram of a simplified feedforward control system. If we change the order of the plant and control filter in the block diagram, the response will be unchanged as long as the system is linear. The resulting block diagram is shown Fig. 18.35b. The result is a new reference signal u that is the original reference signal filtered by the plant transfer function P. The residual of this new system can be written as

$$e(n) = d(n) - \sum_{n=0}^{N_w-1} w(k)u(n-k) = d(n) - \mathbf{w}^T\mathbf{u} \qquad (18.34)$$

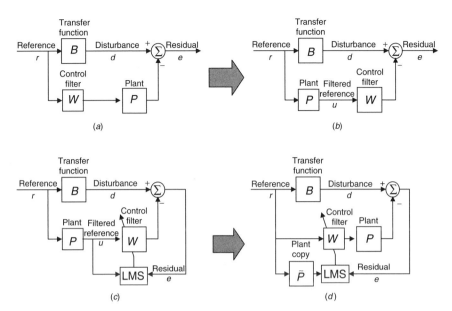

FIGURE 18.35 Feedforward block diagram with and without adaptation showing the progression to the LMS filtered-x algorithm: (a) original diagram; (b) changing the order of plant and control filter; (c) LMS algorithm; (d) filtered-x algorithm.

where $e(n)$ is the nth element of a time sequence e, $d(n)$ is similarly defined, $\mathbf{w}^T$ is the transpose of a vector, the N_w elements of which are the control filter tap coefficients, and $\mathbf{u}$ is a column vector of the last N_w samples in the filtered reference, which is given by

$$u(n) = \sum_{m=0}^{N_w-1} p(m)r(n-m) \tag{18.35}$$

where p is the impulse response of the system estimate of the plant filter P and r is the reference signal time sequence. The block diagram showing the adaptation process is shown in Fig. 18.35c, and a slight rearrangement of the blocks in the block diagram in Fig. 18.35d results in the specialized LMS algorithm called the filtered-x LMS algorithm.

Unlike the direct-estimation approach in the previous section in which the filter coefficients are estimated in a single calculation, the filtered-x algorithm uses an iterative approach which updates the estimate at each time step. The update equation for the control filter is derived similarly to that for the LMS algorithm (see Adaptive Digital Filter Design in Section 18.3). The resulting equation to estimate the coefficients of w is typically written as

$$\mathbf{w}(n+1) = \mathbf{w}(n) + 2\mu e(n)\mathbf{u}(n) \tag{18.36}$$

where

$$\mathbf{w}(n) = \begin{Bmatrix} w(0) \\ w(1) \\ \vdots \\ w(N_w - 1) \end{Bmatrix}$$

is a column vector of the nth estimate of the N_w control filter coefficients, $w(0)$ is the first filter coefficient, $w(1)$ is the second, and so on,

$$\mathbf{u}(n) = \begin{Bmatrix} u(n) \\ u(n-1) \\ \vdots \\ u(n - N_w + 1) \end{Bmatrix}$$

is a column vector of the last N_w samples in the filtered reference, and μ is a scalar convergence coefficient set by the user to ensure stable convergence to the optimum filter coefficients. This equation is very similar to that derived for the LMS algorithm, the only difference being that the filtered reference $\mathbf{u}(n)$ is used in the update equation rather than the reference itself.

In Eq. (18.36) the larger is μ, the faster the algorithm will converge to the optimum filter coefficients. However, if μ is too large, the algorithm will go unstable and will not converge. The convergence coefficient can be selected using the following rule of thumb[58]:

$$0 < \mu \leq \frac{1}{N_w E\{u^2\}} \qquad (18.37)$$

where N_w is the number of control filter coefficients and $E\{u^2\}$ is the mean-square value of the filtered reference. This is very similar to the rule of thumb for the LMS algorithm presented in Section 18.3 except that the mean-square value of the filtered reference, $E\{u^2\}$, is substituted for mean-square value of the reference itself, $E\{r^2\}$.

Equation (18.36) along with Eq. (18.35) states the SISO filtered-x algorithm. This algorithm has enjoyed a great deal of popularity because it is very easy to implement and functions well in the presence of errors in the plant estimate used to filter the reference. If the convergence rate is kept slow enough, the algorithm will still find the optimum filter coefficients even in the presence of plant phase errors of up to 90°.[64] In addition, the algorithm can be very effective in maintaining control even if the reference is nonstationary. Finally, it is possible to introduce something comparable to control effort weighting as discussed in the previous section. In the filtered-x algorithm this is referred to as *leakage*. By reducing the control filter coefficients by a small percentage at each time step,[59] the effect is similar to broadband control effort weighting.

FEEDFORWARD CONTROL SYSTEMS 763

Example. To illustrate the application of this algorithm, we will utilize it to design an optimum filter for the system of the previous section. To do so, we will require the impulse response function (IRF) associated with the cross-spectrum between disturbance and reference, S_{rd}, and the IRF of the plant, P. We obtain these by simply inverse Fourier transforming the FRF functions in Eqs. (18.32) and (18.33), respectively, and retaining a sufficient number of terms to ensure that we capture all of the impulse response. Figure 18.36 illustrates the two IRFs. The plant IRF is a decaying sinusoid with a frequency of 25 Hz, the natural frequency of the plant. The cross-spectrum IRF shows the delay of 0.02 s introduced into the model. We will retain the full 1 s duration (200 taps at a sampling rate of 200 Hz) of the IRFs in the figure, although it would probably be sufficient to retain substantially less. The filter coefficients obtained using the direct [Eq. (18.30)] and filtered-x [Eqs. (18.35) and (18.36)] estimation procedures are compared in Fig. 18.37, and the frequency response functions are compared in Fig. 18.38. There are clear differences between the two control filters that may be a consequence of the fact that the filtered-x procedure is still converging. That the procedure is still converging can be seen in Fig. 18.39, which shows a time history of the residual, or controlled disturbance. Even after over 80 s the residual still shows a tendency to decrease. Figure 18.40 shows the ratio of the spectrum of the controlled disturbance to the spectrum of the uncontrolled disturbance as a function of frequency at the end of the time interval in Fig. 18.39. Also shown in the figure is the same ratio for the direct-estimation procedure. In this case the two procedures give comparable results despite the differences in the control filters.

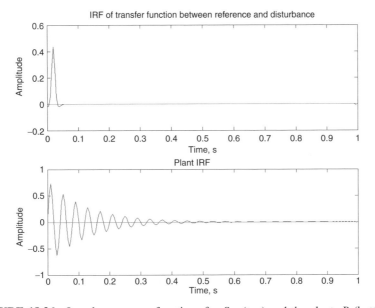

FIGURE 18.36 Impulse response functions for S_{rd} (top) and the plant, P (bottom).

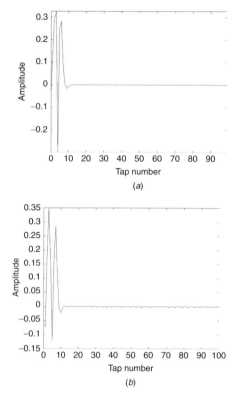

FIGURE 18.37 (a) Direct-estimation control filter coefficient compared to (b) filtered-x control filter coefficients.

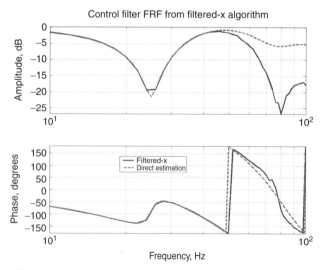

FIGURE 18.38 Filtered-x control filter frequency response compared to the direct estimation control filter frequency response.

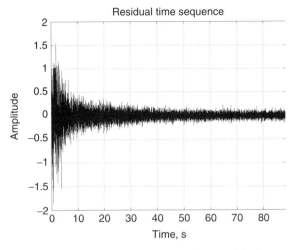

FIGURE 18.39 Time history of the residual.

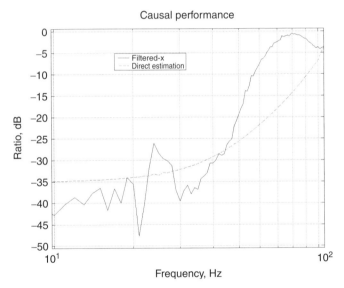

FIGURE 18.40 Ratio of controlled to uncontrolled disturbance.

Control of Aliasing Effects

In the previous sections we have focused on control of the disturbance below the Nyquist frequency. In fact, because digital filters are periodic in the frequency domain with a period equal to the sampling frequency, we must also concern ourselves with the performance above the Nyquist frequency, especially if the disturbance has significant amplitude there. Figure 18.41 shows the ratio

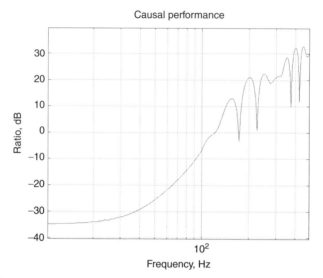

FIGURE 18.41 Change in the ratio of controlled to uncontrolled disturbance beyond the Nyquist frequency.

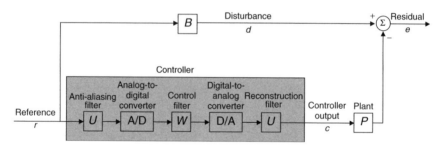

FIGURE 18.42 Feedforward block diagram further simplified (adaptation blocks not shown).

of the controlled-to-uncontrolled disturbance beyond the Nyquist frequency for the direct-estimation case in the previous section on optimal control filter estimation. Above ~125 Hz the figure shows that the control system will amplify the disturbance. To control this out-of-band amplification, one usually employs anti-aliasing filters, low-pass filters that minimize energy above the Nyquist frequency entering the controller. In addition, there are other low-pass filters and delays that should be included in a more realistic model of the active control system. To that end Fig. 18.42 shows the controller expanded into five blocks. First the analog reference signal needs to be digitized before the controller can utilize it. That is the function of the A/D converter shown as the second block in the controller. The first block is a low-pass filter called the anti-aliasing filter. Its cutoff frequency is usually set slightly below the Nyquist frequency in order to reduce aliasing in the digital controller, as described above. The third block is the digital control filter, the output of which is fed to a D/A converter to convert the digital signals to analog signals. The D/A converter is usually a sample-and-hold device that produces a signal with discrete steps or jumps at the sampling rate. Consequently, there is usually a low-pass filter at the output of the D/A converter, typically called a reconstruction filter, to smooth or reconstruct the signal. This filter is typically the same as the anti-aliasing filter, although there is no requirement that the two filters be the same. The output of the reconstruction filter is an analog signal, c, that is applied to power electronics that will in turn drive a speaker or other actuator. The plant, P, accounts for all of the dynamics relating the controller output to the control signal, which cancels the disturbance d. In designing the control filter there may be significant delays associated with the anti-aliasing filter, the reconstruction filter, and the D/A converter, which need to be included in designing and evaluating the control filter. The delays associated with the A/D converter are usually small and can, in most cases, be neglected. The delays associated with the D/A converter, on the other hand, are on the order of half a sample period and usually need to be included in the design and evaluation process. To carry out the analysis, the anti-aliasing filter, reconstruction filter, and sample-and-hold delay (D/A converter) are usually lumped in with the plant transfer function. The D/A converter is typically modeled by a simple delay of one-half of the sample period. The total plant then becomes

$$U^2(\omega)P(\omega)e^{-j\omega\Delta T/2} \qquad (18.38)$$

where ΔT is the sample period and $U^2(\omega)$ is the product of the frequency response functions of the anti-aliasing and reconstruction filters. To illustrate the impact on the design process and on the resulting performance, we will select an anti-aliasing filter and go through the design process once again using the same plant and disturbance-to-residual transfer function as used in Section 18.4 and given by Eqs. (18.32) and (18.33). As will be discussed in Section 18.6 under Hardware Selection, there is a wide variety of low-pass filter types that are commonly used for anti-aliasing and reconstruction, such as Butterworth, Bessel, and elliptical (Cauer), to name just a few. In this example we will use the third-order elliptical filter whose frequency response function is shown in Fig. 18.43. The filter has a cutoff frequency of 70 Hz, slightly below the Nyquist frequency, and the ratio of passband to stop-band amplitude is nominally 50 dB.

Using the direct-estimation procedure outlined in Section 18.4 under Optimal Control Filter Estimation, we obtain the causal control filter whose frequency response is shown in Fig. 18.44. Also shown in the figure is the noncausal control filter given by $B(\omega)\{U^2(\omega)P(\omega)\}^{-1}e^{j\omega\Delta T/2}$. The frequency response functions of the two filters are clearly quite different, but the difference is more clearly seen in Figs. 18.45a and b, where the impulse response function of the noncausal control filter and the filter coefficients of the causal control filter, respectively, are compared. The noncausal filter IRF shows a strong increase at the end of the time interval, which, as explained earlier, indicates that the filter responds before

768   ACTIVE CONTROL OF NOISE AND VIBRATION

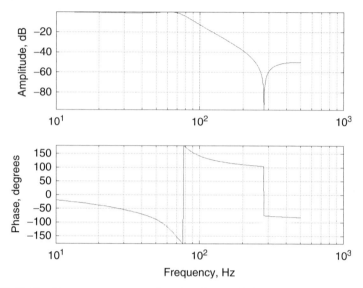

FIGURE 18.43 Frequency response function for a third-order elliptical filter with a cutoff frequency of 70 Hz and a passband–stop band ratio of 50 dB.

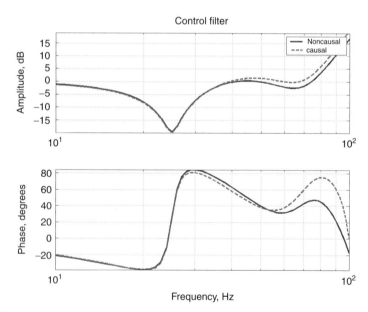

FIGURE 18.44 Comparison of the causal and noncausal control filter frequency response functions.

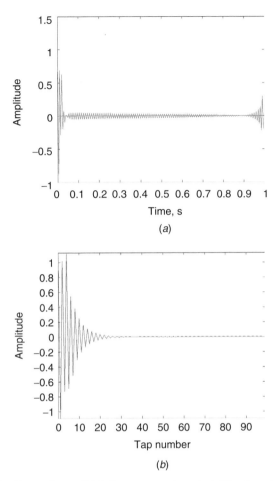

FIGURE 18.45 Comparison of (a) the noncausal control filter impulse response function and (b) the filter coefficients of the causal control filter.

it receives an input signal and is clearly noncausal. The filter coefficients of the causal filter, on the other hand, decay away to zero. The performance of the active control system is shown in Fig. 18.46. The figure shows the ratio of the controlled disturbance spectrum to the uncontrolled spectrum. The performance has clearly been degraded, especially in the vicinity of the Nyquist frequency. Nevertheless, the out-of-band amplification clearly visible in Fig. 18.41 has been completely removed through the use of the anti-aliasing and reconstruction filters.

System Identification

The control algorithms discussed in the previous sections require internal models of the transfer functions between the outputs of the controller (i.e., the signals sent

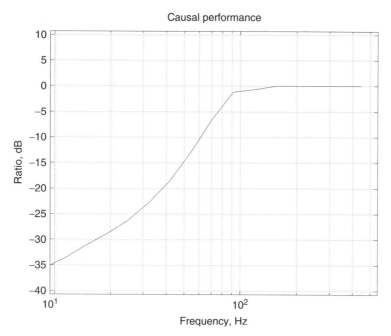

FIGURE 18.46 Ratio of controlled to uncontrolled disturbance with anti-aliasing filters, reconstruction filters, and D/A converter delays.

to the D/A converters) and the inputs to the controller from the residual sensors (i.e., outputs from the A/D converters). These transfer functions are commonly referred to as *plant* transfer functions. The procedure used to measure these transfer functions (or equivalently impulse response functions) is referred to as *system identification*. In this section, we discussed system identification approaches that apply to feedforward controllers.

Since most control algorithms require plant models for the control algorithms to converge to (or solve directly for) the optimal control filters,[58] system identification must be performed prior to any adjustments to the control filters. As such, when system identification is performed with control filters set to zero (i.e., all coefficients are equal to zero), it is referred to as *"open-loop"* system identification. It derives this name because the system identification algorithms are run when the control loop is open (i.e., no signals are allowed to pass through the control filters). The plant estimates obtained during open-loop system identification are used to initialize the plant models within the controller to support the control algorithms that subsequently adjust the control filter coefficients.

For systems where the plant transfer functions are expected to be time invariant (i.e., they do not change characteristics as a function of time), the plant estimates obtained during open-loop system identification can be held constant as the control filter coefficients are adjusted. It is more common, however, to expect that these transfer functions will change over time. These changes may occur because of changes in environmental conditions (e.g., temperature, pressure), changes in the operating conditions of the system to be controlled (e.g., speed, load), or the presence of external factors (e.g., number and movement of passengers in an automobile).

There are two basic approaches to account for time-varying transfer functions that may be encountered in ANVC systems: the use of robust control algorithms and periodic system identification. Control algorithms are often formulated to allow for certain levels of uncertainty associated with the internal plant models. For example, the magnitude and phase response of the plant at a given frequency may only be known to within, say, ± 2 dB in magnitude and $\pm 10°$ in phase. In effect, these algorithms recognize that the internal plant model is "close" to the actual plant transfer function but not exact. Control algorithms that are stable and provide performance in the presence of this plant uncertainty are called "robust." For example, the filtered-x LMS algorithm is robust to errors in the plant model of up to $90°$.[68] The second approach to account for time-varying transfer functions is to perform system identification at the same time that control filter coefficients are being adjusted. This approach provides *periodically updated* plant models to the control algorithms that track changes in the plant over time. System identification performed in this manner is referred to as *closed-loop* or *concurrent* since it occurs while the control loop is closed.

In practice, a combination of both approaches is often required to account for time-varying plants while at the same time maximizing achievable system performance. Control algorithms that rely on exact models of the plants to be controlled can offer maximum performance but are also very sensitive to slight changes in the plant transfer function. As such, small changes in the plant can lead to instability, and for this reason such systems are rarely used in practice. On the other hand, algorithms that make do with very inaccurate plant models will often provide very little performance. As such, there is a trade-off between control system robustness and performance. These trade-offs ultimately influence the approach chosen for system identification.

We now discuss procedures for open- and closed-loop system identification that are applicable to feedforward control systems. Consider first the feedforward system of Fig. 18.47. For this system, the plant, P, is the transfer function between the output of the control filter, y, and the response of the residual sensor, e. Though not shown explicitly, the plant transfer function P is assumed to include the following: D/A sample and hold, smoothing filters, transfer function through the structure to be controlled, anti-alias filters, and A/D converters. An open-loop estimate of P can be performed by setting the coefficients of the control filter, W, to zero and injecting a band-limited probe signal v into the output of the control filter. The probe signal is therefore a digital signal that adds to the output of the control filter y to produce a net signal c at the output of the controller.

The bandwidth of the probe signal is chosen to span the frequency range over which control performance is desired. For a tonal control problem, where the tone is nominally fixed in frequency, the probe signal may span only a few hertz.

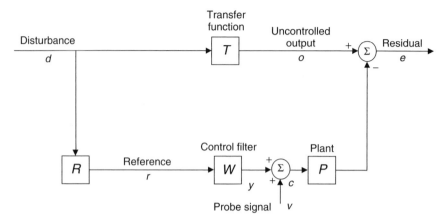

FIGURE 18.47 Probe signal injection for system identification in a feedforward system with no coupling between actuators and reference.

For broadband control, the probe signal bandwidth is often chosen to extend marginally beyond the control bandwidth to ensure an accurate plant estimate throughout the bandwidth for control.

The transfer function between the probe, v, and the residual, e, can be estimated in the frequency domain using the expression

$$\hat{P} = -S_{ve}S_{vv}^{-1} \qquad (18.39)$$

where $\hat{P}$ = estimate of plant transfer function P
S_{ve} = cross-spectrum between probe and residual
S_{vv} = autospectrum of probe

This expression will provide an unbiased estimate of the plant transfer function for $W = 0$ (i.e., open-loop system identification) as well as for $W \neq 0$ (i.e., closed-loop system identification) provided the probe signal is uncorrelated with the disturbance and sufficient averaging time is allowed to obtain good estimates of S_{ve}.

An estimate of the plant transfer function in Fig. 18.47 can be made in the frequency domain using Fourier transforms of the probe and residual signals as outlined above. Alternatively, an estimate of the plant *impulse response* can be obtained using the optimal or adaptive FIR filter design procedures discussed in Section 18.3.

An example of system identification using the LMS algorithm integrated into a filtered-x control algorithm is depicted in Fig. 18.48. The filtered-x algorithm adapts the coefficients of the control filter W using the filtered reference signal and a modified residual signal (see later) as inputs. A separate LMS algorithm is used to identify the plant impulse response as shown in the lower portion of the figure. A probe signal v is injected into the control loop and sums with

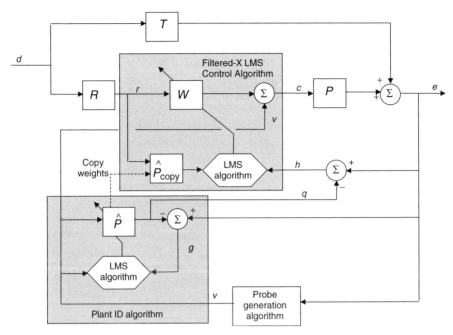

FIGURE 18.48 System identification using the LMS algorithm, embedded within filtered-x LMS controller.

the output of the control filter. The probe signal is also filtered by the adaptive FIR filter that is labeled $\hat{P}$. Finally, the probe signal is an input signal to the LMS algorithm that adapts the coefficients of $\hat{P}$. The output of $\hat{P}$ is summed with the residual signal e to form the error signal g, which is the second input to the LMS algorithm adapting the coefficients of $\hat{P}$. The coefficients of $\hat{P}$ are adapted to minimize the contribution in g that is correlated with the probe, v, in a least-squares sense. An optimal estimate of $\hat{P}$ is achieved when the LMS error signal g becomes uncorrelated with the probe, v. The coefficients of $\hat{P}$ are copied periodically into the filter, $\hat{P}_{\text{copy}}$, which filters the reference signal r, as required by the filtered-x LMS control algorithm.

Following the development of Section 18.3 under Adaptive Digital Filter Design, the update equation for the plant estimate is

$$\hat{\mathbf{p}}(n+1) = \hat{\mathbf{p}}(n) + 2\mu \mathbf{v}(n) g(n) \tag{18.40}$$

where

$$\hat{\mathbf{p}}^T = [\hat{p}(0), \hat{p}(1), \hat{p}(2), \ldots, \hat{p}(I-1)] \tag{18.41}$$
$$\mathbf{v}^T(n) = [v(n), v(n-1), \ldots, v(n-I+1)]$$

We note here that the convergence behavior of the LMS algorithm of Eq. (18.40) is improved if the portion of the residual signal correlated with the reference is removed before being used as input for the plant estimate. This can be achieved by including a third LMS algorithm with the reference and residual signals as inputs, and the modified residual signal as output. The update equation is similar to that in Eq. (18.40). Though not shown explicitly in Fig. 18.84 to preserve clarity, this approach was used for the adaptive feedforward control example presented near the end of this chapter (MIMO Feedforward Active Locomotive Exhaust Noise Control System with Passive Component) and is illustrated there in Fig. 18.93.

A consequence of being able to periodically update the model of the plant using the system identification procedure outlined in Fig. 18.48 is that the injected probe signal will increase the residual signal e. For purposes of system identification, it is desirable to used high-level probe signals because the probe signal will dominate the response at the residual sensor, and consequently accurate plant estimates can be obtained in a shorter length of time. During open-loop identification of ANVC systems, it is often acceptable to use high levels of probe signals. During closed-loop operation, however, the use of high-level probe signals is not acceptable since performance gains achieved as a result of the control filter will be overwhelmed by the presence of the probe signals reaching the residual. As a consequence, low-level (i.e., *covert*) probe signals are required during closed-loop operation.* The drawback to using low-level probes is that the plant estimate will adapt slowly and will not be able to track changes in the plant that occur over relatively short time frames. When this is the case, the control algorithms must be adjusted to deal with larger levels of uncertainty in the plant models or alternative strategies involving gain scheduling[69] must be considered.

For systems where the time constants of plant changes are *large* relative to the adaptation time of the LMS algorithm when using low-level probes, the system identification approach of Fig. 18.48 has been used effectively. For these cases, the probe signals can be generated using low-level band-limited "white" noise. However, fixed-level, white-noise probe signals are not necessarily the best choice. In particular, drawbacks of using white-noise probes include the following:

- The magnitude of the probe signal is held constant. Therefore, as the magnitude of the disturbance increases, the effective convergence rate for the plant filter will decrease. Alternatively, as the disturbance decreases relative to the probe, the convergence rate will increase at the expense of increasing the level of the closed-loop residual signals.
- The spectral shape of the probe signal is independent of the spectral shape of the residual signal and plant transfer function. Consequently, the quality (e.g., estimation errors) of the plant estimate will vary as a function of

---

*Typically the probe signal is sized such that the resulting signal at the residual sensors is nominally 6 dB below the closed-loop signal (absent the probe) in those sensors.

frequency. This can lead to temporary losses in system performance for control of slewing tonals and nonuniform control of broadband noise.

For ANVC applications where white-noise probes may not be appropriate, probe-shaping algorithms can be employed to generate probe signals that provide plant estimates with nominally uniform estimation errors as a function of frequency. Further, these algorithms can account for changes in the disturbance spectrum and plant transfer functions over time.[70–72]. Probe generation approaches are depicted collectively by the *probe generation* block in Fig. 18.48.

The final aspect to discuss concerning Fig. 18.48 is possible interaction between the filtered-x LMS algorithm adapting the control filter and the LMS algorithm adapting the plant estimate. In particular, we consider the effects of probe signal injection on convergence of the control filters. Note that the injected probe signal used for system identification contributes to the net residual signal e, which is typically used to adapt the control filter coefficients using the filtered-x LMS algorithm. The contribution from the probe signal is, by design, uncorrelated with the reference signal r. However, its presence in the residual signal will impact the instantaneous residual signal used in the gradient estimate and reduce the convergence speed of the filtered-x algorithm. As a consequence, it is beneficial to remove an estimate of the probe's contribution to the residual signal by subtracting the output of the filter, $\hat{P}$ (labeled q in Fig. 18.48), from the residual, e. This modified residual signal (labeled h in Fig. 18.48) is then used to adapt the control filter coefficients using the filtered-x LMS algorithm.

Up to this point, we have assumed that any coupling from the actuators back to the reference sensors is negligible (i.e., sufficiently small that it can be ignored with respect to system performance and stability). This is not always the case. In general, there exists physical coupling which must be accounted for by both the control algorithm and the system identification algorithms. Control algorithms that account for this path include *feedback neutralization* and Eriksson's *filtered-U* algorithm, both of which were discussed in the previous section. A fully adaptive filtered-U algorithm that includes concurrent system identification is discussed by Eriksson and Allie.[61] In those instances where the physical coupling to the references becomes so strong that electronic cancellation of the coupling may be ineffective, other approaches must be employed to minimize the physical coupling, such as the use of directional reference sensors and selection of alternate sensors.

Figure 18.49 shows the injection of a probe signal in a feedforward system that contains coupling C from the controller output to the reference input and neutralization filter $\tilde{C}$ in the controller to minimize the effects of this feedback path. The procedures for open-loop identification of the coupling transfer function C follows from the discussion above for identifying P, provided the reference signals are used in place of the residual signals. As such, unbiased estimates of the coupling transfer function can be obtained during open-loop operation, provided that the probe signal is uncorrelated with the disturbance. During closed-loop operation, however, transfer functions from the probe signal to the residual and

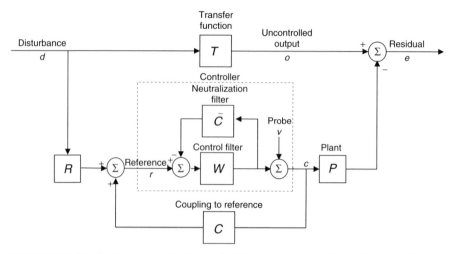

FIGURE 18.49 Probe signal injection in feedforward system with coupling to reference sensor and feedback neutralization filter.

reference sensors no longer provide unbiased estimates of P and C. Instead, these transfer functions include the loop gain associated with the feedback loop comprised of the coupling transfer function, neutralization filter, and control filter.

Following the nomenclature of Fig. 18.49, these transfer functions are given by

$$G_e = \frac{S_{ve}}{S_{vv}} = \frac{-P}{1-(C-\tilde{C})W} \qquad G_r = \frac{S_{vr}}{S_{vv}} = \frac{C}{1-(C-\tilde{C})W} \qquad (18.42)$$

These expressions for G_e and G_c contain the desired transfer functions for P and C in the numerator, but both contain the impact of the feedback loop in the denominator. The denominator term can be isolated by estimating the transfer function from the probe signal to the controller output signal c. This transfer function is given by

$$G_c = \frac{S_{vc}}{S_{vv}} = \frac{1}{1-(C-\tilde{C})W} \qquad (18.43)$$

Asymptotically unbiased estimates for the plant and coupling transfer functions can then be obtained by combining Eqs. (18.42) and (18.43) such that

$$\hat{P} = \frac{-G_e}{G_c} = P \qquad \tilde{C} = \frac{G_r}{G_c} = C \qquad (18.44)$$

This approach for closed-loop identification is known as the joint input–output method.[73] Estimates of P and C can be made in the frequency domain by estimating the cross-spectra in Eqs. (18.42) and (18.43) and taking the ratios as indicated in Eq. (18.44). Alternatively, multiple-embedded LMS algorithms could be employed to adapt estimates of P and C on a sample-by-sample basis.

Although we have discussed the joint input–output method applied to a feedforward system with feedback to the reference sensors, it is a general approach that is applicable to traditional feedback systems. As such, this method of concurrent system identification can be used to implement fully adaptive feedforward and feedback control systems. Further, the method is valid for use during open- and closed-loop operation. An example of a fully adaptive feedback control system using this approach to system identification when applied to an active machinery mount can be found in Berkman et al.,[74] Curtis et al.,[75] and Berkman and Bender[76] and is included here as an example in Section 18.7 under Active Machinery Isolation.

Extension to MIMO Systems

The equations and examples so far have considered only SISO controllers. In many instances there will be the need to design controllers that can handle multiple inputs and/or multiple outputs. While a complete consideration of the issues involved in the design of MIMO systems is beyond the scope of this chapter, we provide here a brief look at the algorithms for the design of optimum MIMO control filters.

Let us begin by defining a notional MIMO system like that shown in Fig. 18.50, where K reference signals, **r**, are applied to a control filter which in turn generates M output signals, **u**, to drive actuators on the system to be controlled. The system has L control sensors to monitor the system performance. Since there are multiple inputs and multiple outputs to each of the blocks in Fig. 18.50, the variables in the blocks $B(\omega)$, $W(\omega)$, and $P(\omega)$ no longer represent scalar transfer functions but must be transfer function matrices. For example, if there are M actuators driving the plant and L control sensors monitoring the system response, the plant transfer function $P(\omega)$ between the sensors and the

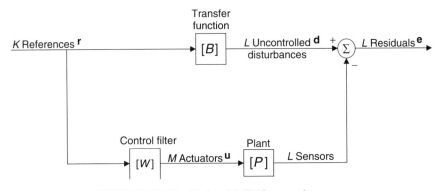

FIGURE 18.50 Notional MIMO control system.

actuators becomes the matrix of transfer functions $[P(\omega)]$:

$$[P(\omega)] = \begin{bmatrix} P_{11}(\omega) & P_{12}(\omega) & \cdots & P_{1M}(\omega) \\ P_{21}(\omega) & & & \vdots \\ \vdots & \ddots & \ddots & \\ P_{L1}(\omega) & \cdots & \cdots & P_{LM}(\omega) \end{bmatrix} \quad (18.45)$$

where the $P_{nm}(\omega)$ are each frequency response functions relating sensor output n to actuator input m.

In Eq. (18.24) we defined an optimum SISO control filter for tonal disturbances given by $W(\omega) = P(\omega)^{-1} B(\omega)$. It turns out that the optimal MIMO control filter is given by a very similar expression if the plant is a square transfer function matrix, that is, $L = M$:

$$[W(\omega)] = [P(\omega)]^{-1}[B(\omega)] \quad (18.46)$$

In the equation the square brackets indicate that the variables are transfer function matrices and not scalar transfer functions. For the more general case of a nonsquare plant matrix of K rows and L columns the same expression applies except that the pseudoinverse of the plant matrix is substituted for the normal inverse:

$$[W(\omega)] = [P(\omega)]^{\#}[B(\omega)] \quad (18.47)$$

where the pseudoinverse of the plant, $[P(\omega)]^{\#}$, is given by

$$[P]^{\#} = \begin{cases} \{[P]^H[P]\}^{-1}[P]^H & \text{for } K \leq L \\ [P]^H\{[P][P]^H\}^{-1} & \text{for } K > L \end{cases}$$

where $[\cdot]^H$ means the complex conjugate of the transpose of the matrix. Since for broadband disturbances Eq. (18.47) can result in noncausal control filters, there is a need to apply causality constraints to the MIMO filter design process similar to those applied to the SISO case in Section 18.4 under Optimal Control Filter Estimation. Unfortunately the derivation of the direct-estimation equations for the optimal MIMO control filter is very involved and beyond the scope of this chapter. The interested reader is referred to reference 77.

Fortunately the MIMO filtered-x algorithm is much simpler than the MIMO direct-estimation algorithm, and the equations have been developed by Nelson and Elliott.[16,67] The residual that we are seeking to minimize is given by

$$e_l(n) = d_l(n) - \sum_{m=1}^{M} \sum_{j=0}^{J-1} P_{lmj} u_m(n-j) \quad (18.48)$$

where P_{lmj} is the jth filter coefficient of the plant transfer function relating the ith sensor output to the mth actuator input, M is the number of actuators, J is the length of the digital representation of the plant transfer functions, and u_m is the mth output of the control filter driven by the K references as indicated in the equation

$$u_m(n) = \sum_{k=1}^{K} \sum_{i=0}^{I-1} w_{mki} r_k(n-i) \qquad (18.49)$$

In this equation w_{mki} is the ith coefficient of the control filter relating the mth actuator drive signal to the kth reference, I is the length of each control filter, and r_k is the kth reference signal input to the control filters. The MIMO filtered-x update equation for each control filter coefficient is given by

$$w_{mki}(n+1) = w_{mki}(n) + 2\mu \sum_{l=1}^{L} e_l(n) q_{lmk}(n-i) \qquad (18.50)$$

where $w_{mki}(n)$ is the existing estimate of the ith filter coefficient for the control filter connecting reference k to actuator m, $w_{mki}(n+1)$ is the new estimate, and

$$q_{lmk}(n) = \sum_{j=0}^{J-1} P_{lmj} r_k(n-j) \qquad (18.51)$$

Equation (18.50) is the MIMO filtered-x control filter update equation for multiple references, multiple sensors, and multiple actuators. It is similar in form (except for the sum over control sensors) to the SISO case, since the coefficients are updated by the product of the residual and the reference filtered by the plant transfer function. Its simple form allows for easy, computationally efficient implementation and, as a consequence, has become a very popular MIMO algorithm for feedforward applications. Furthermore, as we shall see in the next section, it can also be applied to feedback systems.

18.5 FEEDBACK CONTROL SYSTEMS

Basic Architecture

The block diagram of Fig. 18.51 shows the basic architecture of a SISO feedback control system. Such a block diagram might arise from the notional physical system shown in Fig. 18.52. The physical system consists of a second-order oscillator with a control sensor, an accelerometer for example, to monitor the motion of the mass. In addition there is a control actuator which can apply forces to the mass to control its motion. In the block diagram the residual, e, is the output signal from the residual sensor. That output is fed to a control filter,

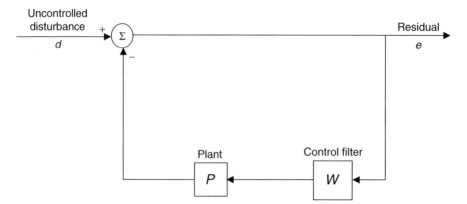

FIGURE 18.51 Basic feedback architecture.

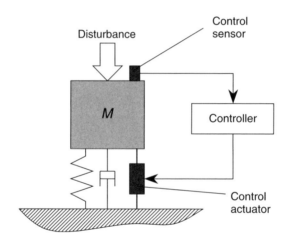

FIGURE 18.52 Notional system with a feedback controller.

which is shown as W in the block diagram. The control filter output drives an actuator that is used to control the motion of the mass.

The plant transfer function P in the block diagram relates the control filter output signal, which drives the actuator to the residual sensor output. In Fig. 18.51 the control filter drives the actuator in such a way as to reduce the input to the control filter. This is in contrast to feedforward systems in which the input to the control filter is a reference signal correlated with the disturbance. In one sense, feedback control systems can be thought of as feedforward systems in which the reference and control (residual) sensors are the same.

The solution for the closed-loop response is given by

$$e(\omega) = \{1 + P(\omega)W(\omega)\}^{-1} d(\omega) \qquad (18.52)$$

Clearly the residual will be small if $PW \gg 1$, and in the limit the residual amplitude will become

$$e(\omega) = \{P(\omega)W(\omega)\}^{-1}d(\omega) \tag{18.53}$$

The quantity PW is referred to as the loop gain and the residual amplitude will be inversely proportional to the loop gain. Consequently, good performance from a feedback controller will be achieved if the loop gain is made sufficiently large. In addition, when the loop gain is large, the phase of PW is immaterial to the performance such that if the control filter deviates somewhat from its design goals the performance will suffer only minimally.

The above discussion makes the design of a feedback control filter seem easy. Unfortunately, stability considerations limit our ability to increase the loop gain without limit. Stability issues will be discussed more fully below under Alternate Suboptimum Control Filter Estimation, where the compensator-regulator approach to feedback control filter design is discussed.

Optimal Control Filter Estimation

The literature dealing with the design of optimal feedback controllers is vast and beyond the scope of what can be reasonably included in this chapter. Instead we will focus on two approaches—one optimal and one suboptimal—that each provide tools for the design of SISO or MIMO controllers. The collection of methods often referred to as *modern optimal control theory* will not be dealt with here. Those methods, while providing powerful tools for control filter design when an analytical model of the plant dynamics is available, are not especially well suited to the types of problems where only measured data on the plant dynamics are available. While curve-fitting techniques can be utilized to obtain an analytical model that approximates the measured data, many of those techniques begin to founder if the order of the system model (number of states) becomes too large. For a sampling of the technology in this field the interested reader is referred to references[78-81].

In this section, we focus on what has come to be called the Youla transform or inner model formulation. Through selection of a particular feedback architecture, the Youla transform allows the feedback problem to be transformed into a feedforward problem, making it possible for the feedforward control filter design techniques in Section 18.4 to be used for feedback. In addition, because the system is effectively transformed to a feedforward system, the resulting system is guaranteed to be stable, provided an accurate digital model of the plant has been utilized.

Figure 18.53 shows a simplified block diagram of the Youla transform. In the figure a second feedback loop has been inserted around the control filter. In that second loop a digital filter approximating the plant transfer function has been inserted. The term *inner model* is often used when referring to this approach, precisely because a digital model of the plant is inserted into the control filter.

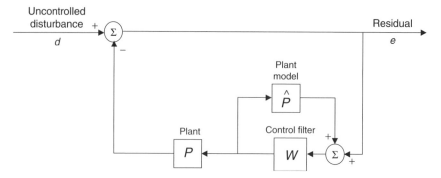

FIGURE 18.53 Youla transform architecture.

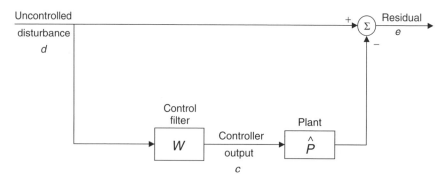

FIGURE 18.54 Feedforward system resulting from the Youla transform.

By making this change in the architecture of the feedback system, we have modified the transfer function from that shown in Eq. (18.52) to the following:

$$e(\omega) = \{1 - \hat{P}(\omega)W(\omega)\}\{1 + (P(\omega) - \hat{P}(\omega))W(\omega)\}^{-1}d(\omega) \quad (18.54)$$

Note that if $\hat{P}$ is a good match to P, then Eq. (18.54) simplifies to

$$e(\omega) = \{1 - \hat{P}(\omega)W(\omega)\}d(\omega) \quad (18.55)$$

Equation (18.55) is the equation for the residual of a feedforward system, the block diagram for which is shown in Fig. 18.54. The resulting block diagram is identical to the feedforward block diagram in Fig. 18.31 if the transfer function $B(\omega)$ is set to unity. Consequently, virtually all of the control filter design equations in Section 18.4 can be applied to a feedback system to which the Youla transform has been applied provided $B(\omega) = 1$. From Eq. (18.24) the noncausal control filter becomes

$$W(\omega) = \hat{P}^{-1}(\omega) \quad (18.56)$$

For the filtered-x algorithm Eqs. (18.35) and (18.36) can be applied to the Youla transform feedback system by simply changing the reference time sequence $r(n)$ in the equations to the disturbance time sequence $d(n)$ in Fig. 18.54:

$$\mathbf{w}(n+1) = \mathbf{w}(n) + \mu e(n)\mathbf{u}(n) \tag{18.57}$$

where $\mathbf{w}(n+1)$ is the new vector of filter coefficients, $\mathbf{w}(n)$ is the old vector,

$$\mathbf{u}(n) = \begin{Bmatrix} u(n) \\ u(n-1) \\ \vdots \\ u(n-N_w+1) \end{Bmatrix} \qquad u(n) = \sum_{m=0}^{N_w-1} \hat{p}(m)d(n-m)$$

and $\hat{p}(m)$ is the mth filter coefficient of the FIR filter representation of the plant frequency response function $\hat{P}(\omega)$.

The equations for determining the optimum feedforward control filter with causality constraints using the direct-estimation approach can be similarly employed by simply interchanging the disturbance d for the reference r. In this case what are affected are the autospectra and cross-spectra in the equations. Modifying Eqs. (18.30) and (18.31), we obtain for a control filter with N taps

$$\begin{Bmatrix} W_0 \\ W_1 \\ \vdots \\ W_{N-1} \end{Bmatrix} = \begin{bmatrix} C_0 & C_{-1} & \cdots & C_{1-N} \\ C_1 & C_0 & \cdots & C_{2-N} \\ \vdots & \vdots & \ddots & \vdots \\ C_{N-1} & C_{N-2} & \cdots & C_0 \end{bmatrix}^{-1} \begin{Bmatrix} A_0 \\ A_1 \\ \vdots \\ A_{N-1} \end{Bmatrix} \tag{18.58}$$

where the coefficients in the matrix equation become

$$A_p = \mathrm{Re}\left\{\int_{\omega_1}^{\omega_2} P S_{dd} e^{-jp\omega \Delta t} d\omega \right\}$$

$$C_{(p-m)} = \mathrm{Re}\left\{\int_{\omega_1}^{\omega_2} \{|P^2| + \alpha^2\} S_{dd} e^{-j(p-m)\omega \Delta t} d\omega \right\} \tag{18.59}$$

where $p, m = 0, 1, \ldots N - 1$, S_{dd} is the autospectrum of the disturbance, and α^2 is the control effort weighting factor.

For MIMO systems the equations of Section 18.4 under Extension to MIMO Systems can be similarly adjusted for the design of a MIMO controller in the Youla feedback architecture.

The transformation of Eq. (18.54), a feedback system, into Eq. (18.55), a feedforward system, only occurs if $\hat{P} = P$. Consequently, one would expect that the digital representation of the plant transfer function would have to be fairly accurate. If it is not, the term

$$\{1 + (P(\omega) - \hat{P}(\omega))W(\omega)\}^{-1} \tag{18.60}$$

may indeed lead to instability if the error in the plant model is large enough. If we substitute $E(\omega)$ for the difference between the true plant and the digital representation, Eq. (18.60) becomes

$$\{1 + E(\omega)W(\omega)\}^{-1} \tag{18.61}$$

A simple criterion to ensure that the feedback system remains stable is to require for all frequencies where the controller is functioning that

$$|E(\omega)W(\omega)| < 1 \tag{18.62}$$

where $|\cdot|$ means the absolute value. Furthermore, it can be easily shown that $|E(\omega)W(\omega)| = |E(\omega)||W(\omega)|$. Consequently, we can write a simple criterion for the maximum allowable error in the plant model:

$$|E(\omega)| = |\hat{P}(\omega) - P(\omega)| \leq \frac{\beta}{|W(\omega)|} \qquad 0 < \beta < 1 \tag{18.63}$$

where the criterion applies to all frequencies where the control system is functioning and β provides some stability margin. For example, if $\beta = 0.1$, Eq. (18.63) will provide, at least, 20 dB of stability margin. This is a very simple criterion requiring that we know only the amplitude of the control filter as a function of frequency. No phase information is required. However, it is also very conservative, and systems that violate it may not necessarily be unstable. A more precise method for determining stability is to employ the Nyquist criterion, which we will discuss in the next section. Note that a similar criterion can be fashioned for MIMO systems based on the maximum singular values of the matrices $[E(\omega)]$ and $[\hat{P}(\omega)]$. The interested reader is referred to references 80 and 81.

Alternate Suboptimum Control Filter Estimation

Compensator-Regulator Architecture. The compensator-regulator approach to feedback controller design results in a suboptimal system but provides useful insights into functions performed by the controller by breaking the control filter up into a cascade of two filters, as illustrated in Fig. 18.55. The compensation filter is designed to approximate the plant inverse and *compensate* for its amplitude and phase over a range of frequency, which we will refer to as the compensation band. Consequently, if the compensation filter is given by

$$G(\omega) \approx P(\omega)^{-1} \tag{18.64}$$

then we conclude from Eq. (18.52) that the closed-loop residual response $e(\omega)$ is related to the disturbance $d(\omega)$ by

$$\frac{e(\omega)}{d(\omega)} = \frac{1}{1 + H(\omega)} \tag{18.65}$$

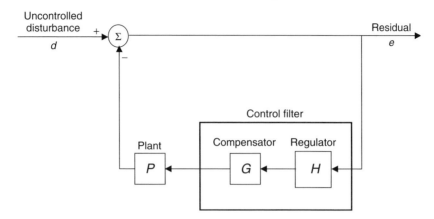

FIGURE 18.55 Compensator-regulator controller architecture.

where $H(\omega)$ is the regulation filter. As this equation shows, to provide reduction in the residual, $|H(\omega)| \gg 1$. When $|H(\omega)| \gg 1$,

$$\frac{e}{d} = \frac{1}{H(\omega)} \ll 1$$

and when $|H(\omega)| \ll 1$,

$$\frac{e}{d} = 1$$

As a result, the design of the regulation filter involves making its magnitude or loop gain large at those frequencies where control is desired (the regulation band) and small where control is not desired. As we shall see in the next section, the trick is to reduce the magnitude and phase of $H(\omega)$ in such a way as to maintain stability and avoid excessive noise amplification outside of the regulation band.

Regulation Filter Design. The range of frequency over which $|H(\omega)| \gg 1$ is referred to as the regulation band. To ensure stability as the magnitude of $H(\omega)$ is reduced outside of the regulation band, the phase of $H(\omega)$ should not exceed 180° until the magnitude is less than 1. This requirement is a consequence of the Nyquist stability criterion.

To employ the Nyquist criterion, one must plot in the phase plane the loop gain, or the change in amplitude and phase as a signal propagates around the feedback loop. In Eq. (18.65) $H(\omega)$ is the loop gain. In the phase plane, the real part of the loop gain is plotted on the horizontal axis and the imaginary part is plotted on the vertical axis such that as the frequency changes, a curve is traced out in two dimensions, as illustrated in Fig. 18.56. In the phase plane, the radius from the origin to a point on the curve represents the magnitude of the loop gain at a particular frequency, and the angle the radius makes with the positive horizontal axis is the phase angle. This is shown in Fig. 18.56 for

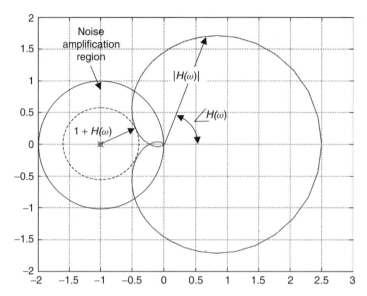

FIGURE 18.56 Stability and noise amplification requirement in the phase plane.

the case where the loop gain is $H(\omega)$. We will not present any proof here of the Nyquist criterion but will simply illustrate its application. For the interested reader, detailed discussions of the Nyquist criterion can be found in a number of texts on control theory.[56,82]

To apply the Nyquist criterion, one simply plots $H(\omega)$ on the phase plane as the frequency ω varies from $-\infty$ to ∞ and counts the number of times that -1 is encircled. That number is equal for most cases of interest to the number of unstable poles of $\{1 + H(\omega)\}^{-1}$. Figure 18.56 shows a Nyquist plot of a notional $H(\omega)$ in the phase plane. The amplitude and phase of $H(\omega)$ at a particular frequency are shown in the figure when the phase is less than 180° but the amplitude is greater than 1. As the frequency increases and the phase of $H(\omega)$ approaches 180° in the figure, the amplitude decreases to less than 1, satisfying the criterion. As the figure shows, this ensures that that the curve of $H(\omega)$ will not encircle -1 and that the system is stable.

In the figure, when the phase of $H(\omega)$ is 180°, the amplitude is ~ 0.2, which gives a modest gain margin of ~ 14 dB. This means that the amplitude of $H(\omega)$ would have to be 14 dB higher before the feedback control system would go unstable. Similarly, when the amplitude of $H(\omega)$ is unity, the phase is approximately 120°, giving a phase margin of 60°, meaning that the phase would have to increase by approximately 60° before the system would go unstable.

To ensure that there will be no noise amplification as the magnitude of $H(\omega)$ is reduced outside of the regulation band, the curve traced out by $H(\omega)$ in the phase plane should not enter a circle of unit radius centered on -1. This last requirement is also illustrated in Fig. 18.56. There, the trajectory of $H(\omega)$ is seen

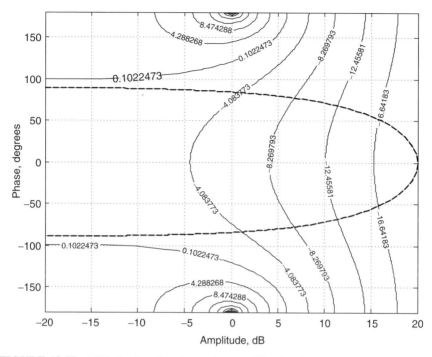

FIGURE 18.57 Nichols plot showing noise amplification in a feedback system as a function of the amplitude and phase of the loop gain.

entering the unit circle, indicating that some degree of noise amplification will occur.

While the loop gain plot in the phase plane can be used to estimate the noise amplification, the so-called Nichols plot provides that information more directly. Figure 18.57 shows a Nichols plot. In the figure, the phase of the loop gain is plotted on the vertical axis and the amplitude is plotted on the horizontal axis. The dashed curve in the figure is the loop gain of a second-order filter.* The contour lines in the plot show the noise amplification. Whenever the dashed curve crosses one of the contour lines, the feedback system with that loop gain will have the noise amplification in decibels indicated by the number on the contour line. For example, Fig. 18.57 shows that the second-order filter with the loop gain given by the dashed curve will never have a noise amplification greater than 0 dB. In fact, a simple examination of the Nyquist and Nichols plots will show that *any*

---

*The form of the filter is given by

$$H(\omega) = K \left\{ \frac{j(\omega/\omega_0)}{(1 - (\omega/\omega_0)^2 + j\eta\omega/\omega_0)} \right\}$$

where ω is the frequency and the other terms are constants.

filter whose phase is bounded by $\pm 90°$, as the second-order filter is, will always be stable with 0 dB noise amplification when placed in a feedback loop.

All of the above discussion assumes that $P(\omega)$ has been measured accurately and that a good causal representation of $P(\omega)^{-1}$ exists. If, however, the causal representation of $P(\omega)^{-1}$ is imperfect, then the loop gain in the controller will not be just $H(\omega)$ but will be $P(\omega)G(\omega)H(\omega)$. The same criteria for stability and noise amplification apply to this modified loop gain. The consequence is that $H(\omega)$ may need to be designed with larger gain and/or phase margins to maintain stability, and this may result in decreased performance. In addition, the imperfectly compensated system may also have increased noise amplification.

Another consideration is the frequency band over which the compensation filter, $G(\omega)$, must be a good representation of $P(\omega)^{-1}$. In general, the compensation bandwidth must be much greater than the regulation bandwidth to ensure stability and to control noise amplification. To determine the required compensation bandwidth, we first consider noise amplification. The basic approach is to compensate the plant until $H(\omega)$ is small enough so that uncompensated variations in the plant, $P(\omega)$, times the aliased compensation filter, $G(\omega)$, will not lead to noise amplification greater than that specified. If there are measurements of $P(\omega)$ up to frequencies well beyond even the compensation bandwidth, then it is possible to construct the Nyquist plot similar to that shown in Fig. 18.56 for $P(\omega)G(\omega)H(\omega)$ and determine the noise amplification for candidate compensation bandwidths in the same way as illustrated in the figure. Unfortunately, information on the plant is rarely available over such a broad bandwidth, and one must design in robustness by utilizing margins of safety. For example, let us assume that 2 dB of noise amplification is all that is allowed. For this to be true for any phase angle, the amplitude of the regulation filter must satisfy $|H(\omega)| < 0.206$ (-14 dB). If we have a margin of safety of, say, 6 dB to allow for uncompensated plant variations, we must compensate the plant until the regulation filter has declined to -20 dB. If we design the regulation filter to provide 10 dB of performance reduction in the residual, then $H(\omega)$ must have an amplitude of at least 10 dB in the regulation band. Consequently, we will need to compensate the plant out to frequencies beyond the regulation frequency until the amplitude of $H(\omega)$ has decreased by 30 dB from its maximum. Since we must not allow $H(\omega)$ to decrease too rapidly with increasing frequency or instability will result, the resulting compensation bandwidth can easily exceed the regulation bandwidth by a factor of 10–100.

To illustrate these concepts, we will carry out an example for narrow-band control and broadband control, in both cases assuming perfect compensation. For the narrow-band case we choose

$$H(\omega) = K \frac{j\omega/\omega_0}{1 - (\omega/\omega_0)^2 + j\eta(\omega/\omega_0)} \qquad (18.66)$$

where K and η are constant, $\omega = 2\pi f$, and f is the frequency in hertz. Equation (18.66) is a narrow-band regulation filter with high gain near $\omega = \omega_0$. The frequency response is shown in Fig. 18.58 for $\eta = 0.1$ and $K = 1$. This filter will provide approximately 20 dB of loop gain at the center frequency and

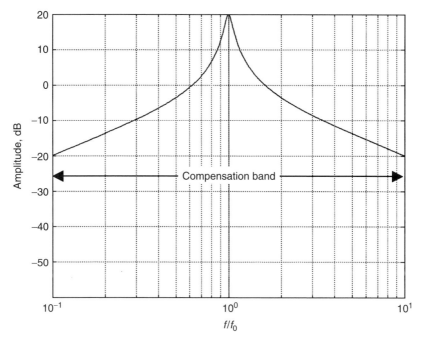

FIGURE 18.58 Narrow-band regulation filter.

consequently we would expect about 20 dB reduction in the residual. In addition, as mentioned above, the phase of the filter is bounded by $\pm 90°$ and consequently, when placed in the feedback loop, will be stable with 0 dB noise amplification.

Figure 18.59 shows the reduction in the residual with this regulation filter in the feedback loop. The system frequency response looks like a notch filter with approximately 20 dB of noise reduction over a bandwidth of about 5% of the center frequency f_0. It shows no out-of-band noise amplification, as expected.

These estimates have assumed that the compensation for the plant is perfect and that $G(\omega) = P(\omega)^{-1}$ for all frequencies. In reality we need to define a realistic range over which the plant will be measured and the compensation filter will be designed to work. The compensation band should be extended out far enough in frequency such that the regulation filter gain is small enough to ensure that no matter what uncompensated plant phase is introduced, the noise amplification will not exceed a design value (e.g., 2 dB). Earlier we determined that that criterion would be satisfied if $|H(\omega)| < 0.206$ (-14 dB). To allow for possible increases in loop gain as the compensation is gradually turned off, we introduce some margin by reducing this value by an additional factor of 2 (6 dB). Consequently, the plant will be compensated until the regulation filter reduces its amplitude to -20 dB. From Fig. 18.58 we can see that the compensation bandwidth must extend from $\sim 0.1 f_0$ to $\sim 10 f_0$, or 2 decades. This is substantially broader than the control bandwidth.

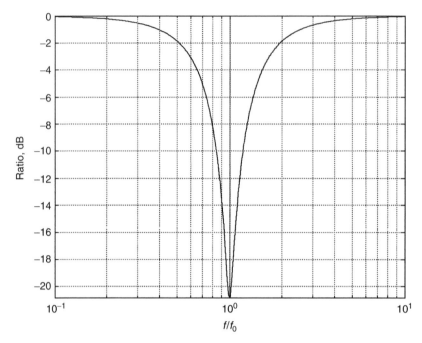

FIGURE 18.59 Ratio of controlled to uncontrolled residual for the narrow-band regulation filter.

Note that the bandwidth and noise reduction in Fig. 18.59 can be adjusted by varying the parameters of the regulation filter. For example, increasing η will increase the bandwidth but decrease the noise reduction performance of the feedback system, essentially broadening the notch and decreasing its depth in Fig. 18.59. Increasing K will increase the noise reduction performance, essentially deepening the notch in the figure but requiring that we compensate the plant out to still higher frequency to control noise amplification.

For broadband control we will use a second-order bandpass filter for the regulation filter given by

$$H(\omega) = Kb \left\{ \frac{j(\omega/\omega_0)}{(j\omega/\omega_0 + a)(j\omega/\omega_0 + b)} \right\} \quad (18.67)$$

where ω_0 is the center frequency of the filter.

The frequency response of the filter with $K = 3$, $a = 0.1$, and $b = 10$ is shown in Fig. 18.60. The phase of this filter (not shown) is also bounded by $\pm 90°$ and hence the system would be expected to be stable with no noise amplification, provided the plant is properly compensated. The noise reduction performance of the feedback system employing this regulation filter is shown in Fig. 18.61. The figure shows about 10 dB or more of noise reduction for $0.12 f_0 < f < 8 f_0$. It

FEEDBACK CONTROL SYSTEMS   791

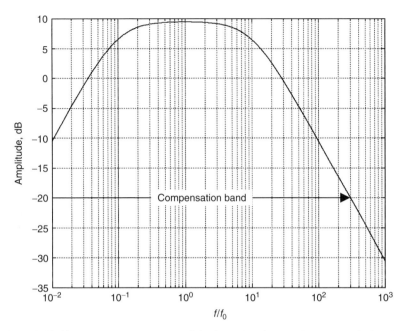

FIGURE 18.60 Frequency response of the broadband second-order regulation filter.

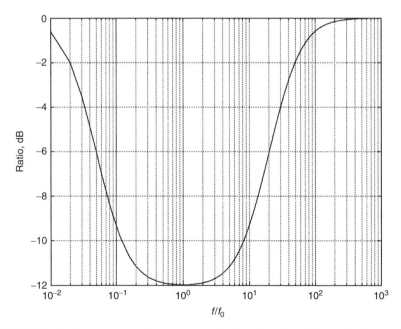

FIGURE 18.61 Ratio of controlled to uncontrolled residual for the broadband regulation filter.

also shows that there is no out-of-band noise amplification above and below the regulation frequency band, as expected.

The above calculations assume that the plant has been perfectly compensated. As for the narrow-band case, the compensation band should be extended out at least far enough in frequency such that the regulation filter gain is small enough to ensure that no matter what the phase, the noise amplification will not exceed 2 dB. Earlier we determined that that criterion would be satisfied if $|H(\omega)| < 0.206$ (-14 dB). Again, to allow for possible increases in loop gain as the compensation is gradually turned off, we introduce some margin by reducing this value by a factor of 2 (6 dB). Consequently, the plant will be compensated until the regulation filter reduces its amplitude to -20 dB. From Fig. 18.60 we can determine that the compensation band must extend out to $300 f_0$, or almost 40 times the highest frequency in the regulation frequency band. Since this is a fairly broad band over which to provide good compensation, it would be interesting to examine what would happen if we used a higher order filter. Since the loop gain associated with a higher order regulation filter would decrease more rapidly with frequency, the compensation band should be narrower. For example, if we simply square the frequency response function of the second-order filter in Eq. (18.67), we would have a fourth-order filter given by

$$H(\omega) = Kb \left\{ \frac{j(\omega/\omega_0)}{(j\omega/\omega_0 + a)(j\omega/\omega_0 + b)} \right\}^2 \qquad (18.68)$$

The frequency response function of this filter is shown in Fig. 18.62, which indicates that this filter rolls off much more quickly than the second-order filter. The compensation band now extends out to $\sim 50 f_0$ rather than $300 f_0$. There is, however, a penalty for this higher rolloff. That penalty is illustrated in the Nyquist and Nichols plots of Fig. 18.63 and the noise reduction performance plot of Fig. 18.64. The Nyquist plot shows that the system is stable but that the curve formed by $H(\omega)$ enters the noise amplification region. The Nichols plot shows that the expected noise amplification should be about 2.5 dB, which is confirmed in Fig. 18.64. Finally the regulation frequency band over which the noise reduction is 10 dB or more has been reduced slightly due to the more rapid rolloff of this higher order filter. The noise reduction performance and regulation bandwidth issues can be dealt with by changing the parameter values in the filter frequency response function (with concomitant increases in the compensation bandwidth and noise amplification). However, *the noise amplification generated by this higher order filter occurs even with perfect compensation and is the real reason the low-order filter is preferred.*

Compensation Filter. There are a number of approaches that can be applied to designing the compensation filter. Here we will examine the use of feedforward control filter design techniques, as discussed in Section 18.4, to design a control filter that will converge to the inverse of the plant. The notion for this approach is illustrated in the feedforward control architecture of Fig. 18.65. There the plant,

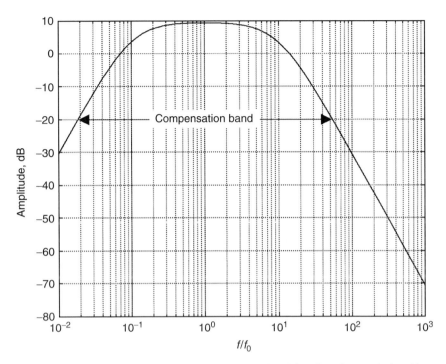

FIGURE 18.62 Frequency response of the broadband fourth-order regulation filter.

$P(\omega)$, and the compensation filter, $G(\omega)$, are connected in series and the disturbance and the residual are the same function, that is, the transfer function $B(\omega)$ in the simplified feedforward architecture of Fig. 18.31 is unity. With this arrangement, the control filter design algorithms will attempt to make $G(\omega)P(\omega) = 1$ or $G(\omega) \approx P(\omega)^{-1}$. Both the direct-estimation and filtered-x design tools can be applied to this problem and the inverses of both MIMO and SISO plants can be determined.

Example. To illustrate this approach, we define the notional position-keeping system for the suspended mass shown in Fig. 18.66. The mass might be machinery that is vibration isolated from the deck of a ship. For frequencies well above ω_0 the suspension system under the mass will reduce the excitation of the deck due to machine vibration from imbalance, for example. Our position-keeping system prevents the machinery from striking the deck when the ship encounters rough seas. In this system the relative displacement between the machine and the deck is measured and used as the input to a controller. The controller in turn generates a control force to minimize that relative deflection so that the machine and the deck do not come into contact. The control force could be generated by an inertial shaker or an interstructural shaker between the deck and the machine.

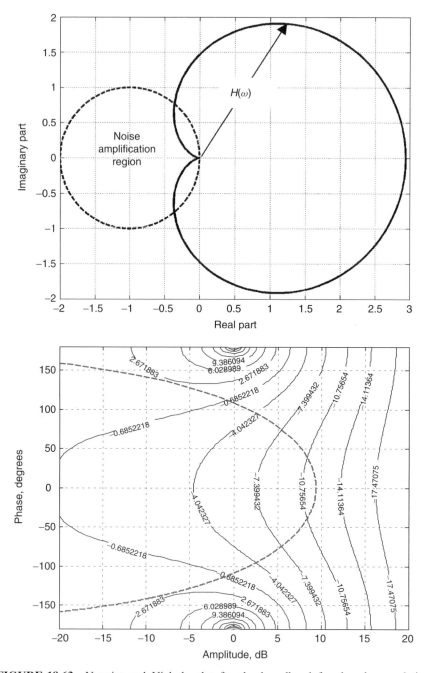

FIGURE 18.63 Nyquist and Nichols plot for the broadband fourth-order regulation filter.

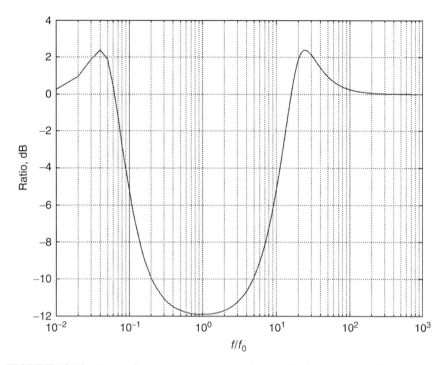

FIGURE 18.64 Ratio of controlled to uncontrolled residual for the broadband fourth-order regulation filter.

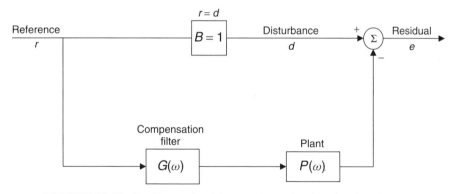

FIGURE 18.65 Feedforward architecture for estimating the plant inverse.

The plant for this system, which is the ratio of the relative deflection between deck and machine to actuator force is given by

$$P(\omega) = \frac{1}{M\omega_0^2\{1 - (\omega/\omega_0)^2 + j(\omega/\omega_0)\eta\}} \qquad (18.69)$$

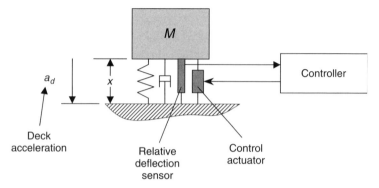

FIGURE 18.66 Notional feedback position-keeping system.

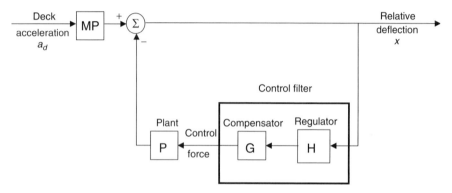

FIGURE 18.67 Block diagram of the notional feedback position-keeping system.

where ω_0 is the natural frequency of the mass on its suspension and η is the loss factor for the system. The feedback control system block diagram for our notional position-keeping system is shown in Fig. 18.67.

The parameter values that we will use are

$$f_o = \frac{\omega_0}{2\pi} = 25 \text{ Hz} \qquad \eta = 0.1$$

For this calculation we will not use any anti-aliasing or reconstruction filters but will rely on the plant itself to provide low-pass filtering. If we were to use any additional low-pass filtering, the frequency response of those filters would simply be included with the plant, since the compensation filter will have to compensate for the amplitude and phase variations of those components as well. In a real system with more complex high-frequency response than our simple notional system we would probably not be able to eliminate anti-aliasing and reconstruction filters, but eliminating them here simplifies the example. Finally, we must account for the half-sample delay introduced by the sample-and-hold D/A converters. We simply include that delay in the plant frequency response function.

For a regulation filter we will use the second-order broadband filter of Eq. (18.67) with the following parameters:

$$K = 3 \quad a = 0 \quad b = 10$$
$$f_o = \frac{\omega_0}{2\pi} = 1$$

These parameters make the regulation filter have the low-pass characteristics illustrated in Fig. 18.68. The filter has a nominal loop gain of $\sim$10 dB out to $\sim$10 Hz.

Based on the results in Fig. 18.60, we know that the compensation band will need to extend out to about 40 times the highest frequency in the regulation band or out to 400 Hz. Since we cannot expect to design a compensation filter that will be effective right out to the Nyquist frequency (one-half the sampling rate), we will put in a margin of safety and use a sampling rate of 1000 Hz, which gives a Nyquist frequency of 500 Hz. Using the direct-estimation filter design technique

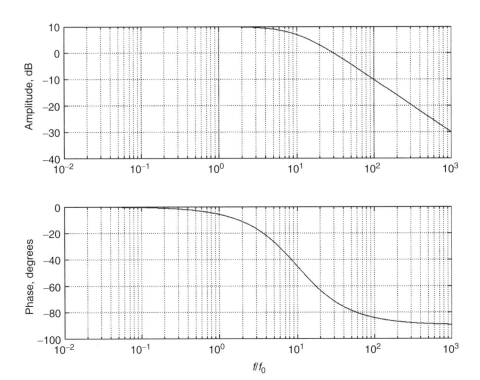

FIGURE 18.68 Amplitude and phase of the regulation filter.

of Section 18.4 for the system in Fig. 18.65, we obtain a 50-tap compensation filter $G(\omega)$ that when multiplied by the plant transfer function gives the result shown in Fig. 18.69. If the filter were a perfect compensator, the amplitude of $G(\omega)P(\omega)$ would be unity and the phase would be zero.

The figure shows that the amplitude is very close to unity almost all the way to the Nyquist frequency.* The phase, however, reaches almost 180° at 300 Hz, which indicates that we should expect some noise amplification due to imperfect compensation. That expectation is confirmed in the Nichols plot of Fig. 18.70 and the noise reduction performance plot of Fig. 18.71, which each slow a little over 2 dB of noise amplification. The Nyquist plot also in Fig. 18.70 shows that the system is stable. The noise reduction performance in Fig. 18.71 shows that 10 dB of noise reduction has been achieved up to more than 10 Hz.

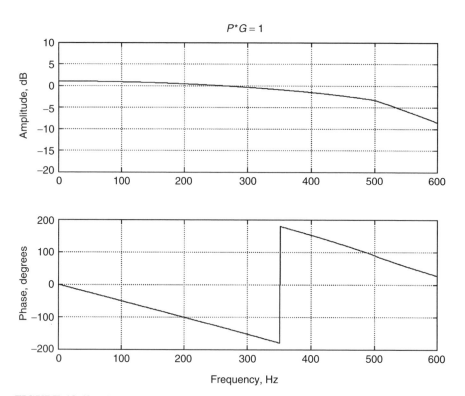

FIGURE 18.69 Amplitude and phase of the product of the compensation filter and the plant transfer function $\{G(\omega)P(\omega)\}$ as a function of frequency.

---

*Note that in computing the compensation filter using the direct-estimation approach it is sometimes necessary to introduce a small delay in the block diagram of Fig. 18.65 to deal with uncompensatable delays in the plant. Instead of setting $B(\omega) = 1$, we set it to $B(\omega) = e^{-j\omega\lambda T}$, where T is the sample period and λ a constant representing the length of the delay in sample periods. While introducing some phase error in the compensator, this approach often results in a better plant compensator than if no delay were introduced. For the case examined here best results were obtained with $\lambda \sim 1.5$.

FEEDBACK CONTROL SYSTEMS **799**

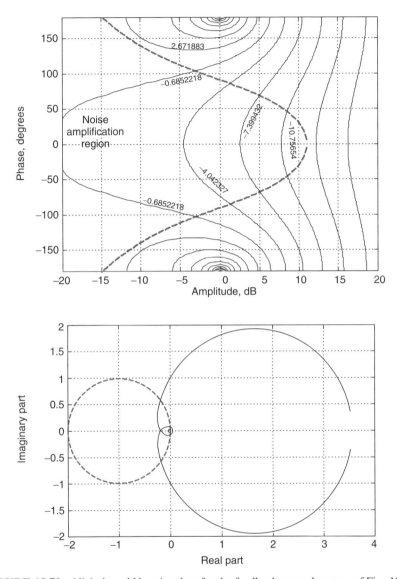

FIGURE 18.70 Nichols and Nyquist plots for the feedback control system of Fig. 18.67.

This example points out how feedback control systems can make severe demands on hardware, typically much more severe than feedforward systems. To achieve 10 dB of noise reduction out to 10 Hz required in this case a sampling rate of 1 kHz, 100 times higher than the regulation bandwidth. *Such a huge factor between regulation bandwidth and compensation bandwidth is not uncommon and is one of the reasons why feedforward systems are often favored over feedback.*

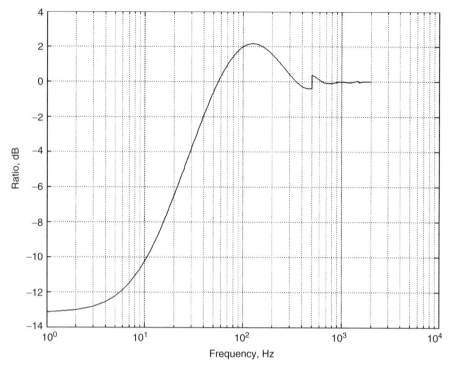

FIGURE 18.71 Ratio of controlled to uncontrolled residual for the feedback control system of Fig. 18.67.

18.6 CONTROL SYSTEM DESIGN CONSIDERATIONS

In this section, we discuss the process and steps involved in designing and implementing a successful ANVC system. This process is illustrated in the flow chart of Fig. 18.72 As shown, the first step is to establish design goals and requirements, which include performance goals and objectives. These goals, together with experimental data or structural–acoustic models, are used to conduct simulations to predict achievable system performance. These performance simulations typically involve trade-off studies of system performance as a function of actuator output levels, sensor dynamic ranges and noise floors, control algorithms, architecture, and controller parameters (e.g., sample rate, analog filter characteristics, digital filter sizes, computational load, and memory requirements). These simulations ultimately guide decisions concerning hardware selection, such as analog filtering, A/D and D/A converters, and digital-controller hardware. The output of this simulation step is a set of design specifications for the sensor and actuator subsystems as well as for the controller (i.e., control algorithm and architecture). Preliminary and detailed designs are then performed to produce sets of component specifications for each subsystem as well as interface control documentation between each subsystem and with the system to be controlled. At this

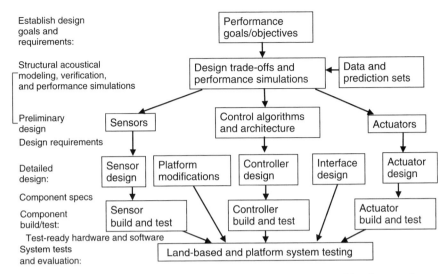

FIGURE 18.72 Flow chart for ANVC system development and implementation.

point, the subsystems are acquired if they are commercially available and fabricated if not. Each is then tested to ensure that both hardware and software meet specifications. The components are then integrated with the system to be controlled. Typically, a series of tests is conducted to validate system performance. If successful, more extensive in-service testing is usually performed to determine long-term performance and reliability before the prototype system is transformed into a commercial product or deployable system. That transformation, which can be an extensive and costly process, is not discussed here.

In the sections that follow, we discuss the major steps in this development process, including identification of performance goals, specification of actuator and sensor requirements, and selection of control architecture. Also discussed in what follows are the benefits of conducting both non causal and causal performance simulations; practical considerations concerning hardware selection of analog filters, D/A and A/D converters, and digital signal processors; and an overview of issues related to control system user interface, operating modes of the system, and system testing guidelines.

Identifying Performance Goals

Identifying performance goals is an important, yet often overlooked, first step in developing an ANVC system. Performance goals state what effect the control system should have on a particular physical quantity as a function of frequency. A schematic that is useful for discussing performance goals is shown in Fig. 18.73. This figure shows two conceptual curves as a function of frequency. The solid curve is representative of the uncontrolled response (e.g., mean-square vibration level on a particular component). The dashed curve is the desired mean-square

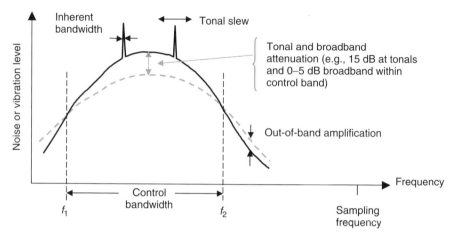

FIGURE 18.73 Schematic of performance goal specification, uncontrolled response (solid line), controlled response (dashed line).

response as a result of implementing an ANVC system. This desired response may be required to meet product specifications or may be related to meeting certain radiated noise specifications. In either case, the difference between these two curves identifies the reductions, as a function of frequency, that are required. As an example, the performance goals for the case illustrated in Fig. 18.73 might include the following:

- Reduce tonal components in mean-square vibration response by at least 15 dB.
- Maintain tonal performance in the presence of slewing tonals (e.g., where the tonal frequency changes over time at a rate of less than 3 Hz/s).
- Reduce broadband mean-square response in the control band (i.e., between frequencies f_1 and f_2) by 0–5 dB.
- Maintain performance in the presence of changes in the dynamics of the system (i.e., plant) that will occur over time.
- Limit out-of-band noise amplification to less than 2 dB at any frequencies outside the control band.

Identifying performance goals is important for several reasons. First, the expectations for the ANVC system are clearly stated and can be evaluated versus practically achievable ANVC performance. For example, typical ANVC systems can provide 15–20 dB of tonal reductions, depending on the slew rate of the tonals. Further, broadband performance in excess of 10 dB is typically associated with some level of out-of-band amplification. Second, these specifications will directly impact implementation aspects of the ANVC system. With regard to the above example, we can make the following initial observations. The broadband performance requirement will require a feedback implementation

CONTROL SYSTEM DESIGN CONSIDERATIONS **803**

or a feedforward implementation with reference sensors that are well correlated with the mean-square response to be controlled. The need to provide additional reductions at slewing tonal frequencies will require control filters that adapt relatively quickly in response to changing tonal frequency. In addition, since the plant transfer functions are expected to change over time, the control system will need to support concurrent system identification and adaptation to track relatively slower changes in the system dynamics. Finally, the constraint on noise amplification will impact the bandwidth of a feedback implementation (as discussed in Section 18.5) and probe signal design (as discussed in Section 18.4 under System Identification).

In general, the performance goals will have a direct impact on *all* aspects of the ANVC system design (e.g., sensors, actuators, and controller). In the next sections, we discuss how the performance goals impact each of these subsystems. In particular, we discuss procedures to answer the following questions:

- What number of actuator channels is required?
- Where should the actuators typically be located?
- What are the output drive requirements (force, volume velocity, etc.) for the actuators?
- What is the number of sensor channels required?
- Where should sensors typically be located?
- What control algorithm/architecture can be used to design control filters that will maximize achievable system performance?
- Is it likely that the program goals can be met (at least throughout the frequency range of the available models) with a reasonable number of actuator and sensor channels and with a causal controller?

These questions will be addressed within the context of the technical approach illustrated in the flow diagram of Fig. 18.74. As shown along the left side, the approach involves three sequential studies:

- Controllability: where to place the actuators.
- Observability: where to place the sensors.
- Realizability: how to design causal control filters.

Number, Location, and Sizing of Actuators

Actuator and sensor placement studies require either a model of the system to be controlled [e.g., finite-element model (FEM), lumped-parameter model] or access to experimental data from the system itself. The first questions to be addressed are where actuators should be located, how many are necessary as a function of achievable performance, and what size of actuator is needed (shaker maximum force output, speaker maximum volume velocity, etc.) to effectively drive the system and provide control.

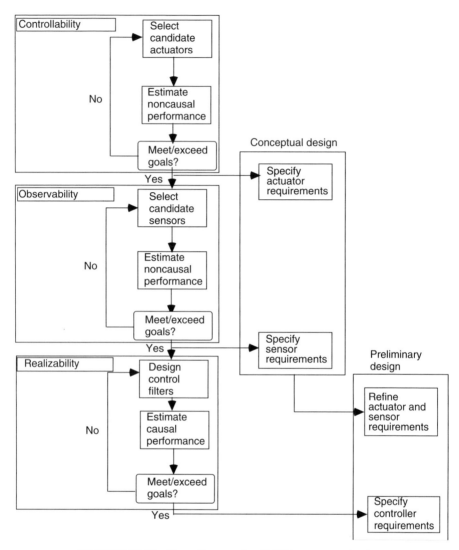

FIGURE 18.74 Flow diagram for ANVC strategy selection.

Number of Actuators. The question of how many actuator channels are required can be roughly estimated from experimental data or model results of the response to be controlled. For example, suppose that the response to be controlled is the mean-square vibration response on a component that is characterized by L point response measurements. The minimum number of actuator channels is equal to the number of independent ways the system is responding at the L measurement locations (referred to as *response dimensionality*) on a frequency-by-frequency basis. The response dimensionality can be estimated by computing the singular-valued decomposition (SVD) of the cross-spectral density (CSD) matrix of the L responses. We can express this CSD matrix as

$$[S_{ee}] = [T][S_{dd}][T]^H \qquad (18.70)$$

where $[S_{dd}]$ = cross-spectral density matrix of disturbance
$[T]$ = transfer function matrix from source(s) to L response locations
$[\cdot]^H$ = conjugate transpose of matrix

Note that $[S_{ee}]$ can be evaluated directly by measuring the CSD matrix of a number of sensors at candidate locations. As an alternate approach analytical/numerical models of the system to be controlled can be used to estimate $[T]$ along with approximations to $[S_{dd}]^*$ in Eq. (18.70) to estimate $[S_{ee}]$.

The SVD of a matrix $[A]$ is defined as

$$\mathrm{svd}\{[A]\} = [U_A][\lambda_A][V_A]^T \qquad (18.71)$$

where $[U_A]$ and $[V_A]$ are unitary matrices containing the left and right singular vectors of $[A]$ and $[\lambda_A]$ is a diagonal matrix containing the singular values for $[A]$ ranked from largest to smallest.[60] A typical plot of the singular values of $[S_{ee}]$ (i.e., diagonal elements of $[\lambda_e]$) at a particular frequency will reveal several large (i.e., significant) singular values followed by a transition to a region containing smaller (i.e., less significant) singular values. An example of this type of plot, where the singular values are normalized by the largest, is shown in Fig. 18.75. These types of plots can be conveniently generated using the svd function in MATLAB.

The number of large singular values of $[S_{ee}]$ at a given frequency indicates the number of significant and independent ways in which the disturbances on the system express themselves at the L measurement locations. At any frequency, the number of significant singular values is equal to the minimum of

- the number of independent disturbances acting on the structure (termed *source dimensionality*),
- the number of structural degrees of freedom (i.e., modes),
- the number of measurement responses (ie, sensors), and
- degrees of freedom of spectral estimates (i.e., number of ensemble averages in estimating the CSD matrix from experimental data).

As such, care must be taken to ensure that the estimated response dimensionality is not limited by using too few sensors, by insufficient averaging if $[S_{ee}]$ is measured, or by too few sources and sensors if analytical models are used to estimate $[S_{ee}]$. We note that controlling the response characterized by the L measurement

---

*For analytical/numerical simulations, where the CSD of the disturbances may be unknown, it is often sufficient to assume a large number of candidate disturbance locations and use a disturbance CSD matrix in Eq. (18.70) equal to the identity matrix at all frequencies of interest. This corresponds to the case where the sources are statistically independent with unit variance at each frequency.

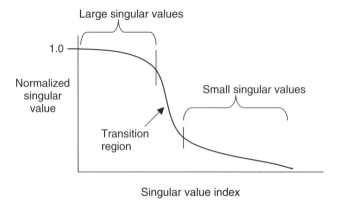

FIGURE 18.75 Typical decay of singular values of a sensor CSD matrix, normalized by the largest singular value.

locations will require at least as many actuator *channels** as significant singular values of $[S_{ee}]$. Thus, this approach establishes a lower bound on the number of required actuator channels.

Finally, we make two remarks concerning the use of SVD spectra to determine lower bounds on the number of actuator channels. First, since the number of significant singular values will typically increase as a function of frequency (because the number of modes supported by the system increases with frequency), the number of required actuator channels is often dictated by the singular-value spectra at high frequency. Second, the transition region in Fig. 18.75 may be gradual rather than showing a distinct "knee." For those cases, a useful "rule of thumb" is that the number of significant singular values at a given frequency is equal to the number of singular values whose magnitude is within, say, 30 dB of the largest singular value.

Actuator Locations. The selection of specific actuator locations is typically the result of an optimization procedure to select the best K locations from a set of M possible locations ($M > K$). This procedure requires the response CSD discussed above as well as the $L \times M$ transfer function matrix $[\mathbf{P}]$ between the M possible actuator locations and the L sensor responses. An exhaustive search over all combinations to determine the optimal set of K actuator locations would require a search of $\dfrac{M!}{[K!(M-K)!]}$ possibilities. This can be a daunting task for even modest values of K and M. When this is the case, it is often useful to invoke suboptimal techniques to choose a "good" set of actuators.

One suboptimal approach that has been used successfully is based on Gram–Schmidt orthogonalization.[60] Given M candidate actuator locations,

---

*The term *actuator channels* is not necessarily the same as the number of actuators. For instance, a group of 10 actuators driven coherently requires only a single output channel from a controller.

choose the one that maximizes reductions in the trace of [S_{ee}]. In this context, the best candidate is the column of [**P**] that is "most parallel" to the L sensor responses caused by the disturbances, integrated across frequency. Once an actuator (i.e., column of [**P**]) is chosen, components of [S_{ee}] and remaining columns of [**P**] in the direction of the chosen column of [**P**] are removed. The process is repeated K times, where the value of K is estimated from SVD spectra of [S_{ee}]. Variants of this procedure are discussed by LePage et al.[42] and Heck et al.[83] An interesting comparison of alternative procedures for selecting source locations based on using genetic algorithms and simulated annealing is discussed by Elliott.[58]

Regardless of the selection procedure, the mean-square response of the L sensors can be estimated assuming all actuators chosen so far can be used in an optimal, unconstrained sense. That is, we assume a controller can be developed to produce the required magnitude and phase spectra to drive each output channel. At each frequency, the L vector of residual responses ε can be expressed in terms of the L vector of uncontrolled responses **e** and the k vector of actuator drive signals **a** (for $k \leq K$) acting though the $L \times k$ plant transfer function matrix [**P**]:

$$\varepsilon = [\mathbf{T}]\mathbf{d} + [\mathbf{P}]\mathbf{a} = \mathbf{e} + [\mathbf{P}]\mathbf{a} \qquad (18.72)$$

The optimal solution for **a** to reduce the residual mean-square response tr{[$S_{\varepsilon\varepsilon}$]} at each frequency is

$$\mathbf{a} = -[\mathbf{P}]^{\#}\mathbf{e} \qquad (18.73)$$

where # signifies the pseudoinverse, which can be conveniently computed using the pinv function in MATLAB or by the expression defined in Section 18.4 under Extension to MIMO Systems.

As a consequence, the optimal noise reduction (NR, in decibels) of the mean-square response using the selected actuators is

$$\mathrm{NR} = -10 \log_{10}\left(\frac{\mathrm{tr}\{[\mathbf{S}_{\varepsilon\varepsilon}]\}}{\mathrm{tr}\{[\mathbf{S}_{ee}]\}}\right) \qquad (18.74)$$

where

$$[\mathbf{S}_{ee}] = E\{\mathbf{e}\mathbf{e}^{H}\} = [\mathbf{T}][\mathbf{S}_{dd}][\mathbf{T}]^{H}$$
$$[\mathbf{S}_{\varepsilon\varepsilon}] = [\mathbf{I} - \mathbf{P}\mathbf{P}^{\#}][\mathbf{S}_{ee}][\mathbf{I} - \mathbf{P}\mathbf{P}^{\#}]^{H} \qquad (18.75)$$

In Eq. (18.75), [S_{ee}] is the CSD matrix of the *uncontrolled* responses at the L residual sensor locations. This can be estimated from experimental data (i.e., $E\{\mathbf{e}\mathbf{e}^{H}\}$) or can be computed using transfer functions from source-to-residual locations, [**T**], based on analytical models (i.e., lumped-parameter or FEM), and an assumed CSD matrix of the sources, [S_{dd}], for example, the unit matrix as described above.

Equation (18.74) can be plotted frequency by frequency (or integrated across the frequency bandwidth of concern) to determine an upper bound on system performance, parameterized by the number of assumed actuators, as shown in Fig. 18.76. These performance predictions represent upper bounds because at this point we have not placed any causality constraints on the control filter, and we have not identified what sensor signals will be used as inputs to the controller. As such, these predictions are referred to as noncausal predictions, corresponding to those identified in Fig. (18.74) pertaining to controllability. Since Eq. (18.74) represents an upper bound on achievable performance as a function of frequency for the selected actuator set, it provides a first-order assessment as to whether the identified performance goals are achievable.

Actuator Sizing. A critical issue in the design of an active control system is the sizing of the actuators. Actuators with too little drive capacity will result in an underperforming control system, while actuators with too much capacity may impose excessive weight or cost penalties. A rough guide is that the actuators must be able to create response levels at the residual sensors that are comparable to the levels observed there due to the disturbance. A more precise estimate of the required drive levels from the actuators can be determined from the diagonal elements of $[S_{aa}]=[P]^{\#}[S_{ee}][P^{\#}]^{H}$, which correspond to the power spectra of the required drive signals to each actuator. For this calculation $[S_{ee}]$ should be based on measurements unless good analytical models of the disturbances and their CSD matrix, $[S_{dd}]$, are available. The square root of the integrated mean-square levels of each diagonal term across frequency provides an estimate of the rms levels for each drive signal. Peak drive requirements can be estimated from these

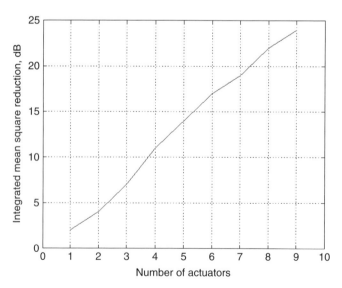

FIGURE 18.76 Example of noncausal performance versus number of actuator channels.

rms levels by applying an appropriate crest factor (i.e., ratio of the peak value of a waveform to its rms value). For pure-tone cases, the crest factor is 1.41; for broadband noise, the crest factor will depend on the statistics of the disturbance. A useful approximation to the crest factor for random noise is 4.

Note that if the plant [**P**] used in the equations above relates actuator output (force, volume velocity, etc.) to the residual channel responses, then the diagonal elements of [**S_{aa}**] (and the associated rms and peak levels) correspond to the actual actuator drive level (e.g., pounds, cubic feet per minute) that must be delivered to the structure by the actuators to produce the predicted performance. As such, these predictions provide initial actuator sizing information.

Number and Location of Sensors

The goals and requirements for selecting residual and reference channels are summarized below.

Residual Sensor Channel Selections for Feedforward and Feedback Strategies. The goal is to determine a minimal set of measurable responses that are

- well correlated with the radiated noise or vibration response to be controlled,
- nominally colocated with actuators for feedback control, and
- consistent with satisfying "obvious" causal constraints*: at or "downstream" of actuators for feedforward control strategy.

Reference Sensor Channel Selection (for Only Feedforward Strategy). The goal is to determine a minimal set of measurable responses that are

- well correlated with the disturbance sources,
- well correlated with candidate residual channels,
- consistent with satisfying obvious causal constraints: with sufficient "lead time" to implement an effective causal controller (i.e., located sufficiently "upstream" of the actuators).

Approaches for determining sensor channels that meet these goals and requirements are presented below.

Number of Sensors. If the number of residual sensors is equal to the number of control actuators, it is theoretically possible to achieve infinite noise reduction in a feedforward system. Of course, very large noise reduction at residual

---

*By obvious causal constraints, we rule out consideration of residual sensors that are closer to the sources than the selected actuator locations (i.e., upstream of the actuator locations). Similarly, we rule out consideration of reference sensor locations that are downstream of (farther away from the sources than) the actuator locations.

sensors does not necessarily mean that the system will also provide good noise reduction globally. Alternatively, if the number of control actuators exceeds the number of residual sensors, the system is overdetermined and the control effort may be excessive with some actuators competing against others unless control effort weighting is carefully included in the control filter design algorithm. Most practical systems have more residual sensors than control actuators so that the noise reduction is less concentrated at the residual sensor locations, resulting in more uniform global control.

The selection of residual and reference sensors (when required or available) can be determined following similar procedures to those outlined for actuator selection. As before, it is advisable to perform SVD analyses of candidate residual and reference CSD matrices to determine the minimum number of channels necessary to characterize the response caused by the disturbances at those locations. Since response dimensionality will never increase as measurement locations move away from the sources, it is common that the response dimensionality at the reference sensor locations will exceed that at the residual sensor locations. We note, however, that the number of sensors necessary to characterize the response dimensionality may exceed the number of sensor channels determined by the dimensionality studies, as, for example, when arrays of sensors are processed to characterize modal responses of a system.

If the response to be controlled is already known, the choice of residual sensing may be relatively straightforward. For example, if the mean-square response of a plate structure on a machine is to be reduced, it may be sufficient to instrument that plate with L point sensors, where L exceeds the response dimensionality of the plate (e.g., number of significant modes). Subsets of these L sensors can also be considered to determine the sets that when minimized in an optimal sense will produce the largest mean-square reductions of the full set of L sensors. Again, either exhaustive or suboptimal search techniques can be used.

For cases where reducing radiated noise is the main objective, it may not be possible to use residual sensor inputs to the controller that directly measure the noise field. Instead, for these cases, residual sensors must be located on the machine or device to be controlled. Selection of these on-board sensors is typically the result of using detailed radiation models to determine which on-board sensor responses are highly correlated with the radiated noise. If models or experimental data that include radiated noise information are available for this purpose, then candidate residual sensors can be selected using the exhaustive or suboptimal search techniques discussed above.

Residual Sensor Evaluation. As was done when considering actuator selection, the resulting sets of residual sensors obtained at each step can be evaluated to assess the maximum achievable performance as a function of the number of residual channels. In particular, reductions in the mean-square response of the selected residual channels can be determined using Eq. (18.74), where the optimal drive vector **a** in Eq. (18.73) is chosen to minimize only the candidate set of residual channels. These drives are then used in Eq. (18.72) to estimate the reductions in mean-square performance across a larger and more complete set of residual channels, or the radiated noise response. Again, these predictions represent upper bounds on achievable noncausal performance assuming fixed sets of actuators and residual sensors. That is, we again assume that a controller can be developed to produce the spectral magnitudes and phases that are necessary to drive each output channel to reduce the mean-square residual response in an optimal sense.

Alternatively, residual sensors can be selected simultaneously with implementing the procedure for estimating performance. A similar approach can be used to select reference sensors for a feedforward system. Details of this approach are discussed below with respect to evaluating reference and residual sensors for the case where residual sensors cannot directly measure radiated noise.

Assuming statistical quantities, we define the following frequency domain relationships between the vectors of radiated pressures (or some other desired residual which cannot be measured directly by the control system) **p**, residual channels **e**, and reference channels **r** and the vector of disturbances **d**,

$$\mathbf{p} = [\mathbf{Q}]\mathbf{d} \qquad \mathbf{e} = [\mathbf{T}]\mathbf{d} \qquad \mathbf{r} = [\mathbf{R}]\mathbf{d} \qquad (18.76)$$

where $[\mathbf{Q}]$, $[\mathbf{T}]$, and $[\mathbf{R}]$ are the transfer function matrices relating the disturbances to the radiated pressure, residual channels, and reference channels, respectively. For purposes of the discussion below, we assume that either the transfer functions $[\mathbf{Q}]$, $[\mathbf{T}]$, and $[\mathbf{R}]$ are available from analytical models (e.g., lumped-parameter or FEM) or the response vectors **p**, **e**, and **r** are available from experimental data.

The radiated pressure can be related to the residual channels as

$$\mathbf{p} = [\mathbf{H_e}]\mathbf{e} + \mathbf{n} \qquad (18.77)$$

where **n** represents the radiated pressure response that is uncorrelated with (or unobserved by) the residual channels. Assuming that **e** and **n** are independent statistical responses, we can obtain an expression for $[\mathbf{H_e}]$ in terms of the CSD matrix of the uncontrolled residual $[\mathbf{S_{ee}}]$ and the CSD matrix between the pressure and residual channels $[\mathbf{S_{ep}}]$:

$$[\mathbf{H_e}] = [\mathbf{S_{ep}}][\mathbf{S_{ee}}]^{-1}. \qquad (18.78)$$

As discussed earlier with regards to Eq. (18.75), $[\mathbf{S_{ee}}]$ is the CSD matrix of the *uncontrolled* responses at the L residual sensor locations and can be estimated from experimental data (i.e., $E\{\mathbf{ee^H}\}$) or can be computed by using transfer functions from source-to-residual locations, $[\mathbf{T}]$, based on analytical models (i.e., lumped-parameter or FEM), and an assumed CSD matrix of the sources, $[\mathbf{S_{dd}}]$. Similarly, $[\mathbf{S_{ep}}]$ is the CSD matrix of the uncontrolled responses of the pressure and residual sensors. It can be estimated from experimental data (i.e., $E\{\mathbf{pe^H}\}$) or be computed from analytical models for [**T**] and [**Q**] together with an assumed CSD matrix of the sources, [S_{dd}]:

$$[S_{ep}] = [Q][S_{dd}][T^H] \tag{18.79}$$

We can now write an expression for the maximum noise reduction, at each frequency, that can be achieved by eliminating the response in the radiated pressure that is correlated with the chosen set of residual channels:

$$NR_p = 10 \log_{10}\left(\frac{\text{tr}\{[S_{pp}] - [H_e][S_{ee}][H_e^H]\}}{\text{tr}\{[S_{pp}]\}}\right) \tag{18.80}$$

where

$$[S_{pp}] = [Q][S_{dd}][Q^H] \qquad [S_{ee}] = [T][S_{dd}][T^H] \tag{18.81}$$

Note that if there is no correlation between the residual sensors and radiated pressure, then NR_p will equal 0 dB. If the output is entirely due to the input (i.e., the residual sensors are perfectly correlated with the radiated pressure, then NR_p will be infinite (i.e., $-\infty$ decibels).

The evaluation of Eq. (18.80) requires either measured data or analytical/numerical models of the system to be controlled. If *simultaneous* measurements of the residual sensors, **e**, and the desired residual, **p**, are available, then the data can be processed to generate the cross-spectra density matrices [S_{ep}], [S_{ee}], and [S_{pp}]. Equation (18.78) can then be used to determine [H_e], which can then be substituted along with [S_{ee}] and [S_{pp}] into Eq. (18.80). If, on the other hand, analytical/numerical models are available, they can be used to estimate the transfer function matrices [**Q**] and [**T**]. If the CSD of the disturbance [S_{dd}] is then taken to be the unit matrix as described earlier, then [S_{ep}], [S_{ee}], and [S_{pp}] can be easily estimated from the equations given above, allowing the evaluation of Eq. (18.80) as if measured data were available.

Reference Sensor Evaluation. In a similar fashion, the procedure outlined above can be used to select reference sensors that are highly correlated with the residual and/or radiated pressure. The residual channels can be related to the reference channels as

$$\mathbf{e} = [H_r]\mathbf{r} + \mathbf{m} \tag{18.82}$$

where **m** represents the response in the residual channels that is uncorrelated with (or unobserved by) the reference channels. Following the procedure outlined above, the maximum reduction in the mean-square response of the residual channels, **e**, assuming a set of reference sensors **r** is given by

$$NR_e = 10 \log_{10}\left(\frac{\text{tr}([S_{ee}] - [H_r][S_{rr}][H_r^H])}{\text{tr}([S_{ee}])}\right) \tag{18.83}$$

where

$$[S_{ee}] = [T][S_{dd}][T^H] \qquad [S_{rr}] = [R][S_{dd}][R^H] \qquad (18.84)$$

and

$$[H_r] = S_{re}[S_{rr}]^{-1} \qquad (18.85)$$

The expression given in Eq. (18.83) represents the maximum reduction, at each frequency, that can be achieved by eliminating the response in the residual channels that is correlated with the chosen reference channels. If there is no correlation between the residual sensors and radiated pressure, then NR_e will equal 0 dB. If the output is entirely due to the input (i.e., the residual sensors are perfectly correlated with the radiated pressure), then NR_e will be infinite.

As with the residual sensors the evaluation of Eq. (18.83) requires either measured data or analytical/numerical models of the system to be controlled. If simultaneous measurements of the residual sensors, **e**, and the reference sensors, **r**, are available, then the data can be processed to generate the cross-spectra density matrices $[S_{er}]$, $[S_{ee}]$, and $[S_{rr}]$. Equation (18.85) can then be used to determine $[H_r]$, which can then be substituted along with $[S_{ee}]$ and $[S_{rr}]$ into Eq. (18.83). If, on the other hand, analytical/numerical models are available, they can be used to estimate the transfer function matrices $[R]$ and $[T]$. If the CSD of the disturbance $[S_{dd}]$ is then taken to be the unit matrix as described earlier, then $[S_{er}]$, $[S_{ee}]$, and $[S_{rr}]$ can be easily estimated from the equations given above, allowing the evaluation of Eq. (18.83) as if measured data were available.

Sensor Selection and Noncausal Performance. Choices for candidate residual and reference sensors are typically guided by an understanding of the physics of energy propagation for the system to be controlled as well as guidance on the required number of channels required based on SVD analyses. Assuming candidate actuator, reference, and residual locations have been identified using the procedures discussed above, an upper bound on achievable performance for this set of transducers can be obtained by estimating the noncausal performance. These predictions correspond to those identified in Fig. 18.74 under the observability phase. For a feedforward system, this means that we assume the controller can implement whatever filter is necessary to relate the reference sensor responses to the required actuator drive signals to minimize the mean-square residual response. For this case, the uncontrolled and controlled mean-square residual responses are given by

$$\text{tr}([S_{ee}]) = \text{tr}([T][S_{dd}][T^H]) \qquad (18.86)$$

and

$$\text{tr}([S_{\varepsilon\varepsilon}]) = \text{tr}(\{I-[P][P^\#]\}[H_r][S_{rr}][H_r]^H\{I-[P][P^\#]\}^H) \qquad (18.87)$$

where $[\mathbf{S_{ee}}]$ and $[\mathbf{S_{rr}}]$ are defined in Eq. (18.84), $[\mathbf{H_r}]$ is defined in Eq. (18.85), and the noise reduction is given by

$$\text{NR} = 10 \log_{10}\left(\frac{\text{tr}([\mathbf{S}_{\varepsilon\varepsilon}])}{\text{tr}([\mathbf{S_{ee}}])}\right) \quad (18.88)$$

As described in the above discussion for the residual and reference sensors, Eq. (18.88) can be evaluated using either measured data or analytical/numerical models. The one difference here is that measurements or analytical/numerical predictions of the plant transfer function matrix $[\mathbf{P}]$ are also required.

On a final note, we remark that while the required numbers of reference and residual channels are typically determined by the system response at the highest frequency of interest, they may be more than is necessary at lower frequencies. As a consequence, the CSD matrices of the reference and residual sensors (and the plant matrix) may be ill-conditioned (i.e., noninvertible) at low frequencies, which may impact the matrix inverses in Eqs. (18.78), (18.85), and (18.87). For these situations, it may be necessary to add small values to the diagonal elements of these matrices before computing the inverse. This procedure is referred to as "regularization." The values added are ideally a function of frequency and will have a larger influence at low frequencies (where the matrices are ill-conditioned) than at high (where the matrices are not ill-conditioned). One procedure that satisfies this requirement (and has been used successfully in many ANVC applications) is to add a scalar times the maximum singular value of the matrix to each diagonal element of the matrix on a frequency-by-frequency basis.

Controller Architecture and Performance Simulations

The upper bound predictions given in the sections above on the number, location, and sizing of sensors and actuators provide a series of tools for bounding system performance as more of the control system is defined. As more realism is included (e.g., identifying both actuator and sensor numbers and locations), the bounds on achievable performance typically get smaller but also more realistic. Comparisons of these bounds with performance goals can be made at each step to determine if the control system under consideration is likely to meet its objectives. Should the system still appear feasible after selecting both actuator and sensor locations [i.e., from Eqs. (18.86) and (18.87)], the next step is to add more realism related to the implementation of the controller.

In this section, we consider the impact of controller architecture on achievable performance. In particular, we discuss procedures to include effects of analog filtering, sample rate, and length of control filter.

As an introduction to these issues, consider the signal path from sensor inputs to actuator drive signals for a conventional digital controller as shown in Fig. 18.77. With reference to this figure, a digital controller requires that the analog signals from the sensors be sampled and digitized at the sample rate of the controller. The sample rate of the controller is usually expressed in terms of a sampling rate or frequency (f_s), which implies that the signals are sampled

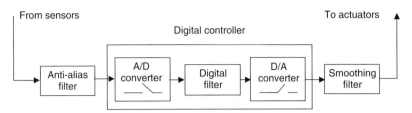

FIGURE 18.77 Signal path from sensor input to actuator drive signal for a digital controller.

at a time interval (T_s) equal to the inverse of the sample rate. For example, a sample rate $f_s = 1000$ Hz corresponds to a sample interval $T_s = 0.001$ s. To satisfy Nyquist's sampling theorem,[57] the sample rate of the controller must be at least twice the highest frequency of interest for the controller. For feedforward systems, the sample rate may be a factor of 4 above the highest frequency to be controlled. For broadband feedback systems, as discussed in Section 18.5 under Alternate Suboptimum Control Filter Estimation, it is not uncommon for the controller sample rate to be a factor of 100 above the highest frequency to be controlled (to minimize the effects of time delay through the controller and quantization noise on system performance).

Once the sample rate has been chosen, analog filters must be used to limit the frequency content of the analog sensor signals to below the Nyquist frequency of the controller, where

$$f_{\text{nyq}} = \tfrac{1}{2} f_s \tag{18.89}$$

These filters are called "anti-alias" filters, and their output is passed to an A/D converter, which in turn samples and quantizes the signal. Similarly, the output of the digital controller is converted to an analog signal by a D/A converter. Before these output signals are sent to the actuators, however, they must be filtered to band limit the signals below the Nyquist frequency; otherwise signals above the Nyquist frequency will be sent to the actuators. These filters are referred to as "reconstruction" or "smoothing" filters

Although issues regarding the selection of anti-alias and smoothing filters are discussed in the next section, we note here that a major consequence of filtering is that it introduces a delay in the signal path, which is referred to as the "group delay" of the filter. This delay is related to the derivative of the phase response of the filter:

$$\tau_g(\omega) = \frac{d\varphi(\omega)}{d\omega} \tag{18.90}$$

In addition to the filters, delays in the signal path of Fig. 18.77 are also associated with the A/D converter, sample-and-hold functionality of the D/A, and the processing time associated with implementing the digital filter. With regards to the A/D converter, delays can be kept sufficiently small by choosing successive-approximation A/D converters. Issues related to A/D selection for applications where time delay is not a limiting factor in controller performance (e.g., tonal control systems or for residual sensing in some feedforward applications) are discussed in the next section. The delay associated with the sample and hold of the D/A converter $\tau_{D/A}$ is directly related to the sample rate of the controller:

$$\tau_{D/A} = \frac{1}{2f_s} \qquad (18.91)$$

For efficient implementations, the delay (τ_{DSP}) associated with implementing the digital filter within a DSP chip can be limited typically to a few tens of microseconds. These implementations require that the sample clock controlling the D/A converters be offset relative to the sample clock for the A/D converters. As such, output samples can be sent to the D/A converters as soon as they are computed so that the processing delay can be held to a fraction of a sample interval. This sampling strategy is in contrast to one that waits until the next time step to output a sample. The latter approach imposes a minimum one-sample delay in the processing, which can be unacceptable for certain applications.

The total delay from analog signal input to analog signal output is termed the *latency* of the controller. Where appropriate, any delay associated with the amplifiers and actuators will also contribute to the latency. Latency is an important parameter since it can directly limit achievable performance of broadband feedforward and feedback systems. As such, it is important to limit overall latency in ANVC control systems consistent with achieving performance goals.

At this point, specific characteristics of the analog filtering, sampling, and control filter parameters can be merged with model or experimental results to make causal performance predictions. The steps involved are as follows:

- Specify a system sample rate based on control bandwidth and architecture (feedforward or feedback).
- Modify the transfer function matrices involving sensor inputs and actuator outputs to include
  — transfer functions of anti-alias (AA) and smoothing filters (SF),
  — delay associated with D/A converter sample and hold ($\tau_{D/A}$),
  — controller processing delay (τ_{DSP}), and
  — transfer functions of amplifiers and actuator (SH).
- Specify cost function (including control effort and robustness constraints) to be minimized.
- Specify number of control filter coefficients.
- Solve for control filter coefficients using the procedures outlined in Sections 18.4 and 18.5.
- Estimate causal mean-square residual performance and actuator output drive requirements: Compare estimated performance to performance goals.

If transfer functions of the filters, amplifiers, and actuators are not available, initial causal performance estimates can be made by assuming flat magnitude spectra with linear phase spectra, which correspond to pure time delays that bound expected delays for those components. When this is the case, Elliott[58] suggests approximating the group delay of the anti-alias or smoothing filters as

$$\tau_g \approx \frac{n}{8 f_c} \quad (18.92)$$

where n is the order of the filter and f_c is the cutoff frequency.

Hardware Selection

In this section, we discuss some of the practical issues related to selecting analog filters, A/D converters, and DSPs. Although a complete discussion of these issues is beyond the scope of this book, we present general guidelines and considerations. Detailed discussions of each of these issues can also be found in reference 58.

Anti-Alias and Reconstruction Filters. As discussed earlier, the main purpose of anti-alias and reconstruction filters is to attenuate signals above the Nyquist frequency. The anti-alias filters minimize contamination of frequencies below the Nyquist frequency by attenuating the spectral content of the signals above the Nyquist frequency. These higher frequencies are interpreted as lower frequencies because of the nonlinear sampling process of the A/D converter. This phenomenon is referred to as *aliasing*. Similarly, reconstruction filters attenuate components of the drive above the Nyquist frequency, which occur as a consequence of sampled signals at the output of the D/A converter. The benefits of using anti-alias and reconstruction filters in combination with A/D and D/A converters are discussed below

The amount of attenuation required from an anti-alias filter depends on the spectral shape of the sensor response and the desired bandwidth over which the controller needs an accurate representation of the signal. A sensor response that is inherently band limited below the Nyquist frequency may require only a low-order anti-alias filter or possibly none. Alternatively, a sensor response that is flat or increasing with frequency will require a higher order anti-alias filter. For purposes here, we assume that the sensor response is flat versus frequency. In addition, we impose a requirement of at least 40 dB attenuation of contributions from aliasing frequencies at the highest frequency that is important to the controller (f_a). From the context of controller implementation, however, we wish to achieve this attenuation while minimizing the group delay of the filter to acceptable levels.

There are two basic approaches to selecting the filter. The first is to use a low-order filter with a relatively low cutoff frequency. The second is to use a high-order filter with a higher cutoff frequency. Figure 18.78 compares the magnitude spectra and group delays for a fourth-order Butterworth filter and a sixth-order Cauer filter, both of which were designed to provide at least 40 dB of attenuation above $f_{40\ dB} = 600$ Hz. Both filters will meet the attenuation requirements for

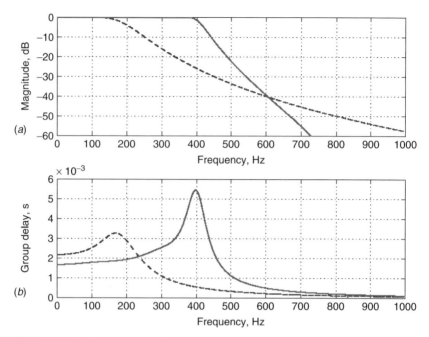

FIGURE 18.78 Comparison of (*a*) magnitude spectra and (*b*) group delay for fourth-order Butterworth (dashed curve) and sixth-order Cauer (solid curve) filters.

frequencies below their respective cutoff frequencies (i.e., $f_a < f_c$) provided that the system sample rate is chosen such that

$$f_s \geq f_a + f_{40\text{dB}} \tag{18.93}$$

Consider first the case with f_a less than the cutoff frequency of the Butterworth filter (nominally 200 Hz) and a sample rate of 800 Hz. For this case, attenuation of aliasing components is assessed by "folding" the magnitude response of the filters about the Nyquist frequency of $800/2 = 400$ Hz. As such, both filters provide the required attenuation of aliasing components in the frequency band below f_a. However, the group delay associated with the Cauer filter is less than that for the Butterworth for all frequencies below f_a, particularly near 200 Hz.

Now consider the case with $f_a = 350$ Hz and a sample rate of 950 Hz. For this case, attenuation of aliasing components is assessed by folding the magnitude response of the filters about the Nyquist frequency of $950/2 = 475$ Hz. The Cauer filter again satisfies the attenuation requirement because the sample rate satisfies Eq. (18.93). However, the attenuation requirement is not met by the Butterworth filter. In fact, less than 20 dB attenuation of aliasing components is achieved at 350 Hz (i.e., the difference in the magnitude responses at 350 Hz and $950 - 350 = 600$ Hz. Further, note that the maximum group delay of both filters in the frequency band below f_a is approximately the same.

The above comparisons suggest that using higher order filters with high cutoff frequencies is preferable to using low-order filters with low cutoff frequencies. Although the preceding discussion centered on selecting anti-alias filters, similar conclusions apply to reconstruction filters provided that the control filters are designed to reduce output levels above f_a. Unfortunately, the cost of implementing high-order filters will exceed the cost of using low-order filters. As a consequence, the performance benefits must be balanced against considerations of additional cost, as will be discussed further in the example ANVC system in Section 18.7 under MIMO Feedforward Active Locomotive Exhaust Noise Control System with Passive Component.

Analog-to-Digital and Digital-to-Analog Converters. A typical data path from sensor input to actuator drive signal was shown in Fig. 18.77. That figure illustrates the use of both types of data converters required for implementing digital controllers, namely A/D and D/A converters. The purpose of A/D converters is to sample the continuous-time signals from the sensors at uniform time intervals ($T_s = 1/f_s$) and quantize the amplitude to a discrete set of amplitude levels. Because the sampling process is a nonlinear operation, anti-alias filters are typically used to band limit the sensor signals below the Nyquist frequency ($f_s/2$) before sending the signals to the A/D converter. This minimizes the possibility that frequencies above Nyquist will contaminate the sensor response below Nyquist, which is referred to as aliasing.

Figure 18.79 shows a frequency domain schematic of how anti-aliasing filters are used in combination with A/D converters to minimize aliasing effects. The upper curve in Fig. 18.79*a* represents the spectrum of an unfiltered continuous-time signal presented at the input of an A/D converter. For an A/D converter operating at a sample rate of f_s, signals above the Nyquist frequency ($F_{\text{nyq}} = f_s/2$) will be interpreted as frequency below Nyquist. Fig. 18.79*a* illustrates schematically how frequencies above the Nyquist frequency will fold down into the frequency region below Nyquist. The estimated spectrum below Nyquist will be the sum of the true spectrum below Nyquist plus any aliasing components resulting from the frequency content of the signal that is above the Nyquist frequency. As a consequence, aliasing components can cause the estimated spectrum to differ from the true spectrum of the signal in the frequency region below Nyquist. Anti-alias filters are typically used to band limit the signals prior to the conversion process to reduce the magnitude of any aliasing components. Figure 18.79*b* is a plot of the magnitude response of an anti-alias filter designed to reject frequencies above the Nyquist frequency. When this filter is applied to the unfiltered signal of Fig. 18.79*a*, the resulting spectrum of the signal presented at the input to the A/D converter is that shown in Fig. 18.79*c*. As shown in this figure, much of the effects of aliasing components on the estimated spectrum are removed, thus ensuring that the estimated spectrum is a more accurate representation of the true spectrum of the signal below the Nyquist frequency.

Digital-to-analog converters are used to convert the digital sequences from the controller into analog signals that can ultimately be used as drive signals to actuator electronics. In the frequency domain, the sample output sequence

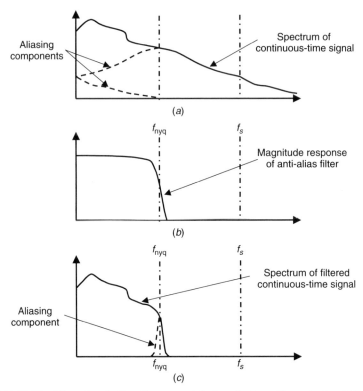

FIGURE 18.79 Use of anti-alias filters to minimize aliasing effects during A/D conversion: (a) aliasing components of an unfiltered continuous-time signal digitized at sample rate f_s; (b) magnitude spectrum of anti-alias filter; (c) aliasing components of filtered continuous-time signal digitized at sample rate f_s.

produces a spectrum that contains "images" of the signal spectrum below the Nyquist frequency, which occur at frequencies above the Nyquist frequency. These image spectra are shown schematically in Fig. 18.80. The effects of the images must be minimized. Otherwise, the controller will drive the actuators at frequencies above the Nyquist frequency, resulting in increased out-of-band noise amplification. These out-of-band effects are reduced to some extent by the sample-and-hold (S&H) functionality that is integral to most D/A converters. As discussed earlier, the S&H introduces a delay equal to one-half the sample interval $(T_s/2)$. In addition, it provides some filtering of the output spectrum, as illustrated in Figs. 18.80b and 18.80c. To suppress the remaining signals above the Nyquist frequency, D/A converters are typically followed by smoothing or reconstruction filters. The effect of applying reconstruction filters on the D/A output signal is shown in Fig. 18.80e.

Two basic types of A/D converters are applicable for use in ANVC systems. The first type is referred to as "successive-approximation" (SA) A/D converters. The second type is referred to as "sigma–delta" ($\Sigma\Delta$) A/D converters. Table 18.1

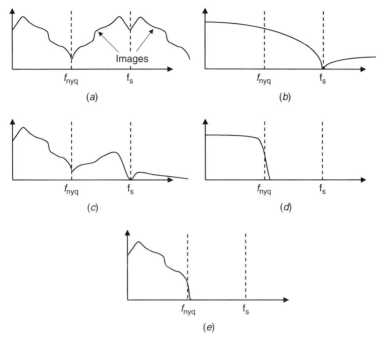

FIGURE 18.80 Use of reconstruction filters to reduce image spectra at the output of D/A converters: (a) spectrum of digitized output signal at sample rate f_s; (b) magnitude response of zero-order sample-and-hold integral to D/A converter; (c) spectrum of analog output of D/A converter prior to reconstruction filter; (d) magnitude response of reconstruction filter; (e) spectrum of analog signal at output of reconstruction filter.

compares some of the important features and differences between these two types of converters.

The choice of A/D converter for an ANVC application will depend on the control architecture (i.e., feedforward or feedback), the type of sensor signals (i.e., reference or residual sensors), and the performance goals. For broadband ANVC applications that require minimal latency through the controller (e.g., reference sensors in a feedforward system or residual sensors in a feedback system), the delay associated with $\Sigma\Delta$ A/D converters is unacceptable. As a consequence, SA A/D converters should be used for those applications, and the extra cost of the converters and anti-alias filtering should be included in the total cost of the ANVC system. We note, however, that low latency is not a necessary requirement for the residual sensors of a feedforward implementation. These sensor signals are used in the design of the control filters but are not themselves filtered by the control filter to produce output signals to the actuators. However, the latency associated with $\Sigma\Delta$ A/D converters used on the residual signals will be seen as a delay in the plant transfer functions. This added delay will in turn reduce allowable convergence coefficients for LMS-based algorithms[58] but does not preclude their use.

TABLE 18.1   Comparison of Successive-Approximation and Sigma-Delta A/D Converters

Successive-Approximation Converters	Sigma-Delta Converters
Sample input signal at rate f_s	Over sample input signal nominally at 64 times desired sample rate (i.e., $f_{\Sigma\Delta} = 64 f_s$)
Require use of relatively high-order anti alias filters to bandlimit signals below Nyquist frequency	Only low-order (if any) anti-alias filters required to bandlimit signals below $f_{\Sigma\Delta}/2$. Integral high-order (linear-phase) digital filters bandlimit signal below Nyquist frequency ($f_s/2$). Output is decimated (i.e., keep only every 64th sample) to produce sampled sequence at f_s sample rate.
Low latency (on the order of a few μs)	High throughput latency (typically $32T_s$, where $T_s = 1/f_s$)
Quantize signal at multiple bit levels (e.g., 12 or 16 bits)	One-bit quantization at oversampled rate to provide comparable multi-bit quantization (e.g., 12 or 16 bits) at f_s sample rate
Relatively expensive compared to $\Sigma\Delta$ converters	Relatively inexpensive compared to SA converters

When low latency is not required, the use of $\Sigma\Delta$ converters can significantly reduce the costs associated with digitizing the sensor signals. For example, a tonal feedforward ANVC system can use a $\Sigma\Delta$ A/D converter to digitize the reference signal. As before, digitization of the residual signals for this example can be done using $\Sigma\Delta$ A/D converters, provided that the extra delay is acceptable in terms of convergence rate (if adaptive algorithms are used). The choice of D/A converter is governed by similar arguments concerning allowable latency. For instance, when low latency is not required, $\Sigma\Delta$ D/A converters can be used. Alternatively, when latency must be minimized, conventional D/A converters should be used.

For most ANVC applications, it is desirable to initiate the sampling across all A/D converters using a common clock signal. In this way, all A/D inputs will be synchronously sampled. Similarly, it is advantageous to sample all the D/A converters using a common clock signal as well. The D/A converters could be clocked using the same clock pulse as for the A/D converters; however, to minimize latency through the controller, it is useful to offset the D/A clock relative to the A/D clock. In this way, subsample latency can be achieved by sending signals to the D/A when they are ready, as opposed to waiting for a full sample period.

The final aspect of converter selection is quantization noise. When analog signals are quantized to a finite number of amplitude values (e.g., 16-bit A/D converter), the errors in the conversion process can be thought of as noise. For the signals of interest for ANVC, this noise is modeled as uniformly distributed from—LSB/2 to LSB/2. Here, LSB stands for the *least-significant bit* and is given by

$$\text{LSB} = \frac{V}{2^{M-1}} \quad (18.94)$$

where V corresponds to the voltage range of the converter (i.e., $\pm V$) and M is the number of resolution bits (e.g., 16 for a 16-bit converter). For the assumed uniform probability distribution of the noise, the rms of the quantization noise (QN_{rms}) is equal to

$$QN_{\text{rms}} = \frac{LSB}{\sqrt{12}} \quad (18.95)$$

For a sample rate f_s, the spectrum of the quantization noise ($QN_{\text{psd}}(f)$) is white (i.e., flat) with a power spectral density (PSD) amplitude given by

$$QN_{\text{psd}}(f) = \frac{LSB^2/12}{fs/2} \quad (18.96)$$

As an example, for a 16-bit converter with a voltage range of ± 10 V and a sample rate of 1 kHz, the rms quantization noise [from Eq. (18.95)] is 88 μV, and the PSD level is -108 dB re 1 V^2/Hz.

At this point, it is useful to compare plots of

- the expected signal levels (in volts from the sensor),
- electrical noise floors (associated with the sensors, analog filters, and gain), and
- quantization noise.

To support a common comparison, all of these voltage levels should be referenced to a common point in the signal path (e.g., the amplifier input). These plots should be generated corresponding to both open- and closed-loop sensor responses. Sensor sensitivities, signal gain, and quantization noise can then be assessed within a common framework. The goal is to choose each of these to ensure that sensor signals will have sufficient signal-to-noise ratio (SNR) throughout the bandwidth of interest.

To illustrate some of the issues related to A/D converter selection, consider the example shown in Fig. 18.81. The various curves grouped in the center of this plot correspond to the expected signal levels from multiple residual sensors, expressed in dB re V^2/Hz, referenced to the input of a bank of sixth-order Cauer filters. The filters are used to provide anti-alias filtering as well as programmable gain for these channels. The cutoff frequency is approximately 800 Hz. The power spectral density of the noise from the anti-aliasing filters at their input (labeled PFI *noise* in the figure) falls between -155 dB and -150 dB re 1 V^2/Hz. The spectral density of the sensor noise floor reference to the filter input is

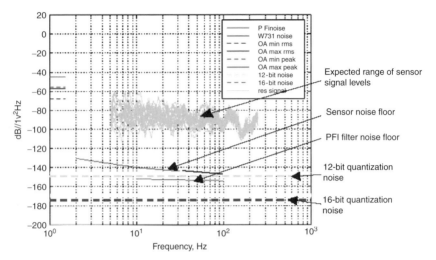

FIGURE 18.81 Comparison of signal level versus electrical and quantization noise.

approximately -130 dB re 1 V^2/Hz at 2 Hz and drops to -145 dB re 1 V^2/Hz at 100 Hz.

The rms levels for the sensor signals (needed for quantization noise estimates as indicated below) are determined by integrating the mean-square voltage responses across frequency and taking the square root. For these spectra, the largest rms level is approximately -57 dB re 1 V. Assuming the statistics of these responses is random (i.e., with a crest factor of 4 or 12 dB), a peak voltage response for this group of sensors is estimated to be approximately -45 dB re 1 V. The spectrum level of the digitization noise (at the filter input) can be estimated from Eq. (18.96), where the voltage V in Eq. (18.94) is taken (at the moment) to be the peak voltage of the sensor signals (i.e., -45 dB re 1 V = 5.5 mV). As such, the spectral levels of digitization noise for 12- and 16-bit converters, assuming a sample rate of 2 kHz, are -152 dB and -176 dB re 1 V^2/Hz, respectively. The SNR for these sensor measurements is a function of frequency and corresponds to the spread in decibels between the estimated signal responses and the maximum of the electrical noise (from filters and sensors) or the digitization noise. The SNR is nearly 30 dB throughout the frequency band of concern.

For the example shown in Fig. 18.81, it is sufficient to use a 12-bit converter since the SNR is limited by sensor noise as opposed to digitization noise up to about 100 Hz. Above that frequency, the SNR will be limited by digitization noise if a 12-bit converter is used or by PFI noise if a 16-bit converter is used. Further, if the converter (12- or 16-bit) has a voltage range of ±1 V, the programmable gain should be set at approximately 33 dB to allow 1–2 bits of headroom to avoid clipping.

The example case in Fig. 18.81 illustrates the relatively complex interdependencies that must be considered when selecting ANVC hardware to ensure adequate SNR, which in general will be a function of sensor noise and sensitivity, anti-alias filter noise, A/D converter quantization noise and voltage range, system gain, and sample rate.

Digital Signal Processors. As discussed earlier, the rapid growth in ANVC systems and applications over the past 25 years can be traced in large part to the advent of DSP chips. These chips are optimized to perform operations on a sample-by-sample basis. That is, these chips read in data one sample at a time, perform large numbers of computations, and then output a sample. This is in contrast to other central-processing units (CPUs) that read in groups (i.e., buffers or blocks) of data, operate on the entire buffer, and then output buffers of data. These later processors can often perform larger numbers of computations than DSP chips, but their latency is on the order of the block size.

Digital signal processors are ideally suited to perform the low-latency digital filtering required by ANVC algorithms. As such, they are widely used in the design of adaptive control systems based on the algorithms presented in this chapter. In addition, current DSP technology supports high-speed communication between DSPs to support applications with large numbers of inputs and outputs (e.g., many tens of inputs and outputs) as well as large digital filter sizes (e.g., IIR filters with over 5000 taps per filter). For these large-scale problems, efficient low-latency implementation of the digital filters can be achieved by combining DSPs with high-powered CPU compute engines such as PowerPC (PPC) chips. Further, the relatively large latency associated with the PPC chips may be acceptable to perform much of the "off-line" algorithms associated with system identification and control filter design when using direct estimation as opposed to adaptive filter algorithms.

Once the control algorithms have been selected, the computation and memory requirements can be determined for all "in-line" filtering and "off-line" processes. These requirements should then be mapped to DSPs and PPC chips, as necessary. Floating-point DSPs such as the TMS320C6701 are rated at 1 Gflop and include up to 128 MB of fast-access memory. Compute engines like the PPC7410 are rated at 2 Gflops with up to 512 MB of local memory. In addition, PC or VME (workstation) boards are currently available with up to four of each type of processor per board and support high-speed communications between processors on a given board as well as processors located on other boards. These processor-to-processor communications are supported by high-speed interconnect architectures such as Mercury RACE^{++}, Spectrum DSPLink, and SKYchannel.

As an example, Fig. 18.82 shows the hardware architecture for a large-scale ANVC system incorporating both DSPs and PPC chips, including high-speed communication over a Mercury RACEWAY interconnect. The signal path from the reference sensors to the actuator drive signals includes high-order anti-alias and reconstruction filters with cutoff frequencies near the Nyquist frequency. Successive-approximation converters are used to minimize latency associated with the conversion. A combination of DSPs and PPC chips are used to efficiently implement the digital control filters to minimize latency.[84] The residual sensor

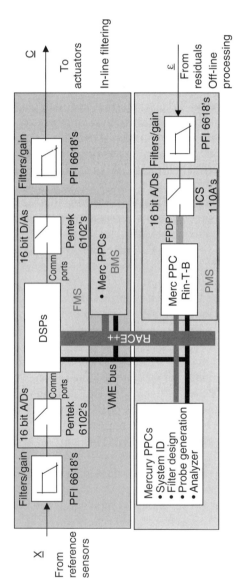

FIGURE 18.82 Control architecture using combination of DSPs and PPCs.

signals pass through anti-alias filters and are digitized using sigma–delta converters. These data support off-line functionality associated with system identification, control filter design, and probe-signal injection, which are performed on multiple PPC chips. Communications between the DSPs and PPC chips are primarily carried over Mercury's RACE^{++} interconnect bus, with additional communications occurring over the VME bus. In this way, the compute power associated with PPC chips can be used to efficiently perform off-line functions (e.g., direct estimation of plant and control filters), which can be passed to the in-line processors (DSPs and PPC chips) to support adaptation of the control filters.

Finally, we note that fixed-point DSPs as well as floating-point DSP devices are available. The floating-point devices are much easier to program and are often chosen for proof-of-principle systems or for final systems that are not highly cost sensitive. When cost is a driving issue, fixed-point devices should be considered. The final choice must balance the higher nonrecurring cost associated with programming fixed-point devices and the potential increased CPU overhead due to implementing floating-point arithmetic on a fixed-point device against the lower recurring per-unit cost

Control System Implementation and Testing

Once the simulations and hardware selections have been made, the control system is implemented and performance tests are conducted. In this section, we discuss the basic operating modes and features of a controller that should be considered during implementation to support the subsequent testing phase.

For each of the modes discussed below, parameters that can be changed during system operation (referred to as "soft" parameters) and those that cannot be changed during operation (referred to as "hard" parameters) must be identified. Soft parameters provide the flexibility to modify and tune certain parameters during system operation, which is often an invaluable feature in prototype systems. Allowing parameters to be changed "on-the-fly," however, adds complexity to the implementation. As such, lists of desired soft and hard parameters should be identified early on in the implementation phase so that the appropriate communications between a user interface and the embedded controller codes can be included. The typical operating modes and features of a controller are summarized in Table 18.2, including references to soft parameters that are useful in supporting each mode.

As indicated, an ANVC controller will typically have two primary operating modes, namely system identification and control. The purpose of system identification is to estimate the transfer functions (or equivalently impulse response functions) from the actuator control signals to sensor responses (both references and residuals). This identification is initially performed with the control filter set to zeros but can subsequently be performed during closed-loop operation, as discussed in Section 18.4 under System Identification. Once an initial plant estimate has been obtained, operational measurements of the system responses are obtained to characterize the "uncontrolled" (i.e., open-loop) response. These

TABLE 18.2 Controller Operating Modes and Features

Operating Mode	Description and Features
Open-loop characterization of system response	Collect measurement of residual sensors during normal "uncontrolled" (i.e., open-loop operation)
Open-loop system identification	Estimate plant models between actuator drive signals and sensors. Soft parameters include probe strength, adaptation coefficients, leakage coefficients, regularization parameters, and actuator channel selection.
Closed-loop operation with filter design based on open-loop response and system identification	Collect measurements of residual sensors during "controlled" (i.e., closed-loop) operation. Soft parameters include adaptation and leakage parameters, control effort and robustness weighting parameters, and regularization parameters.
Concurrent closed-loop system identification and control-filter adaptation	Collect measurements of residual sensors during "controlled" (i.e., closed-loop) operation, while probe signals are injected for the purpose of closed-loop system identification. Soft parameters include those cited above for system identification and closed-loop operation. In addition, software flags should be included to initiate use of updated plant models in the control filter design algorithm.
Save operational "state"	All controller information necessary to restart the controller should be saved. This includes, plant and control filter coefficients, all hard and soft parameters, operating mode, and probe signal path parameters.
Load saved "state" and restart controller	Load in a saved "state" and start operation using saved parameters and filters.

responses are also used to design control filters to support operation in the control mode. Once in the control mode, probe signals can be injected to support closed-loop system identification and to support adaptation of control filter coefficients. Soft parameters can be adjusted to optimize performance. Closed-loop system responses should then be collected and compared with open-loop responses to evaluate system performance versus performance objectives and goals.

The final two rows of Table 18.2 indicate the desire to be able to save the "state" of the controller at any point in time. By state, we mean all controller parameter values and filter coefficients are saved for the purpose of restarting the system at that state in the future. As such, the controller can be started from a closed-loop operating state or from some initial operating state (e.g., open loop). It is convenient for prototype systems to include a graphical user interface (GUI) through which operation mode, soft parameter values, performance assessments, and save/load state functionality can be monitored and controlled. Once an ANVC system has been successfully tested, values of soft parameters and logic for progressing from open-loop to closed-loop operation can be automated within the software, thus resulting in a stand-alone ANVC system.

18.7 EXAMPLES OF ANVC SYSTEMS

In this section we present three prototype active systems that illustrate the application of the principles outlined in the previous sections of this chapter. All of these systems were developed to demonstrate the technology and one is currently operational on a complete class of U.S. Navy ships. All three demonstrate the effectiveness of active noise and vibration control technology when appropriate design procedures are followed. The first example involves the synergistic application of a combination of active and passive noise reduction treatments. The example demonstrates the successful control of a source of noise that would have been difficult using only passive or only active approaches. The second example demonstrates the application of feedback technology to the control of both broadband and narrow-band vibration transmission through a vibration isolation mount. The final example illustrates the use of active noise control technology to generate a zone of silence in a noisy environment.

MIMO Feedforward Active Locomotive Exhaust Noise Control System with Passive Component

Problem Description. When operated at full power diesel-electric locomotives generate significant noise and can have a significant adverse impact on the quality of life near major railroad lines. The sources of noise are shown in Fig. 18.83.[85] As indicated in the figure, the primary sources of diesel-electric locomotive noise are the engine exhaust and the cooling fans, both of which must be reduced before significant noise reductions can be realized. The active noise control system developed for this application focused on just the exhaust noise, fully recognizing that later efforts would have to attack cooling-fan noise if significant overall locomotive noise reductions were to be achieved. Active technology was considered for this application because locomotive exhaust noise is significant at very low frequencies (below 40 Hz), and passive noise control treatments (e.g., dissipative or reactive mufflers) that could achieve the desired noise reduction would simply be too large to fit in the space available.

The basic concept for the system[86,87] is illustrated in Fig. 18.84, where a plan view of the top of the locomotive hood is shown near the exhaust stack. The figure shows a number of loudspeakers, the control actuators, surrounding the exhaust

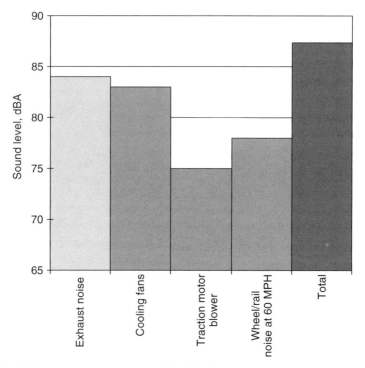

FIGURE 18.83 Noise sources on an SD40-2 diesel electric locomotive measured at 100 ft with the locomotive running at throttle 8 at full load.

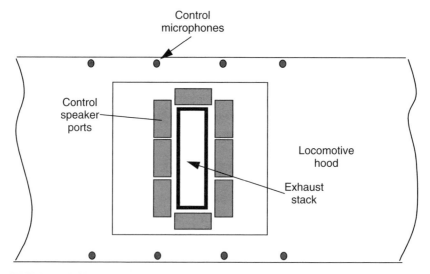

FIGURE 18.84 Plan view of the locomotive hood with the basic system concept.

stack and a number of control microphones located near the edge of the hood that will act as the residual sensors. It was determined based on exhaust noise measurements that at low frequencies where the active system would be designed to operate, the noise was primarily tonal and so a feedforward architecture was selected that used a tachometer on the locomotive diesel engine as the reference signal.

Performance Goals. Based on the source information in Fig. 18.83, it was decided that 10 dBA of overall reduction in exhaust noise would be desirable to have a significant impact on community noise (assuming of course that cooling-fan noise would eventually be similarly reduced). After examining the locomotive exhaust noise spectra we determined that to achieve 10 dBA of overall noise reduction would require that the exhaust noise be controlled out to at least 5 kHz. Since extending the bandwidth of the active system out to so high a frequency and requiring broadband control would place excessive demands on the technology, we decided on a hybrid approach. At low frequency where the noise is primarily tonal we decided to employ an active system. At high frequency where the noise is primarily broadband in character we determined that a passive silencer would be most advantageous. Such an approach is desirable because active technology is well developed for the control of low-frequency tonal noise and at high frequency passive silencers can be effective without having to be large in size. Both technologies were required because we found that

- active control of tones in the exhaust below 250 Hz with no broadband control at the higher frequencies would result in less than 1 dBA of noise reduction and
- no control of tones in the 0–250-Hz band would limit the maximum reduction of exhaust noise to ∼5 dBA.

We ultimately decided that 10 dBA of exhaust noise reduction could be achieved with

- an active system providing 10 dB reduction of exhaust tones below 250 Hz along with
- a passive silencer providing 5 dB of broadband noise reduction from 250 to 500 Hz and 15 dB reduction from 500 to 5500 Hz.

Figure 18.85 shows the silencer designed to provide the above noise reduction performance. It is a very compact design with the center body and side chambers designed to provide the necessary insertion loss while maintaining the back pressure low enough to meet locomotive diesel engine specifications. The silencer was designed to allow the speaker enclosures to surround it, with the whole assembly fitting within a protective enclosure in the engine compartment. While the passive silencer was a critical component in the entire system design,

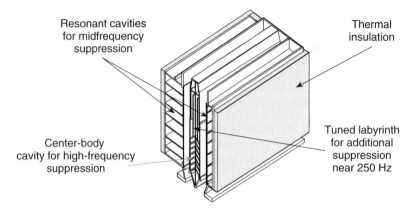

FIGURE 18.85 Compact passive silencer design.

we focus here only on the active system. Details on the passive silencer can be found in references 86 and 87.

Number and Location of Actuators. We began the design process by determining the number and location of control actuators (loudspeakers). Figure 18.86 shows a typical arrangement of actuators that we examined in simulations in which the exhaust stack and control actuators were treated as point sources. In the simulations we formed a matrix of transfer functions relating the sound pressure at 90 locations in the horizontal plane 30 m from the exhaust stack to the control speaker volume velocities at 32 speaker locations around the exhaust stack, as illustrated in Fig. 18.86. Figure 18.87 shows the ratio of the significant singular values to the largest singular value as a function of frequency for that transfer function matrix. In principle the number of significant singular values tells us the minimum number of actuators needed to provide significant control of the source. We found (not surprisingly), while carrying out calculations of this type, that the minimum number of singular values (control sources) was achieved when the control sources were placed as close as possible to the exhaust stack. Consequently, we carried out a preliminary design of the control speaker enclosures to determine a realistic minimum spacing between the exhaust stack and the enclosure outlets. The calculations in Fig. 18.87 were carried out for that minimum spacing. The figure shows that at 250 Hz eight singular values (including the largest) lie within 20 dB of the maximum. Since we are looking for only 10 dB reduction in noise, the use of eight actuators arranged as shown in Fig. 18.85 seemed to be a conservative choice. The performance predicted by this arrangement is shown in Fig. 18.88. While the noncausal noise reduction predicted by this calculation is much larger than we would expect to achieve, it was comforting to see that eight control actuators seemed to be more than adequate.

Number and Location of Control Sensors. The evaluations in the previous section placed the control sensors in the far field. In reality the microphones

EXAMPLES OF ANVC SYSTEMS   **833**

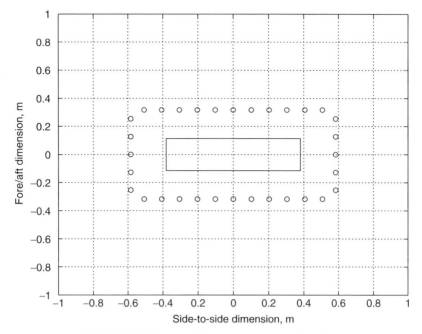

FIGURE 18.86   Candidate control source locations.

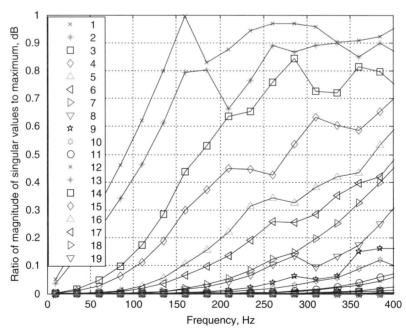

FIGURE 18.87   Ratio of each singular value to the largest singular value as a function of frequency.

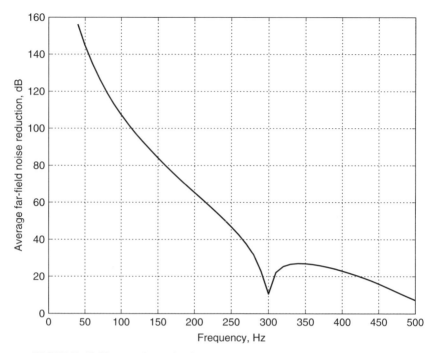

FIGURE 18.88 Predicted far-field performance of eight control actuators.

will have to be placed somewhere on the locomotive hood. Consequently, we developed an additional set of transfer function matrices relating the sound pressure at candidate control microphone locations on the locomotive hood to control speaker volume velocity at the previous 32 locations around the exhaust stack. We then determined the volume velocity required to minimize the pressure at the control microphones and then used those values with the previous transfer function matrix to predict the reduction in far-field pressure. Figure 18.89 shows the results of that calculation for a number of different control sensor locations. The solid curve in the figure is for the final control sensor placement, a line of four sensors along the two edges of the locomotive hood, similar to that shown in Fig. 18.84. A number of calculations of this type were used to help in the selection of control sensor locations. Other issues that came into play in the selection included the degree of dominance of exhaust noise over other sources at the candidate sensor locations, interference with other locomotive components, heat, and routing of cabling.

Control Actuator Design. The simulations for control sensor placement can provide, as part of the calculation, the estimated volume velocities required from the control speakers, provided realistic values of the uncontrolled sound pressure at the control sensors are used in the calculations. Measurements of the sound pressure at various locations on the hood of a test locomotive were acquired to

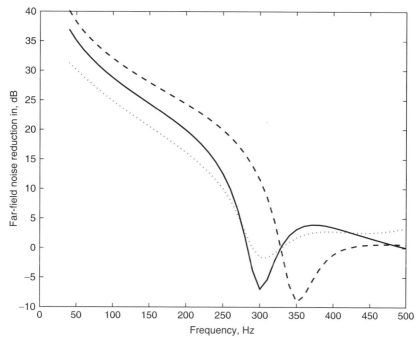

FIGURE 18.89 Predicted far-field noise reduction of eight control actuators with various configurations of eight control sensors.

provide those data. Simulations were then performed to determine the optimum control speaker volume velocities. That information was used in the selection of the control speakers and the design of the speaker enclosures. The enclosure design is shown in Fig. 18.90. Two different enclosure geometries were required to allow the necessary number of enclosures to fit in the space available around the exhaust duct. Each enclosure contains two 12-in.-diameter high-fidelity speakers and is designed to provide bandpass frequency response, enhancing the volume velocity in the 40–250-Hz frequency range. The enclosure–speaker system was designed using a commercially available computer program.

Figure 18.91 shows the arrangement of the control speaker enclosures around the exhaust stack. By careful design we were able to fit 10 enclosures in the space available. However, we retained only eight independent channels to drive the speakers. Because the simulations indicated that very high volume velocity would be required of the speakers on the centerline of the locomotive, we placed two speaker enclosures on each side of the centerline, as indicated in Fig. 18.91, rather than a single one on the centerline. We then drove each of the two pairs of enclosures with a single controller output channel.

Control Architecture. As indicated above, we decided early in the design process to use a feedforward control architecture since the technology is well

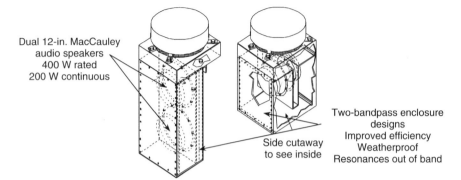

FIGURE 18.90 Control speaker enclosure design.

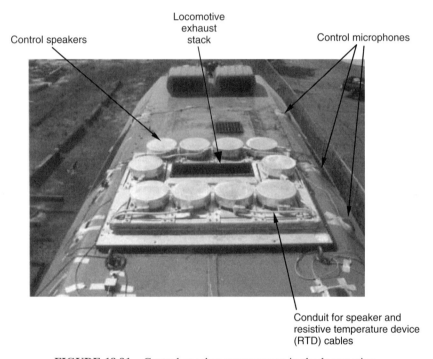

FIGURE 18.91 Control speaker arrangement in the locomotive.

developed for tonal control problems. A simplified control block diagram is shown in Fig. 18.92. The figure shows the typical adaptive MIMO LMS filtered-x architecture with eight control filters and an 8×8 plant transfer function matrix. A more detailed block diagram is shown in Fig. 18.93, where probe injection is shown for plant identification along with the additional LMS blocks for estimating the plant while minimizing the noise in the estimate.

EXAMPLES OF ANVC SYSTEMS   837

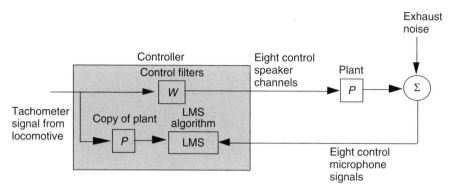

FIGURE 18.92 Simplified basic block diagram of the control architecture.

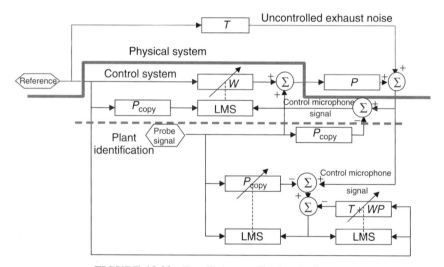

FIGURE 18.93 Detailed controller block diagram.

Hardware Selection. We decided early in the design process to use 12-bit successive-approximation A/D converters, since the lower discretization noise associated with 16-bit converters was not needed in this application. A commercially available input–output (I/O) board from Loughborough Sound Images was found that provided 16 input channels (A/D converters) and 8 output channels (sample-and-hold D/A converters) and in addition provided third-order low-pass Butterworth filters for anti-aliasing and reconstruction. We decided to use the on-board filters because providing a separate set of filters would have been too costly. Since these filters roll off very slowly with increasing frequency, we needed to use a sampling frequency much higher than would be needed to achieve the desired control bandwidth. Consequently, we sampled at 2000 Hz and set the cutoff frequency of the filters to 720 Hz; however, in the control computations we downsampled the digital signals by a factor of 4. This gave us an effective sampling rate of 500 Hz and a control bandwidth on the order of 250 Hz.

Estimates of computational load and memory requirements for the application are shown in Table 18.3. These estimates were based on up to 200 taps for each of the plant filters and 180 taps for each control filter. The table is divided into two functions: in-line control and system identification. In-line control is the function associated with implementing the control algorithm and system identification is the measurement of the plant transfer function, which must be updated periodically to maintain good control performance. It was decided to use two Texas Instruments TMS320C44 DSP chips available on a commercially available carrier board from Loughborough Sound Images. The board provided more than sufficient memory for the two operations. Each DSP chip is clocked at 60 MHz and is capable of up to 30 million floating-point operations per second (Mflops). It was decided to separate the two functions mentioned above with the primary DSP providing all of the control processing and the secondary DSP carrying out all of the system identifications tasks.

System Performance. The control system described above was implemented on an F40PH passenger locomotive operated by Chicago Metra, a commuter rail line in the Chicago metropolitan area. Testing was carried out at the 51st St. Rail Yard in Chicago. The microphone locations for the evaluation are shown in Fig. 18.94. The number of available microphone locations was limited because of the presence of other equipment and structures in the yard that would have interfered with the acoustical evaluation. Uncontrolled measurements were made before installation of the passive silencer. The performance of the system was then measured with the passive silencer in place and with the active system turned on and turned off.

Figure 18.95 shows the reduction of the tonal noise at microphone 5 on the roof of the locomotive due to the use of the active system for the locomotive operating loaded at throttle 4.* The figure shows significant reduction of all of the important tones with some amplification of the low-amplitude tones.

TABLE 18.3 DSP Computation Requirements for the Active System

Operation	MFLOPS	Memory (Kbytes)
Control	22.4	107
System identification	9.6	10
Total	32.0	117.0

*The locomotive diesel engine was loaded by passing the power from the alternator driven by the engine through the locomotive dynamic brake grids (large resistors cooled by the dynamic brake fan). The locomotive can be operated unloaded at idle and can be operated loaded or unloaded in any one of eight throttle setting. Throttle 1 corresponds to the lowest speed and lowest power and throttle 8 is the highest speed at full power.

EXAMPLES OF ANVC SYSTEMS **839**

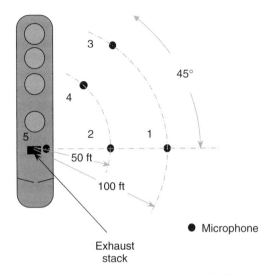

FIGURE 18.94 Microphone locations for the evaluation measurements.

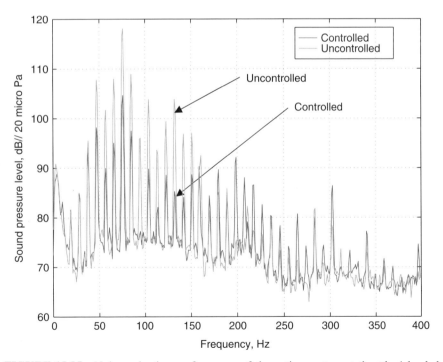

FIGURE 18.95 Noise reduction performance of the active system at throttle 4 loaded as measured at a microphone on the roof of the locomotive.

TABLE 18.4 Estimated Noise Reduction at the Far-Field Microphones

Throttle	Load	mic 5	mic 1	mic 2	mic 3	mic 4
idle	unloaded	5.1	5.2	6.5	6.1	6.4
hi idle	unloaded	8.7	6.5	8.9	5.8	7.4
t4	unloaded	6.9	5.2	4.1	5.6	5.4
t6	unloaded	7.7	6.6	5.6	6.4	6.0
t8	unloaded	5.8	−0.1	3.7	6.2	4.7
t4	loaded	6.4	5.2	4.1	4.7	4.6
t6	loaded	4.3	−1.9	1.2	2.8	1.7
t8	loaded	6.9	2.9	6.5	6.6	6.6

The reduction of the overall sound level below 250 Hz is in excess of 12 dB. Table 18.4 shows the overall A-weighted noise reduction due to the passive silencer and active system operating together. The reductions are somewhat less that the 10-dBA goal but are still significant for most operating conditions and for most microphone locations, showing that the hybrid active–passive system has provided significant broadband global noise reduction.

Active Machinery Isolation

Problem Description. It is difficult to achieve desired vibration isolation at low frequency using passive machinery mounts. Often, to achieve low-frequency isolation, two-stage isolators are employed, which incurs a very large weight penalty for the intermediate mass between the two isolators. In some applications, such as marine vessels or aircraft, the weight of the intermediate mass has an adverse economic and vehicle performance impact.

The forces transmitted into machinery foundations generally contain narrow-band and broadband components. Reduction of the continuous or broadband spectral components as well as the discrete or narrow-band spectral components is often necessary to meet noise or vibration goals. In addition, the tonal frequencies of the narrow-band excitation can change quite rapidly with time due to changes in operating speed. Further, the plant transfer function (i.e., the transfer function between the drive signal to the actuator and the sensor response) is expected to vary with time.

Finally, the machinery foundations of interest are often complex, large, distributed mechanical structures that are lightly damped and have a large number of resonant modes in the frequency range of interest. Thus, typically, plant transfer functions are of high order with high-Q response components. Moreover, the order (e.g., complexity) of the plant increases as the bandwidth of the controller is increased.

Based on the above observations, the control problem is defined as follows:

- Provide both narrow-band and broadband reduction of forces transmitted into the machinery foundation structures.

- Adapt rapidly to variations in the tonal frequencies of narrow-band excitations. Adapt at a relatively slower rate to variations in the plant.
- Minimize controller bandwidth to minimize complexity of the plant transfer function and consequently the controller.
- Provide the desired performance while avoiding out-of-band vibration amplification.

The final item is based on our experience with applications which require increased isolation performance within a certain frequency range but which will not tolerate significant degradation in isolation system performance outside the regulation bandwidth of the active system. Acceptable levels of out-of-band enhancement (i.e., noise amplification) are typically 2–3 dB.

In the following section, we describe an active isolation system designed to meet these requirements. We consider a feedback algorithm, because in general there is not a suitable reference sensor available to achieve broadband control using feedforward control. Although this discussion pertains to SISO control, the algorithms and control architecture are extensible to MIMO control.

Description of Control Strategy.

Real-Time Control Processing The basic real-time controller structure is shown in Fig. 18.96. This is the compensator-regulator architecture that was introduced in Section 18.5, where the plant, designated as $P(\omega)$ in Fig. 18.96, is the transfer function between the output of the controller and the net force transmitted to the foundation.

The controller is implemented as a cascade of two filters because we wish to adapt the low-order regulation filter coefficients quickly to track changes in the center frequency of narrow-band components of the disturbance. At the same time, we choose to adapt the relatively high-order compensation filter at a slower rate to track changes in the plant transfer function.

Adaptation Processing The full functionality of the adaptive controller is shown in Fig. 18.97. The "concurrent adaptation processing" block provides two types of adaptation:

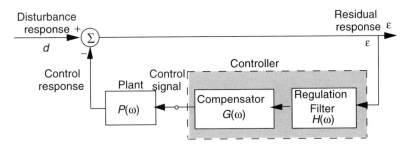

FIGURE 18.96 Feedback compensator regulator.

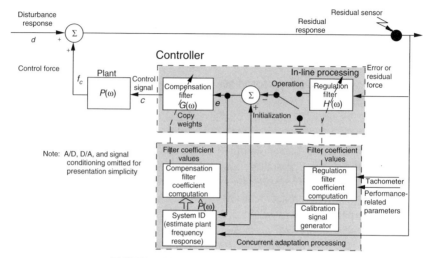

FIGURE 18.97 Controller functionality.

1. Variable narrow-band center frequency adaptation:
   - measurement via a tachometer (or other distinct speed or repetition rate measurement sensor) and a tracking filter of the center frequency of narrow-band response due to periodic excitation and
   - computation and updating of filter coefficients of the digital narrow-band regulation filters based on the currently measured center frequency and rate of change of the center frequency.
2. Variable plant adaptation:
   - system identification in the sense of plant frequency response estimation and
   - computation and updating of filter coefficients of the digital broadband compensation filter.

System identification is performed during closed-loop operation using the procedure outlined in Section 18.4. The major aspects of this procedure include (a) inserting a low-level calibration signal, which is uncorrelated with the external disturbance, into the compensation filter; (b) estimating the cross-spectra of the calibration signal with both the plant input and output; and (c) estimating the plant transfer function as the ratio of these two cross-spectra. Details of the processing of cross-spectra to estimate the plant as well as the design and injection of a covert probe signal are also discussed in Section 18.4 under System Identification.

The probe signal is sufficiently low in level that it does not add appreciably to the residual, and its bandwidth is matched to the full compensation band. Although it is low level, a good-quality estimate of the plant transfer function can be obtained with sufficient averaging time. In the performance plots that follow, estimates of the plant were updated approximately every 2 min.

Since we are concerned with a lightly damped mechanical structure and relatively large compensation bandwidths (e.g., 800 Hz), a FIR filter implementation of the compensation filters would have required a large number of coefficients. To reduce both the off-line and on-line computation load, we opted to use IIR filters designed by a multistep Yule–Walker method.

Hardware Description. Figure 18.98 presents a block diagram of the controller hardware. The algorithmic functionality of the controller is performed by four TMS320C30 DSP chips operating in parallel. The allocation of controller functionality to the individual DSP processors can be summarized as follows:

In-Line Processor. Performs the digital filtering of the regulation and compensation filters for the in-line data path.

Desamp Processor. Responsible for all off-line tasks associated with generating the desampled data buffers presented to the system identification algorithm. Replication filters and reference signal processing are also performed in this processor.

Sys_ID Processor. Performs the system identification and compensation filter weight algorithms. The compensation filter weights are then copied to the in-line processor.

Monitor Processor. Collects data from the other processors and uploads data to the host for monitoring and evaluating system performance.

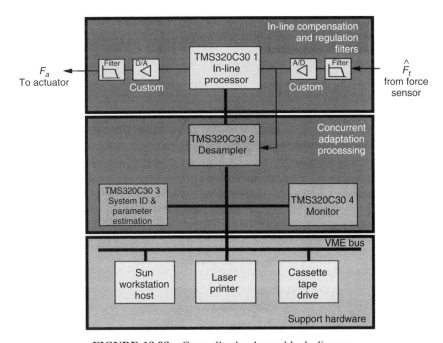

FIGURE 18.98 Controller hardware block diagram.

844 ACTIVE CONTROL OF NOISE AND VIBRATION

The controller implementation included custom-built A/D and D/A converter boards plus a logic controller board. The controller board provided a high-speed two-way interface between the DSP processors and the A/D and D/A boards. Two Sky Challenger boards, each of which contain two TMS320C30 chips per board, communicated with each other and the host computer over the VME bus and provide up to 132 Mflop computation rate. A SPARC workstation was chosen as the host computer.

Performance Example. The controller was connected to a prototype active/passive machinery mount located between one leg of a 1300-lb, 140 brake horsepower (BHP) at 2800 rpm, Detroit Diesel model 4–53 engine and a representative complex foundation structure. A picture of the active/passive mount and a schematic of the system are shown in Fig. 18.99. A detailed discussion of the engine, active/passive mount design, and test rig can be found in reference 88.

The test objectives were to provide at least 15 dB of narrow-band regulation at the fundamental and next four harmonics of the piston-firing frequency. In

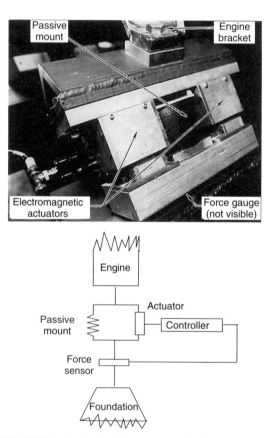

FIGURE 18.99 Active/passive mount and schematic of system.

addition, 10 dB of broadband regulation was desired over the frequency interval from 10 to 80 Hz while maintaining noise amplification below 5 dB outside the regulation bandwidth. To meet these objectives, a compensation bandwidth of 833 Hz and the system sample rate of 10 kHz were chosen. Latency of the digital system (including delay associated with the sample-and-hold on the D/A converter) was 60 μs.

The closed-loop performance of the active isolation system is shown in Fig. 18.100. This plot presents the ratio of the open- to closed-loop residual force transmitted into the foundation structure. As shown in this figure, narrow-band reductions in excess of 15 dB are achieved at the first five piston-firing tonals. In addition, approximately 10 dB of broadband regulation is achieved from about 15 to 80 Hz, which spans most of the frequency interval containing the tonals. Finally, this performance is achieved while limiting noise amplification to approximately 5 dB outside the regulation bandwidth. The very fine structure at higher frequencies of the measured ratio of open- to closed-loop forces, which sometimes goes below −5 dB, is due to harmonic distortion of the actuators (which were being driven very hard), rather than any shortcoming of the control processor.

Active Control of Airborne Noise in a High-Speed Patrol Craft

The Navy's new high-speed patrol crafts are powered by four main propulsion diesel engines that deliver a total of 13,000 shaft horsepower to a set of four propulsion shafts. The operation of these engines produces propeller blade rate tonal noise and broadband cavitation noise that propagate through the hull, producing high noise levels in the aft-crew berthing compartment, which is located

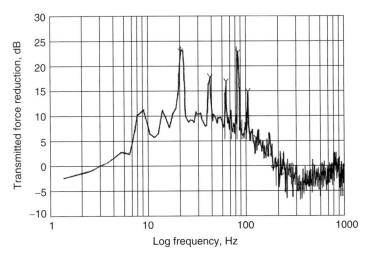

FIGURE 18.100 Reduction of transmitted force into foundation for active/passive mount relative to passive mount.

just forward of the propellers. Although extensive passive noise control treatments have been applied to reduce the noise in this compartment (including floating floors, double bulkheads, and constrained-layer damping), the 63- and 125-Hz octave-band and overall (dBA) airborne noise levels exceeded the Navy's acoustical habitability specifications. Table 18.5 compares measured third-octave-band and overall sound levels in the aft-crew berthing compartment for the first three ships of the class to the noise specification limits.[89] These measurements indicated that reductions in the 63- and 125-Hz octave bands were necessary to reduce overall A-weighted noise to acceptable levels.

To address this problem, several options were explored for reducing the low-frequency noise in the aft-crew compartment. Because passive treatments are least effective in this frequency region, several active noise and vibration control concepts were evaluated. These concepts included methods for both global and local control, using feedforward and feedback control strategies, and with digital and analog control hardware. To assess the benefits of each approach, a set of acoustical trials were conducted initially to measure the operational noise characteristics, path transfer functions, and compartment acoustical characteristics. These data were used to simulate performance of the various control options.

The approach chosen for implementation was local control of the sound field near the head of each bunk using a SISO feedback control strategy. This approach creates an effective "zone of silence" in the immediate vicinity of the occupant's head. This approach is shown schematically in Fig. 18.101. Laboratory simulations using measured data and a mock-up of one rack of bunks suggested that the necessary reductions at the occupant's head location could be achieved using this approach.

The active noise control system for a single bunk is contained within a single prismatic-shaped enclosure that is located in the upper corner of the bunk above and behind the occupant's head. Each unit contains a loudspeaker to create the cancellation noise, a pair of microphones to sense the noise field to be controlled, and a microprocessor to compute the cancellation signal in real time. The control algorithm implemented on the microprocessor was an adaptive feedback algorithm (based on the adaptive Youla transform discussed in Section 18.5) to reduce both narrow-band and broadband noise. The reading light originally placed in this location was incorporated into the active noise control (ANC) enclosure design.

TABLE 18.5 Average Aft Crew Berthing Octave-Band and A-Weighted Noise Levels at Full-Speed[89]

	31.5 Hz	63 Hz	125 Hz	250 Hz	A-Weighted
Noise Specification	105	100	95	90	82
Ship 1	98	113	100	88	87
Ship 2	97	104	94	85	85
Ship 3	95	111	96	86	86

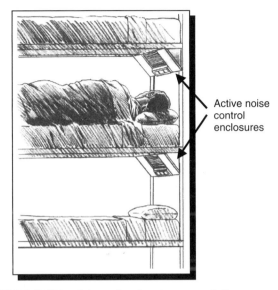

FIGURE 18.101 Schematic of active zone-of-silence approach.

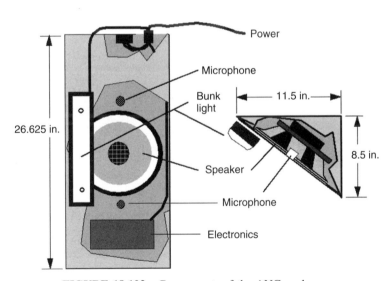

FIGURE 18.102 Components of the ANC enclosure.

The components of the ANC enclosure are illustrated in Fig. 18.102. Pictures of the actual implementation are contained in Figs. 18.103 and 18.104.

The prototype control system described above was evaluated during underway testing. A narrow-band plot of the measured noise reduction at the occupant's head location (not at the control microphones, where greater reductions were achieved) is shown in Fig. 18.105. This plot compares the noise spectra obtained

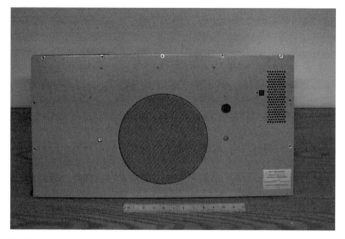

FIGURE 18.103 Faceplate of the ANC enclosure (integrated reading light not shown).

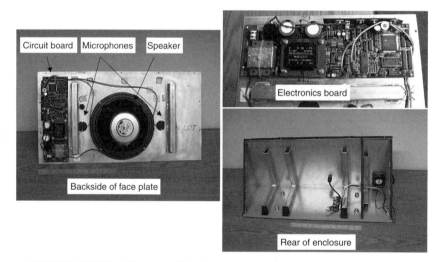

FIGURE 18.104 Pictures of the internal components of the ANC enclosure.

during full-speed operation when the control system was turned *off* to that when the control system was turned *on*. As shown in this plot, reductions at the dominant blade-rate tonals (at approximately 60 Hz) were greater than 15 dB, while 7–10 dB of broadband noise reduction was achieved between 30 and 85 Hz. Third-octave-band performance is summarized in Fig. 18.106. This summary shows that the system reduced the third-octave-band levels to below the limits of the Navy's acoustical habitability specifications, the goal of the program. Based on extensive testing of the prototype system on one patrol craft, the Navy contracted for production units for the entire class of PC1 patrol craft.

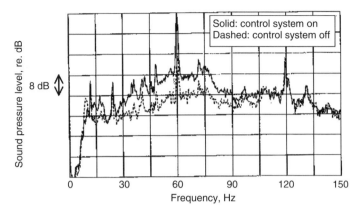

FIGURE 18.105 Narrow-band plot of noise reduction at occupant's head location during full-speed operation.

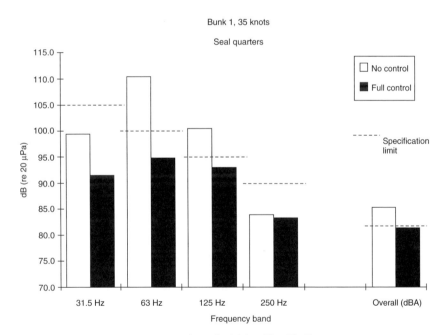

Acoustical data, dB re 20 µPa

Octave band	No control	Control	Attenuation
31.5	99.5	91.5	8
63	110.5	95	15.5
125	100.5	93	7.5
250	84	83.5	0.5
Overall dBA	85.5	81.5	4

FIGURE 18.106 Summary of third-octave-band performance at occupant's head location.

REFERENCES

1. W. B. Conover, "Fighting Noise with Noise," *Noise Control*, **2**, 78–82 (1956).
2. W. B. Conover and R. J. Ringlee, "Recent Contributions to Transformer Audible Noise Control," *Trans. AIEE Pt. III Power Apparatus and Systems*, **74**, 77–90 (1955).
3. K. Kido, "Reduction of Noise by Use of Additional Noise Sources," *Proc. Internoise 75*, 1975, pp. 647–650.
4. G. B. B. Chaplin and R. A. Smith, "The Sound of Silence," *Engineering*, **218**, 672–673 (1978).
5. L. A. Blondel and S. J. Elliott, "Tuned Loudspeaker Enclosures for Active Noise Control," paper presented at Internoise 96, International Congress on Noise Control Engineering, Liverpool, United Kingdom, July 30–August 2, 1996. Proceedings, pp. 1105–1108; Abstract, p. 67.
6. G. J. Stein, "A Driver's Seat with Active Suspension of Electro-pneumatic Type," *ASME J. Vib. Acoust.*, **119**, 230–235 (1997).
7. J. der Hagopian, L. Gaudiller, and B. Maillard, "Hierarchical Control of Hydraulic Active Suspensions of a Fast All-Terrain Military Vehicle," *J. Sound Vib.*, **222**, 723–752 (1999).
8. A. K. Abu-Akeel, "The Electrodynamic Vibration Absorber as a Passive or Active Device," ASME Paper 67-Vibr-18, *ASME J. Eng. Ind.*, **89**, 741–753 (1967).
9. S. Ikai; K. Ohsawa, K. Nagaya, and H. Kashimoto, "Electromagnetic Actuator and Stacked Piezoelectric Sensors for Controlling Vibrations of a Motor on a Flexible Structure," *J. Sound Vib.*, **231**, 393–409 (2000).
10. S. W. Sirlin and R. A. Laskin, "Sizing of Active Piezoelectric Struts for Vibration Suppression on a Space-Based Interferometer," paper presented at the First Joint U.S./Japan Conference on Adaptive Structures, Maui, HI, November 13–15, 1990. Proceedings, pp. 47–63.
11. P. K. C. Wang, "Feedback Control of Vibrations in a Micromachined Cantilever Beam with Electrostatic Actuators," *J. Sound Vib.*, **213**, 537–550 (1998).
12. D. L. Hall and A. B. Flatau, "Broadband Performance of a Magnetostrictive Shaker," in C. J. Radcliffe, K.-W. Wang, H. S. Tzou, and E. W. Hendricks (Eds.), *Active Control of Noise and Vibration*. Papers presented at the ASME Winter Annual Meeting, Anaheim, CA, November 8–13, 1992.*ASME Book DSC-Vol. 38*, pp. 95–104, American Society of Mechanical Engineers, New York.
13. B-T. Wang, R. A. Burdisso, and C. R. Fuller, "Optimal Placement of Piezoelectric Actuators for Active Structural Acoustic Control," *J. Intelligent Material Syst. Struct.*, **5**, 67–77 (1994).
14. G. E. Warnaka, "Active Attenuation of Noise—The State-of-the-Art," *Noise Control Eng.*, May/June 1982, pp. 100–110.
15. D. Guicking, *Active Noise and Vibration Control Reference Bibliography*, 3rd ed., Dritte Physicalisches Institut, University of Göttingen, Göttingerm 1988 and 1991 supplement.
16. S. J. Elliott and P. A. Nelson, "Active Noise Control," *IEEE Signal Process. Mag.*, October 1993, pp. 12–35.
17. C. Fuller, "Active Control of Sound and Vibration," tutorial lecture presented at the 120th Acoustical Society of America Meeting, San Diego, November 26, 1990.

18. H. Coanda, "Procédé de Protection Contre les Bruits," French Patent No. 722274, filed October 21, 1930, patented December 29, 1931, published March 15, 1932.
19. P. Lueg, "Process of Silencing Sound Oscillations," U. S. Patent 2,043,416, June 19, 1936.
20. M. A. Swinbanks, "The Active Control of Sound Propagation in Long Ducts," *J. Sound Vib.*, **27**(3), 411–436 (1973).
21. M. J. M. Jessel, "Sur les absorbeur actif," in *Proceedings of the sixth International Congress on Acoustics*, Paper F 5–6 82, Tokyo, 1968.
22. M. J. M. Jessel and G. Magiante, "Active Sound Absorbers in an Air Duct," *J. Sound Vib.*, **23**, 383–390 (1972).
23. H. G. Leventhall, "Developments in Active Attenuators," in *Proceedings of the 76 Noise Control Conference*, Warsaw, Poland, 1976, pp. 175–180.
24. G. Canevet, "Active Sound Absorption in an Air Conditioning Duct," *J. Sound Vib.*, **58**, 333–345 (1978).
25. L. J. Eriksson, M. C. Allie, and R. A. Greiner, "The Selection and Application of IIR Adaptive Filter for Use in Active Sound Attenuation," *IEEE Trans. Acoust. Speech Signal Process.*, **ASSP-35**, 433–437 (1987).
26. L. J. Eriksson and M. C. Allie, "A Digital Sound Control System for Use in Turbulent Flows," Paper presented at NOISE-CON 87, National Conference on Noise Control Engineering, State College, PA, June 8–10, 1987. Proceedings, pp. 365–370.
27. L. J. Eriksson and M. C. Allie, "A Practical System for Active Attenuation in Ducts," *Sound Vib.*, **22**(2), 30–34 (February 1988).
28. L. J. Eriksson, "The Continuing Evolution of Active Noise Control with Special Emphasis on Ductborne Noise," in C. A. Rogers and C. R. Fuller (Eds.), *Proceedings of the First Conference on Recent Advances in Active Control of Sound and Vibration*, Blacksburg, VA, April 15–17, 1991, pp. 237–245.
29. H. F. Olsen and E. G. May, "Electronic Sound Absorber," *J. Acoust. Soc. Am.*, **25**, 1130–1136 (1953).
30. D. C. Perry, S, J, Elliott, I. M. Stothers, and S. J. Oxley, "Adaptive Noise Cancellation for Road Vehicles," in *Proceedings of the Institution of Mechanical Engineers Conference on Automotive Electronics*, 1989, pp. 150–163.
31. S. Hasegawa, T. Tabata, A. Kinsohita, and H. Hyodo, "The Development of an Active Noise Control System for Automobiles," SAE Technical Paper Series, Paper No. 922087, Society of Automotive Engineers, Warrendale, PA. 1992.
32. R. J. Bernhard, "Active Control of Road Noise Inside Automobiles," in *Proceedings of ACTIVE 95, The 1995 International Symposium on Active Control of Sound and Vibration*, Newport Beach, CA, July 6–8, 1995, pp. 21–32
33. S. J. Elliott, P. A. Nelson, T. J. Sutton, A. M. McDonald, D. C. Quinn, I. M. Stothers, and I. Moore, "The Active Control of Low Frequency Engine and Road Noise Inside Automotive Interiors," Paper presented at ASME Winter Annual Meeting, Session NCA-8, Dallas, TX, Nov. 25–30, 1990. Proceedings, pp. 125–129.
34. W. Dehandschutter, R. Van Cauter, and P. Sas, "Active Structural Acoustic Control of Structure Borne Road Noise: Theory, Simulations, and Experiments," in S. D. Sommerfeldt and H. Hamada (Eds.), *Proceedings of ACTIVE 95, The 1995 International Symposium on Active Control of Sound and Vibration*, Newport Beach, CA, July 6–8, 1995, pp. 735–746.

35. S. J. Elliott, P. A. Nelson, and I. M. Stothers, "In Flight Experiments on the Active Control of Propeller Induced Cabin Noise," *J. Sound Vib.*, **140**, 219–238 (1990).
36. C. F. Ross and M. R. J. Purver, "Active Cabin Noise Control," in S. J. Elliott and G. Horváth (Eds.), *Proceedings of ACTIVE 97, The 1997 International Symposium on Active Control of Sound and Vibration*, Budapest, Hungary, August 21–23, 1997, pp. 39–46.
37. G. P. Mathur, B. N. Tran, and M. A. Simpson, "Active Structural Acoustic Control of Aircraft Cabin Noise Using Optimized Actuator Arrays—Laboratory Tests," AIAA Paper 95-082, presented at the First Joint CEAS/AIAA Aeroacoustics Conference (AIAA 16th Aeroacoustics Conference), München, DE, June 12–15, 1995.
38. M. A. Simpson, T. M. Luong, M. A. Swinbanks, M. A. Russell, and H. G. Leventhall, "Full Scale Demonstration Tests of Cabin Noise Reduction Using Active Noise Control," Paper presented at Internoise 89, International Congress on Noise Control Engineering, Newport Beach, CA, December 4–6, 1989. Proceedings, pp. 459–462.
39. P. D. Wheeler, "The Role of Noise Cancellation Techniques in Aircrew Voice Communications Systems," in *Proceedings of the Royal Aeronautical Society Symposium on Helmets and Helmet Mounted Devices*, 1987.
40. I. Veit, "Noise Gard—An Active Noise Compensation System for Headphones and Headsets," NATO Research Study Group (RSG) 11, Panel 3: Proceedings of the Workshop on Active Cancellation of Sound and Vibration, 22nd Meeting, Vicksburg, MS, October 31, 1990, and 23rd Meeting, Bremen, DE, June 5, 1991, pp. 164–167.
41. B. Rafaely and M. Jones, "Combined Feedback-Feedforward Active Noise-Reducing Headset—The Effect of the Acoustics on Broadband Performance," *J. Acous. Soc. Am.*, **112**, 981–989 (2002).
42. K. D. LePage, P. J. Remington, and A. R. D. Curtis, "Reference Sensor Selection for Automotive Active Noise Control Applications," Paper presented at 130th Meeting of the Acoustical Society of America, Adam's Mark Hotel, St. Louis, MO, November 27–December 1, 1995.
43. P. Remington, D. Sutliff, and S. Sommerfeldt, "Active Control of Low Speed Fan Tonal Noise Using Actuators Mounted in Stator Vanes Part 1: Control System Design and Implementation," Paper No. AIAA-2003-3191, presented at the AIAA/CEAS Aeroacoustics Conference, Hilton Head, SC, 2003.
44. D. Sutliff, P. Remington, and B. Walker, "Active Control of Low Speed Fan Tonal Noise Using Actuators Mounted in Stator Vanes Part 3: Results," Paper No. AIAA-2003-3193, presented at the AIAA/CEAS Aeroacoustics Conference, Hilton Head, SC, 2003.
45. A. R. D. Curtis, "Active Control of Fan Noise by Vane Actuators," BBN Report 1193, BBN Technologies, Cambridge, MA, 1998.
46. P. Joseph, P. A. Nelson, and M. J. Fisher, "Active Control of Turbofan Radiation Using an In-Duct Error Sensor Array," in S. J. Elliott and G. Horvath (Eds.), *Proceedings of ACTIVE 97, the 1997 International Symposium on Active Control of Sound and Vibration*, Active 97, Budapest, Hungary, August 21–23, 1997, pp. 273–286.
47. R. H. Thomas, R. A., Burdisso, C. R. Fuller, and W. F. O'Brien, "Active Control of Fan Noise from a Turbofan Engine," *AIAA J.*, **32**(1), 23–30 (1994).
48. R. A. Burdisso, R. H. Thomas, C. R. Fuller, and W. F. O'Brien, "Active Control of Radiated Inlet Noise from Turbofan Engines," in *Proceedings of the 2nd Conference*

on Recent Advances in Active Control of Sound and Vibration, Blacksburg, VA, April 28–30, 1993, pp. 848–860.
49. R. G. Gibson, J. P. Smith, R. A. Burdisso, and C. R. Fuller, "Active Reduction of Jet Engine Exhaust Noise," Paper presented at Internoise 95, International Congress on Noise Control Engineering, Newport Beach, CA, July 10–12, 1995. Proceedings, pp. 518–520.
50. J. P. Smith, R. A. Burdisso, and C. R. Fuller, "Experiments on the Active Control of Inlet Noise from a Turbofan Jet Engine Using Multiple Circumferential Control Arrays," AIAA Paper 96–1792, presented at the 2nd AIAA/CEAS Aeroacoustics Conference, State College, PA, May 6–8, 1996.
51. H. F. Olsen, "Electronic Control of Noise, Vibration and Reverberation," *J. Acoust. Soc. Am.*, **28**, 966–972 (1956).
52. E. F. Berkman, "An Example of a Fully Adaptive Integrated Narrowband/Broadband SISO Feedback Controller for Active Vibration Isolation of Complex Structures," Paper presented at the ASME Annual Meeting, November 1992.
53. J. Scheuren, "Principles, Implementation and Application of Active Vibration Isolation," Paper C3, presented at the International Workshop on Active Control of Noise and Vibration in Industrial Applications, CETIM, Senlis, France, April 9–12, 1996.
54. P. Gardonio and S. J. Elliott, "Active Control of Structural Vibration Transmission between Two Plates Connected by a Set of Active Mounts," Proceedings of ACTIVE 99, Paper presented at the 1999 International Symposium on Active Control of Sound and Vibration, Ft. Lauderdale, FL, December 2–4, 1999, pp. 118–128
55. A. H. von Flotow, "An Expository Overview of Active Control of Machinery Mounts," Paper presented at the 27th IEEE Conference on Decision and Control (CDC), Austin, TX, December 7–9, 1988. Proceedings, Vol. 3, pp. 2029–2032.
56. P. A. Nelson and S. J. Elliott, *Active Control of Sound*, Academic, San Diego, CA, 1992.
57. A. V. Oppenheim and R. W. Schafer, *Digital Signal Processing*, Prentice-Hall, Englewood Cliffs, NJ, 1975.
58. S. J. Elliott, *Signal Processing for Active Control*, Academic, San Diego, CA, 2001.
59. B. Widrow and S. D. Stearns, *Adaptive Signal Processing*, Prentice-Hall, Englewood Cliffs, NJ, 1985.
60. G. H. Golob and C. F. Van Loan, *Matrix Computations*, Johns Hopkins University Press, Baltimore, MD, 1983.
61. L. J. Eriksson and M. J. Allie, "Use of Random Noise for On-Line Transducer Modeling in an Adaptive Active Attenuation System," *J. Acoust. Soc. Am.*, **85**, 797–802 (1989).
62. L. J. Eriksson, "Active Sound Attenuation System with On-Line Adaptive Feedback Cancellation," U. S. Patent 4,677,677, June 30, 1987.
63. L. J. Eriksson, "Active Attenuation System with On-Line Modeling of Speaker, Error Path and Feedback Path," U. S. Patent 4,677,676, June 30, 1987.
64. D. R. Morgan, "An Analysis of Multiple Correlation Cancellation Loops with a Filter in the Auxiliary Path," *IEE Trans. Acoust. Speech Signal Process.*, **ASSP-28**, 454–467 (1980).

65. B. Widrow, D. Shur, and S. Shaffer, "On Adaptive Inverse Control," in *Proceeding of the 15th ASILOMAR Conference on Circuits, Systems and Computers*, Pacific Grove, CA, November 9–11, 1981, pp. 185–195.
66. J. C. Burgess, "Active Adaptive Sound Control in a Duct: A Computer Simulation," *J. Acoust. Soc. Am.*, **70**, 715–726 (1981).
67. S. J. Elliott and P. A. Nelson, "Algorithm for Multi-Channel LMS Adaptive Filtering," *Electron. Lett.*, **21**, 978–981 (1985).
68. C. C. Boucher, S. J. Elliott, and P. A. Nelson, "The Effects of Errors in the Plant Model on the Performance of Algorithms for Adaptive Feedforward Control," *Proc. IEE-F*, **138**, 313–319 (1991).
69. K. J. Astrom and B. Wittenmark, *Adaptive Control*, Addison-Wesley, Reading, MA, 1989.
70. R. B. Coleman, E. F. Berkman, and B. G. Watters, "Optimal Probe Signal Generation for On-Line Plant Identification within Filtered-X LMS Controllers," paper presented at the ASME Winter Annual Meeting 1994, Chicago, IL, November 1994.
71. R. B. Coleman and E. F. Berkman, "Probe Shaping for On-line Plant Identification," in *Proceedings of ACTIVE 95, the 1995 International Symposium on Active Control of Sound and Vibration*, Newport Beach, CA, July 6–8, 1995.
72. R. B. Coleman, B. G. Watters, and R. A. Westerberg, "Active Noise and Vibration Control System Accounting for Time Varying Plant, Using Residual Signal to Create Probe Signal," U.S. Patent 5,796,849, August 18, 1998.
73. I. Gustavsson, L. Lyung, and T. Soderstrom, "Survey Paper: Identification of Processes in Closed Loop—Identifiability and Accuracy Aspects," *Automatica*, **13**, 59–75 (1977).
74. E. F. Berkman, R. B. Coleman, B. Watters, R. Preuss, and N. Lapidot, "An Example of a Fully Adaptive Integrated Narrowband/Broadband SISO Feedback Controller for Active Vibration Isolation of Complex Structures," paper presented at the ASME Annual Meeting, New York November 1992.
75. A. R. D. Curtis, E. F. Berkman, R. B. Coleman, and R. Preuss, "Controller Strategies for Fully Adaptive Integrated Narrowband and Broadband Feedback Control for Active Vibration Isolation of Complex Structures," in *Active Control of Noise and Vibrations in Industrial Applications*, CETIM, Senlis, France, April 1996.
76. E. F. Berkman and E. K. Bender, "Perspectives on Active Noise and Vibration Control," *Sound Vib.*, 30th anniversary issue, January 1997.
77. R. D. Preuss, "Methods for Apparatus Designing a System Using the Tensor Convolution Block Toeplitz Preconditioned Conjugate Gradient Method," U. S. Patent 6,487,524 B1, November 26, 2002.
78. M. Athans and P. L. Falb, *Optimal Control: An Introduction to the Theory and Its Application*, McGraw-Hill, New York, 1966.
79. H. Kwakernaak and R. Sivan, *Linear Optimal Control Systems*, Wiley-Interscience, New York, 1972.
80. M. A. Dahleh and I. J. Diaz-Bobillo, *Control of Uncertain Systems*, Prentice-Hall, Englewood Cliffs, NJ, 1995.
81. K. Zhou and J. C. Doyle, *Essentials of Robust Control*, Prentice-Hall, Upper Saddle River, NJ, 1998.
82. J. E. Gibson, *Nonlinear Automatic Control*, McGraw-Hill, New York, 1963.

83. L. P. Heck, J. A. Olkin, and H. Naghshineh, "Transducer Placement for Broadband Active Vibration Control Using a Novel Multidimensional QR Factorization," *ASME J. Vib. Acoust.*, **120**, 663–670 (1997).
84. R. D. Preuss and B. Musicus, "Digital Filter and Control System Employing Same," U.S. Patent 5,983,255, awarded November 9, 1999.
85. P. J. Remington and M. J. Rudd. "An Assessment of Railroad Locomotive Noise," Report No. DOT-TSC-OST-76-4/FRA-OR&D-76-142, U.S. Department of Transportation, Washington, DC, August 1976.
86. P. J. Remington, S. Knight, D. Hanna, and C. Rowley, "A Hybrid Active-Passive System for Controlling Locomotive Exhaust Noise," Paper No. IN2000/0035, presented at Internoise 2000, Nice, France, 2000.
87. P. J. Remington, S. Knight, D. Hanna, and C. Rowley, "A Hybrid Active/Passive Exhaust Noise Control System (APECS) for Locomotives," BBN Report No. 8302, BBN Technologies, Cambridge, MA, March 29, 2001.
88. B. G. Watters et al., "A Perspective on Active Machinery Isolation," in *Proceedings of 27th Conference on Decision and Control*, Austin, TX, December 1988.
89. M. Dignan et al., "The Active Control of Airborne Noise in a High Speed Patrol Craft," paper presented at Noise-Con 94, Ft. Lauderdale, FL, May 1994.

CHAPTER 19

Damage Risk Criteria for Hearing and Human Body Vibration

SUZANNE D. SMITH
Air Force Research Laboratory
Wright-Patterson Air Force Base, Ohio

CHARLES W. NIXON
Consultant
Kettering, Ohio

HENNING E. VON GIERKE
Consultant
Yellow Springs, Ohio

19.1 INTRODUCTION

Noise and vibration are closely related biodynamic environments in terms of their origins, manifestations, and effects on people. Those effects on people that are undesirable are often threatening, induce fatigue, compromise working performance, modify physiological responses, and harm human systems. At the present time, avoidance of excessive noise and vibration exposure is the only assured way to prevent these major hazardous effects.

The practical alternative to complete avoidance is to limit exposures in these environments to those defined as acceptable by appropriate standards, guidelines, and damage risk criteria. Scientific exposure guidelines and criteria established to curtail these effects are vital parts of comprehensive protection programs of concern to governments, industry, and affected personnel. Most adverse effects of these commonly encountered mechanical forces on human systems can be minimized and controlled through engineering and design efforts. Central to these programs and actions are guidelines and criteria that describe potential damage risk and/or establish acceptable exposure limits.

Criteria and limits define conditions above which the risk of damage due to an exposure is considered substantial or unacceptable. Noise and vibration criteria describe exposure characteristics and corresponding undesirable effects such as noise-induced hearing loss, vibration-induced hand–arm vibration syndrome

(HAVS), and vibration-related spinal injury. Damage risk criteria have been developed and implemented worldwide. The basic exposure–effects relationships that underlie these criteria are reasonably well understood and are derived from observations and experience as well as good laboratory and field studies. The various criteria contain different limiting values because of variations in interpretations of the basic data and in the rationale used to establish them. The rationale may include practical, legal, and economic considerations as well as humanitarian concerns. It is very important that the rationale underlying damage risk criteria is fully understood by the user to ensure that the application is justified and accurate.

Estimates of noise and vibration exposures and of their probable effects on populations are usually expressed in terms of population distribution statistics. These group population effects are not precise descriptors, and they are not appropriate for evaluating an individual. Nevertheless, they are adopted by nations and incorporated into national regulations and laws. Many become mandatory requirements included in governmental and industrial activities involving exposure of people to noise and vibration environments.

This chapter presents contemporary regulatory and voluntary noise and vibration exposure standards and criteria along with background information that will facilitate their understanding and application in engineering control and design.

19.2 DAMAGE RISK CRITERIA FOR THE AUDITORY RANGE

Noise Factor

Permanent hearing loss and its associated problems are clearly the most critical and widespread of the various consequences of excessive noise exposure. The extent of damage to the hearing mechanism caused by noise is related to the amount of acoustical energy reaching the hearing mechanism. Such damage cannot be estimated accurately for an individual because of the variability of the noise and the susceptibility of the exposed ears. The primary factors in noise-induced hearing loss are the level of the noise, the frequency content or spectrum of the noise, the duration or time course of the noise exposure, and the susceptibility of the ear.

Exposure limits are defined in terms of level, spectrum, and duration of the noise. A-weighted sound energy of an exposure is directly related to noise-induced hearing loss. No other measure of noise exposure provides a better *cause–effect* relationship with hearing loss.[1] Impulse noise is also included in this measure for many criteria, as was concluded by a special workshop on impulse noise.[2] It was agreed that there is no convincing evidence against acceptance of A-weighting measurement of all noises from 20 to 20,000 Hz in determining their permanent threshold shift (noise-induced hearing loss that does not recover to preexposure levels after cessation of exposures) hazard except when their unweighted, instantaneous, peak-sound-pressure levels exceed approximately 145 dB. Consequently, exposures are typically described in terms of the average A-weighted levels, or the equivalent continuous A-weighted sound pressure level (L_{eq}), over an average workday.

Practical measures for the prevention of noise-induced hearing loss are centered in hearing conservation programs. These programs involve definitions of acceptable noise exposure, personal hearing protection, monitoring the hearing of the affected personnel, and appropriate administrative actions to minimize and eliminate identified temporary hearing problems before they become permanent. The basis of a hearing conservation program is the definition of acceptable noise exposure or exposure criteria that specify the acceptable exposure limits and the proportion of the population to be protected. The criteria of various hearing conservation programs and applications differ in their limiting values.

The parameters of particular exposure criteria are selected to satisfy the needs of the user. Factors that may influence these selections are various interpretations of available data, policies or requirements of organizations, and the remaining uncertainty in the noise exposure–hearing loss databases. Consequently, criteria may differ in such features as estimates of beginning hearing loss, corrections for nonnoise effects such as aging, percentage of the population to be protected, and the extent of protection to be provided.

The most obvious differences among noise exposure criteria are the sound level at which the implementation occurs and how the duration of the exposure and the sound level of the noise are combined.[3] The duration–sound level relationships are referred to as time–intensity trading rules, which assume that damage to hearing is related to total A-weighted sound level and the duration of exposure time. The equal energy relationship between these parameters results in the 3-dB rule. This and other relationships are shown in Table 19.1 which displays permissible noise exposures in A-weighted sound pressure levels for the cited criteria. The permissible A-weighted level for an 8-h exposure ranges from 75 dBA for the Environmental Protection Agency (EPA) to 90 dBA for the Occupational Safety and Health Administration (OSHA).

Most criteria utilize the 3-, 4-, or 5-dB rule. The 3-dB rule is based on the equal-energy concept and is the most conservative or protective of the three rules. The 4- and 5-dB rules assume that intermittency and interruptions of exposures reduce the risk to less than that expected from the total energy. Consequently, a 50% increase in exposure duration corresponds to sound-level decreases of 3 and 5 dB for the respective 3- and 5-dB rules (Table 19.1). Intermittency of exposure is discussed later in the chapter.

Hearing Sensitivity

The human ear is sensitive to a much wider range of sounds than the generally cited 20 Hz–20 kHz audio frequency range. A compilation of independent measurement studies by several investigators using a wide variety of instrumentation and methodologies shows very good agreement and provides confidence in the data summarized in Fig. 19.1 Infrasound (<20 Hz) and ultrasound (traditionally >20,000 Hz but practically above about 12,000 Hz) are normally detected by the ear only at very high sound pressure levels. The traditional audio frequency

TABLE 19.1 Equal Energy and Other Trading Rules Used to Define Permissible Noise Exposures in A-Weighted Sound Pressure Level (dBA) and Exposure Time (h)

Duration of Exposure (h)	Equal Energy[a]	OSHA	EPA	NIOSH[b]	Army	Navy	Air Force	Music
24								
16								
8	90	90	75[c]	85	85	84	85	
4	93	95		88	88	88	88	
2	96	100		91	91	92	91	94[d]
1	99	105		94	94	96	94	
0.5	102	110		97	97	100	97	
0.25	105	115[e]		100	100	104	100	

[a] Equal energy rule of 3-dB decrease for doubling of exposure applied to a basic 8-h criterion of 90 dBA.
[b] National Institute of Occupational Safety and Health.
[c] Threshold for detectable noise-induced permanent threshold shift (NIPTS) at 4000 Hz: exposures exceeding 75 dBA may cause NIPTS exceeding 5 dB in 100% of the population after cumulative noise exposure of 10 years.
[d] Time-averaged A-weighted sound level, in dB, over a 2-h period once a week.
[e] Ceiling on exposure level and duration.

region (20 Hz–20 kHz) is well defined for stimuli including discrete tones, bands of noise, speech materials, loudness, comfort, and acceptability. These databases provide the information necessary to support the development of noise exposure criteria.

The sensitivity of human hearing for high-frequency sounds (3000–4000 Hz and above) gradually decreases with advancing age. This process is called *presbyacusis*. Auditory system components are affected in both the peripheral and the central nervous systems. Although individual patterns of presbyacusis vary widely, normative data describing hearing sensitivity as a function of age have been compiled for various segments of society (some are reported in ref. 5). Loss of sensitivity due to accident, disease, or substances toxic to the auditory system is called *nosoacusis* while that attributed to the noises of everyday living is *sociacusis*. The major high-level noises to which people are exposed are the occupational environments.

Environmental noise occurs over the full spectrum to which the human auditory system is sensitive. Exposure to various segments of this sensory continuum produces differential effects on humans. Limiting levels and durations of acoustical exposure are defined for a number of specific portions of this spectrum, which include infrasound (0.5–20 Hz), audio frequencies (20 to about 12,000 Hz), ultrasound (12,000 to about 40,000 Hz), and impulsive sounds (characterized by rapid onset and durations of less than 1 s) described in terms of peak sound pressure level and duration. Some of these limits are well substantiated by experience and experimental evidence while others remain tentative until more evidence is available.

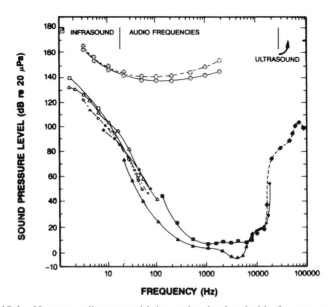

FIGURE 19.1 Human auditory sensitivity and pain thresholds for pure tones, octave bands of noise, and static pressure: (◉) BENOX (1953), pain MAP tones; (◉) pain static pressure; (◉) tickle, pain tones; (□) Bekesy (1960), MAP tones; (▲) ISO R226 (1961), MAF tones; (♦) Corso (1963) bone conduction minus 40-dB tones; (○) Yeowart, Bryan, and Tempest (1960), MAP tones; (×) MAP octave bands of noise; (■) standard reference threshold values (American National Standard on Specifications for Audiometers) (1969), MAP tones; (●) Northern et al. (1972), MAP tones; (△) Whittle, Collins, and Robinson (1972), MAP tones; (★) Yamada et al. (1986), MAF tones. (Data adapted from Ref. 4.) Minimum audible pressure (MAP) indicates that the sound was presented to the ears through earphones and the sound pressure levels were measured in an earphone–microphone coupler that approximated the cavity created by the earphone/pinna. Minimum audible field (MAF) indicates that the sound was presented to the listeners facing a loudspeaker and located in anechoic space. The sound pressure levels for MAF were measured at the location of the center of the head without the listener present. For the same listeners, MAP thresholds are generally several decibels higher than MAF thresholds.

19.3 AUDIO FREQUENCY REGION

CHABA

Noise exposure criteria for the audio frequency region (20–12,000 Hz) were developed by the National Academy of Sciences—National Research Council, Committee on Hearing, Bioacoustics and Biomechanics (CHABA) in 1965.[6] This method described noise exposure in terms of pure tones, third-octave, and òctave bands of noise, and it includes the audio frequencies of 100–7000 Hz. Acceptable exposures to noise can be determined from 11 sets of curves. An environmental noise is considered acceptable if it produces, on average, a NIPTS after 10 years or more of near daily exposure of no more than 10 dB at 1000 Hz and below, 15 dB at 2000 Hz, and no more than 20 dB at 3000 Hz and above. These criteria are based on the assumption that noise exposures producing temporary threshold shifts (TTSs) would eventually produce permanent threshold shifts (PTSs). The possible relationship that TTS is a precursor to PTS is still an open question. The CHABA criteria have been widely used and are an excellent tool. However, they are not simple to use or to relate to current standards, regulations, and guidelines and are less preferred than criteria employing A-weighted sound level.

OSHA

The Occupational Safety and Health Administration has adopted a noise exposure limit of 90 dBA with a 5-dB trading relationship to control excessive noise exposure in industry (Table 19.1). When employees are exposed to noise at different levels during the day, ratios of the actual to the allowed duration for that level are computed and the fractions summed for the day. Total daily exposure calculated from these fractions or ratios must not exceed unity. No other corrections or adjustments are applied to these criteria.

The OSHA noise exposure criteria were verified in 1983 with the publication of "Occupational Noise Exposure; Hearing Conservation Amendment; Final Rule."[7] The basic conditions of the original OSHA noise exposure regulation remain the same with a few exceptions. Continuous A-weighted sound levels are not permitted above 115 dBA regardless of duration. A permissible exposure level (PEL) is defined as that noise dose that would result from a continuous 8-h exposure to a sound level of 90 dBA. The limit of 90 dBA is a dose of 100%, which is the basic criterion level. A *time-weighted average* (TWA) is the sound level that would produce a given noise dose when the employee is exposed to that level continuously over an 8-h workday regardless of the length of the work shift. Workday exposures of 4 h at 90 dB, 8 h at 85 dB, or 12 h at 82 dB all correspond to a TWA of 85 dBA and a noise dose of 50%. The Hearing Conservation Amendment includes computational formulas and tables showing the time–intensity relationships for the 5-dB rule and conversions of dose to time-weighted averages. Guidance is given on calculations of age corrections to audiograms. However, the use of age correction procedures is not required for compliance.

A noise dose of 50% or a TWA of 85 dB is the "action level" at which hearing conservation measures must be implemented. All workers receiving noise doses at or above the action level must be included in a hearing conservation program that requires noise monitoring, audiometric testing, hearing protection, employee training, and record keeping. A baseline audiogram is one taken within six months of the employee's first exposure above the action level, against which subsequent audiograms can be compared. An annual audiogram must be taken for each employee exposed at or above the action level. A *standard threshold shift* (STS) is a change in hearing sensitivity from the baseline audiogram that exceeds an average of 10 dB or more at 2000, 3000, and 4000 Hz in either ear. Appropriate action by the employer must be taken in response to the STS to ensure the continued protection of the hearing of the employee.

Environmental Protection Agency

The EPA published "Information on Levels of Environmental Noise Requisite to Protect Public Health and Welfare with an Adequate Margin of Safety"[1] in 1974 in response to the Noise Control Act of 1972. The objective was to identify levels of environmental noise required to protect the public from adverse health and welfare effects. The levels for noise-induced hearing loss described in this document were based upon reviews and analyses of scientific materials as well as consultations and interpretations of experts. It was concluded that an L_{eq} of 70 dB over a 24-h day (over a 40-year working life) would protect virtually the entire population (96th percentile) for hearing conservation purposes. An $L_{eq(8)}$ limit of 75 dB was considered appropriate protection for the typical 8-h daily work period. This criterion is considered to be very restrictive for most applications, and it has not been incorporated into any DRC for occupational noise exposures.

Air Force

The U.S. Air Force (USAF) hearing conservation criterion is 85 dBA for a maximum allowable 8-h daily exposure. The trading relationship of 3 dB allows such exposures as 16 h at 82 dBA and 4 h at 88 dBA. Higher level continuous exposures for shorter durations, such as 94 dBA for 1 h, are limited to a maximum level of 115 dBA. The ratios of the actual to the allowable daily exposure times are not to exceed unity. Air Force criteria also include limiting conditions for infrasound, ultrasound, and impulse noises.

Army

The U.S. Army (USA) hearing conservation program criterion is 85 dBA as the maximum allowable exposure regardless of duration. Personnel experiencing these exposures must be enrolled in the hearing conservation program. In training and noncombat scenarios single hearing protection will be worn in steady noises of 85–107 dBA and double hearing protection for 108–118 dBA. Protection requirements for unique military noise sources are determined individually. The Army criteria also include limiting exposures for impulsive noises.

Navy

All noise-exposed Navy personnel are required to wear hearing protection when exposed to environmental noise exceeding the criterion of 84 dBA or 140 dB peak, regardless of duration. Personnel are entered into the hearing conservation program based on the 84-dBA damage risk criterion for an 8-h work day, with a 4-dB exchange rate. Double hearing protection is required when sound levels exceed 104 dBA, which administratively assumes 20 dB of single protection from an approved earplug or earmuff.

ISO-1999

The International Organization for Standardization (ISO) standard ISO-1999 (1990), "Acoustics—Determination of Occupational Noise Exposure and Estimation of Noise-Induced Hearing Impairment,"[5] is a landmark document that establishes practical procedures for estimating noise-induced hearing loss in populations. Standard ISO-1999 does not provide a specific formula for assessing risk of hearing handicap, but it specifies uniform methods for the prediction of hearing impairment that can be used for assessment of handicap according to the formula stipulated in a specific nation. The procedures are based on the equal-energy, 3-dB rule (adopted by the ISO and member nations as a conservative criterion) and deal with the measurement and description of noise exposure, the prediction of effects of noise on hearing threshold, and the assessment of risk of noise-induced hearing impairment and handicap. Annexes, which are not part of the standard, provide calculation procedures, examples, tabular data used in the calculations, and a method for relating this information to that of the preceding standard, ISO-1999 (1975). These procedures allow agencies, industries, and governments to select parameters and establish criterion values according to their respective needs. This document will form the basis for legislation in many countries.

The ISO method describes all exposures during an average work day in terms of the A-weighted sound exposure or energy average. The integration period is taken as a working day or a working week. All noises are included in the exposure, ranging from steady state to impulses. Exposures that contain steady tonal noise or impulsive/impact noise are considered about as harmful as the same exposure without these components but about 5 dB higher in level. Exposure can be measured with personal noise dosimeters or an integrating-averaging sound-level meter. Direct and indirect methods for the determination of exposure level are discussed as well as sampling methods.

The only measure of environmental noise needed to calculate hearing impairment or risk of hearing handicap under the following conditions is the energy-averaged daily noise exposure. The maximum instantaneous sound pressure level must be less than 140 dB, the average 8-h daily exposure must not exceed 100 dBA, and the maximum individual daily exposure must not be more than 10 dB above the average of all daily exposures to permit this determination of energy-averaged daily exposure.

Implementation of this standard follows a well-defined series of operations. The first involves determination of the age-related hearing levels of the target population for all test frequencies (e.g., population of 50-year-old males, 90th percentile at 500–6000 Hz). The long-standing difficulties of defining a "normal" population for this purpose were overcome with the utilization of two databases. Database A contains standardized distributions of hearing threshold of an ideal "highly screened" population free from all signs of ear disease, obstructions of wax, and without undue history of noise exposure. Database B can be any carefully collected database covering an *occupationally non-noise-exposed population* considered to be a valid control for the noise-exposed population under consideration. Each user of the standard can select the subpopulation most appropriate for its analysis. As an example, for database B the standard provides the data from the U.S. Public Health Service Surveys reported in 1965.[8]

Next, the predicted NIPTS of the population is calculated for all test frequencies by considering both the number of years of exposure and the average daily noise exposure levels. Data in the standard for calculating NIPTS are valid for frequencies from 500 to 6000 Hz, exposure times of 0–40 years, and average daily noise exposure levels between 75 and 100 dB.

The hearing handicap or risk of hearing handicap may be calculated using the appropriate NIPTS values and a formula selected by the user or a member nation. The document contains nine formulas that are proposed or commonly used among nations for assessing hearing handicap by averaging hearing threshold levels at selected audiometric test frequencies. In the United States, hearing handicap for conversational speech is assessed using the average of the hearing levels at 500, 1000, 2000, and 4000 Hz. Other procedures are available for determining overall percentage of hearing loss for purposes of compensation.

Long-Duration Noise Exposure

Noise exposure durations that exceed 8 h may occur in some work assignments and when substantial nonoccupational noise is added to that received at work. These longer duration exposures are extended by noises from daily living activities, recreation, transportation, proximity to industry, and even other vocational activities. Although the baseline of most criteria is the allowable exposure for an 8-h day, many do extend their time–intensity trading relationship to 16 or 24 h as additional guidance.

Although the 8-h day, 5-day work week is considered standard, numerous variations are employed for many occupations. Some of these are four 10-h days with three days off, three 12-h days with three and four days off, and 12 h on and 12 h off. The noise exposure criteria do not cover these exposures with the same degree of accuracy as with the standard work week. However, it is reasonable to consider the exposure per work week as a basis for calculating noise exposure (i.e., the 3-dB rule).

An important discovery in studies of effects of long-duration exposure to continuous (nonimpulsive) noise of 24 h and longer on human hearing was the phenomenon represented in Fig. 19.2 and called asymptotic threshold shift (ATS).[9] Hearing threshold levels progressively increased with time until the exposure durations reached 8–16 h. Hearing threshold levels reached a plateau or asymptote between 8 and 16 h and did not increase further with continuation of the same exposure levels for 24 and 48 h. Recovery from these asymptotic levels to the preexposure threshold levels was related to exposure time. Even though the asymptotic levels were the same for 24- and 48-h exposures to a particular stimulus, the time needed to recover was significantly longer for the 48- than for the 24-h exposures. The longer period of time to recover is interpreted by

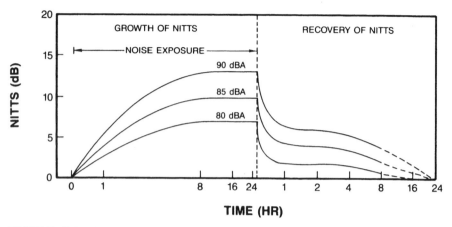

FIGURE 19.2 Growth recovery of noise-induced temporary threshold shift (NITTS) measured during and following the noise exposure at the times marked on the abscissa. The stimulus was a third-octave band of random noise centered on 1000 Hz presented at 80, 85, and 90 dBA. The curves represent the averages of the hearing levels (HLs) for the 1000-, 1500-, and 2000-Hz test frequencies $\frac{1}{3}(HL_{1000} + HL_{1500} + HL_{2000})$.

some as an indication of greater risk to hearing from the same stimulus (same asymptotic threshold level) for 48 than for 24 h. On the basis of analyses of the actual recovery times the guideline was established that the period of time in effective quiet required for recovery should be at least as long as the duration of the exposure.

Audiometric data on the crew members and information on the internal noise levels of the *Voyager* lightweight aircraft (two 110-hp piston engines) during a practice (five days) and a round-the-world flight (nine days) have provided additional important data points. The overall noise environments at the crew locations ranged between 99 and 103 dBA. Crew members wore communications headset and earplug equipments that allowed estimations of the exposures at the ears to be between 84 and 95 dB for the octave band at 500 Hz. Comparisons of preflight and postflight audiograms for both flights revealed substantial shifts of hearing levels. The threshold shifts from the nine-day flight were no greater than those for the five-day flight. One week following the nine-day flight the hearing thresholds of both crew members had returned to their preflight levels.

The hearing level data on the five- and nine-day exposures is consistent with laboratory data derived from both human studies of shorter duration and animal studies of similar duration. It is considered acceptable by some criteria to extend the 8-h limits to as much as 24 h using appropriate trading relationships. Noise exposures of atypical work schedules such as four 10-h days with three days off and 12 h on, 12 h off might be calculated on the bases of work week. Although recovery of the hearing thresholds occurred prior to nine days for these crew members, it remains reasonable to have personnel remain in effective quiet for a period at least as long as the exposure prior to reentering the noise.

Music Exposure Criteria

The USAF has adopted music exposure criteria that consider customers or clients of military "clubs" to be "recreationally exposed" and employees to be "occupationally exposed." The occupationally exposed persons are governed by the same provisions as those of workers exposed in any other occupational noise. A separate set of criteria are used to control or limit the recreational exposures.

An average A-weighted sound level of 94 dB is considered acceptable when it does not exceed 2 h duration once a week. It is important to recognize that the 94-dBA guideline is not a peak of a maximum level value but is the average sound level. The average sound-level concept does not specify a fixed maximum level or eliminate crescendos and special effects or even some selections. It does permit these intermittent high levels of entertainment music to be averages in such a way that the overall performance is acceptable.

19.4 IMPULSE NOISE

Impulse or impact noise is a very brief sound or short burst of acoustical energy with a sound pressure rise of 40 dB in 0.5 s or faster that may occur singly or as a series of events. The noise may be treated as steady state when the repetition rate of a series of impulses exceeds 10 per second and the decay from the individual peaks to minima does not exceed 6 dB.

The effects on the auditory system have been examined for such characteristics of the impulsive stimulus as frequency spectrum, duration, peak pressure level, total energy, type of impulse, and rise time. Although work continues with some of these characteristics, present exposure criteria use only peak pressure level and duration and type of impulse to describe safe impulse exposures.

In 1968, CHABA developed exposure criteria for impulse noise[10] based on extensive work in the United Kingdom on firing small arms.[11] The limiting noise exposure values for the impulsive stimuli are summarized in Fig. 19.3. These criteria define exposures that should produce, on average, no more NITTS than 10 dB at 1000 Hz, 15 dB at 2000 Hz, and 20 dB at 3000 Hz and above in 95% of the exposed ears. The criteria provide for adjustments or corrections for exposure situations that vary from the basic condition. The criteria provide for a daily exposure of 100 impulses during any time period ranging from about 4 min to several hours. The values are increased for fewer and decreased for more than 100 impulses per day by a factor of 1.5 dB for each doubling or halving of the number of impulses. The allowable level must be decreased by 5 dB for impulses that strike the ear at perpendicular incidence.

Simple, nonreverberating impulses that occur in open spaces are evaluated using the A-duration or pressure wave duration, which is the time required for the initial or principal wave to reach peak pressure level and return momentarily to zero (Fig. 19.3). The B-duration or pressure envelope duration is used for impulses that occur under various reverberant conditions and is the total time that the envelope of the pressure fluctuations (positive and negative) is within 20 dB

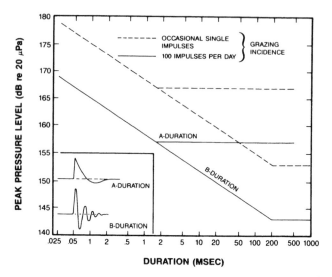

FIGURE 19.3 Proposed damage risk criteria for impulse noise arriving from the front or from behind (grazing incidence) exposed persons. The A-duration curve reflects the simple, nonfluctuating impulse that occurs in open areas. The B-duration curve reflects the pressure fluctuations of impulses that occur under various reverberant conditions.

of the peak pressure level, including reflected waves. The 143-dB floor of the B-duration curve represents the reduction of the energy entering the ear after 200 ms by action of the acoustic reflex (reflex contraction of the middle-ear muscles that reduces the transmission of energy to the inner ear). This comprehensive dose–response curve (DRC) for impulses continues to represent the present data and scientific understanding of impulse noise effects on hearing.

The OSHA amendment, the Air Force, and the Army, all limit exposure to impulse or impact noises to 140 dB peak sound pressure level. The Army has unique equipment that generates high-level impulse noises that are measured and treated individually.

Sonic booms do not constitute a threat of noise-induced hearing loss for human beings. Most of the energy in sonic booms generated by aircraft in supersonic flight is in the low and infrasonic frequency ranges and contributes little to the A-weighted sound level. Field and laboratory investigations with human exposures have revealed no significant effects on hearing from sonic booms at levels typically experienced in the community. One field study involving human exposures to extremely intense sonic booms ranging in level from about 50 to 144 lb/ft^2 observed no changes in the hearing levels of the participants.[12] Naturally, such intense sonic booms may generate higher A-weighted levels indoors than outdoors due to rattling of windows and doors.

Air bag systems are designed to provide crash protection of occupants during side and forward impacts of motor vehicles. These systems generate a loud, impulse noise inside the vehicle upon inflation of the air cushion. In an early study[13] of a prototype system, 91 volunteers experienced this air bag deployment inside a small automobile at a median peak pressure level of 168 dB. Some TTS was experienced by about 50% of the subjects. About 95% of those with TTS recovered preexposure hearing levels on the same day. About 5% required longer times for recovery with one subject showing a gradually returning shift at one frequency that persisted for several months.

19.5 INFRASOUND

Relationships among human exposures to infrasound (0.5–20 Hz) and resulting hearing loss are presently not sufficiently understood for the establishment of national (U.S.) or international standards on exposure limits. Few investigations have been conducted because of difficulties in measuring hearing thresholds for infrasound and in producing infrasound stimuli free from audible overtones that are required for exposure studies. Tentative criteria have been established on the basis of laboratory investigations and field experiences with noises containing intense infrasound components. These criteria have been incorporated in some Department of Defense regulations on hazardous noise exposure.[14]

Human whole-body vibration exposures in intense levels of infrasound that exceeded 150 dB sound pressure level were reported in a classic study.[15] The sample size was small. However, the subjects were highly experienced professionals. Relationships were observed between exposure levels and human tolerance as a function of subjective "symptoms." These symptoms are described in Section 19.8 for airborne vibration. Certain exposures were judged to be very close to tolerance limits. Hearing levels of the subjects were measured 3 min following termination of these very intense exposures and no changes were found in hearing sensitivity.

The absence of hearing sensitivity effects immediately following the intense infrasound exposures verifies laboratory findings that infrasound is not a major threat to hearing. The exposure criteria in Fig. 19.4 have been developed on the basis of data such as that in the cited report and that presented in Fig. 19.5.[16] Numerous experimental subjects experienced exposures to 10 Hz at 144 dB for 8 min with no adverse effects. This set of safe exposure conditions was accepted as a baseline exposure from which 8-h limits (136 dB at 1 Hz and 123 dB at 20 Hz) and 24-h limits (130 dB at 1 Hz and 118 dB at 20 Hz) were extrapolated. (Development of this sound pressure level formulation is described in ref. 16.)

19.6 ULTRASOUND

Ultrasound (~16–40 kHz) is widespread in our society due to sources such as ultrasonic cleaners, measuring devices, drilling and welding processes, animal repellants, alarm systems, and communications control applications as well as a wide range of medical applications. Although the number of people exposed to

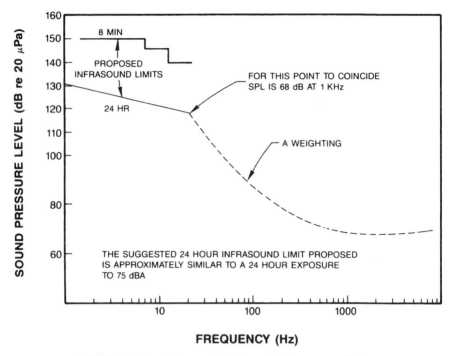

FIGURE 19.4 Infrasound 8-min and 24-h exposure limits.

ultrasound is large, it poses no threat to human hearing because neither temporary nor permanent hearing loss caused by ultrasound has been reported, with one exception. Temporary threshold shift was reported following experimental exposures to discrete tones in the region of 17–37 kHz at levels of 148–154 dB. The TTS occurred at subharmonics of the stimulus frequency and was likely caused by nonlinear distortion of the eardrum.

Nevertheless, ultrasound continues to be viewed as a threat to hearing as well as a cause of other subjective symptoms. Ultrasound is readily absorbed by air and its intensity diminishes rapidly with increasing distance from the source. The impedance match with human flesh is poor, and much energy is reflected away from the surface of the body. Consequently, the ear is the primary channel for transmitting airborne ultrasound to the internal systems.

Ultrasonic energy at frequencies above about 17 kHz and at levels in excess of about 70 dB may produce adverse subjective effects experienced as fullness in the ear, tinnitus, fatigue, headache, and malaise. These subjective effects are mediated through the hearing mechanism and are related to hearing ability. Persons who do not hear in this frequency region do not experience these subjective symptoms. Women experience the symptoms more often than men, and younger individuals report them more often than older ones. This reporting is consistent with the relative hearing abilities of the three groups. Neither disorientation nor loss of balance has been attributed to ordinary exposures to ultrasound.

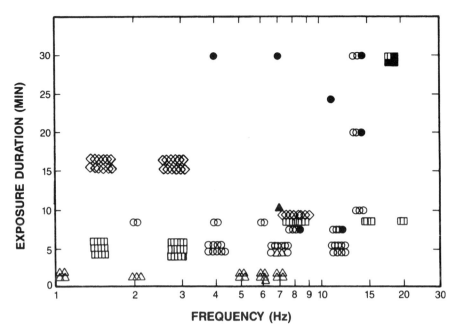

FIGURE 19.5 Infrasound exposure effects on hearing. Exposure levels (filled symbols indicate that some TTS has occurred): (△) > 150 dB; (○) 140–149 dB; (□) 130–139 dB; (◊) 120–129 dB.

Ultrasound exposures usually contain various amounts of high audio frequency energy (5–20 kHz). Subjective effects attributed to the ultrasound are usually caused by the audio frequency energy. Reducing the level of the audio frequency energy in the exposure usually results in the disappearance of the auditory and subjective symptoms.

Criteria for limiting the levels of ultrasound to control auditory and subjective effects are very similar. The data refer to the ultrasound levels at the head of the exposed person. The limiting levels at 20 kHz and above apply to protection from the subjective symptoms described earlier. Representative international and national criteria adopted by the World Health Organization (WHO), Norway, Sweden, and the American Conference of Governmental Industrial Hygienists (ACGIH) are displayed in Fig. 19.6. Norway accepts a level of 120 dB for frequencies higher that 22 kHz.

19.7 HEARING PROTECTION

The damage risk or noise exposure criteria presented above define allowable exposures for unprotected ears. These criteria are also used to identify exposure conditions above which personnel should be included in hearing conservation programs and where hearing protection should be worn. Hearing protection extends

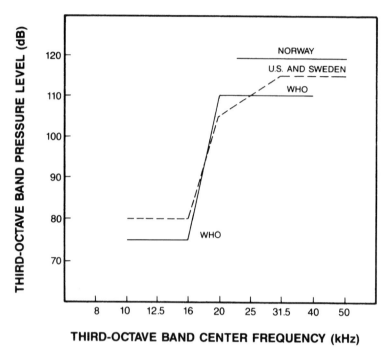

FIGURE 19.6 International and national criteria for exposure to airborne ultrasound.

the limits of allowable exposures by reducing the levels at the ear, allowing personnel to experience more intense and longer duration exposures than with unprotected ears and still remain within established whole-body exposure criteria.

Hearing protector performance is influenced by characteristics of the wearer, the protective device, and the noise exposure. Protection is best with devices that provide good attenuation, are comfortable, fit properly, are easy to use, are in a good state of repair, and are worn. Hearing protector effectiveness is reduced by air leaks, transmission properties of the materials, and noise-induced motion of the device that produces sound under the protector. A limit on performance of ideal devices is imposed by the sound conduction properties of the tissue and bone of the head. In high-level acoustical fields, sound travels "around" the protector to the inner ear through the tissue and bone of the head. The level of the sound reaching the ear by such means is about 50 dB below that of air conduction reaching the ear through an open ear canal. Bone-conducted sound is not a major concern for most noise environments. Total-head enclosures extend the tissue and bone limits by about 10 dB.

Infrasound

Good insert-type hearing protection devices should provide attenuation of infrasound approaching that observed at the 125-Hz third-octave band. Earmuff hearing protectors provide very little protection and may even amplify the sound at some of these infrasonic frequencies. Exposures to levels of infrasound above 150 dB should be avoided even with maximal hearing protection because of possible adverse nonauditory effects.

Audio Frequency Region

Hearing protection performance varies as a function of the frequency spectrum of the noise. In general, conventional earplugs and earmuffs provide good sound protection at higher frequencies (above 1000 Hz). Earplugs provide effective protection at both high and low frequencies. Ample protection is observed across all frequencies with certain foam insert earplugs. Earmuffs provide poor attenuation at the low frequencies.

A combination of earplug and earmuff is required when an individual protector is unable to reduce a noise to an acceptable level. The resulting attenuation is not the sum of the two protectors but an amount determined by the particular combination of devices. The attenuation of the combined units often reaches the bone conduction limits at the high frequencies. At the mid- and lower frequencies the amount of attenuation is determined primarily by the earplug. Selection of a good earplug for use with a muff will provide good double hearing protection at all frequencies.

Special earmuff hearing protectors called nonlinear devices allow face-to-face speech communication at low levels of noise and provide typical amounts of earmuff protection at high levels. Active noise reduction (ANR) earmuffs employ electronic (noise cancellation) and acoustical means of reducing low-frequency noise at the ears. This active earmuff can increase intelligibility and comfort and reportedly reduce fatigue when used with speech communication systems in noise. The ANR earmuffs are widely used in such areas as communications, private aircraft, entertainment, and sport activities and by passengers in commercial aircraft.

Hearing protection devices have the potential capability to reduce and eliminate most noise-induced hearing loss. However, issues such as comfort, selection, fit, training, use, motivation, and others limit the full-performance capability of the devices to be realized in the workplace.

Long-Duration Exposures

Long-duration noise exposures are typically steady state, continuous, and generally within the audio frequency range. The information on hearing protection for audio frequencies is also appropriate for long-duration noises.

Ultrasound

Conventional hearing protection devices, earplugs and earmuffs, provide good protection against airborne ultrasound at frequencies above about 20,000 Hz.

Attenuation exceeding 30 dB is generally provided for frequencies from about 10,000 to 20,000 Hz. Hearing protection is most effective in eliminating subjective symptoms that occur when protectors are not worn. Reduction and elimination of subjective symptoms is also a good indication of the effectiveness of the hearing protector.

Impulses

Earmuff and earplug protectors should provide adequate attenuation for impulses comprised primarily of high-frequency energy, such as small-arms fire. Earmuff attenuation decreases as the concentration of energy in the impulses moves to the lower frequencies, as with large-caliber weapons (the attenuation vs. frequency of the protectors is not changing, only the spectral distribution of the impulse is different). The reduction of the peak sound pressure level of a particular earmuff is about 30 dB for pistol fire, about 18 dB for rifle fire, and as little as 5 dB for cannon fire. The peak level of the impulses of most pistol and standard rifle shots is reduced to less than 140 dB by good earmuffs. A combination of earplugs and earmuffs should be used for impulses requiring good low-frequency attenuation.

The USAF and the USA require single hearing protection when impulses reach 140 dB peak positive pressure and double hearing protection when the levels reach 160 dB (USAF) and 165 dB (USA).

19.8 HUMAN VIBRATION RESPONSE

The human body is a dynamic system, possessing mass and the ability to achieve relative motion between parts of the body (elasticity). It can therefore be affected by exposure to oscillatory motion or vibration. Although the ear is the most sensitive body organ for the reception of vibratory energy in the auditory frequency range, structure-borne and airborne vibration occurring at lower frequencies can be transmitted to a variety of anatomical structures, including skin, bone, muscles, joints, and internal organs. Most of our population is exposed to occasionally moderate levels of vibration with relatively harmless effects. Occupational exposures to vibration are more severe than nonoccupational exposures and have been associated with biological, psychological, and human performance effects.

Vibration is transmitted to the human body via the oscillatory motion from vibrating structures in contact with the body surface (structure borne) or via the transmission of sound pressure waves in high-noise environments (airborne). In structure-borne whole-body vibration, oscillatory motions enter the body usually at the feet (standing) or buttocks (sitting) at the supporting surface and can be transmitted throughout the body to other anatomical structures (such as the head).[17] The major sources of nonoccupational whole-body exposures include transportation vehicles such as automobiles, buses, trains, airplanes, and boats. Whole-body occupational vibration is experienced by operators of agricultural and forestry tractors, earth-moving and construction equipment, all types of trucks, sea-going ships, airplanes and helicopters, as well as mining and factory workers on vibrating platforms. Structure-borne hand-transmitted vibration is isolated to a very specific anatomical region, primarily through the operation of vibrating hand tools such as jackhammers, chipping hammers, chain saws, grinders, and other types of pneumatic or electric hand-held devices. In airborne whole-body vibration, oscillatory motions enter the body via high levels of low-frequency sound pressure waves. Substantial airborne vibration occurs during aircraft engine run-ups and ground-based maneuvering, particularly in military environments where ground crews are required to work close to powerful aircraft in restrictive areas.

The absorption of vibratory energy by the body is determined by the body's characteristics as a mechanical system.[18] Engineering techniques (transfer function, power absorption) have been used and continue to be improved for defining these characteristics and for developing various models of the whole body and its subsystems (such as the hand–arm). These methods and models give insight into the energy transmission through the body and are useful tools for identifying the overall sensitivity of the body as well as for explaining the various effects on specific target organs and structures.

Whole-Body Vibration Effects

For whole-body vibration, the mechanical stresses imposed on the body can potentially lead to interference with bodily functions and tissue damage in practically all parts of the body. Historically, the major concern has focused on whole-body vertical vibration. More recent research includes multiaxis vibration effects. Figure 19.7 illustrates the short-time, 1-min, and 3-min tolerance limits reported by healthy adult male subjects exposed to vertical sinusoidal vibration.[19] Human tolerance to vibration tends to decrease with longer exposure periods. While acute exposures in the vicinity of human tolerance have not resulted in demonstrable harm or injury, prolonged and repeated exposures to these levels are considered to have a high potential for producing bodily damage. For vertical vibration, minimal tolerance occurs between 4 and 8 Hz.[19] This frequency range coincides with the major whole-body resonance observed in biodynamic data. Most physiologic effects in the region of 2–12 Hz are associated with excessive movement of the thoraco-abdominal viscera that can interfere with respiration and cause changes in cardiovascular functions that typically resemble the response to exercise.[20] Prolonged and repeated (chronic) occupational exposures, as they occur in drivers of tractors and earth-moving equipment, are implicated by many investigations in the development of spinal column and other joint disorders and pathologies and of stomach and duodenal diseases.[21,22] However, their causal relationship with vibration stress has not been clearly proven. For example, back pain and back disorders have been reported for other occupations with no vibration. These symptoms have also been associated with poor body posture. Studies, including the early study by Coermann,[23] have shown that body posture can affect whole-body vibration response.

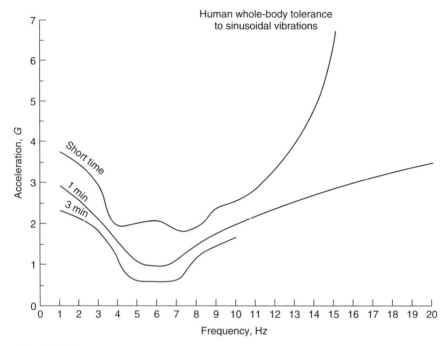

FIGURE 19.7 The short-time, 1-min, and 3-min tolerance limits reported during vertical sinusoidal vibration.[19]

Whole-body vibration can also influence working performance. It can cause involuntary body motions that interfere with the active motor control of the operator with undesirable effects. Relative motion of the eye with respect to an object or target can cause difficulty in reading instruments and performing visual searches.[24] This is particularly a concern at low frequencies where there is a high propensity for head movement in the region of whole-body resonance (below 10 Hz). Below about 20 Hz, compensatory eye movement aids in stabilizing the line of sight to a stationary object during head vibration.[25] When the image moves with the head, as occurs with a helmet-mounted display, compensatory eye movement becomes ineffective, increasing the potential for visual blurring.[26] Eye resonance has been reported to occur in a broad frequency range between 20 and 70 Hz.[25] Relatively high levels of whole-body vibration or direct contact of the head with the vibrating structure are necessary to induce visual blurring due to the body's damping effect at these higher frequencies.

With regard to airborne vibration, the mismatch between the acoustical impedance of air and the human body surfaces prevents significant amounts of acoustical energy from entering the body, particularly at higher frequencies.[27] With decreasing frequencies below 1000 Hz, more acoustical energy is absorbed in the form of transverse shear waves. With exposure to high-intensity noise levels [120 dB sound pressure level (SPL)] between 100 and 1000 Hz, tissue vibration occurs and the noise is felt via the stimulation of somatic mechanoreceptors.[28] Below 100 Hz, intense noise can cause whole-body vibration that not only affects motion in the chest, abdominal wall, viscera, limbs, and head but also can generate motions in the body cavities and air-filled or gas-filled spaces.[28] Von Gierke and Nixon[27] reported that resonance of the chest wall and air-filled lungs occurs around 60 Hz. For noise exposures below 150 dB, the most common symptom reported by subjects was mild to moderate chest vibration.[15] Above 150 dB, symptoms included mild nausea, giddiness, subcostal discomfort, cutaneous flushing and tingling (around 100 Hz, 153 dB); coughing, severe substernal pressure, choking respiration, salivation, pain on swallowing, hypopharyngeal discomfort, and giddiness (60 Hz, 154 dB and 73 Hz, 150 dB); and headache (50 Hz, 153 dB).[15] More recent efforts have been made to measure the biodynamic response of the body to airborne vibration and compare the body acceleration spectra to the noise spectra.[29] The preliminary results for exposures to jet aircraft noise during engine run-ups confirm the upper torso resonance reported by von Gierke and Nixon.[27]

Hand-Transmitted Vibration Effects

Hand–arm vibration syndrome (HAVS) refers to the "complex of peripheral vascular, neurological, and musculoskeletal disorders associated with exposure to hand-transmitted vibration" (ref. 30, p. 11). HAVS is recognized worldwide as a health concern. Prolonged and repeated exposure to hand-transmitted vibration can lead to a very specific disease called vibration-induced white finger (VWF), or Raynaud's phenomenon. This impairment of the blood circulation of the hand progresses with exposure time from intermittent numbness and tingling in selected fingers to extensive blanching of most fingers, first only in combination with cold and finally at all environmental temperatures. In the final stages the disease interferes severely with social activities and continuation of the occupation. According to Taylor and Pelmear,[31] the clinical manifestations of the finger-blanching attacks progress in four stages, with stage 1 signaling the onset of the disease, mainly outdoors in winter. While symptoms of the first stages might still be reversible if vibration exposure is discontinued, the later stages are considered irreversible by most researchers. In later stages, the vascular manifestations are reported to be accompanied by nerve, bone, joint, and muscle involvement.[32] Studies have shown that HAVS, specifically VWF, is influenced by several factors, including the frequency, magnitude, and duration of the exposure, in addition to the type of tool. As with whole-body vibration, the symptoms of VWF appear to increase with exposure duration.

19.9 HUMAN VIBRATION EXPOSURE GUIDELINES

Extensive research over the last four decades has led to a reasonable understanding of the potential pathological and physiological effects of structure-borne whole-body as well as hand-transmitted vibration, although the mechanisms of these effects continue to be topics of research. Frequency-dependent sensitivity curves have been derived for the whole-body and for the hand–arm from both subjective and objective frequency response characteristics. These sensitivity curves form the basis for current whole-body and hand-transmitted vibration standards. There are major national and international standards that provide guidelines and recommended criteria for safe vibration exposure based on these curves and that are generally relied upon in designing vehicles, equipment, tools, and mitigation strategies for use in vibration environments. These standards are periodically revised as vibration research continues and new data become available.

Structure-Borne Whole-Body Vibration Criteria

The International Organization for Standardization (ISO) is the most widely recognized body for providing the basic guidelines for assessing the damage risk of structure-borne whole-body vibration (ISO 2631). As of 2004, the standard entitled "Mechanical Vibration and Shock—Evaluation of Human Exposure to Whole-Body Vibration" included three parts: ISO 2631-1:1997,[33] ISO 2631-2:2003,[34] ISO 2631-4:2001,[35] and ISO 2631-5:2004.[36] Part 1, entitled "General Requirements," describes the general methods for measuring whole-body vibration. Informative annexes are included that give guidance on possible effects of vibration on comfort, perception, health, and motion sickness based on current knowledge in these areas. In 2002, the American National Standards Institute (ANSI) accepted the ISO 2631-1:1997 as a nationally adopted international standard (ANSI S3.18–2002),[37] replacing the long-standing 1979 version. For whole-body vibration, the frequency range considered is 0.5–80 Hz although the range from 1 to 80 Hz can be used if appropriate. When the vibration is transmitted by a resilient structure, such as a seat cushion, a suitably shaped transducer support is interposed between the person and that structure. The standard recommends the collection of acceleration data between the body and three supporting surfaces for the seated occupant, including the seat pan or surface, seat back, and feet. For the recumbent occupant, the supporting surfaces include the pelvis, back, and head. Translational vibrations are typically measured in the three axes defined by an orthogonal coordinate system. For standing, seated, and recumbent whole-body vibration, the x axis is defined in the back-to-chest (fore-and-aft) direction, the y axis for the side-to-side (lateral) direction, and the z axis for the foot- (or buttocks-) to-head (vertical) directions.[33] Rotational vibration may also be measured or estimated. There are three primary frequency-weighting curves (including motion sickness) and three additional frequency-weighting curves that are applied to the acceleration–time histories or frequency response spectra depending on the measurement site and the particular effect being assessed. Figure 19.8 illustrates the two primary frequency weightings that reflect the sensitivity of the body in the respective directions.

The overall weighted root-mean-square (rms) acceleration is determined in the time or frequency domain in each axis. The assessment is made based on

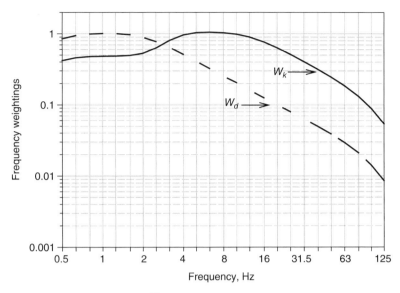

FIGURE 19.8 ISO 2631-1:1997[33] frequency-weighting curves for whole-body vibration in the horizontal (W_d) and vertical (W_k) directions. (Redrawn with permission. Copyright by the International Organization for Standardization. Standard can be obtained from any member body or directly from the ISO Central Secretariat, ISO, Casepostal 56, 1211 Geneva 20, Switzerland.)

the highest weighted acceleration level. If vibration in two or more directions is comparable, then additional direction-dependent multiplying factors are applied and the vibration total value (VTV) or vector sum is calculated.

Other methods are also described, including the running rms method and fourth-power vibration dose method (using the vibration dose value, or VDV). The VDV is more sensitive to peaks in the exposure. Figure 19.9 depicts the Health Guidance Caution Zones provided in the ISO 2631-1:1997 and ANSI S3.18–2002. These guidance zones apply to the overall weighted accelerations determined at the seat pan in the three translational directions. In the figure, the zones defined by the solid lines are applicable to the rms acceleration level. The zones defined by the dotted lines are applicable to the fourth-power vibration dose method. Below the zones, health effects have not been clearly documented or observed. Within the zones, caution is indicated for potential health risks. Above the zones, health risks are likely.[33]

For estimating the effects of structure-borne whole-body vibration on comfort and perception, rotational frequency weightings and multiplying factors are included for the seated occupant.[33] The VTV is recommended for assessing comfort. The standard ISO 2631-1:1997 provides six comfort reactions to vibration and the associated overall weighted acceleration levels ranging from "not uncomfortable" to "extremely uncomfortable." These reactions have no time dependency.

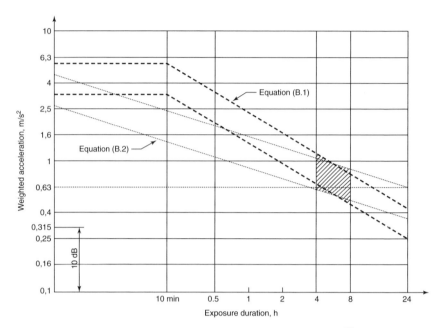

FIGURE 19.9 ISO 2631-1:1997 Health Guidance Caution Zones.[33] Equations (B.1) and (B.2) are defined in the standard. (Reprinted with permission. Copyright by the International Organization for Standardization. Standard can be obtained from any member body or directly from ISO Central Secretariat, ISO, Casepostal 56, 1211 Geneva 20, Switzerland.)

Guidelines on motion sickness are primarily applicable to ships and other sea vessels. The weighted rms acceleration is determined between 0.1 and 0.5 Hz in the vertical direction at the supporting surface. Two methods are given for calculating the motion sickness dose value ($MSDV_z$). An approximation of the percentage of individuals who may vomit is given by $K_m(MSDV_z)$, where $K_m = \frac{1}{3}$ for a population of unadapted male and female adults.[33]

The standard ISO 2631-2:2003, entitled "Part 2: Vibration in Buildings (1 Hz to 80 Hz)," is directed toward human exposure in buildings with respect to comfort and annoyance. A single frequency weighting, W_m, is defined that is related to the combined-response rating curve used in the older 1989 edition. The general guidelines for measuring the vibration are similar to those given in ISO 2631-1:1997. The vibration is measured in the three orthogonal axes with the directions relating to the structure and defined for the standing individual (Part 1). The measurements are made at a location where the highest frequency-weighted vibration occurs. Categories of sources are provided and include (a) continuous or semicontinuous processes (e.g., industry), (b) permanent intermittent activities (e.g., traffic), and (c) limited duration (nonpermanent activities, e.g., construction). The 2003 standard does not include acceptable levels of vibration. The consensus of opinion is that sufficient information was not available at the time for establishing appropriate vibration levels given the complexity of human response to building vibration. The standard does indicate that adverse comments about residential building vibration may occur when the magnitudes or levels are only slightly higher than the perception levels (ISO 2631-1:1997, Annex C).[33] Other complaints may be due to secondary effects such as reradiated noise (ISO 2631-2:2003, Annex B).[34] There should be no reason to consult the Health Guidance Caution Zones in ISO 2631-1:1997, Annex B, except in an extremely rare building vibration environment.

The standard ISO 2631-5:2001,[36] entitled "Part 5: Method for Evaluation of Vibration Containing Multiple Shocks," directs attention to the health of the lumbar spine. Exposures to multiple shocks may occur in the operation of equipment and vehicles over rough terrain or rough seas. A biomechanical model is used to predict the response of the human spine to a given input and to generate an acceleration dose. The standard provides guidelines on using the acceleration dose to assess the health effects of multiple shocks based on the probability of an adverse health effect that depends on the relationship between the ultimate strength of the lumbar spine, the age of the person, and the number of years of exposure.

Airborne Whole-Body Vibration Criteria

Current guidelines for airborne vibration are given in terms of the noise exposure. The Air Force Occupational, Safety, and Health Standard AFOSHSTD 48-19[14] recommends that, for minimizing whole-body vibration effects, no octave- or one-third-octave-band noise level exceed 145 dB for frequencies in the range of 1–40,000 Hz and that the overall A-weighted sound pressure level be below 150 dBA with no time limits. The ACGIH[38] recommends that one-third-octave-band levels between 1 and 80 Hz should not exceed 145 dB, and the overall unweighted SPL should not exceed 150 dB. The assessment of airborne vibration exposure may differ depending on which guideline is used. Minimal research has been conducted on the physiological and pathological effects that may be associated with airborne vibration.

Hand-Transmitted Vibration Criteria

The ISO provides basic guidelines for assessing hand-transmitted vibration in ISO 5349, "Mechanical Vibration—Measurement and Evaluation of Human Exposure to Hand-Transmitted Vibration." The standard consists of two parts: Part 1: General Requirements (ISO 5349-1:2001)[30] and Part 2: Practical Guidance for Measurement at the Workplace (ISO 5349-2:2001).[39] For hand-transmitted vibration, the center of the coordinate system is the head of the third metacarpal with the z axis defined as the longitudinal axis of the third metacarpal, the y axis in the general direction of the tool handle (basicentric coordinate system), and the x axis passing from the top to the bottom of the hand. Figure 19.10 illustrates the frequency weightings or factors associated with the center frequencies of one-third-octave bands for assessing the effects of hand-transmitted vibration.

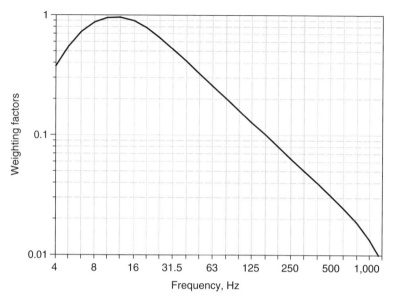

FIGURE 19.10 ISO 5349-1:2001 Frequency-weighting curve W_h for hand-transmitted vibration.[33] (Redrawn with permission. Copyright by the International Organization for Standardization. Standard can be obtained from any member body or directly from the ISO Central Secretariat, ISO, Casepostal 56, 1211 Geneva 20, Switzerland.)

The quantity used to assess health effects is the 8-hr energy-equivalent vibration total value [$A(8)$], or daily vibration exposure. It is assumed that the health risk for hand-transmitted vibration applies to all three orthogonal directions. Therefore, the same weighting curve (Fig. 19.10) is applied to vibration in each of the three orthogonal axes. The vector sum of the overall weighted accelerations estimated for each axis defines the vibration total value (a_{hv}). Figure 19.11 depicts the daily exposure values [$A(8)$] expected to produce vibration-induced white finger in 10% of persons exposed for a given number of years, D_y. The standard ISO 5349-2:2001[39] provides additional guidelines for measuring hand-transmitted vibration in the workplace, determining the daily vibration exposure, and selecting appropriate operations. The measurement of vibration entering the hand may not always be practical. Part 2 includes practical measurement locations for selected power tool types and summarizes the location of accelerometers used in vibration-type test standards for a variety of power tools. These locations are usually close to the hand.

Additional Exposure Standards and Guidelines

Other organizations and countries have generated standards and guidelines that compliment, supplement, or even differ from the human vibration standards described above, particularly the ISO 2631-1:1997. The ACGIH[38] provides guidelines for whole-body as well as hand-transmitted vibration. The European Union

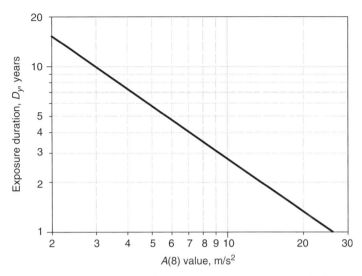

FIGURE 19.11 Vibration exposure for predicted 10% prevalence of vibration-induced white finger in a group of exposed persons. (Redrawn from ISO 5349-1:2001[30] with permission. Copyright by the International Organization for Standardization. Standard can be obtained from any member body or directly from the ISO Central Secretariat, ISO, Casepostal 56, 1211 Geneva 20, Switzerland.)

has recently published a directive on health and safety requirements for exposure to both whole-body and hand-transmitted vibration. It is important that designers and assessors understand the similarities and differences among the various standards and the appropriateness of their applicability. It is also critical to use current editions of these standards since new information could have significant influences on defining health risk.

REFERENCES

1. U.S. Environmental Protection Agency (EPA), "Information on Levels of Environmental Noise Requisite to Protect Public Health and Welfare with an Adequate Margin of Safety," Report No. 550/9-74004, EPA, Washington, DC, March 1974.
2. H. E. Von Gierke, D. Robinson, and S. J. Karmy, "Results of the Workshop on Impulse Noise and Auditory Hazard," ISVR Memorandum 618, Institute of Sound and Vibration Research, Southampton, United Kingdom, October 1981. Also *J. Sound Vib.*, **83**, 579–584 (1982).
3. J. Tonndorf, H. E. Von Gierke, and W. D. Ward, "Criteria for Noise and Vibration Exposure," in C. M. Harris (Ed.), *Handbook of Noise Control*, 2nd ed., McGraw-Hill, New York, 1979.
4. C. W. Nixon, "Excessive Noise Exposure," in S. Singh (Ed.), *Measurement Procedures in Speech, Hearing, and Language*, University Park Press, Baltimore, MD, 1975.

5. International Organization for Standardization (ISO), "Acoustics—Determination of Occupational Noise Exposure and Estimation of Noise-Induced Hearing Impairment," ISO 1999.2, ISO, Geneva, Switzerland, 1990.
6. K. D. Kryter et al., "Hazardous Exposure to Intermittent and Steady-State Noise," paper presented before the NAS-NRC Committee on Hearing, Bioacoustics, and Biomechanics, WG 46, Washington, DC, 1965.
7. Occupational Safety and Health Administration, "Occupational Noise Exposure; Hearing Conservation Amendment," *Fed. Reg.*, **48**(46), 9738–9785 (1983).
8. National Center for Health Statistics, "Hearing Levels of Adults by Age and Sex," Vital Statistics, Public Health Service Publication No. 1000, Series-11-No. 11, U.S. Government Printing Office, Washington, DC, 1965.
9. M. R. Stephenson, C. W. Nixon, and D. L. Johnson, "Long Duration (24–48 Hours) Exposure to Continuous and Intermittent Noise," Joint EPA/USAF Report, AFAMRL Technical Report No. 82–92, Wright Patterson AFB, OH, 1982.
10. W. D. Ward et al., "Proposed Damage Risk Criteria for Impulsive Noise (Gunfire)," paper presented before the NAS-NRC Committee on Hearing, Bioacoustics, and Biomechanics, WG 46, Washington, DC, 1968.
11. R. R. A. Coles, G. R. Garinther, D. C. Hodge, and C. G. Rice, "Hazardous Exposure to Impulse Noise," *J. Acoust. Soc. Am.*, **43**, 336–343 (1968).
12. C. W. Nixon, H. H. Hille, H. C. Sommer, and E. Guild, "Sonic Booms Resulting from Extremely Low Altitude Supersonic Flight: Measurements and Observations on Houses, Livestock, and People," AAMRL Technical Report No. 68-52, Wright Patterson AFB, OH, 1968.
13. C. W. Nixon, "Human Auditory Response to an Air Bag Inflation Noise," PB-184-837, Clearinghouse for Federal Scientific and Technical Information, Arlington, VA, 1969.
14. Office of the Air Force Surgeon General, Air Force Occupational Safety and Health (AFOSH) Program, "Hazardous Noise Program," AFOSHSTD 48-19, March 31, 1994, Washington, DC.
15. G. C. Mohr, J. N. Cole, E. Guild, and H. E. Von Gierke, "Effects of Low Frequency and Infrasonic Noise on Man," *Aerospace Med.*, **36**, 817–824 (1965).
16. C. W. Nixon and D. L. Johnson, "Infrasound and Hearing," in W. D. Ward (Ed.), *International Congress on Noise as a Public Health Problem*, 550/9-73-008, U. S. Environmental Protection Agency, Washington, DC, May 1973.
17. H. E. Von Gierke and D. E. Goldman, "Effects of Shock and Vibration on Man," in C. M. Harris (Ed.), *Shock and Vibration Handbook*, 3rd ed., McGraw-Hill, New York, 1988.
18. H. E. Von Gierke, "To Predict the Body's Strength," *Aviat. Space Environ. Med.*, **59**, A107–A115 (1988).
19. E. B. Magid, R. R. Coermann, and G. H. Ziegenruecker, "Human Tolerance to Whole Body Sinusoidal Vibration," *Aerospace Med.*, **31**, 915–924 (1960).
20. W. B. Hood, R. H. Murray, C. W. Urschel, et al., "Cardiopulmonary Effects of Whole-Body Vibration in Man," *J. Appl. Physiol.*, **21**(6), 1725–1731 (1966).
21. H. Dupuis and G. Zerlett, *The Effects of Whole-Body Vibration*, Springer-Verlag, Berlin, Heidelberg, New York, Tokyo, 1986.

22. H. Seidel and R. Heide, "Long-Term Effects of Whole-Body Vibration: A Critical Survey of the Literature," *Int. Arch. Occup. Environ. Health*, **58**, 1–26 (1986).
23. R. R. Coermann, "The Mechanical Impedance of the Human Body in Sitting and Standing Position at Low Frequencies," ASD Technical Report 61–492, Aeronautical Systems Division, Air Force Systems Command, U. S. Air Force, Wright Patterson AFB, OH, 1961.
24. M. J. Moseley and M. J. Griffin, "Effects of Display Vibration and Whole-Body Vibration on Visual Performance," *Ergonomics*, **29**(8), 977–983 (1986).
25. M. J., Griffin, *Handbook of Human Vibration*, Academic, London, 1990, pp. 128–134.
26. M. J. Wells and M. J. Griffin, "Benefits of Helmet-Mounted Display Image Stabilisation under Whole-Body Vibration," *Aviat. Space Environ. Med.*, **55**(1), 13–18 (1984).
27. H. E. Von Gierke and C. W. Nixon, "Effects of Intense Infrasound on Man," in W. Tempest (Ed.), *Infrasound and Low Frequency Vibration*, Academic, New York, 1976, pp. 115–150.
28. J. C. Guignard and P. F. King, "Aeromedical Aspects of Vibration and Noise," AGARDograph No. 151, NATO Advisory Group for Aerospace Research and Development (AGARD), 1972.
29. S. D. Smith, "Characterizing the Effects of Airborne Vibration on Human Body Vibration Response," *Aviat. Space Environ. Med.*, **73**(1), 36–45 (2001).
30. International Organization for Standardization (ISO), "Mechanical Vibration and Shock—Measurement and Evaluation of Human Exposure to Hand-Transmitted Vibration—Part 1: General Requirements," ISO 5349-1: 2001, ISO, Geneva, Switzerland.
31. W. Taylor and P. L. Pelmear (Eds.), *Vibration White Fingers in Industry*, Academic, London, 1975.
32. U.S. Department of Health and Human Service, National Institute for Occupational Safety and Health (NIOSH), "Occupational Exposure to Hand-Arm Vibration," DHHS Publication No. 89–106, NIOSH, Washington, DC, 1989.
33. International Organization for Standardization (ISO), "Mechanical Vibration and Shock—Evaluation of Human Exposure to Whole-Body Vibration—Part 1: General Requirements," ISO 2631-1:1997, ISO, Geneva, Switzerland.
34. International Organization for Standardization (ISO), "Mechanical Vibration and Shock—Evaluation of Human Exposure to Whole-Body Vibration—Part 2: Vibration in Buildings (1–80 Hz)," ISO 2631-2:2003, ISO, Geneva, Switzerland.
35. International Organization for Standardization (ISO), "Mechanical Vibration and Shock—Evaluation of Human Exposure to Whole-Body Vibration—Part 4: Guidelines for the Evaluation of the Effects of Vibration and Rotational Motion on Passenger and Crew Comfort in Fixed-Guideway Transport Systems," ISO 2631-4:2001, ISO, Geneva, Switzerland.
36. International Organization for Standardization (ISO), "Mechanical Vibration and Shock—Evaluation of Human Exposure to Whole-Body Vibration—Part 5: Method for Evaluation of Vibration Containing Multiple Shocks," ISO 2631-5:2004, ISO, Geneva, Switzerland.
37. American National Standards Institute, "Guide for the Evaluation of Human Exposure to Whole-Body Vibration," ANSI S3.18–2002, Acoustical Society of America, Melville, NY, 2002.

38. American Conference of Governmental Industrial Hygienists (ACGIH), "2003 TLVs® and BEIs®, Based on the Documentation of the Threshold Limit Values for Chemical Substances and Physical Agents & Biological Exposure Indices," ACGIH, Cincinnati, OH, 2003.
39. International Organization for Standardization (ISO), "Mechanical Vibration and Shock—Measurement and Evaluation of Human Exposure to Hand-Transmitted Vibration—Part 2: Practical Guidance for Measurement at the Workplace," ISO 5349-2:2001, ISO, Geneva, Switzerland.

CHAPTER 20

Criteria for Noise in Buildings and Communities

LEO L. BERANEK

Consultant
Cambridge, Massachusetts

One of the early steps in the architectural design of a building is the specification of acceptable noise levels in the spaces for occupancy by human beings. Noise sources in rooms include the air-handling and conditioning system (HVAC), the occupants themselves, machinery in the room, and vibration of the surfaces of the room by machinery outside. In this chapter, the sources of the noise in a space are, in general, not discussed. However, in many cases, such as in offices, dwellings, schools, auditoriums, and studios, it is generally true that the most likely annoying noise is created by the HVAC system. In those cases the criteria presented here apply to the HVAC noise alone as measured without the presence of people. In shops, manufacturing plants, and restaurants, the principal noise sources are machinery or large numbers of people. In such spaces the designer only has to make sure that the HVAC levels do not exceed the liberal limits suggested in this chapter.

20.1 SOUND LEVELS: DEFINITIONS

The sound (noise) levels, all in decibels, that are used in this chapter are taken from Chapter 1, which, in turn, were taken from U.S. and international standards:

- Sound pressure level (SPL) L_p [see Eq. (1.21)].
- A-weighted sound pressure level L_A [see Eq. (1.22)].
- The equivalent sound level L_{eq}, which is broadly defined by

$$L_{eq} = L_{av,T} = 10 \log(1/T) \frac{\int_0^T p_A^2(t)\,dt}{p_{ref}^2} \quad \text{dB} \qquad (20.1)$$

where T is the time over which the averaging takes place and must be specified [see Eq. (1.23)].

- A-weighted equivalent sound level $L_{eq,A}$ is that given by Eq. (20.1), but with the A-weighted frequency response in place in the measuring instrument. Often this quantity is simply labeled L_{eq} and the A weighting is known through its expression in dBA [see Eq. (1.24)].
- The band average sound level is that given in Eq. (19.1), but with the band width specified.
- The averaging time T can be a matter of fractional seconds, minutes, hours, days, or even a year. Always, it must be specified.
- Day–night sound level L_{dn} [see Eq. (1.25)]. Note, in some nations, the 24-h period is divided into three intervals, day, evening, and night, for which Eq. (1.25) must have three integrals with the interval times specified.
- A-weighted sound (noise) exposure $E_{A,T}$ (proportional to energy flow, intensity times time, in a sound wave in the time period T) [see Eq. (1.26)].
- A-weighted sound (noise) exposure level L_{EAT} [see Eq. (1.27)].
- Speech interference level SIL is the average of the sound pressure levels in the four octave bands, whose midfrequencies are 500, 1000, 2000, and 4000 Hz.

20.2 EVALUATION OF ROOM NOISE: SURVEY METHOD, ENGINEERING METHOD, AND PRECISION METHOD[1]

Current-Day Sound-Level Meters

Current-day sound-level meters are used to determine the sound levels in some or all of the ten octave bands with midfrequencies from 16 to 8000 Hz. The meter "speed" can be set to slow, fast, or impulse. When the meter speed is set to "fast," a running integration time of about 125 ms is achieved. The output of the meter can be set to read or store the following levels: L_{eq}, the equivalent noise level over the measurement time period; the peak level L_{max}; and the statistical levels L_{10}, L_{50}, L_{90}, and L_{95}, where the subscript means that level is exceeded 10, 50, 90, or 95% of the time. The 50% level is the median level. When measuring A-weighted sound levels, the meter speed is set at "slow." For monitoring, the meters usually have storage capacity from 1 s up to 24 h. The stored data can be sampled as often as every 100 ms. Acoustical consultants say that their measuring time of HVAC noise at a particular location in a room is usually under a minute.

Survey Method — L_A

An A-weighted sound level L_A is often used for surveying whether a measured noise level satisfies a specified noise criterion. It is readily measured with the simplest sound-level meters. The measurement of L_A is the average of the maximum meter readings, with the meter response set at slow, and obtained by moving the microphone about the room, avoiding positions near the surfaces or the geometric center. It is assumed that the measured noise is free of pure tones and HVAC surging. The presence of HVAC surging can usually be detected by listening or by observing the meter swings using the "flat" (or "C") frequency weighting and the fast meter response.

Because L_A lacks specific spectral information, it can be misleading in evaluating noise other than that arising from reasonably well-designed HVAC systems. The values given in Table 20.1 are recommended for estimating whether measured A-weighted levels may be suitable for the purposes indicated. The numbers presented in that table are largely influenced by the results of a request to a number of consulting firms for examples of excessive noise situations.[2] Fourteen firms responded. The acceptable and unacceptable octave-band levels in school rooms, auditoria, offices, and hotel rooms resulted. Those data indicated that the best method to use in writing specifications, where HVAC surging and large sound fluctuations are not expected, is to use noise criterion (NC) curves, described in the next section. (*Note*: The acceptable NC curves for various occupancies of the spaces are given in Table 20.2. Table 20.1 was determined from Table 20.2 and the NC curves of Fig. 20.1 as follows: For each of the NC curves the speech interference level (SIL) and the A-weighted effective level (L_A) were calculated and L_A − SIL found. These differences were added to the levels in Table 20.2 to obtain the levels of Table 20.1. For example, L_A − SIL for the NC-40 criterion curve is 8 dB, which means that where 40 dB appears in Table 20.2, 48 dB appears in Table 20.1. Also in reference 2, the octave-band levels for 14 cases of excessive noise at low frequencies revealed that L_A − SIL = 14 dB is clearly unacceptable, while in offices with acceptable noise levels L_A − SIL did not exceed approximately 7.5 dB.)

Engineering Method: NC Curves

The most widely used method today for evaluating the suitability of the noise in a space for human occupancy and for writing noise specifications employs the NC curves shown in Fig. 20.1.[3,4] The numerical values from which Fig. 20.1 was plotted are given in Table 20.2 (rounded to the nearest decibel). In measuring a noise the sound-level meter (with octave-band filtering) is set to yield L_{eq} with a slow meter, flat response. The integration time T can be of any desired length—seconds, hours, or months. If each octave-band meter reading slowly fluctuates over a small range, the average value is used.

For writing specifications, the suggested NC ratings for various occupancies are shown in Table 20.2. For example, for small offices, NC-35 to NC-40 is listed.

When evaluating a measured noise spectrum, the octave-band levels are plotted on the set of curves in Fig. 20.1. The rating given to the noise is set by the band level that "touches" the highest NC curve (interpolated to the nearest decibel). For example, assume that the levels in the eight bands from 31.5 to 4000 Hz were 68, 65, 61, 58, 50, 42, 37, and 30 dB, respectively. The highest NC rating

TABLE 20.1 Recommended A-Weighted Sound-Level Criteria for HVAC Systems for Rooms (Unoccupied) of Various Uses

Occupancy	A-Weighted Sound Level L_A in dBA
Small auditoriums	35–39
Large auditoriums, large drama theaters, and large churches (for very good speech articulation)	30–35
TV and broadcast studios (close microphone pickup only)	16–35
Legitimate theaters	30–35
Private residences	
Bedrooms	35–39
Apartments	39–48
Family rooms and living rooms	39–48
Schools	
Lecture and classrooms	
With areas less than 70 m²	44–48
With areas greater than 70 m²	39–44
Open-plan classrooms	44–48
Hotels/motels	
Individual rooms or suites	39–44
Meeting/banquet rooms	35–44
Service support areas	48–57
Office buildings	
Offices	
Executive	35–44
Small, private	44–48
Large, with conference tables	39–44
Conference rooms	
Large	35–39
Small	39–44
Open-plan areas	44–48
Business machines, computers	48–53
Public circulation	48–57
Hospitals and clinics	
Private rooms	35–39
Wards	39–44
Operating rooms	35–44
Laboratories	44–53
Corridors	44–53
Public areas	48–52
Movie theaters	39–48
Churches, small	39–44
Courtrooms	39–44
Libraries	44–48
Restaurants	48–52
Light maintenance shops, industrial plant control rooms, kitchens and laundries	52–62
Shops and garages	57–67

Note: Data based on L_A − SIL for NC curves of Table 20.2.

TABLE 20.2 Recommended NC and RNC Noise Criteria for HVAC Systems for Rooms (Unoccupied) of Various Uses

Occupancy	NC and RNC Recommended Criterion Curve
Recording studios	Lowest curve of Fig. 20.1
Broadcast studios (distant microphone pickup used)	10
Concert halls, opera houses, and recital halls (listening to faint musical sounds)	15–18
Small auditoriums	25–30
Large auditoriums, large drama theaters, and large churches (for very good speech articulation)	20–25
TV and broadcast studios (close microphone pickup only)	15–25
Legitimate theaters	20–25
Private residences	
Bedrooms	25–30
Apartments	30–40
Family rooms and living rooms	30–40
Schools	
Lecture and classrooms	
With areas less than 70 m^2	35–40
With areas greater than 70 m^2	30–35
Open-plan classrooms	35–40
Hotels/motels	
Individual rooms or suites	30–35
Meeting/banquet rooms	25–35
Service support areas	40–50
Office buildings	
Offices	
Executive	25–35
Small, private	35–40
Large, with conference tables	30–35
Conference rooms	
Large	25–30
Small	30–35
Open-plan areas	35–40
Business machines, computers	40–45
Public circulation	40–50
Hospitals and clinics	
Private rooms	25–30
Wards	30–35
Operating rooms	25–35
Laboratories	35–45
Corridors	35–45
Public areas	40–45
Movie theaters	30–40
Churches, small	30–35
Courtrooms	30–35
Libraries	35–40
Restaurants	40–45
Light maintenance shops, industrial plant control rooms, kitchens and laundries	45–55
Shops and garages	50–60

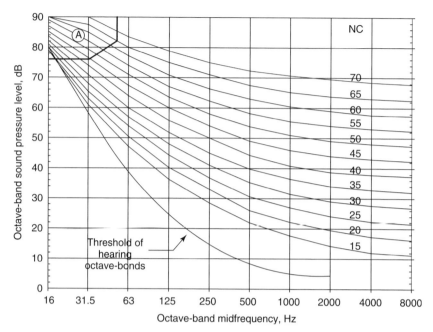

FIGURE 20.1 Noise criteria curves derived from interviews and simultaneous noise measurements extended to low frequencies by comparison with balanced NC (NCB) curves. These NC curves are used to rate noises by the "tangency method."[2,3,5,10] Region A indicates low-frequency levels that may produce feelable vibrations and rattles from lighting fixtures in lightweight constructions if fluctuations or surging is present. The threshold of hearing is taken from ANSI S12.2–1995.

curve "touched" by this spectrum is determined by the 58-dB level in the 250-Hz band and equals NC-50. This is the so-called tangency method of rating.

The NC curves were originally derived from a survey of noise in a large number of offices and spaces in a military Air Force base[5] and in several office buildings.[6] The surveys were conducted by asking occupants of rooms how they rated the noise both generally and at that instant (on a scale with six divisions from "very quiet" to "intolerably noisy") and then measuring the noise at that instant. It was determined that the activity most affected by the noise was speech communication both person to person and by telephone. However, it was also found that even though speech communication might be satisfactory in a space, the low-frequency noise levels could be so high as to be annoying. How much higher can the low-frequency noise be compared to the midfrequency noise? It was found that the overall loudness of the noise had to be taken into consideration. To answer this question, two quantities were measured, the SIL and the loudness level of the noise.

The SIL is a measure of the degree of interference of a noise with speech communication. It is standardized as the average sound pressure level, in decibels, in the four octave bands 500, 1000, 2000, and 4000 Hz.[7]

TABLE 20.3 Noise Criteria Curves from Fig. 20.1 to Nearest Decibel

NC CURVE	By Octave-Band Center Frequencies, Hz								
	16	31.5	63	125	250	500	1000	2000	4000	8000
NC-70	90	90	84	79	75	72	71	70	68	68
NC-65	90	88	80	75	71	68	65	64	63	62
NC-60	90	85	77	71	66	63	60	59	58	57
NC-55	89	82	74	67	62	58	56	54	53	52
NC-50	87	79	71	64	58	54	51	49	48	47
NC-45	85	76	67	60	54	49	46	44	43	42
NC-40	84	74	64	56	50	44	41	39	38	37
NC-35	82	71	60	52	45	40	36	34	33	32
NC-30	81	68	57	48	41	35	32	29	28	27
NC-25	80	65	54	44	37	31	27	24	22	22
NC-20	79	63	50	40	33	26	22	20	17	16
NC-15	78	61	47	36	28	22	18	14	12	11

The loudness level is determined by a standardized computational procedure which yields results in units called phons.[8] Needed are measured octave- or one-third-octave-band levels in all the frequency bands.

The studies revealed that the loudness level (in phons) should not exceed the SIL (in decibels) by more than about 24 units if the space is not to be annoying to the occupants. Because low-frequency sounds are less loud than high-frequency sounds, their levels can be higher. Thus with increasing frequency the NC curves slope downward monotonically and their shape is determined to resemble that of well-known "equal-loudness" curves.[9] The amount of slope and the shape of the spectrum are determined by the 24-unit difference just mentioned. (*Note*: The average of the levels in the four speech interference bands for each NC curve in Fig. 20.1 is between 0.25 and 1 dB higher than the designating number shown for the curve. The reason is that a shift in the frequency limits for the octave bands was standardized some years after these curves were developed. However, the decibel values for the points and the designating numbers on these curves are exactly the same as originally presented.)

It must be emphasized that the NC curves so derived give the upper permissible limit of the low-frequency noise once the SIL is known. Thus, the shape of a NC curve is not the shape of an ideal noise spectrum, but rather is the shape of a "not-to-be-exceeded" spectrum.

Two basic assumptions concerning the curves of Fig. 20.1 are that the noise contains no pure tones and there is no surging of the noise from fans that drive the HVAC system.

Concerning the measurement of sound levels at each point on a NC curve, the meter response is set to suit the desired result. If the equivalent noise level over a period of hours or days is decided, the slow meter response is selected. For short samples of the measured noise, the fast response is used. For example, a series of 100–1000 samples each 100 ms long requires fast response. The meter can show the equivalent level L_{eq} in decibels for the sequence, or the maximum level L_{max}, or the level above which 10% of the interval levels are found, called L_{10}, or the L_{50} level, which is the mean level, or the L_{90} level. Some consultants believe the L_{50} level correlates well with people's judgment of the acceptability of their noise environment, provided that the speech intelligibility requirement is met.

Comments about NC Curves. The NC curves as originally derived did not extend down to the 31.5- and 63-Hz octave bands. These two bands have been added by the author to the original curves, based on the studies that led to the balanced noise criterion (NCB) curves that are briefly discussed below. The revised curves of Fig. 20.1 have been shown in two studies to predict successfully all known cases of sizable complaints by occupants of existing building spaces.[10,11]* In none of these cases was surging detected.

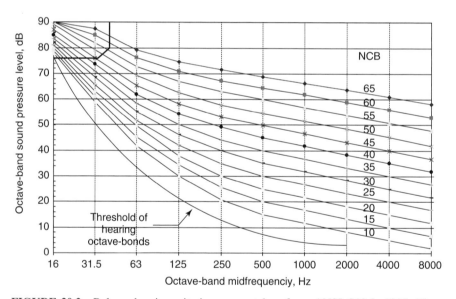

FIGURE 20.2 Balanced noise criteria curves, taken from ANSI S12.2–1995. These curves are used to rate noises with a procedure described in the standard and ref. 10.

*In ref. 11, 238 measured sound-pressure-level spectra were obtained from the consulting files of Cavanaugh Tocci Associates. Plots were made of the relation between NC (tangency method) and NCB (tangency method). The two were closely related with an $R^2 = 0.97$. Similarly, the NCB rating was plotted in relation to the RC or RC II rating, with a correlation of $R^2 = 0.98$, except that the RC values are about 2 dB greater than the NC values because the NC curves are based on the average of the levels in the 500-, 1000-, 2000-, and 4000-Hz bands, while the RC curves are labeled based on the average of the levels in the 500-, 1000-, and 2000-Hz bands.

NCB Curves

The NCB curves shown in Fig. 20.2[7] were derived from the following premises: (1) the rating number for each curve should equal its four-band SIL; (2) the calculated octave-band loudness for those bands from 125 to 8000 Hz should contain the same number of critical (hearing) bands, and for the two lower bands, the loudnesses should be weighted downward in proportion to the fraction of critical bands contained; and (3) the difference between the SIL (in decibels) and the loudness level (in phons) for a criterion curve should not exceed 24 units. The detailed method of applying NCB curves to the evaluation of HVAC noise is not discussed here, because they have not found widespread use, in spite of their being contained in ANSI S12.2–1995. As mentioned above, the NC curves have been extended to the 16- and 31.5-Hz bands by comparison with the rationale that led to the NCB curves. Except for the method of applying them to the rating of a noise, the NCB curves differ little from the NC curves of Fig. 20.1.

The levels taken from the NCB curves (rounded to the nearest decibel) are listed in Table 20.4.

RC (Originally Called "Revised Criteria") Curves

The RC criteria curves for evaluating noise in a room or for specifications were built around the point of view that sound levels lower than those shown by the NC curves of Fig. 20.1 should always be specified for the bottom three-octave frequency bands (31.5, 63, and 125 Hz). The primary reason given is that if there are strong fluctuations or surging in the HVAC noise, the specified low-frequency octave-band levels for the RC curves ought to be low enough so that the noise is not disturbing.

The basis for the RC curves was a set of measurements made in the early 1970s of the noise in 68 offices where the HVAC noise levels were judged satisfactory.

TABLE 20.4 Balanced Noise Criteria Curves from Fig. 20.2 to Nearest Decibel

NCB CURVE	By Octave-Band Center Frequencies, Hz									
	16	31.5	63	125	250	500	1000	2000	4000	8000
NCB-65	97	88	79	75	72	69	66	64	61	58
NCB-60	94	85	76	71	67	64	62	59	56	53
NCB-55	92	82	72	67	63	60	57	54	51	48
NCB-50	89	79	69	62	58	55	52	49	45	42
NCB-45	87	76	65	58	53	50	47	43	40	37
NCB-40	85	73	62	54	49	45	42	38	35	32
NCB-35	84	71	58	50	44	40	37	33	30	27
NCB-30	82	68	55	46	40	35	32	28	25	22
NCB-25	81	66	52	42	35	30	27	23	20	17
NCB-20	80	63	49	38	30	25	22	18	15	12
NCB-15	79	61	45	34	26	20	17	13	10	7
NCB-10	78	59	43	30	21	15	12	8	5	2

When plotted, those noise-level spectra, on average, sloped off linearly with frequency at the rate of 5 dB/octave in the bands from 63 to 4000 Hz. It was concluded that if the measured spectra in satisfactory office environments had that characteristic, the −5 dB/octave line is "optimum," but no comparisons with other spectra were made.

The RC curves are shown in Fig. 20.3 and the values are listed in Table 20.5.[12]

In this chapter, only the Mark II version of the RC method is employed, which differs from the original version in that the levels in the 16- and 31.5-Hz bands are equal instead of the 16-Hz band being 5 dB higher than the 31.5-Hz band. Also the procedure for evaluating measured data is different.

An RC curve suitable for a particular type of room may be determined from Table 20.6. The values there were chosen to permit satisfactory speech intelligibility, where that is important, or to provide noise levels that are not above normal activity sounds.

When evaluating a noise spectrum using the RC method, the first step is to determine the average of the measured sound levels in the 500-, 1000-, and 2000-Hz octave bands, the *midfrequency average*, which Reference 11 calls *level at midfrequencies* (LMF). This number identifies the level of the measured spectrum in the frequency region most important to speech communication, and it selects the reference RC curve used as a starting point for the analysis that follows. For example, if the LMF is 36 dB, the RC-36 becomes the reference curve. If that LMF is suited to the activities that take place in the space in which

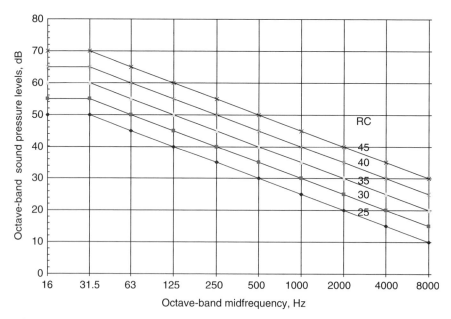

FIGURE 20.3 Noise criteria RC Mark II curves, derived from measurements of HVAC noise in offices. These RC curves are used to rate noises by a procedure detailed in ref. 12.

TABLE 20.5 Noise Criteria RC Curves from Fig. 20.3 to Nearest Decibel

RC CURVE	By Octave-Band Center Frequencies, Hz								
	16	31.5	63	125	250	500	1000	2000	4000	8000
RC-25	50	50	45	40	35	30	25	20	15	10
RC-26	51	51	46	41	36	31	26	21	16	11
RC-27	52	52	47	42	37	32	27	22	17	12
RC-28	53	53	48	43	38	33	28	23	18	13
RC-29	54	54	49	44	39	34	29	24	19	14
RC-30	55	55	50	45	40	35	30	25	20	15
RC-31	56	56	51	46	41	36	31	26	21	16
RC-32	57	57	52	47	42	37	32	27	22	17
RC-33	58	58	53	48	43	38	33	28	23	18
RC-34	59	59	54	49	44	39	34	29	24	19
RC-35	60	60	55	50	45	40	35	30	25	20
RC-36	61	61	56	51	46	41	36	31	26	21
RC-37	62	62	57	52	47	42	37	32	27	22
RC-38	63	63	58	53	48	43	38	33	28	23
RC-39	64	64	59	54	49	44	39	34	29	24
RC-40	65	65	60	55	50	45	40	35	30	25
RC-41	66	66	61	56	51	46	41	36	31	26
RC-42	67	67	62	57	52	47	42	37	32	27
RC-43	68	68	63	58	53	48	43	38	33	28
RC-44	69	69	64	59	54	49	44	39	34	29
RC-45	70	70	65	60	55	50	45	40	35	30

the measurements were made (from Table 20.6), and if the spectrum closely approximates the reference curve, the noise is given a rating of "N," neutral.

If the measured spectrum is not "neutral," the RC (latest version, Mark II) rating procedure begins with the reference curve derived from it. Next, determination of three quantities is calculated from the measured spectrum, each called, *energy-average spectral deviation factor*. Three frequency regions are chosen for this analysis—low (16–63 Hz), middle (125–500 Hz), and high (1000–4000 Hz)—dubbed the LF, or "rumble," the MF, or "roar," and the HF, or "hiss," regions.

The procedure is as follows: First, for each octave-frequency band in the measured spectrum, the difference in decibels between that spectrum and the reference RC curve is found. The sign may be either positive or negative. For each of the three frequency regions, LF, MF and HF, the three octave-band differences therein are summed on an *energy basis*, which yields an energy-average spectral deviation factor for that region. For example, assume the three-octave-band differences in the LF region are 4, 5, and 9 dB, giving an averaged energy level of 4.54, and $10 \log 4.54 = 6.57$ dB. This factor is also determined for the MF and HF regions, which, as an example here, we shall assume are equal to 4.82 and -0.92 dB. Next, the *quality assessment index* QAI is defined

TABLE 20.6 Recommended RC Criteria for HVAC Systems for Rooms (Unoccupied) of Various Uses

Occupancy	RC Criterion Curve[a]
Private residences	
Bedrooms	25–30 (N)
Apartments	30–35 (N)
Family rooms and living rooms	30–40 (N)
Schools	
Lecture and classrooms	30–35 (N)
Open-plan classrooms	35–40 (N)
Hotels/motels	
Individual rooms or suites	30–35 (N)
Meeting/banquet rooms	30–35 (N)
Halls, corridors, lobbies	35–40 (N)
Service support areas	40–50 (N)
Office buildings	
Executive offices	25–35 (N)
Private offices	30–35 (N)
Private offices with conference tables	25–30 (N)
Conference rooms	25–30 (N)
Open-plan areas	35–40 (N)
Business machines/computers	40–45 (N)
Public circulation	40–50 (N)
Hospitals and clinics	
Private rooms	25–30 (N)
Wards	30–35 (N)
Operating rooms	25–35 (N)
Laboratories	35–45 (N)
Corridors	35–45 (N)
Public areas	40–45 (N)
Churches	30–35 (N)
Libraries	35–40 (N)
Courtrooms	30–40 (N)
Restaurants	40–45 (N)

[a] N = neutral spectrum.

as the difference (in decibels) between the maximum and minimum values of the three spectral deviation factors just determined. In our example QAI = 6.57 − (−0.92) = 7.5 dB.

By the RC Mark II procedure, a *neutral* spectrum is one for which QAI is less than 5 dB and L_{16} and $L_{31.5}$ are at most 65 dB. If the highest *positive* spectral deviation factor occurs in the LF region, the spectrum is evaluated as having (i) a "marginal rumble" if 5 dB < QAI ≤ 10 dB or (ii) an "objectionable rumble" if QAI > 10 dB. If the highest positive factor falls in the MF or the HF regions, the same decibel ranges for QAI are listed for evaluating a spectrum as having "marginal or objectionable roar or hiss." For example, a measured spectrum might be designated as RC-45 (i.e., reference curve), roar (i.e., MF region), and probably objectionable (i.e., QAI exceeds 10 dB).[12]

The RC procedure would seem to mean that in writing a specification, a particular RC curve would be specified, but with the understanding that the measured levels in *any one* of the three frequency regions can be exceeded by about 5 dB without penalty.

It must be noted that in the 68 offices originally studied, the A-weighted slow meter sound level only covered a range of 40 to 50 dB, and thus none of the sound levels were considered that are found in spaces like concert halls at one extreme or factory spaces at the other. Second, the RC spectra are typical of HVAC practice in quality offices of the 1970 period and may not be typical at any future time.

The difficulty with the undisciplined use of the RC curves is that they penalize those HVAC installations that are free of surging or of large low-frequency fluctuations. In concert halls or other sites where very low background noise is required, the HVAC systems must be of highest quality, free of surging or significant low-frequency fluctuations. The measured NC noise spectra for concert halls usually are specified to be in the range of NC-15 to NC-18 (See Fig. 20.1 and Table 20.2). The levels in the completed halls usually follow a NC curve down to the 63-Hz band or lower. If RC curves equal to RC-15 to RC-20 were to be specified at such a site, 15 or more decibels of noise reduction would be required in the lowest bands. The cost of such an unnecessary reduction at low frequencies would be prohibitive. Of course, the perceived danger, from the specification point of view, is that a supplier might be told to meet, say, the NC-20 criterion curve, but that he or she might, in fact, install a low-cost HVAC system whose air supply source exhibited surging and then insist that the system meets the specification based on measurements in the octave bands with the usual sound-level meter set to slow response. Advanced techniques for evaluating spectra to determine whether the HVAC noise exhibits surging or large random fluctuations are presented in the next section.

Precision Method — RNC Curves[13,14]

There is a need to evaluate HVAC systems in which surging from the driving fan(s) takes place and to write specifications for the design of HVAC systems to avoid the chance of surging. To listeners, surging is very apparent and disturbing, and it must be addressed both in the writing of specifications and when diagnosing the seriousness of the noise afterward. Surging cannot be determined by a conventional sound-level meter with the slow meter speed and "A" frequency-weighted response in place, partly because surging primarily affects the levels in the low-frequency bands and partly because surging may be averaged out if the slow speed is used.

HVAC: Well Behaved or Surging? Surging usually can be heard aurally or it can be seen visually on a standard sound-level meter by observing the levels in the 16-, 31-, and 63-Hz octave bands. The sound-level meter must be set for fast response and flat frequency characteristic (or C weighting). If surging is found from listening or observing the levels in the octave bands above, then the RNC method of evaluation should be used to determine the seriousness of the noise condition.

First, Some Psychoacoustics. In devising a more precise method of evaluating whether the HVAC noise in a room is intrusive for its primary use, one must turn to the field of psychoacoustics for help. First, the ear only integrates (lumps together) significant variations in the level of a sound provided they occur within a short time period, that is, if the variations occur within a time period that is less than about 125 ms. That is, any fluctuations in HVAC noise that occur within a time interval of about 100 ms are perceived as though the noise were continuous (with the same energy) in that period without fluctuations.

A second reality is that the ear integrates the sound energy that falls within frequency bands called *critical bands*. Critical bands increase in width at about the same rate as the widths of the octave frequency bands increase except at low frequencies. In the frequency region below 125 Hz, the critical bands are about 100 Hz wide. Hence, the 16-, 31.5-, and 63-Hz octave-band levels must be combined in some manner into one band centered, say, at 31.5 Hz, in order that it can be treated in the same way as the higher frequency bands.

A third fact arising from the properties of hearing involves the loudness of sounds at low frequencies, that is, at frequencies below about 200 Hz. For the three octave bands 16, 31.5, and 63 Hz, a change of approximately 5 dB in level creates a doubling in the subjective loudness of a sound. For the 125-Hz band, a change of about 8 dB causes a doubling of loudness. For octave bands with midfrequencies of 250 Hz and above, a change of 10 dB is necessary to create a doubling in loudness. Thus, variations in level of the successive 100-ms intervals for the four lowest frequency bands must be treated differently from variations in level of the 100-ms intervals for the higher frequency bands. The RNC procedures, about to be described, are designed to take into account these three characteristics of the hearing system when evaluating the noise of a space.

Large Random Fluctuations. Even though surging is not observed orally or visually, the presence of large random fluctuations at low frequencies is still a possibility. The simplest test for this requires the observation of the differences between the L_{max} and L_{eq} levels or the L_{10} and the L_{eq} levels in the 16-, 31-, 63-, and 125-Hz bands with the meter set on fast. If the difference between L_{max} and L_{eq} for any one of the three lowest bands (16-, 31-, and 63-Hz octave bands) is greater than 7 dB, or $L_{10} - L_{eq} > 4$ dB, or if the difference between L_{max} and L_{eq} for the 125-Hz band is greater than 6 dB, or $L_{10} - L_{eq} > 3$ dB, large random fluctuations are indicated and the full RNC method should be used.

Well-Behaved HVAC Noise — Use NC Curves. When no surging or large-scale fluctuations are observed, L_{eq} measurements (slow scale) can be made in all the octave bands for any length of time T (fractions of minutes or longer) and the results compared with the NC curves of Fig. 20.1. The microphone used for the decibel measurements can be at one place in the room if the noise field is uniform throughout, or by measuring at a number of points and averaging the results, or simply by rotating the microphone at arm's length (2-m-diameter circle) and taking the average of the results.

Again, it must be borne in mind that all of the methods described in this chapter assume that there are no pure tones audible in the noise. If a pure tone(s) is suspected, narrow-band measurements (one-third octave or less) should be made to determine its frequency and intensity. Then, the pure-tone source should be located and steps taken to reduce its level before evaluating the remaining HVAC noise.

Found or Suspected Surging or Large Low-Frequency Fluctuations in HVAC Noise — Use RNC Curves.
When surging or large low-frequency fluctuations are found or suspected, the RNC procedure is recommended to determine the penalty. It requires the use of a contemporary sound-level meter and a computer. The method incorporates the psychoacoustic facts discussed above. The RNC curves are shown in Fig. 20.4. Each RNC-X curve (e.g., RNC-40, RNC-50)

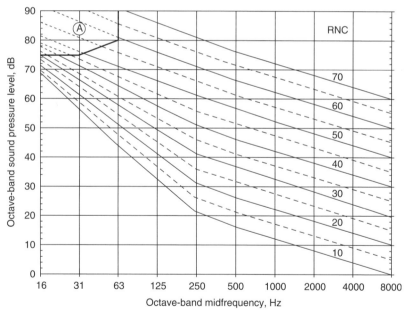

FIGURE 20.4 Balanced Noise criteria curves, devised in refs. 1, 13, and 14 to rate noises that exhibit HVAC surging or large random fluctuations. In well-behaved HVAC systems the curves yield the same ratings as NC curves. When there are clearly audible HVAC surging or random vibrations, the curves yield ratings more like those of the RC curves — a penalty is imposed for surging that requires lower levels as measured with a "slow," sound-level meter setting.

TABLE 20.7 Coefficients Used in Eqs. (20.2) and (20.3) to Calculate RNC-X Curves of Fig. 20.4

Octave Band, Hz	Sound-Level Range, dB	K_{1ob}[a]	K_{2ob}
16	≤81	64.3333	3
	>81	31	1
31	≤76	51	2
	>76	26	1
63	≤71	37.6667	1.5
	>71	21	1
125	≤66	24.3333	1.2
	>66	16	1
250	All	11	1
500	All	6	1
1000	All	2	1
2000	All	−2	1
4000	All	−6	1
8000	All	−10	1

[a] The four-decimal figures are only to give calculated RNC-X even-decibel values.

can be derived from

$$L_{ob} = \frac{\text{RNC-}X}{K_{2ob}} + K_{1ob} \quad (20.2)$$

where RNC-X is the number on the curve, L_{ob} is the x-axis level for the particular octave band, and the constants K_{2ob} and K_{1ob} are taken from Table 20.7 for the particular octave bands.

Example 20.1. Calculate the SPL of the 63-Hz band for the RNC-20 curve of Fig. 20.4:

$$L_{ob} = (20/1.5) + 37.6667 = 13.3333 + 37.6667 = 51 \text{ dB}.$$

Conversely, the nearest RNC evaluation curve for a measured octave-band level can be obtained from

$$\text{RNC-}X = (L_{ob} - K_{1ob}) \times K_{2ob} \quad (20.3)$$

Example 20.2. Assume the measured level in the 125-Hz octave band L_{ob} is 55 dB. Determine which RNC curve (to nearest decibel) that measurement touches:

$$\text{RNC-}X = (55 - 24.3333) \times 1.2 = \text{RNC-37}$$

Recommended RNC Curves. The recommended RNC curves for spaces are the same as those for the NC curves and are given in Table 20.2.

If there is no surging or large scale-fluctuations, so that L_{eq} levels can be measured with the slow meter and flat frequency response, evaluation by the new RNC curves will yield results that are approximately equal to those from the NC curves. Therefore, the permissible L_{eq} levels at low frequencies compared to the SIL at midfrequencies will be as large as those shown in Fig. 20.1.

If surging and fluctuations exist, the RNC method "penalizes" the L_{eq} levels in the octave frequency bands below 300 Hz, so that if the disturbance in those frequency bands is large, the permissible L_{eq} levels may approach the low-frequency levels of the RC curves, as in Fig. 20.3.

Corrections to Measured Levels in Octave Bands at 16, 31.5, and 63 Hz When Surging or Large Fluctuations Exist.

If surging and fluctuations exist, the RNC method penalizes the L_{eq} levels in the octave frequency bands below 300 Hz, so that if the disturbance in those frequency bands is large, the permissible L_{eq} levels may approach the low-frequency levels of the RC curves, as in Fig. 20.3. Contemporary sound-level meters can store up to 24 h of data for later downloading to a computer. The data should be recorded with a fast meter setting and a flat frequency response. The stored data (of whatever total time length) in bands or overall is sampled every 100 ms, forming the required octave-band time series for the RNC method. For the total time period, L_{peak}, L_{10}, and L_{50} are also available.

Each of the successive 100-ms samples is labeled i, where the ultimate i should be between 150 and 1000 depending on the suspected magnitude of the surging.

Because the loudness of a sound, at a given level, is considerably less at 16 Hz than at 31.5 Hz and, in turn, the loudness is considerably less at 31.5 Hz than that at 63 Hz, in the RNC method the three octave bands with midfrequencies at 16, 31.5, and 63 Hz are combined and replaced by a "three-band-sum" at 31.5 Hz. To do this, all of the measured 100-ms samples for the 16-Hz octave band are reduced by 14 dB and the samples for the 63-Hz octave band are increased by 14 dB. These adjusted levels for the 16- and 63-Hz bands are combined with the measured levels in the 31.5-Hz band on *an energy basis* and the new level (in decibels) is called a three-band-sum.

The ith 100-ms sample of this three-band-sum is called L_{LFi}. Calculate $L_{LF,eq}$ for the three-band-sum by summing the energies of the samples and dividing by the number of samples and taking 10 log of the result. Also, calculate L_{LFm}, the mean level of the set of three-band-sum samples.

The object now is to determine a correction K_{LFC} to be added to the *measured* L_{eq} at 31.5 Hz (not the three-band-sum) that takes into consideration both the L_{LFi} values and the fact that a change of only 5 dB is needed for a doubling of loudness at 31.5 Hz—and at 125 Hz, 8 dB is needed. Let δ stand for this change.

First calculate the correction at 31.5 Hz, K_{LFC}:

$$K_{LFC} = K_{LF\delta} - (L_{LF,eq} - L_{LFm}) \quad (20.4)$$

where

$$K_{LF\delta} = 10 \log \left(\frac{1}{N} \sum_{i=1}^{N} 10^{L\alpha/10} \right) \quad (20.5)$$

and where, at 31.5 Hz ($\delta = 5$ dB),

$$L\alpha = \frac{10}{\delta}(L_{LFi} - L_{LFm}) = 2(L_{LFi} - L_{LFm}) \qquad (20.6)$$

The quantity K_{LFC} is added to the *measured* equivalent level L_{eq} (long time energy average) in the 31.5-Hz octave band and is plotted on Fig. 20.4 (no numbers are plotted in the 16- and 63-Hz bands). If data are not available for the 16-Hz band, the three-band-sum is determined for the 31.5- and 63-Hz bands.

Correction to Measured Level in Octave Band at 125 Hz If Surging Is Observed or Suspected. The level of each element of the 100-ms series at 125 Hz is called L_{125i}. Calculate L_{125eq} at 125 Hz by combining the energies of the samples and dividing by the number of samples and taking 10 log of the result. Also, calculate L_{125m}, the mean level of the set of samples at 31.5 Hz. The correction to be added to the *measured* 125-Hz octave-band L_{eq} is called K_{125C} and is given by

$$K_{125C} = K_{125\delta} - (L_{125eq} - L_{125m}) \qquad (20.7)$$

where

$$K_{125\delta} = 10 \log \left(\frac{1}{N} \sum_{i=1}^{N} 10^{L\beta/10} \right) \qquad (20.8)$$

and where, at 125 Hz ($\delta = 8$ dB),

$$L\beta = \frac{10}{\delta}(L_{125i} - L_{125m}) = 1.25(L_{125i} - L_{125m}) \qquad (20.9)$$

Example for Surging. Used is a set of random data having a Gaussian distribution with a standard deviation of about 3 dB for the combined 16–31.5–63-Hz band. Superimposed is a sinusoidal variation with a peak-to-peak amplitude of 15 dB. Shown in Table 20.8 are 10 of the 1000 measured octave-band levels each 100 ms in length. The three-band-sum defined above is given in the last column.

The summary data for 1000 samples are listed in Table 20.9. The correction K_{LFC} to be added to the L_{eq} at 31.5 Hz comes from Eq. (20.4) and equals 11.5 dB, yielding the adjusted 31.5-dB band level equal to 73.1 dB. From Eq. (20.7), the correction K_{125C} is found to be 1.6 dB, yielding 44.7 dB. The corrected spectrum is shown in the last line of Table 20.9. When the corrected spectrum is plotted in Fig. 20.4 or is determined from Eq. (20.3), it is found that the highest RNC curve touched is RNC-44 (at 31.5 Hz). If there had been no correction, the highest curve touched would have been RNC-22. The strong surging and turbulence has changed the rating from RNC-22 to RNC-44.

TABLE 20.8 Ten 100-ms Samples from a Total of 1000 Samples[a]

ith Band Numbers	By Octave-Band Mid frequencies, Hz									L_{Lfi} Three-Band-Sum, (dB)
	16	31.5	63	125	250	500	1000	2000	4000	8000	
1	41.3	52.0	32.4	33.8	26.7	21.7	14.2	10.7	7.4	3.4	53.1
2	63.8	53.8	35.3	35.8	28.5	24.7	17.7	14.1	8.3	4.0	56.3
3	48.8	46.9	41.9	32.5	24.8	22.2	16.1	10.0	5.7	2.5	56.4
4	69.6	54.8	39.6	42.3	31.8	28.2	21.1	19.4	12.9	9.3	59.5
5	67.3	55.7	45.8	46.4	41.0	36.7	29.2	25.2	22.3	17.4	61.9
6	75.5	52.3	44.6	39.0	32.3	26.8	22.7	19.2	12.7	8.7	63.6
7	59.7	52.8	50.0	50.3	37.8	33.8	28.5	25.5	19.5	16.8	64.4
8	61.8	63.0	46.8	47.8	41.4	35.9	27.6	25.2	18.7	16.1	65.1
9	65.2	65.2	46.6	44.3	39.5	34.1	26.7	23.9	19.1	14.3	66.6
10	64.8	66.5	44.2	42.7	34.2	30.0	24.0	19.2	15.3	12.4	67.2
Mean level L_{LFm} of these 10 samples.											61.4
Standard deviation σ of these 10 samples.											4.8
Equivalent sound level L_{eq} of these 10 samples.											63.4

[a]Data Represent a Gaussian noise with a superimposed surge implemented by a sine wave with a 2-s period and a 15-dB peak-to-peak amplitude.

Simplified Method: Corrections to Measured Levels in 16–125-Hz Octave Bands When There Are Large Random Reflections and Not Strong Surging. By this method, the same equations are used to determine the corrections to the adjusted 31- and 125-Hz bands, that is, Eqs. (20.4) and (20.7), but the quantities $K_{LF\delta}$ and $K_{125\delta}$ are calculated from

$$K_{LF\delta} = 0.115 \left(\frac{10}{\delta}\right)^2 \sigma_{LF}^2 = 0.115(2)^2 \sigma_{LF}^2 = 0.46 \sigma_{LF}^2 \qquad (20.10)$$

where σ_{LF} is the standard deviation of the L_{LFi} (three-band-sum) series, and

$$K_{125\delta} = 0.115 \left(\frac{10}{\delta}\right)^2 \sigma_{125}^2 = 0.115(1.2)^2 \sigma_{125}^2 = 0.166 \sigma_{125}^2 \qquad (20.11)$$

where σ_{125} is the standard deviation of the L_{125i} series.

20.3 ACOUSTICALLY INDUCED VIBRATIONS AND RATTLES

From experience, primarily on the West Coast of the United States, an assessment has been made of the probability of acoustically induced vibration and rattles in lightweight wall and ceiling constructions (including light fixtures, windows, and some furnishings). Obviously, significant fluctuations or surging must be present in the HVAC noise. The region of likely probability where such vibrations in lightweight construction can be clearly felt is shown as A in Figs 20.1–20.4.[11] In

TABLE 20.9 Summary Data for RNC Example[a]

	Three-Band-Sum	\multicolumn{9}{c}{Octave-Band Mid frequencies, Hz}
		16	31.5	63	125	250	500	1000	2000	4000	8000
L_{eq}	64.5	68.6	61.8	46.0	43.2	35.9	30.9	24.9	20.9	17.0	12.9
L_m (mean)	60.3	60.2	62.0	54.9	42.1	39.9	32.9	28.0	22.0	18.0	14.0
$L_{eq} - L_m$	4.2	6.6	6.8	3.9	3.2	2.9	2.9	2.9	2.9	2.9	2.9
Eqs. (20.5) and (20.8)	15.5				4.8						
Eqs. (20.4) and (20.7)											
Band corrections			11.3		1.6						
Corrected spectrum			73.1		44.7	35.9	30.9	24.9	20.9	17.0	12.9
Highest RNC curve "touched"			44		25	25	25	23	23	23	22

[a] The L_{eq} is the energy average level in each octave band for all 1000 of the 100-ms samples. The three-band-sum correction of 11.3 dB is added to the L_{eq} of the 31.5-Hz octave band to give an adjusted (for surging) level of 73.1 dB. At 125 Hz the adjusted level is 44.8 dB, 1.6 dB above the L_{eq} level.

very well built structures, such as are expected in concert halls and opera houses, the sound levels can extend into region A without expectation of feelable surface vibrations or audible rattles. In office and school buildings attention should be paid to L_{eq} levels that reach 70 dB or more in the 16- and 31.5-Hz octave bands because of current trends in HVAC design and installation.

20.4 CRITERIA FOR NOISE ANNOYANCE IN COMMUNITIES

The major sources of noise in communities are road, rail and air traffic, industries, construction, and public works. The most common metric used in the United States for estimating the annoyance or disturbance of community noise is the A-weighted day–night sound level (L_{dn}) [see Eq. (1.25)] with slow meter response. The A weighting discriminates against low-frequency sound energy, similar to the way that the hearing apparatus judges loudness. The sound energy of each noise-producing event is summed separately into the total. All of the energy of a noise that rises and falls as a source approaches and departs is summed into the total. The sound energy of each event at night is multiplied by 10 before it is added into the total. This treatment means that a few aircraft fly-bys at night may be as annoying to residents as a large number in the daytime. The European Union countries, for the most part, use a day–evening–night level.

The day–night sound level (L_{dn}) that is agreed on by nearly all agencies, boards, and standards-setting bodies to be the threshold for noise impact in urban residential areas is $L_{dn} = 55$ dBA.[15] In other words, people will begin to complain seriously about the noise if the level is higher. Some agencies speak of this energy-average level as being determined over a period of 12 months. Only the U.S. Federal Aviation Administration (FAA) and the Department of Defense (DOD) speak of a L_{dn} equal to 65 dBA as being the level at which noise begins to impact unfavorably on people in residential areas. Some studies indicate that in urban communities just under half of the U.S. population is exposed to L_{dn} levels of 55 dBA or greater. The World Health Organization (WHO, 1999) has stated that serious annoyance may occur for a L_{dn} of approximately 55 dBA, while moderate annoyance will have a threshold of about 50 dBA. It further recommends that the maximum level L_{max} at night should not exceed 40 dBA.

A possible reason for the 65-dBA level used by the FAA and DOD is that people near airports, on average, have learned to accept higher noise levels or, if not, they have moved away. This also assumes that the airport's relations with the communities surrounding it are good. Quiet rural communities seem to demand lower (5–10 dBA) noise levels.

20.5 TYPICAL URBAN NOISE

A summary of A-weighted noise levels measured in U.S. city areas, daytime and nighttime, is given in Fig. 20.5. Trucks and buses are principal sources of

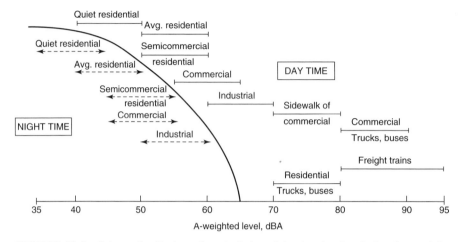

FIGURE 20.5 Schematic display of typical A-weighted noise levels by day and by night in different U.S. urban areas.[4] Noise around industrial areas is frequently caused by trucks.

noise. Parenthetically, the levels in this figure are not L_{dn}, but rather A-weighted levels in a residential area that vehicles produce when they travel on the streets surrounding that neighborhood. In urban residential areas, the noise is frequently the roar of traffic on thoroughfares that may be up to a kilometer away. Traffic noise is likely to be relatively steady and go on all night at a reduced level. If thoroughfares exist on two or more sides of an area, the noise is often nearly uniform over the area.

REFERENCES

1. American National (Draft) Standard ANSI S12.2-200x, "Criteria for Evaluating Room Noise" (not yet subject to extended comment or vote, 2004).
2. L. L. Beranek, "Applications of NCB and RC noise criterion curves for specification and evaluation of noise in buildings," *Noise Control Eng. J.*, **45**, 209–216 (1997).
3. L. L. Beranek, "Revised Criteria for Noise Control in Buildings," *Noise Control*, **3**, 19–27 (1957).
4. L. L. Beranek, *Noise and Vibration Control*, rev. ed., Institute of Noise Control Engineering, Marstun Itall, Iowa State University, Ames, IA, 1988.
5. L. L. Beranek, "Criteria for Office Quieting Based on Questionnaire rating Studies," *J. Acoust. Soc. Am.*, **28**, 833–852 (1956).
6. L. L. Beranek, *Transactions Bulletin No. 18*, Industrial Hygiene Foundation, Pittsburgh, PA, 1950.
7. ANSI S3.14–1977 (R-1986), "*Rating Noise with Respect to Speech Interference,*" Acoustical Society of America, Melville, NY, 1986.
8. ANSI S3.3–1960 (R 1992), "*Procedure for the Computation of the Loudness of Noise,*" Acoustical Society of America, Melville, NY, 1992.

9. D. W. Robinson, and L. S. Whittle, "The Loudness of Octave Bands of Noise," *Acustica*, **14**, 24–35 (1964).
10. L. L. Beranek, "Balanced Noise Criterion (NCB) Curves," *J. Acoust. Soc. Am.*, **86**, 650–664 (1989); "Application of NCB Noise Criterion Curves," *Noise Control Eng. J.*, **33**, 45–56 (1989).
11. G. C. Tocci, "Room Noise Criteria—The State of the Art in the Year 2000," *NOISE/NEWS International*, **8**, 106–119 (2000).
12. W. E. Blazier, "RC Mark II: A Refined Procedure for Rating the Noise of Heating, Ventilating and Air-Conditioning (HVAC) Systems in Buildings," *Noise Control Eng. J.*, **45**, 243–250 (1997).
13. P. D. Schomer, "Proposed Revisions to Room Noise Criteria," *Noise Control Eng. J.*, **48**, 85–96 (2000).
14. P. D. Schomer, and J. S. Bradley, "A Test of Proposed Revisions to Room Noise Criteria Curves," *Noise Control Eng. J.*, **48**, 124–129 (2000).
15. P. D. Schomer, *A White Paper: ASSESSMENT OF NOISE ANNOYANCE*, Schomer and Associates, Champaign, IL, April 22, 2001.

CHAPTER 21

Acoustical Standards for Noise and Vibration Control

ANGELO CAMPANELLA
Consultant
Columbus, Ohio

PAUL SCHOMER
Consultant
Champaign, Illinois

LAURA ANN WILBER
Consultant
Wilmette, Illinois

21.1 WHERE TO FIND AND HOW TO SELECT U.S. NATIONAL AND ISO STANDARDS

This chapter is arranged to provide the titles of North American and international (ISO) standards of interest to noise control engineers, according to specific noise control technology categories such as absorption and insulation (isolation). Acoustical terminology may be found in ANSI S1.1-1994 (R1999), *Acoustical Terminology* and ASTM C634, *Standard Terminology Relating to Environmental Acoustics*. The titles of some standards listed here are truncated to save space. Verbal descriptions, if offered, are brief. As with any technology, state-of-the-art advances bring updated versions of many of these documents. These standards are periodically reviewed and possibly revised by the responsible committees, often on a five-year cycle. A few ANSI standards have become an adoption of the corresponding ISO document, indicated as "Adoption" or "NAIS" (Nationally Adopted International Standard). The reader is encouraged to search anew for recent revisions of these documents. Be advised that some engineering applications may specify past versions, especially as part of a long procurement program or arbitrated contract. It is up to the reader to determine, in participation with interested parties as necessary, which version to use. For the latest standards version, often listed according to the standard's numeral title (e.g., ANSI S12.19 can be found by searching www.ansi.org for S12.19), see the public Internet web page listing by the providing agencies. Some of these agency pages are:

www.ANSI.org (ANSI)
www.ari.org (ARI—free downloads)
www.asastore.aip.org (ANSI + ISO standards available)
www.ashrae.com (ASHRAE)
www.asme.org (ASME)
www.astm.org (ASTM)
www.iec.ch (IEC)
www. iso.ch (ISO)
www.nssn.org (National Standards Systems Network, all standards available)
www.sae.org (SAE)

21.2 SOUND LEVEL MEASUREMENTS

Instrumentation

Standards for noise-measuring instruments, generally measuring sound pressure level (SPL) or "sound level," are listed separate from hearing-measuring instruments (audiometry). Common sound-measuring instruments include sound-level meters, calibrators, filters, and microphones. The most common sound measurement instruments for noise control are specified in ANSI S1.4. Such an instrument comprises a pressure sensor such as a condenser microphone, an amplifier with special wide-band and narrow-band filters, time-averaging circuits, a display, and some memory functions. The precision varies as type 0 (most accurate) to type 2 (least accurate). This entire system can be calibrated at any time with an appropriate calibrator. The acoustical velocity is rarely measured in practice, except for the use of dynamic (velocity) microphones for some audio purposes. The "intensity" method combines SPL and acoustical velocity measurements at a point to compute the intensity value at that point. Applicable standards are as follows:

ANSI S1.4-1983 (R2001), *Specification for Sound Level Meters.* Conforms closely to the IEC standard for sound-level meters, Publication 651, first edition, issued in 1979, but improved for measurement of transient sound signals and it permits the use of digital displays, rigorous definition of the fast and slow exponential time averaging, increase in the crest factor requirement to 10 for type 1 instruments, specification of a type 0 laboratory instrument with generally smaller tolerance limits than specified for type 1, and deletion of the type 3 survey instrument. The corresponding IEC standard (IEC 61672-1:2002 and IEC 61672-2:2003) contains more stringent environmental requirements. In 2003, ANSI S1.4 was under revision to better correspond.

ANSI S1.6-1984 (R2001), *Preferred Frequencies, Frequency Levels, and Band Numbers for Acoustical Measurements.*

ANSI S1.8-1989 (R2001), *Reference Quantities for Acoustical Levels.*

ANSI S1.9-1996 (R2001), *Instruments for the Measurement of Sound Intensity.* Requirements for instruments to measure sound intensity employing the two-microphone technique. Similar to IEC 1043.

ANSI S1.10-1966 (R2001), *Method for the Calibration of Microphones.*

ANSI S1.11-1986 (R1998), *Specification for Octave-Band and Fractional-Octave-Band Analog and Digital Filters.* Performance requirements for fractional-octave-band bandpass filters, in particular octave-band and one-third-octave-band filters, applicable to passive or active analog filters that operate on continuous-time signals to analog and digital filters that operate on discrete-time signals and to fractional-octave-band analyses synthesized from narrow-band spectral components. Four accuracy grades are allowed: the most accurate for precise analog and digital filters. The two least accurate grades meet the requirements of S1.11-1966. Also see IEC 61260:1995.

ANSI S1.14-1998, *Recommendations for Specifying and Testing the Susceptibility of Acoustical Instruments to Radiated Radio-Frequency Electromagnetic Fields, 25 MHz to 1 gHz.*

ANSI S1.15-1997/Part 1 (R2001), *Measurement Microphones.* Part 1: *Specifications for Laboratory Standard Microphones.* This is comparable to the international standard IEC 1094-1:1992, "Measurement Microphones—Part 1: Specifications for Laboratory Standard Microphones."

ANSI S1.16-2000, *Method for Measuring the Performance of Noise Discriminating and Noise Canceling Microphones.*

ANSI S1.17-2004/Part 1, *Measurement and Specifications of Insertion Loss of Wind Screens in Still or Slightly Moving Air.* Determines the insertion loss of wind screens when installed on measuring microphones over a defined frequency range.

ANSI S1.40-1984 (R2001), *Specification for Acoustical Calibrators.* Requirements for coupler-type acoustical calibrators, including the SPL in the coupler, the frequency of the sound, and the determination of the influence of atmospheric pressure, temperature, humidity, and magnetic fields on the pressure level and frequency of the sound produced by the calibrator. (Also see IEC 60942:2003.)

ANSI S1.42-2001, *Design Response of Weighting Networks for Acoustical Measurements.*

ANSI S1.43-1997 (R2002), *Specifications for Integrating Averaging Sound Level Meters.* Consistent with the relevant requirements of ANSI S1.4-1983 (R1997) but specifies additional characteristics needed to measure the time-average SPL of steady, intermittent, fluctuating, and impulsive sounds. (See also IEC 61672-1:2002 and IEC 61672-2:2003.)

ANSI S12.3-1985 (R2001), *Statistical Methods for Determining and Verifying Stated Noise Emission Values of Machinery and Equipment.* Preferred methods for determining and verifying noise emission values for machinery and equipment which are stated in product literature or labeled by other means.

SPL Measurement Techniques

ANSI S1.13-1995 (R1999), *Measurement of Sound Pressure Levels in Air.* Procedures for the measurement of sound levels in air at a single point indoors but may be utilized in outdoor measurements under specified conditions. Includes identification of prominent discrete tones.

ASTM E1014-2000. *Measurement of Outdoor A-Weighted Sound Levels*. A guide for measuring outdoor sound levels over various time periods, intended to be used by practitioners in the field as well as by members of the general public who have little or no special technical training in areas relating to acoustics.

ANSI S1.25-1991 (R2002), *Specification for Personal Noise Dosimeters*. Makes provision for three exchange rates: 3, 4, and 5 dB per doubling of exposure time. Provides tolerances for the entire instruments, parameters, including frequency response, exponential averaging employing slow and fast transient time response, threshold, and dynamic range, for SPL measurement in a random-incidence sound field without the presence of a person wearing the instrument.

ANSI S12.19-1996 (R2001), *Measurement of Occupational Noise Exposure*. Methods that can be used to measure a person's noise exposure received in a workplace. Provides uniform procedures and repeatable results for the measurement of occupational noise exposure.

ISO 11204:1995, *Noise Emitted by Machinery and Equipment—Measurement of Emission Sound Pressure Levels at a Work Station and at Other Specified Positions—Method Requiring Environmental Corrections*.

Intensity Instruments and Measurement Techniques

ANSI S1.9-1996 (R2001), *Instruments for the Measurement of Sound Intensity*. Requirements for instruments to measure sound intensity employing the two-microphone technique. Similar to IEC 1043.

ANSI S12.12-1992 (R2002), *Standard Engineering Method for the Determination of Sound Power Levels of Noise Sources Using Sound Intensity*. Measures the sound power level of noise sources in indoor or outdoor environments using sound intensity instrumentation.

ISO 9614-1:1993, *Determination of Sound Power Levels of Noise Sources Using Sound Intensity*. Part 1: *Measurement at Discrete Points*. Part 2: *Measurement by Scanning*.

21.3 SOUND POWER MEASUREMENT TECHNIQUES

ANSI S12.5-1990 (R1997), *Requirements for the Performance and Calibration of Reference Sound Sources*. Requirements, including free-field laboratory calibration above a reflecting plane, for reference sound sources used in the substitution measurement of sound power.

ANSI S12.11-1987 (R1997), *Method for the Measurement of Noise Emitted by Small Air-Moving Devices*. Measures sound power emitted by small air-moving devices, reported as the noise power emission level in bels. Includes device-mounting methods, test environmental conditions, and device operation method during the tests.

ISO 10302:1996, *Method for the Measurement of Airborne Noise Emitted by Small Air-Moving Devices*.

ANSI S12.15-1992 (R2002), *Measurement of Sound Emitted by Portable Electric Power Tools, Stationary and Fixed Electric Power Tools, and Gardening Appliances*. Methods for SPL measurement and the subsequent calculation of sound power levels.

ANSI S12.23-1989 (R2001), *Method for the Designation of Sound Power Emitted by Machinery and Equipment*. Expressing the noise emission of machinery and equipment as overall A-weighted sound power on labels or other noise emission documentation.

ANSI S12.30-1990 (R2002), *Guidelines for the Use of Sound Power Standards and for the Preparation of Noise Test Codes*. Six standards for determining the sound power levels of equipment for sound test codes for machines and equipment.

ANSI S12.35-1990 (R2001), *Precision Methods for the Determination of Sound Power Levels of Noise Sources in Anechoic and Hemi-Anechoic Rooms*. Determination of the sound power levels of noise sources in laboratory anechoic or hemi-anechoic room, the instrumentation, installation, source operation, SPL measurement surface; calculation of sound power level, directivity index, and directivity factor.

ANSI S12.44-1997 (R2002), *Methods for Calculation of Sound Emitted by Machinery and Equipment at Workstations and Other Specified Positions from Sound Power Level*. Determining emission SPLs from the sound power level values of machinery and equipment at workstations and other locations.

ANSI S12.50-2002 National Adopted International Standard (NAIS), *Determination of Sound Power Levels of Noise Sources—Guidelines for the Use of Basic Standards*. Adoption of ISO 3740:2000. Summaries of ISO 3741–3747 and selection of one or more standards appropriate to any particular type (Clause 6 and Annex D). Used in the preparation of noise test codes (see ISO 12001) and noise testing where no specific noise test code exists.

ANSI S12.51-2002 (including Corrigendum 1:2001) NAIS standard, *Acoustics—Determination of Sound Power Levels of Noise Sources Using Sound Pressure—Precision Method for Reverberation Rooms*. Adoption of ISO 3741:1999, a direct method and a comparison method for determining the sound power level produced by a source in a standard environment.

ANSI S12.53/1-1999 ISO 3743-1:1994, *Determination of Sound Power Levels of Noise Sources—Engineering Methods for Small, Movable Sources in Reverberant Fields*. Part 1: *Comparison Method for Hard-Walled Test rooms. (Adoption of ISO 3743-1.)* Part 2: *Methods of Special Reverberation Test Rooms. (Adoption of ISO 3743-2.)*

ANSI S12.54-1999 ISO 3744:1994, *Determination of Sound Power Levels of Noise Sources Using Sound Pressure—Engineering Method in an Essentially Free Field over a Reflecting Plane*. Adoption of ISO 3744, sound power engineering accuracy, measurement by the comparison method in an anechoic environment.

ANSI S12.56-1999 ISO 3746:1995, *Determination of Sound Power Levels of Noise Sources Using Sound Pressure—Survey Method Using an Enveloping Measurement Surface over a Reflecting Plane*. Adoption of ISO 3746, determining the sound power level of sound sources In Situ, especially if nonmovable, in octave bands from which the A-weighted sound power level is calculated. The accuracy will be that of either an engineering method or a survey method.

ARI 250-2001, *Performance and Calibration of Reference Sound Sources.*

ARI 280-95, *Requirements for the Qualification of Reverberant Rooms in the 63 Hz Octave Band.*

ASTM E1124-97, *Standard Test Method for Field Measurement of Sound Power Level by the Two-Surface Method.* An In Situ survey measurement of sound power level with sound pressure measurements at two concentric surfaces.

ISO 3740:2000, *Determination of Sound Power Levels of Noise Sources—Guidelines for the Use of Basic Standards.*

ISO 3741:1999, *Determination of Sound Power Levels of Noise Sources Using Sound Pressure—Precision Methods for Reverberation Rooms.*

ISO 3743-1:1994, *Determination of Sound Power Levels of Noise Sources—Engineering Methods for Small, Movable Sources in Reverberant Fields.* Part 1: *Comparison Method for Hard-Walled Test Rooms.* Part 2: *Methods for Special Reverberation Test Rooms.*

ISO 3744:1994, *Determination of Sound Power Levels of Noise Sources Using Sound Pressure—Engineering Method in an Essentially Free Field over a Reflecting Plane.*

ISO 3745:1977, *Determination of Sound Power Levels of Noise Sources—Precision Methods for Anechoic and Semi-Anechoic Rooms.*

ISO 3746:1995, *Determination of Sound Power Levels of Noise Sources Using Sound Pressure—Survey Method Using an Enveloping Measurement Surface over a Reflecting Plane.*

ISO 3747:2000, *Determination of Sound Power Levels of Noise Sources Using Sound Pressure—Comparison Method In Situ.*

ISO 5135:1997, *Determination of Sound Power Levels of Noise from Air-Terminal Devices, Air-Terminal Units, Dampers and Valves by Measurement in a Reverberation Room.*

ISO 5136:1990, *Determination of Sound Power Radiated into a Duct by Fans—In-Duct Method.*

ISO 6926:1999, *Requirements for the Performance and Calibration of Reference Sound Sources Used for the Determination of Sound Power Levels.*

21.4 MACHINERY NOISE EMISSION MEASUREMENT TECHNIQUES

ANSI S12.1-1983 (R2001), *Guidelines for the Preparation of Standard Procedures to Determine the Noise Emission from Sources.* Included are the general questions to be asked during development of a measurement procedure. Included are prefatory material, measurement conditions, measurement operations, data reduction, preparation of a test report, and guidelines for the selection of a descriptor for noise emission.

ANSI S12.10-2002 ISO 7779:1999 NAIS standard, *Measurement of Airborne Noise Emitted by Information Technology and Telecommunications Equipment.* Adoption of ISO 7779:1999 and its amendment ISO 7779: 1999/FDAM 1.

ANSI S12.16-1992 (R2002), *Guidelines for the Specification of Noise of New Machinery*. Methods to calculate workplace noise SPL from test data produced by manufacturers of stationary equipment. This references existing ANSI, trade, and professional association measurement standards and techniques to be used by manufacturers to produce individual raw machine sound power or SPL test data.

ANSI S12.43-1997 (R2002), *Methods for the Measurement of Sound Emitted by Machinery and Equipment at Workstations and Other Specified Positions*. Three methods to measure SPLs: (1) in a free field over a reflecting plane, (2) in normal operating environments, and (3) when operating in their normal environments when less accurate measurements are acceptable.

ARI 530-95, *Method of Measuring Sound and Vibration of Refrigerant Compressors*.

ARI 575-94, *Method of Measuring Machinery Sound within an Equipment Space*.

ISO 4412-1:1991, *Hydraulic Fluid Power—Test Code for Determination of Airborne Noise Levels*. Part 1: *Pumps*. Part 2: *Motors*. Part 3: *Pumps—Method Using a Parallelepiped Microphone Array*.

ISO 4871:1996, *Declaration and Verification of Noise Emission Values of Machinery and Equipment*.

ISO 7574-1:1985, *Statistical Methods for Determining and Verifying Stated Noise Emission Values of Machinery and Equipment*. Part 1: *General Considerations and Definitions*. Part 2: *Methods for Stated Values for Individual Machines*. Part 3: *Simple (Transition) Method for Stated Values for Batches of Machines*. Part 4: *Methods for Stated Values for Batches of Machines*.

ISO 7779:1999, *Measurement of Airborne Noise Emitted by Information Technology and Telecommunications Equipment*.

ISO 9295:1988, *Measurement of High-Frequency Noise Emitted by Computer and Business Equipment*.

ISO 9296:1988, *Declared Noise Emission Values of Computer and Business Equipment*.

ISO 9611:1996, *Characterization of Sources of Structure-Borne Sound with Respect to Sound Radiation from Connected Structures—Measurement of Velocity at the Contact Points of Machinery When Resiliently Mounted*.

ISO 11200:1995, *Noise Emitted by Machinery and Equipment—Guidelines for the Use of Basic Standards for the Determination of Emission Sound Pressure Levels at a Work Station and at Other Specified Positions*.

ISO 11201:1995, *Noise Emitted by Machinery and Equipment—Measurement of Emission Sound Pressure Levels at a Work Station and at Other Specified Positions—Engineering Method in an Essentially Free Field over a Reflecting Plane*.

ISO 11202:1995, *Noise Emitted by Machinery and Equipment—Measurement of Emission Sound Pressure Levels at a Work Station and at Other Specified Positions—Survey Method* In Situ.

ISO 11203:1995, *Noise Emitted by Machinery and Equipment—Determination of Emission Sound Pressure Levels at a Work Station and at Other Specified Positions from the Sound Power Level.*

ISO 10846, *Acoustics and Vibration—Laboratory Measurement of Vibro-Acoustic Transfer Properties of Resilient Elements.* Part 1—1997: *Principles and Guidelines.* Part 2—1997: *Dynamic Stiffness of Elastic Supports for Translatory Motion—Direct Method.* Part 3—2002: *Indirect Method for Determination of the Dynamic Stiffness of Resilient Supports for Translatory Motion.*

ISO 11690-1, *Recommended Practice for the Design of Low-Noise Workplaces Containing Machinery.* Part 1: *Noise Control Strategies.* Part 2: *Noise Control Measures.* Part 3: *Sound Propagation and Noise Prediction in Workrooms.*

21.5 IMPACT OF NOISE IN WORKPLACE

Two qualities of workplace noise are of concern as they affect damage to hearing and speech communication. Environmental noise includes vibration and infrasound. Hearing conservation measurements are of two forms: measurement of hearing acuity and measurement of noise levels in work and recreational environments.

Hearing Testing (Audiometry) Equipment and Procedures

ANSI S3.1-1999, *Maximum Permissible Ambient Noise Levels for Audiometric Test Rooms.* Specifies maximum permissible ambient noise levels (MPANLs) allowed in an audiometric test room.

ANSI S3.6-1996, *Specification for Audiometers.* The audiometers covered in this specification are devices designed for use in determining the hearing threshold of an individual in comparison with a chosen standard reference threshold level.

ANSI S3.21-1978 (R1997), *Methods for Manual Pure-Tone Threshold Audiometry.* Outlines the procedure for pure-tone threshold audiometry used in the assessment of an individual's threshold of hearing for pure tones.

ISO 226:1987, *Normal Equal-Loudness Level Contours.* Data in this standard formed the basis for the dBA and dBC curves used in noise measurement.

ISO 389, 1964 Acoustics, *Standard Reference Zero for the Calibration of Pure-Tone Audiometers.* Part 1: *Reference Equivalent Sound Pressure Levels for Pure Tones and Supra-Aural Earphones.* Part 2: *Reference Equivalent Threshold Sound Pressure Levels for Pure Tones and Insert Earphones.* These data are also shown in ANSI S3.6-1996.

ISO 6189:1983, *Pure Tone Air Conduction Threshold Audiometry for Hearing Conservation Purposes.* Outlines the procedure for testing hearing thresholds as well as describes the conditions of the testing. Part of this standard is similar to information found in ANSI S3.21-1978 (R1997).

ISO 7029:2000, *Statistical Distribution of Hearing Thresholds as a Function of Age.* Gives expected threshold values for men and women without a history of noise exposure. Data are utilized in ISO 1999 and ANSI S3.44-1996.

ISO 8253-1:1989, *Audiometric Test Methods*—Part 1: *Basic Pure Tone Air and Bone Conduction Threshold Audiometry.* Part 2: *Sound Field Audiometry with Pure Tone and Narrow-Band Test Signals.* Part 3: *Speech Audiometry.*

ISO 11904-1:2002, *Determination of Sound Immission from Sound Sources Placed Close to the Ear.* Part 1: *Technique Using a Microphone in a Real Ear (MIRE Technique).* Some of the data in this standard are similar to that in ANSI S12. 42-1995 (R1999).

IEC 60645-1 Ed 2.0 b: 2001, *Electroacoustics—Audiological Equipment.* Part I: *Pure-Tone Audiometers.* Has material also found in ANSI S3.6-1996.

Hearing Conservation Programs

ANSI S3.44-1996 (R2001), *Determination of Occupational Noise Exposure and Estimation of Noise-Induced Hearing Impairment.* Adoption of ISO 1999:1990(E) of the same name; however, S3.44 allows assessment of noise exposure using a time–intensity trading relation other than a 3-dB increase per halving of exposure time.

ANSI S12.6-1997 (R2002), *Methods for Measuring the Real-Ear Attenuation of Hearing Protectors.* Specifies laboratory-based procedures for measuring, analyzing, and reporting the noise-reducing capabilities of conventional passive hearing protection devices.

ANSI Technical Report S12.13 TR-2002, *Evaluating the Effectiveness of Hearing Conservation Programs through Audiometric Data Base Analysis.* Describes methods for evaluating the effectiveness of hearing conservation programs in preventing occupational noise-induced hearing loss by using techniques for audiometric database analysis.

ANSI S12.42-1995 (R1999), *Microphone-in-Real-Ear and Acoustic Test Fixture Methods for the Measurement of Insertion Loss of Circumaural Hearing Protection Devices.* Describes the microphone-in-real-ear and the acoustical test fixture methods for the measurement of the insertion loss of circumaural earmuffs, helmets, and communications headsets.

ISO 1999:1999, *Acoustics—Determination of Occupational Noise Exposure and Estimation of Noise-Induced Hearing Impairment.* Similar to ANSI S3.44-1996 except that ISO 1999 is more restrictive.

ISO 4869-1:1990, *Hearing Protectors*—Part 1: *Subjective Method for the Measurement of Sound Attenuation.* Part 2: *Estimation of Effective A-Weighted Sound Pressure Levels When Hearing Protectors Are Worn.* Part 3: *Simplified Method for the Measurement of Insertion Loss of Ear-Muff Type Protectors for Quality Inspection Purposes.* Part 4: *Measurement of Effective Sound Pressure Levels for Level-Dependent Sound-Restoration Ear-Muffs.*

ISO 7029:2000, *Statistical Distribution of Hearing Thresholds as a Function of Age.*

ISO 9612:1997, *Guidelines for the Measurement and Assessment of Exposure to Noise in a Working Environment.*

Speech Communication Metrics

ANSI S3.2-1989 (R1999), *Method for Measuring the Intelligibility of Speech over Communication Systems*. A revision of ANSI S3.2-1960 (R1982). Provides three alternative sets of lists of English words to be spoken by trained talkers over the speech communication system to be evaluated.

ANSI S3.4-1980 (R1997), *Procedure for the Computation of Loudness of Noise*. Specifies a procedure for calculating the loudness of certain classes of noise.

ANSI S3.5-1997 (R2002), *Methods of the Calculation of the Speech Intelligibility Index (SII)*. Computation of the SII based on the intelligibility of speech as evaluated by speech perception tests given a group of talkers and listeners.

ASTM 1130-2002, *Objective Measurement of Speech Privacy in Open Offices Using Articulation Index (AI)*. Computation of the Articulation Index (AI) from standard speech levels, ambient noise, and articulation weighting factors.

ANSI S3.14-1977 (R1997), *Rating Noise with Respect to Speech Interference*. Defines a simple numerical method for rating the expected speech-interfering aspect (Speech Interference Level, SIL) of noise using acoustical measurements of the noise.

ISO/TR 3352, 1974, *Acoustics—Assessment of Noise with Respect to Its Effect on the Intelligibility of Speech*. This technical report has never been approved as a standard in part because of the concern over whether the impact of noise is similar for each language.

ISO/TR 4870:1991, *The Construction and Calibration of Speech Intelligibility Tests*.

Environmental Noise SPL Measurement Methods

Environmental noise SPL measurements are typically made outdoors to mirror most regulations. Sound measurement locations are often near houses or in empty fields and near transportation and industrial noise sources. In addition to meeting all the requirements of ANSI S1.4, this acoustical equipment and its application must be tolerant of wind, rain, birds, insects, and mechanical abuse. Primary noise sources include transportation (aircraft, road traffic, trains, etc.), industrial facilities, neighborhood noise (pets, heating, ventilation, and air conditioning), recreational noise (shooting ranges, races, and music venues).

ANSI S12.18-1994 (R1999), *Procedures for Outdoor Measurement of Sound Pressure Level*. Procedures for the measurement of SPLs outdoors, considering ground effects and refraction due to wind and temperature gradients and to turbulence. Measurement of SPLs produced by specific sources outdoors. Method 1: general method, outlines conditions for routine measurements. Method 2: precision method, describes strict conditions for more accurate measurements, providing short-term A-weighted SPL or time-averaged SPL, A-weighted or in octave- or in one-third-octave or narrow-band SPL, but does not preclude determination of other sound descriptors.

ASTM E1014-84(2000), *Standard Guide for Measurement of Outdoor A-Weighted Sound Levels*. Basic technique for performing a reliable A-weighted sound-level measurement outdoors.

ASTM E1503-97, *Standard Test Method for Conducting Outdoor Sound Measurements Using a Digital Statistical Analysis System*. Covers the measurement of outdoor sound levels at specific locations using a digital statistical analyzer and a formal measurement plan.

ISO 7196: 1995, *Frequency-Weighting Characteristic for Infrasound Measurements*.

ISO 10843:1997, *Methods for the Description and Physical Measurement of Single Impulses or Series of Impulses*.

Environmental Noise Measurement Applications

ANSI S12.7-1986 (R1998), *Methods for Measurement of Impulse Noise*. Measures impulse noise from discrete events, such as quarry and mining explosions or sonic booms, or from multiple-event sources such as pile drivers, riveting, or machine-gun firing. Data may be reported as time variation of the sound pressure, with or without frequency weighting, and sound exposure level.

ANSI S12.8-1998 (R2003), *Methods for Determining the Insertion Loss of Outdoor Noise Barriers*. Determine the insertion loss of outdoor noise barriers, including direct before and after measurements, indirect before measurements at an "equivalent" site, and indirect predictions of "before" sound levels. Indirect before measurements and indirect before prediction methods require direct measurements of "after" sound levels. May use sound sources naturally present at a site, controlled natural sound sources, or controlled artificial sound sources. Receiver location and atmospheric, ground, and terrain conditions may be chosen. Worksheets are provided

ANSI S12.9-1988 (R2003), *Quantities and Procedures for Description and Measurement of Environmental Sound*. Part 1: *Basic Quantities for Description of Sound in Community Environments and General Procedures for Measurement of These Quantities*. Part 2: *Measurement of Long-Term, Wide Area Sound*. Part 3: *Short-Term Measurements with an Observer Present*. Part 4: *Noise Assessment and Prediction of Long-Term Community Response*. Part 5: *Sound Descriptors for Determination of Compatible Land Use*. Part 6: *Methods for Estimation of Awakenings Associated with Aircraft Noise Events Heard in Homes*.

ANSI/ASME PTC 36-1985 (R1998), *Measurement of Industrial Sound*. Procedures for measuring and reporting airborne sound emissions from mechanical equipment.

ARI 260-2001, *Sound Rating of Ducted Air Moving and Conditioning Equipment*.

ARI 270-95, *Sound Rating of Outdoor Unitary Equipment*.

ARI 275-97, *Application of Sound Rating Levels of Outdoor Unitary Equipment*.

ARI 300-2000, *Sound Rating and Sound Transmission Loss of Packaged Terminal Equipment*.

ARI 350-2000, *Sound Rating of Non-Ducted Indoor Air-Conditioning Equipment*.

ARI 370-2001, *Sound Rating of Large Outdoor Refrigerating and Air-Conditioning Equipment.*

ISO 1996, *Description and Measurement of Environmental Noise.* Part 1-2003: *Basic Quantities and Assessment Procedures.* Part 2-1987: *Acquisition of Data Pertinent to Land Use.* Part 3-1987: *Application to Noise Limits.*

ISO 3891:1978, *Procedure for Describing Aircraft Noise Heard on the Ground.*

ISO 8297:1994, *Determination of Sound Power Levels of Multisource Industrial Plants for Evaluation of Sound Pressure Levels in the Environment—Engineering Method.*

ISO 10847:1997, *In-Situ Determination of Insertion Loss of Outdoor Noise Barriers of All Types.* See also ANSI S12.8.

ISO/TS 13474:2003, *Impulse Sound Propagation for Environmental Noise Assessment.*

SAE J1075 (revised June 2000), *Surface Vehicle Standard (R) Sound Measurement—Construction Site.* Procedures and instrumentation to be used for determining a representative sound level during a representative time period at selected measurement locations on a construction site boundary.

Environmental Sound Propagation Outdoors

ANSI S1.18-1999, *Template Method for Ground Impedance.* Procedures for obtaining the real and imaginary parts of the specific acoustical impedance of natural ground surface outdoors.

ANSI S1.26-1995 (R1999), *Method for Calculation of the Absorption of Sound by the Atmosphere.* Still-atmosphere absorption losses of sound for a wide range of meteorological conditions.

ANSI S12.17-1996 (R2001), *Impulse Sound Propagation for Environmental Noise Assessment Response.* Engineering methods to calculate the propagation and attenuation of high-energy impulsive sounds through the atmosphere. It estimates the mean C-weighted sound exposure level of impulsive sound at distances ranging from 1 to 30 km, applicable for explosive masses between 50 and 1000 kg.

ISO 9613, *Attenuation of Sound during Propagation Outdoors.* Part 1-1993: *Calculation of the Absorption of Sound by the Atmosphere. Attenuation of Sound During Propagation Outdoors.* Part 2-1996: *General Method of Calculation.* Propagation attenuation, including barrier attenuation calculations, excess attenuation by hard and soft ground surfaces, and attenuation of sound by rows of buildings and by trees and shrubs.

ISO/TS 13474:2003, *Impulse Sound Propagation for Environmental Noise Assessment.*

Environmental Vibrations

ANSI S3.18-2002, *NAIS Standard Mechanical Vibration and Shock—Evaluation of Human Exposure to Whole-Body Vibration.* Part 1: *General Requirements.* Part of ISO 2631 defining methods for the measurement of periodic, random, and transient whole-body vibration. It indicates the principal factors that combine to determine the degree to which a vibration exposure will be acceptable.

ANSI S3.29-1983 (R2001), *Guide to the Evaluation of Human Exposure to Vibration in Buildings*. Reactions of humans to vibrations of 1–80 Hz inside buildings are assessed as degrees of perception and associated vibration levels and duration.

ANSI S3.34-1986 (R1997), *Guide for the Measurement and Evaluation of Human Exposure to Vibration Transmitted to the Hand*. The recommended method for the measurement, data analysis, and reporting of human exposure to hand-transmitted vibration.

ANSI S3.40-2002, *NAIS Standard Mechanical Vibration and Shock—Hand-Arm Vibration—Method for the Measurement and Evaluation of the Vibration Transmissibility of Gloves at the Palm of the Hand*. Adoption of ISO 10819:1996. A method for the laboratory measurement, data analysis, and reporting of the vibration transmissibility of gloves of vibrations from a handle to the palm of the hand in the frequency range 31.5–1250 Hz.

ISO 2631, *Mechanical Vibration and Shock—Evaluation of Human Exposure to Whole-Body Vibration*. Part 1-1997: *General Requirements*. Part 2-1989: *Continuous and Shock-Induced Vibrations in Buildings (1 to 80 Hz)*. Part 4-2001: *Guidelines for The evaluation of the Effects of Vibration and Rotational Motion on Passenger and Crew Comfort in Transport Systems*.

ISO 2671:1982, *Environmental Tests for Aircraft Equipment—Part 3.4: Acoustic Vibration*.

ISO 4866:1990, *Mechanical Vibration and Shock—Vibration of Buildings—Guidelines for the Measurement of Vibrations and Evaluation of Their Effects on Building*.

ISO 4867:1984, *Code for the Measurement and Reporting of Shipboard Vibration Data*.

ISO 5007:2003, *Agricultural Wheeled Tractors—Operator's Seat—Laboratory Measurement of Transmitted Vibration*.

ISO 5008:2002, *Agricultural Wheeled Tractors and Field Machinery—Measurement of Whole-Body Vibration of the Operator*.

ISO 5347-3:1993, *Calibration of Vibration and Shock Pick-ups*. Part 3: *Secondary Vibration Calibration*. Part 4: *Secondary Shock Calibration*. Part 5: *Calibration by Earth's Gravitation*. Part 6: *Primary Vibration Calibration at low Frequencies*. Part 7: *Primary Calibration by Centrifuge*. Part 8: *Primary Calibration by Dual Centrifuge*. Part 10: *Primary Calibration by High Impact Shocks*. Part 11: *Testing of Transverse Vibration Sensitivity*. Part 12: *Testing of Transverse Shock Sensitivity*. Part 13: *Testing of Base Strain Sensitivity*. Part 14: *Resonance Frequency Testing of Undamped Accelerometers on a Steel Block*. Part 15: *Testing of Acoustic Sensitivity*. Part 16: *Testing of Mounting Torque Sensitivity*. Part 17: *Testing of Fixed Temperature Sensitivity*. Part 18: *Testing of Transient Temperature Sensitivity*. Part 19: *Testing of Magnetic Field Sensitivity*. Part 22: *Accelerometer Resonance Testing—General Methods*.

ISO 5348:1998, *Mechanical Mounting of Accelerometers*.

ISO 5349-1:2001, *Mechanical Vibration—Guidelines for the Measurement and the Assessment of Human Exposure to Hand-Transmitted Vibration.* Parts of this standard are similar to ANSI S3.34-1986 (R1997). Part 2: *Practical Guidance for Measurement at the Workplace.*

ISO 5805:1997, *Mechanical Vibration and Shock—Human Exposure—Vocabulary.*

ISO 8041:1990, *Human Response to Vibration—Measuring Instrumentation.*

ISO 13091-1:2001, *Mechanical Vibration—Vibrotactile Perception Thresholds for the Assessment of Nerve Dysfunction.* Part 1: *Methods of Measurements at the Fingertips.*

ISO 8042:1988, *Characteristics to be Specified for Seismic Pick-Ups.*

ISO 9022-10:1998, *Optics and Optical Instruments—Environmental Test Methods.* Part 10: *Combined Sinusoidal Vibration and Dry Heat or Cold.* Part 15: *Combined Digitally Controlled Broad-Band Random Vibration and Dry Heat or Cold.*

21.6 VEHICLE EXTERIOR AND INTERIOR NOISE

There are many SAE and ISO standards dealing with the source of noise emissions of various motor vehicles, including road vehicle (automobiles, trucks, buses, and motorcycles), trains, boats, aircraft, construction equipment (e.g., dozer), agricultural equipment (e.g., tractor), and small-engine equipment (e.g., lawn edger). Full listings of these standards are maintained at the SAE and ISO websites. The following is a sampling of these standards. Quite a few are identically SAE and ISO standards.

Vehicle Noise — Interior Noise Measurement Techniques

ISO 2923:1996, *Measurement of Noise on Board Vessels.*

ISO 3095:1975, *Measurement of Noise Emitted by Railbound Vehicles.*

ISO 3381:1976, *Measurement of Noise Inside Railbound Vehicles.*

ISO 5128:1980, *Measurement of Noise Inside Motor Vehicles.*

ISO 5129:2001, *Measurement of Sound Pressure Levels in the Interior of Aircraft During Flight.*

ISO 5130:1982, *Measurement of Noise Emitted by Stationary Road Vehicles—Survey Method.*

ISO 5131:1996, *Tractors and Machinery for Agriculture and Forestry—Measurement of Noise at the Operator's Position—Survey method.*

ISO 7188:1994, *Measurement of Noise Emitted by Passenger Cars under Conditions Representative of Urban Driving.*

ISO 11819-1:1997, *Measurement of the Influence of Road Surfaces on Traffic Noise.* Part 1: *Statistical Pass-By Method.*

Road Surface Sound Absorption

See ISO 13472 below under Sound Absorption.

Vehicle Noise — Exterior Noise Measurement (Noise Emission) Techniques

SAE J366, *Exterior Sound Level for Heavy Trucks and Buses (APR 2001)*. Test procedure, environment, and instrumentation for determining the maximum exterior sound level for highway motor trucks, truck tractors, and buses.

ISO 3095:1975, *Measurement of Noise Emitted by Railbound Vehicles*. Obtaining reproducible and comparable measurement results of levels and spectra of noise emitted by all kinds of vehicles operating on rails or other types of fixed track except for track maintenance vehicles in operation.

SAE J34 (June 2001), *Exterior Sound Level Measurement Procedure for Pleasure Motorboats*. Procedure for measuring the maximum exterior sound level of pleasure motorboats while being operated under wide open-throttle conditions.

ISO 362:1998, *Measurement of Noise Emitted by Accelerating Road Vehicles—Engineering Method*. (Available in English only.)

SAE J16395 (ISO 6395) (February 2003), *Measurement of Exterior Noise Emitted by Earth-Moving Machinery—Dynamic Test Conditions*. Determining the noise emitted to the environment by earth-moving machinery in terms of the A-weighted sound power level while the machine is working under dynamic test conditions.

SAE J17216 (ISO 7216) (February 2003), *Agricultural and Forestry Wheeled Tractors and Self-Propelled Machines—Measurement of Noise Emitted When in Motion*. Measuring the A-weighted SPL in an extensive open space of the noise emitted by agricultural and forestry wheeled tractors and self-propelled machines fitted with elastic tires while the vehicle is in motion. It is not applicable to special forestry machinery, for example, forwarders, skidders, etc., as defined in ISO 6814.

ISO 4872:1978, *Measurement of Airborne Noise Emitted by Construction Equipment Intended for Outdoor Use*. Method for determining compliance with noise limits

ISO 6393:1998, *Measurement of Exterior Noise Emitted by Earth-Moving Machinery—Stationary Test Conditions*. (Available in English only.)

ISO 6394:1998, *Measurement at the Operator's Position of Noise Emitted by Earth-Moving Machinery—Stationary Test Conditions*. (Available in English only.)

ISO 6395:1988, *Measurement of Exterior Noise Emitted by Earth-Moving Machinery—Dynamic Test Conditions*.

ISO 6396:1992, *Measurement at the Operator's Position of Noise Emitted by Earth-Moving Machinery—Dynamic Test Conditions*.

ISO 6798:1995, *Reciprocating Internal Combustion Engines—Measurement of Emitted Airborne Noise—Engineering Method and Survey Method*.

ISO 7216:1992, *Agricultural and Forestry Wheeled Tractors and Self-Propelled Machines—Measurement of Noise Emitted When in Motion*.

ISO 7182:1984, *Measurement at the Operator's Position of Airborne Noise Emitted by Chain Saws*.

ISO 9645:1990, *Measurement of Noise Emitted by Two-Wheeled Mopeds in Motion—Engineering Method.*

ISO 11094:1991, *Test Code for the Measurement of Airborne Noise Emitted by Power Lawn Mowers, Lawn Tractors, Lawn and Garden Tractors, Professional Mowers, and Lawn and Garden Tractors with Mowing Attachments.*

21.7 ARCHITECTURAL NOISE CONTROL IN BUILDINGS

There are three categories of sound control standards for buildings. The first is for testing materials that limit sound transmission from one room to another. The second is to test the sound absorption capability of materials. The third tests the general capabilities of rooms to cope with sounds within it for favorable (offices, classrooms, and auditoria) and unfavorable or noisy situations.

Sound Transmission

ASTM E90-02, *Standard Test Method for Laboratory Measurement of Airborne Sound Transmission Loss of Building Partitions and Elements*. Laboratory measurement of airborne sound transmission loss of building partitions such as walls of all kinds, operable partitions, floor–ceiling assemblies, doors, windows, roofs, panels, and other space-dividing elements.

ASTM E336-97, *Standard Test Method for Measurement of Airborne Sound Insulation in Buildings*. Determining the sound insulation in the field between two rooms in a building. The evaluation may be made including all paths by which sound is transmitted or attention may be focused only on the dividing partition. The word "partition" in this test method includes all types of walls, floors, or any other boundaries separating two spaces. The boundaries may be permanent, operable, or movable.

ASTM E413-87 (1999), *Classification for Rating Sound Insulation*. The method of calculating, from laboratory or field sound insulation measurements at frequencies from 125 to 4000 Hz the single-number acoustical rating called sound transmission class (STC) or field sound transmission class (FSTC) and the single-number rating between two spaces called the noise isolation class (NIC).

ASTM E497, *Practice for Installing Sound-Isolating Lightweight Partitions*. Preferred design methods for constructing high-noise-insulation walls and floor/ceilings.

ASTM E557, *Practice for Architectural Application and Installation of Operable Partitions*. Preferred design and installation methods for constructing high-noise-insulation operable partitions in the field.

ASTM E596-96 (2002), *Standard Test Method for Laboratory Measurement of Noise Reduction of Sound-Isolating Enclosures*. The reverberation room measurement of the noise reduction of sound-isolating enclosures.

ASTM E597, *Practice for Determining a Single-Number Rating of Airborne Sound Isolation for Use in Multi-Unit Building Specifications*. A short test method for field use where a sound source of a standard spectrum is used to test the noise insulation between two existing rooms in a building. Broadband A-weighted sound levels are measured. Computation provides a single number sound insulation value analogous to FSTC.

ASTM E966-02, *Standard Guide for Field Measurements of Airborne Sound Insulation of Building Facades and Facade Elements*. Procedures for measuring the sound insulation of an installed building facade or facade element (window, door). These values may be used separately to predict interior levels or combined into a single number such as by classification E 413 (STC with precautions) or classification E 1332 (outside–inside transmission class, OITC) for sound insulation against transportation noises.

ASTM E492-90 (1996), *Standard Test Method for Laboratory Measurement of Impact Sound Transmission through Floor-Ceiling Assemblies Using the Tapping Machine*. Measurement of impact sound levels (e.g., footfalls) produced by the ISO tapping machine. Results are used to determine the impact isolation class (IIC) by E989.

ASTM E989-89 (1999), *Standard Classification for Determination of Impact Insulation Class (IIC)*. A single-number rating of data from ASTM E492 and E1007 for comparing floor–ceiling assemblies for general building design purposes. The rating is called an impact insulation class (IIC).

ASTM E1007-97, *Standard Test Method for Field Measurement of Tapping Machine Impact Sound Transmission through Floor-Ceiling Assemblies and Associated Support Structures*. A field measurement analogous to E492.

ASTM E1123-86 (1998), *Standard Practices for Mounting Test Specimens for Sound Transmission Loss Testing of Naval and Marine Ship Bulkhead Treatment Materials*. Describes test specimen mountings for test method E90, to be used for naval and marine ship bulkhead noise insulation measurement.

ASTM E1222-90 (2002), *Standard Test Method for Laboratory Measurement of the Insertion Loss of Pipe Lagging Systems*. Covers the measurement of the insertion loss of pipe lagging systems under laboratory conditions.

ASTM E1289-97, *Standard Specification for Reference Specimen for Sound Transmission Loss*. Construction and installation of a standard reference specimen for interlaboratory sound transmission loss measurement evaluation using test method E90.

ASTM E1332-90 (1998), *Standard Classification for Determination of Outdoor-Indoor Transmission Class*. A single-number rating for exterior doors, windows, and walls for their noise insulation against ground and air transportation noise, including aircraft roadway vehicles and trains.

ASTM E1408-91 (2000), *Standard Test Method for Laboratory Measurement of the Sound Transmission Loss of Door Panels and Door Systems*. Laboratory measurement of the sound transmission loss for door panels and door systems. It includes specimen-mounting instructions and the force required to close, latch, unlatch, and open a door.

ASTM E1414-00, *Standard Test Method for Airborne Sound Attenuation between Rooms Sharing a Common Ceiling Plenum*. Uses a special laboratory space to simulate a pair of adjacent offices or rooms separated by a partition and sharing a common plenum space (a common economical construction method used for walls between offices and classrooms). The only significant laboratory sound transmission path is by way of the ceiling structure and the plenum space.

ISO 140-1:1997, *Measurement of Sound Insulation in Buildings and of Building Elements*. Part 1: *Requirements for Laboratory Test Facilities with Suppressed Flanking Transmission*. Part 2: *Determination, Verification and Application of Precision Data*. Part 3: *Laboratory Measurements of Airborne Sound Insulation of Building Elements*. Part 4: *Field Measurements of Airborne Sound Insulation between Rooms*. Part 5: *Field Measurements of Airborne Sound Insulation of Façade Elements and Façades*. Part 6: *Laboratory Measurements of Impact Sound Insulation of Floors*. Part 7: *Field Measurements of Impact Sound Insulation of Floors*. Part 8: *Laboratory Measurements of the Reduction of Transmitted Impact Noise by Floor Coverings on a Heavyweight Standard Floor*. Part 9: *Laboratory Measurement of Room-to-Room Airborne Sound Insulation of a Suspended Ceiling with a Plenum Above It*. Part 10: *Laboratory Measurement of Airborne Sound Insulation of Small Building Elements*. Part 12: *Laboratory Measurement of Room-to-Room Airborne and Impact Sound Insulation of an Access Floor*. Part 13: *Guidelines*. (Available in English only.)

ISO 717-1:1996, *Rating of Sound Insulation in Buildings and of Building Elements*. Part 1: *Airborne Sound Insulation*. Part 2: *Impact Sound Insulation*.

ISO 9052-1:1989, *Determination of Dynamic Stiffness*. Part 1: *Materials Used under Floating Floors in dwellings*.

ISO 3822, *Laboratory Tests on Noise Emission from Appliances and Equipment Used in Water Supply Installations*. Part 1-1999: *Method of Measurement*. Part 2-1995: *Mounting and Operating Conditions for Draw-Off Taps and Mixing Valves*. Part 3-1997: *Mounting and Operating Conditions for In-Line Valves and Appliances*. Part 4-1997: *Mounting and Operating Conditions for Special Appliances*.

ISO 11546-1:1995, *Determination of Sound Insulation Performances of Enclosures*. Part 1: *Measurements under Laboratory Conditions (for Declaration Purposes)*. Part 2: *Measurements In Situ (for Acceptance and Verification Purposes)*.

ISO 11957:1996, *Determination of Sound Insulation Performance of Cabins—Laboratory and In Situ Measurements*.

ISO 15186-1:2000, *Measurement of Sound Insulation in Buildings and of Building Elements Using Sound Intensity*. Part 1: *Laboratory Measurements*.

Sound Absorption

ASTM C384-98, *Standard Test Method for Impedance and Absorption of Acoustical Materials by the Impedance Tube Method*. The use of an impedance tube (standing-wave apparatus) for the measurement of impedance ratios and the normal-incidence sound absorption coefficients of acoustical materials.

ASTM C423-02, *Standard Test Method for Sound Absorption and Sound Absorption Coefficients by the Reverberation Room Method*. Measures the random-incidence sound absorption coefficients of material test specimens, mounted according to ASTM E795, in a reverberation room by measuring sound decay rate.

ASTM E477, *Test Method for Measuring Acoustical and Airflow Performance of Duct Liner Materials and Prefabricated Silencers*. Laboratory measurement method for the sound insertion loss provided by duct sound attenuators.

ISO 7235:1991, *Measurement Procedures for Ducted Silencers—Insertion Loss, Flow Noise and Total Pressure Loss*.

ISO 11691:1995, *Measurement of Insertion Loss of Ducted Silencers without Flow—Laboratory Survey Method*.

ISO 11820:1996, *Measurements on Silencers In Situ*.

ISO 11821:1997, *Measurement of the In Situ Sound Attenuation of a Removable Screen*.

ASTM C522-87(1997), *Standard Test Method for Airflow Resistance of Acoustical Materials*. The measurement of airflow resistance, specific airflow resistance, and airflow resistivity of porous materials. Materials may be thick boards or blankets to thin mats, fabrics, papers, and screens.

ISO 9053:1991, *Materials for Acoustical Applications—Determination of Airflow Resistance*.

ASTM E756-98, *Standard Test Method for Measuring Vibration-Damping Properties of Materials*. Measures the vibration-damping properties—loss factor η, Young's modulus E, and shear modulus G—of materials over a frequency range of 50 Hz to 5 kHz and over the useful temperature range of the material. This method tests materials that have application in structural vibration, building acoustics, and the control of audible noise. Test materials include metal, enamel, ceramics, rubber, plastic, reinforced epoxy matrix, and wood that can be formed to the test specimen bar configuration.

ASTM E795-00, *Standard Practices for Mounting Test Specimens during Sound Absorption Tests*. Standard specimen mountings for tests performed in accordance with C423.

ASTM E1042, *Classification for Acoustically Absorptive Materials Applied by Trowel or Spray*.

ASTM E1050-98, *Standard Test Method for Impedance and Absorption of Acoustical Materials Using a Tube, Two Microphones, and a Digital Frequency Analysis System*. The two-microphone impedance tube method, including a digital frequency analysis system, to measure the normal-incidence sound absorption coefficients and normal specific acoustical impedance ratios of materials.

ISO 354. 2003, *Acoustics—Measurement of Sound Absorption in a Reverberation Room*. Similar to ASTM C423.

ISO 10534-1:1996, *Determination of Sound Absorption Coefficient and Impedance in Impedance Tubes*. Part 1: *Method Using Standing Wave Ratio*. Similar to ASTM C384.

ISO 11654:1997, *Sound Absorbers for Use in Buildings—Rating of Sound Absorption*.

ISO 10844:1994, *Specification of Test Tracks for the Purpose of Measuring Noise Emitted by Road Vehicles.*

ISO 13472-1:2002, *Measurement of Sound Absorption Properties of Road Surfaces In Situ.* Part 1: *Extended Surface Method.* A vertical impedance tube test method for measuring In Situ the sound absorption coefficient of road surfaces as a function of frequency in the range from 250 Hz to 4 kHz.

Architectural Acoustics, Reverberation, and Noise Control Design

ANSI S12.2-1995 (R1999), *Criteria for Evaluating Room Noise.* Defines NC, NCB, RC, and perceptible acoustically induced low-frequency vibration criterion curves for the octave-band SPL spectrum of noise and gives rules for using them to evaluate room background noise.

ANSI S12.60-2002, *Acoustical Performance Criteria, Design Requirements, and Guidelines for Schools.* Acoustical performance criteria, design requirements, and design guidelines for new school classrooms and other learning spaces to achieve a high degree of speech intelligibility in learning spaces. Conformance test procedures are provided.

ASTM E1041, *Guide for Measurement of Masking Sound in Open Offices.*

ASTM E1110-01, *Standard Classification for Determination of Articulation Class.* Provides a single figure rating that can be used for comparing building systems and subsystems for speech privacy purposes. The rating is designed to correlate with transmitted speech intelligence between office spaces.

ASTM E1111-02, *Standard Test Method for Measuring the Interzone Attenuation of Ceiling Systems.* Measures the sound reflective characteristics of ceiling systems with partial-height space dividers, used in offices and sometimes in schools to achieve speech privacy between work zones in the absence of full-height partitions. Restricted to a fixed space divider height of 5 ft, a ceiling height of nominally 9 ft, a source height of 4 ft, and microphone positions at 4 ft height.

ASTM E1130-02, *Standard Test Method for Objective Measurement of Speech Privacy in Open Offices Using Articulation Index (AI).* Measuring speech privacy objectively between existing locations in open offices. Uses acoustical measurements, published information on speech levels, and speech intelligibility to compute the Articulation Index (AI).

ASTM E1179-87 (1998), *Standard Specification for Sound Sources Used for Testing Open Office Components and Systems.* Specifies the sound source used for measuring the speech privacy between open offices or for measuring the laboratory performance of acoustical components (see E1111 and E1130).

ASTM E1375-90 (2002), *Standard Test Method for Measuring the Interzone Attenuation of Furniture Panels Used as Acoustical Barriers.* Measurement of the interzone attenuation of furniture panels used as acoustical barriers in open-plan spaces to provide speech privacy or sound isolation between working positions.

ASTM E1376-90 (2002), *Standard Test Method for Measuring the Interzone Attenuation of Sound Reflected by Wall Finishes and Furniture Panels.* Measures the degree to which reflected sound is attenuated by vertical surfaces in open-plan spaces.

ASTM E1573-02, *Standard Test Method for Evaluating Masking Sound in Open Offices Using A-Weighted and One-Third Octave Band Sound Pressure Levels*. Procedures to evaluate the spatial and temporal uniformity of masking sound in open offices using A-weighted or one-third-octave-band SPLs.

ASTM E1574-98, *Standard Test Method for Measurement of Sound in Residential Spaces*. Practical measurement of residual building interior noise SPLs.

ASTM E2235, *Standard Test Method for the Measurement of Decay Rates for Use in Sound Insulation Test Methods*. Measurement of sound decay rate ($60/T$) in any room, where T is the reverberation time, seconds.

ISO 3382:1997, *Measurement of the Reverberation Time of Rooms with Reference to Other Acoustical Parameters*. Measurement of reverberation time T in performance spaces. Specifies additional measures of auditorium gain (G), early decay (EDT), clarity (C80), Deutlichkeit (D50), central time (CT) lateral energy fraction (LF) and interaural cross correlation (IACC) and of T in any room.

ISO 10053-1991, *Measurement of Office Screen Sound Attenuation under Specific Laboratory Conditions*. Similar to ASTM E1375.

APPENDIX A
General References

M. Belanger, *Digital Processing of Signals: Theory and Practice*, Wiley, New York (1985).

J. S. Bendat and A. G. Piersol, *Random Data, Analysis and Measurement Procedures*, 2nd ed., Wiley, New York (1986).

L. L. Beranek, *Acoustics*, reprinted with changes by the Acoustical Society of America, New York (1986).

L. L. Beranek, *Acoustical Measurements*, rev. ed., Acoustical Society of America, New York (1988).

D. A. Bies and C. H. Hansen, *Engineering Noise Control*, Unwin Hyman, London (1988).

M. P. Blake and W. S. Mitchell, *Vibration and Acoustic Measurement Handbook*, Spartan Books, New York (1972).

R. N. Bracewell, *The Fast Fourier Transform and Its Applications*, 2nd ed., McGraw-Hill, New York (1986).

L. Cremer and H. A. Mueller, *Principals and Applications of Room Acoustics*, Vols. 1 and 2, translated by T. J. Schultz, Applied Science, London and New York (1982).

M. J. Crocker, *Noise Control*, Van Nostrand Reinhold, New York (1984).

J. D. Erwin and E. R. Graf, *Industrial Noise and Vibration Control*, Prentice-Hall, Englewood Cliffs, NJ (1979).

F. Fahy, *Sound and Structural Vibration—Radiation, Transmission and Response*, Academic, New York (1985).

C. M. Harris, Ed., *Shock and Vibration Handbook*, 3rd ed., McGraw-Hill, New York (1988).

M. C. Junger and D. Feit, *Sound, Structures, and Their Interaction*, 2nd ed., MIT Press, Cambridge, MA (1986).

L. E. Kinsler, A. R. Frey, A. B. Coppens, and J. V. Sanders, *Fundamentals of Acoustics*, 3rd ed., Wiley, New York (1982).

H. Kuttruff, *Room Acoustics*, Applied Science Publishers, London (1973).

R. L. Lyon, *Statistical Energy Analysis of Dynamical Systems*, MIT Press, Cambridge, MA (1975).

A. V. Oppenheim and R. W. Schafer, *Digital Signal Processing*, Prentice-Hall, Englewood Cliffs, NJ (1975).

R. B. Randall, *Frequency Analysis*, 3rd ed., Bruel & Kjaer, Naerum, Denmark (1987).

APPENDIX B

American System of Units

The American system of units [which used to be referred to as the English system of units until England converted to the meter-kilogram-second (mks) system] is inherently confusing. In everyday American life, the *pound* (abbreviated lb) is used either as a force or as a weight, both having the units of force. To further complicate matters, some technical writers seek to provide a system parallel to the metric system by using either of two quantities for force and either of two corresponding quantities for mass. Thus, some adopt the pound as the unit of force and define a *slug* as the unit of mass. Others define a *poundal* as the unit of force and adopt the pound as the unit of mass. Neither the slug nor the poundal has found general acceptance in the literature, although we use the former in this text.

As one well-traveled acoustician said, "I have determined that 1 kg of butter bought in Zurich is exactly the same amount as 2.2 lb bought in New York. Whenever I wish to solve a technical problem in America without confusion, I immediately divide the number of pounds by 2.2 to obtain the equivalent number of kilograms. Then I work in the mks system, where force and mass are clearly distinguished." Let us take a moment to distinguish further between mass and force.

The mass of a body is defined as $m = F/\ddot{x}$, where F is the vector sum of all forces acting on the center of gravity of the unrestrained body and $\ddot{x}$ is the acceleration produced in the direction of the force F.

The weight of a body is defined as $w = mg$, where g is the acceleration of gravity (9.81 m/s^2 on earth) and w is the force that must act on the otherwise unrestrained body to keep it stationary when exposed to the gravitational field. Consequently, a body has the same mass but different weight on the moon than on earth. On earth the weight of a 1-kg mass, which is designated as one kilopond (1 kp), is 1 kp weight = 1 kg mass × 9.81 m/s^2 = 9.81 newtons.

CONSISTENT SYSTEMS OF UNITS USED IN THIS TEXT

Two consistent systems of units are used in this text, the *mks* and the *fss* systems. To describe them, let us start with Newton's second law:

$$\text{Force} = \text{mass} \times \text{acceleration} \tag{B.1}$$

In the *meter-kilogram-second* system

$$\text{No. of newtons} = \text{no. of kg} \times \text{no. of m/s}^2 \qquad (B.2)$$

In the *foot-slug-second* system

$$\text{No. of pounds force (lb)} = \text{no. of slugs} \times \text{no. of ft/s}^2 \qquad (B.3)$$

The relations among the magnitudes of the units are

$$1\text{kp} = 2.205\text{lb weight}$$
$$1\text{ kg} = 0.0685\text{slug}$$
$$1\text{slug} = 14.59\text{ kg}$$
$$1\text{ newton} = 0.225\text{lb force}$$
$$1\text{lb force} = 4.448\text{ newtons}$$
$$1\text{slug} = 32.17\text{lb weight}$$
$$1\text{lb weight} = 0.03108\text{slug} = 0.454\text{ kg}$$

Example. If 1 kg mass is to be accelerated 1 m/s², we see, by Eq. (B.2), that a force of 1 newton is required. How many pounds (force) is required for the same result?

Solution 1 kg equals (2.205/32.17) slug and 1 m/s² = 3.28 ft/s². Thus, by Eq. (B.3), 0.225 lb (force) is required.

INCONSISTENT SYSTEMS OF AMERICAN UNITS USED IN THE LITERATURE

Two inconsistent systems of American units are commonly encountered, the *fps* and the *ips* systems.

In the *foot-pound-second* system (inconsistent system)

$$\text{No. of pounds force (lb)} = \frac{\text{no. of pounds weight (lb)}}{g} \times \text{no. of ft/s}^2 \qquad (B.4)$$

where g is the acceleration due to gravity in units of ft/s², that is, 32.17 ft/s².

In the *inch-pound-second* system (inconsistent system)

$$\text{No. of pounds force (lb)} = \frac{\text{no. of pounds weight (lb)}}{g} \times \text{no. of in./s}^2 \qquad (B.5)$$

where g is the acceleration due to gravity in units of in./s^2, that is, 3.86 in./s^2. Mechanical engineers often use the *in.-lb-s* system in the field of shock and vibration.

Example. A 1-kg mass is accelerated 5 m/s^2. Find the force necessary to do this in newtons and pounds (force).

Solution

$$1 \text{ kg} = 2.2 \text{ lb weight} = 0.0685 \text{ slug}$$
$$5 \text{ m/s}^2 = 16.4 \text{ ft/s}^2$$
$$F \text{ (newtons)} = 1 \times 5 = 5 \text{ newtons}$$
$$F \text{ (lb)} = 0.0685 \times 16.4 = 1.124 \text{ lb (force)}$$

APPENDIX C

Conversion Factors

The following values for the fundamental constants were used in the preparation of the factors:

$$1 \text{ m} = 39.37 \text{ in.} = 3.281 \text{ ft}$$
$$1 \text{ lb (weight)} = 0.4536 \text{ kp} = 0.03108 \text{ slug}$$
$$1 \text{ slug} = 14.594 \text{ kg}$$
$$1 \text{ lb (force)} = 4.448 \text{ newtons}$$
$$\text{Acceleration due to gravity} = 9.807 \text{ m/s}^2$$
$$= 32.174 \text{ ft/s}^2$$
$$\text{Density of H}_2\text{O at 4°C} = 10^3 \text{ kg/m}^3$$
$$\text{Density of Hg at 0°C} = 1.3595 \times 10^4 \text{ kg/m}^3$$
$$1 \text{ U.S. lb} = 1 \text{ British lb}$$
$$1 \text{ U.S. gallon} = 0.83267 \text{ British gallon}$$
$$°F = (\tfrac{9}{5})°C + 32$$
$$°C = (\tfrac{5}{9})(°F - 32)$$

TABLE C.1 Conversion Factors

To convert	Into	Multiply by	Conversely, multiply by
acres	ft^2	4.356×10^4	2.296×10^{-5}
	miles^2 (statute)	1.562×10^{-3}	640
	m^2	4,047	2.471×10^{-4}
	hectare (10^4 m^2)	0.4047	2.471
atm	in. H$_2$O at 4°C	406.80	2.458×10^{-3}
	in. Hg at 0°C	29.92	3.342×10^{-2}
	ft H$_2$O at 4°C	33.90	2.950×10^{-2}
	mm Hg at 0°C	760	1.316×10^{-3}
	lb/in.^{2}	14.70	6.805×10^{-2}

(continued overleaf)

TABLE C.1 (*continued*)

To convert	Into	Multiply by	Conversely, multiply by
	newtons/m^2	1.0132×10^5	9.872×10^{-6}
	kp/m^2	1.033×10^4	9.681×10^{-5}
°C	°F	(°C $\times \frac{9}{5}$) + 32	(°F − 32) $\times \frac{5}{9}$
cm	in.	0.3937	2.540
	ft	3.281×10^{-2}	30.48
	m	10^{-2}	10^2
circular mils	in.^{2}	7.85×10^{-7}	1.274×10^6
	cm^2	5.067×10^{-6}	1.974×10^5
cm^2	in.^{2}	0.1550	6.452
	ft^2	1.0764×10^{-3}	929
	m^2	10^{-4}	10^4
cm^3	in.^{3}	0.06102	16.387
	ft^3	3.531×10^{-5}	2.832×10^4
	m^3	10^{-6}	10^6
deg (angle)	radians	1.745×10^{-2}	57.30
dynes	lb (force)	2.248×10^{-6}	4.448×10^5
	newtons	10^{-5}	10^5
dynes/cm^2	lb/ft^2 (force)	2.090×10^{-3}	478.5
	newtons/m^2	10^{-1}	10
1 in. H$_2$O	N/m^2	249.18	4.013×10^{-3}
ergs	ft-lb (force)	7.376×10^{-8}	1.356×10^7
	joules	10^{-7}	10^7
ergs/cm^3	ft-lb/ft^3	2.089×10^{-3}	478.7
ergs/s	watts	10^{-7}	10^7
	ft-lb/s	7.376×10^{-8}	1.356×10^7
ergs/s-cm^2	ft-lb/s-ft^2	6.847×10^{-5}	1.4605×10^4
fathoms	ft	6	0.16667
ft	in.	12	0.08333
	cm	30.48	3.281×10^{-2}
	m	0.3048	3.281
ft^2	in.^{2}	144	6.945×10^{-3}
	cm^2	9.290×10^2	0.010764
	m^2	9.290×10^{-2}	10.764
ft^3	in.^{3}	1728	5.787×10^{-4}
	cm^3	2.832×10^4	3.531×10^{-5}
	m^3	2.832×10^{-2}	35.31
	liters	28.32	3.531×10^{-2}
ft H$_2$O at 4°C	in. Hg at 0°C	0.8826	1.133
	lb/in.^{2}	0.4335	2.307
	lb/ft^2	62.43	1.602×10^{-2}
	newtons/m^2	2989	3.345×10^{-4}
gal (liquid U.S.)	gal (liquid Brit. Imp.)	0.8327	1.2010
	liters	3.785	0.2642
	m^3	3.785×10^{-3}	264.2
gm	oz (weight)	3.527×10^{-2}	28.35

TABLE C.1 (*continued*)

To convert	Into	Multiply by	Conversely, multiply by
	lb (weight)	2.205×10^{-3}	453.6
hp (550 ft-lb/s)	ft-lb/min	3.3×10^4	3.030×10^{-5}
	watts	745.7	1.341×10^{-3}
	kW	0.7457	1.341
in.	ft	0.0833	12
	cm	2.540	0.3937
	m	0.0254	39.37
in.^{2}	ft^2	0.006945	144
	cm^2	6.452	0.1550
	m^2	6.452×10^{-4}	1550
in.^{3}	ft^3	5.787×10^{-4}	1.728×10^3
	cm^3	16.387	6.102×10^{-2}
	m^3	1.639×10^{-5}	6.102×10^4
kg	lb (weight)	2.2046	0.4536
	slug	0.06852	14.594
	g	10^3	10^{-3}
kg/m^2	lb/in.^{2} (weight)	0.001422	703.0
	lb/ft^2 (weight)	0.2048	4.882
	g/cm^2	10^{-1}	10
kg/m^3	lb/in.^{3} (weight)	3.613×10^{-5}	2.768×10^4
	lb/ft^3 (weight)	6.243×10^{-2}	16.02
liters	in.^{3}	61.03	1.639×10^{-2}
	ft^3	0.03532	28.32
	pints (liquid U.S.)	2.1134	0.47318
	quarts (liquid U.S.)	1.0567	0.94636
	gal (liquid U.S.)	0.2642	3.785
	cm^3	1000	0.001
	m^3	0.001	1000
$\log_e n$, or $\ln n$	$\log_{10} n$	0.4343	2.303
m	in.	39.371	0.02540
	ft	3.2808	0.30481
	yd	1.0936	0.9144
	cm	10^2	10^{-2}
m^2	in.^{2}	1550	6.452×10^{-4}
	ft^2	10.764	9.290×10^{-2}
	yd^2	1.196	0.8362
	cm^2	10^4	10^{-4}
m^3	in.^{3}	6.102×10^4	1.639×10^{-5}
	ft^3	35.31	2.832×10^{-2}
	yd^3	1.3080	0.7646
	cm^3	10^6	10^{-6}
microbars (dynes/cm^2)	lb/in.^{2}	1.4513×10^{-5}	6.890×10^4
lb/ft^2		2.090×10^{-3}	478.5

(*continued overleaf*)

TABLE C.1 (*continued*)

To convert	Into	Multiply by	Conversely, multiply by
	newtons/m^2	10^{-1}	10
miles (nautical)	ft	6080	1.645×10^{-4}
	km	1.852	0.5400
miles (statute)	ft	5280	1.894×01^{-4}
	km	1.6093	0.6214
miles^2 (statute)	ft^2	2.788×10^7	3.587×10^{-8}
	km^2	2.590	0.3861
	acres	640	1.5625×10^{-3}
mph	ft/min	88	1.136×10^{-2}
	km/min	2.682×10^{-2}	37.28
	km/h	1.6093	0.6214
nepers	db	8.686	0.1151
newtons	lb (force)	0.2248	4.448
	dynes	10^5	10^{-5}
newtons/m^2	lb/in.^{2} (force)	1.4513×10^{-4}	6.890×10^3
	lb/ft^2 (force)	2.090×10^{-2}	47.85
	dynes/cm^2	10	10^{-1}
lb (force)	newtons	4.448	0.2248
lb (weight)	slugs	0.03108	32.17
	kg	0.4536	2.2046
lb H$_2$O (distilled)	ft^3	1.602×10^{-2}	62.43
	gal (liquid U.S.)	0.1198	8.346
lb/in.^{2} (weight)	lb/ft^2 (weight)	144	6.945×10^{-3}
	kg/m^2	703	1.422×10^{-3}
lb/in.^{2} (force)	lb/ft^2 (force)	144	6.945×10^{-3}
	N/m^2	6894	1.4506×10^{-4}
lb/ft^2 (weight)	lb/in.^{2} (weight)	6.945×10^{-3}	144
	gm/cm^2	0.4882	2.0482
	kg/m^2	4.882	0.2048
lb/ft^2 (force)	lb/in.^{2} (force)	6.945×10^{-3}	144
	N/m^2	47.85	2.090×10^{-2}
lb/ft^3 (weight)	lb/in.^{3} (weight)	5.787×10^{-4}	1728
	kg/m^3	16.02	6.243×10^{-2}
poundals	lb (force)	3.108×10^{-2}	32.17
	dynes	1.383×10^4	7.233×10^{-5}
	newtons	0.1382	7.232
slugs	lb (weight)	32.17	3.108×10^{-2}
	kg	14.594	0.06852
slugs/ft^2	kg/m^2	157.2	6.361×10^{-3}
tons, short (2,000 lb)	tonnes (1000 kg)	0.9075	1.102
watts	ergs/s	10^7	10^{-7}
	hp (550 ft-lb/s)	1.341×10^{-3}	745.7

INDEX

A

Absorption of sound, *see* Sound absorption
Acceleration level, 20
ACGIH, *see* American Conference of Governmental Industrial Hygienists
Acoustical efficiency (throttling valves), 647–649
Acoustical enclosures, 521–552. *See also* Small enclosures, sound in
  close-fitting, sealed enclosures, 531–537
  for cooling towers, 667
  for feed pumps, 672
  for industrial gas turbines, 673
  intermediate-size enclosures, 537–538
  large, with interior sound-absorbing treatment, 540–551
    effect of sound-absorbing treatment, 547, 548
    flanking transmission through floor, 549, 550
    inside-outside vs. outside-inside transmission, 549–550
    key parameters influencing insertion loss, 544–545
    leaks, 547, 548
    machine position, 549
    machine vibration pattern, 549
    model for insertion loss at high frequencies, 541–544
    wall panel parameters, 545–547
  large, without internal sound-absorbing treatment, 538–540
  partial, 551–552
  performance of:
    insertion loss as measure for, 522–523
    measures for, 518–519
    qualitative description, 523–525
  size of, 521
  small, sealed enclosures, 525–531
  for transformers, 679
Acoustical impedance, 39–41
  characteristic 40
  complex 39–40
  specific 40
Acoustically induced vibrations and rattles, 905, 907
Acoustical levels, 12–24
Acoustical modal response, 149–151
Acoustical privacy, 203–204. *See also* Speech privacy
Acoustical standards, *see* Standards
Acoustical terminology, 1–41, 911
Acoustical wave equation, 25–34, 149
Active machinery isolation, 840–845
Active noise and vibration control (ANVC), 721–849
  actuators:
    locations of, 806–808
    number of, 804–806
    placement and selection of, 722–730
    sizing of, 808–809
  controller architecture and performance simulations, 814–817
  control sensors/architectures, 730–732
  design considerations, 800–801
  digital filters, 740–751
    adaptive design for, 749–751
    description of, 741–747
    optimal design for, 747–749
  feedback control systems, 779–800
    alternate suboptimum control filter estimation, 784–800
    basic architecture, 779–781
    optimal control filter estimation, 781–784
  feedforward control systems, 751–779
    adaptive control, 760–765
    basic architecture, 751–754
    control of aliasing effects, 765–769
    extension to MIMO systems, 777–779
    optimal control filter estimation, 754–760
    system identification, 769–777
  identifying performance goals for, 801–804
  implementation and testing of, 827–829
  performance expectations, 732–733
  placement and selection of control sources/actuators, 722–730
  prototype ANVC systems, 733–740
    active machinery isolation, 840–845
    airborne noise in high-speed patrol craft, 845–849
    MIMO feedforward active locomotive exhaust with passive component, 829–840

943

944 INDEX

Active noise and vibration control (ANVC) (*continued*)
  sensors:
    control sensors/architectures, 730–732
    number/location of, 809–814
    performance of, 813–814
    reference, 809, 812–813
    residual, 809–812
Actuators (ANVC):
  locations of, 806–808
  number of, 804–806
  placement and selection of, 722–730
  sizing of, 808–809
Actuator channels, 806
Adaptive control algorithm, 760–765
A/D converters, *see* Analog-to-digital converters
Added mass, 348, 355, 383–384
Added mass coefficient, 348
Added mass/fluid density, 348
Added volume, 348, 355
Admittance, 41
Aeroacoustical sources, 611–616
  aerodynamic dipoles, 612–613
  aerodynamic monopoles, 611–612
  aerodynamic quadrupoles, 613–614
  of fractional orders, 614
  influence of source motion, 615–616
Aerodynamic dipoles, 614
Aerodynamic monopoles, 614
Aerodynamic sound, 611, 643–656. *See also* Gas flow noise
AFOSHSTD (Air Force Occupational, Safety, and Health Standard), 881
AI, *see* Articulation index
Air bag deployment noise, 869
Airborne excitation, simultaneous dynamic excitation and, 462–465
Airborne sound (noise):
  outdoors, 121
Airborne whole-body vibration criteria, 881
Air compressors, predicting noise from, 660–662
Air-Conditioning and Refrigeration Institute (ARI), 911
Air-cooled condensers, predicting noise from, 667–668
Aircraft interior noise control, 736
Airflow velocity (HVAC systems), 701–703
  fans, 707
  near grilles, 701–702
  terminal boxes/valves, 708–709
Airfoils:
  "singing" of, 640
  sound generation by, 639–640
Air Force Occupational, Safety, and Health Standard (AFOSHSTD), 881
Air mounts, 739
Air Movement and Control Association (AMCA), 670

Air pressure ratios, 648
Air spring isolators, 577
Aliasing, 817
  active control of, 765–769
  in spectral analysis, 54
All zero filters, 743
AMCA (Air Movement and Control Association), 670
American Conference of Governmental Industrial Hygienists (ACGIH), 871, 881, 882
American National Standards Institute (ANSI), 878, 911
American Society of Heating, Refrigerating, and Air-Conditioning Engineers (ASHRAE), 706, 711, 911
American Society of Mechanical Engineers (ASME), 911
American Society of Testing and Materials (ASTM), 911
American system of units, 935–937
Amplification at resonance, 583
Analog-to-digital (A/D) converters, 741–742, 766, 767
  for ANVC systems, 819–825
  sampling process of, 817
Anechoic spaces/rooms, 82, 97–98, 211
ANSI, *see* American National Standards Institute
Antialiasing filters, 54, 57, 765–769
  and ANVC performance, 815
  selection of, 817–819
ANVC, *see* Active noise and vibration control
Applications Handbook (ASHRAE), 711
Architectural acoustics, standards for, 930–931
Architectural noise control in buildings, standards for, 926–931
Area sources, 122
ARI (Air-Conditioning and Refrigeration Institute), 911
ARMA filters, 744
Armed forces sound exposure criteria, 863, 867–868, 874
Articulation index (AI), 203–204
ASHRAE, *see* American Society of Heating, Refrigerating, and Air-Conditioning Engineers
ASHRAE handbooks, 706, 711
ASME (American Society of Mechanical Engineers), 911
ASTM (American Society of Testing and Materials), 911
Asymptotic threshold shift (ATS), 865, 866
Atmospheric absorption, 136
Atmospheric pressure, 1
Attenuation:
  active noise control, 721–849
  ducts, lined, 311–313

INDEX

of sound:
   by atmosphere, 136
   by barriers, 132–135, 137–139
   by ground cover and trees, 137
   total, 137
Audio frequency region, 859–867, 873
Auralization, 190
Autocorrelation function, 60
Automotive interior noise control system, 735–736
Autoregressive moving average filters, 744
Autospectral density functions, 55–59
Averages, 8, 19:
   linear, 48
   running, 49–50
   synchronous averaging, 51–52
   unweighted (linear), 48, 49
   weighted, 48–49
Average A-weighted sound level, 17
Average diffuse-field surface absorption coefficient, 184–185
Average sound level, 16
A-weighted sound level, 17, 887–891
A-weighted sound (noise) exposure level, 18, 858, 859
   coal car shakers, 664
   diesel-engine-powered equipment, 668
   transformers, 677–678
A-weighted sound power level, 73–75
   air compressors, 661
   in diffuse field, 81
   estimates of, 659
   feed pumps, 673
   steam turbines, 676
   wind turbines, 680
A-weighted sound pressure level, 15–16
   long-term, 139
   for noise exposure, 858, 859
   for outdoor sound, 121
   overall, 142–143

B

Background noise:
   in classrooms, 196–197
   for speech intelligibility, 192–193
Baffled pistons, sound power output of, 35, 36
Balanced noise criterion (NCB) curves, 894, 895
Bandwidths:
   of continuous-spectrum sound, 7–9
   conversion of, 10–11
   half-power, 165
   with tapering, 57
Barriers, 132
   cooling towers, 667
   effectiveness of, 345
   for open-plan offices, 202–203
   for outdoor sound, 120, 132–135
      close to trees, with gaps and slots, 138–139
      interaction of ground and, 137–138
      and uncertainly of attenuation, 141–142
   with transformers, 679
Beams:
   damping due to reinforcements, 591–592
   infinite:
      effective length connecting force and moment impedance, 502
      transmission through, 395–397
   power transmission to plate from, 410–413
   viscoelastic damping of:
      three-component beams, 602–606
      two-component beams, 599–600
Beam-tracing techniques, 189–190
BEM (boundary element model), 121
Bending waves:
   complete power transmission for, 412–413
   reflection loss of, 407–410
      in beams, 411–412
      and change in cross-sectional area, 407
      free bending waves at L-junctions, 408
      in plates with vibration break, 415–416
      through cross junctions and T-junctions, 409–410
   through infinite plates, 417–418
Bias error, 59
   in coherent output power calculations, 65
   in gain factor estimates, 62
   in intensity measurement, 105, 106
Blocked pressure, 626, 629
Block's formula for shallow cavities, 641–642
Boilers, predicting noise from, 662–663
Boundaries, damping due to, 591
Boundary conditions, 150, 151
   forced sound pressure response, 159
Boundary element method, 156
Boundary element model (BEM), 121
Breakin noise, HVAC, 703–705
Breakin sound transmission loss, 450–451
Breakout noise, HVAC, 703–705
Breakout sound transmission loss, 448–450
Broadband disturbance control, 756–760
Broadband regulation filter, 790–795, 797
Bucket ladder unloaders, 664
Butterworth filters, 817–818

C

Cabins:
   acoustical performance measures of, 520
   defined, 517
Cauer filters, 817–818
Cavities, flows past, 640–643
Ceilings, for mechanical rooms, 710
Central-processing units (CPUs), 825, 827
CHABA, *see* National Academy of Sciences–National Research Council, Committee on Hearing, Bioacoustics and Biomechanics
Characteristic resistance, 40

Choked jets, 621
Circular effects (with FFT), 50
Clamshell bucket unloaders, 664
Classrooms, 191–197
 controlling sound in, 196–197
 predicting acoustical quality of, 193–196
 speech intelligibility in, 192–193
Closed-loop system identification, 771, 772, 774–776, 827, 828
Closed offices, sound pressure levels in, 190–191
Close-fitting enclosures:
 defined, 522
 sealed, 531–537
  free standing, 531–534
  machine-mounted, 535–537
 wrappings vs., 552
Coincidence frequency, 432–434, 436. See also Critical frequency for orthotropic plates, 437
Coal car shakers, 663–664
Coal crushers, 665
Coal-handling equipment:
 diesel-powered, 668
 predicting noise from, 663–665
Coal mills and pulverizers, 665
Coal transfer towers, 665
Coefficient of variation, 50
Coherence functions (coherency squared), 58–59
Coherent output power relationship, 64–65
Coil spring isolators, 576
Combination mufflers, 335–338
Combustion noise (gas turbine engines), 624–626
Community noise, 907
Comparison method (sound power output), 81, 88–89
Compensation filter (feedback control), 792–800
Compensator-regulator architecture (feedback control), 784–785
Complex admittance, 41
Complex impedance, 38–40, 588
Complex sound spectra, 12
Composite partitions, sound transmission through, 451
Composite structures (statistical energy analysis), 456
Composite transmission factor, 461, 462
Comprehensive models (noise prediction), 198
Compression waves, 6, 390–391
 and change in cross-sectional area, 407
 in plates with vibration break, 414–415
Compressive force, 347–348
Concentric-tube resonator (CTR), 304–311
Concurrent system identification, 771
Conference rooms, 191, 890, 891, 898
Construction equipment, diesel-powered, 668

Continuous spectra, 2–3, 7–11
Control authority (ANVC), 730
Conversion factors, 939–942
Cooley-Tukey algorithm, 53–54
Cooling towers, predicting noise from, 666–667
Core noise, 624–625, 679
Correlation functions, 60–61
Cosine-squared taper, 55, 57
Coulomb damping, 588
Coupling loss factor, 456
CPUs, *see* Central-processing units
Criteria for noise limits in buildings, 886–907
 survey method, L_A, 888–889
 engineering method, NC, 889, 892–894
 precision method, RNC, 899–905
Critical bands, 900
Critical frequency, 432–433, 436. See also Coincidence frequency
Critical damping coefficient, 559, 581
Critical damping ratio, 162
Cross-correlation function, 67–69
Cross junctions, reflection loss of bending waves through, 409–410
Cross-sectional area:
 of HVAC ducts, 687, 691
 power transmission owing to change in, 406–407
Cross-spectral density (CSD) matrix, 804–808, 810–814
Cross-spectral density functions, 57–59
CSD matrix, *see* Cross-spectral density matrix
CTR, *see* Concentric-tube resonator
Cylindrical array (microphones), 93, 94, 96
Cylindrical sound sources, 4

D

D/A converters, *see* Digital-to-analog converters
Damage risk criteria:
 for hearing, 858–874
  impulse noise, 867–869
  infrasound exposure, 869, 870
  noise exposure criteria for audio frequency region, 861–867
  protection of, 871–874
  ultrasound exposure, 869–871
 human vibration response, 874–883
  exposure guidelines, 877–883
  hand-transmitted vibration effects, 877
  whole-body vibration effects, 875–877
Damping, 174–176
 critical damping ratio, 162
 effect of, 561
 in HVAC ducts, 701, 703
 structural, 561, 579–607
  analytical models of, 588
  due to boundaries and reinforcements, 591–592

due to energy transport, 592–594
  effects of, 579–580
  energy dissipation and conversion, 589–590
  measurement of, 586–588
  measures of, 580–586
  models of, 588–589
  viscoelastic, 594–607
 in two-stage isolation systems, 573, 574
 viscous, 561
Damping capacity, 584
Damping ratio, 559, 561, 581
Dashpots, 395
 in mass-spring-dashpot system, 557–560
 point force impedance for, 394
Data analysis, 43–69
 analog, 52
 correlation functions, 60–61
 deterministic data, 44–45
 mean-square values, 3, 47–48
 mean values, 47
 for periodic excitation source identification, 63–64
 for propagation path identification, 66–69
 random data, 45–47
 for random excitation source identification, 64–66
 running averages, 49–50
 spectral functions, 52–59
  auto (power) spectral density functions, 55–58
  coherence functions, 58–59
  FFT algorithm, 52–54
  line and Fourier spectral functions, 54–55
 statistical sampling errors, 50–51, 59
 synchronous averaging, 51–52
 for system response properties identification, 62–63
 types of data signals, 43
 weighted averages, 48–49
Data signals, 43
 deterministic, 44–45
 random, 45–47
Day-night sound (noise) level, 17, 888, 907
dB, *see* Decibels
Decay, *see* Sound decay
Decay distance, 603
Decay rates, 582, 587. *See also* Reverberation time
Decibels (dB):
 fractions of, 24
 and reference quantities, 12
 sound power expression, 75
 sound pressure level expression, 15, 17
Deflection, 599
Deformation of solids, 492
Department of Defense (DOD), 907

Deterministic data signals, 44–45
 spectral computations for, 55
 statistical sampling errors with, 59
DI, *see* Directivity index
Diesel-engine-powered equipment, predicting noise from, 668–669
Diffuse-field theory, 184–187
Diffuse (reverberant) field, 76–77
 control of, 200–201
 driving freely hung panel, 457–459
 measurement in, 82
 sound power determination in, 81
 sound power in, 81, 85–91
Diffusers:
 HVAC, 701–703
 in reverberations rooms, 209–210
Digital data:
 mean-square value of, 47–48
 mean values of, 47
Digital filters (ANVC), 740–751
 adaptive design for, 749–751
 advantages of, 740–741
 description of, 741–747
 optimal design for, 747–749
Digital signal processor (DSP) chips, 740–741, 825–827
Digital signal processors, 825–827
Digital-to-analog (D/A) converters, 767, 770, 815–816
 for ANVC systems, 819–825
 sampling process of, 817
Dilatation resonance, 439
Dipoles:
 aerodynamic, 614, 633
 sound power output of, 34, 35
Direct-estimation algorithm, MIMO, 778
Direct field, control of, 200
Directional sources, 72, 73, 78–79
Directivity:
 defined, 72
 determination of, 113–117
Directivity factor, 113–116
Directivity index (DI), 115–117, 123
Directivity pattern, 113, 114
Direct method (sound power output), 81, 89–91
Displacement level, 582
Displacement ventilation systems, 703
Dissipation, 165–166
Dissipative silencers, 311–335
 economic considerations, 335
 effect of flow on silencer attenuation, 329–331
 factors in acoustical performance of, 282–283
 flow-generated noise, 331–333
 key performance parameters, 313–316
 lined ducts, 281–313
 parallel-baffle silencers, 316–325
 pod silencers, 328–329

948 INDEX

Dissipative silencers (*continued*)
   prediction of silencer pressure drop, 333–335
   round silencers, 325–328
Distributed sound-masking systems, 206
Doppler factor, 615
Dose-response curve (DRC), 868
Double-layer partitions, sound transmission through, 443–449
Double-tuned expansion chamber (DTEC), 297–300
Drag, 633
DRC (dose-response curve), 868
Driving point force impedance, 496–500
Dry cooling systems, *see* Air-cooled condensers
Dry friction damping, 588
DSP chips, *see* Digital signal processor chips
DTEC, *see* Double-tuned expansion chamber
Ducts:
   flow noise in, 635–638
   HVAC, noise transmission attentuation for, 685–699
      by cross-sectional area changes, 687, 691
      by divisions, 687
      by elbows, 691–693
      end-reflection loss, 697–698
      by plenums, 696–697
      prefabricated sound attenuators, 693–696
      room effect, 698–699
      in straight ducts, 686–691
   lined, 280–313
   for rooftop air conditioning units, 715–716
   sound power determination in, 112–113
   transmission loss of, 448–451
Dynamic absorbers, 593
Dynamic capability (of instruments), 105–107
Dynamic excitation:
   by point force, 393–394
   simultaneous airborne excitation and, 462–465

E

Early decay time (EDT), 183
ECMA International, 112
EDT (early decay time), 183
Effective value, 2, 31. *See also* Root-mean-square sound pressure
Eigenfrequency, 400
Elastic surface layer, improvement of impact noise isolation with, 479–483
Elastomeric isolators, 576–577
Elbows, HVAC ducts, 691–693
Elementary radiators, sound power output of, 35–36
Emission, 71
   HVAC outdoor noise, 717–718
   immission vs., 73
   machinery noise measurement standards, 916–918
   noise power emission level, 75n.
   strength descriptors for, 71
Empirical models of sound in rooms, 190
Enclosures, *see* Acoustical enclosures
End-reflection loss, HVAC, 697–698
Energy, dissipated, *see* Damping
Energy-average spectral deviation factor, 897
Energy speed, 391, 393
English system of units, 835
Entrance loss (dissipative mufflers), 313–314
Environmental corrections, 80, 98–101
Environmental Protection Agency (EPA), 863
Environmental sound (noise):
   measurement application standards, 921–922
   measurement method standards, 920–921
   outdoor propagation of, 922. *See also* Outdoor sound propagation
Environmental vibrations standards, 922–924
EPA (Environmental Protection Agency), 863
Equipment (machine) mounted enclosures, 522, 535–537
Equivalent continuous A-weighted noise level, 17
Equivalent sound absorption area, 99
Equivalent sound power level, estimating, 659–660
Equivalent viscous damping, 589n.
Excitation:
   of freely hung panel, 457–459
   by incident waves vs. other means, 427
   by point force, 393–394
   of solid structures:
      extension of reciprocity to, 473–476
      with sound field vs. point force, 462–465
   sources of:
      and gain factors, 62–63
      periodic, 63–64
      random, 64–66
Exit loss (dissipative mufflers), 314
Expansion chamber mufflers, 293–300
   double-tuned, 297–299
   extended-outlet, 294–297
   general design guidelines for, 299–300
   simple, 293–294
Extended-outlet muffler, 294–297
Eyring approach (diffuse-field theory), 185

F

FAA (Federal Aviation Administration), 907
Fans:
   air-cooled condensers, 667
   cooling towers, 667
   HVAC, 705–707
   industrial, 669–672
   industrial gas turbines, 674
   for mechanical equipment, 718
   transformers, 677–679

Far field, 5, 6, 76–77
  cooling towers, 667
  transformers, 678–679
Fast-field program (FFP), 121
Fast Fourier transform (FFT) algorithms, 50, 52–55
Federal Aviation Administration (FAA), 907
Federal Highway Administration (FHWA), 132
Feedback control systems, 779–800
  alternate suboptimum control filter estimation, 784–800
  architecture for, 731–732, 779–781
  optimal control filter estimation, 781–784
  residual sensor channels for, 809
  system identification in, 771–777
Feedback neutralization, 775
Feedforward control systems, 751–779
  adaptive control, 760–765
  architecture for, 731–732
  basic architecture, 751–754
  control of aliasing effects, 765–769
  extension to MIMO systems, 777–779
  MIMO active locomotive exhaust with passive component, 829–840
  optimal control filter estimation, 754–760
  reference sensor channels for, 809
  residual sensor channels for, 809
  system identification, 769–777
Feed pumps, predicting noise from, 672–673
FFP (fast-field program), 121
FFT algorithms, *see* Fast Fourier transform algorithms
FFT analyzer, 104
FHWA (Federal Highway Administration), 132
Field-incidence mass law, 435
Field-incidence sound transmission, 435–437
Field-incidence transmission loss, 546
Filtered-U algorithm, 775
Filtered-x LMS algorithm, 761–765, 771, 775, 778, 779
Finite-element method, 156–159
Finite impulse response (FIR) filters, 743, 744, 746–749
Finite panels, sound transmission through, 439–443
FIR filters, *see* Finite impulse response filters
5-dB rule, 859
Fixed diffusers, 209
Fixed -point DSPs, 827
Flanking sound transmission, 451–453, 549, 550
Flexible-wall effect on sound pressure, 166–171
  coupled structural–acoustical response, 169–170
  wall as noise source, 166–168
  wall as reactive impedance, 167, 169

Floating floors:
  checking performance of, 486
  impact noise isolation with, 482
  locally reacting, 482–382
  for mechanical rooms, 710
  resonantly reacting, 483
Floating-point DSPs, 825, 827
Flow(s):
  past cavities, 640–643
  in pipes, 630–631
  separated, 627
  and silencer attenuation, 329–331
Flow-generated noise:
  gas flow noise, 611–656
    aeroacoustical sources, 611–616
    aerodynamic noise of throttling valves, 643–656
    airfoils and struts sound generation, 638–640
    in flows past cavities, 640–643
    from fluid flow in pipes, 630–631
    gas jet noise, 616–624
    gas turbine engine combustion noise, 624–626
    grid or grille noise, 635–638
    spoiler noise, 631–635
    turbulent boundary layer noise, 626–630
  HVAC:
    in duct systems, 699–700
    flow-generated sound power, 694
  of silencers, 331–333
Flow resistance, flow resistivity, 235–237
Fluctuating drag (flow spoilers), 633
Fluctuating lift forces (flow spoilers), 633
Fluids, propagation speed of sound in, 351–352
Foot-pound-second system of units, 936
Foot-slug-second system of units, 936
Forced sound pressure response, 159–161
Formstiff small enclosures, 527–530
4-dB rule, 859
Fourier spectral functions, 54–55
Fraction of critical damping, 581. *See also* Damping ratio
Free field, 4, 32, 76
  measurement in, 82
  outdoor sound propagation in, 126–128
  sound power determination in, 91–98
Free-field approximation, for sound intensity, 79–81
Free-field radiation, from jet in ideal acoustical medium, 616
Free standing enclosures, 522, 531–534
Frequency response function, 62, 161
Frequency spectral density, 638–639
Fresnel number, 132
Fresnel zones, 134
Friction force, 347
Furnishings, sound in rooms and, 184

G

Gain factor estimates, 62–63
Gas flow noise, 611–656
 aeroacoustical sources, 611–616
 aerodynamic noise of throttling valves, 643–656
  acoustical efficiency, 647–649
  aerodynamic noise, 643–647
  due to high velocities in valve outlet, 652
  methods of valve noise reduction, 652–656
  pipe transmission loss coefficient, 649–651
 airfoils and struts sound generation, 638–640
 in flows past cavities, 640–643
 from fluid flow in pipes, 630–631
 gas jet noise, 616–624
 gas turbine engine combustion noise, 624–626
 grid or grille noise, 635–638
 spoiler noise, 631–635
 turbulent boundary layer noise, 626–630
Gas jet noise, 616–624
 flight effects, 624
 from imperfectly expanded jets, 621–623
 jet mixing noise, 616–621
Gas turbines:
 engine combustion noise, 624–626
 industrial, predicting noise from, 673–675
Gauges of metal plates, 513–514
Generalized-terrain PE, 121
Gradient, 25, 26, 103
Gradient descent algorithms, 749–750
Graphical user interface (GUI), 829
Grid noise, 635–638
Grilles:
 HVAC, 701–703
 noise from, 635–638
Ground cover, outdoor sound propagation and, 137
Group speed (energy speed), 391, 393
GUI (graphical user interface), 829

H

Half-power bandwidth, 165
Half-power points, 583n.
Hand-arm vibration syndrome (HAVS), 857, 877
Hand-transmitted vibration criteria, 881–882
Hand-transmitted vibration effects, 877
Hanning window, 55, 57
HAVS, see Hand-arm vibration syndrome
Head-related transfer functions (HRTFs), 190
Hearing, 858–874
 damage risk criteria, 858–861
 impulse noise, 867–869
 infrasound exposure, 869, 870
 noise exposure criteria for audio frequency region, 861–867
 protection of, 871–874
 psychoacoustics, 900
 and sound pressure, 1
 standards for testing equipment/procedures, 918–919
 ultrasound exposure, 869–871
Hearing conservation program standards, 919
Hearing impairment (hearing loss), 18
Hearing threshold, 18, 19, 894
Hearing threshold levels, 18
Heating, ventilating, and airconditioning (HVAC) systems, 685–718
 airflow velocity, 701–703
 A-weighted sound-level criteria for, 890–891
 diffuser selection, 703
 duct-borne noise transmission attentuation, 685–699
  by cross-sectional area changes, 687, 691
  by divisions, 687
  by elbows, 691–693
  end-reflection loss, 697–698
  by plenums, 696–697
  prefabricated sound attenuators, 693–696
  room effect, 698–699
  in straight ducts, 686–691
 duct sizes, 700
 for especially quiet spaces, 700–703
 fans, 705–707
 flow noise in ducted systems, 699–700
 grille selection, 703
 mechanical plant room sound isolation and noise control, 709–710
 NCB criteria curves for, 901
 NC criteria curves for, 901
 noise breakout/break-in, 703–705
 outdoor noise emissions, 717–718
 RC criteria curves for, 898
 RNC criteria curves for, 900–902
 rooftop air conditioning units, 715–717
 terminal boxes/valves, 707–709
 vibration isolation considerations, 711–715
Heat recovery steam generators (HRSGs), 674
Helmholtz equation, 126
Helmholtz resonators, 149, 384, 640, 642–643
Hemi-anechoic spaces/rooms, 82, 92–97, 211
Hemispherical array (microphones), 92–94, 100–101
Hemispherical spaces, directivity index in, 116
Horizontal translational motions, stiffness of isolators and, 564–565
HRSGs (heat recovery steam generators), 674
HRTFs (head-related transfer functions), 190
Human-occupancy areas, noise criteria for, 887–908
 acoustically induced vibrations and rattles, 905, 907
 evaluation methods for, 888–906

A-weighted sound level, 888–891
  current-day sound-level meters, 888
  NCB curves, 895
  NC curves, 889, 892–894
  RC curves, 895–899
  RNC curves, 899–906
  noise annoyance in communities, 907
  sound level definitions, 887–888
  typical urban noise, 907–908
Human systems, effect of noise/vibration on, 857–858. *See also* Hearing
Human vibration response, 874–883
  exposure guidelines, 877–883
  hand-transmitted vibration effects, 877
  whole-body vibration effects, 875–877
Huygens' model for wave fields, 133–134
HVAC systems, *see* Heating, ventilating, and airconditioning systems

I

IEC, *see* International Electrotechnical Commission
IIR filters, *see* Infinite impulse response filters
IL, *see* Insertion loss
Image sources, method of, 188–189
Immission, 71
  emission vs., 73
  preferred descriptor of, 71
Impact dampers, 590
Impact noise, 477–487
  coal-handling equipment:
    car shakers, 664
    rotary car dumpers, 664
  with elastic surface layer, 479–483
  noise isolation vs. sound transmission loss, 483–486
  standard tapping machine, 477–479
Impedance, 38–41
  complex, 39, 588
  equivalent lumped, 397–398, 496
  of infinite plates and beams, 393–397
  mechanical, 40, 394–395
  moment, 501
  plane-wave, 39, 352–354
  point force, 394–396, 496–500
  separation, 430–434
Impulse noise. *See also* Impact noise
  exposure to, 860
  and hearing, 867–869
  protection against, 874
Impulse response (digital systems), 742–746
Inch-pound-second system of units, 936
Incident sound waves:
  excitation of structures by, 427
  sound absorption:
    normal incidence on porous layer in front of rigid wall, 248–251
    oblique incidence, 251–254
  sound transmission through infinite plate:
    normal-incidence waves, 427–430
    oblique-incidence waves, 430–433
    random-incidence (diffuse) sound, 433–439
Industrial fans, predicting noise from, 669–672
Industrial gas turbines, predicting noise from, 673–675
Industrial workshops, 197–201
  noise control for, 200–201
  noise prediction for, 198–200
Inertia bases, 561–562, 712
Inertia force, 347
Infinite impulse response (IIR) filters, 744, 746, 750
Infinite rigid pistons, 417
Infrasound exposure, 859, 860
  and hearing, 869, 870
  protection against, 872–873
Inhomogeneous atmosphere, refraction of sound in, 129–132
Inner model transform, 781. *See also* Youla transform
Insertion loss (IL), 170, 284–286
  as acoustical performance measure, 522–523
  air compressors, 662
  close-fitting enclosures:
    free standing, 532–534
    machine-mounted, 535–537
  HVAC ducts, 690–693
  large enclosures:
    and flanking, 549, 550
    at high frequencies, 541–544
    inside-outside vs. outside-inside transmission, 549–550
    key parameters influencing, 544–545
    and leaks, 547, 548
    and machine position, 549
    and machine vibration pattern, 549
    and sound-absorbing treatment, 547, 548
    and wall panel thickness, 545–547
    without internal sound-absorbing treatment, 538–540
  small enclosures:
    leaky, 530–531
    sealed, 525–530
Instrumentation standards:
  intensity instruments, 914
  noise-measuring instruments, 912–913
Insulation, thermal-acoustical blanket, 672
Intensity, *see* Sound intensity
Intermediate office speech level (IOSL), 205
Intermediate-size enclosures, 537–538
Internal friction, 589
International Electrotechnical Commission (IEC), 104, 911
International Organization for Standardization (ISO), 82, 878, 911
Inverse square law, 4–5

IOSL (intermediate office speech level), 205
ISO, *see* International Organization for Standardization
Isolation effectiveness, 568–569
  and isolator mass effects, 570–571
  with two-stage isolation, 573–575
Isolation efficiency, 560
Isolation range, 559

J

Jet mixing noise, 616
Joint input-output method, 776–777

K

Kronecker delta function, 742
Kuttruff model, 198–199

L

Large enclosures:
  with interior sound-absorbing treatment, 540–551
    analytical model for insertion loss at high frequencies, 541–544
    effect of sound-absorbing treatment, 547, 548
    flanking transmission through floor, 549, 550
    inside-outside vs. outside-inside transmission, 549–550
    key parameters influencing insertion loss, 544–545
    leaks, 547, 548
    machine position, 549
    machine vibration pattern, 549
    wall panel parameters, 545–547
  without interior sound-absorbing treatment, 538–540
Large partitions, sound transmission of, 425–427
Lateral quadrupole, 349–350
Leakage, in filtered-x algorithm, 762
Leakage errors, 54
Leaky enclosures, 522
  large, 547, 548
  small, 530–531
Lecture rooms, 191. *See also* Classrooms
Levels, 12–18
  average A-weighted sound level, 17
  average sound level, 16
  A-weighted sound (noise) exposure level, 18
  A-weighted sound pressure level, 15–16
  day–night, 17
  equivalent continuous A-weighted, 17
  hearing impairment (hearing loss), 18
  hearing threshold levels:
    associated with age, 18
    associated with age and noise, 18
    for setting "zero" at each frequency on a pure-tone audiometer, 18
  noise-induced permanent threshold shift, 18
  overall levels, determining, 22–24
  sound intensity level, 14–15
  sound power level, 13–14
  sound pressure level, 15, 17
Level at mid-frequencies (LMF), 896, 897
Linear average, 48
Linear isolators, massless, 569–570
Lined ducts, 280–313
Line sources, 122
Line spectral function, 54–55
Line spectrum, 2
L-junctions, reflection loss of free bending waves at, 408
LMF, *see* Level at mid-frequencies
LMS algorithm, 750–751
  adaptive control based on, 760
  filtered-x, 761–762, 771, 775
Logarithmic decrement, 581, 587
Longitudinal quadrupole, 350
Longitudinal waves, 6
  in beams, reflection loss for, 410–411
  resonance of, 640–641
Long-term A-weighted sound pressure level, 139
Loss factor:
  damping, 584
  for panels/uniform plates, 591
  in piping, 630–632
  structural damping, 561
  of three-component beam, 604–606
  of viscoelastic material, 595–596
Loudness, perceived, 900
Loudspeakers, 729
Low-noise compressors, 660

M

Machine-mounted enclosures, *see* Equipment mounted enclosures
Machinery noise:
  active machinery isolation, 840–845
  emission measurement standards, 916–918
  inertia bases, 561–562
  prediction of, *see* Prediction of machinery noise
Machinery vibration, active control of, 739–740
Mach number, 304
Masking of sound, 205, 206
Mass, 394, 579
Mass-controlled boundaries, 169
Mass law, 443
Massless linear isolators, 569–570
Massless springs, 395
  in mass-spring-dashpot system, 557–560
  point force impedance for, 394
Mass-spring-dashpot system, 557–560, 580
  energy dissipation in, 585

INDEX **953**

steady-state response of, 584
  time variation of, 581
Material damping, 589
Mean-square error (FIR filters), 748–749
Mean-square (ms) value (variance), 47–48
  in data analysis, 47–48
  normalized random error of, 50, 51
Mean-square sound pressure:
  and bandwidth conversion, 10–11
  for complex spectra, 12
  for contiguous frequency bands, 8, 9
  of continuous-spectrum sound, 7, 11
Measurement environments, 82
Measurement surfaces, 92
Mechanical hysteresis, 589
Mechanical impedance, 40, 394–395
Mechanical induced-draft cooling towers, 666–667
Mechanical plant rooms, sound isolation/noise control in, 709–710
Mercury RACE++, 825, 827
Metallic isolators, 576
Meteo-BEM, 121
Meteorology, outdoor sound propagation and, 139–140
Meter-kilogram-second (mks) system of units, 936
Method of image sources, 188–189
Micrometeorology, 130
Microphones (for measurements), 102
  for hemi-anechoic spaces, 92–97
  for reverberation rooms, 87–88
Midfrequency average, 896
MIMO systems, *see* Multiple-input, multiple-output systems
mks system of units, 936
Mobility, (admittance, 43), 396
  measurement of, 588
  receiver, 569
  in two-stage isolation systems, 575–576
Modal analysis technique, 159
Modal damping, 163, 164
Modal damping constant (decrement), 159
Modal density, 400–402, 454, 507–508
Modal energy, 453–455
Modal parameter extraction instrumentation, 588
Modal testing, 588
Modern optimal control theory, 781
Mode shapes, 150, 152–155, 400, 507–508
Modes of vibration:
  equal energy of, 455
  modal energy, 453–455
  system of modal groups, 453–454
Monin-Obukhov boundary layer theory, 121
Monopoles:
  aerodynamic, 611–612, 614
  sound power output of, 34, 35
Motion sickness guidelines, 880
Moving-average filters, 743

Moving sound sources:
  outdoor, 124–126
  reciprocity in moving, 476–477
ms, *see* Mean-square value
Mufflers. *See also* Silencers
  combination, 335–338
  expansion chamber, 293–300
  feed pumps, 674
  for industrial fans, 670
  perforated-element, 300–311
  steam vents, 677
  transformers, 679
Multilayer partitions, sound transmission through, 443–448
Multiple-input, multiple-output (MIMO) systems, 740
  feedforward, 754
  feedforward active locomotive exhaust with passive component, 829–840
    actuator number/location, 832–834
    control actuator design, 834–836
    control architecture, 835–836
    hardware selection, 837–838
    performance goals, 831–832
    problem description, 829–830
    sensor number/location, 832, 834, 835
    system performance, 838–840
  LMS algorithms for, 751
  optimal design of, 777–779
  Youla feedback architecture in, 783
Music, exposure criteria for, *867*
Music waves, superposition principle and, 36

N

National Academy of Sciences–National Research Council, Committee on Hearing, Bioacoustics and Biomechanics (CHABA), 861–862, 867
Natural-draft cooling towers, 666
Natural frequencies, 150, 152–155
  of finite structures, 400
  and horizontal stiffness of isolators, 565
  of mechanical equipment isolators, 711
  for pure vertical vibration, 563
  of rigid and nonrigid masses, 563
  for rocking, 564
  in two-stage isolation systems, 572–573
NCB curves, *see* Balanced noise criterion curves
NC curves, *see* Noise criterion curves
Near field, 76
NEMA values (transformer noise level), 678, 679
Net oscillatory force, 349
Neutralizers, 593
Newton's second law of motion, 25, 26
Nichols plot, 787–788, 794

954 INDEX

NIPTS, see Noise-induced permanent threshold shift
NITTS, see Noise-induced temporary threshold shift
Noise annoyance (in communities), 907
Noise-cancellation headsets, 736–738
Noise control, 345
 active, see Active noise and vibration control
 in buildings, acoustical standards for, 926–931
 in classrooms, 196–197
 design standards for, 930–931
 for industrial workshops, 200–201
 in mechanical plant rooms, 709–710
 in solid structures, see Solid structures, interaction of sound waves with as system problem, 81
 transformers, 679
 in very small enclosures, 145–146
Noise criterion (NC) curves, 889, 892–894, 901
Noise exposure:
 A-weighted, 858
 criteria for audio frequency region, 861–867, 886–906
 duration of, 865–866
Noise exposure level, A-weighted, 18
Noise-induced hearing loss, 857, 859
Noise-induced permanent threshold shift (NIPTS), 861–862, 864, 865
Noise-induced temporary threshold shift (NITTS), 866, 867
Noise level, perceived, 4
Noise mapping, computer software for, 139
Noise power emission level (NPEL), 75n.
Noise reduction (NR), 284, 286, 345, 382–387
 active, 345
 actuators, 807
 air compressors, 662
 coal-handling equipment:
  car shakers, 664
  crushers, 665
  rotary car dumpers, 664
  transfer towers, 665
  unloaders, 664
 cooling towers, 666, 667
 feed pumps, 673
 HVAC systems:
  criteria for noise control in, 886–906
  by duct cross-sectional area changes, 687, 691
  by duct divisions, 687
  by elbows, 691–693
  for especially quiet spaces, 700–703
  by plenums, 696–697
  for straight ducts, 686–691
 industrial fans, 669–670
 by minimizing added mass, 383–384
 passive, 345–346
 at specific frequencies, 384–387
 steam turbines, 676
 throttling valves, 652–656
 wind turbines, 680
Noise reduction coefficient (NRC), 207
Noise sources, see Sound (noise) sources
Nondirectional sources, 72, 78
Nondispersive waves, 391–392
Nonperiodic steady-state signals, 44
Nonrecursive filters, 743
Nonrigid masses:
 natural frequencies of, 563
 transmissibility of, 565–567
Nonstationary random data, 46–47, 49–50
Non-volume-displacing sound sources, radiation by, 361–377
 effect of surrounding point force by rigid pipe, 375–377
 force acting on fluid, 371–375
 response of bounded fluid to point force excitation, 375
 response of unbounded fluid to excitation by oscillating small rigid body, 363
 in response to excitation of fluid by oscillating small rigid sphere, 363–371
Normal-incidence sound transmission, 427–430
Normalized LMS algorithm, 751
Normalized random error, 50, 65
Normal modes, 150
Normal-mode expansion, 159
Normal mode of vibration, see Resonance
Normal specific acoustical impedance, 41
Nosoacusis, 860
NPEL (noise power emission level), 75n.
NR, see Noise reduction
NRC (noise reduction coefficient), 207
Nyquist frequency, 53, 54, 57
Nyquist plots, 583, 585, 786–788, 794, 798, 799
Nyquist's sampling theorem, 815
Nyquist stability criterion, 785

O

OAPWL, see Overall sound power level
OASPL, see Overall sound pressure level
Oblique-incidence sound transmission, 430–433
Occupational Safety and Health Administration (OSHA), 862
Octave-band sound power level:
 air compressors, 661
 estimates of, 659
 feed pumps, 673
 steam turbines, 676
 with surging or large fluctuations, 903–905
Octave-band sound pressure level, 15, 678
Octave-band source spectra, 122–123

Offices:
  closed, sound pressure levels in, 190–191
  criteria for noise control in, 890, 891
  open-plan, 201–208
One-dimensional wave equation:
  backward-traveling plane wave, 32
  general solution to, 27–28
  intensity, 31–32
  outwardly traveling plane wave, 28–31
  particle velocity, 31
  root-mean-square sound pressure, 31
  solutions to, 27–34
  spherical wave, 32–34
Open area ratio, 306, 307
Open-loop system identification, 770–772, 774, 775, 827, 828
Open-plan offices, 201–208
  barriers for, 202–203
  design parameters for, 206–208
  rating acoustical privacy of, 203–204
  speech and noise levels in, 205–206
  workstation acoustical design, 204–205
Optimal control theory, 781
Oscillating rigid bodies, radiation efficiencies of, 418–419
Oscillating small rigid bodies:
  fluid excitation, 363–371
    force acting on fluid, 371–375
    harmonic excitation, 363–365
    predicting body length, 366–370
    predicting radiated sound power, 366, 371
    random excitation, 371
    sound radiation of nonspherical bodies, 365
    unbounded fluid, 363
  point force excitation, 375–377
  radiation by, 348–349
Oscillating spheres, sound power output of, 34, 35
Oscillating three-dimensional bodies, radiation efficiencies of, 418–419
OSHA noise exposure criteria, 862
OSHA (Occupational Safety and Health Administration), 862
Outdoor noise emissions, from HVAC systems, 717–718
Outdoor sound propagation, 119–143
  accounting for meteorology, 139–140
  airborne, 121
  atmospheric absorption, 136. *See also* Air absorption
  barriers, 132–135
  computer software for noise mapping, 139
  effects of ground cover and trees, 137
  in homogeneous free space over ground, 120–121, 126–128
  inhomogeneity of atmosphere, 129–132
  interaction of barriers and ground, 137–138
  moving sources, 124–126
  point sources at rest, 120–124
  reflectors and reverberation, 135–136
  refraction of sound, 129–132
  relevance of criteria for, 119–120
  standard regulations for interactions, 137
  standards for, 922
  uncertainties of, 140–143
Outwardly traveling plane wave, 28–31
Overall A-weighted sound pressure level, 17, 142–143
Overall level (OA), 22–24
Overall sound power level (OAPWL):
  air compressors, 661
  air-cooled condensers, 667
  boilers, 662
  coal-handling equipment:
    car shakers, 663
    crushers, 665
    mills and pulverizers, 665
    rotary car dumpers, 664
    transfer towers, 665
    unloaders, 664
  diesel-engine-powered equipment, 668–669
  feed pumps, 672
  for gas turbine engines, 625
  industrial fans, 670–672
  industrial gas turbines, 675
  steam turbines, 676
  steam vents, 676–677
Overall sound power output (cooling towers), 666
Overall sound pressure level (OASPL), 618–621, 624
Overlapped processing, 57

P

Parabolic equation (PE), 121
Parallel-baffle silencers, 311, 312, 316–325
  baffle thickness, 324–325
  cross-sectional area, 322
  effect of temperature on, 322–324
  predicting acoustical performance of, 316–322
  quantitative considerations with, 317
Parallelepiped array (microphones), 93, 95, 100
Particle velocity, 5–6
  measurement of, 102–103
  for outwardly traveling plane wave, 31
  for spherical waves, 33–34
Partitions, sound transmission by:
  composite partitions, 451
  double-layer partitions, 443–449
  large partitions, 425–427
  multilayer partitions, 443–447
  small partitions, 421–425
Passive silencers, 279–338
  combination mufflers, 335–338
  dissipative, 311–335
    economic considerations, 335

956 INDEX

Passive silencers (*continued*)
    effect of flow on silencer attenuation, 329–331
    flow-generated noise, 331–333
    key performance parameters, 313–316
    lined ducts, 281–313
    parallel-baffle silencers, 316–325
    pod silencers, 328–329
    prediction of silencer pressure drop, 333–335
    round silencers, 325–328
  expansion chamber mufflers, 293–300
    double-tuned expansion chamber, 297–299
    extended-outlet muffler, 294–297
    general design guidelines for, 299–300
    simple expansion chamber muffler, 293–294
  perforated-element mufflers, 300–311
    acoustical performance, 303–310
    back pressure, 310–311
    range of variables for, 300–303
  performance metrics for, 281–286
  reactive, 286–293
    representation by basic silencer elements, 287–288
    transfer matrices for, 288–293
Pendulum isolators, 577–578
Perceived noise level, 49
Perforated-element mufflers, 300–311
  acoustical performance, 303–310
  back pressure, 310–311
  range of variables for, 300–303
Perforated plates, open area of, 515
Performance:
  ANVC systems:
    and controller architecture, 814–817
    goals for, 801–804
    and doors/windows/ventilation openings, 660
    insertion loss as measure of, 522–523
    measures of, 518–519
    qualitative description of, 523–525
Periodic excitation sources, identification of, 63–64
Periodic signals, 44
  mean-square value for, 48
  spectral computations for, 54
  statistical sampling errors with, 59
Permanent threshold shifts (PTSs), 862
Permissible exposure level (PEL), 862
Phase speed, 391–393
Pipes:
  fluid flow in, 630–631
  to isolated mechanical equipment, 714
  with open side branch, sound radiation in, 355
  transmission loss of, 448–451
Pipe transmission loss coefficient (throttling valves), 649–651

Pipe wrappings, 554–555
Piston-in-cylinder isolators, 577
Plane boundaries, sound radiation and, 377–382
  oscillating moment in infinite liquid, 381–382
  pressure-release boundaries, 379–381
  rigid boundaries, 378–379
Plane waves, 6
  backward-traveling, 32
  outwardly traveling, 28–31
  reflection and transmission at interfaces of, 403–406
  sound pressure and particle velocity relationship, 38
Plane-wave impedance, 352–354
Plane-wave transfer matrix method, 280–281
Plant transfer functions, 770–771
Plateau method (sound transmission), 442–443
Plates:
  finite, point-excited, 420–421
  infinite, 395–397, 417–418
    effective length connecting force and moment impedance, 502
    normal-incidence plane waves, 427–430
    oblique-incidence plane waves, 430–433
    point-excited, 420–421
    random-incidence sound, 433–439
  inhomogeneous, 438–439
  orthotropic, 437–438
  perforated, open area of, 515
  power transmission from beam to, 410–413
  steel, gauges and weights of, 513–514
  thin isotropic, field-incidence transmission for, 435–437
  transmission through, 395–397
    normal-incidence plane waves, 427–430
    oblique-incidence plane waves, 430–433
    random-incidence sound, 433–439
    with vibration break, power transmission between, 413–416
  with viscoelastic coatings, 601–602
  with viscoelastic interlayers, 606–607
Plenums:
  HVAC ducts, 696–697
  for rooftop air conditioning units, 715
Plug muffler, 304–311
Pneumatic isolators, 577
Pod silencers, 328–329
Point force:
  active control of disturbance, 725–727
  dynamic excitation by, 393–394
  excitation by sound field vs., 462–465
Point force impedance, 394–396, 496–500
Point input admittance (mobility), 396
Point moment, dynamic excitation by, 393–394
Point moment impedance, 394–395, 501
Point sound sources, 122–124, 361

Poisson's ratio, 391, 493
Pole-zero filters, 744
Position-keeping system, 739, 796
Positive definite (quadratic function), 748
Power, see Sound power
Power balance:
  and response of finite structures, 399–403
  in two-structure system, 456–457
Power balance equation, 406
Power law damping, 588
PowerPC (PPC) chips, 825–827
Power spectral density (PSD) functions, 55–58, 172–174
Power spectral density spectrum, 9
Power transmission between structural elements, 406–416
  from beam to plate, 410–413
  and change in cross-sectional area, 406–407
  plates separated by thin resilient layer, 413–416
  reflection loss:
    of bending waves through cross junctions/T-junctions, 409–410
    of free bending waves at L-junctions, 408
PPC chips, see PowerPC chips
Prediction of machinery noise, 659–681
  air compressors, 660–662
  air-cooled condensers, 667–668
  boilers, 662–663
  coal-handling equipment, 663–665
  cooling towers, 666–667
  diesel-engine-powered equipment, 668–669
  feed pumps, 672–673
  industrial fans, 669–672
  industrial gas turbines, 673–675
  steam turbines, 675–676
  steam vents, 676–677
  transformers, 677–679
  wind turbines, 679–681
Prefabricated sound attenuators, HVAC, 693–696
Presbyacusis, 860
Pressure, see Sound pressure
Pressure-residual intensity index, 106
Privacy, in open-plan offices, 201–204
Processor-to-processor communications, 825
Propagation of sound:
  in fluids, 351–352
    identifying propagation paths, 66–69
  outdoor, see Outdoor sound propagation
  in rooms, 182
Propagation speed:
  in solids, 492
  velocity vs., 391
PSD functions, see Power spectral density functions
PTSs (permanent threshold shifts), 862
Pulsating small rigid bodies, radiation by, 348

Pulsating sphere:
  maximum achievable volume velocity of, 359–361
  sound radiation of, 355
Pumps:
  feed, predicting noise from, 672–673
  at transformers, 677
Pure tones, 2
  in continuous spectrum, 7, 8
  and hearing thresholds, 18
  setting pure-tone audiometers, 18
  and spectral density, 9

Q

Quadrupoles, aerodynamic, 613–614
Quality assessment index (QAI), 897–899
Quantization noise, 822–823
Quarter-spherical spaces, directivity index in, 116–117

R

Radiation efficiency, 418–420, 509–511
Radiation fields of sources, 75–77
Radiation impedance of sphere, 364
Radiation of sound:
  by thin plates, 417, 420
  effect of plane boundaries on, 377–382
    oscillating moment in infinite liquid, 381–382
    pressure-release boundaries, 379–381
    rigid boundaries, 378–379
  global cancellation, 722
  by non-volume-displacing sound sources, 361–377
    effect of surrounding point force by rigid pipe, 375–377
    force acting on fluid, 371–375
    response of bounded fluid to point force excitation, 375
    response of unbounded fluid to excitation by oscillating small rigid body, 363
    in response to excitation of fluid by oscillating small rigid sphere, 363–371
  by small rigid bodies, 346–355
    acoustical parameters, 351–354
    lateral quadrupole, 349–350
    longitudinal quadrupole, 350
    oscillating small rigid bodies, 348–349
    by pipe with open side branch representing pulsating sphere, 355
    by piston in rigid tube with open side branch, 354–355
    pulsating small rigid bodies, 348
    radiated sound power, 351
  in solid structures, 416–421
  tonal components in, 635
  by vibrating structures, 593–594
  by volume-displacing sound sources, 356–361

Radiators, elementary, 35–36
Random-analysis theory, 172–174
Random data signals, 45–47
    cross-spectral density function, 57–58
    mean values for, 47
    statistical sampling errors with, 59
    tapering windows with, 57
Random error, 62
Random excitation sources, identification of, 64–66
Random-incidence (diffuse) sound transmission, 433–439
Random-incidence mass law, 435
Random sound pressure response, 170, 172–175
Rattles, acoustically induced, 905, 907
Rayleigh, Lord, 119, 129
Raynaud's phenomenon, 877
Ray theory, 121
Ray-tracing techniques, 189–190, 198
RC curves, 895–899
RC filter, 48
Reactive silencers, 286–293
    factors in acoustical performance of, 282–283
    representation by basic silencer elements, 287–288
    transfer matrices for, 288–293
Receiver mobility, 569
Reciprocity, 162, 465–477
    extension to sound excitation of structures, 473–476
    for higher order excitation sources/responses, 512
    in moving media, 476–477
    prediction of noise from multiple correlated forces, 470–471
    source strength identification by, 471–473
    theorem, 162
Reconstruction filters, 817–819
Recursive filters, 744
Reference quantities, 12, 19–22
"Reference Quantities for Acoustical Levels" (ANSI S1.9–1989, Reaffirmed 2001), 19, 20
Reference sensors (ANVC), 809, 812–813
Reflection loss:
    for bending waves:
        in beams, 411–412
        and change in cross-sectional area, 407
        at interfaces, 404–405
        free bending waves at *L*-junctions, 408
        in plates with vibration break, 415–416
        through cross junctions and T-junctions, 409–410
    for compression waves:
        and change in cross-sectional area, 407
        in plates with vibration break, 414–415
    for longitudinal waves in beams, 410–411

Reflection of sound:
    at plane interfaces, 37, 403–406
    by room surfaces, 183
Reflectors, outdoor, 135–136
Refraction of sound, 129–132
Regularization (ANVC sensors), 814
Regulation filter design (feedback control), 785–792
Reinforcements, damping due to, 591–592
Relative bandwidth, 583
Residual sensors (ANVC), 809–812
Resonance, 38, 150, 640–643
Resonance frequencies, 165
    of boilers, 663
    of finite structures, 400, 401, 507–508
    standing-wave, 570
Resonance transmission loss, 460
Reverberant field, *see* Diffuse (reverberant) field
Reverberation:
    in classrooms, 197
    in industrial workshops, 197–198
    outdoor sound propagation, 135–136
    for speech intelligibility, 193
    standards for, 930–931
    with steam turbines, 676
Reverberation rooms, 82, 209–210
    characteristics of, 85–86
    experimental setup for, 87–88
    qualification requirements for, 86–87
Reverberation time, 175–176, 582
Reynolds stresses, 119, 613
Rigid masses, 395, 563. *See also* Three-dimensional masses
Rigid tube with open side branch, sound generation in, 354–355
Ring frequency, 500
Rms sound pressure, *see* Root-mean-square sound pressure
RNC curves, 899–906
Road surface sound absorption standards, 924
Rocking motions, vertical motions coupled with, 563–564
Rooftop air conditioning units, 715–717
Rooms, sound in, 181–211
    air absorption, 183
    alternative prediction approaches, 187–188
    anechoic and hemi-anechoic chambers, 211
    auralization, 190
    classrooms, 191–197
    criteria for noise control in, 886–907
    diffuse-field theory, 184–187
    domestic rooms and closed offices, 190–191
    empirical models, 190
    and furnishings, 184
    from HVAC ducts, 698–699
    industrial workshops, 197–201
    method of image sources, 188–189
    open-plan offices, 201–208

INDEX   **959**

propagation of sound, 182
ray and beam tracing, 189–190
reverberation rooms, 209–210
sound decay, 182–183
surface absorption and reflection, 183
Room noise evaluation:
  engineering method, 889, 892–894
  precision method, 899–905
  survey method, 888–889
Root-mean-square (rms) sound pressure, 2
  for contiguous frequency bands, 8, 9
  of continuous-spectrum sound, 7, 11
  normalized random error of, 50, 51
  for outwardly traveling plane wave, 31
Rotary coal car dumpers, 664
Rotating diffusers, 209–210
Round silencers, 312, 325–328
Running averages, 49–50

S

SA A/D converters, *see* Successive-approximation A/D converters
Sabine approach (diffuse-field theory), 185–186
Sampling errors, *see* Statistical sampling errors
Schroeder frequency, 209
Screech tones, 621, 622
SEA, *see* Statistical energy analysis
Seals, for mechanical rooms, 710
Sealed enclosures:
  close-fitting, 531–537
  defined, 522
  small, 525–531
SECM, *see* Simple expansion chamber muffler
Seismic restraint of equipment, 713
SEL, *see* Sound exposure level
Self-noise, 639
Sensors (ANVC), 730–732
  performance of, 813–814
  reference, 809, 812–813
  residual, 809–812
Separation impedance, 430–434
Shakers (ANVC), 730
Shear modulus (viscoelastic material), 391, 595–596
Shear parameter (3-component beam), 603
Sheet Metal and Air Conditioning National Association (SMACNA), 686
Shock-associated noise, 616, 621–623
Shock response spectrum, 55
Side-lobe leakage, 54
Sigma-delta A/D converters, 820–822
Signal-to-noise ratio (SNR):
  ANVC converters, 823–825
  for classrooms, 196
  for open-plan offices, 203–204
  for synchronous averaged signal, 52

SII, *see* Speech intelligibility index
SIL, *see* Speech interference level
Silencers. *See also* Mufflers
  active noise control, 280
  defined, 280
  dissipative mufflers, 314, 315
  downstream of valves, 654, 655
  effect of flow on attenuation, 329–331
  flow-generated noise of, 331–333
  for HVAC ducts, 693–696, 705
  modeling, 283–286
  passive, *see* Passive silencers
  for rooftop air conditioning units, 716
  selection factors for, 281–282
Simple expansion chamber muffler (SECM), 293–294, 299–300
Simplified models (noise prediction), 198–199
Single-input, single-output (SISO) systems, 733
  feedback, 779–781
  feedforward, 754
  filtered-x algorithm for, 762
  LMD algorithm for, 751
Singular-valued decomposition (SVD), 804–807, 810, 813
SISO systems, *see* Single-input, single-output systems
SMACNA (Sheet Metal and Air Conditioning National Association), 686
Small enclosures, sound in, 145–178. *See also* Rooms, sound in
  acoustical modal response, 149–151
  enclosure driven at resonance, 164–166
  flexible-wall effect on sound pressure, 166–171
  forced sound pressure response, 159–161
  natural frequencies and mode shapes, 150, 152–155
  numerical methods for acoustical analysis, 156–159
  random sound pressure response, 170, 172–175
  sealed enclosures, 525–531
  steady-state sound pressure response, 161–164
  transient sound pressure response, 174–178
  very small enclosures, 145–149
Small partitions, sound transmission of, 421–425
Small rigid bodies:
  oscillating, *see* Oscillating small rigid bodies
  radiation by, 346–355
    acoustical parameters of, 351–354
    lateral quadrupole, 349–350
    longitudinal quadrupole, 350
    pipe with open side branch representing pulsating sphere, 355
    piston in rigid tube with open side branch, 354–355

960 INDEX

Small rigid bodies: (*continued*)
  pulsating bodies, 348
  radiated sound power, 351
Smoothing filters (ANVC) 815
Snell's law, 405
Sociacusis, 860
Soft parameters, 827, 828
Solids:
  key acoustical parameters of, 493–495
  wave motion in, 390–393
Solid structures, interaction of sound waves with, 389–515
  impact noise, 477–487
    impact noise isolation vs. sound transmission loss, 483–486
    improvement of impact noise isolation by elastic surface layer, 479–483
    standard tapping machine, 477–479
  mechanical impedance, 394–395
  power balance and response of finite structures, 399–403
  power input, 395–399
  power transmission between structural elements, 406–416
    from beam to plate, 410–413
    and change in cross-sectional area, 406–407
    plates separated by thin resilient layer, 413–416
    reflection loss of bending waves through cross junctions/T-junctions, 409–410
    reflection loss of free bending waves at L-junctions, 408
  reciprocity, 465–477
    in moving media, 476–477
    prediction of noise from multiple correlated forces, 470–471
    and sound excitation of structures, 473–476
    source strength identification by, 471–473
  reflection and transmission at plane interfaces, 403–406
  and simultaneous airborne/dynamic excitation, 462–465
  sound radiation, 416–421
  sound transmission:
    flanking, 451–453
    of large partitions, 425–427
    loss in ducts and pipes, 448–451
    of normal-incidence waves through infinite plate, 427–430
    of oblique-incidence waves through infinite plate, 430–433
    of random-incidence (diffuse) sound through infinite plate, 433–439
    of small partitions, 421–425
    through composite partitions, 451
    through double- and multilayer partitions, 443–449
    through finite-size panel, 439–443
  statistical energy analyses, 453–462
    composite structures, 456
    diffuse sound field driving freely hung panel, 457–459
    equal energy of modes of vibration, 455
    modal energy, 454–455
    noncorrelation between waves in two systems, 455
    power balance in two-structure system, 456–457
    realization of equal coupling loss factor, 456
    SEA method, 459–462
    system of modal groups, 453–454
    transmission loss through simple homogeneous structure, 459–462
  superposition, 465–467
  wave motion in solids, 390–393
Sound absorption, 215–274
  by air in rooms, 183
  atmospheric, 136
  design charts for fibrous sound-absorbing layers, 254–264
    monolayer absorbers, 255–259
    multilayer absorbers, 261
    thin porous surface layers, 261–264
    two-layer absorbers, 259–261
  for enclosures, 521
  equivalent sound absorption area, 99
  large flat absorbers, 246–265
    normal incidence on porous layer in front of rigid wall, 248–251
    oblique sound incidence, 251–254
    plane sound waves at normal incidence, 247–248
  limp porous layer, 225–229
  multiple limp porous layers, 229
  by non-sound absorbers, 218–221
  in open-plan offices, 206–207
  by plate and foil absorbers, 271–274
  by porous bulk materials and absorbers, 229–246
    acoustical properties of porous materials, 238–239
    analytical characterization of porous granular/fibrous materials, 238
    empirical prediction of flow resistivity, 237
    empirical predictions from regression analyses of measured data, 239–243
    flow resistance and flow resistivity, 235–237
    plastic foams, 244–245
    polyester fiber materials, 241, 243–244
    porosity, 231–233
    process of, 231
    and temperature, 245–246

theoretical prediction of flow resistivity, 237–238
tortuosity, 232, 234
process of, 216
by resonance absorbers, 264–271
  absorption cross section of individual resonators, 266–268
  acoustical impedance of resonators, 265–266
  internal resistance of resonators, 268–271
  nonlinearity and grazing flow, 268
  resonance frequency, 266
  spatial average impedance of resonator arrays, 271
by rigid porous layer, 221–225
by room surfaces, 183
sound absorption coefficients, 216–218
standards for, 928–930
by thin-flow resistive layer in front of rigid wall, 221
waveguide absorbers, 593
for wrappings, 521
Sound absorption average (SAA), 206–208
Sound attenuation, see Noise reduction; Silencers
Sound barriers, see Barriers
Sound decay, 182–183, 186
Sound energy density, 7
Sound exposure, A-weighted, 18
Sound exposure level (SEL), 125–126
Sound generation, 345–387
  effect of nearby plane boundaries on, 377–382
  non-volume-displacing sound sources, 361–377
    effect of surrounding point force by rigid pipe, 375–377
    force acting on fluid, 371–375
    response of bounded fluid to point force excitation, 375
    response of unbounded fluid to excitation by oscillating small rigid body, 363
    in response to excitation of fluid by oscillating small rigid sphere, 363–371
  reducing sound radiation, 382–387
  small rigid bodies, 346–355
    acoustical parameters of, 351–354
    lateral quadrupole, 349–350
    longitudinal quadrupole, 350
    oscillating small rigid bodies, 348–349
    by pipe with open side branch representing pulsating sphere, 355
    by piston in rigid tube with open side branch, 354–355
    pulsating small rigid bodies, 348
    radiated sound power, 351
  volume-displacing sound sources, 356–361

Sound intensity, 3–4
  for cylindrical sound source, 4
  defined, 3
  in determining sound power, 101–112
  free-field approximation for, 79–81
  instrumentation for measuring, 104–105
  for outwardly traveling plane wave, 31–32
  and sound power, 22
  sound power vs., 14–15
  and sound pressure, 4
  for spherical sound source, 4
  for spherical waves, 34
  standards for measurement instruments/techniques, 914
Sound intensity analyzer, 103–104
Sound intensity level, 14–15
  reference quantities for, 19–22
  and sound power level, 22
Sound isolation, in mechanical plant rooms, 709–710
Sound levels, see Levels
Sound level measurement standards, 912–914
Sound (noise) sources, 71–117, 887
  aeroacoustical, 611–616
  in classrooms, 192, 196
  cylindrical, 4
  dipole-type, 638
  directional, 72, 73
  directivity of, 113–117
  excitation sources:
    and gain factors, 62–63
    periodic, 63–64
    random, 64–66
  measurement environments for, 82
  nondirectional (monopole), 34, 72
  outdoor:
    moving sources, 124–126
    point sources at rest, 122–124
  point (monopole), 34, 361
  radiation by, 35, 361–377
  radiation field of, 75–77
  sound intensity:
    free-field approximation for, 14, 35, 79–81
    and sound power, 14, 77–79
  sound power levels, 14, 73–74
    determination of, 82–85
    in diffuse field, 81, 85–91
    in ducts, 112–113
    environmental corrections in determining, 98–101
    in free field, 14, 91–98
    and sound intensity, 14, 77–79
    sound intensity in determining, 14, 101–112
  spherical, 4, 5, 35
  volume-displacing:
    radiation by, 35, 356–361
  in workshops, 199

Sound power, 4, 5
  of baffled pistons, 36
  defined, 4, 14
  in diffuse field, 81, 85–91
  of dipoles, 34, 35
  in ducts, 112–113
  in free field, measurement of, 91–98
    anechoic spaces, 97–98
    hemi-anechoic spaces, 92–97
    measurement surfaces, 92
  of monopoles, 34
  of oscillating spheres, 34, 35
  radiated, 351
  and sound intensity, 4, 22, 77–79
Sound power input:
  in solid structures, 395–399
  to structures, 503–506
Sound power level, 13–14, 73–74
  defined, 72
  and environmental corrections, 98–101
  environmental noise:
    measurement application standards, 921–922
    measurement method standards, 920–921
  expression of, 75
  HVAC duct-borne noise, 685–686
  with impact noise, 479
  ISO standards for determining, 82–85
  reference quantities for, 19–22
  and sound intensity level, 22
Sound power measurement standards, 914–916
Sound power output:
  air-cooled condensers, 667
  cooling towers, 666
  small boilers, 662
Sound pressure, 1
  in determination of sound power, 4, 91–98
  flexible-wall effect on, 166–171
  forced sound pressure response, 159–161
  and particle velocity, 5–6
  in plane waves, 32, 38
  random sound pressure response, 170, 172–175
  and sound intensity, 4
  sound power vs., 4, 14–15
  space-averaged squared, 7
  for spherical waves, 32–33
  steady-state sound pressure response, 161–164
  transient sound pressure response, 174–178
Sound pressure level, 15, 17
  A-weighted, 15–16
  calculation of, 18, 72, 73
  defined, 18, 71
  expression of, 75
  long-term A-weighted, 16, 139
  overall A-weighted, 142–143
  reference pressure for, 19
  reference quantities for, 19–22
Sound propagation:
  in industrial workshops, 197, 198
  between office workstations, 202, 204–205
  underwater, 121
Sound radiation, see Radiation of sound
Sound spectra, 7–12
  complex, 12
  continuous, 7–11
Sound transmission, see Transmission
Sound waves, 1–7
  compressional, 6
  interference of, 36–37
  inverse square law for, 4–5
  longitudinal, 6
  and noise, 1
  and particle velocity, 5–6
  period of, 2
  plane, 6, 28–31
  reflection, 36–38, 404–405
  resonant, 38
  root-mean-square amplitude of, 2
  sound energy density, 7
  and sound spectra, 2–3. See also Sound spectra
  speed of, 6
  spherical, 32–34
  standing, 37–38
  transverse, 6
  traveling, 37
  wave equation, 25–34
  wavelength, 7
Source dimensionality, 805
Space-averaged sound energy density, 7
Space-averaged squared sound pressure, 7
Spacers, 600n.
Spatial decay rate, 582
Specific acoustical impedance, 40
Spectral density function, 50
Spectral functions:
  data analysis, 52–59
    auto (power) spectral density functions, 55–58
    coherence functions, 58–59
    FFT algorithm, 52–54
    line and Fourier spectral functions, 54–55
    statistical sampling errors, 59
  time domain procedures for, 60
Spectral method (underwater sound), 121
Spectral (spectrum) density, 9
Speech communication metrics standards, 920
Speech intelligibility, in classrooms, 192–193
Speech intelligibility index (SII), 204, 206–208
Speech interference level (SIL), 888, 889, 892, 893

Speech privacy, 203–204
  and ambient noise levels, 206
  and voice levels, 205
Speech waves, superposition principle and, 36
Speed of sound, 6
  group (energy) speed, 391, 393
  phase speed, 391–393
  propagation speed vs. velocity, 391
  in solids, 492
Spherical sound sources, 4, 5, 116
Spherical waves, 32–34
Spinal injury, vibration-related, 857
Spoiler noise, 631–635
Spring isolators, 712–713
Squared correlation coefficient function, 61
Square law damping, 588
Standards, 911–931
  for architectural noise control in buildings, 926–931
    acoustics, reverberation, and noise control design, 930–931
    sound absorption, 928–930
    sound transmission, 926–928
  for criteria for noise in rooms, 886–906
  for environmental correction, 98–101
  finding/selecting, 911–912
  hand-transmitted vibration criteria, 881–882
  for HVAC silencers, 694
  for impact noise isolation, 486
  for impact of noise in workplace, 918–924
    environmental noise measurement applications, 921–922
    environmental noise SPL measurement methods, 920–921
    environmental sound propagation outdoors, 922
    environmental vibrations, 922–924
    hearing conservation programs, 919
    hearing test equipment/procedures, 918–919
    speech communication metrics, 920
  for industrial workshop noise, 201
  for intensity measurement, 108, 109–112
  for intensity measurement instrumentation, 104, 108
  for machinery noise emission measurements, 916–918
  for noise control by enclosures and cabins, 517–521
  noise-induced hearing impairment, 864–865
  occupational noise exposure, 864–865
  for outdoor sound, 131, 132, 134–135, 140
  for qualification of reverberation rooms, 86–90
  for sound level measurements, 912–914
    intensity instruments and measuring techniques, 914
    noise-measuring instrumentation, 912–914

SPL measurement techniques, 913–914
for sound power, 82–85
  in anechoic space, 97–98
  in ducts, 112–113
  in hemi-anechoic space, 92–96
  in reverberation rooms, 209–210
for sound power measurements, 914–916
for speech intelligibility, 193
structure-borne whole-body vibration, 878–881
for vehicle exterior/interior noise, 924–926
  exterior noise measurement techniques, 925–926
  interior noise measurement techniques, 924
  road surface sound absorption, 924
for wind turbines, 681
Standard tapping machine, 477–479
Standard threshold shift (STS), 862
Standing-wave resonance frequencies, 570
Standing waves, 37–38
Stationary random data, 45–46
Stationary signals, ms value of, 47–48
Statistical energy analysis (SEA), 453–462
  composite structures, 456
  diffuse sound field driving freely hung panel, 457–459
  equal energy of modes of vibration, 455
  modal energy, 454–455
  noncorrelation between waves in two systems, 455
  power balance in two-structure system, 456–457
  principle of, 455
  realization of equal coupling loss factor, 456
  system of modal groups, 453–454
  transmission loss through simple homogeneous structure, 459–462
Statistical sampling errors:
  with correlation functions, 61
  with mean and ms value calculation, 50–51
  with spectra calculation, 59
Steady state, diffuse-field theory prediction of, 186
Steady-state drag, 633
Steady-state particle velocity, 38
Steady-state response:
  damping, 587
  forced vibrations, 582–584
Steady-state signals:
  ms value of, 47–48
  nonperiodic, 44
Steady-state sound pressure, 38
Steady-state sound pressure level, 15, 187
Steady-state sound pressure response, 161–164, 167
Steam turbines, predicting noise from, 675–676
Steam vents, predicting noise from, 676–677

Steel plates, gauges and weights of, 513–514
Stiffness-controlled boundaries, 169
Stochastic transients, 46–47
Strouhal number, 619, 620, 640, 641
Structural–acoustical analogy, 156, 158
Structural–acoustical coupling, 169–170
Structural acoustics, 389
Structural damping, 561, 579–607
　analytical models of, 588
　due to boundaries and reinforcements, 591–592
　due to energy transport, 592–594
　effects of, 579–580
　energy dissipation and conversion, 589–590
　measurement of, 586–588
　measures of, 580–586
　　complex stiffness, 584–586
　　decay of unforced vibrations with viscous damping, 580–582
　　interrelation of, 586
　　steady forced vibrations, 582–584
　models of, 588–589
　retarding force in, 588
　viscoelastic, 594–607
　　materials and material combinations, 594–595
　　mechanical properties of viscoelastic materials, 595–599
　　plates with viscoelastic coatings, 601–602
　　plates with viscoelastic interlayers, 606–607
　　structures with viscoelastic layers, 599
　　three-component beams with viscoelastic interlayers, 602–606
　　two-component beams, 599–600
Structural dynamics, 389
Structure-borne sound, 611
Structure-borne whole-body vibration criteria, 878–881
Struts, sound generation by, 638–639
Successive-approximation (SA) A/D converters, 820–822, 825
Superposition, 465–467
　with multiple correlated force input, 470
　in source strength identification, 471–473
Supersonic jets, 621
Surging in HVAC noise, 903–905
SVD, *see* Singular-valued decomposition
Synchronous averaging, 51–52
Systems of units, 20, 935
System identification, 769–777, 827
　closed-loop, 771, 772, 774–776, 827, 828
　in feedback control systems, 771–777
　open-loop, 771, 772, 774, 775, 827, 828
System response properties, identification of, 62–63

T

Tapering functions and windows, 54–57
Tapped delay line filters, 743
Tapping machine, *see* Standard tapping machine
Temperature, refraction of outdoor sound and, 129–130
Temperature-frequency equivalence (viscoelastic materials), 597
Temporary threshold shifts (TTSs), 862, 869, 870
Terminal boxes/valves, HVAC, 707–709
Thermal-acoustical blanket insulation, 672
3-dB rule, 859
Three-dimensional masses, vibration isolation for, 563–567
Three-dimensional silencer analysis, 280
Three-duct muffler, 304–311
Throttling valves, aerodynamic noise of, 643–656
　acoustical efficiency, 647–649
　aerodynamic noise, 643–647
　due to high velocities in valve outlet, 652
　methods of valve noise reduction, 652–656
　pipe transmission loss coefficient, 649–651
Time constant, 48
Time history signals, 43–44
Time-varying signals, 44–45
Time-varying transfer functions (ANVC systems), 771
Time-weighted average (TWA), 862
T-junctions, reflection loss of bending waves through, 409–410
TL, *see* Transmission loss
TNM (Traffic Noise Model), 132
Tonal disturbance control, 754–760
Tonal noise:
　boilers, 663
　industrial fans, 669–671
　steam turbines, 676
　transformers, 677, 679
　from turbofan engines, 737–738
Tonal oscillations, 635
Tones:
　in continuous spectrum, 7, 8
　pure, 2
　in sound spectra, 2–3
Traffic noise, 124–126, 132
Traffic Noise Model (TNM), 132
Trailing-edge noise, 640
Transfer function, 62
Transfer matrices (silencers), 280–281, 288–293
　cross-sectional discontinuities, 289–291
　pipe with uniform cross section, 289
　resonators, 291–293
Transformers:
　active noise control for, 723, 725
　predicting noise from, 677–679
Transient signals, 45, 55

Transient sound pressure response, 174–178
Transmissibility, 557–560
  and damping, 561
  and inertia base, 561–562
  and isolation effectiveness, 569
  and isolator mass effects, 570
  and machine speed, 562
  of nonrigid masses, 565–567
  with two-stage isolation, 572–574
Transmission, sound transmission, 71
  flanking, 451–453
  of large partitions, 425–427
  of small partitions, 421–424
  loss of ducts and pipes, 448–451
  at plane interfaces, 403–406
  of small partitions, 421–425
  standards for, 926–928
  structural, 592–593
  through composite partitions, 451
  through double- and multilayer partitions, 443–449
  through finite-size panel, 439–443
  through infinite plate:
    normal-incidence plane sound waves, 427–430
    oblique-incidence plane sound waves, 430–433
    random-incidence (diffuse) sound, 433–439
Transmission coefficient, 425
Transmission loss (TL), 284, 286
  breakin, 450–451
  breakout, 448–450
  classical definition of, 439
  for double partitions, 447–449
  for ducts, 448–451
  field-incidence, 435–436
  of finite panels, 439–443
  flanking, 439–443
  impact noise isolation vs., 483–486
  for inhomogeneous plates, 438–439
  and mass law barrier, 443
  for multilayered partitions, 443–447
  normal-incidence, 427–430
  oblique-incidence, 430–433
  for orthotropic plates, 437
  for pipes, 448–451
  pipe transmission loss coefficient, 649–651
  random-incidence, 434–435
  resonance, 460
  for single partitions, 442–443
  through partitions, 425–427
  through simple homogeneous structure, 459–462
Transversal filters, 743
Transverse waves, 6
Traveling waves, 37
Travel-limited isolators, 713
Trees, outdoor sound propagation and, 137–139

Trigger signal (synchronous averaging), 52
TTS, *see* Temporary threshold shifts
Tuned dampers, 593
Turbines:
  industrial gas turbines, 673–675
  steam turbines, 675–676
  wind turbines, 679–681
Turbofan engine noise control, 737–738
Turbulent boundary layer noise, 626–630
Two-stage vibration isolation, 572–576
Two-structure system:
  noncorrelation between waves in, 455
  power balance in, 456–457

U

Ultrasonic waves, 1
Ultrasound exposure, 859, 860, 869–871, 873–874
Uncertainty, in outdoor sound propagation, 140–143
Underexpanded jets, 621
Underwater sound propagation, 121
Uniform pressure mode, 154, 162
Unit-area acoustical impedance, 41
U.S. armed forces hearing conservation criterion, 863, 867, 868, 874
Unit-sample sequence, 742
Unweighted (linear) averages, 48, 49
Urban noise, 907–908

V

Vacuum bubbles, 385–387
Vacuum pumps, 660
Variation, coefficient of, 50
VDV (vibration dose value), 879
Vehicle exterior/interior noise standards, 924–926
Velocity, propagation speed vs., 391
Vents, steam, predicting noise from, 676–677
Vertical motions, rocking motions coupled with, 563–564
Vibration break, 413
Vibration data, 43
Vibration dose value (VDV), 879
Vibration-induced white finger (VWF), 877
Vibration isolation, 557–578
  active, *see* Active noise and vibration control
  active machinery isolation, 840–845
  for building mechanical systems, 711–715
  classical model, 557–563
    damping effect, 561
    inertia base effects, 561–562
    isolation efficiency, 560
    limitations of, 562–563
    machine speed effect, 562
    mass-spring-dashpot system, 557–560
    transmissibility, 557–560
  high-frequency considerations in, 567–572

966 INDEX

Vibration isolation (*continued*)
  practical isolators, 576–578
  for rooftop air conditioning units, 716
  for three-dimensional masses, 563–567
  two-stage, 572–576
  uses of, 557
Vibration sources:
  acoustic, 905, 907
  loading of, 567–568
Vibration total value (VTV), 879
Viscoelastic damping, 594–607
  beams:
    three-component, with viscoelastic interlayers, 602–606
    two-component, 599–600
  materials and material combinations, 594–595
  mechanical properties of viscoelastic materials, 595–599
  plates:
    with viscoelastic coatings, 601–602
    with viscoelastic interlayers, 606–607
  structures with viscoelastic layers, 599
Viscous damping, 561, 589n.
Viscous damping coefficient, 580
Viscous damping ratio, 583
Volume-displacing sound sources, radiation by, 356–361
Volume velocity, 146–147, 359–361
Voyager lightweight aircraft, 866
VTV (vibration total value), 879
VWF (vibration-induced white finger), 877

W

Walls:
  flexible-wall effect on sound pressure, 166–171
  for mechanical rooms, 710
Wall pressure wavenumber-frequency spectrum, 627, 629–630
Wave equation, 25–34, 149
  continuity equation in, 27
  equation of motion in, 25–26
  gas law in, 26–27
  one-dimensional:
    backward-traveling plane wave, 32
    general solution to, 27–28
    intensity, 31–32
    outwardly traveling plane wave, 28–31
    particle velocity, 31
    root-mean-square sound pressure, 31
    solutions to, 27–34
    spherical wave, 32–34
  in rectangular coordinates, 27
Waveguide absorber, 593
Wavelength, 7
Wave motion (in solids), 390–393
Wavenumber, 29
Weighted averages, 48–49
WHO, *see* World Health Organization
Whole-body vibration effects, 875–877
Wiener filter, 749
Wiener–Khinchine relationship, 50
Wind turbines, predicting noise from, 679–681
Workplace noise standards, 918–924
Workshops, *see* Industrial workshops
Workstations, sound propagation between, 202, 204–205
World Health Organization (WHO), 871, 907
Wrappings, 552–555
  acoustical enclosures vs., 521
  close-fitting enclosures vs., 552
  defined, 517

Y

Youla transform, 781–784
Young's modulus, 391, 493, 595

Z

"Zone of silence," 345, 846, 847
Zoom transforms, 57